HUMANOID ROBOTS

HUMANOID ROBOTS

Modeling and Control

Dragomir N. Nenchev
Tokyo City University
Tokyo, Japan

Atsushi Konno
Hokkaido University
Sapporo, Japan

with contribution by

Teppei Tsujita
National Defense Academy
Yokosuka, Japan

ELSEVIER
Butterworth-Heinemann
An imprint of Elsevier

Butterworth-Heinemann is an imprint of Elsevier
The Boulevard, Langford Lane, Kidlington, Oxford OX5 1GB, United Kingdom
50 Hampshire Street, 5th Floor, Cambridge, MA 02139, United States

Library of Congress Cataloging-in-Publication Data
A catalog record for this book is available from the Library of Congress

British Library Cataloguing-in-Publication Data
A catalogue record for this book is available from the British Library

ISBN: 978-0-12-804560-2

Publisher: Mara Conner
Acquisition Editor: Sonnini R. Yura
Editorial Project Manager: Ana Claudia A. Garcia
Production Project Manager: Bharatwaj Varatharajan
Designer: Mark Rogers

Typeset by VTeX

Dedication

To my loving wife Mariana and my grandchildren Nika, Ruka and Dominic.

DNN

Contents

5. Balance Control

6. Cooperative Object Manipulation and Control

7. Motion Generation and Control: Selected Topics With Applications

The companion materials for this book are available at https://www.elsevier.com/books-and-journals/book-companion/9780128045602

Preface

A robot may not injure a human being or, through inaction, allow a human being to come to harm.

– Isaac Asimov

Humanoid robots are the most universal machines designed so far. Their human-like appearance presumes that one day humanoid robots will become the ubiquitous helpers of mankind as an embodiment of artificial intelligence (AI). Although still in its initial phase, research in the field of humanoid robotics is advancing at a fast pace, covering thereby a diverse set of problems. The research is benefiting from but is also contributing to technological achievements in such areas as self-driving transportation means in a complex environment (e.g. in relation to sensing, perception, and motion planning), natural language communication (e.g. in relation to personal assistance), and AI in general. This is also true for the areas in the field of the mechanical sciences and control.

Characterized as universal, a humanoid robot is also an inherently complex machine. Its control architecture comprises a hierarchical structure. At the middle levels of the hierarchy, kinematic, kinetostatic, and dynamic models have to be employed to ensure appropriate control of the motion and the flow of forces. This is a challenging problem since the models have to account for the relatively large number of actuated joints and the best way to simultaneously control them while performing a varying number of tasks, with a kinematic structure that varies frequently by forming closed kinematic chains and unforming them while the robot establishes new contacts and breaks old ones. The models also have to account for the "floating base" and its "underactuation," as well as for the varying environment conditions.

The aim of this book is the in-depth covering of a core set of problems related to the modeling and the model-based control of humanoid robots. A large body of research exists in the robotics field that can support this goal. Kinematically redundant manipulators and the related problem of multi-degree-of-freedom (DoF) resource allocation to manage a number of robot tasks have been studied from the mid-1970s. A number of studies exist on robots with closed kinematic chains, referred to as parallel-link manipulators. Studies on structure-varying robotic mechanisms, such as multifingered hands, multilegged robots, and dual-/multiarm manipulators, are also abundant. Underactuated, articulated multibody systems on a floating base, such as free-floating space robots, manipulators on flexible bases, and macro-mini manipulators (i.e. a small manipulator attached to the end of a larger arm) have been studied from the mid-1980s. There are a number of studies in the closely related field of constrained multibody systems. Contact modeling is also a well-established area of research.

The ultimate goal is to design a humanoid robot controller that can guarantee the performance of a broad variety of motion and force control tasks. This can only be done when a whole-body model of the robot is employed. With a whole-body model, the structure of the controller becomes quite complex, though, as already noted. The control inputs then have to be derived with the help of an optimization approach. A number of humanoid robot controllers that appeared recently in the literature have been based on this approach, while taking advantage of the existing general optimization software packages. Because of the complexity, however, real-time control may not be feasible with this approach. On the other hand, with a simplified model it is possible to derive the optimal control solution analytically. Analytical solutions yield the advantage of shorter computational cycles. The simplest possible model is the (linear) inverted pendulum model. It was proposed some time ago to deal with the task of balance stabilization on a flat ground. Later it was shown that the model can be enhanced by adding a reaction wheel assembly component. A centroidal moment can then be generated that plays an important role when dealing with unknown disturbances applied to the body of the robot or that appear while the robot steps over irregular terrain, for example. Analytical optimal solutions for a whole-body model are quite rare, though. At present, the analytical approach is indispensable for motion generation and control in real-time.

It is quite exciting that the research in the field of humanoid robotics can contribute to other areas as well. Such areas include the biomechanics and motor control of human movement, physical therapy, the sport sciences, and physics-based animation of articulated figures. Researchers in these fields could benefit from the results described in this book as well.

The authors of this work have been involved for decades in the areas of research that constitute the foundation of humanoid robotics. It was a challenging task to draw on the past results, to organize and reinterpret them in an attempt to exemplify their role when applied to humanoid robots. This work also includes many important and up-to-date results reported by other researchers. From this point of view, the style resembles somewhat that of a reference. What makes it different from a handbook, though, is the revealing of some important interrelations between the results in the different areas of research and their contribution to the main goal of this work.

Tokyo, July 2018 The authors

Acknowledgments

The following students from the Robotics Life Support laboratory at Tokyo City University have contributed to this work.

Shoichi Taguchi, Hiroki Sone, Ryosuke Mitsuhira, Masahiro Hosokawa, Sho Miyahara, Takashi Shibuya, and Ryotaro Hinata derived some of the equations.

Takuya Mimura, Ryosuke Mitsuhira, Hironori Yoshino, Kengo Tamoto, Masahiro Hosokawa, Satoshi Shirai, Sho Miyahara, Takashi Shibuya, and Ryotaro Hinata helped to improve the readability of the text. They also provided data from simulations and experiments, along with Yuki Yoshida, Shogo Onuma, Kajun Nishizawa, Takuma Nakamura, Ryo Iizuka, and Ryota Yui.

Yuki Yoshida, Shogo Onuma, Shoichi Taguchi, Jie Chen, Keita Sagara, Tomonori Tani, Ryosuke Mitsuhira, Takahide Hamano, Masamitsu Umekage, Kajun Nishizawa, Kei Kobayashi, Takashi Shibuya, Ryota Yui, Yukiya Endo, Shingo Sakaguchi, Takayuki Inoue, and Yuki Hidaka contributed with figure preparation.

I appreciate the enormous patience of my wife to go carefully through the text and to considerably improve the English text.

DNN

CHAPTER

1

Introduction

1.1 HISTORICAL DEVELOPMENT

For centuries, imagination has been the driving force behind creativity. This is true especially for humanoid robots with their origins being traceable back to ancient Greek mythology. The drive to create human-like artificial machines has never ceased during the centuries since then [32]. In modern times, the technological advancements achieved by engineers have empowered them to bring into life the creations of the finest dreamers and science-fiction writers.

The development of humanoid robots was pioneered in Japan with WABOT-1, designed in 1973 by the late professor Ichiro Kato and his students at Waseda university [97]. For years, Japanese engineers remained the sole pursuers of the humanoid dream. Their efforts eventually paid off; the fascinating humanoid platform developed throughout the years by Honda Motor Co. and culminated in the world-renowned ASIMO humanoid [94]. In 1996, the impact of unveiling ASIMO's predecessor P2 to the public was enormous. This eventually led to the accelerating development of humanoid robot technology we are witnessing nowadays. Governments in the developed countries are readily financing robot technology. This will inevitably lead to an ever-accelerating cycle in synchrony with the existing exponential advancements in computing machinery, as revealed by Moore's law.

Humanoid robots are complex machines. It took Honda's engineers a decade to develop a number of prototypes and to arrive at the P2 in 1996 [49,48]. It took another 15 years of development to arrive at the "all-new" ASIMO in 2011 [99]. This robot is considered to be the world's most advanced humanoid robot, capable of running fast, very human-like. The robot can also climb stairs, run backwards, hop on one or two legs continuously, and even walk over uneven terrain. These physical capabilities of the robot have been achieved with perfectionist mechanical design, advanced sensors and actuator technology, and dynamic motion control. Besides improving ASIMO's physical capabilities, Honda's engineers also developed the robot's artificial intelligence abilities such as decision making for motion replanning based on sensor data fusion, natural language, gesture-based communication, and others [118].

Following Honda's pioneering efforts, since the end of the 20th century a number of humanoid robots have been built [33]. Humanoid robots have demonstrated such abilities as driving a lift truck [43] or a backhoe [44], pushing [39], lifting [93,83] or moving by pivoting [137] various heavy objects, opening and closing doors [110], pulling drawers [59], nailing [127], lifting and carrying objects in cooperation with humans [27,16,2], and cooking [34].

Humanoid Robots
https://doi.org/10.1016/B978-0-12-804560-2.00008-0

The robots developed so far, however, are only prototypes. They lack sufficient robustness to function in the real-life environment. The DARPA Robotics Challenge (DRC) competition [100] was conceived to address some aspects of this problem, in particular with regard to an extreme environment such as a disaster zone. Indeed, walking through uneven terrain and rubble and driving a car and getting out of it proved to be too difficult tasks for most of the eighteen biped robots that took part in the competition. Seven of the biped teams used the Atlas robot [95], but only one of them, the runner-up Running Man of team IHMC [96], could complete all of the tasks. With nearly identical hardware (the Atlas robot teams used their own lower-arm/hand designs), the robustness of balance maintenance can be identified as being mainly a control issue. The competition outcome has clearly demonstrated, though, that environment-specific design can also play an important role. The winner and the third-place team (DRC-Hubo of team KAIST [61] and CHIMP of CMU team Tartan Rescue [98]) incorporated design elements that deviated from human-like forms, such as a biped-plus-wheel and caterpillar-based locomotion, respectively.

Biped humanoid robots cannot be expected to be commercialized in the near future. In fact, Honda has announced the retirement of ASIMO [103]. The company is developing a new humanoid robot with a design to solve specific tasks in a disaster-related environment [138]. The company also revealed its plan to apply the know-how, accumulated throughout the years of research, in such areas as physical therapy and self-driving vehicles. Without a doubt, pursuing mankind's dream with continuing research efforts in the field of humanoid robotics will inevitably pay off some day.

1.2 TRENDS IN HUMANOID ROBOT DESIGN

1.2.1 Human Likeness of a Humanoid Robot

The design of a universal humanoid robot capable of performing a variety of tasks within different environments remains an open issue. The common "form follows function" design principle can lead to both advantages and disadvantages in humanoid robot design, as discussed by Stanford university professor Bernard Roth [117]. The external appearance of a humanoid robot, including the way it moves, plays an important role for its acceptance in society, as noted by professor Masahiro Mori of Tokyo Institute of Technology [79]. Attempts of quantifying the human-likeness of a humanoid robot are presented in [104,146].

As an example, consider one of the main functions of a biped robot: walking on a level ground. Walking has been realized with the help of a 3D linear inverted pendulum (LIP) model [62]. This, however, has resulted in "crouched" gaits without straightening the knees as in the typical erect gait of the modern-times human. Besides the external appearance problem of the "crouched" robot gait, there is also a functional problem related to gait efficiency. As noted in a study on the bipedal walking of the early hominid *Australopithecus afarensis* [20], net energy absorption is predicted for the "bent" joints, which would have resulted in increased heat load. Indeed, erect (straight-leg) walking is a characteristic of the most energy-efficient machine gait: that of a powerless biped descending a slope, known as passive dynamic walk (PDW) [74]. The bent-knee gait in powered bipeds has been identified as a problem in [119,92] and tackled later on in [70,80,64,124,63,38,71,41,139]. Improved humanoid designs that can circumvent the problematic gait have been demonstrated with WABIAN-2/LL [91], WABIAN-2R [67], and HRP-4C [65,76] and the robot described in [12]. This has been achieved with mechanisms and control methods for ensuring human-like heel-contact and/or toe-off gait phases. A result reported in [35] shows that a straightened-knee gait can be achieved with the Atlas humanoid robot through appropriate control, without involving any special design.

1.2.2 Trade-Offs in Humanoid Robot Design

A design approach aiming at harmony between form and function might not always be possible. Trade-offs between form and function resulting in "environment-specific" designs seem inevitable at present. Consider again the level-ground mobility function. From the viewpoint of stability and safety, a wheel base is definitely preferable instead of bipedalism. It is also preferable in terms of cost effectiveness. These issues are important when bringing humanoids to markets. Hence, commercialized humanoids such as MHI's Wakamaru [101], Hitachi's EMIEW [55] or Softbank's Pepper [102] are all wheel-based. There is also a considerable number of such humanoids that have been designed as research platforms, e.g. [75,24,58,30,123,85,121]. Some of them have been subsequently redesigned and refurbished with a bipedal lower body for increased mobility, e.g. DLR's Rollin' Justin/TORO [30,25] and NASA's Robonaut [23,23,22].

Robonaut-2 is a special case of a "biped" design that does not conform to the human-like form factor; the robot comprises a nice-looking upper body; its legs, however, have a "creepy" appearance. Nevertheless, the leg design seems to best suit the environment on the International Space Station [22,60]. Another example of biped design that differs from the human-like form is the design with legs without feet, i.e. legs that establish point contacts. Such legs are suitable when negotiating highly irregular terrain [106,143]. Another type of environment-specific design is Honda's new prototype, the E2-DR disaster response robot [138].

The existence of the abovementioned form/function design trade-offs has been identified, in fact, as the main reason for the lack of a commonly acceptable definition of the term "humanoid robot" [4].

1.2.3 Human-Friendly Humanoid Robot Design

A class of humanoid robots have been designed especially to support studies in the field of behavioral science, physical embodiment, and social interaction. Various designs have been made, e.g. with a humanoid upper body on a wheel base such as WENDY [81] and TWENDY-ONE [58] by Waseda University, the ARMAR-family humanoids of Karlsruhe Institute of Technology [24], and DB [66,7] and Robovie [75] of the Advanced Telecommunication Research (ATR) Institute in Japan. Other designs include COG of the Massachusetts Institute of Technology (MIT), comprising a fixed-base humanoid upper body [15], the full-body hydraulically driven CB of Sarcos [7,18], and the iCub robot of the Italian Institute of Technology (IIT) [107]. Being equipped with multiple sensory and motor systems and advanced control algorithms, these robots can mimic human abilities quite well. From the viewpoint of this work, the joint torque sensing and control capability of the latter robots deserve special attention. Torque-based control can ensure compliant behaviors of the robot in response to the external force inputs. Such behaviors are useful in exploring physical human–robot interaction paradigms. Compliant behavior is conceived as a necessary condition for robots to function alongside humans. Robot compliance in response to unexpected physical inputs is needed to guarantee the safety of operations [145,3,37,122]. Generally, force/torque control can handle smooth external forces; when dealing with impacts, however, this type of control may not ensure the necessary time of response due to the inherent bandwidth limitations. This led to the design of the intrinsically compliant biped Lucy, powered by pneumatic muscles [128]. More recently, advanced humanoids have been designed that comprise the so-called series-elastic actuators (SEAs), i.e. actuators with embedded passive mechanical elements (springs/dampers) [129]. Such robots are the lower-body humanoid robot M2V2 [112] and the humanoid robots COMAN [126,144] and WALK-MAN [125] developed at IIT, DLR's TORO [25], NASA's humanoid robot Valkyrie [113,105], and others. Other advanced designs for compliant behaviors are those mimicking the human musculoskeletal system [87,77,57,5].

1.3 CHARACTERISTICS OF HUMANOID ROBOTS

There is no commonly accepted definition of the term humanoid robot, as already noted. Nevertheless, the generic characteristics of a humanoid robot can be derived based on the following assumptions. Humanoid robots are designed:

- to operate autonomously in various environments such as dwelling, office, factory, disaster zones;
- to perform a broad spectrum of physical tasks;
- to communicate with humans;
- to come in physical contact with humans without endangering them;
- to operate tools and manipulate objects designed for humans.

From the design viewpoint, these assumptions imply a human-like physical appearance, i.e. a torso, a head, two legs, and two arms with multifingered hands. Only articulated joints are used. From the control viewpoint, these assumptions imply a hierarchical controller structure and a sensor subsystem that are needed to realize:

- perception and cognition,
- learning,
- task sequence planning,
- locomotion trajectory (gait) planning and generation,
- walking control,
- whole-body manipulation planning with motion/force components,
- end-link motion/force trajectory generation, transformation, and tracking control,
- balance and posture control with optimal force distribution in the presence of external disturbances,
- low-level actuator and joint space control.

The upper-level functions, i.e. perception, cognition, learning, and task sequence and motion planning, are related to the advancements in the field of artificial intelligence; they are beyond the scope of the present work. Readers interested in this field are referred to [17] for a treatment from the viewpoint of neuroscience and to [73] and [42] for issues related to human–humanoid communication using natural speech and gestures, respectively. Task sequence and motion planning are currently under extensive development. Some of the problems are covered in [40].

The main focus in this work is on the derivation of kinematic, kinetostatic, and dynamic models of a humanoid robot and the usage of these models in motion/force trajectory generation and control of humanoid robots.

1.4 AREAS OF RESEARCH RELATED TO HUMANOID ROBOTS

Model-based trajectory generation and control design requires an in-depth understanding of the kinematics, kinetostatics, and dynamics of humanoid robots. The following areas of research are well established in the field of robotics and can serve as a basis.

1.4.1 Kinematic Redundancy, Task Constraints, and Optimal Inverse Kinematics Solutions

A humanoid robot comprises a relatively large number of degrees-of-freedom (DoFs). This is the reason why sometimes humanoid robots are characterized as being kinematically redundant. A kinematically redundant robot is modeled as an *underdetermined* system. Such characterization, however, depends on the number of tasks that the robot is supposed to perform. For example, the desired hand position/orientation of the humanoid robot might be reachable in an infinite number of ways when the motion of the trunk is taken under consideration in addition to that of the arm. The motion of the trunk, however, is preferably used in balance control rather than in a reaching task. This simple example demonstrates the fact that when the robot is required to perform a number of tasks *simultaneously*, the DoFs of the robot may not suffice. In this case, the robot should be modeled as an *overconstrained* system rather than as an underconstrained, i.e. as a kinematically redundant one.

From a review on the whole-body control methods used by the teams participating in the DRC [52], it becomes apparent that at present, inverse kinematics–based motion generation and control is the prevailing technique.

Problems related to kinematic redundancy, singularities, and the optimal inverse kinematics solutions of underconstrained and overconstrained systems are discussed in Chapter 2.

1.4.2 Constrained Multibody Systems and Contact Modeling

A humanoid robot is characterized as a *multibody system*. With the exception of jumping or the flight phase of running, a humanoid robot is always in contact with the environment with one or more of its links, e.g. the feet, the hands, the trunk, or the elbows. Existing contacts can be broken and new contacts established at any time instant. The contacts that are formed depend on the geometry of the contacting bodies. The contacts constrain the motion of the generic (i.e. the unconstrained) kinematic chain of the robot. Since the kinematic chain has a tree–like structure, one or more closed kinematic loops will be inevitably formed via the contacts. Thus, a more precise characterization of a humanoid robot would be as a *structure-varying constrained multibody system* [86].

The modeling of the contacts and the instantaneous motion kinematics of a constrained multibody system with closed loops is discussed in Chapter 2. The flow (or distribution) of forces within the closed loops is explained in Chapter 3.

1.4.3 Multifingered Hands and Dual-Arm Object Manipulation

A multifingered hand itself represents a multibody system. When an object is grasped by a multifingered hand, kinematic closed loops are formed. The theory of constrained multibody systems applies in this case as well. Multifingered hands and dual-arm object manipulation have been studied thoroughly throughout the years. An excellent reference text is [84].

Cooperative object manipulation with a multifingered hand, with the two arms of a humanoid robot, and with multiple humanoid robots is discussed in Chapter 6.

1.4.4 Underactuated Systems on a Floating Base

A humanoid robot is characterized as an *underactuated system* since it comprises more DoFs than the number of its actuators. As already noted, the generic kinematic chain of a humanoid robot is structured as a tree. The *root link* of the robot is free to move in 3D space. Such motion implies six nonactuated DoFs.

The root link is quite often referred to as the *floating base* of the robot. The class of underactuated multibody systems on a floating base includes *flexible-base manipulators*, the so-called *macro-micro manipulators* (i.e. a small manipulator mounted at the tip of a larger one), and *free-floating space robots* [90]. Such systems were studied intensively in the late 1980s and 1990s. The studies have contributed to the in-depth understanding of the roles of inertial coupling, angular momentum, and reactionless manipulation. These problems play an important role in balance control of a humanoid robot as well.

The dynamics of floating-base systems are introduced in Chapter 4. Balance control methods are discussed in detail in Chapter 5.

1.4.5 Other Related Areas of Research

Single-Leg, Multilegged, and Multilimb Robots

The development of robots hopping on a single leg has been pioneered by Marc Raibert at MIT [114]. He also developed running bipeds and quadruped robots. At Boston Dynamics, the company Marc Raibert founded after leaving MIT, he used the technology as a basis to design a number of interesting legged robots, including the two biped humanoids Petman and Atlas [95]. Marc Raibert's robots are capable of running fast, negotiating very rough terrain, robust balance control under strong disturbances, and jumping.

Quite similar to humanoid robots, multilegged robots are modeled as constrained multibody systems with a varying kinematic structure. The modeling and control approaches developed for humanoid robots can be directly applied to multilegged robots, and vice versa

[53,56,120,131]. This is also true in the more general case of bio-inspired multilimb robots that provide new ways of locomotion [31,142].

Physics-Based Animation of Articulated Figures

Physics-based animation of articulated figures is a field closely related to humanoid robotics. There are a number of common problems, the main one being motion generation under space-time constraints [135]. The respective methods make use of optimized inverse kinematics solutions [140,141,10,1]. These methods have also been adopted in the field of humanoid robotics [36,68]. Contact modeling, the computation of reaction forces and force-based motion control, is another area of common interest [11,108,82]. This is also the case for interactive motion generation in response to the external perturbations of the character [136, 72,21]. Methods developed for physics-based character animation are also being adopted for motion generation and control in the field of humanoid robotics and vice versa. The key-framing technique, for example, is used in animation to generate the motion of a character by interpolating key postures over time [10]. This technique is also used in the field of humanoid robotics for motion generation [45,46,13,14,26].

Studies on the Biomechanics of Human Movement

In the fields of biomechanics and motor control of human movement [134], physical therapy, and sport sciences there is a large body of research on the mechanism of human balance [88,89,54,29,47], on the ways how the brain controls the balance when perturbations are applied [116,115,69], on the balance during human walking [133,132,9], on the role of arm movements in balance [78,19], and so on. The research results were accumulated well before the advancement of humanoid robotics. Not surprisingly, the human body models used in the research, such as simple inverted pendulums on foot models, were in fact very similar to those that later appeared in the humanoid robotics field (e.g. the LIP model [62]). Two identical concepts, that of the *extrapolated CoM* [51] and that of the *capture point* [111], were derived independently in the biomechanics and robotics fields. Rapid advancements in humanoid robotics have also contributed to the deeper understanding of the human motion control [66, 6], especially by clarifying the role of the angular momentum in balance control [109,50].

With the development of wearable robotics (exoskeletons), a considerable contribution to the field of physical therapy can be attributed to the research in the humanoid robotics field [8].

1.5 PREREQUISITE AND STRUCTURE

This work introduces advanced methods for modeling, motion generation, and control of humanoid robots. It is assumed that readers have a solid general robotics background. A number of excellent textbooks do exist, such as [84] for example. The textbook [28] on rigid-body dynamics would be very helpful. Also, the understanding of the basic concepts in the humanoid robotics field is highly recommended, as presented in [62]. Such concepts include the Zero-Moment Point (ZMP) [130], the Center of Pressure (CoP), the ground reaction forces and moments, the various LIP models in 2D and 3D, the cart-table model and ZMP-based walking pattern generation with preview control, whole-body motion generation, and methods of simulation.

This text is organized as follows. The kinematics, kinetostatics, and dynamics of humanoid robots are discussed in Chapters 2, 3, and 4, respectively. Balance control plays a major role in humanoid robotics; the details are presented in Chapter 5. Chapter 6 explains cooperative object manipulation and control with multifingered hands, with two arms of a robot, and with multiple robots. The research area of motion generation is quite wide. Selected topics with applications are discussed in Chapter 7. Finally, Chapter 8 highlights the importance of simulators and provides step-by-step instructions for implementing a MATLAB-based simulator.

The authors' contribution is as follows. Atsushi Konno contributed Section 5.4.5, Sections 6.1 through 6.4, Section 7.2, Section 7.3, and Section 7.8. Teppei Tsujita wrote Chapter 8. The rest has been written by Dragomir N. Nenchev (Yoshikazu Kanamiya).

References

[1] Y. Abe, M. Da Silva, J. Popović, Multiobjective control with frictional contacts, in: ACM SIGGRAPH/Eurographics Symposium on Computer Animation, 2007, pp. 249–258.
[2] D.J. Agravante, A. Cherubini, A. Bussy, A. Kheddar, Human-humanoid joint haptic table carrying task with height stabilization using vision, in: IEEE/RSJ International Conference on Intelligent Robots and Systems, Tokyo, Japan, 2013, pp. 4609–4614.
[3] A. Albu-Schaffer, O. Eiberger, M. Grebenstein, S. Haddadin, C. Ott, T. Wimbock, S. Wolf, G. Hirzinger, Soft robotics, IEEE Robotics & Automation Magazine 15 (2008) 20–30.
[4] R. Ambrose, Y. Zheng, B. Wilcox, Assessment of International Research and Development in Robotics, Ch. 4 – Humanoids, Technical report, 2004.
[5] Y. Asano, H. Mizoguchi, T. Kozuki, Y. Motegi, J. Urata, Y. Nakanishi, K. Okada, M. Inaba, Achievement of twist squat by musculoskeletal humanoid with screw-home mechanism, in: IEEE International Conference on Intelligent Robots and Systems, 2013, pp. 4649–4654.
[6] C. Atkeson, J. Hale, F. Pollick, M. Riley, S. Kotosaka, S. Schaul, T. Shibata, G. Tevatia, A. Ude, S. Vijayakumar, E. Kawato, M. Kawato, Using humanoid robots to study human behavior, IEEE Intelligent Systems 15 (2000) 46–56.
[7] C.G. Atkeson, J. Hale, F. Pollick, M. Riley, Using a humanoid robot to explore human behavior and communication, Journal of the Robotics Society of Japan 19 (2001) 584–589.
[8] K. Ayusawa, E. Yoshida, Y. Imamura, T. Tanaka, New evaluation framework for human-assistive devices based on humanoid robotics, Advanced Robotics 30 (2016) 519–534.
[9] C. Azevedo, P. Poignet, B. Espiau, Artificial locomotion control: from human to robots, Robotics and Autonomous Systems 47 (2004) 203–223.
[10] P. Baerlocher, R. Boulic, An inverse kinematics architecture enforcing an arbitrary number of strict priority levels, The Visual Computer 20 (2004) 402–417.
[11] D. Baraff, Fast contact force computation for nonpenetrating rigid bodies, in: 21st Annual Conference on Computer Graphics and Interactive Techniques – SIGGRAPH '94, ACM Press, New York, New York, USA, 1994, pp. 23–34.
[12] S. Behnke, Human-like walking using toes joint and straight stance leg, in: International Symposium on Adaptive Motion in Animals and Machines, 2005, pp. 1–6.
[13] K. Bouyarmane, A. Kheddar, Static multi-contact inverse problem for multiple humanoid robots and manipulated objects, in: IEEE-RAS International Conference on Humanoid Robots, Nashville, TN, USA, 2010, pp. 8–13.
[14] K. Bouyarmane, A. Kheddar, Using a multi-objective controller to synthesize simulated humanoid robot motion with changing contact configurations, in: IEEE/RSJ International Conference on Intelligent Robots and Systems, 2011, pp. 4414–4419.
[15] R.A. Brooks, C. Breazeal, M. Marjanović, B. Scassellati, M.M. Williamson, The Cog project: building a humanoid robot, in: C.L. Nehaniv (Ed.), Computation for Metaphors, Analogy, and Agents, in: Lecture Notes in Computer Science, vol. 1562, Springer, Berlin, Heidelberg, 1999, pp. 52–87.

[16] A. Bussy, P. Gergondet, A. Kheddar, F. Keith, A. Crosnier, Proactive behavior of a humanoid robot in a haptic transportation task with a human partner, in: IEEE International Symposium on Robot and Human Interactive Communication, 2012, pp. 962–967.

[17] G. Cheng (Ed.), Humanoid Robotics and Neuroscience: Science, Engineering and Society, CRC Press Book, 2014.

[18] G. Cheng, S.-H. Hyon, J. Morimoto, A. Ude, J.G. Hale, G. Colvin, W. Scroggin, S.C. Jacobsen, CB: a humanoid research platform for exploring neuroscience, Advanced Robotics 21 (2007) 1097–1114.

[19] P.J. Cordo, L.M. Nashner, Properties of postural adjustments associated with rapid arm movements, Journal of Neurophysiology 47 (1982) 287–302.

[20] R.H. Crompton, L. Yu, W. Weijie, M. Günther, R. Savage, The mechanical effectiveness of erect and "bent-hip, bent-knee" bipedal walking in Australopithecus afarensis, Journal of Human Evolution 35 (1998) 55–74.

[21] M. De Lasa, I. Mordatch, A. Hertzmann, Feature-based locomotion controllers, ACM Transactions on Graphics 29 (2010) 131.

[22] M.A. Diftler, T.D. Ahlstrom, R.O. Ambrose, N.A. Radford, C.A. Joyce, N. De La Pena, A.H. Parsons, A.L. Noblitt, Robonaut 2 — initial activities on-board the ISS, in: 2012 IEEE Aerospace Conference, 2012, pp. 1–12.

[23] M.A. Diftler, R.O. Ambrose, S.M. Goza, K.S. Tyree, E.L. Huber, Robonaut mobile autonomy: initial experiments, in: Proceedings – IEEE International Conference on Robotics and Automation, 2005, pp. 1425–1430.

[24] M. Do, P. Azad, T. Asfour, R. Dillmann, Imitation of human motion on a humanoid robot using non-linear optimization, in: IEEE-RAS International Conference on Humanoid Robots, 2008, pp. 545–552.

[25] J. Englsberger, A. Werner, C. Ott, B. Henze, M.A. Roa, G. Garofalo, R. Burger, A. Beyer, O. Eiberger, K. Schmid, A. Albu-Schaffer, Overview of the torque-controlled humanoid robot TORO, in: IEEE-RAS International Conference on Humanoid Robots, IEEE, Madrid, Spain, 2014, pp. 916–923.

[26] A. Escande, A. Kheddar, S. Miossec, Planning contact points for humanoid robots, Robotics and Autonomous Systems 61 (2013) 428–442.

[27] P. Evrard, E. Gribovskaya, S. Calinon, A. Billard, A. Kheddar, Teaching physical collaborative tasks: object-lifting case study with a humanoid, in: IEEE-RAS International Conference on Humanoid Robots, Paris, France, 2009, pp. 399–404.

[28] R. Featherstone, Rigid Body Dynamics Algorithms, Springer Science+Business Media, LLC, Boston, MA, 2008.

[29] J.S. Frank, M. Earl, Coordination of posture and movement, Physical Therapy 70 (1990) 855–863.

[30] M. Fuchs, C. Borst, P. Giordano, A. Baumann, E. Kraemer, J. Langwald, R. Gruber, N. Seitz, G. Plank, K. Kunze, R. Burger, F. Schmidt, T. Wimboeck, G. Hirzinger, Rollin' Justin – design considerations and realization of a mobile platform for a humanoid upper body, in: IEEE International Conference on Robotics and Automation, IEEE, 2009, pp. 4131–4137.

[31] T. Fukuda, Y. Hasegawa, K. Sekiyama, T. Aoyama, Multi-Locomotion Robotic Systems: New Concepts of Bio-Inspired Robotics, 1st edition, Springer, 2012.

[32] T. Fukuda, R. Michelini, V. Potkonjak, S. Tzafestas, K. Valavanis, M. Vukobratovic, How far away is "artificial man", IEEE Robotics & Automation Magazine 8 (2001) 66–73.

[33] A. Goswami, P. Vadakkepat (Eds.), Humanoid Robotics: A Reference, Springer, Netherlands, Dordrecht, 2018.

[34] F. Gravot, A. Haneda, K. Okada, M. Inaba, Cooking for humanoid robot, a task that needs symbolic and geometric reasonings, in: Proceedings 2006 IEEE International Conference on Robotics and Automation, 2006, pp. 462–467.

[35] R.J. Griffin, G. Wiedebach, S. Bertrand, A. Leonessa, J. Pratt, Straight-leg walking through underconstrained whole-body control, in: IEEE International Conference on Robotics and Automation (ICRA), 2018, pp. 5747–5754.

[36] M. Guihard, P. Gorce, Dynamic control of bipeds using ankle and hip strategies, in: IEEE/RSJ Int. Conf. on Intelligent Robots and Systems, Lausanne, Switzerland, 2002, pp. 2587–2592.

[37] S. Haddadin, A. Albu-Schaffer, G. Hirzinger, Requirements for safe robots: measurements, analysis and new insights, The International Journal of Robotics Research 28 (2009) 1507–1527.

[38] N. Handharu, J. Yoon, G. Kim, Gait pattern generation with knee stretch motion for biped robot using toe and heel joints, in: IEEE-RAS International Conference on Humanoid Robots, Daejeon, Korea, 2008, pp. 265–270.

[39] K. Harada, S. Kajita, H. Saito, F. Kanehiro, H. Hirukawa, Integration of manipulation and locomotion by a humanoid robot, in: M.H. Ang, O. Khatib (Eds.), Experimental Robotics IX, in: Springer Tracts in Advanced Robotics, vol. 21, Springer Berlin Heidelberg, Berlin, Heidelberg, 2006, pp. 187–197.

[40] K. Harada, E. Yoshida, K. Yokoi (Eds.), Motion Planning for Humanoid Robots, Springer London, London, 2010.
[41] Y. Harada, J. Takahashi, D.N. Nenchev, D. Sato, Limit cycle based walk of a powered 7DOF 3D biped with flat feet, in: IEEE/RSJ International Conference on Intelligent Robots and Systems, 2010, pp. 3623–3628.
[42] M. Hasanuzzaman, H. Ueno, User, gesture and robot behaviour adaptation for human–robot interaction, in: R. Zaier (Ed.), The Future of Humanoid Robots – Research and Applications, InTech, 2012, pp. 229–256.
[43] H. Hasunuma, M. Kobayashi, H. Moriyama, T. Itoko, Y. Yanagihara, T. Ueno, K. Ohya, K. Yokoi, A tele-operated humanoid robot drives a lift truck, in: Proceedings 2002 IEEE International Conference on Robotics and Automation, 2002, pp. 2246–2252.
[44] H. Hasunuma, K. Nakashima, M. Kobayashi, F. Mifune, Y. Yanagihara, T. Ueno, K. Ohya, K. Yokoi, A tele-operated humanoid robot drives a backhoe, in: IEEE International Conference on Robotics and Automation, 2003, pp. 2998–3004.
[45] K. Hauser, T. Bretl, J.C. Latombe, Non-gaited humanoid locomotion planning, in: IEEE-RAS International Conference on Humanoid Robots, 2005, pp. 7–12.
[46] K. Hauser, T. Bretl, J.-C. Latombe, K. Harada, B. Wilcox, Motion planning for legged robots on varied terrain, The International Journal of Robotics Research 27 (2008) 1325–1349.
[47] J. He, W. Levine, G. Loeb, Feedback gains for correcting small perturbations to standing posture, IEEE Transactions on Automatic Control 36 (1991) 322–332.
[48] K. Hirai, M. Hirose, Y. Haikawa, T. Takenaka, The development of Honda humanoid robot, in: IEEE Int. Conf. on Robotics and Automation, Leuven, Belgium, 1998, pp. 1321–1326.
[49] M. Hirose, T. Takenaka, H. Gomi, N. Ozawa, Humanoid robot, Journal of the Robotics Society of Japan 15 (1997) 983–985 (in Japanese).
[50] A.L. Hof, The equations of motion for a standing human reveal three mechanisms for balance, Journal of Biomechanics 40 (2007) 451–457.
[51] A.L. Hof, M.G.J. Gazendam, W.E. Sinke, The condition for dynamic stability, Journal of Biomechanics 38 (2005) 1–8.
[52] E.M. Hoffman, Simulation and Control of Humanoid Robots for Disaster Scenarios, Ph.D. thesis, Genoa University and Italian Institute of Technology, 2016.
[53] P. Holmes, R.J. Full, D. Koditschek, J. Guckenheimer, The dynamics of legged locomotion: models, analyses, and challenges, SIAM Review 48 (2006) 207–304.
[54] F.B. Horak, L.M. Nashner, Central programming of postural movements: adaptation to altered support-surface configurations, Journal of Neurophysiology 55 (1986) 1369–1381.
[55] Y. Hosoda, S. Egawa, J. Tamamoto, K. Yamamoto, Development of human-symbiotic robot "EMIEW" – Design concept and system construction –, Journal of Robotics and Mechatronics 18 (2006) 195–202.
[56] M. Hutter, H. Sommer, C. Gehring, M. Hoepflinger, M. Bloesch, R. Siegwart, Quadrupedal locomotion using hierarchical operational space control, The International Journal of Robotics Research 33 (2014) 1047–1062.
[57] S. Ikemoto, Y. Nishigori, K. Hosoda, Advantages of flexible musculoskeletal robot structure in sensory acquisition, Artificial Life and Robotics 17 (2012) 63–69.
[58] H. Iwata, S. Sugano, Design of human symbiotic robot TWENDY-ONE, in: IEEE International Conference on Robotics and Automation, 2009, pp. 580–586.
[59] A. Jain, C.C. Kemp, Pulling open novel doors and drawers with equilibrium point control, in: IEEE-RAS International Conference on Humanoid Robots, Paris, France, 2009, pp. 498–505.
[60] C. Joyce, J. Badger, M. Diftler, E. Potter, L. Pike, Robonaut 2 – building a robot on the international space station, 2014.
[61] T. Jung, J. Lim, H. Bae, K.K. Lee, H.M. Joe, J.H. Oh, Development of the Humanoid Disaster Response Platform DRC-HUBO+, IEEE Transactions on Robotics 34 (2018) 1–17.
[62] S. Kajita, H. Hirukawa, K. Harada, K. Yokoi, Introduction to Humanoid Robotics, Springer Verlag, Berlin, Heidelberg, 2014.
[63] K. Kameta, A. Sekiguchi, Y. Tsumaki, Y. Kanamiya, Walking control around singularity using a spherical inverted pendulum with an underfloor pivot, in: IEEE-RAS International Conference on Humanoid Robots, 2007, pp. 210–215.
[64] K. Kameta, A. Sekiguchi, Y. Tsumaki, D. Nenchev, Walking control using the SC approach for humanoid robot, in: IEEE-RAS International Conference on Humanoid Robots, 2005, pp. 289–294.

[65] K. Kaneko, F. Kanehiro, M. Morisawa, K. Miura, S. Nakaoka, S. Kajita, Cybernetic human HRP-4C, in: IEEE-RAS International Conference on Humanoid Robots, Paris, France, 2009, pp. 7–14.

[66] S. Kotosaka, Synchronized robot drumming with neural oscillators, in: International Symposium on Adaptive Motion of Animals and Machines, 2000.

[67] P. Kryczka, K. Hashimoto, H. Kondo, A. Omer, H.-o. Lim, A. Takanishi, Stretched knee walking with novel inverse kinematics for humanoid robots, in: IEEE International Conference on Intelligent Robots and Systems, 2011, pp. 3221–3226.

[68] S. Kudoh, T. Komura, K. Ikeuchi, Stepping motion for a human-like character to maintain balance against large perturbations, in: IEEE International Conference on Robotics and Automation, 2006, pp. 2661–2666.

[69] A.D. Kuo, F.E. Zajac, Human standing posture multi-joint movement strategies based on biomechanical constraints, Progress in Brain Research 97 (1993) 349–358.

[70] R. Kurazume, S. Tanaka, M. Yamashita, T. Hasegawa, K. Yoneda, Straight legged walking of a biped robot, in: IEEE/RSJ International Conference on Intelligent Robots and Systems, 2005, pp. 337–343.

[71] Z. Li, N.G. Tsagarikis, D.G. Caldwell, B. Vanderborght, Trajectory generation of straightened knee walking for humanoid robot iCub, in: IEEE International Conference on Control, Automation, Robotics and Vision, 2010, pp. 2355–2360.

[72] A. Macchietto, V. Zordan, C.R. Shelton, Momentum control for balance, ACM Transactions on Graphics 28 (2009) 80.

[73] Y. Matsusaka, Speech communication with humanoids: how people react and how we can build the system, in: R. Zaier (Ed.), The Future of Humanoid Robots – Research and Applications, InTech, 2012, pp. 165–188.

[74] T. McGeer, Passive walking with knees, in: IEEE International Conference on Robotics and Automation, 1990.

[75] N. Mitsunaga, T. Miyashita, H. Ishiguro, K. Kogure, N. Hagita, Robovie-IV: a communication robot interacting with people daily in an office, in: IEEE International Conference on Intelligent Robots and Systems, 2006, pp. 5066–5072.

[76] K. Miura, M. Morisawa, F. Kanehiro, S. Kajita, K. Kaneko, K. Yokoi, Human-like walking with toe supporting for humanoids, in: IEEE/RSJ International Conference on Intelligent Robots and Systems, San Francisco, CA, USA, 2011, pp. 4428–4435.

[77] I. Mizuuchi, M. Kawamura, T. Asaoka, S. Kumakura, Design and development of a musculoskeletal humanoid, in: IEEE-RAS International Conference on Humanoid Robots, Osaka, Japan, 2012, pp. 811–816.

[78] P. Morasso, Spatial control of arm movements, Experimental Brain Research 42 (1981) 223–227.

[79] M. Mori, The uncanny valley, IEEE Robotics & Automation Magazine 19 (2012) 98–100.

[80] M. Morisawa, S. Kajita, K. Kaneko, K. Harada, F. Kanehiro, K. Fujiwara, H. Hirukawa, Pattern generation of biped walking constrained on parametric surface, in: IEEE International Conference on Robotics and Automation, 2005, pp. 2405–2410.

[81] T. Morita, H. Iwata, S. Sugano, Development of human symbiotic robot: WENDY, in: 1999 IEEE International Conference on Robotics and Automation, vol. 4, 1999, pp. 83–88.

[82] U. Muico, Y. Lee, J. Popović, Z. Popović, Contact-aware nonlinear control of dynamic characters, ACM Transactions on Graphics 28 (2009) 81.

[83] M. Murooka, S. Noda, S. Nozawa, Y. Kakiuchi, K. Okada, M. Inaba, Manipulation strategy decision and execution based on strategy proving operation for carrying large and heavy objects, in: International Conference on Robotics and Automation, 2014, pp. 3425–3432.

[84] R.M. Murray, Z. Li, S.S. Sastry, A Mathematical Introduction to Robotic Manipulation, CRC Press, 1994.

[85] K. Nagasaka, Y. Kawanami, S. Shimizu, T. Kito, T. Tsuboi, A. Miyamoto, T. Fukushima, H. Shimomura, Whole-body cooperative force control for a two-armed and two-wheeled mobile robot using generalized inverse dynamics and idealized joint units, in: IEEE International Conference on Robotics and Automation, Anchorage, AK, USA, 2010, pp. 3377–3383.

[86] Y. Nakamura, K. Yamane, Dynamics computation of structure-varying kinematic chains and its application to human figures, IEEE Transactions on Robotics and Automation 16 (2000) 124–134.

[87] Y. Nakanishi, T. Izawa, M. Osada, N. Ito, S. Ohta, J. Urata, M. Inaba, Development of musculoskeletal humanoid Kenzoh with mechanical compliance changeable tendons by nonlinear spring unit, in: IEEE International Conference on Robotics and Biomimetics, 2011, pp. 2384–2389.

[88] L.M. Nashner, Brain adapting reflexes controlling the human posture, Experimental Brain Research 72 (1976) 59–72.

[89] L.M. Nashner, G. McCollum, The organization of human postural movements: a formal basis and experimental synthesis, Behavioral and Brain Sciences 8 (1985) 135–150.

[90] D.N. Nenchev, Reaction null-space of a multibody system with applications in robotics, Mechanical Sciences 4 (2013) 97–112.

[91] Y. Ogura, T. Kataoka, H. Aikawa, K. Shimomura, H.-o. Lim, A. Takanishi, Evaluation of various walking patterns of biped humanoid robot, in: IEEE International Conference on Robotics and Automation, 2005, pp. 603–608.

[92] Y. Ogura, H.-o. Lim, A. Takanishi, Stretch walking pattern generation for a biped humanoid robot, in: IEEE/RSJ International Conference on Intelligent Robots and Systems, 2003, pp. 352–357.

[93] Y. Ohmura, Y. Kuniyoshi, Humanoid robot which can lift a 30kg box by whole body contact and tactile feedback, in: IEEE/RSJ International Conference on Intelligent Robots and Systems, IEEE, San Diego, CA, USA, 2007, pp. 1136–1141.

[94] ASIMO by Honda: the world's most advanced humanoid robot, http://asimo.honda.com [online].

[95] Boston dynamics: dedicated to the science and art of how things move, https://www.bostondynamics.com/atlas [online].

[96] DARPA robotics challenge — IHMC robotics lab, http://robots.ihmc.us/drc [online].

[97] Development of Waseda robot, http://www.humanoid.waseda.ac.jp/booklet/katobook.html [online].

[98] DRC Tartan rescue team, https://www.nrec.ri.cmu.edu/solutions/defense/other-projects/tartan-rescue-team.html [online].

[99] Honda worldwide, Honda unveils all-new ASIMO with significant advancements, http://hondanews.com/releases/honda-unveils-all-new-asimo-with-significant-advancements/videos/all-new-asimo, November 8, 2011 [online].

[100] The DARPA robotics challenge, http://archive.darpa.mil/roboticschallenge [online].

[101] Wakamaru, Mitsubishi Heavy Industries, Ltd., https://en.wikipedia.org/wiki/Wakamaru [online].

[102] Who is Pepper, https://www.softbankrobotics.com/emea/en/robots/pepper [online].

[103] Honda's Asimo robot bows out but finds new life – Nikkei Asian Review, https://asia.nikkei.com/Business/Companies/Honda-s-Asimo-robot-bows-out-but-finds-new-life [online].

[104] E. Oztop, T. Chaminade, D. Franklin, Human-humanoid interaction: is a humanoid robot perceived as a human?, in: IEEE/RAS International Conference on Humanoid Robots, 2004, pp. 830–841.

[105] N. Paine, J.S. Mehling, J. Holley, N.A. Radford, G. Johnson, C.-L. Fok, L. Sentis, Actuator control for the NASA-JSC Valkyrie humanoid robot: a decoupled dynamics approach for torque control of series elastic robots, Journal of Field Robotics 32 (2015) 378–396.

[106] H.-W. Park, A. Ramezani, J. Grizzle, A finite-state machine for accommodating unexpected large ground-height variations in bipedal robot walking, IEEE Transactions on Robotics 29 (2013) 331–345.

[107] A. Parmiggiani, M. Maggiali, L. Natale, F. Nori, A. Schmitz, N.G. Tsagarakis, J.S. Victor, F. Becchi, G. Sandini, G. Metta, The design of the iCub humanoid robot, International Journal of Humanoid Robotics 09 (2012) 1250027.

[108] N. Pollard, P. Reitsma, Animation of humanlike characters: dynamic motion filtering with a physically plausible contact model, in: Yale Workshop on Adaptive and Learning Systems, 2001.

[109] M. Popovic, A. Hofmann, H. Herr, Angular momentum regulation during human walking: biomechanics and control, in: IEEE International Conference on Robotics and Automation, 2004, pp. 2405–2411.

[110] M. Prats, S. Wieland, T. Asfour, A. del Pobil, R. Dillmann, Compliant interaction in household environments by the Armar-III humanoid robot, in: IEEE-RAS International Conference on Humanoid Robots, 2008, pp. 475–480.

[111] J. Pratt, J. Carff, S. Drakunov, A. Goswami, Capture point: a step toward humanoid push recovery, in: IEEE-RAS International Conference on Humanoid Robots, Genoa, Italy, 2006, pp. 200–207.

[112] J. Pratt, B. Krupp, Design of a bipedal walking robot, in: G.R. Gerhart, D.W. Gage, C.M. Shoemaker (Eds.), Proceedings of SPIE – The International Society for Optical Engineering, 2008.

[113] N.A. Radford, P. Strawser, K. Hambuchen, J.S. Mehling, W.K. Verdeyen, A.S. Donnan, J. Holley, J. Sanchez, V. Nguyen, L. Bridgwater, R. Berka, R. Ambrose, M. Myles Markee, N.J. Fraser-Chanpong, C. McQuin, J.D. Yamokoski, S. Hart, R. Guo, A. Parsons, B. Wightman, P. Dinh, B. Ames, C. Blakely, C. Edmondson, B. Sommers, R. Rea, C. Tobler, H. Bibby, B. Howard, L. Niu, A. Lee, M. Conover, L. Truong, R. Reed, D. Chesney, R. Platt, G. Johnson, C.-L. Fok, N. Paine, L. Sentis, E. Cousineau, R. Sinnet, J. Lack, M. Powell, B. Morris, A. Ames, J. Akinyode, Valkyrie: NASA's first bipedal humanoid robot, Journal of Field Robotics 32 (2015) 397–419.

[114] M.H. Raibert, Legged Robots that Balance, MIT Press, 1986.

[115] C.F. Ramos, L.W. Stark, Postural maintenance during fast forward bending: a model simulation experiment determines the "reduced trajectory", Experimental Brain Research 82 (1990) 651–657.
[116] C.F. Ramos, L.W. Stark, Postural maintenance during movement: simulations of a two joint model, Biological Cybernetics 63 (1990) 363–375.
[117] B. Roth, How can robots look like human beings, in: IEEE International Conference on Robotics and Automation, IEEE, 1995, p. 1.
[118] Y. Sakagami, R. Watanabe, C. Aoyama, S. Matsunaga, N. Higaki, K. Fujimura, The intelligent ASIMO: system overview and integration, in: IEEE/RSJ International Conference on Intelligent Robots and System, 2002, pp. 2478–2483.
[119] A. Sekiguchi, Y. Atobe, K. Kameta, D. Nenchev, Y. Tsumaki, On motion generation for humanoid robot by the SC approach, in: Annual Conference of the Robotics Society of Japan, 2003, 2A27 (in Japanese).
[120] C. Semini, V. Barasuol, T. Boaventura, M. Frigerio, M. Focchi, D.G. Caldwell, J. Buchli, Towards versatile legged robots through active impedance control, The International Journal of Robotics Research 34 (2015) 1003–1020.
[121] L. Sentis, J. Petersen, R. Philippsen, Experiments with balancing on irregular terrains using the Dreamer mobile humanoid robot, in: Robotic Science and Systems, Sydney, 2012, pp. 1–8.
[122] D. Shin, I. Sardellitti, Y.L. Park, O. Khatib, M. Cutkosky, Design and control of a bio-inspired human-friendly robot, The International Journal of Robotics Research 29 (2009) 571–584.
[123] M. Stilman, J. Wang, K. Teeyapan, R. Marceau, Optimized control strategies for wheeled humanoids and mobile manipulators, in: IEEE-RAS International Conference on Humanoid Robots, IEEE, Paris, France, 2009, pp. 568–573.
[124] K. Takahashi, M. Noda, D. Nenchev, Y. Tsumaki, A. Sekiguchi, Static walk of a humanoid robot based the singularity-consistent method, in: IEEE/RSJ International Conference on Intelligent Robots and Systems, 2006, pp. 5484–5489.
[125] N.G. Tsagarakis, D.G. Caldwell, F. Negrello, W. Choi, L. Baccelliere, V.G. Loc, J. Noorden, L. Muratore, A. Margan, A. Cardellino, L. Natale, E. Mingo Hoffman, H. Dallali, N. Kashiri, J. Malzahn, J. Lee, P. Kryczka, D. Kanoulas, M. Garabini, M. Catalano, M. Ferrati, V. Varricchio, L. Pallottino, C. Pavan, A. Bicchi, A. Settimi, A. Rocchi, A. Ajoudani, WALK-MAN: a high-performance humanoid platform for realistic environments, Journal of Field Robotics 34 (2017) 1225–1259.
[126] N.G. Tsagarakis, S. Morfey, G. Medrano Cerda, Z. Li, D.G. Caldwell, COMpliant huMANoid COMAN: optimal joint stiffness tuning for modal frequency control, in: IEEE International Conference on Robotics and Automation, 2013, pp. 673–678.
[127] T. Tsujita, A. Konno, S. Komizunai, Y. Nomura, T. Owa, T. Myojin, Y. Ayaz, M. Uchiyama, Humanoid robot motion generation for nailing task, in: IEEE/ASME International Conference on Advanced Intelligent Mechatronics, 2008, pp. 1024–1029.
[128] B. Vanderborght, Dynamic Stabilisation of the Biped Lucy Powered by Actuators with Controllable Stiffness, Springer, 2010.
[129] B. Vanderborght, A. Albu-Schaeffer, A. Bicchi, E. Burdet, D.G. Caldwell, R. Carloni, M. Catalano, O. Eiberger, W. Friedl, G. Ganesh, M. Garabini, M. Grebenstein, G. Grioli, S. Haddadin, H. Hoppner, A. Jafari, M. Laffranchi, D. Lefeber, F. Petit, S. Stramigioli, N. Tsagarakis, M. Van Damme, R. Van Ham, L.C. Visser, S. Wolf, Variable impedance actuators: a review, Robotics and Autonomous Systems 61 (2013) 1601–1614.
[130] M. Vukobratovic, B. Borovac, Zero-moment point — thirty five years of its life, International Journal of Humanoid Robotics 01 (2004) 157–173.
[131] P.M. Wensing, L.R. Palmer, D.E. Orin, Efficient recursive dynamics algorithms for operational-space control with application to legged locomotion, Autonomous Robots 38 (2015) 363–381.
[132] D. Winter, Human balance and posture control during standing and walking, Gait & Posture 3 (1995) 193–214.
[133] D.A. Winter, Foot trajectory in human gait: a precise and multifactorial motor control task, Physical Therapy 72 (1992) 45–53, discussion 54–56.
[134] D.A. Winter, Biomechanics and Motor Control of Human Movement, Wiley, 2009.
[135] A. Witkin, M. Kass, Spacetime constraints, Computer Graphics 22 (1988) 159–168.
[136] K. Yin, D. Pai, M.V.D. Panne, Data-driven interactive balancing behaviors, in: Pacific Graphics, 2005, pp. 1–9.
[137] E. Yoshida, M. Poirier, J.-P. Laumond, O. Kanoun, F. Lamiraux, R. Alami, K. Yokoi, Pivoting based manipulation by a humanoid robot, Autonomous Robots 28 (2009) 77–88.
[138] T. Yoshiike, M. Kuroda, R. Ujino, H. Kaneko, H. Higuchi, S. Iwasaki, Y. Kanemoto, M. Asatani, T. Koshiishi, Development of experimental legged robot for inspection and disaster response in plants, in: IEEE/RSJ International Conference on Intelligent Robots and Systems (IROS), 2017, pp. 4869–4876.

[139] Y. You, S. Xin, C. Zhou, N. Tsagarakis, Straight leg walking strategy for torque-controlled humanoid robots, in: IEEE International Conference on Robotics and Biomimetics (ROBIO), 2016, pp. 2014–2019.
[140] J. Zhao, N.I. Badler, Real-Time Inverse Kinematics with Joint Limits and Spatial Constraints, Technical Report MS-CIS-89-09, University of Pennsylvania, 1989.
[141] J. Zhao, N.I. Badler, Inverse kinematics positioning using nonlinear programming for highly articulated figures, ACM Transactions on Graphics 13 (1994) 313–336.
[142] Y. Zhao, A Planning and Control Framework for Humanoid Systems: Robust, Optimal, and Real-Time Performance, Ph.D. thesis, The University of Texas at Austin, 2016.
[143] Y. Zhao, B.R. Fernandez, L. Sentis, Robust optimal planning and control of non-periodic bipedal locomotion with a centroidal momentum model, The International Journal of Robotics Research 36 (2017) 1211–1243.
[144] C. Zhou, Z. Li, X. Wang, N. Tsagarakis, D. Caldwell, Stabilization of bipedal walking based on compliance control, Autonomous Robots 40 (2016) 1041–1057.
[145] M. Zinn, O. Khatib, B. Roth, J. Salisbury, Playing it safe, IEEE Robotics & Automation Magazine 11 (2004) 12–21.
[146] J. von Zitzewitz, P.M. Boesch, P. Wolf, R. Riener, Quantifying the human likeness of a humanoid robot, International Journal of Social Robotics 5 (2013) 263–276.

CHAPTER

2

Kinematics

2.1 INTRODUCTION

This chapter discusses the kinematics of humanoid robots. Kinematics plays a basic role in motion analysis, generation, and control. Kinematics-based motion control is quite often employed in control algorithms for hand/feet movements.

The chapter is divided in eleven sections. In Section 2.2, the kinematic structure of a generic humanoid robot is presented and main coordinate frames of interest are defined. In Section 2.3, the forward and inverse kinematics problems are addressed via notions for rigid-body position/orientation. Section 2.4 focuses on the forward and inverse differential kinematic relations, including spatial velocity and spatial transformations. The special case of differential kinematic relations at singular configurations is discussed in Section 2.5. Generic singular configurations for robots with and without kinematically redundant limbs are highlighted. Section 2.6 introduces the manipulability ellipsoid. Further on, a number of humanoids comprise kinematically redundant arms and/or legs. Kinematic redundancy resolution is discussed in Section 2.7, focusing on the limbs. Section 2.8 analyzes the important problem of redundancy resolution under motion constraints imposed by multiple motion tasks. This problem is related to whole-body motion control problem and is tackled in Chapter 4 and Chapter 5. Section 2.9 introduces motion constraints due to physical contacts between the robot links and the environment. Differential motion relationships within closed kinematic loops resulting from the contacts are discussed in Section 2.10. Finally, the differential motion relationships of a humanoid robot are derived in Section 2.11.

It is assumed that readers are familiar with the basic concepts concerning systems of rigid bodies, such as representations of positions and orientations, joint models, coordinate frame assignment techniques for serial-link and parallel-link mechanisms, coordinate transformations, and the forward and inverse kinematics problems.

2.2 KINEMATIC STRUCTURE

The skeleton figure of a generic humanoid robot is shown in Fig. 2.1A. The anthropomorphic structure of the robot becomes apparent from the figure: there are six-degree-of-freedom (6-DoF) legs, 7-DoF arms, a 2-DoF torso, and a 2-DoF head. In the real robot, all joints are single-DoF rotational R joints. Robot models, on the other hand, quite often make use of

Humanoid Robots
https://doi.org/10.1016/B978-0-12-804560-2.00009-2

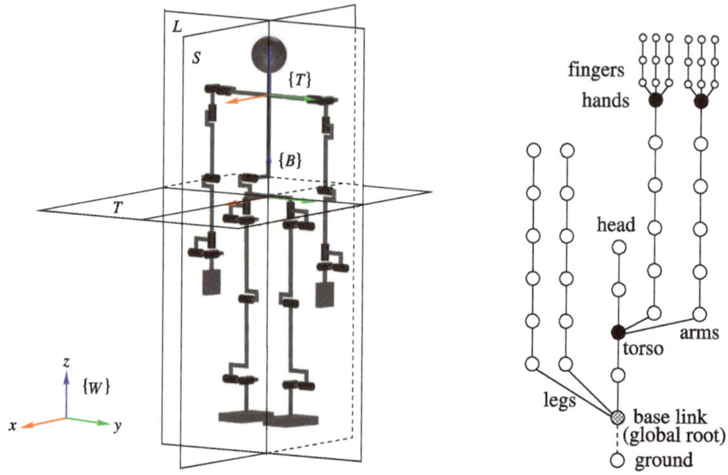

FIGURE 2.1 (A) Left: Skeleton figure of a humanoid robot with three basic reference planes derived from human anatomy and three reference frames. The coloring of the coordinate axes applies henceforth throughout the text. (B) Right: Tree connectivity structure of the kinematic chain. The black circles denote local root links for the branches. The white circles denote the joints.

equivalent multi-DoF joints. In this case, in resemblance to the human skeleton structure, the 3R joint assemblies at the hips, shoulders, and wrists are represented as equivalent spherical joints (3-DoF). The 2R joint assemblies at the ankles, torso, and neck, on the other hand, constitute equivalent universal joints (2-DoF). All robot links in the skeleton figure are assumed to represent rigid bodies. The position/orientation of each link can then be assessed via the position/orientation of a coordinate frame embedded into the link in an appropriate way. For example, with single-DoF models, the Denavit and Hartenberg notation [26] is frequently used. With multi-DoF joint models, on the other hand, the link coordinate frame is attached to one of the connecting joints. Details will be given at a later point.

One of the links in the kinematic chain, the "pelvis" link, plays a special role. Note that the kinematic chain of a humanoid robot can be expressed in the form of a tree, as shown with the connectivity graph in Fig. 2.1B [36]. The pelvis link is the global root link of the tree. It is connected via a virtual 6-DoF joint to the ground. This connection expresses the fact that the robot has a "floating base," i.e. the base link moves in 3D space like a free rigid body [164]. The root link then branches out into the two legs and the torso. The torso itself represents a local root that branches out into the two arms and the head. The arms terminate with the hands that branch out into the fingers (not shown in the skeleton figure, Fig. 2.1A). The figure shows the so-called *spanning tree* of a humanoid robot. It signifies the case when only the global root link is connected to the ground. The usual situation, however, is that one or more of the other links are also connected. With a quiet stance, for example, the two feet connect to the ground via temporary *contact joints*. In this way, closed loops are formed within the chain. Due to this reason, the kinematic chain of a humanoid robot is characterized as a structurally varying one [107].

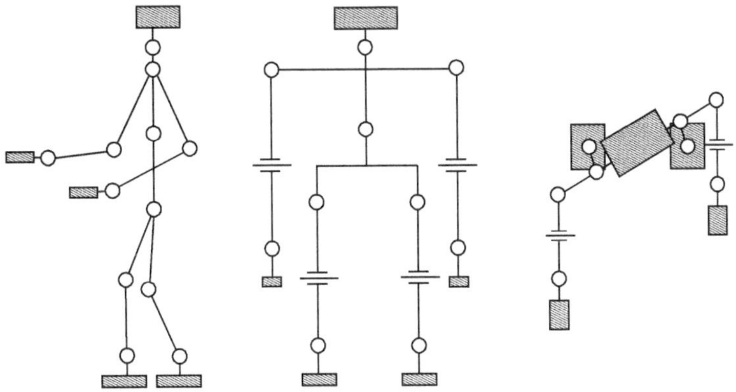

FIGURE 2.2 Planar models in the three planes. (A) Left: sagittal plane. (B) Middle: lateral plane. (C) Right: transverse plane.

In Fig. 2.1A, three reference frames are shown that play special roles. Frame $\{W\}$ is attached to the fixed link (the ground) and hence represents the inertial frame. Frame $\{B\}$ is fixed at the base (root) link. Frame $\{T\}$ is attached to the torso link; it acts as the "local" root for the arm and head branches. There is also another class of coordinate frames with a special role: those attached to the terminal links in the tree structure (the end links). Indeed, the robot interacts with the environment mainly via its end links. For instance, the fingers need to be controlled for grasping and manipulating small objects. The hands, considered as the end links of the arm branches, likewise need to be controlled for fetching/placing small objects and for manipulating larger objects. The feet, on the other hand, are controlled for locomotion. Finally, the head is controlled for an appropriate gaze. In this work, the main focus is on locomotion and arm manipulation, with special emphasis on motion analysis and control of the feet (F) and hands (H). The respective coordinate frames will be denoted as $\{e_j\}$, $e \in \{F, H\}$, $j \in \{r, l\}$, "r" and "l" standing for "right" and "left."

Planar models can be quite helpful in motion analysis and motion generation. Three basic planes adopted from human anatomy are used to devise such planar models, as shown in Fig. 2.1A. First, the sagittal plane (S) is an x-z plane perpendicular to the ground. It passes through the "head" and the "spinal cord" and divides the body into left and right (sinister and dexter) portions. Second, the lateral plane (L), also called a frontal or coronal plane, is an y-z plane perpendicular to the ground. It divides the body into back and front (dorsal and ventral, or posterior and anterior) portions. Finally, the transverse plane (T), also referred to as a cross-section, is an x-y horizontal plane parallel to the ground. It separates the head from the feet (or the superior from the inferior). A 3D motion pattern, observed e.g. while walking or while balancing in response to an external disturbance, is often decomposed into motion patterns within the planes. Planar models on the three planes are shown in Fig. 2.2. In the sagittal plane, all joints are 1-DoF rotational joints. In the frontal and transverse planes, on the other hand, the elbow and knee joints are represented as translational joints since the arm and leg lengths appear to vary. This feature is used in motion analysis and generation via simplified models, such as the inverted linear pendulum model described in Chapter 4.

2.3 FORWARD AND INVERSE KINEMATIC PROBLEMS

A rigid-body position/orientation in 3D space, henceforth referred to as *6D position*, is represented as an element of the special Euclidean group $SE(3) = \Re^3 \times SO(3)$. Note that the parametrization of the 3D rotation group $SO(3)$ is not unique [148]. In robotics, most used are 3×3 rotation matrices and the minimal parametrization via Euler angles, as well as Euler parameters (the unit quaternion) [169]. When parametrized by a rotation matrix, for example, bold font notation $\boldsymbol{X} \in SE(3)$ will be used to denote a 4×4 homogeneous matrix (see e.g. [104]). A trailing subscript denotes the link frame, a leading superscript the reference frame. For instance, ${}^B\boldsymbol{X}_{F_j}$ expresses the 6D position of the feet w.r.t. the base frame, ${}^T\boldsymbol{X}_{H_j}$ gives the 6D position of the hands w.r.t. the torso frame. Furthermore, it is important to note that the elements $\boldsymbol{X}$ signify a coordinate transformation (also called a *rigid-body motion*) between two frames. As an example, ${}^B\boldsymbol{X}_T$ denotes the transformation from the arm local-root (or torso) frame $\{T\}$ to the base frame $\{B\}$. This transformation is used to express the 6D position of the hand ${}^T\boldsymbol{X}_{H_j}$ in base coordinates, i.e.

$$\boldsymbol{X}_{H_j} = {}^B\boldsymbol{X}_T {}^T\boldsymbol{X}_{H_j}. \tag{2.1}$$

The two rigid-body motion components, ${}^B\boldsymbol{X}_T$ and ${}^T\boldsymbol{X}_{H_j}$, are obtained from the trunk/limb joint angle data sets via the *forward kinematics functions* for the branches. Denote by $\boldsymbol{\theta}_{e_j} \in \Re^{n_{e_j}}$ the joint angle vector w.r.t. the local-root link, where n_{e_j} stands for the number of joints in the limb. The *forward kinematics problem* is defined as

$${}^r\mathcal{X}_{e_j} = {}^r\boldsymbol{\varphi}(\boldsymbol{\theta}_{e_j}),$$

where ${}^r\boldsymbol{\varphi}(\boldsymbol{\theta}_{e_j})$ is the forward kinematics function and ${}^r\mathcal{X}_{e_j} \in WS_r \subset \Re^6$ denotes the 6D position of the end link w.r.t. reference frame $\{r\}$. The 6D position is expressed in *local coordinates*, obtained by employing the minimal parametrization of $SO(3)$ (and hence of $SE(3)$). Local coordinates are used since the minimal parametrization cannot be global [148]. Care is then needed to deal with singularities of the chosen set of Euler angles. WS_r denotes the *workspace* of the limb w.r.t. reference frame $\{r\}$. The derivation of the forward kinematics function for serial-link manipulators is straightforward; see e.g. [104].

Robot tasks are defined w.r.t. the inertial frame embedded into the environment (frame $\{W\}$ in Fig. 2.1A). Then, the 6D position of the controlled bodies (the end links of the limbs) has to be expressed in the inertial frame. This is done via the rigid-body motion transform ${}^W\boldsymbol{X}_B$. For example, to obtain the 6D position of the hands in $\{W\}$, premultiply (2.1) from the left by ${}^W\boldsymbol{X}_B$, i.e.

$${}^W\boldsymbol{X}_{H_j} = {}^W\boldsymbol{X}_B {}^B\boldsymbol{X}_T {}^T\boldsymbol{X}_{H_j}.$$

When the robot is in a single or double stance posture, the rigid-body motion component ${}^W\boldsymbol{X}_B$ is calculated from the leg(s) joint angle data. When the robot is airborne or tipping over, the base-link localization sensor data obtained via sensor fusion (e.g. from a LIDAR, an IMU, and/or stereo vision) can be used.

For control purposes, the *inverse kinematics problem* has to be solved. That is, "given the 6D position of a specific link (e.g. an end link) in the kinematic chain w.r.t. the branch root frame or the inertial frame, find the respective joint angles". This problem may have a closed-form solution. An example with humanoid robot HUBO is presented in [126]. Whenever a closed-form solution cannot be found, numerical solutions using differential kinematic relationships have to be used. A closed-form solution is preferable since the differential inverse kinematics have inherent problems, as will be clarified in what follows.

2.4 DIFFERENTIAL KINEMATICS

Differential kinematics are essential for instantaneous motion analysis and motion generation. Since the robot links are represented as rigid bodies, the instantaneous motion of the robot is uniquely characterized by the motion rates in the joints, under the assumption that the instantaneous motion of the base link is known. Information about the motion rates is derived from sensors in the joints (e.g. optical encoders). Using this information, it is possible to calculate the instantaneous motion of any link of interest, e.g. that of the end links. Proper assessment of the instantaneous motion state is essential for activities like walking, manipulating objects with the hands or whole-body reconfiguration to avoid collisions. Further on, differential kinematics also play an important role in control. Task space–based kinematic feedforward/feedback control schemes [136] make use of control commands specified in terms of instantaneous motion of the end links of interest. These control commands are then transformed into joint motion control commands via (inverse) differential kinematics. The differential kinematics are as well essential in deriving dynamical models, wherein both first-order (velocity level) and second-order (acceleration level) relationships are taken into consideration.

2.4.1 Twist, Spatial Velocity, and Spatial Transform

The instantaneous motion of a link is completely characterized by the velocity of a characteristic point (point P) on the link and the angular velocity of the link. These are vector quantities of well known geometrical origin that are subjected to vector operations such as inner and outer product. In what follows, the following coordinate form representation will be employed: $\boldsymbol{v}_P \in \Re^3$ and $\boldsymbol{\omega} \in \Re^3$, for velocity and angular velocity, respectively. The above vector operations will also be represented in coordinate form. Given two coordinate vectors $\boldsymbol{a},\ \boldsymbol{b} \in \Re^3$, their inner and outer products are denoted as $\boldsymbol{a}^T\boldsymbol{b}$ and $[\boldsymbol{a}^\times]\boldsymbol{b}$, respectively. If $\boldsymbol{a} = \begin{bmatrix} a_x & a_y & a_z \end{bmatrix}^T$, then

$$[\boldsymbol{a}^\times] = \begin{bmatrix} 0 & -a_z & a_y \\ a_z & 0 & -a_x \\ -a_y & a_x & 0 \end{bmatrix}$$

is the respective outer-product operator, represented as a skew-symmetric matrix. Note that skew-symmetric matrices are characterized by the relation $[\boldsymbol{a}^\times] = -[\boldsymbol{a}^\times]^T$.

Further on, the velocity and the skew-symmetric representation of the angular velocity, $[\boldsymbol{\omega}^\times] \in so(3)$, constitute an element of $se(3)$, the infinitesimal generator of the special Euclidean group $SE(3)$. This element can be parametrized by a 6D vector, $\mathcal{V} \in \Re^6$, composed of two components representing the velocity and the angular velocity vectors. Given a 6D position in homogeneous coordinates, $\boldsymbol{X}$, describing the motion of a body w.r.t. the inertial frame, the elements of $\mathcal{V}$ can be extracted from $\dot{\boldsymbol{X}}\boldsymbol{X}^{-1}$ or $\boldsymbol{X}^{-1}\dot{\boldsymbol{X}}$. In the former case, the instantaneous motion is described in inertial coordinates and referred to as the *spatial velocity*. In the latter case, the instantaneous motion is described in body coordinates; it will be referred to as the *body velocity*. Sometimes, it might be preferable to express the velocity in a way independent of the particular choice of the characteristic point P, i.e. in the form of a *vector field*. Denote $\mathcal{V} = \begin{bmatrix} \boldsymbol{v}_O^T & \boldsymbol{\omega}^T \end{bmatrix}^T \in \Re^6$. With this parametrization, $\boldsymbol{v}_O$ is interpreted as the velocity of a (fictitious) point on the link that coincides instantaneously with the origin O of an arbitrarily chosen fixed coordinate frame; $\mathcal{V}$, on the other hand, can be interpreted as an operator that, given a point on the link, extracts its velocity in spatial coordinates, i.e.

$$\mathcal{V}(P) \equiv \mathcal{V}_P = \begin{bmatrix} \boldsymbol{v}_P^T & \boldsymbol{\omega}^T \end{bmatrix}^T,$$

where

$$\boldsymbol{v}_P = \boldsymbol{v}_O - [\boldsymbol{r}^{\times}_{\overleftarrow{PO}}]\boldsymbol{\omega}. \tag{2.2}$$

Here $\boldsymbol{r}_{\overleftarrow{PO}}$ denotes the vector pointing from O to P, as apparent from the subscript with the over-arrow notation. The spatial velocity, along with other "spatial" quantities like spatial transform, spatial force, and spatial inertia, constitute the elements of "spatial algebra" [35]. This notation has been widely accepted due to its compactness. Historically, 6D velocity/angular velocity vector pairs (also called bivectors) appeared first in the theory of screws under the name *twist* [9]. A twist can be equated with a spatial velocity vector [104,22]; henceforth these two terms will be used interchangeably. It should be pointed out that implementations based on the spatial notation require great care because of the nonuniform dimensions of the bivectors [31,30].

The instantaneous translation and rotation of the link can be expressed in Cartesian coordinates. Six basis vectors can be defined that constitute a *Plücker coordinate system*, as known from the theory of screws; $\mathcal{V}_P$ then represents the Plücker coordinates of the link spatial velocity for point P. The notation introduced for rigid-body motion objects applies. For instance, the spatial velocity of the hands w.r.t. the base coordinate frame will be expressed as ${}^{B}\mathcal{V}_{H_j}$. Furthermore, for the purposes of analysis and control it is often required to express the spatial velocity of a point (e.g. point P) on a given link represented in a given coordinate frame (e.g. base frame $\{B\}$), w.r.t. a different coordinate frame (e.g. world frame $\{W\}$). This is accomplished with the following relation:

$${}^{W}\mathcal{V}_P = {}^{W}\mathbb{R}_B\, {}^{B}\mathcal{V}_P,$$

where

$${}^{W}\mathbb{R}_B = \begin{bmatrix} {}^{W}\boldsymbol{R}_B & \boldsymbol{0}_3 \\ \boldsymbol{0}_3 & {}^{W}\boldsymbol{R}_B \end{bmatrix} \in \Re^{6\times 6}. \tag{2.3}$$

The leading superscript identifies the reference frame. Notation $\boldsymbol{0}_3$ stands for the 3×3 null matrix, ${}^{W}\boldsymbol{R}_B \in \Re^{3\times 3}$ denotes the rotation matrix that transforms vectors from the base frame to the world frame, and $\mathbb{R}_B$ will be referred to as the *spatial rotation transform*.

Another operation that is frequently required is the following: given a spatial velocity at point O, find the spatial velocity at another point, P. To this end, the following relation can be employed:

$$ {}^{B}\mathcal{V}_{P} = {}^{B}\mathbb{T}_{\overleftarrow{PO}}\, {}^{B}\mathcal{V}_{O}, \tag{2.4} $$

where

$$ {}^{B}\mathbb{T}_{\overleftarrow{PO}} = \begin{bmatrix} \boldsymbol{E}_3 & -[{}^{B}\boldsymbol{r}^{\times}_{\overleftarrow{PO}}] \\ \boldsymbol{0}_3 & \boldsymbol{E}_3 \end{bmatrix} \in \Re^{6\times 6}. \tag{2.5} $$

Here $\boldsymbol{E}_3$ stands for the 3×3 identity matrix.[1] The validity of (2.2) can be confirmed from the above relation; $\mathbb{T}$ will be referred to as the *spatial translation transform*. The role of this transform is to account for the contribution of the angular velocity of coordinate frame $\{B\}$ to the linear velocity at point P. Note that the translation transform does not change the coordinate frame, as apparent from the same leading superscripts in (2.4). In the particular case when all quantities are represented w.r.t. the world coordinate frame, the leading superscripts will be omitted.

Consecutive spatial translation and rotation can be applied to a twist acting at some given point on the body and expressed in a given coordinate frame to obtain the twist acting at some different point and expressed in a different coordinate frame, as follows:

$$ \begin{aligned} {}^{W}\mathcal{V}_{P} &= {}^{W}\mathbb{R}_{B}\, {}^{B}\mathbb{T}_{\overleftarrow{PO}}\, {}^{B}\mathcal{V}_{O} \\ &= {}^{W}\mathbb{X}_{B\overleftarrow{PO}}\, {}^{B}\mathcal{V}_{P}. \end{aligned} \tag{2.6} $$

The combined spatial transform, $\mathbb{X} : \Re^6 \to \Re^6$, is expressed in this case as

$$ \begin{aligned} {}^{W}\mathbb{X}_{B\overleftarrow{PO}} &= \begin{bmatrix} {}^{W}\boldsymbol{R}_{B} & \boldsymbol{0}_3 \\ \boldsymbol{0}_3 & {}^{W}\boldsymbol{R}_{B} \end{bmatrix} \begin{bmatrix} \boldsymbol{E}_3 & -[{}^{B}\boldsymbol{r}^{\times}_{\overleftarrow{PO}}] \\ \boldsymbol{0}_3 & \boldsymbol{E}_3 \end{bmatrix} \\ &= \begin{bmatrix} {}^{W}\boldsymbol{R}_{B} & -{}^{W}\boldsymbol{R}_{B}[{}^{B}\boldsymbol{r}^{\times}_{\overleftarrow{PO}}] \\ \boldsymbol{0}_3 & {}^{W}\boldsymbol{R}_{B} \end{bmatrix}. \end{aligned} \tag{2.7} $$

In [104], this transform is called the adjoint transformation associated with the rigid body motion $\mathcal{X} \in SE(3)$, denoted as $\mathrm{Ad}_{\mathcal{X}}$. In [35], the term "Plücker transform" is used to express a connection with the theory of screws. Note that the inverse of the spatial transform,

$$ \begin{aligned} \mathbb{X}^{-1} &= \mathbb{T}^{-1}\mathbb{R}^{-1} \\ &= \begin{bmatrix} \boldsymbol{E} & [\boldsymbol{r}^{\times}] \\ \boldsymbol{0} & \boldsymbol{E} \end{bmatrix} \begin{bmatrix} \boldsymbol{R}^{T} & \boldsymbol{0} \\ \boldsymbol{0} & \boldsymbol{R}^{T} \end{bmatrix} \\ &= \begin{bmatrix} \boldsymbol{R}^{T} & [\boldsymbol{r}^{\times}]\boldsymbol{R}^{T} \\ \boldsymbol{0} & \boldsymbol{R}^{T} \end{bmatrix}, \end{aligned} \tag{2.8} $$

can be used to exchange the coordinate frames. In the above example, ${}^{W}\mathbb{X}_{B}^{-1} \equiv {}^{B}\mathbb{X}_{W}$.

[1] The subscripts in $\boldsymbol{0}_3$ and $\boldsymbol{E}_3$ will be dropped when there is no ambiguity.

2.4.2 Forward Differential Kinematic Relations

The *forward kinematics problem* for a branch of the kinematic chain (e.g. arm or leg), is formulated as follows: "Given the joint angles and the motion rates in the joints, find the end-link spatial velocity." The instantaneous motion of the end link is expressed in the coordinate frame at the local root of the respective branch. The motion depends on the motion rate in each kinematic joint.

Jacobian Matrix

First, consider a real robot that comprises only single-DoF (rotational) joints. To keep the notation simple, for the time being assume a "generic" branch (or limb) of the kinematic chain composed of n joints. The joint angles are the elements of vector $\boldsymbol{\theta} \in \Re^n$. This vector specifies the *limb configuration*. For a given limb configuration $\boldsymbol{\theta}$, the end-link spatial velocity is determined as a linear combination of the joint motion rates $\dot{\theta}_i$, i.e.

$$\mathcal{V}_n = \sum_{i=1}^{n} \mathcal{V}_i, \quad \mathcal{V}_i = \mathcal{J}_i(\boldsymbol{\theta})\dot{\theta}_i. \tag{2.9}$$

Here $\mathcal{J}_i(\boldsymbol{\theta})$ denotes the end-link spatial velocity obtained when the joint rate in the ith joint is set at 1 rad/s and all the other joints are locked. This vector can be determined from the following geometrical relation:

$$\mathcal{J}_i(\boldsymbol{\theta}) = \begin{bmatrix} \boldsymbol{e}_i(\boldsymbol{\theta}) \times \Delta\boldsymbol{r}_i(\boldsymbol{\theta}) \\ \boldsymbol{e}_i(\boldsymbol{\theta}) \end{bmatrix}. \tag{2.10}$$

Here $\boldsymbol{e}_i(\boldsymbol{\theta})$ denotes the unit vector along the axis of rotation, $\Delta\boldsymbol{r}_i(\boldsymbol{\theta}) = \boldsymbol{r}_E(\boldsymbol{\theta}) - \boldsymbol{r}_i(\boldsymbol{\theta})$ stands for the position vector of the characteristic point on the end link w.r.t. the link frame chosen, e.g. according to the Denavit and Hartenberg notation [26]. In Fig. 2.3, an example is shown wherein the above vectors are given for the first axis of rotation ($i = 1$) of the right arm. The torso reference frame for the right arm, $\{T_R\}$, is obtained by a translation of the torso root frame $\{T\}$. Formula (2.9) can be used to calculate the spatial velocity of any link from the given branch. Note that when calculating spatial velocity $\mathcal{V}_k$ of Link k, $k < n$, all spatial velocities $\mathcal{V}_j$, $k < j < n$, are zero.

Usually, relation (2.9) is represented in a compact form as follows:

$$\mathcal{V}_n = \boldsymbol{J}(\boldsymbol{\theta})\dot{\boldsymbol{\theta}}, \tag{2.11}$$

where matrix $\boldsymbol{J}(\boldsymbol{\theta}) = \begin{bmatrix} \mathcal{J}_1 & \mathcal{J}_2 & \ldots & \mathcal{J}_n \end{bmatrix} \in \Re^{6\times n}$ denotes the *Jacobian matrix* of the limb. Eq. (2.11) represents the solution to the forward kinematics problem at the velocity level. As apparent from the equation, the solution is unique for the given branch configuration. The forward kinematic problem plays an important role in motion analysis and motion control, especially in feedback kinematic control [136].

Multi-DoF Joint Models

In motion analysis, planning, and simulation, quite often models with multi-DoF kinematic joints are employed, like spherical (S), universal (U), and the 6-DoF rigid-body (RB)

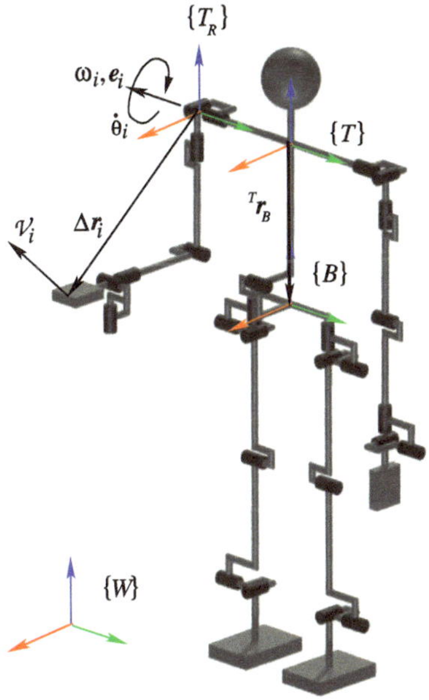

FIGURE 2.3 End-link spatial velocity $\mathcal{J}_i = [(\boldsymbol{e}_i \times \boldsymbol{r}_i)^T \;\; \boldsymbol{e}_i^T]^T$ is obtained with joint rate $\dot{\theta}_i = 1$ rad/s. Vector $\boldsymbol{e}_i$ signifies the joint axis of rotation. The position $\boldsymbol{r}_i$ of the characteristic point on the end link is determined w.r.t. reference frame $\{T_R\}$, obtained by translating the common root frame for the arms, $\{T\}$, to a suitably chosen point on the joint axis, e.g. according to the Denavit and Hartenberg notation [26].

joint connecting the base link to the ground. Then, to express the differential kinematic relations between two neighboring links, Link i and its predecessor Link $p(i)$, the following joint model is used. Denote by $\boldsymbol{\vartheta}_i \in \Re^{\eta_i}$ the joint coordinate vector, where η_i is the joint DoF number. The joint DoF is determined as $\eta_i = 6 - c_i$, where c_i is the number of constraints imposed by the joint. The body velocity of Link i is

$$ {}^i\mathcal{V} = {}^i\mathbb{X}_{p(i)} {}^{p(i)}\mathcal{V} + {}^i\mathbb{B}_m(\boldsymbol{\vartheta}_i)\dot{\boldsymbol{\vartheta}}_i. \tag{2.12}$$

Here $\mathbb{B}_m(\boldsymbol{\vartheta}_i) \in \Re^{6\times\eta_i}$ is composed of basis vectors that determine the possible motion directions across the joint. In general, the basis vectors depend on the *joint configuration* $\boldsymbol{\vartheta}_i$. When represented in Link i coordinates though, i.e. as in the above equation, $\mathbb{B}_m$ is constant. For the joints mentioned above, we have

$$ {}^i\mathbb{B}_m(\boldsymbol{\vartheta}_S) = \begin{bmatrix} 0 & 0 & 0 \\ 0 & 0 & 0 \\ 0 & 0 & 0 \\ 1 & 0 & 0 \\ 0 & 1 & 0 \\ 0 & 0 & 1 \end{bmatrix}, \; {}^i\mathbb{B}_m(\boldsymbol{\vartheta}_U) = \begin{bmatrix} 0 & 0 \\ 0 & 0 \\ 0 & 0 \\ 1 & 0 \\ 0 & 1 \\ 0 & 0 \end{bmatrix}, \; {}^i\mathbb{B}_m(\boldsymbol{\vartheta}_{RB}) = \begin{bmatrix} 1 & 0 & 0 & 0 & 0 & 0 \\ 0 & 1 & 0 & 0 & 0 & 0 \\ 0 & 0 & 1 & 0 & 0 & 0 \\ 0 & 0 & 0 & 1 & 0 & 0 \\ 0 & 0 & 0 & 0 & 1 & 0 \\ 0 & 0 & 0 & 0 & 0 & 1 \end{bmatrix}. \tag{2.13}$$

Models of other types of joints are described in [35], Table 4.1. The above notations imply that rotations in 3D are defined w.r.t. the (body-)fixed axes. Another possibility is to employ rotations w.r.t. the relative coordinate axes. In this case, the joint velocity transform $\mathbb{B}(\boldsymbol{\vartheta})$ is not constant anymore and the underlying expressions are not so simple. The joint model, however, completely matches the physical system, where equivalent spherical joint motion, for instance, is realized by compound 3R joints. Readers interested in detailed analysis of joint models are referred to [73].

Using (2.12), the end-link spatial velocity can be represented in the torso coordinate frame as ${}^T\mathcal{V} = \sum_i^n ({}^T\mathbb{X}_i {}^i\mathcal{V})$.

Parametrization of Instantaneous Rotation

For the purpose of instantaneous motion analysis it is sometimes necessary to express the end-link spatial velocity as time derivative of a physically meaningful quantity. Note that the spatial velocity $\mathcal{V}$ does not comply with this requirement since the angular velocity vector component does not have an integral. This problem is alleviated by involving a transformation that relates the angular velocity vector to the time differential of the chosen local $SO(3)$ parametrization. For example, assume a minimal parametrization, by employing the ZYX set of Euler angles, (ϕ, θ, ψ), defined w.r.t. the inertial frame $\{W\}$. Under this parametrization, the angular velocity is represented as

$$\boldsymbol{\omega} = \boldsymbol{A}_{ZYX}(\phi, \theta)\begin{bmatrix}\dot{\phi} & \dot{\theta} & \dot{\psi}\end{bmatrix}^T, \tag{2.14}$$

$$\boldsymbol{A}_{ZYX}(\phi, \theta) = \begin{bmatrix} 0 & -\sin\phi & \cos\phi\cos\theta \\ 0 & \cos\phi & \sin\phi\cos\theta \\ 1 & 0 & -\sin\theta \end{bmatrix}.$$

The 6D position of the end-link is then expressed by $\mathcal{X} \in \Re^6$ (cf. Section 2.3), while its first-order time differential assumes the form

$$\dot{\mathcal{X}} = \mathbb{X}_{ZYX}(\phi, \theta)\mathcal{V} = \mathbb{X}_{ZYX}(\phi, \theta)\boldsymbol{J}(\boldsymbol{\theta})\dot{\boldsymbol{\theta}}, \tag{2.15}$$

where $\mathbb{X}_{ZYX}(\phi, \theta) = \text{diag}\begin{bmatrix}\boldsymbol{E}_3 & \boldsymbol{A}_{ZYX}^{-1}(\phi, \theta)\end{bmatrix} \in \Re^{6\times 6}$. It is important to note that under the above parametrization, there are specific limb configurations s.t. $\det \boldsymbol{A}_{ZYX} = \cos\theta = 0$, and hence, the inverse transform $\boldsymbol{A}_{ZYX}^{-1}(\phi, \theta)$ does not exist. This is a well-known problem inherent to the minimal parametrization of $SO(3)$, via Euler angles. The problem can be alleviated by employing another type of parametrization, e.g. Euler parameters (the unit quaternion) [169]. The same problem exists for the spherical and rigid-body joint models that are based on relative rotations.

By introducing a local parametrization of $so(3)$, as above, the Jacobian matrix can be expressed as the partial derivative of the forward kinematic function, i.e.

$$\boldsymbol{J}_a(\boldsymbol{\theta}) = \frac{\partial \boldsymbol{\varphi}(\boldsymbol{\theta})}{\partial \boldsymbol{\theta}}.$$

Here $\boldsymbol{J}_a$ and $\boldsymbol{J}$ are referred to as the *analytical* and *geometrical* Jacobians, respectively. Note that $\boldsymbol{J}_a \neq \boldsymbol{J}$. For example, under the parametrization used in (2.15), $\boldsymbol{J}_a = \mathbb{X}_{ZYX}\boldsymbol{J}$. Thus, it can be expected that the use of the geometrical Jacobian in kinematic, kinetostatic, and dynamic analysis yields simpler notations. In fact, it would be more appropriate to refer to $\boldsymbol{J}$ as a joint-to-spatial velocity transform, rather than as a (geometrical) "Jacobian."

2.4.3 Inverse Differential Kinematic Relations

Kinematic control is quite often used with humanoid robots. This type of control is based on differential kinematic relations for the forward and inverse kinematics problems. The later is defined as follows: "Given the joint angles and the end-link spatial velocity, find the motion rates in the joints." In order to find a solution in a straightforward manner, the following two conditions have to be satisfied:

1. the Jacobian matrix at branch configuration $\boldsymbol{\theta}$ should be of full rank;
2. the number of joints of the branch should be equal to the DoF of the end link.

These conditions imply that the inverse of the Jacobian matrix exists. For example, the maximum DoF of each end link of the limb branches (hands and feet) is six. Therefore, according to the above conditions, $\text{rank}\boldsymbol{J}(\boldsymbol{\theta}) = 6$ and $n = 6$. For finger branches used in point-contact grasps, the conditions imply $\text{rank}\boldsymbol{J}(\boldsymbol{\theta}) = 3$ and hence, $n = 3$. Henceforth, without loss of generality, the former (general) case will be considered.

When the conditions are satisfied, solving (2.11) for the joint rates yields the following solution to the inverse kinematics problem:

$$\dot{\boldsymbol{\theta}} = \boldsymbol{J}(\boldsymbol{\theta})^{-1}\mathcal{V}. \tag{2.16}$$

A branch configuration yielding a full-rank Jacobian is called a *nonsingular configuration.* A branch with a number of joints that conforms to the second condition is called a *kinematically nonredundant* branch.

Whenever any of the above two conditions cannot be met, the inverse problem needs to be handled with care. Indeed, there are special branch configurations, $\boldsymbol{\theta}_s$, where the Jacobian matrix loses rank: $\text{rank}\boldsymbol{J}(\boldsymbol{\theta}_s) < 6$. Such configurations are called *singular.* The branch can attain a singular configuration irrespective of the number of its joints. Further on, when the branch comprises more joints than the DoF of its end link ($n > 6$), then (2.11) is underdetermined. This implies the existence of an infinite set of inverse kinematics solutions for the joint rates [147]. In this case, the branch is referred to as a *kinematically redundant* branch. Singular configurations and kinematically redundant branches will be discussed further in detail in the following subsections.

The second-order (acceleration level) differential kinematics for the forward problem are obtained through differentiating (2.11) w.r.t. time, i.e.

$$\dot{\mathcal{V}} = \boldsymbol{J}(\boldsymbol{\theta})\ddot{\boldsymbol{\theta}} + \dot{\boldsymbol{J}}(\boldsymbol{\theta})\dot{\boldsymbol{\theta}}. \tag{2.17}$$

The solution to the respective inverse problem is then represented as

$$\ddot{\boldsymbol{\theta}} = \boldsymbol{J}(\boldsymbol{\theta})^{-1}\left(\dot{\mathcal{V}} - \dot{\boldsymbol{J}}(\boldsymbol{\theta})\dot{\boldsymbol{\theta}}\right). \tag{2.18}$$

The forward and inverse differential relationships examined so far were "generic," derived w.r.t. a suitably chosen coordinate frame at the local root of the branch. Referring to Fig. 2.3, the hand spatial velocities are conveniently expressed in the common arm frame $\{T\}$ attached at the torso. We introduce now a coordinate frame-dependent notation for the first-order differential relationship (2.11). We have

$$^{T}\mathcal{V}_{H_j} = \boldsymbol{J}(\boldsymbol{\theta}_{H_j})\dot{\boldsymbol{\theta}}_{H_j}. \tag{2.19}$$

The differential relationships for the feet, on the other hand, are more conveniently represented in the base frame $\{B\}$. This is expressed as

$$^{B}\mathcal{V}_{F_j} = \boldsymbol{J}(\boldsymbol{\theta}_{F_j})\dot{\boldsymbol{\theta}}_{F_j}. \tag{2.20}$$

Further on, as already noted, for the purposes of analysis and control it is often required to express all these relationships in the same frame, e.g. in the base frame $\{B\}$ or the inertial frame $\{W\}$. Using rotational operator (2.3), it is easy to obtain the hand spatial velocities in base coordinates as

$$^{B}\mathcal{V}_{H_j} = {}^{B}\mathcal{V}_T + {}^{B}\mathbb{R}_T{}^{T}\mathcal{V}_{H_j} = {}^{B}\mathcal{V}_T + {}^{B}\mathbb{R}_T\boldsymbol{J}(\boldsymbol{\theta}_{H_j})\dot{\boldsymbol{\theta}}_{H_j}, \tag{2.21}$$

where $^{B}\mathcal{V}_T$ stands for the spatial velocity of the torso frame. This velocity is a function of the joint angles/rates in the torso joints. Likewise, the spatial velocity of the hand can be obtained in inertial coordinates as

$$^{W}\mathcal{V}_{H_j} = {}^{W}\mathcal{V}_B + {}^{W}\mathbb{R}_B{}^{B}\mathcal{V}_T + {}^{W}\mathbb{R}_T{}^{T}\mathcal{V}_{H_j} = {}^{W}\mathcal{V}_B + {}^{W}\mathbb{R}_B{}^{B}\mathcal{V}_T + {}^{W}\mathbb{R}_T\boldsymbol{J}(\boldsymbol{\theta}_{H_j})\dot{\boldsymbol{\theta}}_{H_j}, \tag{2.22}$$

where $^{W}\mathcal{V}_B$ denotes the spatial velocity of the base frame. This velocity can be derived from the leg joint angles and rates, when the robot is standing or walking. In the special cases when the humanoid robot is jumping or running, both legs loose contact with the ground. The base velocity is then determined via the IMU (gyro) sensor of the robot [54], if needed.

2.5 DIFFERENTIAL KINEMATICS AT SINGULAR CONFIGURATIONS

There are certain limb configurations where the end link looses mobility, i.e. the ability of instantaneous motion in one or more directions. Fig. 2.4A shows a robot standing upright with stretched legs and arms hanging at the sides of the body. This posture, though commonly used by humans, is usually avoided with a humanoid robot. The reason is that all four limbs are in singular configurations w.r.t. their root frames ($\{T\}$ and $\{B\}$ for arms and legs, respectively). Indeed, since the arms are fully extended, the hands cannot move in the downward direction w.r.t. the $\{T\}$ frame. Similarly, since the legs are stretched, the $\{B\}$ frame cannot be moved in the upward direction. These singular configurations of the arms/legs are called *elbow/knee singularities*, respectively. They are characterized as *unavoidable* singular configurations: there are no alternative nonsingular configurations that would place the end links at the same locations, at the workspace boundaries of each limb. As apparent from the example,

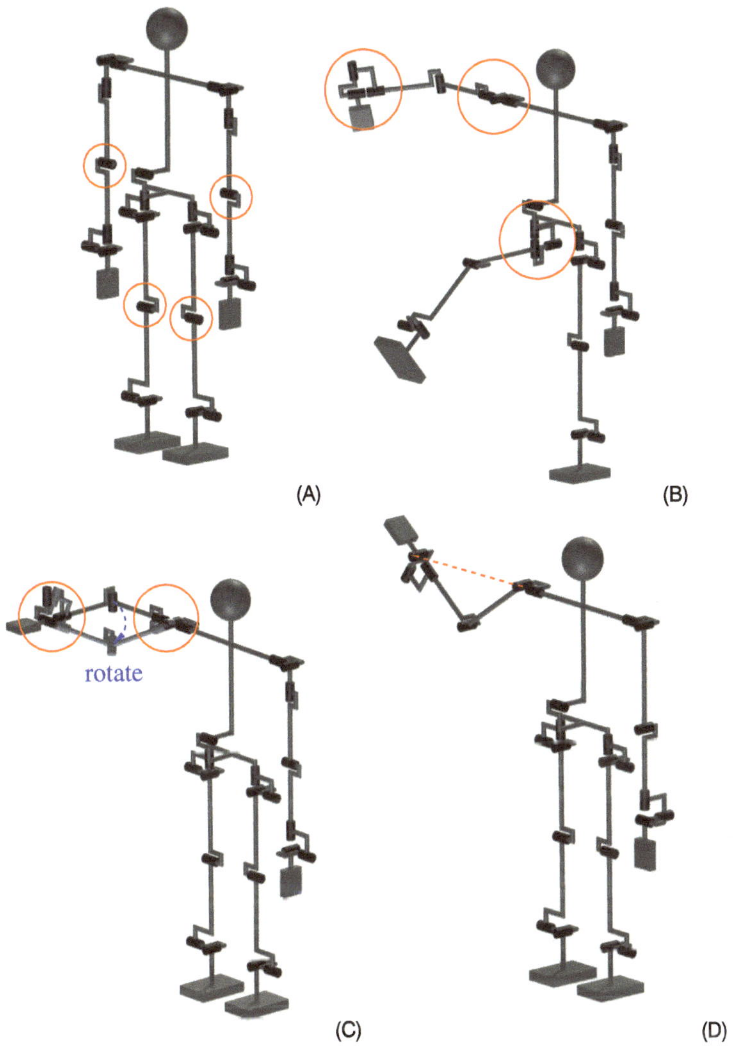

FIGURE 2.4 Singular configurations. (A) Elbow/knee singularities in both arms/legs (unavoidable). (B) Wrist/shoulder singularity in 7-DoF right arm and hip singularity in right leg. (C) Shoulder singularity in 7-DoF right arm (avoidable). (D) Shoulder singularity in 6-DoF right arm (unavoidable).

unavoidable singular configurations are inherent to both redundant and nonredundant limbs (the arms and legs, respectively).

Another singular configuration of the robot is shown in Fig. 2.4B. In addition to the elbow/knee singularities in the left arm/leg, there are singularities associated with the 3-DoF spherical joint–type subchains in the shoulder and wrist joints of the right arm and the hip joint of the right leg, respectively. As apparent from the figure, two out of the three joints axes in each of these joints are aligned, which leads to mobility loss in one direction. The singular configuration of the arm is called *wrist/shoulder singularity* [71], that for the leg *hip singularity*.

Fig. 2.4C shows another type of singular configuration for the right arm. The singularity is due to the alignment of the two axes in the shoulder joint, the elbow joint being at 90 degrees. The end link loses mobility in the translational direction of the lower-arm link. This configuration is called *shoulder singularity*. Note that the end link is placed within the workspace; it is not on the boundary. In this case, the self-motion of the arm, i.e. a motion whereby the end link is fixed (cf. Section 2.7.1), yields a transition to a nonsingular configuration, as shown in Fig. 2.4C. Such types of singular configurations are characterized as *avoidable*.

A singular configuration for a kinematically nonredundant arm is shown in Fig. 2.4D. It occurs when the wrist center is placed on the axis of the first (root) joint at the shoulder. The translational mobility of the end link is then constrained within the arm plane. This type of configuration is referred to as the *shoulder singularity*. Confusion with the shoulder singularity of the redundant arm should be avoided.

The singular configurations discussed so far are characterized by loss of end-link mobility in a single direction. As already noted, there are singular configurations when the end link can lose mobility in more than one direction. This happens, for example, when the hip and knee singularities in the leg or the wrist/shoulder and elbow singularities in the arm occur simultaneously. Other combinations, as well as other types of singular configurations do exist. Readers interested in in-depth analysis are referred to [71].

When a limb is in a singular configuration, the respective Jacobian matrix becomes rank-deficient, reflecting the loss of mobility. In Fig. 2.4A for example, all four limb Jacobians are rank-deficient; $\operatorname{rank} \boldsymbol{J}_k(\boldsymbol{\theta}_s) = 5,\ k \in \{H_r, H_l, F_r, F_l\}$. The rank is five since each limb has lost mobility in one direction. Multiple singularities, e.g. an instantaneous elbow and wrist singularity, lead to a further decrease in the rank of the Jacobian. Rank deficiency implies that the inverse kinematics solutions cannot be found with the formulas introduced in the previous subsections. The problem can be alleviated in a straightforward manner by avoiding singular configurations, e.g. by using self-motion (kinematically redundant limb) and/or by proper motion planning.

On the other hand, it is important to note that singular configurations can be useful. Indeed, humans use such configurations to resist external forces with minimal load in the joints. For example, during normal walk the support leg is almost fully stretched. This is a configuration in the vicinity of the knee singularity. With this configuration, the ground reaction force imposes a minimal load on the knee joint. Another example is shown in Fig. 2.5: when pushing a heavy object, the arms are almost fully stretched (elbow singularity), as is the pushing leg (knee singularity). Not much of the research so far has addressed the singularity issue: the stretched-knee walking problem for humanoids has been identified in [137,121] and tackled later on in [72,101,56,152,55,46,79,47]. Improved humanoid designs that can circumvent the problematic bent-knee gait have been demonstrated with WABIAN-2/LL, WABIAN-2R, HRP-4C, and other humanoid robots. The use of upper-limb singular configurations is discussed in [70,3].

There are two main approaches to deal with the singularity problem: algebraic and geometric. The algebraic approach is based on the least-squares method for obtaining an approximate solution with a lack of mobility: minimize $\|\boldsymbol{J}\dot{\boldsymbol{\theta}} - \hat{\mathcal{V}}\|^2$, where $\hat{\mathcal{V}}$ is the desired spatial velocity of the limb. At this point it is important to note that the singularity problem should

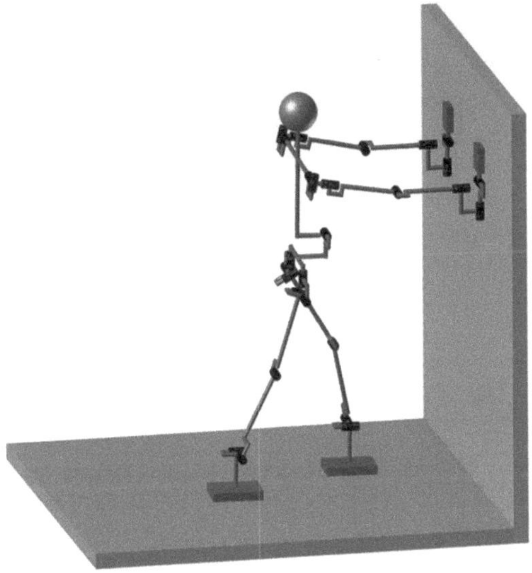

FIGURE 2.5 Pushing with almost singular configurations in the arms and the pushing leg.

not be considered only a pointwise problem. In other words, when the limb is within the vicinity of (but not exactly at) a singular configuration, the Jacobian matrix is ill-conditioned. Then, a high-norm joint velocity will be obtained as a solution to the inverse kinematics problem. This is highly undesirable and therefore, pure least-squares minimization does not work. The problem is alleviated by augmenting the objective with a "damping" term for the joint velocity norm: minimize $\|\boldsymbol{J}\dot{\boldsymbol{\theta}} - \hat{\mathcal{V}}\|^2 + \alpha^2\|\dot{\boldsymbol{\theta}}\|^2$, where α is a weighting factor for the damping. The solution is

$$\dot{\boldsymbol{\theta}} = (\boldsymbol{J}^T\boldsymbol{J} + \alpha^2\boldsymbol{E}_n)^{-1}\boldsymbol{J}^T\hat{\mathcal{V}}. \tag{2.23}$$

The method is known as the *Levenberg–Marquardt* or *Damped Least-Squares* method [78]. Its application in the field of robotics has been pioneered in [106,158]. The method has also been applied to solve the inverse kinematics problem of humanoids when their motion is constrained by multiple tasks. There are, however, a few problems associated with the damping factor α, e.g. the errors it introduces into the solution, its unintuitive nature, and the difficulty of determination. Details will be given in Section 2.8. An attempt to tackle some of these problems has been presented in [149].

Next, consider the geometric approach. The *Singularity-Consistent* (SC) method [111,114] is constraint-based, wherein the differential inverse kinematics are represented in the form of an autonomous dynamical system. The end link is required to follow the desired path (the constraint) exactly, even if it arrives at or passes through a *singular point* in the workspace. Such a point represents the spatial position of the end link obtained as the image of a singular configuration of the limb under the forward kinematics map. A path passing through a singular point is called a *singular path*. Following a singular path is possible since the magnitude of the desired spatial velocity along the path can be set appropriately to comply with the decreased

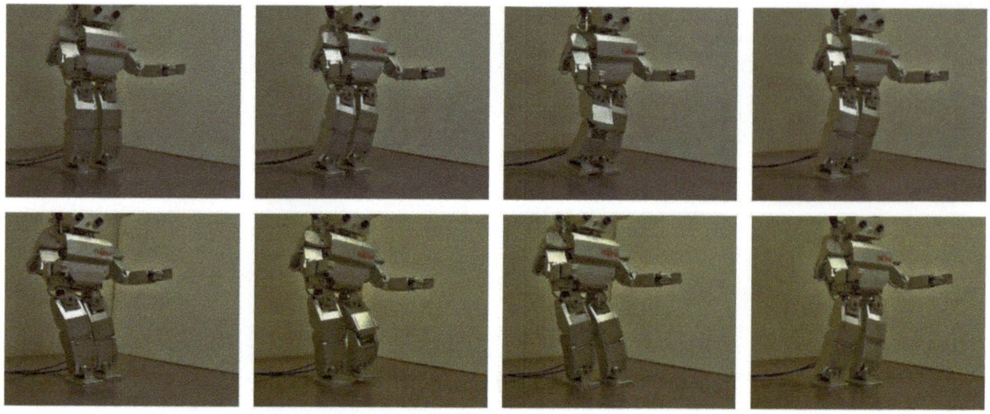

FIGURE 2.6 Implementation of the Singularity-Consistent inverse kinematics approach: stretched-knee walking [152].

mobility of the end link at the singular point. The differential kinematic relations are derived as follows. First, consider a nonredundant limb $n = 6$. The forward kinematic relation (2.11) is rewritten as

$$\boldsymbol{J}(\boldsymbol{\theta})\dot{\boldsymbol{\theta}} - \hat{\mathcal{S}}(\boldsymbol{\theta})\dot{s} = \boldsymbol{0}. \tag{2.24}$$

The rationale behind this notation is that, according to Chasles' theorem [104], the path constraint can be regarded as a virtual screw motion that closes the kinematic chain. In terms of instantaneous motion, the closure is represented by the normalized twist, $\hat{\mathcal{S}}(\boldsymbol{\theta})$. The scalar $\dot{s}$ denotes the joint rate in the virtual screw joint. It is assumed that $\dot{s}$ is unknown as are the rates in the limb joints. Hence, the above equation is underdetermined. The general solution can be written as

$$\begin{bmatrix} \dot{\boldsymbol{\theta}} \\ \dot{s} \end{bmatrix} = b\boldsymbol{n}(\boldsymbol{\theta}) = b\begin{bmatrix} \{\operatorname{adj}\boldsymbol{J}(\boldsymbol{\theta})\}\hat{\mathcal{S}}(\boldsymbol{\theta}) \\ \det \boldsymbol{J}(\boldsymbol{\theta}) \end{bmatrix}, \tag{2.25}$$

where $\boldsymbol{n}(\boldsymbol{\theta})$ denotes a vector spanning the null space of the *column-augmented Jacobian*, $\begin{bmatrix} \boldsymbol{J}(\boldsymbol{\theta})^T & -\hat{\mathcal{S}}^T \end{bmatrix}^T$. Scalar b can assume an arbitrary value to scale the joint motion rates in an appropriate way. The notation adj($\circ$) denotes the adjoint matrix. With $b = 1/\det \boldsymbol{J}(\boldsymbol{\theta})$, solution (2.16) is obtained. To arrive at a singular point on the workspace boundary along a constrained direction (unavoidable singularity, e.g. the elbow or knee singularity), set b as a constant within a suitably chosen vicinity of the singularity, e.g. determined from the maximum joint speed [113]. This implies that the end link approaches the singular point with a speed proportional to the determinant. This type of motion is referred to as a *natural motion* [114]. It is apparent that the end link's speed will be zero upon arrival at the singular point and, hence, consistent with the singularity constraint. Erect walking patterns (static walk) based on this notation are reported in [56] and [152]. Snapshots from the latter realization are shown in Fig. 2.6.

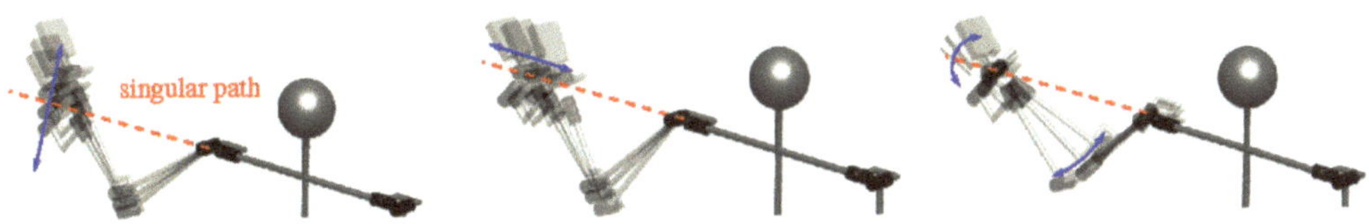

FIGURE 2.7 Implementation of the Singularity-Consistent inverse kinematics approach: three motion patterns at the shoulder singularity of a nonredundant limb. Left (A): motion through the singular path. Middle (B): motion along the singular path. Right (C): rotation around the singular path [156].

In addition to arrival at the singularity, as shown in Video 2.5-1 [98], the SC method also ensures departure from and motion through a singular point. Motion through a singular point on the workspace boundary results in reversal of motion along the path: the end link "reflects" from the boundary, with nonzero joint velocity. This implies a limb reconfiguration, as shown in Video 2.5-2 [99] (i.e. the elbow/knee joint angle changes sign).

Further on, the end-link motion direction along the desired path at the singular point may coincide with the unconstrained motion direction. From the algebraic viewpoint, this means that the linear system, though rank-deficient, will be consistent. Hence, the end link can approach, pass, and depart from the singular point with any speed. In this case, the above differential relations cannot be used since the vector field (2.25) vanishes. This problem can be alleviated via an appropriate coordinate transform: the end-link differential relationships are then expressed w.r.t. a rotating coordinate frame, one of the axes being always aligned with the singular (constrained) direction. Interested readers are referred to [153], where an implementation with a kinematically redundant limb is also explained.

At the (avoidable) shoulder singularity of a nonredundant limb, on the other hand, the SC method can be applied to control instantaneous motions resulting in three motion patterns [156]. These are obtained with the first three joints (the positioning subchain) of a 6R limb, as shown in Fig. 2.7. The dashed (red) line represents the singular path at the shoulder singularity. Whenever the end link is on that path, motion is constrained in the direction parallel to the arm-plane normal. Motion along the unconstrained directions is shown in Figs. 2.7A and B. In Fig. 2.7A, the end link moves through the singularity in the direction transverse to the singular path. In Fig. 2.7B, the end link moves in the direction parallel to the singular path, whereby the singularity will not be escaped. Note that the end link may reach the workspace boundary along the singular path, thus arriving at a double shoulder-elbow singular configuration. Finally, Fig. 2.7C displays a self-motion pattern resulting from a commanded end-link velocity having a nonzero component in the constrained direction (parallel to the arm-plane normal). The joint at the shoulder base rotates the limb and hence the arm plane, until the commanded end-link velocity component along the constrained direction is nullified. Thereafter, the end link will leave the singularity along the unconstrained directions.

As already mentioned, the inverse kinematics solution (2.25) is in the form of an autonomous dynamical system. In [51], it was shown that the nonlinear dynamical system approach can also be used to create attractors for the learning process of a humanoid robot, the so-called "dynamic movement primitives" (DMPs). Recently, DMPs have been implemented in a number of motion generation algorithms.

2.6 MANIPULABILITY ELLIPSOID

From the differential kinematics relation (2.11), it is apparent that the ability of the end link to move instantaneously along a given spatial (rigid-body motion) direction will depend on the current limb configuration. In particular, as already clarified, at a singular configuration the ability to move along the singular directions becomes zero, and hence, mobility is lost in these directions. To facilitate instantaneous motion analysis and control, it is quite desirable to quantify the mobility in a given direction, at any given configuration. This can be done via *Singular-Value Decomposition* (SVD) [42,147,90] of the Jacobian matrix. For the general case of an n-DoF kinematically redundant limb, we have

$$\boldsymbol{J}(\boldsymbol{\theta}) = \boldsymbol{U}(\boldsymbol{\theta})\boldsymbol{\Sigma}(\boldsymbol{\theta})\boldsymbol{V}(\boldsymbol{\theta})^T, \tag{2.26}$$

where $\boldsymbol{U}(\boldsymbol{\theta}) \in \Re^{6\times 6}$ and $\boldsymbol{V}(\boldsymbol{\theta}) \in \Re^{n\times n}$ are orthonormal matrices and

$$\boldsymbol{\Sigma}(\boldsymbol{\theta}) = \left[\operatorname{diag}\{\sigma_1(\boldsymbol{\theta}),\ \sigma_2(\boldsymbol{\theta}), ..., \sigma_6(\boldsymbol{\theta})\} \ \middle|\ \boldsymbol{0}\right] \in \Re^{6\times n}. \tag{2.27}$$

Here $\sigma_1 \geq \sigma_2 \geq, ..., \geq \sigma_6 \geq 0$ are the singular values of the Jacobian. The columns of matrix $\boldsymbol{U}(\boldsymbol{\theta}), \boldsymbol{u}_i,\ i = 1, ..., 6$, provide a basis for the instantaneous motion space of the end link at the given limb configuration. At a nonsingular limb configuration, all singular values are positive. At a singular configuration of corank $6-\rho$ $(\rho = \operatorname{rank}\boldsymbol{J})$, $6-\rho$ of the singular values become zeros, i.e. $\sigma_1 \geq \sigma_2 \geq, ..., \geq \sigma_\rho > 0,\ \sigma_{\rho+1} = ... = \sigma_6 = 0$. The singular value σ_i quantifies the instantaneous mobility of the end link along the instantaneous motion direction $\boldsymbol{u}_i$. Assuming that the magnitude of the joint rate vector is limited at each limb configuration as $\|\dot{\boldsymbol{\theta}}\| \leq 1$, the highest mobility is along the direction corresponding to the maximum singular value. At a singular configuration of corank 1, $\sigma_{min} = 0$ and the respective direction $\boldsymbol{u}_{min}$ becomes a singular direction. Vectors $\sigma_i \boldsymbol{u}_i$ constitute the principal axis of an ellipsoid—a useful graphic tool for visualizing the instantaneous mobility along each possible motion direction. The dimension of the ellipsoid is determined by the rank of the Jacobian. Fig. 2.8 shows a robot configuration wherein the right arm is at a nonsingular configuration, whereas the left one is at the elbow singularity. The two ellipsoids at the end links visualize the instantaneous translational motion abilities. The ellipsoid for the right arm is 3D (full translational mobility), while that for the left arm is flat (an ellipse). The ellipse lies in a plane parallel to the floor since translational mobility in the vertical direction is nil at the singularity. The ellipsoid-based instantaneous mobility analysis has been introduced in [166]; the ellipsoid is referred to as the *manipulability ellipsoid*.

2.7 KINEMATIC REDUNDANCY

Whenever the limb mobility, determined by the number of limb joints n, exceeds the DoFs of the end link (six), the limb is characterized as kinematically redundant. Some humanoid robots are equipped with kinematically redundant, 7-DoF arms [120,161,57,127,173]. Such robots can control the position of their elbows without affecting thereby the instantaneous

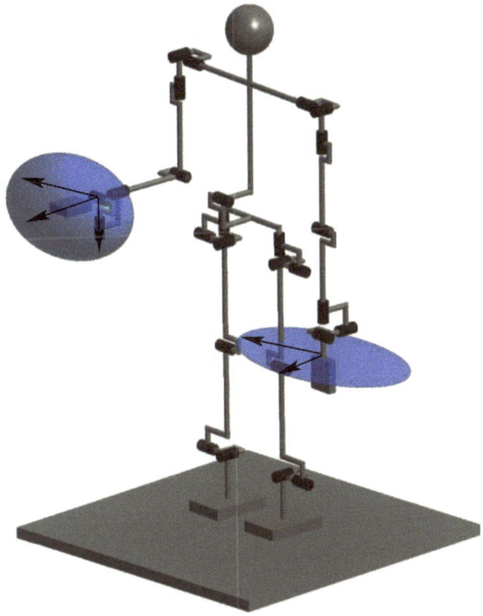

FIGURE 2.8 Manipulability ellipsoid for translational motion. The right arm is in a nonsingular configuration and the respective ellipsoid is 3D, with principal axes $\sigma_1 \boldsymbol{u}_1$, $\sigma_2 \boldsymbol{u}_2$, and $\sigma_3 \boldsymbol{u}_3$. The left arm is at a singular configuration: the downward translational mobility has been lost, and therefore, the manipulability ellipsoid is only 2D. The principal axes are $\sigma_1 \boldsymbol{u}_1$ and $\sigma_2 \boldsymbol{u}_2$.

motion of the hands. Thus, they attain the capability to perform tasks in clustered environments avoiding collisions with their elbows, similar to humans. Also, there are humanoids that comprise 7-DoF legs. With proper control, their gait appears more human-like than that of robots with 6-DoF legs [121,19]. The difference $r = n - 6$ is referred to as the *degree of redundancy* (DoR).

2.7.1 Self-Motion

In contrast to a nonredundant limb, a kinematically redundant limb can move even when its end link is immobilized ($\mathcal{V} = \mathbf{0}$). Such motion is shown in Fig. 2.9 for the arm; the hand remains fixed w.r.t. the arm root frame while the elbow rotates around the line connecting the shoulder and wrist joints. Such type of motion is known as *self-motion*, *internal motion*, or *null motion*.

Self-motion is generated by the joint velocity obtained from the following homogeneous differential relation:

$$\boldsymbol{J}(\boldsymbol{\theta})\dot{\boldsymbol{\theta}} = \mathbf{0}, \quad \dot{\boldsymbol{\theta}} \neq \mathbf{0}. \tag{2.28}$$

Since $n > 6$, the Jacobian is nonsquare ($6 \times n$) and the above equation is characterized as an underdetermined linear system. Hence, there is an infinite set of solutions, each nontrivial

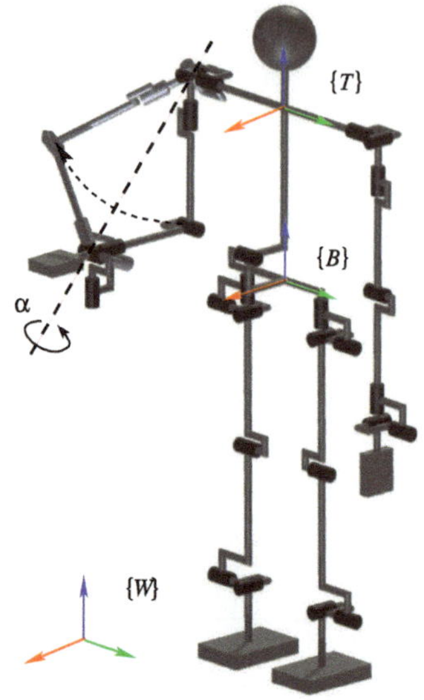

FIGURE 2.9 The self-motion of the arm is shown as a rotation of the arm plane, determined by the upper/lower arm links, around the line connecting the shoulder and wrist joints. The rotation angle α can be associated with parameter b_v in (2.35).

solution representing a self-motion joint velocity:

$$\{\dot{\boldsymbol{\theta}}_h : \dot{\boldsymbol{\theta}} = \boldsymbol{N}(\boldsymbol{J}(\boldsymbol{\theta}))\dot{\boldsymbol{\theta}}_a,\ \forall\dot{\boldsymbol{\theta}}_a\}. \tag{2.29}$$

Matrix $\boldsymbol{N}(\boldsymbol{J}(\boldsymbol{\theta})) \in \Re^{n\times n}$, a projection matrix onto the null space of the limb Jacobian, $\mathcal{N}(\boldsymbol{J}(\boldsymbol{\theta}))$, can be expressed as

$$\boldsymbol{N}(\boldsymbol{J}(\boldsymbol{\theta})) \equiv \left(\boldsymbol{E}_n - \boldsymbol{J}^{\#}(\boldsymbol{\theta})\boldsymbol{J}(\boldsymbol{\theta})\right). \tag{2.30}$$

Matrix $\boldsymbol{J}^{\#} \in \Re^{n\times 6}$ is a generalized inverse of the Jacobian, i.e. a matrix such that $\boldsymbol{J}\boldsymbol{J}^{\#}\boldsymbol{J} = \boldsymbol{J}$. The properties of the generalized inverses are discussed in [14,105]. It is important to note that for a given matrix $\boldsymbol{J}$, there is an infinite number of generalized inverses. Note also that $\boldsymbol{N}(\boldsymbol{J}(\boldsymbol{\theta}))$ has special properties. They are:

- a singular matrix of rank r (at a nonsingular configuration);
- a symmetric matrix $\boldsymbol{N}^T(\boldsymbol{J}(\boldsymbol{\theta})) = \boldsymbol{N}(\boldsymbol{J}(\boldsymbol{\theta}))$;
- an idempotent matrix $\boldsymbol{N}^2(\boldsymbol{J}(\boldsymbol{\theta})) = \boldsymbol{N}(\boldsymbol{J}(\boldsymbol{\theta}))$.

Further on, vector $\dot{\boldsymbol{\theta}}_a$ in (2.29) is an arbitrary joint velocity vector that parametrizes the null space. Because of the rank deficiency of $\boldsymbol{N}$, this parametrization is referred to as the *nonminimal*. A minimal parametrization can be obtained from the SVD of the Jacobian, (2.26), as

$$\dot{\boldsymbol{\theta}}_h = \boldsymbol{V}_r(\boldsymbol{\theta})\boldsymbol{b}. \tag{2.31}$$

Here $\boldsymbol{V}_r(\boldsymbol{\theta}) \in \Re^{n \times r}$ is formed by extracting the last r rows from the orthonormal matrix $\boldsymbol{V}^T(\boldsymbol{\theta})$ in the SVD formula; $\boldsymbol{b} \in \Re^r$ is an arbitrary vector parameter with the dimension of (truncated) joint velocity. The columns of $\boldsymbol{V}_r$ span the null space. Hence, the following alternative representation of the null-space projector can be obtained [69]:

$$\boldsymbol{N}(\boldsymbol{J}(\boldsymbol{\theta})) = \boldsymbol{V}_r \boldsymbol{V}_r^T. \tag{2.32}$$

In the case of a single DoR, $r = 1$, the minimal parametrization assumes the simple form

$$\dot{\boldsymbol{\theta}}_n = b\boldsymbol{n}(\boldsymbol{\theta}), \tag{2.33}$$

where b is an arbitrary scalar with the dimension of joint rate. Vector $\boldsymbol{n} = \begin{bmatrix} n_1 & n_2 & \dots & n_n \end{bmatrix}^T$ is nondimensional; it is the sole nonzero vector spanning the null space. Its components can usually be obtained in closed form, or calculated from cofactors [7,13]: $n_i = (-1)^{i+1} \det \boldsymbol{J}_i$, where $\boldsymbol{J}_i$, $i = 1, 2, .., n$, is derived from the Jacobian matrix by removing the ith column.

It should be noted that a continuous self-motion, $\dot{\boldsymbol{\theta}}_h(t)$, signifies a motion along the integral curves of (2.31). The integral curve for a particular end-link spatial position is defined by $\boldsymbol{f}(\boldsymbol{\theta}) = \text{const}$. The curve will be henceforth referred to as the *self-motion manifold* [18]. The dimension of this manifold equals the rank of the null-space projector $\boldsymbol{N}$ (or $\boldsymbol{V}_r$). The curve can be closed or open, depending on the specified 6D position of the end link [18].

2.7.2 General Solution to the Inverse Kinematics Problem

In the case of a kinematically redundant manipulator, the instantaneous motion relation (2.11) becomes underdetermined. The general solution of an underdetermined linear system can be represented as the sum of a particular solution, $\dot{\boldsymbol{\theta}}_p$, and a solution to the homogeneous equation (2.28), i.e. a self-motion joint velocity $\dot{\boldsymbol{\theta}}_h$ [14]. We have

$$\dot{\boldsymbol{\theta}} = \boldsymbol{J}(\boldsymbol{\theta})^{\#}\mathcal{V} + \left(\boldsymbol{E}_n - \boldsymbol{J}(\boldsymbol{\theta})^{\#}\boldsymbol{J}(\boldsymbol{\theta})\right)\dot{\boldsymbol{\theta}}_a. \tag{2.34}$$

Apparently, the particular solution $\dot{\boldsymbol{\theta}}_p = \boldsymbol{J}(\boldsymbol{\theta})^{\#}\mathcal{V}$ is parametrized with the choice of a specific generalized inverse. The choice of $\dot{\boldsymbol{\theta}}_a$, on the other hand, parametrizes the infinite set of solutions. In the case of a single DoR, the above general solution can be rewritten, with the help of (2.33), as

$$\dot{\boldsymbol{\theta}} = \boldsymbol{J}(\boldsymbol{\theta})^{+}\mathcal{V} + b\boldsymbol{n}(\boldsymbol{\theta}). \tag{2.35}$$

Notation $(\circ)^+$ stands for the Moore–Penrose generalized inverse [14]. This generalized inverse is quite often preferred instead of other generalized inverses because it endows the IK solution for the joint velocity with two desirable properties: minimum norm ($b = 0 \rightarrow \min \|\dot{\boldsymbol{\theta}}\|_2^2$) and orthogonality of the two solution components ($b \neq 0 \rightarrow \dot{\boldsymbol{\theta}}_p \perp \dot{\boldsymbol{\theta}}_h$). The Moore–Penrose generalized inverse is often referred to as the *pseudoinverse*. At a nonsingular configuration, where $\boldsymbol{J}(\boldsymbol{\theta})$ has a full row rank, the (right) pseudoinverse is used. It is expressed with the following formula:

$$\boldsymbol{J}^+ = \boldsymbol{J}^T(\boldsymbol{J}\boldsymbol{J}^T)^{-1}. \tag{2.36}$$

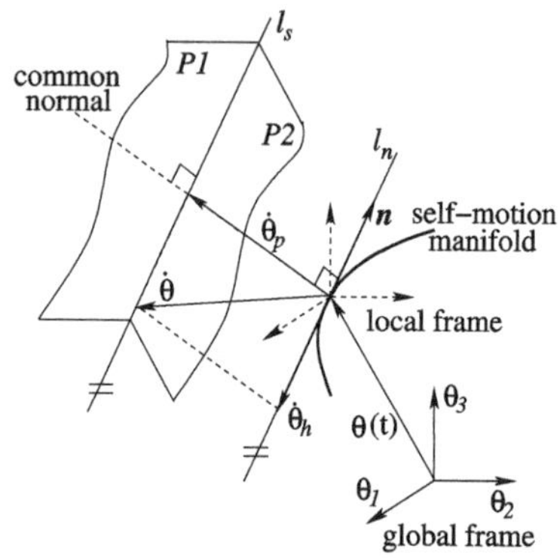

FIGURE 2.10 Geometrical representation of local decomposition of the joint space of a planar 3R redundant manipulator.

The pseudoinverse determines quite a specific behavior of the limb in the joint space. First, note that the induced minimum-norm constraint is of a *nonholonomic* nature, rendering the pseudoinverse-based particular solution nonintegrable. As a consequence, a desired cyclic end-link motion encoded with spatial velocity commands may yield a noncyclic path in the joint space, characterized by a drift [67,85,24]. Second, note that motion can be easily destabilized in the vicinity of a kinematic singularity. This is clearly seen from the following representation of the pseudoinverse, obtained via SVD of the Jacobian (cf. (2.26)):

$$\boldsymbol{J}^{+} = \boldsymbol{V}\boldsymbol{\Sigma}^{+}\boldsymbol{U}^{T}, \tag{2.37}$$

where $\boldsymbol{\Sigma}^{+} = \left[\text{diag}\{\frac{1}{\sigma_1}, \frac{1}{\sigma_2}, \dots, \frac{1}{\sigma_6}\} \;\middle|\; \boldsymbol{0}^{T}\right]^{T}$. When a singular value σ_i approaches zero, the respective component of the pseudoinverse, and hence of the inverse kinematics solution, tends to infinity, which causes the instability.

Finally, it is important to note that the Jacobian matrix induces a local decomposition in the joint space, whereby the normal and tangential subspaces at the self-motion manifold determine minimum-norm motion and self-motion, respectively, i.e.

$$\dot{\boldsymbol{\theta}} = \boldsymbol{J}(\boldsymbol{\theta})^{+}\boldsymbol{J}(\boldsymbol{\theta})\dot{\boldsymbol{\theta}}_a + \left(\boldsymbol{E}_n - \boldsymbol{J}(\boldsymbol{\theta})^{+}\boldsymbol{J}(\boldsymbol{\theta})\right)\dot{\boldsymbol{\theta}}_a. \tag{2.38}$$

Consider as an example a planar 3R redundant manipulator, its end tip following a 2D path in workspace. The end-tip velocity $\mathcal{V}$ is along the tangent at the current point on this path. The respective local decomposition of the joint space is presented graphically in Fig. 2.10. Each of the two equations in the forward kinematics relation $\boldsymbol{J}\dot{\boldsymbol{\theta}} = \mathcal{V}$ determines a plane in 3D joint space (planes $P1$ and $P2$). The infinite set of solutions to the inverse kinematics problem is determined by the intersection of the two planes, i.e. line l_s in the figure. Since the DoR is one, (2.35) holds. In the figure, $\dot{\boldsymbol{\theta}}_p = \boldsymbol{J}^{+}\mathcal{V}$ denotes the minimum-norm particular solution, $\dot{\boldsymbol{\theta}}_h = b\boldsymbol{n}$ describes the infinite set of solutions to homogeneous equation $\boldsymbol{J}\dot{\boldsymbol{\theta}} = \boldsymbol{0}$. This set is represented by line l_n. Note that l_n and l_s are parallel. The minimum-norm character of the particular solution $\dot{\boldsymbol{\theta}}_p$ is apparent from its orthogonality w.r.t. the two lines.

The second-order (acceleration level) differential kinematics for the inverse problem can be obtained as the general solution of the underdetermined system (2.17). Making use again either of the nonminimal or the minimal null-space parametrization, two solutions are obtained, respectively, as

$$\ddot{\boldsymbol{\theta}} = \boldsymbol{J}(\boldsymbol{\theta})^{\#}\left(\dot{\mathcal{V}} - \dot{\boldsymbol{J}}(\boldsymbol{\theta})\dot{\boldsymbol{\theta}}\right) + \left(\boldsymbol{E}_n - \boldsymbol{J}(\boldsymbol{\theta})^{\#}\boldsymbol{J}(\boldsymbol{\theta})\right)\ddot{\boldsymbol{\theta}}_a \tag{2.39}$$

and

$$\ddot{\boldsymbol{\theta}} = \boldsymbol{J}(\boldsymbol{\theta})^{+}\left(\dot{\mathcal{V}} - \dot{\boldsymbol{J}}(\boldsymbol{\theta})\dot{\boldsymbol{\theta}}\right) + b\boldsymbol{n}(\boldsymbol{\theta}). \tag{2.40}$$

Here $\ddot{\boldsymbol{\theta}}_a$ denotes an arbitrary joint acceleration contributing to the self-motion of the limb. Similarly, the arbitrary scalar b has the physical meaning of joint acceleration magnitude that determines the self-motion. Note, however, that in contrast to the velocity-level solution, a zero input for any of these quantities would not terminate the self-motion. This problem can be alleviated by a careful choice of $\ddot{\boldsymbol{\theta}}_a$ or b, such that the respective inverse kinematics solutions are integrable. Further details will be presented in Section 2.11.3. Alternatively, a joint damping term can be added to damp out the conserved self-motion velocity.

Finally, it is worth noting that the inverse kinematics solution for a 7-DoF limb can be obtained in closed form, as an alternative to the above instantaneous-motion solutions [6, 142,84]. Closed-form solutions may be preferable in some cases since they provide a better accuracy.

2.7.3 Weighted Generalized Inverse

Besides the pseudoinverse, there is another type of generalized inverse that is frequently used in inverse kinematics relations: the *weighted generalized inverse*. In the pioneering work on kinematically redundant manipulators [159], the following generalized inverse was employed:

$$\boldsymbol{J}^{-W} = \boldsymbol{W}^{-1}\boldsymbol{J}^T(\boldsymbol{J}\boldsymbol{W}^{-1}\boldsymbol{J}^T)^{-1}, \tag{2.41}$$

where $\boldsymbol{W} \in \Re^{n\times n}$ is a positive definite weighting matrix. This generalized inverse minimizes locally the weighted norm $\|\dot{\boldsymbol{\theta}}\|_W = \sqrt{\dot{\boldsymbol{\theta}}^T\boldsymbol{W}\dot{\boldsymbol{\theta}}} = \|\boldsymbol{W}^{\frac{1}{2}}\dot{\boldsymbol{\theta}}\|_2$. For example, with a proper constant diagonal $\boldsymbol{W}$, the instantaneous motion in the "heavy-duty" joints can be suppressed w.r.t. that in the rest. This method can also be adopted w.r.t. the end-link velocity components, by minimizing $\|\boldsymbol{v}\|_B$. This implies a variable weighting matrix, $\boldsymbol{W}(\boldsymbol{\theta}) = \boldsymbol{J}_v^T(\boldsymbol{\theta})\boldsymbol{B}\boldsymbol{J}_v(\boldsymbol{\theta})$, matrix $\boldsymbol{J}_v(\boldsymbol{\theta})$ denoting the Jacobian for the end-link velocity. Another example of a variable weighting matrix is $\boldsymbol{W}(\boldsymbol{\theta}) = \boldsymbol{M}(\boldsymbol{\theta})$, where $\boldsymbol{M}(\boldsymbol{\theta})$ is the link inertia matrix. In this case, the weighted generalized inverse is denoted as $\boldsymbol{J}^{-M(\boldsymbol{\theta})}$, since $\boldsymbol{W} = \boldsymbol{M}(\boldsymbol{\theta})$. This generalized inverse plays an important role in inverse dynamics. Details will be discussed in Chapter 4. It should also be apparent that the weighted generalized inverse is a generalization of pseudoinverse (2.36), that is, $\boldsymbol{J}^{+} = \boldsymbol{J}^{-E}$.

2.7.4 Redundancy Resolution via Gradient Projection

With the help of self-motion, it is possible to realize tasks subordinated to the main end-link motion task. Such tasks are referred to as *additional tasks* or *subtasks*. They impose a motion constraint in the joint space that determines the self-motion. Often used additional tasks for a limb are those of the avoidance type; the limb has to avoid joint limits, singular configurations, and collisions with an external obstacle or with other links of the robot (self-collision) [109,143,105].

Joint-Limit Avoidance Subtask

Joint-limit avoidance was tackled in the pioneering work [80]. The method is based on local minimization of the following performance criterion:

$$h(\boldsymbol{\theta}) = \frac{1}{n}\sum_{i=1}^{n}\left(\frac{\theta_i - \theta_{mid_i}}{\theta_{mid_i} - \theta_{max_i}}\right)^2, \tag{2.42}$$

where $\theta_{mid_i} = (\theta_{max_i} + \theta_{min_i})/2$, θ_{max_i} and θ_{min_i} denote the maximum and minimum angular limits in joint i. The minimization is achieved by selecting the arbitrary vector in (2.34) as

$$\dot{\boldsymbol{\theta}}_a = -\beta\frac{\partial h(\boldsymbol{\theta})}{\partial \boldsymbol{\theta}}, \tag{2.43}$$

where β is a positive scalar determining the speed of optimization. This method is known as the *gradient projection method.*

Singularity Avoidance Subtask via the Manipulability Measure

Assume a nonsingular limb configuration, $\sigma_i \neq 0, i \in \{1, 2, ..., 6\}$, and denote the product of the singular values as

$$w(\boldsymbol{\theta}) \equiv \sigma_1 \cdot \sigma_2 \cdot, ..., \cdot\sigma_6 = \sqrt{\det(\boldsymbol{J}\boldsymbol{J}^T)}. \tag{2.44}$$

When approaching a singular point in the workspace, scalar $w(\boldsymbol{\theta})$ decreases gradually to become zero at the singularity. In [166], it was suggested to use this scalar as a nondirectional "measure of distance" to the singularity. The maximization of this measure via self-motion leads to the avoidance of (avoidable) singular configurations. As in the above joint-limit avoidance subtask, the maximization is achieved via gradient projection, i.e.

$$\dot{\boldsymbol{\theta}}_a = \beta\frac{\partial w(\boldsymbol{\theta})}{\partial \boldsymbol{\theta}}, \tag{2.45}$$

where scalar $\beta > 0$ determines again the speed of optimization. This scalar can also be used to suppress excessive joint rates, when needed. Further details can be found in [168]. The scalar function $w(\boldsymbol{\theta})$ is referred to as the *manipulability measure* [166].

When using the manipulability as a "distance" measure to singular points, a problem arises due to its nonlinear nature. Appropriate alternatives are the minimum singular value

$\sigma_{min}(\boldsymbol{\theta})$ or the *condition number* $\kappa(\boldsymbol{\theta}) = \sigma_{max}(\boldsymbol{\theta})/\sigma_{min}(\boldsymbol{\theta})$ [68]. Some authors have proposed vector norm–based measures, which however might lead to problems related to the nonuniform dimensions of the two vector components of the spatial end-link velocity [30].

2.7.5 Redundancy Resolution via the Extended Jacobian Technique

The gradient projection method introduced above is based on the nonminimal parametrization of self-motion. An alternative approach is proposed in [8]. It makes use of the fact that the gradient projection term

$$N(\boldsymbol{\theta})\frac{\partial g(\boldsymbol{\theta})}{\partial \boldsymbol{\theta}}$$

that optimizes locally a given scalar function (additional task) $g(\boldsymbol{\theta})$ via self-motion determines $r = n - 6$ independent constraints at a nonsingular configuration. These constraints can be expressed via an r-dimensional vector function $\boldsymbol{g}(\boldsymbol{\theta})$. The relation $\boldsymbol{g}(\boldsymbol{\theta}) = \boldsymbol{0}$ implies that when $g(\boldsymbol{\theta})$ is at its optimum under the forward kinematics constraint, there will be no self-motion.

The additional constraints are represented in terms of instantaneous motion as

$$\boldsymbol{J}_g(\boldsymbol{\theta})\dot{\boldsymbol{\theta}} = \boldsymbol{0},$$

where $\boldsymbol{J}_g(\boldsymbol{\theta}) = \partial \boldsymbol{g}(\boldsymbol{\theta})/\partial \boldsymbol{\theta} \in \Re^{r \times n}$ is the additional task Jacobian. Adjoining the above additional task constraints to the instantaneous forward kinematics relations yields

$$\begin{bmatrix} \boldsymbol{J}(\boldsymbol{\theta}) \\ \boldsymbol{J}_g(\boldsymbol{\theta}) \end{bmatrix} \dot{\boldsymbol{\theta}} = \begin{bmatrix} \mathcal{V} \\ \boldsymbol{0} \end{bmatrix}. \tag{2.46}$$

Matrix $\boldsymbol{J}_e(\boldsymbol{\theta}) \equiv \begin{bmatrix} \boldsymbol{J}^T(\boldsymbol{\theta}) & \boldsymbol{J}_g^T(\boldsymbol{\theta}) \end{bmatrix}^T \in \Re^{n \times n}$ is called the *extended Jacobian matrix*. Its inverse can be used to find the unique solution to the instantaneous inverse kinematics problem, provided $\boldsymbol{J}_e(\boldsymbol{\theta})$ is nonsingular.

Further on, the extended Jacobian approach can be generalized by assuming a time-varying additional task, i.e. $\boldsymbol{h}(\boldsymbol{\theta}) = \boldsymbol{\gamma}(t) \in \Re^r$ [141]. In terms of instantaneous motion, the task is

$$\boldsymbol{J}_h(\boldsymbol{\theta})\dot{\boldsymbol{\theta}} = \dot{\boldsymbol{\gamma}},$$

where $\boldsymbol{J}_h(\boldsymbol{\theta}) = \partial \boldsymbol{h}(\boldsymbol{\theta})/\partial \boldsymbol{\theta} \in \Re^{r \times n}$ is the additional task Jacobian. The extended Jacobian relation becomes

$$\boldsymbol{J}_e(\boldsymbol{\theta})\dot{\boldsymbol{\theta}} = \begin{bmatrix} \mathcal{V} \\ \dot{\boldsymbol{\gamma}} \end{bmatrix}, \tag{2.47}$$

where $\boldsymbol{J}_e(\boldsymbol{\theta}) \equiv \begin{bmatrix} \boldsymbol{J}^T(\boldsymbol{\theta}) & \boldsymbol{J}_h^T(\boldsymbol{\theta}) \end{bmatrix}^T \in \Re^{n \times n}$ is the extended Jacobian (referred to as the "augmented Jacobian matrix" in [141]).

The drawback of the extended Jacobian approach is the introduction of *algorithmic singularities*. Matrix $\boldsymbol{J}_e(\boldsymbol{\theta})$ can become singular while the manipulator Jacobian $\boldsymbol{J}(\boldsymbol{\theta})$ and the additional task Jacobian $\boldsymbol{J}_h(\boldsymbol{\theta})$ are full row rank. The problem can be alleviated via *kinematic decoupling* [129]. The components of the vector on the r.h.s. of (2.47) are said to be kinematically decoupled, if the following relations hold:

$$J_h(\theta)J_e^{-1}(\theta)\begin{bmatrix}\mathcal{V}\\ 0\end{bmatrix}=0, \forall\mathcal{V}$$

and

$$J(\theta)J_e^{-1}(\theta)\begin{bmatrix}0\\ \dot{y}\end{bmatrix}=0, \forall\dot{y}.$$

These relations are valid when the Jacobian of the additional constraint is in the form $J_h(\theta) = Z(\theta)W(\theta)$. Matrix $Z(\theta) \in \Re^{r\times n}$ is a full row-rank matrix providing a minimal set of basis vectors for null space $\mathcal{N}(J)$, i.e. $JZ^T = 0$; $W(\theta)$ is a positive-definite weighting matrix. With these definitions, the "asymmetric" weighted generalized inverse

$$Z_W^{\#}(\theta) = Z^T(\theta)\left(Z(\theta)W(\theta)Z^T(\theta)\right)^{-1}$$

will represent a minimum-parametrization projector onto the null space of the Jacobian, i.e.

$$\{\dot{\theta} : \dot{\theta}_h = Z_W^{\#}(\theta)b\}.$$

Vector $b = Z(\theta)W(\theta)\dot{\theta}_a \in \Re^r, \forall\dot{\theta}_a \in \Re^n$. The infinite set of inverse kinematics solutions is determined by

$$\dot{\theta} = \left[J^{-W}(\theta) \quad Z_W^{\#}(\theta)\right]\begin{bmatrix}\mathcal{V}\\ b\end{bmatrix}. \tag{2.48}$$

The two components of the solution are kinematically decoupled and free of algorithmic singularities, for any $W(\theta)$ [122]. These important properties are quite useful not only in inverse kinematics but also in inverse dynamics and passivity control. The method is referred to as the *kinematically decoupled joint-space decomposition* (KD-JSD) method.

2.8 INVERSE KINEMATICS SOLUTION UNDER MULTIPLE TASK CONSTRAINTS

Single-limb differential kinematic relations play an important role for tasks requiring a predetermined posture of the robot. On the other hand, there is a class of tasks wherein the posture of the humanoid is allowed to vary, i.e. a *whole-body* motion is admissible and also desirable. One such task is arm reach, quite often used as a benchmark task; see e.g. [6,59, 81,16]. Thereby, the whole-body posture variation is employed to enlarge the workspace of the arm. With this task, as well as with other similar motion tasks, the motion of the hand(s) is specified w.r.t. the inertial frame $\{W\}$. This means that motions in the joints within the entire kinematic chain will contribute to the hand motion. In the example in Fig. 2.9, the total number of joints becomes $n_{total} = n_{leg} + n_{torso} + n_{arm} = 6 + 1 + 7 = 17$. Thus, the degree of *kinematic* redundancy is $17 - 6 = 11$. With such a high degree of redundancy, it is possible to realize multiple "additional" tasks.

2.8.1 Motion-Task Constraints

Indeed, besides the hand motion tasks, the robot has to simultaneously perform a number of other motion tasks, e.g. to keep balance while avoiding obstacles, self-collisions, singularities, and joint limits and to follow a moving object visually: the so-called *gaze* task. All these tasks impose constraints on the motion, referred to as the *motion-task constraints*. These constraints are helpful in resolving the kinematic redundancy. Imposing multiple task constraints should be done with great care since the resulting motion may lead to an *overconstrained* state, where no solution to the inverse kinematics problem can be found. Such states are sometimes referred to as a *task conflict*; they occur in fact quite often. Therefore, some means of resolving task conflicts should be embedded into the inverse kinematics solver.

It should be noted that besides motion task constraints, there are also constraints expressed in terms of forces and moments. Such constraints are represented in a subspace dual to that of the motion-task constraints. Force-type constraints may stem from a specific force task (e.g. a joint torque minimization task) or they may stem from the physical contacts occurring when the robot links interfere with objects from the environment. Contact and static force task constraints will be discussed in Sections 2.9 and 3.4, respectively. Force-task constraints within dynamic models, on the other hand, will be treated in Chapter 4.

The problem of determining the most appropriate joint motions under a given set of motion task constraints is nontrivial. A possible approach is to formulate the inverse kinematics problem as a multiobjective optimization problem and to take advantage of readily available numerical optimization methods, e.g. quadratic programming (QP) or differential dynamic programming [38]. An alternative approach is to use an inverse kinematics solution wherein the task-motion constraints are handled within a hierarchical structure derived with the help of null-space projections. This approach was devised for redundant manipulators over three decades ago. The aim was to avoid optimization via general solvers, which was quite time consuming at that time due to the iterative nature of the solver. Recently, it was shown that such solvers can provide optimized solutions in real-time [58]. Both approaches are now under development for use with humanoid robots, on a competitive basis. Proponents of numerical optimization methods refer to the difficulty of incorporating inequality constraints (stemming e.g. from unilateral constraints) with null-space projection methods. Proponents of null-space projection methods, on the other hand, argue that under numerical optimization it is difficult to guarantee control stability. A detailed discussion about these two alternative approaches will be presented in the two following sections.

When tackling motion-task constraint–based inverse kinematics optimization, it is important to distinguish between the following types of constraints:

- link-motion and joint-motion constraints;
- equality- and inequality (unilateral)-type constraints;
- permanently active and temporal constraints;
- high-priority and low-priority constraints.

First, in the case of link-motion constraints, typically the motion of the end links is constrained. For example, the hand moves along a specific path when approaching an object to be grasped, the foot of the swing leg moves along a desired path when taking a step, or the head movement is constrained while tracking an object under a gaze task. Likewise, a link motion-type constraint is imposed by the balance task on the movement of the center of mass (CoM), as will be discussed in detail in Chapter 5. Furthermore, the motion of intermittent links (e.g. related to the spatial motion of the elbow) may be constrained by an obstacle avoidance task.

Second, joint-motion constraints refer to joint angle and rate limits. The joint-limit and the singularity avoidance tasks, discussed in 2.7.4, represent such type of constraints. These tasks are also an example of inequality-type constraints. Obstacle avoidance is another such example, as will be demonstrated in what follows.

Third, to avoid overconstrained states (task conflicts), the number of active tasks should be minimized. This is the so-called *active set* method [83]. Obstacle avoidance is an example of a temporal constraint that should be activated only when an obstacle enters the workspace. Singularity avoidance can also be considered a temporal constraint, to be activated when the measure of distance to the singularity (e.g. manipulability, condition number, minimum singular value) passes a predetermined threshold. Joint limits, on the other hand, are an example of a permanently active constraint.

Finally, introducing priorities between the tasks is helpful to decide whether a temporal constraint should be deactivated to avoid a task conflict. The priority assignment can be either fixed or variable, as will be shown in what follows.

2.8.2 Redundancy Resolution Methods for Multiple Tasks

Motion-task constraints imposed on the motion of specified links are most often resolved via velocity-/acceleration-level inverse kinematics relations. This is because a whole-body motion implies multi-DoF kinematic chains that do not comprise a closed-form solution to the inverse kinematics. As already discussed, kinematic redundancy implies the existence of an infinite set of solutions to the instantaneous inverse kinematics problem. The choice of an appropriate solution is still an open research problem.

Restricted Generalized Inverse and Task Prioritization

Local linear optimization based on the *constrained least-squares* method ([1], Chapter VII) was adopted as a solution in the 1980s [69,45]. The simplest example concerns two tasks: an end-link motion task (the main constraint), expressed with the differential kinematic relation $\boldsymbol{J}_1\dot{\boldsymbol{\theta}} = \mathcal{V}_1$, and an additional motion task (the additional constraint), described by $\boldsymbol{J}_2\dot{\boldsymbol{\theta}} = \mathcal{V}_2$. The objective is to minimize $\|\boldsymbol{J}_2\dot{\boldsymbol{\theta}} - \mathcal{V}_2\|^2$ over all postures, such that the main constraint $\boldsymbol{J}_1\dot{\boldsymbol{\theta}} = \mathcal{V}_1$ is satisfied. The general solution is

$$\dot{\boldsymbol{\theta}} = \boldsymbol{J}_1^+\mathcal{V}_1 + \overline{\boldsymbol{J}}_2^+\overline{\mathcal{V}}_2 + (\boldsymbol{E} - \boldsymbol{J}_1^+\boldsymbol{J}_1)(\boldsymbol{E} - \overline{\boldsymbol{J}}_2^+\overline{\boldsymbol{J}}_2)\dot{\boldsymbol{\theta}}_a, \tag{2.49}$$

where $\overline{\mathcal{V}}_2 = \mathcal{V}_2 - \boldsymbol{J}_2\boldsymbol{J}_1^+\mathcal{V}_1$. Pseudoinverse $\overline{\boldsymbol{J}}_2^+$ is called a "restricted generalized inverse" ([14], p. 88; see also [97,61]). The overbar notation $\overline{\boldsymbol{J}}_2 = \boldsymbol{J}_2\boldsymbol{N}(\boldsymbol{J}_1)$ is used to denote a *restricted Jacobian* [110]: the range of Jacobian $\boldsymbol{J}_2$ is restricted by the null space of another transform (in this case, by the null space projector $N(J_1)$). In Chapter 4, this notation will also be applied to other types of transforms, e.g. the joint-space inertia matrix. The above solution introduces an "order of priority" [45,106] among the tasks, the highest priority being assigned to the main task and the second-highest priority to the additional task. Arbitrary joint velocity vector $\dot{\theta}_a$ parametrizes the remaining DoFs, within the intersection $\mathcal{N}(J_1) \cap \mathcal{N}(J_2)$.

It is easy to confirm that when the underdetermined linear system

$$\begin{bmatrix} J_1 \\ J_2 \end{bmatrix} \dot{\theta} = \begin{bmatrix} \mathcal{V}_1 \\ \mathcal{V}_2 \end{bmatrix} \tag{2.50}$$

has full (row) rank, its pseudoinverse solution coincides with the minimum-norm solution derived from (2.49), i.e. with $\dot{\theta}_a = 0$. This means that the above task-prioritization scheme makes sense only when system (2.50) is rank-deficient. But then, either J_1 and/or J_2 or $\overline{J}_2$ is rank-deficient, and hence, at least one of the (right) pseudoinverses in (2.49) ceases to exist. Therefore, when implementing (2.49), some means of handling singularities should be provided. The rank deficiencies of J_1, J_2, and $\overline{J}_2$ are referred to as kinematic, *task*, and *algorithmic* singularity [8,110], respectively. An algorithmic singularity indicates linear dependency among the tasks, such that $\text{rank}\begin{bmatrix} J_1^T & J_2^T \end{bmatrix}^T < \text{rank} J_1 + \text{rank} J_2$. This type of rank deficiency is the essence of a *task conflict*.

Multiple Tasks With Fixed Priorities

With the help of the arbitrary vector parameter $\dot{\theta}_a$ in (2.49), it is possible to extend the hierarchical inverse kinematics structure to cope with more than two subtasks. The following recursive scheme has been proposed for m subtasks [109,115]:

$$\begin{aligned} \dot{\theta}_k &= \dot{\theta}_{k-1} + \overline{J}_k^+ \overline{\mathcal{V}}_k, \\ \overline{\mathcal{V}}_k &= \mathcal{V}_k - J_k \dot{\theta}_{k-1}, \\ \overline{J}_k &= J_k N_{k-1}, \\ N_k &= \prod_{i=1}^{k} (E - \overline{J}_i^+ \overline{J}_i), \\ k &= 1, 2, \ldots, m, \quad \dot{\theta}_0 = 0, \quad N_0 = E. \end{aligned} \tag{2.51}$$

This scheme is characterized by the restriction of Jacobian J_k by the intersection of null spaces $\mathcal{N}(J_1) \cap \mathcal{N}(J_2) \cap \ldots \cap \mathcal{N}(J_{k-1})$.

Another recursive formulation has been presented in [144], i.e.

$$\begin{aligned} \dot{\theta}_k &= \dot{\theta}_{k-1} + \overline{J}_k^{\#} \overline{\mathcal{V}}_k, \\ \overline{\mathcal{V}}_k &= \mathcal{V}_k - J_k \dot{\theta}_{k-1}, \\ \overline{J}_k &= J_k N_{C_{k-1}}, \\ N_{C_{k-1}} &= (E - J_{C_k}^{\#} J_{C_k}), \end{aligned} \tag{2.52}$$

$$J_{C_k} = [J_1^T \ \ J_2^T \ \ .... \ \ J_k^T]^T,$$
$$k = 1, 2, ..., m, \ \ \dot{\theta}_0 = 0, \ \ N_{C_0} = E.$$

This scheme is characterized by the restriction of Jacobian J_k by the null space $\mathcal{N}(J_{C_k})$. The Lyapunov-based stability analysis in [2] has clarified that, to guarantee stability, task relations in the first scheme should be defined in a more conservative way than those in the second one. As shown in [115], both schemes are characterized by the following: (1) the kth task is of lower priority than the first $k-1$ tasks due to the fixed priority and (2) the kth task cannot be realized in two cases: either due to task singularity, i.e. when the first $k-1$ tasks are all well conditioned but matrix J_k is not, or due to an algorithmic singularity (J_k is well conditioned but the restricted Jacobian $\overline{J}_k$ is not). The latter case means however that the first $k-1$ tasks definitely influence the executability of task k. The two schemes can be easily extended to second-order differential relations [135].

In the two schemes (2.51) and (2.52), the task priorities are fixed. Based on the above analysis, it is straightforward to conclude that the way of allocating priorities to the tasks will be crucial for the performance. In humanoid robotics, fixed priorities have been introduced in [138], where three main priority levels were suggested: highest priority for joint-motion constraints, mid-level priority for link-motion constraints, and lowest priority for postural-variation constraints. Indeed, joint-motion constraints cannot be violated since at the extreme, they become physical constraints, e.g. hitting a joint limit or saturating a joint rate. At the mid-level priority level, link-motion constraints are further hierarchically substructured to account for the need of a higher priority, e.g. for balance (the CoM control subtask) w.r.t. the hand position control. At the lowest level, postural-variation constraints refer to constraints on the remaining DoFs (if any), imposed in joint space, such as trying to pull all joint angles toward their center values, i.e., $\theta_i \rightarrow \theta_{mid_i}$.

A pioneering example of redundancy resolution with hand motion control as the main task and *obstacle avoidance* as a low-priority (additional) task is presented in [89]. It is assumed that the shortest distance d between the obstacle and a specific point on the (intermittent) link closest to the obstacle is available from sensor data. The instantaneous motion of this point, expressed in terms of spatial velocity $\mathcal{V}_2$, is to be controlled for obstacle avoidance. Based on (2.49), the obstacle avoidance solution was derived as

$$\dot{\theta} = J_1^+ \mathcal{V}_1 + \alpha_\eta \overline{J}_2^+ (\alpha_2 \mathcal{V}_2 - J_2 J_1^+ \mathcal{V}_1), \tag{2.53}$$

where subscripts "1" and "2" refer to the main (hand motion) and additional (obstacle avoidance) tasks, respectively. Two scalar variables, $\alpha_\eta(d)$ and $\alpha_2(d)$, were introduced to determine the behavior of the manipulator in the vicinity of an obstacle. Three specific distances around the obstacle were determined as the task-abort distance, the unity-gain distance, and the sphere-of-influence distance. The latter is used to activate self-motion via α_η, when approaching the obstacle. Outside the sphere of influence, self-motion can be used for other subtasks, thus avoiding unnecessary overconstraints. Once activated, the rate of self-motion gradually increases, i.e. to ensure an appropriate repelling behavior via velocity $\alpha_2 \mathcal{V}_2$. The value of α_2 increases quadratically when the link comes closer to the obstacle. A collision is imminent when the task-abort distance is reached.

The above example clearly demonstrates a major problem inherent to the multitask null-space projection methods: the *difficulty in dealing with unilateral (inequality-type) constraints.*

The scalar $\alpha_2(d)$ determines in fact a repelling potential that represents a unilateral constraint. Potential functions are typically employed in obstacle avoidance tasks [8,65,41,168], but also in joint-limit and singularity avoidance subtasks. Apparently, expressions (2.42) and (2.44) represent such functions. Note that these potentials remain activated throughout the entire motion, as their gradients are projected onto the null space. In contrast, the potential in the obstacle avoidance example is activated/deactivated via the sphere-of-influence threshold. This approach, known as the *active set* method [83], is preferable in lieu of permanently active potentials, since in the latter case the system may easily get overconstrained. On the other hand, the nature of activation/deactivation is discrete, and therefore, special means for smoothing are needed to avoid discontinuity in the joint velocity solution. In the obstacle-avoidance example, this was achieved via $\alpha_\eta(d)$, defined as a smooth polynomial of d, with boundary values "0" (no self-motion) and "1" (self-motion active).

Variable Task Priorities With Smooth Task Transitions

Another possible approach for dealing with algorithmic singularities (overconstrained system) and the related discontinuity was reported in [115,112]. The main idea is to *dynamically allocate the priority among the tasks*. The method makes use of the fact that when the first $k-1$ tasks have a consistent solution, they can be reordered arbitrarily without affecting the solution. According to the method, a multidimensional task is *fully decomposed* into single-dimensional components. The priority of each component is determined dynamically, at each time step, the best-conditioned task component receiving thereby the highest priority. Less well-conditioned components have lower priority, and ill-conditioned components, if any, have the lowest priority. The latter are damped in a smooth way, in accordance with the idea behind the damped least-squares method described in Section 2.5. Thereby, because of the full task decomposition approach, the time consuming SVD is avoided.

Others have also considered variable task-priority allocation. In [17], the idea of exchanging the priorities among the main task and an obstacle avoidance subtask was proposed. The reasoning is that if a collision becomes inevitable under low priority obstacle avoidance, then by exchanging the priorities, there will be a possibility to avoid the collision. In [29], it was argued that the priority of unilateral constraints (joint limits typically) does not have to be always fixed at the highest level. With regard to variable-priority tasks, in [63,75,131] methods for smooth task activation/deactivation have been developed.

Algorithmic singularities are inherent to the above prioritization schemes. As already noted, an algorithmic singularity signifies an overconstrained system due to numerous task constraints. Given a desired set of tasks, this is an unavoidable problem. There are two possible approaches to deal with this problem. First, a "damping" or "regularization" term can be added to the solution to suppress the excessive joint velocity; in essence, this is the damped least-squares approach outlined in Section 2.5. Second, the number of tasks can be decreased, e.g. via removal of the "least important" ones. The former method has been criticized by a number of researchers for introducing errors into the solution, disturbing the order of priority, the unintuitive character of the damping factor, and the difficulty of its tuning [116,93,63, 29]. Therefore, the latter method is preferable. Note though that "simple" removal (or deactivation) would imply an abrupt change in system dimensions and hence discontinuity in the solution. The easiest way to alleviate this problem is by introducing a variable scalar function at each level of the hierarchy. The role of this variable is similar to that of α_η in the obstacle avoidance solution (2.53). As an example, consider the approach proposed in [75]. In the case of two tasks, the priority-based solution discussed above, is modified as follows:

$$\dot{\boldsymbol{\theta}} = \boldsymbol{J}_1^+ \mathcal{V}_1^{int} + \overline{\boldsymbol{J}}_2^+ (\mathcal{V}_2^{int} - \boldsymbol{J}_2 \boldsymbol{J}_1^+ \mathcal{V}_1^{int}), \tag{2.54}$$

where

$$\mathcal{V}_1^{int} = \alpha_1 \mathcal{V}_1 + (1 - \alpha_1) \boldsymbol{J}_1 \boldsymbol{J}_2^+ \alpha_2 \mathcal{V}_2,$$

$$\mathcal{V}_2^{int} = \alpha_2 \mathcal{V}_2 + (1 - \alpha_2) \boldsymbol{J}_2 \boldsymbol{J}_1^+ \alpha_1 \mathcal{V}_1$$

are intermediate values that ensure smooth transitions via appropriate choices of scalars $\alpha_1, \alpha_2 \in [0, 1]$. It is straightforward to rewrite the scheme for m tasks. The number of scalar functions increases accordingly, i.e. $\alpha_k, k = 1, 2, ..., m$. These functions are called *activation variables*.

Another possible approach is to modify the null-space projectors $\boldsymbol{N}_k$ appearing in (2.51) for smooth task transitions as follows [131,130]:

$$\boldsymbol{E} - \alpha_k \overline{\boldsymbol{J}}_k^+ \overline{\boldsymbol{J}}_k. \tag{2.55}$$

Scalars α_k have the meaning of activation variables, as above. Their definition plays a crucial role in the behavior of the robot and should be done in a task-dependent way. In [130], for instance, exponential functions of the zero-moment point (ZMP) position were involved to ensure "reflexive" balance control of a humanoid. Other examples of activation variables can be found in [17,63].

In [28], the so-called "null space projection shaping" method was proposed for a self-collision avoidance task of m DoFs. Note that self-collision is handled in the same way as obstacle avoidance, i.e. via potential functions, since the constraints are again unilateral [150, 145]. According to the method, first the null space projector (2.30) is recast via SVD as follows:

$$\boldsymbol{N} = (\boldsymbol{E}_n - \boldsymbol{J}^+ \boldsymbol{J}) = (\boldsymbol{E}_n - \boldsymbol{V} \boldsymbol{\Sigma}^+ \boldsymbol{U}^T \boldsymbol{U} \boldsymbol{\Sigma} \boldsymbol{V}^T) = (\boldsymbol{E}_n - \boldsymbol{V} \overline{\boldsymbol{E}} \boldsymbol{V}^T), \tag{2.56}$$

where $\overline{\boldsymbol{E}} = \text{diag}(\boldsymbol{E}_m, \boldsymbol{0}_{n-m})$. The m unit diagonal elements in $\overline{\boldsymbol{E}}$ indicate that all task components are active. A task component is deactivated by simply replacing the respective "1" on the diagonal with a "0." For smooth transitions, however, the m diagonal elements are redefined as differentiable scalar functions, having the meaning of activation variables as in the above schemes.

2.8.3 Iterative Optimization Methods

When looking for a solution to the inverse kinematics problem under a set of motion-task constraints, the first thought is to apply a readily available optimization method like linear-quadratic or differential dynamic programming. This approach was first used for animation of human figures [172] whereby the main goal was to decrease the animator's workload. Tasks in 3D space, such as reaching a point with the hand, were formulated as potential functions. Multiple tasks were incorporated then as a sum of weighted potentials. The objective potential function was subjected to joint limits, using a nonlinear programming approach with linear equality- and inequality-type constraints. This method, however, turns out to be quite cumbersome since weight tuning is empirical. Moreover, with multiple tasks, the system may easily become overconstrained, or it may get stuck at a suboptimal solution.

Most used are convex optimization methods, a subclass of nonlinear programming methods. Especially, the following quadratic programming (QP) task formulation is suitable:

$$\min_{\boldsymbol{x}} \frac{1}{2}\boldsymbol{x}^T \boldsymbol{G}\boldsymbol{x} + \boldsymbol{g}^T \boldsymbol{x}, \tag{2.57}$$

subject to

$$\begin{aligned}\boldsymbol{A}\boldsymbol{x} + \boldsymbol{a} &= \boldsymbol{0},\\ \boldsymbol{B}\boldsymbol{x} + \boldsymbol{b} &\leq \boldsymbol{0}.\end{aligned}$$

Usually, the minimization problem is set in the form $0.5\|\boldsymbol{C}\boldsymbol{x} - \boldsymbol{c}\|^2$, thus $\boldsymbol{G} = \boldsymbol{C}^T\boldsymbol{C}$ and $\boldsymbol{g} = -\boldsymbol{C}^T\boldsymbol{c}$ [38]. Note that inequality constraints can be handled with this formulation, which is quite suitable, e.g. to ensure the joint-limit constraints.

Introducing a Hierarchical Structure With Fixed Task Priorities

It is desirable to avoid an over-constrained system. One possible approach is to arrange the tasks within the optimization framework in a hierarchical structure [96]. In [23], a recursive prioritization scheme was devised wherein the constrained optimization process at the current priority level remained decoupled from all high-priority processes. Decoupling was achieved via null-space projection operators, in a manner similar to the redundancy resolution methods for multiple tasks. The optimization objective was defined as follows.

For a set of tasks $T_i(\boldsymbol{x}_i)$, $i = 1, 2, ..., m$, find

$$\begin{gathered} h_i = \min\nolimits_{\boldsymbol{x}_i \in S_i} T_i(\boldsymbol{x}_i),\\ \text{subject to}\\ T_k(\boldsymbol{x}_i) = h_k, \forall k < i, \end{gathered}$$

where S_i is a nonempty convex set. The tasks were determined as positive semidefinite quadratic forms of linear equality constraints, i.e.

$$T_i(\boldsymbol{x}_i) = \|\boldsymbol{A}_i\boldsymbol{x}_i - \boldsymbol{b}_i\|^2,$$

for given $\boldsymbol{A}_i$, $\boldsymbol{b}_i$. For example, in the case of instantaneous motion tasks, the following replacements hold: $\boldsymbol{x}_i \to \dot{\boldsymbol{\theta}}_i$, $\boldsymbol{A}_i \to \boldsymbol{J}_i$, $\boldsymbol{b}_i \to \mathcal{V}_i$. The solution to this problem at priority level k is based on appropriate parametrization of null spaces of restricted Jacobians at all preceding $k - 1$ levels.

It turns out that a method called *lexicographic optimization* [52] solves the same problem; a finite number of objective functions are to be optimized on a feasible set in a lexicographic order, i.e. low-priority tasks are optimized as far as they do not interfere with the optimization of higher-priority ones. This method was not well known in the robotics community until recently; it first appeared in relation to a rigid-body contact modeling [146,163]. The lexicographic order is denoted as follows:

$$T_1(\boldsymbol{x}) \succ T_2(\boldsymbol{x}) \succ \ldots \succ T_m(\boldsymbol{x}).$$

This notation signifies hierarchical structuration within the set of tasks.

Variable Task Priorities With Smooth Task Transitions

The importance of the task activation/deactivation feature in combination with reordering the task priorities has been discussed in [91,145,63]. The respective control architecture comprises a low-level controller that ensures a "reflexive" or "reactive"-type task execution with priority determined by the sequence of the tasks defined within a so-called "stack of tasks." The high-level controller makes the decisions about the activation/deactivation (inserting/removing from the stack) of specific tasks and their priority allocation within the stack. In [92,94], the problem of discontinuity in the solution is addressed via activation variables α_k, as already introduced above. These variables are placed on the diagonal of a matrix $\mathbb{A}$ referred to as the *activation matrix*. The inverse kinematics solution is derived in the form

$$\dot{\boldsymbol{\theta}} = \lambda(\mathbb{A}\boldsymbol{J})^{+}\mathbb{A}\mathcal{V}, \tag{2.58}$$

where scalar λ determines the convergence speed of the iterative solution. Further on, it was confirmed that when dealing with the discontinuity of the above pseudoinverse, the damped least-squares method is inappropriate, due to the reasons already discussed. Instead, a new inverse operator was proposed. For the simplest case of two subtasks, the inverse is derived via the following decomposition of the pseudoinverse:

$$\boldsymbol{J}^{+} = \begin{bmatrix} \boldsymbol{J}_1 \\ \boldsymbol{J}_2 \end{bmatrix}^{+} = [\boldsymbol{J}_1^{+} \quad \boldsymbol{J}_2^{+}] + \boldsymbol{C}_{12}, \tag{2.59}$$

where $\boldsymbol{C}_{12}$ accounts for the *coupling* between the two subtask Jacobians $\boldsymbol{J}_1$ and $\boldsymbol{J}_2$. Then, the solution (2.58) is rewritten as

$$\dot{\boldsymbol{\theta}} = \lambda \boldsymbol{J}_c^{\oplus\mathbb{A}}\mathcal{V}, \tag{2.60}$$

where $\boldsymbol{J}_c^{\oplus\mathbb{A}}$ is the new inverse having the property of continuity. The new inverse and the respective solution (2.60) were verified with vision-based servoing experiments. No other reports are available so far with regard to humanoids, though.

Introducing Inequality Constraints

The advantage of iterative optimization methods is that inequality constraints can be embedded in a straightforward manner, via *slack variables*. This approach was implemented for humanoid robots [60,59], as follows. Instead of the above tasks T_i, consider the following linear inequalities:

$$\boldsymbol{b}_i^{min} \leq \boldsymbol{A}_i\boldsymbol{x}_i \leq \boldsymbol{b}_i^{max}.$$

The minimization objective is then

$$\min_{\boldsymbol{x}_i \in S_i, \boldsymbol{w}_i} \|\boldsymbol{w}_i\|, \ \boldsymbol{b}_i^{min} \leq \boldsymbol{A}_i \boldsymbol{x}_i - \boldsymbol{w}_i \leq \boldsymbol{b}_i^{max}, \tag{2.61}$$

subject to

$$\boldsymbol{b}_k^{min} \leq \boldsymbol{A}_k \boldsymbol{x}_i \leq \boldsymbol{b}_k^{max}, \ \forall k < i.$$

The role of the slack variable $\boldsymbol{w}_k$ is to relax the constraints at level k. The norm $\|\boldsymbol{w}_k\|$ can be used as a measure of constraint violation at level k. This violation is propagated to all lower levels.

In these schemes, the particular solutions and the null space basis vectors at each priority level are determined via SVD. The aggregate computational cost of this method, however, is quite high. Recently, alternative approaches have been explored, based on complete orthogonal decomposition [33,34] and QR factorization [58]. Encouraging results that support real-time implementations have been achieved. The QR factorization scheme has been designed to account for the decreasing dimension of the optimization problem when descending through the hierarchical structure. This method was shown to yield the fastest result.

With the computational cost problem being manageable, the remaining problem associated with numerical optimization is the computational stability. Note that in the above schemes, the prioritization is based on restricted Jacobians. As with the redundancy resolution methods for multiple tasks, the inherent algorithmic singularities hinder definitive conclusions about stability. The damped least-squares method was considered in [58] to alleviate the problem. But as already noted, this method suffers from a number of drawbacks. The stability of numerical optimization schemes with task prioritization still remains an open issue.

In a recent work [82], a generalized hierarchical IK algorithm was proposed that claims to avoid the numerical instability problem. The method can handle inequality constraints, as well as fixed and variable task priorities. In the latter case, multiple priority rearrangements can be executed simultaneously with the help of the so-called "generalized null-space projector." It is also claimed that the method is robust w.r.t. both kinematic and algorithmic singularities. Unfortunately, the heavy computational load does not allow for real-time control.

2.8.4 Summary and Discussion

The focus in this section was on motion-task constraints and the respective null-space projections derived through motion-task Jacobians. It was shown that with such projections, a desirable hierarchy among the tasks can be introduced, either fixed or variable. Null-space projections play an important role in the two competing approaches—the classic one, established with studies on kinematically redundant manipulators in the 1980s, and the newer one, based on iterative optimization methods and developed initially to support human figure animation techniques. The desirable features in both approaches are:

- dealing with tasks arranged in a hierarchical structure;
- handling equality- and inequality-type tasks;
- dealing with singularities;
- flexibility through variable priority allocation for the tasks;

- possibility for activation/deactivation of tasks (i.e. the application of the *active set* method [83]);
- ensuring smooth transitions between the tasks;
- stability and passivity of the solution;
- compliance with hard real-time requirements.

Iterative optimization methods are general; they can cope with all types of constraints, including inequality ones. In the past, ensuring real-time performance was problematic. But recently, it was shown that such performance is achievable for frameworks with null space–based task prioritization. The remaining problem is the lack of a proof regarding stability when implemented into robot control.

Not all of the methods discussed here were purely kinematics-based. Dynamical relations need to be taken under consideration when addressing control stability. Asymptotic stability of multitask null space-based redundancy resolution was proven so far in [27,140] only. A related problem is the lack of passivity when using null-space projections, as identified in [133]. In [124], it was shown how to ensure conditional passivity for compliance control.

Switching task priorities and task activation/deactivation is important to avoid an over-constrained system due to task conflicts (or algorithmic singularities). Since these are discrete events, special means have been developed to ensure a smooth solution. In the past, the damped least-squares method was the preferred approach. However, a number of problems arise when implementing the method into hierarchical multitask schemes, as pointed out in [116,93,63,29]:

- the damping factor works against the desirable order of priority;
- the accuracy of the solution decreases not only in the vicinity of a singularity;
- the tuning of the damping factor is empirical and difficult;
- the damping factor is unintuitive and cannot be related to physical parameters of the system.

It is possible to deal with multitask constraints via weighted generalized inverses, instead of null-space projections [172,21,44]. Most researchers agree, however, that this method is not preferable since weight tuning is empirical and lacks rigorousness. Here also, no proof exists with regard to control stability.

The null space–based approaches discussed here can ensure a "reflexive" or "reactive" robot behavior. The nature of this type of behavior is local. Such behavior is considered a necessary but insufficient condition. Global motion planning methods need to be involved as well, as noted e.g. in [63,29,25].

Motion-task constraints only were discussed here. Physical constraints stemming from contacts will be discussed in the following section.

2.9 MOTION CONSTRAINTS THROUGH CONTACTS

A biped humanoid robot moves around by walking, establishing thereby physical contacts between the feet and the ground. Physical contacts also occur when the robot performs manipulation tasks using thereby the hands and fingers to manipulate an object or a tool.

Contacts occurring between one or more of the end links and an object from the environment (the ground, a wall, and so on) are typical cases of physical contacts [107]. Not so common are contacts established between an object and one or more of the robot's intermittent links, e.g. when the robot lies on the ground after a fall [40,66,76,43], while it is sitting in a chair [134,77], or when it is pushing a heavy object with its back [102]. The robot may also contact an object indirectly via a hand-held tool, or when placing an object on a table [20]. Sometimes, a physical contact may occur between the robot links, without involving an object from the environment (e.g. in the case of self-collision).

All these types of physical contacts are modeled via *contact joints*. Modeling and analysis of contact phenomena in multibody systems and robotics have been extensively studied since the 1990s [100,10,154,125,128]. The mobility of the robot is constrained via the contact. Hence, a *constrained multibody system* model becomes appropriate. Research in the area of constrained multibody systems has drawn in fact a lot of attention since the turn of the century [4,12,165, 15,74]. In the field of robotics, the constrained-motion control problem can be traced back to manipulator force [160], hybrid position/force [132], and cooperative multiarm robot control [48,167]; see also [32,171]. Dexterous object manipulation with a multifingered hand is another well-established area of robotics research that refers to constrained motion models. The theoretical base of hand modeling and control, as described in [104] (Chapters 5 and 6), provides a firm foundation for humanoid robot modeling and control, as will be apparent from what follows. In the field of humanoid robots, the importance of motion and force control across contact joints has been recognized e.g. in [128,108,92,139,134].

2.9.1 Contact Joints

A contact joint enforces up to six constraints upon the relative motion between the two bodies in contact. The number of motion constraints enforced depends on the shapes of the bodies and assumptions with regard to friction [100]. These conditions determine the type of the contact joint.

A contact joint with friction constrains the motion in all directions. The focus here will be on *frictionless* contact joints that allow relative motion in one or more directions. Contact joints with friction will be treated in Section 3.3. In the example depicted in Fig. 2.11, the robot stands on the ground and holds a cylindrical object (a rod) with the hands. Contacts occur between the feet and the ground and between the hands and the object. Assuming zero friction at the hand contacts, sliding along and axial rotation around the rod is admissible. The two hand-contact joints can then be characterized as cylindrical joints. The foot contact joints, on the other hand, can be characterized as planar joints since sliding within the ground plane and rotation around the normal is admissible with zero friction. Another often discussed example is object grasping with the finger tips. The contact model includes point-contact joints, without or with friction [104].

The contact joints at the end links will be identified by $k \in \{e_r, e_l\}, e \in \{H, F\}$. The number of motion constraints at each contact joint will be denoted as c_k, $c_k \leq 6$. With $c_k = 6$, the relative motion between the two bodies in contact is fully constrained. With $c_k < 6$, relative motion along $\eta_k = 6 - c_k$ directions is possible. These directions are referred to as *unconstrained*. There are two special types of "contact" joints. A contact joint that enforces constraints along all six directions ($c_k = 6$, $\eta_k = 0$) is called a "welded" joint. Weld-type joints are used to model a

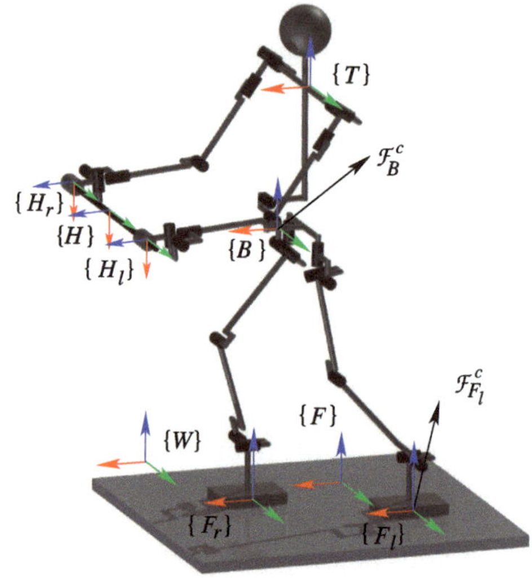

FIGURE 2.11 Closed loops are formed via contact joints at the feet and hands. Contact coordinate frames $\{k\}$, $k \in \{e_r, e_l\}$, $e \in \{H, F\}$ are fixed at the center of pressure (CoP) to the common loop-closure link (floor F and rod H). The z-axes at the feet (shown in blue color) point in a way s.t. the *reaction* force at the contact is always nonnegative. The contact constraints in the vertical direction at the feet are unilateral while those in the angular tangential directions are bilateral, with bounds. All contact constraints at the hands are bilateral.

firm grip of an object or a tool, for example. Next, recall the rigid-body joint introduced in Section 2.4.2. This joint can be characterized as a "contact" joint that does not enforce any constraints, s.t. $c_k = 0$, $\eta_k = 6$.

2.9.2 Contact Coordinate Frames

The constrained/unconstrained motion directions at the contact joint are expressed in an appropriate *contact coordinate frame*. The origin of this frame is fixed at the characteristic contact point, on the site of the loop-closure link. For point contacts, the characteristic point is trivial; for surface or line contacts, the *center of pressure (CoP)* is used as such point [139]. The position of the CoP can be derived through appropriate sensor data, e.g. from force sensing resistors distributed over the soles or from multiaxis force/torque sensors [53]. The z-axis of the contact coordinate frame is along the surface normal of the object that is in contact with the robot link. By convention, the direction of the z-axis is set in a way s.t. the *reaction* force at the contact is always nonnegative [10]. In the case of grasping, for example, such force always results from compression (squeezing the object). On the other hand, the x- and y-axes are chosen by intuition, in dependence with the specific object surface [162]. The commonly accepted contact modeling approach requires also the definition of another coordinate frame. This frame is fixed to the second body taking part in the contact, i.e. to the robot link [100,104, 125].

Consider as an example the contact joints at the hands and the feet displayed in Fig. 2.11. Contact frames $\{e_k\}$ are fixed at the rod and the ground, at the four CoPs. The z_k-coordinate axes are along the object surface normal, which always denotes a motion constraint direction. In the case of the feet contacts, the z_{F_j}-axes point upward so that the reaction contact force along the normal is always positive. The x_{F_j}- and y_{F_j}-coordinate axes are tangent to the contact surface (the ground). The x_{F_j}-axes are selected to be parallel to the inertial frame x_F-axis. In the case of the hand contacts, the y_{H_j}-axes are selected to be parallel to the object's y_H-axis.

Generally, the surfaces coming into contact may have any shape [100,125]. One should bear in mind, then, that there are cases when the selection of the contact frame axes is not as straightforward as in the present example [37].

2.9.3 Kinematic Models of Frictionless Contact Joints

Denote by $\bar{\mathcal{V}}_k^m \in \Re^{\eta_k}$ the first-order instantaneous motion components along the unconstrained-motion directions at contact joint k. These components determine the contact joint twist, i.e.

$$\mathcal{V}_k = {}^k\mathbb{B}_m \bar{\mathcal{V}}_k^m. \tag{2.62}$$

Here ${}^k\mathbb{B}_m \in \Re^{6\times\eta_k}$ is a transform that comprises orthonormal basis vectors for the twist components in the unconstrained motion directions.[2] There is a complementary transform s.t. ${}^k\mathbb{B}_m \oplus {}^k\mathbb{B}_c = \boldsymbol{E}_6$ ($\oplus$ denotes the direct sum operator):

$$\mathcal{V}_k = {}^k\mathbb{B}_c \bar{\mathcal{V}}_k^c. \tag{2.63}$$

Here $\bar{\mathcal{V}}_k^c$ comprises first-order instantaneous motion components in the constrained motion directions. In the above notations (and throughout this text), the overbar notation signifies a restricted quantity, i.e.

$$\bar{\mathcal{V}}_k^m = \boldsymbol{N}({}^k\mathbb{B}_c)\mathcal{V}_k = {}^k\mathbb{B}_m^T \mathcal{V}_k, \tag{2.64}$$

$$\bar{\mathcal{V}}_k^c = \boldsymbol{N}({}^k\mathbb{B}_m)\mathcal{V}_k = {}^k\mathbb{B}_c^T \mathcal{V}_k. \tag{2.65}$$

These relations imply that

$$\begin{bmatrix} \bar{\mathcal{V}}_k^c \\ \bar{\mathcal{V}}_k^m \end{bmatrix} = \begin{bmatrix} {}^k\mathbb{B}_c^T \\ {}^k\mathbb{B}_m^T \end{bmatrix} \mathcal{V}_k, \quad \bar{\mathcal{V}}_k^c \perp \bar{\mathcal{V}}_k^m. \tag{2.66}$$

Note that in the case of *hard* constraints $\bar{\mathcal{V}}_k^c = \mathbf{0}$.

The above transforms are represented in the contact coordinate frame. On the other hand, the motion of objects within the environment that come in contact with the robot links is usually expressed in terms of inertial coordinates. This means that expressions for the transforms in the inertial (world) frame are needed. To this end, use the following relation:

$$\mathbb{B}_{(\circ)}(\boldsymbol{p}_k) = \mathbb{R}_k(\boldsymbol{p}_k){}^k\mathbb{B}_{(\circ)}. \tag{2.67}$$

Here $\mathbb{R}_k$ is the spatial rotation matrix defined in (2.3). Apparently, when expressed in the inertial frame (indicated by a missing lead superscript), $\mathbb{B}_m(\boldsymbol{p}_k)$ and $\mathbb{B}_c(\boldsymbol{p}_k)$ become functions

[2] Recall that such transform was used in the multi-DoF joint models (cf. (2.13)).

of the contact geometry parametrized by the local curvature and torsion of the two surfaces in contact. The respective parameters are collected in the vector parameter $\boldsymbol{p}_k$ [64,100]. This is the reason why $\mathbb{R}_k$ is specified as a function of $\boldsymbol{p}_k$. It should also be noted that the $\mathbb{R}_k(\boldsymbol{p}_k)$ map preserves the complementarity relation between the two transforms.

Examples

In the example in Fig. 2.11, the frictionless cylindrical contact joints at the hands determine

$$
{}^{H_j}\mathbb{B}_m = \begin{bmatrix} 0 & 0 \\ 1 & 0 \\ 0 & 0 \\ 0 & 0 \\ 0 & 1 \\ 0 & 0 \end{bmatrix}, \quad \overline{\mathcal{V}}^m_{H_j} = \begin{bmatrix} v_y \\ \omega_y \end{bmatrix}. \tag{2.68}
$$

The frictionless plane-contact joints at the feet, on the other hand, are modeled with

$$
{}^{F_j}\mathbb{B}_m = \begin{bmatrix} 1 & 0 & 0 \\ 0 & 1 & 0 \\ 0 & 0 & 0 \\ 0 & 0 & 0 \\ 0 & 0 & 0 \\ 0 & 0 & 1 \end{bmatrix}, \quad \overline{\mathcal{V}}^m_{F_j} = \begin{bmatrix} v_x \\ v_y \\ \omega_z \end{bmatrix}. \tag{2.69}
$$

The contact models described so far have been expressed in the contact coordinate frame on the environment/object site. To accomplish tasks like walking and object manipulation, the robot has to control the motion (and force) components across the contact joints. Therefore, contact frames at the robot site have to be introduced into the contact model. This will be done in the following section.

2.10 DIFFERENTIAL KINEMATICS OF CHAINS WITH CLOSED LOOPS

Constraints on the instantaneous motion within the kinematic chain are imposed by the controller (*task-based* constraints) and/or via contacts (*physical* constraints). For example, the additional motion constraints used for redundancy resolution in Section 2.7 are task-based constraints. End-link path following, e.g. according to the SC method (cf. Section 2.5), gives another example. The main focus here will be on physical constraints through contacts.

Through the contact joints, one or more closed loops are formed within the robot kinematic chain. When the robot is in a double stance, for example, a closed loop is formed via the ground/feet contacts. The ground is referred to as the *loop-closure link*. When holding an object with the hands, a second closed loop is formed via the object/hands contacts. In this case, there are two loop-closure links: the ground and the object. Each closed loop is formed

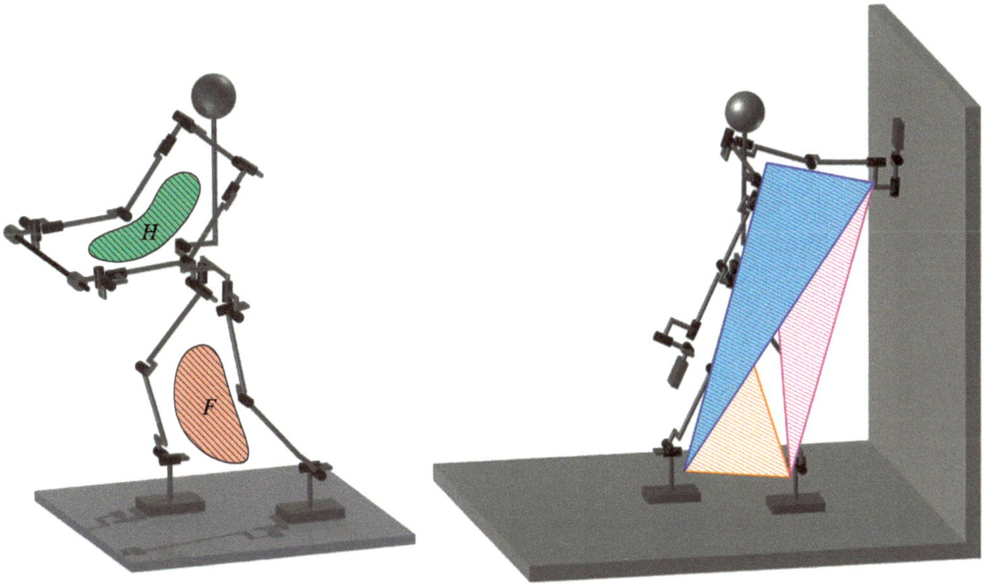

FIGURE 2.12 Closed loops formed via contact joints. Left (A): Two *independent* closed loops are formed with two distinct loop-closure links, floor F and object (rod) H, when the feet rest at the floor and the hands hold an object (a rod). The loops are identified by the respective loop-closure links (floor F and rod H). Each loop comprises two parallel branches composed of the legs and arms, respectively. Right (B): The three colored triangles signify three *interdependent* closed loops formed with three parallel branches: hand/left foot (blue), hand/right foot (red), and foot/foot (yellow). The common loop-closure link is composed of the floor and the wall.

by two parallel branches: the arms and the legs. The two loops are said to be *independent*. This situation is depicted in Fig. 2.12A. With this model, it is possible to design two independent controllers, e.g. a balance controller for the closed loop with the legs and an object manipulation controller for the closed loop with the arms.

Further on, when one of the hands touches a wall while the robot is in a double stance, three closed loops are formed. Note that there is only one common loop-closure link that is composed by the floor and the wall. This link gives rise to interdependent differential motion relations. The three closed loops formed by the three parallel branches, i.e. the two legs and the arm touching the wall, are said to be *interdependent*. Referring to Fig. 2.12B, the three colored triangles and the floor/wall signify the closed loops and the common loop-closure link, F, respectively. Another example where interdependent closed loops are formed is manipulation of an object with the fingers [104]. The fingers represent multiple parallel branches; the grasped object plays the role of the loop-closure link. In this way, multiple interdependent closed loops are formed.

2.10.1 Instantaneous Motion Analysis of Chains With Closed Loops

Closed-loop subchains exhibit differential kinematic relations that reflect the loss of DoF due to the contact constraints within each loop. In this section, the focus will be on the single

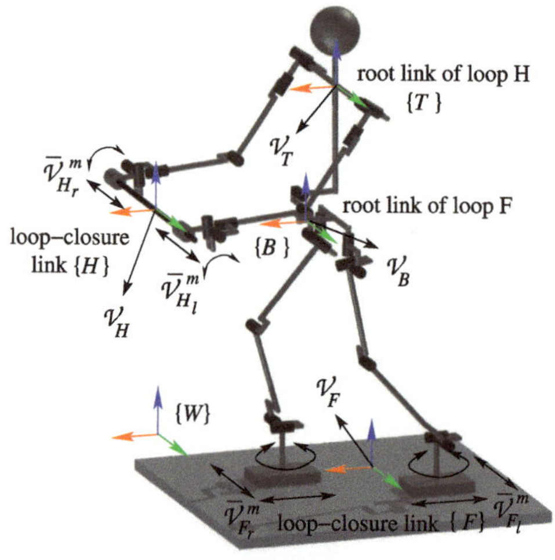

FIGURE 2.13 Humanoid robot with two independent closed loops, F and H, formed by frictionless contact joints at the feet and hands, respectively. The loop-closure link velocities are denoted as $\mathcal{V}_F$ and $\mathcal{V}_H$. The end-link velocities $\bar{\mathcal{V}}_{e_j}^m$ signify instantaneous motions in the unconstrained directions at each contact joint. Each closed loop is composed of two physical parallel branches connecting the loop-root with the loop-closure link, plus one virtual branch (not shown) formed via a virtual 6-DoF joint between the inertial frame and the root link.

closed-loop case, i.e. closed loop F or H, as shown in Fig. 2.12A. For the purpose of instantaneous motion analysis, frictionless contact joints will be assumed unless otherwise specified. Joints with friction will be treated in Section 3.4. Contact coordinate frames $\{k\}$, $k \in \{e_r, e_l\}$, $e \in \{H, F\}$, are assigned according to the rules clarified in Section 2.9.2. The basis vectors along the constrained and unconstrained motion directions are denoted as ${}^k\mathbb{B}_c \in \Re^{6\times c_k}$ and ${}^k\mathbb{B}_m \in \Re^{6\times \eta_k}$, respectively (cf. Section 2.9.3). In the following derivations, all quantities will be expressed in the inertial frame, and therefore, the leading superscript will be omitted. The basis vectors at the contact joints are transformed to the inertial frame in accordance with (2.67). In this way, it becomes possible to handle curved contact surfaces (nonflat floor/curved object). The root and closure links of the closed loop will be denoted as $\{R\}$ and $\{e\}$, respectively. The closed loops formed by the legs and arms are thus characterized by $e = F$, $R = B$ and $e = H$, $R = T$, respectively (cf. Fig. 2.13). The number of limb joints is denoted as $n_k \geq 6$, $n_e = n_{e_r} + n_{e_l} \geq 12$ standing for the total number of actuated joints in the closed loop. The respective joint variables are collected in the vector $\boldsymbol{\theta}_e = \begin{bmatrix}\boldsymbol{\theta}_{e_r}^T & \boldsymbol{\theta}_{e_l}^T\end{bmatrix}^T \in \Re^{n_e}$. Furthermore, the closed-loop generalized coordinate vector is denoted as $\boldsymbol{q}_e = (\mathcal{X}_R, \boldsymbol{\theta}_e) \in \Re^{6+n_e}$, $\mathcal{X}_R$ standing for the 6D position of the root link determined with the preferred choice of local coordinates (e.g. Euler angles, Euler parameters; cf. Section 2.4.2).

The instantaneous motion relations within the closed loop are determined by the twists of the closure and root links, $\mathcal{V}_e$ and $\mathcal{V}_R$, as well as by the loop joint rates, $\dot{\boldsymbol{\theta}}_e$. Referring to Fig. 2.13, the rod held by the hands moves instantaneously with twist $\mathcal{V}_H$. Also, the general case of a support surface, moving instantaneously with twist $\mathcal{V}_F$, can be addressed with this notation.

Limb Velocities

The instantaneous motion of the contact frame k is determined as

$$\mathcal{V}_k(\mathcal{V}_e, \boldsymbol{p}_k) = \mathbb{T}_{\overleftarrow{ke}}(\boldsymbol{p}_k)\mathcal{V}_e + \mathcal{V}_k'(\boldsymbol{p}_k), \tag{2.70}$$

where $\mathcal{V}_k'(\boldsymbol{p}_k)$ denotes relative instantaneous motion w.r.t. the closure link. Recall that the vector parameter $\boldsymbol{p}_k$ parametrizes the contact joint motion in unison with the contact geometry (cf. (2.67)). The instantaneous motion of end link k, on the other hand, can be written as

$$\mathcal{V}_k(\mathcal{V}_R, \boldsymbol{q}_k) = \mathbb{T}_{\overleftarrow{kR}}(\boldsymbol{q}_k)\mathcal{V}_R + \boldsymbol{J}_R(\boldsymbol{q}_k)\dot{\boldsymbol{\theta}}_k, \tag{2.71}$$

where $\boldsymbol{J}_R(\boldsymbol{q}_k) \in \Re^{6\times n_k}$ denotes the limb Jacobian. The closure of the kinematic loop will be ensured when the end links track the motion of the contact frames on the closure link precisely. We have

$$\mathcal{V}_k(\mathcal{V}_e, \boldsymbol{p}_k) = \mathcal{V}_k(\mathcal{V}_R, \boldsymbol{q}_k).$$

Simplifying assumption: From the above equation one can conclude that motion at a contact joint is parametrized by $n_k + 6 + \dim(\boldsymbol{p}_k)$ number of variables. This means that, in general, the instantaneous motion relations in the closed loop will depend on the time derivatives of the $\boldsymbol{p}_k$ vector parameters [64]. To keep the notation simple, henceforth time-invariant contact geometries only will be assumed. Then, $\mathcal{V}_k'(\boldsymbol{p}_k)$, $\mathbb{T}_{\overleftarrow{ke}}(\boldsymbol{p}_k)$, $\mathbb{B}_c(\boldsymbol{p}_k)$, and $\mathbb{B}_m(\boldsymbol{p}_k)$ can be expressed as functions of $\boldsymbol{q}_k$ instead of $\boldsymbol{p}_k$. The twist at the contact joint will be denoted simply as

$$\mathcal{V}_k \equiv \mathcal{V}_k(\mathcal{V}_R, \boldsymbol{q}_k). \tag{2.72}$$

Further on, (2.71) can be projected along the constrained (c) and mobility (m) (i.e. the unconstrained) directions. To this end, premultiply (2.71) first by $\mathbb{B}_c^T(\boldsymbol{q}_k)$ and then by $\mathbb{B}_m^T(\boldsymbol{q}_k)$, to obtain the system

$$\begin{bmatrix} \mathbb{C}_{cR}^T(\boldsymbol{q}_k) \\ \mathbb{C}_{mR}^T(\boldsymbol{q}_k) \end{bmatrix} \mathcal{V}_R + \begin{bmatrix} \mathcal{J}_{cR}(\boldsymbol{q}_k) \\ \mathcal{J}_{mR}(\boldsymbol{q}_k) \end{bmatrix} \dot{\boldsymbol{\theta}}_k = \begin{bmatrix} \mathbb{B}_c^T(\boldsymbol{q}_k) \\ \mathbb{B}_m^T(\boldsymbol{q}_k) \end{bmatrix} \mathcal{V}_k, \tag{2.73}$$

where

$$\mathbb{C}_{cR}^T(\boldsymbol{q}_k) = \mathbb{B}_c^T(\boldsymbol{q}_k)\, \mathbb{T}_{\overleftarrow{kR}}(\boldsymbol{q}_k) \in \Re^{c_k\times 6}, \tag{2.74}$$

$$\mathcal{J}_{cR}(\boldsymbol{q}_k) = \mathbb{B}_c^T(\boldsymbol{q}_k)\boldsymbol{J}_R(\boldsymbol{q}_k) \in \Re^{c_k\times n_k}, \tag{2.75}$$

$$\mathbb{C}_{mR}^T(\boldsymbol{q}_k) = \mathbb{B}_m^T(\boldsymbol{q}_k)\, \mathbb{T}_{\overleftarrow{kR}}(\boldsymbol{q}_k) \in \Re^{\eta_k\times 6},$$

$$\mathcal{J}_{mR}(\boldsymbol{q}_k) = \mathbb{B}_m^T(\boldsymbol{q}_k)\boldsymbol{J}_R(\boldsymbol{q}_k) \in \Re^{\eta_k\times n_k}.$$

Matrices $\mathbb{C}_{cR}(\boldsymbol{q}_k)$ and $\mathbb{C}_{mR}(\boldsymbol{q}_k)$ are always of full column rank, as are $\mathbb{B}_c(\boldsymbol{q}_k)$ and $\mathbb{B}_m(\boldsymbol{q}_k)$. The upper part of (2.73) determines the first-order *differential-motion constraint* of limb e_j, $j \in \{r, l\}$. The two end links in the closed loop move in a synchronous way determined by the two twists $\mathcal{V}_k$. These twists ensure that the closure-link twist is exactly $\mathcal{V}_e$. The lower part of (2.73), on the other hand, determines the instantaneous motion of each end link along the mobility directions. The end links track thereby the instantaneous motion of the respective contact frame, determined by $\mathcal{V}_k'(\boldsymbol{q}_k)$, along the closure link. These tracking movements of the end links are independent.

Further on, referring to (2.70), the r.h.s. of (2.73) can be expressed as

$$\mathbb{C}_e^T(\boldsymbol{q}_k)\mathcal{V}_e + \mathbb{B}^T(\boldsymbol{q}_k)\mathcal{V}_k' \equiv \bar{\mathcal{V}}_k, \tag{2.76}$$

where

$$\begin{aligned}
\mathbb{B}(\boldsymbol{q}_k) &= \begin{bmatrix} \mathbb{B}_c(\boldsymbol{q}_k) & \mathbb{B}_m(\boldsymbol{q}_k) \end{bmatrix} \in \Re^{6\times 6},\\
\mathbb{C}_e^T(\boldsymbol{q}_k) &= \mathbb{B}^T(\boldsymbol{q}_k)\mathbb{T}_{\overleftarrow{ke}}(\boldsymbol{q}_k),\\
&= \begin{bmatrix} \mathbb{C}_{ce}(\boldsymbol{q}_k) & \mathbb{C}_{me}(\boldsymbol{q}_k) \end{bmatrix}^T,\\
\bar{\mathcal{V}}_k &= \begin{bmatrix} (\bar{\mathcal{V}}_k^c)^T & (\bar{\mathcal{V}}_k^m)^T \end{bmatrix}^T.
\end{aligned}$$

As already shown in (2.66), $\bar{\mathcal{V}}_k^c \perp \bar{\mathcal{V}}_k^m$. Also, the range space of the *constraint* Jacobian of the limb, $\mathcal{J}_{cR}(\boldsymbol{q}_k)$, is an orthogonal complement to that of the limb *mobility* Jacobian, $\mathcal{J}_{mR}(\boldsymbol{q}_k)$. With this notation, (2.73) can be written in compact form as

$$\mathbb{C}_R^T(\boldsymbol{q}_k)\mathcal{V}_R + \mathcal{J}_R(\boldsymbol{q}_k)\dot{\boldsymbol{\theta}}_k = \mathbb{C}_e^T(\boldsymbol{q}_k)\mathcal{V}_e + \mathbb{B}^T(\boldsymbol{q}_k)\mathcal{V}_k', \tag{2.77}$$

where

$$\mathbb{C}_R(\boldsymbol{q}_k) = \begin{bmatrix} \mathbb{C}_{cR}(\boldsymbol{q}_k) & \mathbb{C}_{mR}(\boldsymbol{q}_k) \end{bmatrix} \in \Re^{6\times 6} \tag{2.78}$$

and $\mathcal{J}_R(\boldsymbol{q}_k) = \mathbb{B}^T(\boldsymbol{q}_k)\boldsymbol{J}_R(\boldsymbol{q}_k)$; $\mathbb{B}(\boldsymbol{q}_k)$ plays the role of a permutation matrix that reorders the instantaneous motion equation of the limbs in the closed loop. Matrices $\mathbb{C}_R(\boldsymbol{q}_k)$ and $\mathbb{C}_e(\boldsymbol{q}_k)$ are referred to as the *contact maps* (CMs) of the limb for the root and closure links, respectively. They are composed of CMs in the constrained and mobility directions. Note that for a contact joint with friction, $\mathbb{C}_{(\circ)}(\boldsymbol{q}_k) = \mathbb{C}_{c(\circ)}(\boldsymbol{q}_k)$. On the other hand, the CMs of a completely free "contact joint" are $\mathbb{C}_{(\circ)}(\boldsymbol{q}_k) = \mathbb{C}_{m(\circ)}(\boldsymbol{q}_k)$.

Velocities Within the Closed Chain

The instantaneous-motion equations for the closed loop can be written in the same form as (2.73). We have

$$\begin{bmatrix} \mathbb{C}_{cR}^T(\boldsymbol{q}_e) \\ \mathbb{C}_{mR}^T(\boldsymbol{q}_e) \end{bmatrix} \mathcal{V}_R + \begin{bmatrix} \mathcal{J}_{cR}(\boldsymbol{q}_e) \\ \mathcal{J}_{mR}(\boldsymbol{q}_e) \end{bmatrix} \dot{\boldsymbol{\theta}}_e = \begin{bmatrix} \mathbb{B}_c^T(\boldsymbol{q}_e) \\ \mathbb{B}_m^T(\boldsymbol{q}_e) \end{bmatrix} \begin{bmatrix} \mathcal{V}_{e_r} \\ \mathcal{V}_{e_l} \end{bmatrix}, \tag{2.79}$$

where

$$\begin{aligned}
\mathbb{B}_c^T(\boldsymbol{q}_e) &= \operatorname{diag}\left(\mathbb{B}_c^T(\boldsymbol{q}_{e_r}), \mathbb{B}_c^T(\boldsymbol{q}_{e_l})\right) \in \Re^{c_e\times 12},\\
\mathbb{B}_m^T(\boldsymbol{q}_e) &= \operatorname{diag}\left(\mathbb{B}_m^T(\boldsymbol{q}_{e_r}), \mathbb{B}_m^T(\boldsymbol{q}_{e_l})\right) \in \Re^{\eta_e\times 12},
\end{aligned}$$

and $\mathcal{J}_{cR}(\boldsymbol{q}_e) \in \Re^{c_e \times n_e}$, $\mathcal{J}_{mR}(\boldsymbol{q}_e) \in \Re^{\eta_e \times n_e}$ denoting the *joint-space constraint* and *end-link mobility* Jacobians of the closed loop, respectively. The parameters are related as follows: $c_e = c_{e_r} + c_{e_l}$, $\eta_e = \eta_{e_r} + \eta_{e_l} = 12 - c_e$. The r.h.s. of (2.79) can be expressed as

$$\begin{bmatrix} \mathbb{C}_{ce}^T(\boldsymbol{q}_e) \\ \mathbb{C}_{me}^T(\boldsymbol{q}_e) \end{bmatrix} \mathcal{V}_e + \mathbb{B}^T(\boldsymbol{q}_e) \begin{bmatrix} \mathcal{V}'_{e_r} \\ \mathcal{V}'_{e_l} \end{bmatrix} \equiv \begin{bmatrix} \overline{\mathcal{V}}_e^c \\ \overline{\mathcal{V}}_e^m \end{bmatrix}, \quad \overline{\mathcal{V}}_e^c \perp \overline{\mathcal{V}}_e^m, \tag{2.80}$$

where

$$\mathbb{B}^T(\boldsymbol{q}_e) = \text{diag}\left(\mathbb{B}_c^T(\boldsymbol{q}_{e_r}), \mathbb{B}_c^T(\boldsymbol{q}_{e_l}), \mathbb{B}_m^T(\boldsymbol{q}_{e_l}), \mathbb{B}_m^T(\boldsymbol{q}_{e_l})\right) \in \Re^{12\times 12}.$$

The CMs of the closed loop are in stacked form

$$\mathbb{C}_{c(\circ)}(\boldsymbol{q}_e) = \begin{bmatrix} \mathbb{C}_{c(\circ)}(\boldsymbol{q}_{e_r}) & \mathbb{C}_{c(\circ)}(\boldsymbol{q}_{e_l}) \end{bmatrix} \in \Re^{6\times c_e}, \tag{2.81}$$

$$\mathbb{C}_{m(\circ)}(\boldsymbol{q}_e) = \begin{bmatrix} \mathbb{C}_{m(\circ)}(\boldsymbol{q}_{e_r}) & \mathbb{C}_{m(\circ)}(\boldsymbol{q}_{e_l}) \end{bmatrix} \in \Re^{6\times \eta_e}. \tag{2.82}$$

With this notation, (2.79) can be written in compact form as

$$\mathbb{C}_R^T(\boldsymbol{q}_e)\mathcal{V}_R + \mathcal{J}_R(\boldsymbol{q}_e)\dot{\boldsymbol{\theta}}_e = \mathbb{C}_e^T(\boldsymbol{q}_e)\mathcal{V}_e + \mathbb{B}^T(\boldsymbol{q}_e)\mathcal{V}'(\boldsymbol{q}_e), \tag{2.83}$$

where

$$\mathbb{C}_{(\circ)}^T(\boldsymbol{q}_e) = \mathbb{B}^T(\boldsymbol{q}_e)\begin{bmatrix} \mathbb{T}_{\overleftarrow{e_r(\circ)}}^T(\boldsymbol{q}_{e_r}) & \mathbb{T}_{\overleftarrow{e_l(\circ)}}^T(\boldsymbol{q}_{e_l}) \end{bmatrix}^T \in \Re^{12\times 6},$$

$$\mathcal{J}_R(\boldsymbol{q}_e) = \mathbb{B}^T(\boldsymbol{q}_e)\boldsymbol{J}_R(\boldsymbol{q}_e) \in \Re^{12\times n_e},$$

$$\mathcal{V}'(\boldsymbol{q}_e) = [(\mathcal{V}'_{e_r})^T \quad (\mathcal{V}'_{e_l})^T]^T.$$

As an example, consider the frequently occurring case of double stance on stationary ground: $e = F$, $R = B$, $\mathcal{V}_F = \mathbf{0}$. Then, (2.79) becomes

$$\begin{bmatrix} \mathbb{C}_{cR}^T(\boldsymbol{q}_F) \\ \mathbb{C}_{mR}^T(\boldsymbol{q}_F) \end{bmatrix} \mathcal{V}_B + \begin{bmatrix} \mathcal{J}_{cR}(\boldsymbol{q}_F) \\ \mathcal{J}_{mR}(\boldsymbol{q}_F) \end{bmatrix} \dot{\boldsymbol{\theta}}_F = \mathbb{B}^T(\boldsymbol{q}_F) \begin{bmatrix} \mathcal{V}'_{F_r} \\ \mathcal{V}'_{F_l} \end{bmatrix} = \begin{bmatrix} \mathbf{0} \\ \overline{\mathcal{V}}_F^m \end{bmatrix}. \tag{2.84}$$

Note that, since the twist of the loop-closure link is zero ($\mathcal{V}_F = \mathbf{0}$), only relative-motion twists $\mathcal{V}'_{F_j}$ are present. Note also that the projection of these twists along the constrained motion directions results in a zero, under the assumption of *hard constraints*: $\overline{\mathcal{V}}_F^c = \mathbf{0}$. From the upper equation it is apparent that joint rates $\dot{\boldsymbol{\theta}}_F$ ensure the instantaneous motion of the base frame, expressed by twist $\mathcal{V}_B$. From the lower equation, on the other hand, it is apparent that the joint rates also ensure the relative motion of the feet, i.e. sliding on the floor with a twist around the vertical, determined by the composite spatial velocity $\overline{\mathcal{V}}_F^m$. Appropriate choices for the joint rates will be introduced in the following section.

2.10.2 Inverse Kinematics Solution

The inverse kinematics problem for closed loop e can be formulated as follows: "Given the loop-closure and root-link twists $\mathcal{V}_e$ and $\mathcal{V}_R$ and the relative end-link twist $\mathcal{V}_k'$, find the limb joint velocities $\dot{\boldsymbol{\theta}}_k$, $k \in \{e_r, e_l\}$." The solution to the above problem can be derived from (2.73). It would be straightforward to solve this equation for the joint velocity that instantaneously satisfies both constraints determined by the r.h.s., i.e. the loop-closure one and the end-link relative velocity one. It is important to note, however, that the former constraint is a physical one while the latter is a task-induced one. Having in mind that physical constraints should always be satisfied, it would be preferable to resort to a solution with a priority structure such that the loop-closure constraints have the higher priority. To this end, first solve the upper part of (2.73) for the limb joint velocity, i.e.

$$\dot{\boldsymbol{\theta}}_k = \mathcal{J}_{cR}^{+}(\boldsymbol{q}_k)\widetilde{\mathcal{V}}_k^c + \left(\boldsymbol{E} - \mathcal{J}_{cR}^{+}(\boldsymbol{q}_k)\mathcal{J}_{cR}(\boldsymbol{q}_k)\right)\dot{\boldsymbol{\theta}}_{ku}, \tag{2.85}$$

where $\widetilde{\mathcal{V}}_k^c = \overline{\mathcal{V}}_k^c - \mathbb{C}_{cR}^T(\boldsymbol{q}_k)\mathcal{V}_R$. The joint velocity $\dot{\boldsymbol{\theta}}_{ku}$ of the unconstrained limb is then determined to satisfy the end-link relative motion task constraint in the lower part of (2.73). As a result, one obtains the following constrained least-squares solution (cf. (2.49)):

$$\begin{aligned}\dot{\boldsymbol{\theta}}_k &= \mathcal{J}_{cR}^{+}(\boldsymbol{q}_k)\widetilde{\mathcal{V}}_k^c + \overline{\mathcal{J}}_{mR}^{+}(\boldsymbol{q}_k)\widetilde{\mathcal{V}}_k^m + \left(\boldsymbol{E} - \boldsymbol{J}_R^{+}(\boldsymbol{q}_k)\boldsymbol{J}_R(\boldsymbol{q}_k)\right)\dot{\boldsymbol{\theta}}_{ku} \\ &= \dot{\boldsymbol{\theta}}_k^c + \dot{\boldsymbol{\theta}}_k^m + \dot{\boldsymbol{\theta}}_k^n, \quad \text{s.t. } \dot{\boldsymbol{\theta}}_k^c \succ \dot{\boldsymbol{\theta}}_k^m \succ \dot{\boldsymbol{\theta}}_k^n.\end{aligned} \tag{2.86}$$

Here $\overline{\mathcal{J}}_{mR}(\boldsymbol{q}_k) = \mathcal{J}_{mR}(\boldsymbol{q}_k)\boldsymbol{N}(\mathcal{J}_{cR}(\boldsymbol{q}_k))$ is the end-link mobility Jacobian restricted by the null space of the limb constraint Jacobian, $\mathcal{N}\left(\mathcal{J}_{cR}(\boldsymbol{q}_k)\right)$. The end-link velocity $\widetilde{\mathcal{V}}_k^m = \overline{\mathcal{V}}_k^m - \mathcal{J}_{mR}(\boldsymbol{q}_k)\mathcal{J}_{cR}^{+}(\boldsymbol{q}_k)\widetilde{\mathcal{V}}_k^c - \mathbb{C}_{mR}^T(\boldsymbol{q}_k)\mathcal{V}_R$. The joint velocity $\dot{\boldsymbol{\theta}}_{ku}$ can be reused to parametrize any remaining DoFs within the null space of the limb Jacobian, $\mathcal{N}\left(\boldsymbol{J}_R(\boldsymbol{q}_k)\right)$. Such DoFs are available when the limb is kinematically redundant ($n_k - 6 > 0$); they determine its self-motion.

On the other hand, when the limb is nonredundant ($n_k = 6$), the last term in the above equation vanishes. This is the case with the leg branches (cf. Fig. 2.13), i.e. $e = F$ and $R = B$. Recall the example in 2.10.1 where a stationary support with *hard contacts* was assumed. Then, $\mathcal{V}_F = \boldsymbol{0} = \overline{\mathcal{V}}_{F_j}^c$ (cf. (2.84)). The higher-priority component (the first term on the r.h.s. of (2.86)) yields then instantaneous motion in each leg that will only contribute to the base link velocity $\mathcal{V}_B$. The lower-priority component (the second term), on the other hand, will produce a sliding motion within the support plane and/or a rotation around the plane normal. This subtask can be accomplished independently from the respective subtask with the other leg.

Further on, in many cases high-friction contacts at the feet are assumed, s.t. foot velocity $\mathcal{V}_{F_j}$ becomes identically zero (under the assumption of a stationary support surface). This also implies that the constraint basis at the feet $\mathbb{B}_c = \boldsymbol{E}_6$. Hence, $\mathcal{J}_{cB}(\boldsymbol{q}_{F_j})$ becomes identical to the limb Jacobian $\boldsymbol{J}_B(\boldsymbol{q}_{F_j})$ (a square matrix). Since the feet are fixed, the relative twist $\mathcal{V}_F' = \boldsymbol{0}$. Then, from (2.86) the following unique solution is obtained:

$$\dot{\boldsymbol{\theta}}_{F_j} = \boldsymbol{J}_B^{-1}(\boldsymbol{q}_{F_j})\mathcal{V}_B. \tag{2.87}$$

The closed-loop formulas derived above hold even when a contact breaks and the respective closed loop ceases to exist. For instance, consider the change from a double to a single leg stance, e.g. when the left foot lifts off the floor. Then, the last equation expresses the inverse kinematics relationship for a nonredundant serial-link limb (the right leg, $j = r$) on a fixed base, provided the contact is maintained.

The limb joint velocities derived from the above equations will be henceforth referred to as *constraint-compatible* or *constraint-consistent*.

2.10.3 Forward Kinematics Solution

The forward kinematics problem for closed loop e can be formulated as follows: "Given an arbitrary loop-root twist $\mathcal{V}_R$ (or loop-closure twist $\mathcal{V}_e$) and a (constraint-consistent) loop joint velocity $\dot{\boldsymbol{\theta}}_e$, find the loop-closure link twist $\mathcal{V}_e$ (or loop-root twist $\mathcal{V}_R$)." This formulation reflects the relative character of the instantaneous motion relation within the closed loop. The solution, obtained from the upper part of (2.79), will be unique only when the rank of each matrix $\mathbb{B}_c^T(\boldsymbol{q}_e)$, $\mathbb{C}_{cR}^T(\boldsymbol{q}_e) \in \Re^{c_e \times 12}$, is six. This implies six independent loop constraints, i.e. $c_e = 6$. The closed loop state is described then as *fully constrained*. As an example, consider closed loop H formed by the rod, with cylindrical contact joints (cf. Fig. 2.13). Each contact joint imposes four constraints; there are a total of eight constraints within the closed loop. Note, however, that the constraints are *not independent* and hence the rod is not fully constrained: it can rotate and translate between the hands. To obtain a fully constrained object, it would be sufficient to assume that one of the contacts (e.g. at the right hand) is characterized by high friction. Then, the number of independent constraints in the closed loop will be $c_H = 6$. Assume also that the loop-root (torso) twist $\mathcal{V}_T$ is known. The unique solution for the rod twist is then obtained from the upper part of (2.79) as

$$\mathcal{V}_H = \mathbb{B}_c^{-T}(\boldsymbol{q}_{H_r})\left(\mathbb{C}_{cR}^T(\boldsymbol{q}_{H_r})\mathcal{V}_T + \mathcal{J}_{cT}(\boldsymbol{q}_{H_r})\dot{\boldsymbol{\theta}}_{H_r}\right). \tag{2.88}$$

Matrix $\mathbb{B}_c^{-T}(\boldsymbol{q}_{H_r}) \in \Re^{6\times 6}$ represents a spatial transform, regular at all postures. Apparently, the motion of the object is completely determined by the instantaneous motion of the right hand. Arbitrary *constraint-consistent* loop velocities can then be assigned to $\dot{\boldsymbol{\theta}}_H$. The left hand may thereby move relative to the motion of the object, i.e. translating and/or rotating along the unconstrained directions at the left hand contact joint.

In the case $c_e > 6$, matrices $\mathbb{B}_c^T(\boldsymbol{q}_e)$ and $\mathbb{C}_{cR}^T(\boldsymbol{q}_e)$ will have ranks greater than six. The closed loop is then said to be *unilaterally overconstrained*. This situation appears quite frequently in closed loops, $e = F$ (and $R = B$), with a double stance posture. Assuming that the contacts at the feet are characterized by high friction and are *always maintained*, $c_{F_j} = 6$ and $\mathcal{J}_{cB}(\boldsymbol{q}_{F_j}) = \boldsymbol{J}_B(\boldsymbol{q}_{F_j})$. As already clarified above, in this case there is a unique solution to the inverse kinematics problem which would produce the appropriate constraint-consistent joint velocity for the given loop-closure link velocity. This can be verified by inserting the closed-loop joint velocity obtained from (2.87) into the constraint equation (the upper part of (2.79)), under the assumption that $\mathcal{V}_k = \boldsymbol{0}$. Then we have

$$-\begin{bmatrix}\mathbb{C}_{cB}^T(\boldsymbol{q}_{F_r})\\ \mathbb{C}_{cB}^T(\boldsymbol{q}_{F_l})\end{bmatrix}\mathcal{V}_B = \begin{bmatrix}\boldsymbol{J}_B(\boldsymbol{q}_{F_r}) & \boldsymbol{0}\\ \boldsymbol{0} & \boldsymbol{J}_B(\boldsymbol{q}_{F_l})\end{bmatrix}\begin{bmatrix}\boldsymbol{J}_B^{-1}(\boldsymbol{q}_{F_r}) & \boldsymbol{0}\\ \boldsymbol{0} & \boldsymbol{J}_B^{-1}(\boldsymbol{q}_{F_l})\end{bmatrix}\mathcal{V}_B = \begin{bmatrix}\mathcal{V}_B\\ \mathcal{V}_B\end{bmatrix}. \tag{2.89}$$

This implies that $\mathbb{C}_{cB}^T(\boldsymbol{q}_{F_r}) = \mathbb{C}_{cB}^T(\boldsymbol{q}_{F_l}) = -\boldsymbol{E}_6$.

In the case of $c_e < 6$, on the other hand, the rank of matrices $\mathbb{B}_c^T(\boldsymbol{q}_e)$ and $\mathbb{C}_{cR}^T(\boldsymbol{q}_e)$ is less than six and the loop is said to be *underconstrained*. An example of an underconstrained H-loop was mentioned above. Another example are frictionless contact joints at the feet, whereby $c_F = 3$. Then, there will be an infinite number of solutions for the loop-root link velocity $\mathcal{V}_B$. The set of solutions can be expressed as a sum of a particular solution, obtained via a generalized inverse, and a homogeneous solution, i.e.

$$\mathcal{V}_B = -\left(\mathbb{C}_{cB}^T(\boldsymbol{q}_F)\right)^{\#} \mathcal{J}_{cB}(\boldsymbol{q}_F)\dot{\boldsymbol{\theta}}_F + \left(\boldsymbol{E} - \mathbb{C}_{cB}^T(\boldsymbol{q}_F)\left(\mathbb{C}_{cB}^T(\boldsymbol{q}_F)\right)^{\#}\right)\mathcal{V}_{Ba}. \tag{2.90}$$

Here $\mathcal{V}_{Ba}$ denotes an arbitrary base link velocity that parametrizes the infinite set. This velocity is projected onto the null space of $\mathbb{C}_{cB}^T(\boldsymbol{q}_F)$ and accounts for the arbitrary velocities of the base resulting from insufficient loop constraints. This means that the motion of the base will not be controllable.

In the last equation, the generalized inverse of a transposed matrix appeared. This type of generalized inverse is frequently used in kinetostatic relations (cf. Section 3.4). It is straightforward to confirm that the identity $(\boldsymbol{X}^T)^{\#} = (\boldsymbol{X}^{\#})^T$ holds, which implies that the order of the transpose/inverse operations is irrelevant. In addition, the notation will be simplified by dropping the brackets. For example, the generalized inverse in the last expression will be denoted as $\mathbb{C}_{cB}^{\#T}(\boldsymbol{q}_F)$.

2.11 DIFFERENTIAL MOTION RELATIONS OF A HUMANOID ROBOT

2.11.1 Quasivelocity, Holonomic and Nonholonomic Contact Constraints

Quasivelocity, holonomic and nonholonomic constraints are related to the integrability of the first- and second-order differential relations. First, recall that *generalized coordinates* are *physical coordinates* that define the DoFs of the system. *Generalized velocity*, on the other hand, may not necessarily be defined as the time derivative of the generalized coordinates. It is possible to define generalized velocity as a linear combination of the time derivatives of the generalized coordinates. For example, as already clarified in Section 2.4.2, the angular velocity components of a body can be represented as linear combinations of the Euler angle derivatives. No physical coordinates are associated with the integrals of the angular velocity components. Thus, the integrals are referred to as the *quasicoordinates* [95]. Furthermore, the term *quasivelocity* can be introduced to distinguish a generalized velocity that incorporates angular velocity components, from generalized velocities that are time differentials of physical coordinates [49].

The differential motion relations will be expressed as functions of the generalized coordinates of the robot; $\boldsymbol{q} = (\mathcal{X}_B, \boldsymbol{\theta})$, where $\mathcal{X}_B$ denotes the 6D position of the base link derived with the preferred choice of local coordinates (e.g. Euler angles or Euler parameters; cf. Section 2.4.2) and $\boldsymbol{\theta} = \begin{bmatrix}\boldsymbol{\theta}_{F_r}^T & \boldsymbol{\theta}_{F_l}^T & \boldsymbol{\theta}_{H_r}^T & \boldsymbol{\theta}_{H_l}^T\end{bmatrix}^T \in \Re^n$ stands for the joint angle vector, $n \geq 24$ denoting the total number of joints. Further on, a quasivelocity will be denoted with the usual overdot that signifies differentiation w.r.t. time. Different types of quasivelocities will be distinguished via a subscript. For example, the quasivelocity associated with the base-link twist $\mathcal{V}_B$, henceforth referred to as the *base quasivelocity*, will be denoted as $\dot{\boldsymbol{q}}_B = (\mathcal{V}_B, \dot{\boldsymbol{\theta}})$. Another type of quasivelocity, quite frequently used in the field of humanoid robotics, will be introduced in Section 2.11.4. The quasivelocity notation enables the omitting of linear transforms like (2.15) throughout the equations, thus rendering them in simpler form. In simulation and control algorithms, though, this type of transform would be required in the numerical integration procedures.

Next, consider the contact constraints. So far, these constraints were assumed holonomic. This is justified in most situations, when the humanoid robot assumes a single or double leg stance posture with high friction, whereby the base link is (unilaterally) fully constrained or overconstrained. In the case of holonomic constraints, the first-order differential motion constraint (the upper part of (2.79)) is integrable. On the other hand, there are specific postures whereby the base link is characterized as underconstrained. For example, when the robot is on hard ground but the feet are rolling, or when the robot is balancing on soft ground. In this case, the contact constraints are said to be nonholonomic. A nonholonomic contact constraint implies that the time integral of the first-order differential motion constraint has no physical meaning. Nonholonomic contact constraints also appear during multifinger object manipulation with rolling contacts without slipping [103]. Note also that nonholonomic constraints do not always stem from contacts: in Chapter 4 the angular momentum conservation constraint will be introduced to account for a mid-air posture of the robot, e.g. during running or jumping. Nonholonomic motion constraints necessitate a respective nonholonomic motion planning approach to arrive at desired states that are not directly accessible via the differential relation.

In the case of holonomic contact constraints, the 6D position of the floating base can be expressed in terms of the joint variables. Indeed, as already clarified in the last section, for a single or double stance posture with high-friction contacts, there is a unique solution to the forward kinematics problem. This implies that the 6D position of the base link can be obtained from the joint angle sensor data. On the other hand, in the case of nonholonomic contact constraints, the 6D position of the base link cannot be obtained from the joint angle data alone; inertial measurement data will be needed to evaluate the state. This is the reason why the 6D position of the base link should be included into the model as a generalized coordinate. The following discussion will cover both holonomic and nonholonomic constraints.

2.11.2 First-Order Differential Motion Relations Expressed in Terms of Base Quasivelocity

The instantaneous motion relations for independent closed loops derived in Section 2.10.1 apply under the following conditions. First, substitute independent closed-loop generalized coordinates $\boldsymbol{q}_e$ with the generalized coordinates of the robot, $\boldsymbol{q}$ (as defined in 2.11). Next, assume that closed loops are formed by p contact joints. The maximum number of constrained motion directions is then $6p$, i.e. in the case of bilateral contacts. In the general case of mixed unilateral/bilateral constraints, the total number of motion constraints is $c = \sum c_k \leq 6p$. The number of unconstrained motion directions is then determined as $\eta = \sum \eta_k = 6p - c$. For clarity, in the following discussion it will be assumed that contact joints are formed at the end links only ($p = 4$).

Furthermore, the only root link is the base link: $R = B$. The *contact map of the robot w.r.t. the base link* is $\mathbb{C}_B(\boldsymbol{q}) \in \Re^{6\times 24}$, with components $\mathbb{C}_{cB}(\boldsymbol{q}) \in \Re^{6\times c}$ and $\mathbb{C}_{mB}(\boldsymbol{q}) \in \Re^{6\times \eta}$. Jacobian $\boldsymbol{J}_B(\boldsymbol{q}) \in \Re^{24\times n}$ is the *complete Jacobian* of the robot. The twist of the base link, $\mathcal{V}_B$, is determined by the specific task, e.g. to produce a variation in the whole-body posture while maintaining the contacts. The number of closure links, on the other hand, varies with the application task. In the case of a dual-arm manipulation task, two closure links are involved: the floor supporting the feet ($e = F$) and the object ($e = H$). This implies two independent closed loops, as already explained. In the case of a surface cleaning task, on the other hand, the environment (which includes the ground) is the only closure link. Both cases can be handled with the following notation.

The *first-order differential motion constraint* of the humanoid robot can be written in the form of (2.79), i.e.

$$\begin{bmatrix} \mathbb{C}^T_{cB}(\boldsymbol{q}) \\ \mathbb{C}^T_{mB}(\boldsymbol{q}) \end{bmatrix} \mathcal{V}_B + \begin{bmatrix} \mathcal{J}_{cB}(\boldsymbol{q}) \\ \mathcal{J}_{mB}(\boldsymbol{q}) \end{bmatrix} \dot{\boldsymbol{\theta}} = \begin{bmatrix} \bar{\mathcal{V}}^c \\ \bar{\mathcal{V}}^m \end{bmatrix}. \tag{2.91}$$

Note that $\bar{\mathcal{V}}^c \perp \bar{\mathcal{V}}^m$. Jacobians $\mathcal{J}_{cB}(\boldsymbol{q}) \in \Re^{c\times n}$ and $\mathcal{J}_{mB}(\boldsymbol{q}) \in \Re^{\eta\times n}$ will be referred to as the *joint-space constraint* and *mobility* Jacobians of the robot, respectively. The two parts of the above equation have the same inner structure, which is revealed by the following expansion of the upper part:

$$\begin{aligned} \begin{bmatrix} \bar{\mathcal{V}}^c_{F_r} \\ \bar{\mathcal{V}}^c_{F_l} \\ \bar{\mathcal{V}}^c_{H_r} \\ \bar{\mathcal{V}}^c_{H_r} \end{bmatrix} &= \begin{bmatrix} \mathbb{C}^T_{cB}(\boldsymbol{q}_{F_r}) \\ \mathbb{C}^T_{cB}(\boldsymbol{q}_{F_l}) \\ \mathbb{C}^T_{cB}(\boldsymbol{q}_{H_r}) \\ \mathbb{C}^T_{cB}(\boldsymbol{q}_{H_l}) \end{bmatrix} \mathcal{V}_B + \begin{bmatrix} \mathcal{J}_{cB}(\boldsymbol{q}_{F_r}) & \boldsymbol{0} & \boldsymbol{0} & \boldsymbol{0} \\ \boldsymbol{0} & \mathcal{J}_{cB}(\boldsymbol{q}_{F_l}) & \boldsymbol{0} & \boldsymbol{0} \\ \boldsymbol{0} & \boldsymbol{0} & \mathcal{J}_{cB}(\boldsymbol{q}_{H_r}) & \boldsymbol{0} \\ \boldsymbol{0} & \boldsymbol{0} & \boldsymbol{0} & \mathcal{J}_{cB}(\boldsymbol{q}_{H_l}) \end{bmatrix} \begin{bmatrix} \dot{\boldsymbol{\theta}}_{F_r} \\ \dot{\boldsymbol{\theta}}_{F_l} \\ \dot{\boldsymbol{\theta}}_{H_r} \\ \dot{\boldsymbol{\theta}}_{H_l} \end{bmatrix} \\ &= \begin{bmatrix} \mathbb{B}^T_c(\boldsymbol{q}_{F_r}) & \boldsymbol{0} & \boldsymbol{0} & \boldsymbol{0} \\ \boldsymbol{0} & \mathbb{B}^T_c(\boldsymbol{q}_{F_r}) & \boldsymbol{0} & \boldsymbol{0} \\ \boldsymbol{0} & \boldsymbol{0} & \mathbb{B}^T_c(\boldsymbol{q}_{H_r}) & \boldsymbol{0} \\ \boldsymbol{0} & \boldsymbol{0} & \boldsymbol{0} & \mathbb{B}^T_c(\boldsymbol{q}_{H_l}) \end{bmatrix} \begin{bmatrix} \mathcal{V}_{F_r} \\ \mathcal{V}_{F_l} \\ \mathcal{V}_{H_r} \\ \mathcal{V}_{H_l} \end{bmatrix}. \end{aligned} \tag{2.92}$$

The compact-form notation of (2.91) can be written in analogy to (2.83) as

$$\bar{\mathcal{V}} = \mathbb{C}^T_B(\boldsymbol{q})\mathcal{V}_B + \mathcal{J}_B(\boldsymbol{q})\dot{\boldsymbol{\theta}} = \mathbb{B}^T(\boldsymbol{q})\mathcal{V}(\boldsymbol{q}). \tag{2.93}$$

Here $\mathcal{J}_B(\boldsymbol{q}) = \mathbb{B}^T(\boldsymbol{q})\boldsymbol{J}_B(\boldsymbol{q}) \in \Re^{24\times n}$ denotes the permuted Jacobian of the robot. The vector $\mathcal{V}(\boldsymbol{q})$ collects all contact link twists, i.e.

$$\mathcal{V}(\boldsymbol{q}) = \begin{bmatrix} \mathcal{V}(\boldsymbol{q}_F)^T & \mathcal{V}(\boldsymbol{q}_H)^T \end{bmatrix}^T \in \Re^{24}, \tag{2.94}$$

where $\mathcal{V}(\boldsymbol{q}_e) = \begin{bmatrix} \mathcal{V}^T_{e_r} & \mathcal{V}^T_{e_l} \end{bmatrix}^T \in \Re^{12}$.

In the case of holonomic constraints, the first-order differential relations of the humanoid robot, (2.91), can be written in the following compact form:

$$\begin{bmatrix} \boldsymbol{J}_{cB}(\boldsymbol{q}) \\ \boldsymbol{J}_{mB}(\boldsymbol{q}) \end{bmatrix} \dot{\boldsymbol{q}}_B = \begin{bmatrix} \bar{\mathcal{V}}^c \\ \bar{\mathcal{V}}^m \end{bmatrix}. \tag{2.95}$$

Here $J_{cB} \equiv \left[\mathbb{C}_{cB}^T \quad \mathcal{J}_{cB}\right] \in \Re^{c\times(n+6)}$ and $J_{mB} \equiv \left[\mathbb{C}_{mB}^T \quad \mathcal{J}_{mB}\right] \in \Re^{\eta\times(n+6)}$ are referred to as the *constraint* and *mobility Jacobian* of the robot, respectively.

Structural Changes

The above notation is general in the sense that *structural changes* can be treated at ease. Consider first the case of two independent closed loops (F and H). In this case, the end-link twists are determined, in accordance with (2.70), as

$$\begin{aligned} \mathcal{V}_{e_j} &= \mathbb{T}_{\overleftarrow{e_j e}}(\boldsymbol{q}_{e_j})\mathcal{V}_e + \mathcal{V}_{e_j}', \\ & e \in \{F, H\},\ j \in \{r, l\}. \end{aligned}$$

The projection of these twists along the constrained motion directions produces the closure-link twist, i.e.

$$\mathcal{V}_e = \mathbb{C}_{ce}^T(\boldsymbol{q}_e)\mathcal{V}(\boldsymbol{q}_e).$$

As already noted, the constraint bases annihilate the relative motion twists $\mathcal{V}_k'$.

Next, when one of the closed loops or both cease to exist, one or more of the end links will be completely free. The constraint bases of the respective end links change, s.t. ${}^k\mathbb{B}_c = \boldsymbol{0}_6$ and ${}^k\mathbb{B}_m = \boldsymbol{E}_6$. Apparently, the notation directly accounts for the single foot stances.

In multicontact tasks, there is a single immobile loop-closure link (e.g. the floor F). Also, in many cases hard contacts can be assumed. Then, the constraint equation of the humanoid robot (the upper part in (2.95)) can be written in the simple form

$$J_{cB}(\boldsymbol{q})\dot{\boldsymbol{q}}_B = \boldsymbol{0}. \tag{2.96}$$

This equation appears quite frequently in the literature.

Constraint-Consistent Joint Velocity

Usually, it is assumed that the number of constraints is less than the DoFs of the robot ($c < n$). Then, the first-order differential motion constraint in (2.91) will be underdetermined. The general solution for the joint velocity can be derived from this equation as

$$\dot{\boldsymbol{\theta}} = \mathcal{J}_{cB}^+(\boldsymbol{q})\left(\overline{\mathcal{V}}^c - \mathbb{C}_{cB}^T(\boldsymbol{q})\mathcal{V}_B\right) + \left(\boldsymbol{E} - \mathcal{J}_{cB}^+(\boldsymbol{q})\mathcal{J}_{cB}(\boldsymbol{q})\right)\dot{\boldsymbol{\theta}}_u. \tag{2.97}$$

The two components on the r.h.s. denote the particular and homogeneous solutions. Assuming that all limbs are in nonsingular configuration (such posture is henceforth referred to as *regular*), any desired base-link velocity can be achieved with the help of the particular solution. The homogeneous solution, on the other hand, provides an infinite set of joint-velocity vectors that do not affect the state of the base link whatsoever. The null space of the joint-space constraint Jacobian, $\mathcal{N}(\mathcal{J}_{cB})$, is parametrized by the joint velocity $\dot{\boldsymbol{\theta}}_u$ of the unconstrained robot. This velocity can be determined by additional motion constraints, e.g. from the desired motion of a completely free end link, or the desired motion along the unconstrained directions at the contact joints. To this end, make use of the lower part of (2.91). The resultant inverse kinematics solution can be written as

$$\begin{aligned}\dot{\boldsymbol{\theta}} &= \mathcal{J}_{cB}^{+}(\boldsymbol{q})\left(\overline{\mathcal{V}}^{c} - \mathbb{C}_{cB}^{T}(\boldsymbol{q})\mathcal{V}_{B}\right) + \overline{\mathcal{J}}_{mB}^{+} + (\boldsymbol{q})\widetilde{\mathcal{V}}^{m} + \left(\boldsymbol{E} - \boldsymbol{J}_{B}^{+}(\boldsymbol{q})\boldsymbol{J}_{B}(\boldsymbol{q})\right)\dot{\boldsymbol{\theta}}_{u} \\ &= \dot{\boldsymbol{\theta}}^{c} + \dot{\boldsymbol{\theta}}^{m} + \dot{\boldsymbol{\theta}}^{n}, \quad \text{s.t. } \dot{\boldsymbol{\theta}}^{c} \succ \dot{\boldsymbol{\theta}}^{m} \succ \dot{\boldsymbol{\theta}}^{n}.\end{aligned} \tag{2.98}$$

Here $\overline{\mathcal{J}}_{mB}(\boldsymbol{q}) = \mathcal{J}_{mB}(\boldsymbol{q})\boldsymbol{N}(\mathcal{J}_{cB})$ is the mobility Jacobian restricted by $\mathcal{N}(\mathcal{J}_{cB}(\boldsymbol{q}))$. The twist $\widetilde{\mathcal{V}}^{m}$ is defined as $\widetilde{\mathcal{V}}^{m} = \overline{\mathcal{V}}^{m} - \mathcal{J}_{m}(\boldsymbol{q})\mathcal{J}_{cB}^{+}(\boldsymbol{q})(\overline{\mathcal{V}}^{c} - \mathbb{C}_{cB}^{T}(\boldsymbol{q})\mathcal{V}_{B})$. The joint velocity $\dot{\boldsymbol{\theta}}_{u}$ parametrizes now any remaining DoFs within $\mathcal{N}(\boldsymbol{J}_{B}(\boldsymbol{q}))$ that determine the robot's self-motion. With this hierarchical structure, the highest priority is assigned to the constrained base-link motion, ensured via the joint velocity $\dot{\boldsymbol{\theta}}^{c}$. The second component, $\dot{\boldsymbol{\theta}}^{m}$, ensures the motion of the end-links along the unconstrained motion directions. The lowest-priority component, $\dot{\boldsymbol{\theta}}^{n}$, can be used for additional postural adjustments. Since the highest priority is assigned to the motion constraint, joint velocity (2.98) is referred to as the *constraint-compatible* or *constraint-consistent* joint velocity of the robot.

Constraint-Consistent Generalized Velocity

Regular postures of the robot with single/double stance occur quite frequently; they play an essential role in propulsion and balance. As mentioned in Section 2.11.1, the contact constraints are then characterized as holonomic. Such constraints can be represented by a smooth C^2 vector-valued function, $\boldsymbol{\gamma}(\boldsymbol{q}) = \text{const} \in \Re^{c}$. Furthermore, in many practical tasks hard constraints are assumed. In this case, the first-order differential motion constraint of the robot, (2.96), is valid. This is a homogeneous system of c linear equations in $n+6$ unknowns. In the general case, when the constraints are less than the generalized coordinates ($c < n+6$), the system will be underdetermined. This equation is of the same type as the self-motion equation of a redundant manipulator, (2.28). Hence, the discussion in Section 2.7.1 is relevant. Accordingly, the system constraint matrix $\boldsymbol{J}_{cB}$ will induce an orthogonal decomposition of the generalized coordinate space locally. We have

$$\dot{\boldsymbol{q}}_{u} = \dot{\boldsymbol{q}}_{B} + \dot{\boldsymbol{q}}_{c}, \; \dot{\boldsymbol{q}}_{B} \perp \dot{\boldsymbol{q}}_{c}, \tag{2.99}$$

$$\dot{\boldsymbol{q}}_{B} = \boldsymbol{N}(\boldsymbol{J}_{cB})\dot{\boldsymbol{q}}_{u}, \tag{2.100}$$

$$\dot{\boldsymbol{q}}_{c} = (\boldsymbol{E} - \boldsymbol{N}(\boldsymbol{J}_{cB}))\dot{\boldsymbol{q}}_{u} = \boldsymbol{J}_{cB}^{+}\boldsymbol{J}_{cB}\dot{\boldsymbol{q}}_{u}. \tag{2.101}$$

Here $\boldsymbol{N}(\boldsymbol{J}_{cB}) = \left(\boldsymbol{E} - \boldsymbol{J}_{cB}^{+}\boldsymbol{J}_{cB}\right)$ is the null space projector. The generalized velocity of the unconstrained system is denoted as $\dot{\boldsymbol{q}}_{u} \in \Re^{n+6}$. Its projection yields generalized velocity $\dot{\boldsymbol{q}}_{B} \in \mathcal{N}(\boldsymbol{J}_{cB})$, a tangential component at the *constraint manifold* (the self-motion manifold in Section 2.7.1). This velocity, henceforth referred to as the *constraint-compatible* or *constraint-consistent* generalized velocity, ensures motion in harmony with the constraints. The generalized velocity $\dot{\boldsymbol{q}}_{c} \in \mathcal{R}^{T}(\boldsymbol{J}_{cB})$, on the other hand, is normal to the constraint manifold. Whenever the constraints are satisfied, $\dot{\boldsymbol{q}}_{c} = \boldsymbol{0}$. It is interesting to note that when the constraints are not satisfied, a nonzero $\dot{\boldsymbol{q}}_{c}$ will ensure that the constraint manifold behaves as an attractor; $\dot{\boldsymbol{q}}_{c}$ will be henceforth be referred to as the *constraint enforcing* generalized velocity.

From (2.99), it is straightforward to derive the respective joint velocity components by employing the following *underactuation filtering* matrix:

$$\boldsymbol{S} = \begin{bmatrix}\boldsymbol{0}_{n\times 6} & \boldsymbol{E}_{n}\end{bmatrix} \in \Re^{n\times(n+6)}. \tag{2.102}$$

With this matrix, the constraint-consistent joint velocity is obtained as

$$\dot{\boldsymbol{\theta}} = \boldsymbol{S}\dot{\boldsymbol{q}}_B = \overline{\mathbb{S}}(\boldsymbol{J}_{cB})\dot{\boldsymbol{q}}_u, \tag{2.103}$$

where $\overline{\mathbb{S}}(\boldsymbol{J}_{cB}) = \boldsymbol{S}\boldsymbol{N}(\boldsymbol{J}_{cB})$ is the restriction of $\boldsymbol{S}$ by the null space of $\boldsymbol{J}_{cB}$. Similarly, the constraint enforcing joint velocity is obtained as $\dot{\boldsymbol{\theta}}_c = \boldsymbol{S}\dot{\boldsymbol{q}}_c$. The two orthogonal joint velocities, $\dot{\boldsymbol{\theta}}$ and $\dot{\boldsymbol{\theta}}_c$, can be used as *noninterfering* control inputs in kinematics-based motion control; the former will produce the desired constraint-consistent motion along the constraint manifold, the latter can be used to ensure a locally minimized deviation from the manifold during the motion.

Notations based on the system constraint matrix, as above, lead to compactness when used in a general (e.g. QP) solver. In analysis, however, the notation should be used with care since important properties could be easily masked by matrix $\boldsymbol{S}$. Note also that the notation increases the computational burden, which might be critical in some cases (e.g. in the case of real-time control).

2.11.3 Second-Order Differential Motion Constraints and Their Integrability

The differential motion constraints introduced above have to be taken in consideration when deriving the equation of motion of the robot, as will be clarified in Chapter 4. The classical approach is to resort to a second-order representation [12,157]. To this end, differentiate (2.91) w.r.t. time. Then we have

$$\mathbb{C}_{cB}^T(\boldsymbol{q})\dot{\mathcal{V}}_B + \dot{\mathbb{C}}_{cB}^T(\boldsymbol{q})\mathcal{V}_B + \mathcal{J}_{cB}(\boldsymbol{q})\ddot{\boldsymbol{\theta}} + \dot{\mathcal{J}}_{cB}(\boldsymbol{q})\dot{\boldsymbol{\theta}} = \dot{\overline{\mathcal{V}}}^c. \tag{2.104}$$

The compact form of this equation (i.e. the time differential of the upper part in (2.95)) is written as

$$\boldsymbol{J}_{cB}(\boldsymbol{q})\ddot{\boldsymbol{q}}_B + \dot{\boldsymbol{J}}_{cB}(\boldsymbol{q})\dot{\boldsymbol{q}}_B = \dot{\overline{\mathcal{V}}}^c. \tag{2.105}$$

Henceforth $\ddot{\boldsymbol{q}}_B = \frac{d}{dt}\dot{\boldsymbol{q}}_B$ will be referred to as the *constraint-consistent generalized acceleration.*

Determining a constraint-consistent generalized acceleration is an important problem. Note that, since the system is kinematically redundant, there is an infinite number of constraint-consistent generalized accelerations. Not all of these accelerations can ensure contact twists $\overline{\mathcal{V}}^c$ that comply with the constraints. This is a problem of integrability as will be shown in what follows. To focus on the problem, assume for the time being hard constraints, s.t. $\dot{\overline{\mathcal{V}}}^c = \boldsymbol{0} = \overline{\mathcal{V}}^c$. The general solution to the above equation is then obtained as

$$\ddot{\boldsymbol{q}}_B = -\boldsymbol{J}_{cB}^+\dot{\boldsymbol{J}}_{cB}\dot{\boldsymbol{q}}_B + \boldsymbol{N}\boldsymbol{a}. \tag{2.106}$$

Note that vector $\boldsymbol{a}$ is usually assumed arbitrary. Although mathematically correct, this assumption is problematic, however, from the viewpoint of integrability. The integrability of (2.106) ensures that the respective first-order motion constraint, (2.96), is satisfied during the motion. The lack of integrability, on the other hand, would result in velocity drift, and hence in constraint violation. Such velocity drift has been observed and studied in torque minimization–based kinematic redundancy resolution schemes. Interested readers are referred to [151,62,88,86,87,123].

To ensure integrability, vector $\boldsymbol{a}$ in (2.106) has to be specified in the following way. First, observe that the constraint-consistent generalized acceleration can also be derived as the time differential of the generalized velocity (2.100), i.e.

$$\ddot{\boldsymbol{q}}_B = \dot{\boldsymbol{N}}\dot{\boldsymbol{q}}_u + \boldsymbol{N}\ddot{\boldsymbol{q}}_u. \tag{2.107}$$

The time derivative of the projection operator is

$$\begin{aligned}\dot{\boldsymbol{N}} &= -\left(\frac{d}{dt}\boldsymbol{J}_{cB}^{+}\right)\boldsymbol{J}_{cB} - \boldsymbol{J}_{cB}^{+}\dot{\boldsymbol{J}}_{cB} \\ &= \boldsymbol{J}_{cB}^{+}\dot{\boldsymbol{J}}_{cB}\boldsymbol{J}_{cB}^{+}\boldsymbol{J}_{cB} - \boldsymbol{N}\dot{\boldsymbol{J}}_{cB}^{T}(\boldsymbol{J}_{cB}\boldsymbol{J}_{cB}^{T})^{-1}\boldsymbol{J}_{cB} - \boldsymbol{J}_{cB}^{+}\dot{\boldsymbol{J}}_{cB} \\ &= -(\boldsymbol{L}+\boldsymbol{L}^{T}),\end{aligned} \tag{2.108}$$

where $\boldsymbol{L} \equiv \boldsymbol{J}_{cB}^{+}\dot{\boldsymbol{J}}_{cB}\boldsymbol{N}$. In the derivation, the following expression was used[3]:

$$\frac{d}{dt}\boldsymbol{J}_{cB}^{+} = -\boldsymbol{J}_{cB}^{+}\dot{\boldsymbol{J}}_{cB}\boldsymbol{J}_{cB}^{+} + \boldsymbol{N}\dot{\boldsymbol{J}}_{cB}^{T}(\boldsymbol{J}_{cB}\boldsymbol{J}_{cB}^{T})^{-1}. \tag{2.109}$$

Generalized acceleration (2.107) can then be rewritten as

$$\ddot{\boldsymbol{q}}_B = -\boldsymbol{J}_{cB}^{+}\dot{\boldsymbol{J}}_{cB}\boldsymbol{N}\dot{\boldsymbol{q}}_u + \boldsymbol{N}\left(\ddot{\boldsymbol{q}}_u - \dot{\boldsymbol{J}}_{c}^{T}\boldsymbol{J}_{cB}^{+T}\dot{\boldsymbol{q}}_u\right). \tag{2.110}$$

This expression identifies with (2.106) if and only if

$$\boldsymbol{N}\dot{\boldsymbol{q}}_u = \dot{\boldsymbol{q}}_B \tag{2.111}$$

and

$$\left(\ddot{\boldsymbol{q}}_u - \dot{\boldsymbol{J}}_{cB}^{T}\boldsymbol{J}_{cB}^{+T}\dot{\boldsymbol{q}}_u\right) = \boldsymbol{a}. \tag{2.112}$$

Relation (2.111) stands for the constraint-consistent generalized velocity (2.100).

The above result guarantees the integrability of the second-order constraint, in addition to that of the first-order one, via (2.100). Once the initial conditions in the simulation environment are set to satisfy the constraints, i.e. $\boldsymbol{q}(0) \equiv \boldsymbol{q}_0 : \boldsymbol{\gamma}(\boldsymbol{q}_0) = \boldsymbol{0}$ and $\dot{\boldsymbol{q}}(0) \equiv \dot{\boldsymbol{q}}_0 = \boldsymbol{N}\dot{\boldsymbol{q}}_u(0)$, the motion will evolve without constraint violation, provided there are no numerical errors. In practice however, there are difficulties in both aspects, the setting of the initial conditions as well as the presence of numerical errors. Details about the former problem are discussed in [119]. With regard to the latter problem, different approaches have been developed. Baumgarte's method is frequently used [11]; see also [5,155]. Accordingly, correcting terms for the errors in the generalized coordinates and velocities are added to the r.h.s. of the differential

[3] Derived by making use of $\boldsymbol{J}_{cB}^{+} = \boldsymbol{J}_{cB}^{T}(\boldsymbol{J}_{cB}\boldsymbol{J}_{cB}^{T})^{-1}$, $\frac{d}{dt}(\boldsymbol{A}\boldsymbol{A}^{-1} = \boldsymbol{E}) \Rightarrow \frac{d}{dt}\boldsymbol{A}^{-1} = -\boldsymbol{A}^{-1}\dot{\boldsymbol{A}}\boldsymbol{A}^{-1}$ for any square and nonsingular $\boldsymbol{A}$, and set $\boldsymbol{A} \equiv \boldsymbol{J}_{cB}\boldsymbol{J}_{cB}^{T}$. The expression appeared first in [117,118].

constraints, in a fashion resembling PD feedback control. This method may not be efficient, though, when the system constraint matrix is ill-conditioned. The errors are amplified then and result in unreasonably large accelerations and respective forces of constraint. Rank deficiency of the system constraint matrix (i.e. the case of dependent constraints), on the other hand, leads to a singularity problem resembling that discussed in Section 2.5.

The constraint-consistent generalized acceleration plays an important role in the dynamic analysis discussed in Chapter 4. It is therefore important to understand the role of each component in (2.110). First, recall the properties pertinent to the inverse differential kinematic relations of an underdetermined system, as explained in Section 2.7 for the case of a kinematically redundant limb. Note that the form of (2.110) is identical to that of (2.40), when the end-link acceleration is assumed zero and the pseudoinverse is used as a generalized inverse. As already mentioned, the generalized coordinate motion subspace is decomposed locally into two orthogonal subspaces, null space $\mathcal{N}(\boldsymbol{J}_{cB})$ and range space $\mathcal{R}(\boldsymbol{J}_{cB}^T)$. The former contains accelerations that are tangent to the constraint manifold at the given posture. There are two tangential components in (2.110), a linear and a nonlinear one. The acceleration $\ddot{\boldsymbol{q}}_u$, appearing in the linear tangential component, can be set in an appropriate way to achieve the desired acceleration without violating the constraints. The nonlinear tangential component can be represented as $-\dot{\boldsymbol{J}}_{cB}^T \boldsymbol{J}_{cB}^{+T} \dot{\boldsymbol{q}}_u = -\dot{\boldsymbol{J}}_{cB}^T(\boldsymbol{J}_{cB}\boldsymbol{J}_{cB}^T)^{-1}\boldsymbol{J}_{cB}\dot{\boldsymbol{q}}_c$, whereby (2.99) and (2.96) were used. This component will be identically zero under constraint-consistent motion ($\dot{\boldsymbol{q}}_c = \boldsymbol{0}$). Conversely, it can be concluded that the component will play an important role for compensating any drift from the constraint manifold and hence for integrability.

There is also a second nonlinear, state-dependent component: $-\boldsymbol{J}_{cB}^{+}\dot{\boldsymbol{J}}_c\dot{\boldsymbol{q}}_B$. This component is normal to the constraint manifold; it accounts for the centripetal/centrifugal accelerations stemming from the nonlinear geometry of the constraints. It can be shown that this component is in fact a constraint enforcing generalized acceleration. To this end, differentiate (2.101) w.r.t. time. Then we have

$$\ddot{\boldsymbol{q}}_c = (\boldsymbol{E} - \boldsymbol{N})\ddot{\boldsymbol{q}}_u - \dot{\boldsymbol{N}}\dot{\boldsymbol{q}}_u. \tag{2.113}$$

Applying (2.108) for $\dot{\boldsymbol{N}}$, premultiplying first by $\boldsymbol{J}_{cB}$ and then by $\boldsymbol{J}_{cB}^{+}$, and using identities $\boldsymbol{J}_{cB}\boldsymbol{N} = \boldsymbol{0}$ and $\boldsymbol{J}_{cB}\boldsymbol{J}_{cB}^{+} = \boldsymbol{E}$, one arrives at

$$\begin{aligned}\boldsymbol{J}_{cB}^{+}\boldsymbol{J}_{cB}\ddot{\boldsymbol{q}}_c &= \boldsymbol{J}_{cB}^{+}\boldsymbol{J}_{cB}\ddot{\boldsymbol{q}}_u - \boldsymbol{J}_{cB}^{+}\dot{\boldsymbol{J}}_{cB}\boldsymbol{J}_{cB}^{+}\boldsymbol{J}_{cB}\dot{\boldsymbol{q}}_u + \boldsymbol{J}_{cB}^{+}\dot{\boldsymbol{J}}_{cB}\dot{\boldsymbol{q}}_u \\ &= \boldsymbol{J}_{cB}^{+}\boldsymbol{J}_{cB}\ddot{\boldsymbol{q}}_u + \boldsymbol{J}_{cB}^{+}\dot{\boldsymbol{J}}_{cB}\dot{\boldsymbol{q}}_B.\end{aligned} \tag{2.114}$$

Hereby, use was made of $\boldsymbol{J}_{cB}^{+}\boldsymbol{J}_{cB}\dot{\boldsymbol{q}}_u = \dot{\boldsymbol{q}}_c$ (cf. (2.101)) and $\dot{\boldsymbol{q}}_B = \dot{\boldsymbol{q}}_u - \dot{\boldsymbol{q}}_c$. All terms in the above equation are accelerations from $\mathcal{R}(\boldsymbol{J}_{cB}^T)$. Taking the last term on the r.h.s. of (2.114) with minus sign renders its role as a constraint enforcing generalized acceleration.

In the case when motion is confined to the constraint manifold, (2.113) can be rewritten as

$$(\boldsymbol{E} - \boldsymbol{N})\ddot{\boldsymbol{q}}_B = \dot{\boldsymbol{N}}\dot{\boldsymbol{q}}. \tag{2.115}$$

This relation will be used later on. In this case, all terms in (2.114) are rendered identically zero. On the other hand, when motion drifts apart from the manifold, for any $\ddot{\boldsymbol{q}}_u$, (2.114) will ensure that the constraint manifold behaves as an attractor with regard to such drifting motion.

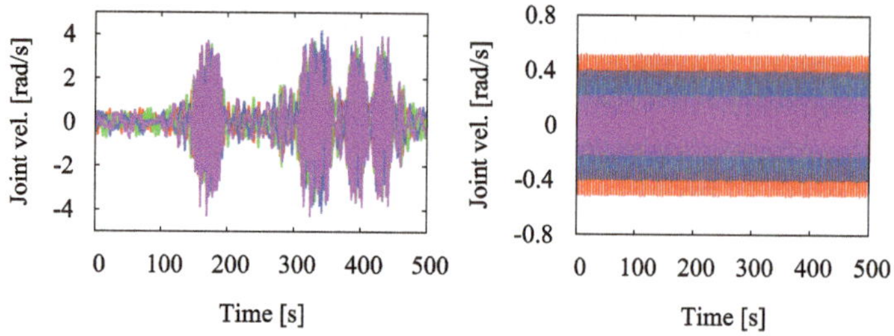

FIGURE 2.14 Acceleration-level redundancy resolution with periodic null-space joint acceleration input. Left (A): the null-space joint velocity obtained via the system-state joint velocity yields aperiodic behavior with large velocity peaks. Right (B): the null-space joint velocity obtained via the integral of the input yields periodic behavior with bounded magnitude.

Example

The goal is to demonstrate the important role of the joint velocity component derived from the null space. To this end, rewrite (2.110) as

$$\ddot{\boldsymbol{q}}_B = -\boldsymbol{J}^+_{cB}\dot{\boldsymbol{J}}_{cB}\dot{\boldsymbol{q}}_B + \boldsymbol{N}\left(\ddot{\boldsymbol{q}}_u - \dot{\boldsymbol{J}}^T_{cB}\boldsymbol{J}^{+T}_{cB}\dot{\boldsymbol{q}}_\zeta\right). \tag{2.116}$$

This equation will be used in the following two numerical simulations: Case (a) $\dot{\boldsymbol{q}}_\zeta = \dot{\boldsymbol{q}}_B$ and Case (b) $\dot{\boldsymbol{q}}_\zeta = \dot{\boldsymbol{q}}_u$. A fixed-base planar four-link manipulator with fixed end tip is employed (self-motion only, two DoRs). Arbitrary joint acceleration $\ddot{\boldsymbol{q}}_u$ is specified via simple harmonic motion components. Case (a) is the "conventional" approach, whereby the system-state joint velocity $\dot{\boldsymbol{q}}_B$ is employed. The results are shown in Fig. 2.14, in terms of system-state joint velocity. The "conventional" approach (on the l.h.s.) yields aperiodic joint velocity behavior with occasional velocity build-up. The integral approach (Case (b)), on the other hand, yields a periodic joint velocity behavior, as expected. This important result should be considered in control design.

2.11.4 First-Order Differential Motion Relations With Mixed Quasivelocity

The equations derived so far were expressed in terms of base quasivelocity. In the field of humanoid robotics, another type of quasivelocity is also used: $\dot{\boldsymbol{q}}_M = \begin{bmatrix}\mathcal{V}^T_M & \dot{\boldsymbol{\theta}}^T\end{bmatrix}^T$, $\mathcal{V}_M = \begin{bmatrix}\boldsymbol{v}^T_C & \boldsymbol{\omega}^T_B\end{bmatrix}^T$. Vector $\boldsymbol{v}_C$ denotes the velocity of the CoM of the robot, also referred to as the *system centroid*. The position and the instantaneous motion of the CoM play an important role in the static, kinematic, and dynamic relations of the humanoid robot. The two components of $\mathcal{V}_M$ are "mixed" in the sense that the CoM velocity is combined with the angular velocity of the base link; hence the "M" subscripts.

Given the quasivelocity in base coordinates, the CoM velocity can be written as[4]

$$\boldsymbol{v}_C = \boldsymbol{v}_B - [\boldsymbol{r}^\times_{\overleftarrow{CB}}]\boldsymbol{\omega}_B + \boldsymbol{J}_{\overleftarrow{CB}}(\boldsymbol{\theta})\dot{\boldsymbol{\theta}}. \tag{2.117}$$

[4] This relation is derived in Section 4.6.4.

Here $\boldsymbol{r}_{\overleftarrow{CB}}$ denotes the vector pointing from the base coordinate frame to the CoM. Matrix $\boldsymbol{J}_{\overleftarrow{CB}}$ is the *CoM Jacobian*:

$$\boldsymbol{J}_{\overleftarrow{CB}}(\boldsymbol{\theta}) = \frac{1}{M}\sum_{i=1}^{n} M_i \boldsymbol{J}_{vi}(\boldsymbol{\theta}) \in \Re^{3\times n}. \tag{2.118}$$

Here M_i and M denote the mass of Link i and the total mass of the robot, respectively;

$$\boldsymbol{J}_{vi} = \begin{bmatrix}[\boldsymbol{e}_1^{\times}]\boldsymbol{r}_{\overleftarrow{i1}} & [\boldsymbol{e}_2^{\times}]\boldsymbol{r}_{\overleftarrow{i2}} & \cdots & [\boldsymbol{e}_j^{\times}]\boldsymbol{r}_{\overleftarrow{ij}} & \boldsymbol{0} & \cdots & \boldsymbol{0}\end{bmatrix} \in \Re^{3\times n}$$

are Jacobians, $\boldsymbol{e}_j = \boldsymbol{R}_j{}^j\boldsymbol{e}_j$, ${}^j\boldsymbol{e}_j = [0\ 0\ 1]^T$, and $\boldsymbol{r}_{\overleftarrow{ij}}$ is the distance vector from the jth joint axis to the ith link CoM ($1 \le j \le i$).

Closed-chain differential motion relations (2.92) can be rewritten in terms of mixed quasivelocity, as follows. First, decompose the contact map in the constrained motion directions into force and moment component maps, i.e.

$$\mathbb{C}_{cB}(\boldsymbol{q}) \equiv \begin{bmatrix}\mathbb{C}_{cB_f}^T(\boldsymbol{q}) & \mathbb{C}_{cB_m}^T(\boldsymbol{q})\end{bmatrix}^T \in \Re^{6\times c}. \tag{2.119}$$

Expressions for the two components can be obtained via definition (2.74) of the respective limb contact map, i.e.

$$\begin{aligned}\mathbb{C}_{cB}(\boldsymbol{q}_k) &= \mathbb{T}_{\overleftarrow{kB}}^T(\boldsymbol{q}_k)\mathbb{B}_c(\boldsymbol{q}_k)\\ &= \begin{bmatrix}\boldsymbol{E} & \boldsymbol{0}\\ [\boldsymbol{r}_{\overleftarrow{kB}}^{\times}] & \boldsymbol{E}\end{bmatrix}\begin{bmatrix}\mathbb{B}_{cf}(\boldsymbol{q}_k) & \mathbb{B}_{cm}(\boldsymbol{q}_k)\end{bmatrix}.\end{aligned}$$

Thus,

$$\begin{aligned}\mathbb{C}_{cB_f}(\boldsymbol{q}_k) &= \mathbb{B}_{cf}(\boldsymbol{q}_k),\\ \mathbb{C}_{cB_m}(\boldsymbol{q}_k) &= \mathbb{B}_{cm}(\boldsymbol{q}_k) + [\boldsymbol{r}_{\overleftarrow{kB}}^{\times}]\mathbb{B}_{cf}(\boldsymbol{q}_k).\end{aligned} \tag{2.120}$$

These components are stacked (as in (2.81)) to obtain an expression for the contact map of the closed chain e in terms of the force $\mathbb{C}_{cB_f}(\boldsymbol{q}_e)$ and moment $\mathbb{C}_{cB_m}(\boldsymbol{q}_e)$ components. These components are in turn stacked to arrive at representation (2.119).

Next, note that via (2.117), the base twist can be expressed in terms of mixed quasivelocity as

$$\mathcal{V}_B = \mathbb{T}_{\overleftarrow{BC}}\mathcal{V}_M - \begin{bmatrix}\boldsymbol{J}_{\overleftarrow{CB}}(\boldsymbol{\theta})\\ \boldsymbol{0}\end{bmatrix}\dot{\boldsymbol{\theta}}. \tag{2.121}$$

Using this relation, the first term on the l.h.s. of (2.92) can be represented as

$$\begin{bmatrix}\mathbb{C}_{cB}^T(\boldsymbol{q}_F)\\ \mathbb{C}_{cB}^T(\boldsymbol{q}_H)\end{bmatrix}\mathcal{V}_B = \begin{bmatrix}\mathbb{C}_{cC}^T(\boldsymbol{q}_F)\\ \mathbb{C}_{cC}^T(\boldsymbol{q}_H)\end{bmatrix}\mathcal{V}_M - \begin{bmatrix}\mathbb{C}_{cB}^T(\boldsymbol{q}_F)\\ \mathbb{C}_{cB}^T(\boldsymbol{q}_H)\end{bmatrix}\begin{bmatrix}\boldsymbol{J}_{\overleftarrow{CB}}(\boldsymbol{\theta})\\ \boldsymbol{0}\end{bmatrix}\dot{\boldsymbol{\theta}}, \tag{2.122}$$

where $\mathbb{C}_{cC}(\boldsymbol{q}_e)$, $e \in \{F, H\}$ comprise stacked components;

$$\begin{aligned}\mathbb{C}_{cC}(\boldsymbol{q}_k) &= \mathbb{T}^T_{\overleftarrow{BC}}\mathbb{C}_{cB}(\boldsymbol{q}_k) \underset{(2.74)}{=} \mathbb{T}^T_{\overleftarrow{BC}}\mathbb{T}^T_{\overleftarrow{kB}}\mathbb{B}_c(\boldsymbol{q}_k)\\ &= \mathbb{T}^T_{\overleftarrow{kC}}\mathbb{B}_c(\boldsymbol{q}_k) \in \Re^{6\times 6}.\end{aligned} \tag{2.123}$$

With this notation, (2.92) can be rewritten as

$$\begin{aligned}&\begin{bmatrix}\mathbb{C}^T_{cC}(\boldsymbol{q}_F)\\ \mathbb{C}^T_{cC}(\boldsymbol{q}_H)\end{bmatrix}\mathcal{V}_M + \left(\begin{bmatrix}\mathcal{J}_{cB}(\boldsymbol{q}_F) & \mathbf{0}\\ \mathbf{0} & \mathcal{J}_{cB}(\boldsymbol{q}_H)\end{bmatrix} - \begin{bmatrix}\mathbb{C}^T_{cB_f}(\boldsymbol{q}_F)\\ \mathbb{C}^T_{cB_f}(\boldsymbol{q}_H)\end{bmatrix}\boldsymbol{J}_{\overleftarrow{CB}}(\boldsymbol{\theta})\right)\dot{\boldsymbol{\theta}}\\ &= \begin{bmatrix}\mathbb{B}^T_c(\boldsymbol{q}_F) & \mathbf{0}\\ \mathbf{0} & \mathbb{B}^T_c(\boldsymbol{q}_H)\end{bmatrix}\begin{bmatrix}\mathcal{V}(\boldsymbol{q}_F)\\ \mathcal{V}(\boldsymbol{q}_H)\end{bmatrix}.\end{aligned} \tag{2.124}$$

The general (e.g. nonholonomic constraints) compact-form representation of this equation is

$$\mathbb{C}^T_{cC}(\boldsymbol{q})\mathcal{V}_M + \mathcal{J}_{cM}(\boldsymbol{q})\dot{\boldsymbol{\theta}} = \mathbb{B}^T_c(\boldsymbol{q})\mathcal{V}(\boldsymbol{q}) \equiv \bar{\mathcal{V}}^c, \tag{2.125}$$

where

$$\mathcal{J}_{cM}(\boldsymbol{q}) = \mathcal{J}_{cB}(\boldsymbol{q}) - \mathbb{C}^T_{cB_f}(\boldsymbol{q})\boldsymbol{J}_{\overleftarrow{CB}}(\boldsymbol{\theta}). \tag{2.126}$$

Matrix $\mathbb{C}_{cC}(\boldsymbol{q}) \in \Re^{6\times c}$ is composed of stacked $\mathbb{C}_{cC}(\boldsymbol{q}_e)$ components. This is the contact map of the humanoid robot in the constrained directions, expressed in mixed quasicoordinates; $\mathbb{C}_{cC}$ can be decomposed into force and moment components, i.e.

$$\mathbb{C}_{cC}(\boldsymbol{q}) = \begin{bmatrix}\mathbb{C}^T_{cC_f}(\boldsymbol{q}) & \mathbb{C}^T_{cC_m}(\boldsymbol{q})\end{bmatrix}^T, \tag{2.127}$$

where $\mathbb{C}_{cC_f}(\boldsymbol{q}) = \mathbb{C}_{cB_f}(\boldsymbol{q}) = \mathbb{B}_{c_f}(\boldsymbol{q})$ and $\mathbb{C}_{cC_m}(\boldsymbol{q})$ is composed of stacked components

$$\mathbb{C}_{cC_m}(\boldsymbol{q}_k) = \mathbb{B}_{c_m}(\boldsymbol{q}_k) + [\boldsymbol{r}^\times_{\overleftarrow{kC}}]\mathbb{B}_{c_f}(\boldsymbol{q}_k).$$

Velocity relations in the mobility directions can be obtained in a similar fashion, resulting in

$$\mathbb{C}^T_{mC}(\boldsymbol{q})\mathcal{V}_M + \mathcal{J}_{mM}(\boldsymbol{q})\dot{\boldsymbol{\theta}} = \mathbb{B}^T_m(\boldsymbol{q})\mathcal{V}(\boldsymbol{q}) \equiv \bar{\mathcal{V}}^m. \tag{2.128}$$

In the special case of holonomic constraints, (2.125) and (2.128) can be written as

$$\boldsymbol{J}_{cM}(\boldsymbol{q})\dot{\boldsymbol{q}}_M = \bar{\mathcal{V}}^c \tag{2.129}$$

and

$$\boldsymbol{J}_{mM}(\boldsymbol{q})\dot{\boldsymbol{q}}_M = \bar{\mathcal{V}}^m, \tag{2.130}$$

respectively, where $\boldsymbol{J}_{cM}(\boldsymbol{q}) = \begin{bmatrix}\mathbb{C}^T_{cC}(\boldsymbol{q}) & \mathcal{J}_{cM}(\boldsymbol{q})\end{bmatrix}$ and $\boldsymbol{J}_{mM}(\boldsymbol{q}) = \begin{bmatrix}\mathbb{C}^T_{mC}(\boldsymbol{q}) & \mathcal{J}_{mM}(\boldsymbol{q})\end{bmatrix}$.

The complete instantaneous motion relation of the humanoid robot in mixed quasicoordinates can then be obtained by stacking (2.125) and (2.128), i.e.

$$\mathbb{C}_C^T(\boldsymbol{q})\mathcal{V}_M + \mathcal{J}_M(\boldsymbol{q})\dot{\boldsymbol{\theta}} = \mathbb{B}^T(\boldsymbol{q})\mathcal{V}(\boldsymbol{q}) \equiv \begin{bmatrix} \overline{\mathcal{V}}^c \\ \overline{\mathcal{V}}^m \end{bmatrix}. \tag{2.131}$$

Here $\mathbb{C}_C^T(\boldsymbol{q}) = \begin{bmatrix} \mathbb{C}_{cC}(\boldsymbol{q}) & \mathbb{C}_{mC}(\boldsymbol{q}) \end{bmatrix}^T$ and $\mathcal{J}_M(\boldsymbol{q}) = \begin{bmatrix} \mathcal{J}_{cM}^T(\boldsymbol{q}) & \mathcal{J}_{mM}^T(\boldsymbol{q}) \end{bmatrix}^T$ denote the complete contact map and the (permuted) joint-space Jacobian matrix of the robot for mixed quasivelocity, respectively.

Implementation Example

The constraint-consistent joint velocity can be expressed in terms of mixed quasivelocity by making use of (2.125). We have

$$\dot{\boldsymbol{\theta}} = \mathcal{J}_{cM}^+ \left(\overline{\mathcal{V}}^c - \mathbb{C}_{cC}^T \mathcal{V}_M \right) + \boldsymbol{N}(\mathcal{J}_{cM})\, \dot{\boldsymbol{\theta}}_u. \tag{2.132}$$

This equation can be used as a control equation for position-controlled robots, as will be shown in the following example.

Assume a double-stance initial posture on a stationary floor s.t. $\overline{\mathcal{V}}^c = \boldsymbol{0}$. The reference input is determined by the two components of twist $\mathcal{V}_M$: upward CoM translation along a vertical line and base rotation regulated at the initial (zero) value.[5] The resolved joint velocity obtained from the above equation is used to control a small-size robot with parameters similar to those of the HOAP-2 robot [39]. For the numbering of the joints and other relevant data, see Section A.1. The simulation result is shown in Video 2.11-1 [170]. Note that initially, the CoM upward motion is achieved with motion in both the legs and the arms. At a given time instant, the legs become fully stretched. This is a singular configuration w.r.t. the motion of the base link, as explained in Section 2.5. With respect to the desired CoM motion, however, the configuration is not singular. The upward CoM motion continues therefore without destabilization, determined now solely by the upward rotation of the arms. The graphs of the simulation and the snapshots are shown in Fig. 2.15. At around 2.5 s, there is an abrupt change in the motion pattern. The reason is that the legs are fully stretched then and the vertical base motion becomes saturated. Nevertheless, the motion remains stable as apparent from the joint velocity graphs. If the motion were to continue, at some point the arms would become fully stretched upwards. This is a singular posture w.r.t. the desired CoM motion; such posture usually yields destabilization. Another example based on the same approach will be presented in Section 7.5.3.

Further on, the additional arbitrary joint velocity input $\dot{\boldsymbol{\theta}}_u$ can be used to implement other subtasks, such as the end-link motion control along the unconstrained motion directions. The constraint-consistent joint velocity can be written in analogy to (2.98) as

$$\begin{aligned} \dot{\boldsymbol{\theta}} &= \mathcal{J}_{cM}^+(\boldsymbol{q}) \left(\overline{\mathcal{V}}^c - \mathbb{C}_{cC}^T(\boldsymbol{q})\mathcal{V}_M \right) + \overline{\mathcal{J}}_{mM}^+(\boldsymbol{q})\widetilde{\mathcal{V}}^m + \left(\boldsymbol{E} - \boldsymbol{J}_M^+(\boldsymbol{q})\boldsymbol{J}_M(\boldsymbol{q}) \right) \dot{\boldsymbol{\theta}}_u \\ &= \dot{\boldsymbol{\theta}}^c + \dot{\boldsymbol{\theta}}^m + \dot{\boldsymbol{\theta}}^n, \quad \text{s.t. } \dot{\boldsymbol{\theta}}^c \succ \dot{\boldsymbol{\theta}}^m \succ \dot{\boldsymbol{\theta}}^n. \end{aligned} \tag{2.133}$$

[5] The additional arbitrary joint velocity input is not used, i.e. $\dot{\boldsymbol{\theta}}_u = \boldsymbol{0}$.

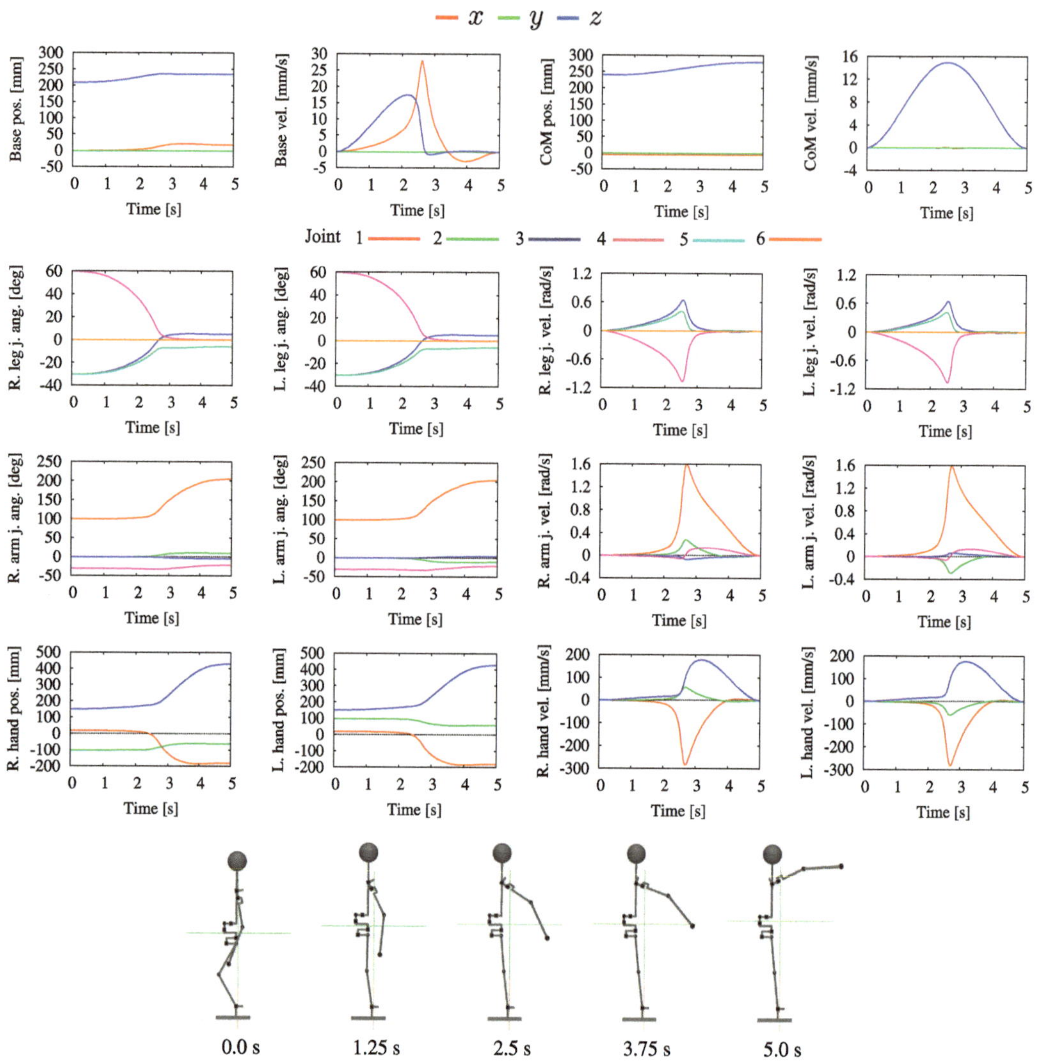

FIGURE 2.15 Simulation of a desired CoM upward motion without hand motion constraints. Initially, the desired motion is achieved with movements in both the legs and the arms. At around 2.5 s, there is an abrupt change in the motion pattern since the legs become fully stretched and the vertical base motion saturates. The CoM motion continues as desired, though, without any disruption.

Here $\overline{\mathcal{J}}_{mM}(\boldsymbol{q}) = \mathcal{J}_{mM}(\boldsymbol{q})\boldsymbol{N}(\mathcal{J}_{cM})$ is the mobility Jacobian restricted by the null space of the constraint Jacobian, $\mathcal{N}(\mathcal{J}_{cM}(\boldsymbol{q}))$. The twist $\widetilde{\mathcal{V}}^m$ is now defined as $\widetilde{\mathcal{V}}^m = \overline{\mathcal{V}}^m - \mathcal{J}_m(\boldsymbol{q})\mathcal{J}_{cM}^{+}(\boldsymbol{q})(\overline{\mathcal{V}}^c - \mathbb{C}_{cC}^T(\boldsymbol{q})\mathcal{V}_M)$. The joint velocity $\dot{\boldsymbol{\theta}}_u$ parametrizes any remaining DoFs within $\mathcal{N}(\boldsymbol{J}_M(\boldsymbol{q}))$. The above equation will be used for synergy-based motion generation in Section 7.5.

2.11.5 Summary and Discussion

The formulation of differential kinematic relations within two parallel branches forming an independent closed loop can be extended in a straightforward manner to deal with the case of interdependent closed loops formed by multiple parallel branches. The interdependency relation in this case results from the fact that all parallel branches have to accomplish a common main task: ensuring the spatial velocity of the loop-closure link in a consistent way. Thereby, the structure of the inverse kinematics solution can be preserved; only the dimension will increase. Multifinger object manipulation is a good representative example, as already pointed out.

Eq. (2.91) represents the first-order instantaneous motion relations in various types of systems with constraints. In multifinger grasping, the equivalent of the contact map $\mathbb{C}_{cB} : \Re^6 \to \Re^c$ is called the *grasp map* [104]. It plays an important role in motion/force analysis, grasp planning, and control. It will be shown in what follows that, likewise, $\mathbb{C}_{cB}$ plays an important role in humanoid robotics. The universality of the motion constraint equation can also be confirmed for the case when one or more of the contacts break and an open-loop structure is obtained. Furthermore, it has been shown that with the same equation it is possible to express both holonomic and nonholonomic constraints. With single and double leg stance, the constraints are said to be holonomic. When the base link is underconstrained, a nonholonomic constraint related to the angular momentum conservation may appear. This is also the case when the robot is in mid-air. Angular momentum–related issues will be discussed in Chapter 4.

Finally, it is interesting to note that the first-order differential motion constraint (the upper part of (2.79)) also describes the differential kinematics of parallel-link manipulators [50]. These manipulators comprise a fixed kinematic structure such that the loop-closure link (often referred to as the "traveling plate") connects to the parallel branches via fixed passive joints. The nature of the constraints is similar to that introduced by contact joints. The only difference is that the fixed passive joints cannot brake and thus they cannot induce a change in the kinematic structure, as is the case with humanoid robots and multifinger grasping.

References

[1] A.E. Albert, Regression and the Moore–Penrose Pseudoinverse, 1st edition, Academic Press, 1972.

[2] G. Antonelli, Stability analysis for prioritized closed-loop inverse kinematic algorithms for redundant robotic systems, IEEE Transactions on Robotics 25 (2009) 985–994.

[3] H. Arisumi, S. Miossec, J.-R. Chardonnet, K. Yokoi, Dynamic lifting by whole body motion of humanoid robots, in: IEEE/RSJ International Conference on Intelligent Robots and Systems, 2008, pp. 668–675.

[4] U.M. Ascher, H. Chin, S. Reich, Stabilization of DAEs and invariant manifolds, Numerische Mathematik 149 (1994) 131–149.

[5] U.M. Ascher, L.R. Petzold, Computer Methods for Ordinary Differential Equations and Differential-Algebraic Equations, Society for Industrial and Applied Mathematics, Philadelphia, PA, USA, 1998.

[6] T. Asfour, R. Dillmann, Human-like motion of a humanoid robot arm based on a closed-form solution of the inverse kinematics problem, in: IEEE/RSJ International Conference on Intelligent Robots and Systems, Las Vegas, Nevada, USA, 2003, pp. 1407–1412.

[7] J. Baillieul, Kinematic programming alternatives for redundant manipulators, in: IEEE International Conference on Robotics and Automation, 1985, pp. 722–728.

[8] J. Baillieul, Avoiding obstacles and resolving kinematic redundancy, in: IEEE International Conference on Robotics and Automation, 1986, pp. 1698–1704.

[9] R. Ball, The Theory of Screws: A Study in the Dynamics of a Rigid Body, Hodges, Foster, and Company, 1876.
[10] D. Baraff, Fast contact force computation for nonpenetrating rigid bodies, in: 21st Annual Conference on Computer Graphics and Interactive Techniques – SIGGRAPH '94, ACM Press, New York, New York, USA, 1994, pp. 23–34.
[11] J. Baumgarte, Stabilization of constraints and integrals of motion in dynamical systems, Computer Methods in Applied Mechanics and Engineering 1 (1972) 1–16.
[12] E. Bayo, S. Barbara, A. Avello, Singularity-free augmented Lagrangian algorithms for constrained multibody dynamics, Nonlinear Dynamics 5 (1994) 209–231.
[13] N. Bedrossian, K. Flueckiger, Characterizing spatial redundant manipulator singularities, in: IEEE International Conference on Robotics and Automation, 1991, pp. 714–719.
[14] A. Ben-Israel, T.N. Greville, Generalized Inverses – Theory and Applications, 2nd edition, CMS Books in Mathematics, Springer-Verlag New York, Inc., 2003.
[15] W. Blajer, A geometric unification of constrained system dynamics, Multibody System Dynamics 1 (1997) 3–21.
[16] M. Brandao, L. Jamone, P. Kryczka, N. Endo, K. Hashimoto, A. Takanishi, Reaching for the unreachable: integration of locomotion and whole-body movements for extended visually guided reaching, in: IEEE-RAS International Conference on Humanoid Robots, 2013, pp. 28–33.
[17] O. Brock, O. Khatib, S. Viji, Task-consistent obstacle avoidance and motion behavior for mobile manipulation, in: IEEE International Conference on Robotics and Automation, 2002, pp. 388–393.
[18] J. Burdick, On the inverse kinematics of redundant manipulators: characterization of the self-motion manifolds, in: IEEE International Conference on Robotics and Automation, Scottsdale, AZ, USA, 1989, pp. 264–270.
[19] T. Buschmann, S. Lohmeier, H. Ulbrich, Humanoid robot Lola: design and walking control, Journal of Physiology Paris 103 (2009) 141–148.
[20] J.-R. Chardonnet, S. Miossec, A. Kheddar, H. Arisumi, H. Hirukawa, F. Pierrot, K. Yokoi, Dynamic simulator for humanoids using constraint-based method with static friction, in: IEEE International Conference on Robotics and Biomimetics, 2006, pp. 1366–1371.
[21] B. Dariush, G.B. Hammam, D. Orin, Constrained resolved acceleration control for humanoids, in: IEEE/RSJ International Conference on Intelligent Robots and Systems, 2010, pp. 710–717.
[22] J.K. Davidson, K.H. Hunt, Robots and Screw Theory: Applications of Kinematics and Statics to Robotics, Oxford University Press, 2004.
[23] M. De Lasa, A. Hertzmann, Prioritized optimization for task-space control, in: IEEE/RSJ International Conference on Intelligent Robots and Systems, 2009, pp. 5755–5762.
[24] A. De Luca, G. Oriolo, Modelling and control of nonholonomic mechanical systems, in: J. Angeles, A. Kecskeméthy (Eds.), Kinematics and Dynamics of Multi-Body Systems, in: CISM International Centre for Mechanical Sciences, vol. 360, Springer, Vienna, 1995, pp. 277–342.
[25] A. Del Prete, F. Romano, L. Natale, G. Metta, G. Sandini, F. Nori, Prioritized optimal control, in: IEEE International Conference on Robotics and Automation, Hong Kong, China, 2014, pp. 2540–2545.
[26] J. Denavit, R. Hartenberg, A kinematic notation for lower-pair mechanisms based on matrices, in: Trans. of the ASME, Journal of Applied Mechanics 22 (1955) 215–221.
[27] A. Dietrich, C. Ott, A. Albu-Schaffer, Multi-objective compliance control of redundant manipulators: hierarchy, control, and stability, in: IEEE/RSJ International Conference on Intelligent Robots and Systems, 2013, pp. 3043–3050.
[28] A. Dietrich, T. Wimbock, A. Albu-Schaffer, G. Hirzinger, Integration of reactive, torque-based self-collision avoidance into a task hierarchy, IEEE Transactions on Robotics 28 (2012) 1278–1293.
[29] A. Dietrich, T. Wimbock, A. Albu-Schaffer, G. Hirzinger, Reactive whole-body control: dynamic mobile manipulation using a large number of actuated degrees of freedom, IEEE Robotics & Automation Magazine 19 (2012) 20–33.
[30] K.L. Doty, C. Melchiorri, C. Bonivento, A theory of generalized inverses applied to robotics, The International Journal of Robotics Research 12 (1993) 1–19.
[31] J. Duffy, The fallacy of modern hybrid control theory that is based on "orthogonal complements" of twist and wrench spaces, Journal of Robotic Systems 7 (1990) 139–144.
[32] R. Ellis, S. Ricker, Two numerical issues in simulating constrained robot dynamics, IEEE Transactions on Systems, Man and Cybernetics 24 (1994) 19–27.
[33] A. Escande, N. Mansard, P.-B. Wieber, Fast resolution of hierarchized inverse kinematics with inequality constraints, in: IEEE International Conference on Robotics and Automation, 2010, pp. 3733–3738.

[34] A. Escande, N. Mansard, P.-B. Wieber, Hierarchical quadratic programming: fast online humanoid-robot motion generation, The International Journal of Robotics Research 33 (2014) 1006–1028.

[35] R. Featherstone, Rigid Body Dynamics Algorithms, Springer Science+Business Media, LLC, Boston, MA, 2008.

[36] R. Featherstone, Exploiting sparsity in operational-space dynamics, The International Journal of Robotics Research 29 (2010) 1353–1368.

[37] R. Featherstone, A. Fijany, A technique for analyzing constrained rigid-body systems, and its application to the constraint force algorithm, IEEE Transactions on Robotics and Automation 15 (1999) 1140–1144.

[38] S. Feng, E. Whitman, X. Xinjilefu, C.G. Atkeson, Optimization-based full body control for the DARPA robotics challenge, Journal of Field Robotics 32 (2015) 293–312.

[39] Fujitsu, Miniature Humanoid Robot HOAP-2 Manual, 1st edition, Fujitsu Automation Co., Ltd, 2004 (in Japanese).

[40] K. Fujiwara, F. Kanehiro, S. Kajita, K. Kaneko, K. Yokoi, H. Hirukawa, UKEMI: falling motion control to minimize damage to biped humanoid robot, in: IEEE/RSJ International Conference on Intelligent Robots and System, 2002, pp. 2521–2526.

[41] K. Glass, R. Colbaugh, D. Lim, H. Seraji, Real-time collision avoidance for redundant manipulators, IEEE Transactions on Robotics and Automation 11 (1995) 448–457.

[42] G.H. Golub, C.F. Van Loan, Matrix Computations, Johns Hopkins University Press, 1996.

[43] A. Goswami, S.-k. Yun, U. Nagarajan, S.-H. Lee, K. Yin, S. Kalyanakrishnan, Direction-changing fall control of humanoid robots: theory and experiments, Autonomous Robots 36 (2013) 199–223.

[44] G.B. Hammam, P.M. Wensing, B. Dariush, D.E. Orin, Kinodynamically consistent motion retargeting for humanoids, International Journal of Humanoid Robotics 12 (2015) 1550017.

[45] H. Hanafusa, T. Yoshikawa, Y. Nakamura, Analysis and control of articulated robot arms with redundancy, in: Prep. of the IFAC '81 World Congress, 1981, pp. 78–83.

[46] N. Handharu, J. Yoon, G. Kim, Gait pattern generation with knee stretch motion for biped robot using toe and heel joints, in: IEEE-RAS International Conference on Humanoid Robots, Daejeon, Korea, 2008, pp. 265–270.

[47] Y. Harada, J. Takahashi, D. Nenchev, D. Sato, Limit cycle based walk of a powered 7DOF 3D biped with flat feet, in: IEEE/RSJ International Conference on Intelligent Robots and Systems, 2010, pp. 3623–3628.

[48] S. Hayati, Hybrid position/force control of multi-arm cooperating robots, in: IEEE International Conference on Robotics and Automation, 1986, pp. 82–89.

[49] P. Herman, K. Kozłowski, A survey of equations of motion in terms of inertial quasi-velocities for serial manipulators, Archive of Applied Mechanics 76 (2006) 579–614.

[50] K.H. Hunt, Structural kinematics of in-parallel-actuated robot-arms, Journal of Mechanisms, Transmissions, and Automation in Design 105 (1983) 705–712.

[51] A. Ijspeert, J. Nakanishi, S. Schaal, Movement imitation with nonlinear dynamical systems in humanoid robots, in: IEEE International Conference on Robotics and Automation, 2002, pp. 1398–1403.

[52] H. Isermann, Linear lexicographic optimization, OR Spektrum 4 (1982) 223–228.

[53] S. Kajita, H. Hirukawa, K. Harada, K. Yokoi, Introduction to Humanoid Robotics, Springer Verlag, Berlin, Heidelberg, 2014.

[54] S. Kajita, T. Nagasaki, K. Kaneko, H. Hirukawa, ZMP-based biped running control, IEEE Robotics & Automation Magazine 14 (2007) 63–72.

[55] K. Kameta, A. Sekiguchi, Y. Tsumaki, Y. Kanamiya, Walking control around singularity using a spherical inverted pendulum with an underfloor pivot, in: IEEE-RAS International Conference on Humanoid Robots, 2007, pp. 210–215.

[56] K. Kameta, A. Sekiguchi, Y. Tsumaki, D. Nenchev, Walking control using the SC approach for humanoid robot, in: IEEE-RAS International Conference on Humanoid Robots, 2005, pp. 289–294.

[57] K. Kaneko, F. Kanehiro, M. Morisawa, K. Akachi, G. Miyamori, A. Hayashi, N. Kanehira, Humanoid robot HRP-4 – humanoid robotics platform with lightweight and slim body, in: IEEE International Conference on Intelligent Robots and Systems, 2011, pp. 4400–4407.

[58] O. Kanoun, Real-time prioritized kinematic control under inequality constraints for redundant manipulators, in: H. Durrant-Whyte, N. Roy, P. Abbeel (Eds.), Robotics: Science and Systems VII, MIT Press, 2012, pp. 145–152.

[59] O. Kanoun, F. Lamiraux, P.-B. Wieber, Kinematic control of redundant manipulators: generalizing the task-priority framework to inequality task, IEEE Transactions on Robotics 27 (2011) 785–792.

[60] O. Kanoun, F. Lamiraux, P.-B. Wieber, F. Kanehiro, E. Yoshida, J.-P. Laumond, Prioritizing linear equality and inequality systems: application to local motion planning for redundant robots, in: IEEE International Conference on Robotics and Automation, 2009, pp. 2939–2944.

[61] V.N. Katsikis, D. Pappas, The restricted weighted generalized inverse of a matrix, The Electronic Journal of Linear Algebra 22 (2011) 1156–1167.
[62] K. Kazerounian, Z. Wang, Global versus local optimization in redundancy resolution of robotic manipulators, The International Journal of Robotics Research 7 (1988) 3–12.
[63] F. Keith, P.-B.B. Wieber, N. Mansard, A. Kheddar, Analysis of the discontinuities in prioritized tasks-space control under discreet task scheduling operations, in: IEEE International Conference on Intelligent Robots and Systems, 2011, pp. 3887–3892.
[64] J. Kerr, B. Roth, Analysis of multifingered hands, The International Journal of Robotics Research 4 (1986) 3–17.
[65] O. Khatib, Real-time obstacle avoidance for manipulators and mobile robots, The International Journal of Robotics Research 5 (1986) 90–98.
[66] O. Khatib, L. Sentis, J. Park, J. Warren, Whole-body dynamic behavior and control of human-like robots, International Journal of Humanoid Robotics 01 (2004) 29–43.
[67] C. Klein, K.-B. Kee, The nature of drift in pseudoinverse control of kinematically redundant manipulators, IEEE Transactions on Robotics and Automation 5 (1989) 231–234.
[68] C.A. Klein, B.E. Blaho, Dexterity measures for the design and control of kinematically redundant manipulators, The International Journal of Robotics Research 6 (1987) 72–83.
[69] M.S. Konstantinov, M.D. Markov, D.N. Nenchev, Kinematic control of redundant manipulators, in: 11th International Symposium on Industrial Robots, Tokyo, Japan, 1981, pp. 561–568.
[70] S. Kotosaka, H. Ohtaki, Selective utilization of actuator for a humanoid robot by singular configuration, Journal of the Robotics Society of Japan 25 (2007) 115–121 (in Japanese).
[71] K. Kreutz-Delgado, M. Long, H. Seraji, Kinematic analysis of 7-DOF manipulators, The International Journal of Robotics Research 11 (1992) 469–481.
[72] R. Kurazume, S. Tanaka, M. Yamashita, T. Hasegawa, K. Yoneda, Straight legged walking of a biped robot, in: IEEE/RSJ International Conference on Intelligent Robots and Systems, 2005, pp. 337–343.
[73] H.G. Kwatny, G. Blankenship, Nonlinear Control and Analytical Mechanics: A Computational Approach, Springer-Verlag, Birkhäuser, New York, 2000.
[74] A. Laulusa, O.A. Bauchau, Review of classical approaches for constraint enforcement in multibody systems, Journal of Computational and Nonlinear Dynamics 3 (2008) 011004.
[75] J. Lee, N. Mansard, J. Park, Intermediate desired value approach for task transition of robots in kinematic control, IEEE Transactions on Robotics 28 (2012) 1260–1277.
[76] S.-H. Lee, A. Goswami, Fall on backpack: damage minimizing humanoid fall on targeted body segment using momentum control, in: ASME 2011 International Design Engineering Technical Conferences and Computers and Information in Engineering Conference, 2011, pp. 703–712.
[77] S. Lengagne, J. Vaillant, E. Yoshida, A. Kheddar, Generation of whole-body optimal dynamic multi-contact motions, The International Journal of Robotics Research 32 (2013) 1104–1119.
[78] K. Levenberg, A method for the solution of certain non-linear problems in least squares, Quarterly of Applied Mathematics II (1944) 164–168.
[79] Z. Li, N.G. Tsagarikis, D.G. Caldwell, B. Vanderborght, Trajectory generation of straightened knee walking for humanoid robot iCub, in: IEEE International Conference on Control, Automation, Robotics and Vision, 2010, pp. 2355–2360.
[80] A. Liegeois, Automatic supervisory control of the configuration and behavior of multibody mechanisms, IEEE Transactions on Systems, Man and Cybernetics 7 (1977) 868–871.
[81] M. Liu, A. Micaelli, P. Evrard, A. Escande, C. Andriot, Interactive virtual humans: a two-level prioritized control framework with wrench bounds, IEEE Transactions on Robotics 28 (2012) 1309–1322.
[82] M. Liu, Y. Tan, V. Padois, Generalized hierarchical control, Autonomous Robots 40 (2016) 17–31.
[83] D.G. Luenberger, Y. Ye, Linear and Nonlinear Programming, 4th edition, Springer International Publishing, 2016.
[84] R.C. Luo, T.-W. Lin, Y.-H. Tsai, Analytical inverse kinematic solution for modularized 7-DoF redundant manipulators with offsets at shoulder and wrist, in: IEEE/RSJ International Conference on Intelligent Robots and Systems, Chicago, Il, USA, 2014, pp. 516–521.
[85] S. Luo, S. Ahmad, Predicting the drift motion for kinematically redundant robots, IEEE Transactions on Systems, Man and Cybernetics 22 (1992) 717–728.
[86] S. Ma, S. Hirose, D.N. Nenchev, Improving local torque optimization techniques for redundant robotic mechanisms, Journal of Robotic Systems 8 (1991) 75–91.

[87] S. Ma, D.N. Nenchev, Local torque minimization for redundant manipulators: a correct formulation, Robotica 14 (1996) 235–239.
[88] A. Maciejewski, Kinetic limitations on the use of redundancy in robotic manipulators, IEEE Transactions on Robotics and Automation 7 (1991) 205–210.
[89] A.A. Maciejewski, C.A. Klein, Obstacle avoidance for kinematically redundant manipulators in dynamically varying environments, The International Journal of Robotics Research 4 (1985) 109–117.
[90] A.A. Maciejewski, C.A. Klein, The singular value decomposition: computation and applications to robotics, The International Journal of Robotics Research 8 (1989) 63–79.
[91] N. Mansard, F. Chaumette, Task sequencing for high-level sensor-based control, IEEE Transactions on Robotics 23 (2007) 60–72.
[92] N. Mansard, O. Khatib, Continuous control law from unilateral constraints, in: IEEE International Conference on Robotics and Automation, 2008, pp. 3359–3364.
[93] N. Mansard, O. Khatib, A. Kheddar, A unified approach to integrate unilateral constraints in the stack of tasks, IEEE Transactions on Robotics 25 (2009) 670–685.
[94] N. Mansard, A. Remazeilles, F. Chaumette, Continuity of varying-feature-set control laws, IEEE Transactions on Automatic Control 54 (2009) 2493–2505.
[95] L. Meirovitch, Methods of Analytical Dynamics, Dover Publications, Inc., Mineola, New York, 2012.
[96] P. Meseguer, N. Bouhmala, T. Bouzoubaa, M. Irgens, M. Sánchez, Current approaches for solving over-constrained problems, Constraints 8 (2003) 9–39.
[97] N. Minamide, K. Nakamura, A restricted pseudoinverse and its application to constrained minima, SIAM Journal on Applied Mathematics 19 (1970) 167–177.
[98] S. Miyahara, D.N. Nenchev, Singularity-consistent approach/departure to/from a singular posture, Robotic Life Support Laboratory, Tokyo City University, 2017 (Video clip), https://doi.org/10.1016/B978-0-12-804560-2.00009-2.
[99] S. Miyahara, D.N. Nenchev, Singularity-consistent motion "through" a singular posture, Robotic Life Support Laboratory, Tokyo City University, 2017 (Video clip), https://doi.org/10.1016/B978-0-12-804560-2.00009-2.
[100] D. Montana, The kinematics of contact and grasp, The International Journal of Robotics Research 7 (1988) 17–32.
[101] M. Morisawa, S. Kajita, K. Kaneko, K. Harada, F. Kanehiro, K. Fujiwara, H. Hirukawa, Pattern generation of biped walking constrained on parametric surface, in: IEEE International Conference on Robotics and Automation, 2005, pp. 2405–2410.
[102] M. Murooka, S. Nozawa, Y. Kakiuchi, K. Okada, M. Inaba, Whole-body pushing manipulation with contact posture planning of large and heavy object for humanoid robot, in: IEEE International Conference on Robotics and Automation, 2015, pp. 5682–5689.
[103] R.M. Murray, Nonlinear control of mechanical systems: a Lagrangian perspective, Annual Reviews in Control 21 (1997) 31–42.
[104] R.M. Murray, Z. Li, S.S. Sastry, A Mathematical Introduction to Robotic Manipulation, CRC Press, 1994.
[105] Y. Nakamura, Advanced Robotics: Redundancy and Optimization, Addison–Wesley Publishing Company, 1991.
[106] Y. Nakamura, H. Hanafusa, Inverse kinematic solutions with singularity robustness for robot manipulator control, Journal of Dynamic Systems, Measurement, and Control 108 (1986) 163.
[107] Y. Nakamura, K. Yamane, Dynamics computation of structure-varying kinematic chains and its application to human figures, IEEE Transactions on Robotics and Automation 16 (2000) 124–134.
[108] S. Nakaoka, S. Hattori, F. Kanehiro, S. Kajita, H. Hirukawa, Constraint-based dynamics simulator for humanoid robots with shock absorbing mechanisms, in: IEEE/RSJ International Conference on Intelligent Robots and Systems, San Diego, CA, USA, 2007, pp. 3641–3647.
[109] D.N. Nenchev, Redundancy resolution through local optimization: a review, Journal of Robotic Systems 6 (1989) 769–798.
[110] D.N. Nenchev, Restricted Jacobian matrices of redundant manipulators in constrained motion tasks, The International Journal of Robotics Research 11 (1992) 584–597.
[111] D.N. Nenchev, Tracking manipulator trajectories with ordinary singularities: a null space-based approach, The International Journal of Robotics Research 14 (1995) 399–404.
[112] D.N. Nenchev, Z.M. Sotirov, Dynamic task-priority allocation for kinematically redundant robotic mechanisms, in: IEEE/RSJ International Conference on Intelligent Robots and Systems, Munich, Germany, 1994, pp. 518–524.

[113] D.N. Nenchev, Y. Tsumaki, M. Uchiyama, Singularity-consistent behavior of telerobots: theory and experiments, The International Journal of Robotics Research 17 (1998) 138–152.

[114] D.N. Nenchev, Y. Tsumaki, M. Uchiyama, Singularity-consistent parameterization of robot motion and control, The International Journal of Robotics Research 19 (2000) 159–182.

[115] D.N. Nenchev, Recursive local kinematic inversion with dynamic task-priority allocation, in: IEEE International Conference on Robotics and Automation, IEEE Comput. Soc. Press, Munich, Germany, 1994, pp. 2698–2703.

[116] D.N. Nenchev, Y. Tsumaki, M. Uchiyama, Real-time motion control in the neighborhood of singularities: a comparative study between the SC and the DLS methods, in: IEEE International Conference on Robotics and Automation, 1999, pp. 506–511.

[117] G. Niemeyer, J.-J. Slotine, Computational algorithms for adaptive compliant motion, in: International Conference on Robotics and Automation, 1989, pp. 566–571.

[118] G. Niemeyer, J.-J.E. Slotine, Adaptive Cartesian control of redundant manipulators, in: American Control Conference, 1990, pp. 234–241.

[119] P.E. Nikravesh, Initial condition correction in multibody dynamics, Multibody System Dynamics 18 (2007) 107–115.

[120] Y. Ogura, H. Aikawa, K. Shimomura, H. Kondo, A. Morishima, H.-O. Lim, A. Takanishi, Development of a new humanoid robot WABIAN-2, in: IEEE International Conference on Robotics and Automation, 2006, pp. 76–81.

[121] Y. Ogura, H.-O. Lim, A. Takanishi, Stretch walking pattern generation for a biped humanoid robot, in: IEEE/RSJ International Conference on Intelligent Robots and Systems, 2003, pp. 352–357.

[122] Y. Oh, W.K. Chung, Y. Youm, I.H. Suh, A passivity-based motion control of redundant manipulators using weighted decomposition of joint space, in: 8th International Conference on Advanced Robotics, 1997, pp. 125–131.

[123] K. O'Neil, Divergence of linear acceleration-based redundancy resolution schemes, IEEE Transactions on Robotics and Automation 18 (2002) 625–631.

[124] C. Ott, A. Dietrich, A. Albu-Schäffer, A. Albu-Schaffer, Prioritized multi-task compliance control of redundant manipulators, Automatica 53 (2015) 416–423.

[125] D. Pai, U.M. Ascher, P. Kry, Forward dynamics algorithms for multibody chains and contact, in: IEEE International Conference on Robotics and Automation, 2000, pp. 857–863.

[126] H.A. Park, M.A. Ali, C. Lee, Closed-form inverse kinematic position solution for humanoid robots, International Journal of Humanoid Robotics 09 (2012) 1250022.

[127] I.-W. Park, J.-Y. Kim, J. Lee, J.-H. Oh, Mechanical design of the humanoid robot platform HUBO, Advanced Robotics 21 (2007) 1305–1322.

[128] J. Park, Control Strategies for Robots in Contact, Ph.D. thesis, Stanford University, USA, 2006.

[129] J. Park, W. Chung, Y. Youm, On dynamical decoupling of kinematically redundant manipulators, in: IEEE/RSJ International Conference on Intelligent Robots and Systems, 1999, pp. 1495–1500.

[130] T. Petrič, A. Gams, J. Babič, L. Žlajpah, Reflexive stability control framework for humanoid robots, Autonomous Robots 34 (2013) 347–361.

[131] T. Petrič, L. Žlajpah, Smooth continuous transition between tasks on a kinematic control level: obstacle avoidance as a control problem, Robotics and Autonomous Systems 61 (2013) 948–959.

[132] M.H. Raibert, J.J. Craig, Hybrid position/force control of manipulators, Journal of Dynamic Systems, Measurement, and Control 103 (1981) 126–133.

[133] A. Rennuit, A. Micaelli, X. Merlhiot, C. Andriot, F. Guillaume, N. Chevassus, D. Chablat, P. Chedmail, Passive control architecture for virtual humans, in: IEEE/RSJ International Conference on Intelligent Robots and Systems, 2005, pp. 1432–1437.

[134] L. Saab, O. Ramos, F. Keith, N. Mansard, P. Soueres, J.Y. Fourquet, Dynamic whole-body motion generation under rigid contacts and other unilateral constraints, IEEE Transactions on Robotics 29 (2013) 346–362.

[135] H. Sadeghian, L. Villani, M. Keshmiri, B. Siciliano, Dynamic multi-priority control in redundant robotic systems, Robotica 31 (2013) 1–13.

[136] L. Sciavicco, B. Siciliano, Modelling and Control of Robot Manipulators, Springer Science & Business Media, 2000.

[137] A. Sekiguchi, Y. Atobe, K. Kameta, D. Nenchev, Y. Tsumaki, On motion generation for humanoid robot by the SC approach, in: Annual Conference of the Robotics Society of Japan, 2003 (in Japanese), p. 2A27.

[138] L. Sentis, O. Khatib, Synthesis of whole-body behaviors through hierarchical control of behavioral primitives, International Journal of Humanoid Robotics 02 (2005) 505–518.

[139] L. Sentis, O. Khatib, Compliant control of multicontact and center-of-mass behaviors in humanoid robots, IEEE Transactions on Robotics 26 (2010) 483–501.

[140] L. Sentis, J. Petersen, R. Philippsen, Implementation and stability analysis of prioritized whole-body compliant controllers on a wheeled humanoid robot in uneven terrains, Autonomous Robots 35 (2013) 301–319.

[141] H. Seraji, Configuration control of redundant manipulators: theory and implementation, IEEE Transactions on Robotics and Automation 5 (1989) 472–490.

[142] M. Shimizu, H. Kakuya, W.-K. Yoon, K. Kitagaki, K. Kosuge, Analytical inverse kinematic computation for 7-DOF redundant manipulators with joint limits and its application to redundancy resolution, IEEE Transactions on Robotics 24 (2008) 1131–1142.

[143] B. Siciliano, Kinematic control of redundant robot manipulators: a tutorial, Journal of Intelligent & Robotic Systems 3 (1990) 201–212.

[144] B. Siciliano, J.-J. Slotine, A general framework for managing multiple tasks in highly redundant robotic systems, in: Fifth International Conference on Advanced Robotics, 1991, pp. 1211–1216.

[145] O. Stasse, A. Escande, N. Mansard, S. Miossec, P. Evrard, A. Kheddar, Real-time (self)-collision avoidance task on a HRP-2 humanoid robot, in: IEEE International Conference on Robotics and Automation, 2008, pp. 3200–3205.

[146] D.E. Stewart, J.C. Trinkle, An implicit time-stepping scheme for rigid body dynamics with inelastic collisions and coulomb friction, International Journal for Numerical Methods in Engineering 39 (1996) 2673–2691.

[147] G. Strang, Linear Algebra and Its Applications, 4th edition, Cengage Learning, July 19, 2005.

[148] J. Stuelpnagel, On the parametrization of the three-dimensional rotation group, SIAM Review 6 (1964) 422–430.

[149] T. Sugihara, Solvability-unconcerned inverse kinematics by the Levenberg–Marquardt method, IEEE Transactions on Robotics 27 (2011) 984–991.

[150] H. Sugiura, M. Gienger, H. Janssen, C. Goerick, Real-time collision avoidance with whole body motion control for humanoid robots, in: IEEE/RSJ International Conference on Intelligent Robots and Systems, 2007, pp. 2053–2058.

[151] K. Suh, J. Hollerbach, Local versus global torque optimization of redundant manipulators, in: IEEE International Conference on Robotics and Automation, 1987, pp. 619–624.

[152] K. Takahashi, M. Noda, D. Nenchev, Y. Tsumaki, A. Sekiguchi, Static walk of a humanoid robot based the singularity-consistent method, in: IEEE/RSJ International Conference on Intelligent Robots and Systems, 2006, pp. 5484–5489.

[153] S. Taki, D.N. Nenchev, A novel singularity-consistent inverse kinematics decomposition for S-R-S type manipulators, in: IEEE International Conference on Robotics and Automation (ICRA), Hong Kong, China, 2014, pp. 5070–5075.

[154] J.C. Trinkle, J.-S. Pang, S. Sudarsky, G. Lo, On dynamic multi-rigid-body contact problems with Coulomb friction, ZAMM – Zeitschrift für Angewandte Mathematik und Mechanik (Journal of Applied Mathematics and Mechanics) 77 (1997) 267–279.

[155] F.-C. Tseng, Z.-D. Ma, G.M. Hulbert, Efficient numerical solution of constrained multibody dynamics systems, Computer Methods in Applied Mechanics and Engineering 192 (2003) 439–472.

[156] Y. Tsumaki, D. Nenchev, S. Kotera, M. Uchiyama, Teleoperation based on the adjoint Jacobian approach, IEEE Control Systems Magazine 17 (1997) 53–62.

[157] F. Udwadia, R.E. Kalaba, What is the general form of the explicit equations of motion for constrained mechanical systems?, Journal of Applied Mechanics 69 (2002) 335.

[158] C.W. Wampler, Manipulator inverse kinematic solutions based on vector formulations and damped least-squares methods, IEEE Transactions on Systems, Man and Cybernetics 16 (1986) 93–101.

[159] D. Whitney, Resolved motion rate control of manipulators and human prostheses, IEEE Transactions on Man-Machine Systems 10 (1969) 47–53.

[160] D. Whitney, Force feedback control of manipulator fine motions, Journal of Dynamic Systems, Measurement, and Control 99 (1977) 91–97.

[161] T. Wimbock, D. Nenchev, A. Albu-Schaffer, G. Hirzinger, Experimental study on dynamic reactionless motions with DLR's humanoid robot Justin, in: IEEE/RSJ International Conference on Intelligent Robots and Systems, St. Louis, USA, 2009, pp. 5481–5486.

[162] K. Yamane, Y. Nakamura, Dynamics filter – concept and implementation of online motion generator for human figures, IEEE Transactions on Robotics and Automation 19 (2003) 421–432.

[163] K. Yamane, Y. Nakamura, A numerically robust LCP solver for simulating articulated rigid bodies in contact, in: Robotics: Science and Systems IV, MIT Press, Zurich, Switzerland, 2008, pp. 89–104.

[164] Y. Yokokohji, T. Toyoshima, T. Yoshikawa, Efficient computational algorithms for trajectory control of free-flying space robots with multiple arms, IEEE Transactions on Robotics and Automation 9 (1993) 571–580.

[165] S. Yoon, R.M. Howe, D.T. Greenwood, Geometric elimination of constraint violations in numerical simulation of Lagrangian equations, Journal of Mechanical Design 116 (1994) 1058–1064.

[166] T. Yoshikawa, Manipulability of robotic mechanisms, The International Journal of Robotics Research 4 (1985) 3–9.

[167] T. Yoshikawa, Dynamic hybrid position/force control of robot manipulators–description of hand constraints and calculation of joint driving force, IEEE Journal on Robotics and Automation 3 (1987) 386–392.

[168] T. Yoshikawa, Analysis and control of robot arms with redundancy, in: First International Symposium on Robotics Research, Pittsburg, Pennsylvania, MIT Press, Cambridge, MA, 1994, pp. 735–747.

[169] J. Yuan, Closed-loop manipulator control using quaternion feedback, IEEE Journal on Robotics and Automation 4 (1988) 434–440.

[170] R. Yui, R. Hinata, D.N. Nenchev, Upward CoM motion of a humanoid robot resolved in terms of mixed quasi-velocity, Robotic Life Support Laboratory, Tokyo City University, 2017 (Video clip), https://doi.org/10.1016/B978-0-12-804560-2.00009-2.

[171] X. Yun, N. Sarkar, Unified formulation of robotic systems with holonomic and nonholonomic constraints, IEEE Transactions on Robotics and Automation 14 (1998) 640–650.

[172] J. Zhao, N.I. Badler, Real-Time Inverse Kinematics With Joint Limits and Spatial Constraints, Technical Report MS-CIS-89-09, University of Pennsylvania, 1989.

[173] M. Zucker, S. Joo, M. Grey, C. Rasmussen, E. Huang, M. Stilman, A. Bobick, A general-purpose system for teleoperation of the DRC-HUBO humanoid robot, Journal of Field Robotics 32 (2015) 336–351.

CHAPTER

3

Statics

3.1 INTRODUCTION

Static and kinetostatic relations play basic role in kinetostatic analysis, optimization, and control. This chapter is divided in eight sections. In the following section, force/moment relations are derived from the principle of virtual work and notations for wrenches and spatial forces are introduced. In Section 3.3, contact joint models with frictional constraints are discussed. Section 3.4 focuses on kinetostatic relations in independent closed-loop chains. Section 3.5 tackles the important problem of body wrench distribution at the contact joints. The kinetostatic relations of a humanoid robot comprising both independent and interdependent closed chains are explained in Section 3.6. Section 3.7 analyzes the stability of a static posture and discusses statics-based optimization. The duality relations of the static and kinematic relations pertinent to a humanoid robot are explained in Section 3.8.

3.2 WRENCH AND SPATIAL FORCE

Let $\boldsymbol{f} \in \Re^3$ denote the force vector and $\boldsymbol{m}_P \in \Re^3$ be the moment vector acting at point P on a robot link. The two vectors are stacked to form a 6D vector that represents a *wrench*. We have

$$\mathcal{F}_P = \begin{bmatrix} \boldsymbol{f}^T & \boldsymbol{m}_P^T \end{bmatrix}^T \in \Re^6.$$

Similarly to the term "twist," the term "wrench" originates from the theory of screws [4]. The two components of a wrench are referred to as bivectors, as it was the case with a twist. The term has also been adopted in the robotics field [49,19]. A wrench is an element of $se^*(3)$, the dual space of $se(3)$. Wrenches and twists form a duality relation, whereby their dot product determines the instantaneous power: $\mathcal{V} \cdot \mathcal{F}$. This quantity can be expressed in coordinate form as $\mathcal{V}_P^T \mathcal{F}_P = \boldsymbol{v}_P^T \boldsymbol{f} + \boldsymbol{\omega}^T \boldsymbol{m}_P$. Similarly to twists, wrenches can be represented either in the inertial or the body frame. In the former case, the term *spatial force* is used [24]. Spatial force can be considered as an operator that, given an arbitrary point P on the rigid body, produces the total moment about that point, i.e.

$$\boldsymbol{m}_O = -[\boldsymbol{r}^{\times}_{\overleftarrow{OP}}]\boldsymbol{f} + \boldsymbol{m}_P, \tag{3.1}$$

where $\boldsymbol{r}_{\overleftarrow{OP}} = \boldsymbol{r}_O - \boldsymbol{r}_P$.

Humanoid Robots
https://doi.org/10.1016/B978-0-12-804560-2.00010-9

Furthermore, it is straightforward to show ([49], p. 62) that a wrench, acting at point P in a given coordinate frame, can be transformed into an equivalent wrench, acting at point O in a different frame, via the transposed spatial transform for twists. Consider, for example, two wrenches ${}^W\mathcal{F}_O$ and ${}^B\mathcal{F}_P$. These wrenches will be equivalent if the following instantaneous power relation holds:

$$({}^W\mathcal{V}_O)^T({}^W\mathcal{F}_O) = ({}^B\mathcal{V}_P)^T({}^B\mathcal{F}_P).$$

Since ${}^W\mathcal{V}_O^T = ({}^B\mathcal{V}_P)^T({}^W\mathbb{X}_{B\overleftarrow{OP}})^T$ (cf. (2.6)), it follows that

$${}^B\mathcal{F}_O = {}^W\mathbb{X}_{B\overleftarrow{OP}}^{-T}({}^W\mathcal{F}_P) = {}^B\mathbb{X}_{W\overleftarrow{OP}}^T({}^W\mathcal{F}_P), \tag{3.2}$$

for any ${}^W\mathcal{V}_O$. Using (2.6), the transpose can be expressed as

$$\begin{aligned} {}^B\mathbb{X}_{W\overleftarrow{OP}}^T &= {}^B\mathbb{T}_{\overleftarrow{OP}}^T\,{}^W\mathbb{R}_B^T \\ &= \begin{bmatrix} \boldsymbol{E} & \boldsymbol{0} \\ -[{}^B\boldsymbol{r}_{\overleftarrow{OP}}^{\times}]^T & \boldsymbol{E} \end{bmatrix} \begin{bmatrix} {}^W\boldsymbol{R}_B^T & \boldsymbol{0} \\ \boldsymbol{0} & {}^W\boldsymbol{R}_B^T \end{bmatrix}. \end{aligned} \tag{3.3}$$

Quite often, spatial forces are subjected to pure translations. From the above relations one obtains

$$\begin{aligned} \mathcal{F}_O &= \mathbb{T}_{\overleftarrow{OP}}^T\mathcal{F}_P \\ &= \begin{bmatrix} \boldsymbol{E} & \boldsymbol{0} \\ -[\boldsymbol{r}_{\overleftarrow{OP}}^{\times}] & \boldsymbol{E} \end{bmatrix}\mathcal{F}_P. \end{aligned} \tag{3.4}$$

Note that, in this transformation rule for wrenches, the overarrow notation in the subscript points in the correct direction. It is straightforward to confirm that the moment component of the resultant wrench is in agreement with (3.1).

The wrench representation leads to compact notations but should be used with care because of the nonuniform dimensions of the bivectors [22,21].

3.3 CONTACT JOINTS: STATIC RELATIONS

Two types of contact joints, with and without friction, were introduced in Section 2.9.1 and examined from the viewpoint of kinematics. It was clarified that only frictionless contact joints enable pure relative motion. In what follows, contact joints will be examined from the viewpoint of statics. Both, contact joints with and without friction, are relevant from this point of view. Indeed, these two types of joints can transmit force/moment components along the constrained motion directions.

3.3.1 Static Models of Frictionless Contact Joints

Denote by $\overline{\mathcal{F}}_k \in \Re^{c_k}$ ($k \in \{e_r, e_l\}$, $e \in \{H, F\}$) the force/moment components that can be transmitted along the constrained motion directions at a frictionless contact joint. These components determine the contact joint wrench, i.e.

$$\mathcal{F}_k = {}^k\mathbb{B}_c \overline{\mathcal{F}}_k. \tag{3.5}$$

Here ${}^k\mathbb{B}_c \in \Re^{6\times c_k}$ is the motion constraint (wrench) subspace basis at contact joint k. Recall that this basis is complementary to the velocity transform basis ${}^k\mathbb{B}_m \in \Re^{6\times \eta_k}$ defined in Section 2.9.3.

Further on, a contact joint may impose *bilateral* and/or *unilateral* motion constraints. For example, when an object with handles is firmly grasped by the hands, bilateral constraints will be imposed. Such contact constraints can also be imposed with a loose grasp. Recall the example depicted in Fig. 2.11; the frictionless cylindrical contact joints at the hands can transmit four force/moment components: two forces in the cross-sectional plane of the rod and two out-of-plane moments. The transmission of these forces and moments is bidirectional; hence, the motion constraints at these two hand contacts are characterized as frictionless and bilateral. Thus:

$${}^{H_j}\mathbb{B}_c = \begin{bmatrix} 1 & 0 & 0 & 0 \\ 0 & 0 & 0 & 0 \\ 0 & 1 & 0 & 0 \\ 0 & 0 & 1 & 0 \\ 0 & 0 & 0 & 0 \\ 0 & 0 & 0 & 1 \end{bmatrix}, \quad \overline{\mathcal{F}}_{H_j} = \begin{bmatrix} f_x \\ f_z \\ m_x \\ m_z \end{bmatrix}. \tag{3.6}$$

Since all constraints are bilateral, no inequalities appear for the magnitudes.

Next, consider the contact joints at the feet in a double stance. These contact joints impose unilateral constraints in the vertical direction: the feet can only push on the ground, they cannot pull it. Contact joints with unilateral constraints are met in fact quite often; besides the feet contacts, such joints also appear in multifinger grasping [49], when bodies impact, and so on. Unilateral constraints involve inequalities and are therefore more difficult to model than the bilateral ones [46]. When the contact joints at the feet are frictionless, they can transmit three force/moment components: one vertical force parallel to the ground normal and two tangential (out-of-plane) moments. The vertical force, however, can be transmitted only downward. Thus,

$${}^{F_j}\mathbb{B}_c = \begin{bmatrix} 0 & 0 & 0 \\ 0 & 0 & 0 \\ 1 & 0 & 0 \\ 0 & 1 & 0 \\ 0 & 0 & 1 \\ 0 & 0 & 0 \end{bmatrix}, \quad \overline{\mathcal{F}}_{F_j} = \begin{bmatrix} f_z \\ m_x \\ m_y \end{bmatrix}. \tag{3.7}$$

Here $f_z \geq 0$ is the normal force magnitude resulting from the unilateral constraint; $\pm m_x$, $\pm m_y$ are tangent moment magnitudes stemming from bilateral constraints. As already mentioned, a positive *reaction* normal force indicates a compressive force at a unilateral contact, by convention.

Further on, note that unilateral constraints are characterized by the following kinematic conditions:

$$v_z = 0$$

and

$$\dot{v}_z \geq 0,$$

v_z standing for the relative speed between the contacting surfaces. This speed is related to the contact state (initial contact condition) and is common for any type of contact joint. The relative acceleration $\dot{v}_z$, on the other hand, is related to the future state and is specific for unilateral contacts. Under this condition, the bodies cannot interpenetrate; it signifies that the contact will be maintained with zero normal acceleration. The main contact condition is expressed via the *complementarity condition* [61],

$$f_z \dot{v}_z = 0. \tag{3.8}$$

Accordingly, since the unilateral contact is characterized by a repulsive normal contact force (by assumption), it can only be maintained with zero normal acceleration [63]. Complementarity then implies that if $\dot{v}_z$ ceases to be zero, the contact is going to brake and, hence, the normal force must become zero ($f_z = 0$) [5].

3.3.2 Models of Contact Joints With Friction

Frictionless contact joint models are used to represent idealized contact conditions. In real life, motion along the "unconstrained" directions is always obstructed by force/moment components stemming from friction. Friction plays an important role in locomotion and dexterous object manipulation with the fingers and hands. Thus, it is important to construct appropriate models for contact joints with friction. Such models will be introduced in what follows.

A contact joint with friction that does not slide can be characterized as a fully constrained joint, i.e. $c_k = 6$, $\eta_k = 0$. Recall that a "welded"-type joint was characterized in the same way in Section 2.9.1. The welded joint does not transmit any motion, whatever the applied external wrench. In the case of a contact joint with friction, on the other hand, motion (sliding or slipping) may occur in the directions of action of friction forces or moments (the tangential directions). These directions will be henceforth referred to as the directions of *soft constraints*. They need to be distinguished from those that do not admit motion, irrespective of the magnitude of the applied friction wrench (i.e. the normal directions). The latter will be referred to as the directions of *hard constraints*. The hard- and soft-constraint directions are determined by the basis vectors ${}^k\mathbb{B}_c \in \Re^{6\times c_k^h}$ and ${}^k\mathbb{B}_m \in \Re^{6\times c_k^s}$, respectively, where $c_k = c_k^h + c_k^s$.

Point-Contact Model

Consider first a point-contact joint model. Without friction, the only force component transmitted across the joint is the force f_{k_z} in the normal direction. This is the hard-constraint direction. With friction, the forces f_{k_x} and f_{k_y} in the tangential directions will be transmitted as well. These are the soft-constraint directions. Point-contact joints are frequently used in multifinger grasping models. They are also quite useful to approximate the flat-surface contacts at the feet (planar contacts), as will be clarified shortly. Motion/force analysis and control at point-contact joints is based on the following *Coulomb friction model* [49]:

$$\sqrt{f_{k_x}^2 + f_{k_y}^2} > \mu_k f_{k_z},$$

where $\mu_k > 0$ denotes the constant static friction coefficient. Hence, the tangential forces that can be used by the robot at the contact must lie within the set

$$\left\{FC_k : \sqrt{f_{k_x}^2 + f_{k_y}^2} \le \mu_k f_{k_z},\ f_{k_z} \ge 0\right\}. \tag{3.9}$$

This set is represented geometrically via the *friction cone*, constructed by rotating a line inclined by the angle $\tan^{-1}\mu_k$ w.r.t. the normal and passing through the contact point. Note that since f_{k_z} is nonnegative (i.e. it is a *reaction* force), only the half-cone in the positive direction is relevant.

When the above friction cone condition cannot be satisfied, the contact joint begins to slide and the friction model changes from a static to a dynamic one. Under dynamic friction, the instantaneous power at the contact obeys the following relation [63]:

$$\mu_k f_{k_z} v_{k_t} + f_{k_t}\sqrt{v_{k_x}^2 + v_{k_y}^2} = 0,\ \ t \in \{x, y\}. \tag{3.10}$$

This means that while the contact joint slides, the contact force stays at the boundary of the friction cone, its direction being thereby opposite to the direction of the velocity of sliding.

When modeling contact joints at the fingertips, contacts with rolling are more realistic than point contacts. While rolling, the contact force changes direction, as does the respective friction cone. The friction cone *inequality* condition applies as long as the contact is a pure rolling one. Thereby, the tangential accelerations $\dot{v}_{k_t} \in \{\dot{v}_{k_x}, \dot{v}_{k_y}\}$ are zero. At the moment when the friction force reaches the boundary of the cone (i.e. when the inequality in (3.9) changes to equality), the contact joint begins to slide (to accelerate in the tangential direction). This condition is expressed as follows [63]:

$$\mu_k f_{k_z} \dot{v}_{k_t} + f_{k_t}\sqrt{\dot{v}_{k_x}^2 + \dot{v}_{k_y}^2} = 0,\ \ t \in \{x, y\}.$$

Soft-Finger Contact Model

In the field of grasping, soft finger tips are used for better manipulability of the grasped object. With a soft fingertip, torsional moments can be applied. In this case, the point-contact model (3.9) is extended as follows [49]:

$$\left\{FC_k : \sqrt{f_{k_x}^2 + f_{k_y}^2} \le \mu_k f_{k_z},\ f_{k_z} \ge 0,\ |m_{k_z}| \le \gamma_k f_{k_z}\right\}, \tag{3.11}$$

where $\gamma > 0$ is the coefficient of torsional friction.

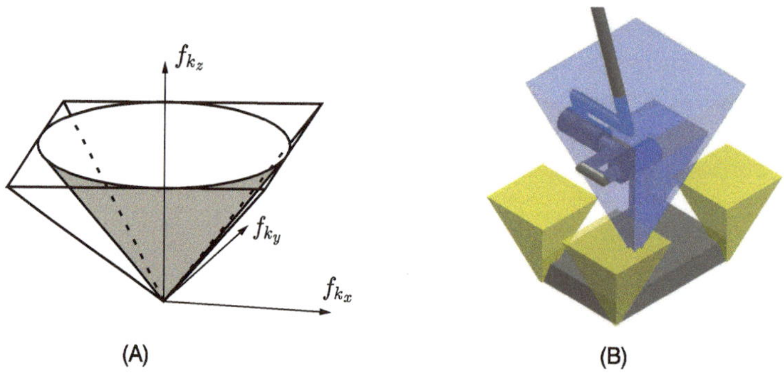

FIGURE 3.1 Left (A): Approximation of the FC with a circumscribed four-sided pyramid. Right (B): Conceptual graphical representation of polyhedral ($N = 4$) point-contact friction cones (FC, colored in yellow) at the vertices of a rectangular foot and the respective polyhedral plane-contact wrench cone (CWC, colored in purple). The CWC is obtained from the four FCs via face/span transforms.

Polyhedral Convex Cone Model

The above representations of the friction cone are nonlinear. To facilitate analysis and the design of faster numerical algorithms, a linearized representation of the friction cone would be more appropriate. In [63], the friction cone was linearized by approximating it with a friction pyramid. In general, an N-sided convex polyhedron (CP) can be employed [64,32,18], i.e.

$$CP_k = \{\boldsymbol{f}_k : \boldsymbol{C}_k \boldsymbol{f}_k \preceq \boldsymbol{0}\}.$$

The curly inequality sign denotes componentwise operation. We have the matrix

$$\boldsymbol{C}_k = \begin{bmatrix} -\sin\alpha_{1_k} & -\cos\alpha_{1_k} & -\mu_k \\ \vdots & \vdots & \vdots \\ -\sin\alpha_{p_k} & -\cos\alpha_{p_k} & -\mu_k \\ \vdots & \vdots & \vdots \\ -\sin\alpha_{N_k} & -\cos\alpha_{N_k} & -\mu_k \end{bmatrix}, \quad \alpha_{p_k} = \frac{2\pi(p_k - 1)}{N_k}.$$

Increasing N improves the accuracy at the expense of the computational load. In the field of humanoid robotics, mostly used are the four-sided pyramid approximations [47,44,42]. The pyramid can be either inscribed (conservative approach) or circumscribed, as shown in Fig. 3.1A. In this text, the latter approach will be adopted. Then

$$\boldsymbol{C}_k = \begin{bmatrix} 0 & -1 & \mu_k \\ -1 & 0 & \mu_k \\ 0 & 1 & \mu_k \\ 1 & 0 & \mu_k \end{bmatrix}. \tag{3.12}$$

An implementation with a three-sided pyramid approximation can be found in [40].

Plane-Contact Model: the Contact Wrench Cone

Consider now the planar-contact joints at the feet. Without friction, only spatial force components f_z, m_x, and m_y can be transmitted across the joints. As already clarified, normal force $f_z \geq 0$ is transmitted unilaterally while tangential moments m_x and m_y are transmitted bilaterally, with bounds (the meaning of the bounds will be clarified in short). In the presence of friction, these three components determine the hard-constraint directions. The remaining three components (i.e. f_x, f_y, and m_z) will be transmitted as well, either completely (no slipping) or partially (with slipping). These components determine the soft-constraint directions. In other words, the contact model should account for all the six contact wrench components. A "friction cone" that accounts for all the contact wrench components in the general case, i.e. for any type of surface contact, is referred to as a "wrench cone" [3]. As noted in [14], a surface contact can be thought of as a continuum of infinitesimal contact forces encoded locally by a scalar normal pressure field and a 2D vector field for mechanical stress in the tangential directions.

In the particular case of plane contact, the general surface model can be simplified by observing only the contact forces at the vertices of the contact polygon. Consider, for example, the four point contacts at the vertices of a rectangular foot sole, conveniently colocated with existing pressure-type sensors. To ensure friction cone conditions (3.9) at all four point contacts, the components of the wrench cone must satisfy the following constraints [15]:

$$|f_t| \leq \mu f_z, \quad t \in \{x, y\}, \tag{3.13}$$

$$|m_x| \leq l_y f_z, \quad |m_y| \leq l_x f_z, \tag{3.14}$$

$$f_z > 0,$$

$$m_z^{min} \leq m_z \leq m_z^{max}.$$

Here l_t denotes the half-distance between two contact points in the tangential direction t. The bounds on yaw moment m_z are determined from

$$m_z^{min} = -\mu(l_x + l_y) f_z + |l_y f_x - \mu m_x| + |l_x f_y - \mu m_y|,$$

$$m_z^{max} \equiv +\mu(l_x + l_y) f_z - |l_y f_x + \mu m_x| - |l_x f_y + \mu m_y|.$$

This model provides valuable information about the admissible bounds on the moment in the vertical direction, which has been shown to play an important role during walking. Note that there is no need to make use of the torsional friction coefficient γ with this model.

An important remark is due at this point. The two motion constraints related to the moment components m_x and m_y have been characterized as bilateral since the moments can be transmitted across the planar-contact joint in both directions. However, as seen from (3.14), there are bounds on the absolute values of these moments. These bounds are needed to ensure that the moments do not grow too large. Note that violating this condition yields a change in the type of the contact joint, from plane-type to line-type or point-type. As a consequence of such a change the foot will begin to roll and the moments would cease to be controllable. The respective dual components (i.e. angular speeds ω_x and/or ω_y) would then become uncontrollable as well, which would in turn cause the robot to lose its stability. Such stability problems will be discussed in more detail in Chapter 5.

The wrench cone for the plane-contact model will be henceforth referred to as the *contact wrench cone* (CWC) [15]. It can be shown that, when the point-contact forces at the vertices are confined within the interior of their friction cones, the contact wrench will be confined within the interior of the CWC [15,14].

Face/Span Polyhedral Convex Cone Double Representations

A general procedure for the construction of the CWC can be obtained with the help of the *face/span polyhedral convex cone double representations* [31,3]. The idea is visualized in Fig. 3.1B. With such procedure, it is possible to construct the CWC at a rectangular foot contact, for example, from the polyhedral cones (PCs) at the four vertices.

The face/span representations of a PC are denoted as

$$\begin{aligned} PC &= \text{face}(\boldsymbol{U}) = \{\boldsymbol{f} : \boldsymbol{U}\boldsymbol{f} \preceq \boldsymbol{0}\} \\ &= \text{span}(\boldsymbol{V}) = \{\boldsymbol{V}\boldsymbol{z} : \boldsymbol{z} \succeq \boldsymbol{0}\}. \end{aligned}$$

The face representation can be used to directly check whether a contact force (or wrench) lies within the interior. The span representation, on the other hand, is useful for linear association. Next, note that for a given $\boldsymbol{U}$ there is a $\boldsymbol{U}^S$, s.t. span$(\boldsymbol{U}^S) =$ face$(\boldsymbol{U})$. Also, for a given $\boldsymbol{V}$ there is a $\boldsymbol{V}^F$, s.t. face$(\boldsymbol{V}^F) =$ span$(\boldsymbol{V})$. Furthermore, denote the force-to-wrench mapping at contact k as

$$\mathcal{F}_k = \mathbb{C}_k \boldsymbol{f}_k.$$

In the case of a rectangular contact area, $\boldsymbol{f}_k \in \Re^{12}$ stands for four stacked point-contact forces and $\mathbb{C}_k \in \Re^{6\times 12}$ is the force-to-wrench map. Assuming that the contact forces are confined within their polyhedral friction cones, i.e. $\boldsymbol{U}_k \boldsymbol{f}_k \preceq \boldsymbol{0}$; it follows then that $\mathcal{F}_k \in$ span$(\boldsymbol{V}_{k_{CWC}})$, where $\boldsymbol{V}_{k_{CWC}} = \mathbb{C}_k \boldsymbol{U}_k^S$. This implies that the contact wrench lies within the interior of the CWC, i.e. $\boldsymbol{U}_{k_{CWC}} \mathcal{F}_k \preceq \boldsymbol{0}$.

It is worth mentioning that plane-contact models derived from reaction forces at the contact area vertices, as above, are used quite frequently [32,1,34,74,73,68]. Note also that there is an open-source software library for face/span transformations (the CDD library [26]).

3.3.3 Motion/Force Duality Relations Across Contact Joints

Contact joints play an important role in humanoid robotics, as already discussed: locomotion depends upon frictional contacts, as does multifinger and dual-arm object manipulation. The motion/force relations across contact joints derived above are based on a fundamental duality property. This property is inherent not only to the contact joints discussed here, but also to any of the robot joints. Moreover, the duality property plays in fact a much broader role and covers motion/force duality relations at the system level.

Consider first the ideal case of a frictionless contact joint. The possible force, $\overline{\mathcal{F}}_k \in \Re^{c_k}$, and the possible motion velocity, $\overline{\mathcal{V}}_k \in \Re^{\eta_k}$, are transmitted across the joint as

$$\mathcal{F}_k = {}^k\mathbb{B}_c \overline{\mathcal{F}}_k \tag{3.15}$$

and

$$\mathcal{V}_k = {}^k\mathbb{B}_m \overline{\mathcal{V}}_k, \tag{3.16}$$

respectively. On the other hand, any spatial force/velocity is suppressed by the joint as

$$^{k}\mathbb{B}_m^T \mathcal{F}_k = \mathbf{0} \tag{3.17}$$

and

$$^{k}\mathbb{B}_c^T \mathcal{V}_k = \mathbf{0}, \tag{3.18}$$

respectively. Since $^{k}\mathbb{B}_c \in \Re^{6\times c_k}$ and $^{k}\mathbb{B}_m \in \Re^{6\times \eta_k}$ are full column rank by definition, there exist respective complementary matrices $^{k}\mathbb{B}_c^{\perp} \in \Re^{6\times \eta_k}$ and $^{k}\mathbb{B}_m^{\perp} \in \Re^{6\times c_k}$, s.t.

$$^{k}\mathbb{B}_c^{\perp} = {}^{k}\mathbb{B}_m, \quad ^{k}\mathbb{B}_m^{\perp} = {}^{k}\mathbb{B}_c, \tag{3.19}$$

where $(\circ)^{\perp} = \boldsymbol{E} - (\circ)$ denotes the orthogonal complement. The pairs $\{^{k}\mathbb{B}_c, {}^{k}\mathbb{B}_c^{\perp}\}$ and $\{^{k}\mathbb{B}_m, {}^{k}\mathbb{B}_m^{\perp}\}$ span the spatial force and motion subspaces, respectively (cf. also [60], Section 2.4.4). Note also that the direct sum $^{k}\mathbb{B}_c \oplus {}^{k}\mathbb{B}_m = \boldsymbol{E}_6$.

Next, consider a joint with friction. Recall that such a joint is fully constrained, whereby the hard constraints (inadmissible motion) are distinguished from the soft ones, the latter admitting motion under friction. The respective constraint directions are determined by the basis vector matrices defined for frictionless joints. Denote by $\overline{\mathcal{F}}_k^h \in \Re^{c_k^h}$ and $\overline{\mathcal{F}}_k^s \in \Re^{c_k^s}$ the friction wrench components transmitted along the hard and soft constraint directions, respectively. Furthermore, a damped velocity component, $\overline{\mathcal{V}}_k^s \in \Re^{c_k^s}$, may be transmitted across the joint while sliding, in accordance with (3.10). From the complementarity relations (3.19) it follows that the friction wrench and the damped velocity components along the soft-constraint directions are transformed within the respective spatial force/velocity domains as

$$\mathcal{F}_k = {}^{k}\mathbb{B}_m \overline{\mathcal{F}}_k^s \tag{3.20}$$

and

$$\mathcal{V}_k = {}^{k}\mathbb{B}_c \overline{\mathcal{V}}_k^s, \tag{3.21}$$

respectively. On the other hand, any wrench acting on the link adjacent to the joint induces the following force component in the soft-constraint directions:

$$\overline{\mathcal{F}}_k^s = {}^{k}\mathbb{B}_m^T \mathcal{F}_k. \tag{3.22}$$

Note that $\overline{\mathcal{F}}_k^h$ and $\overline{\mathcal{F}}_k^s$ have to be evaluated for compliance with the friction cone condition. When the condition is violated, the joint begins to slide and the dynamic friction model (3.10) comes into effect. The components of the velocity in the mobility directions are obtained via the transform

$$\overline{\mathcal{V}}_k^s = {}^{k}\mathbb{B}_c^T \mathcal{V}_k. \tag{3.23}$$

The above duality relations are represented graphically in Fig. 3.2. For a frictionless joint, the joint-motion constraint yields the impossible motion/force subspaces. In the case of a joint with friction, these subspaces become the friction-force and the damped-motion (friction-induced damping) subspaces, respectively.

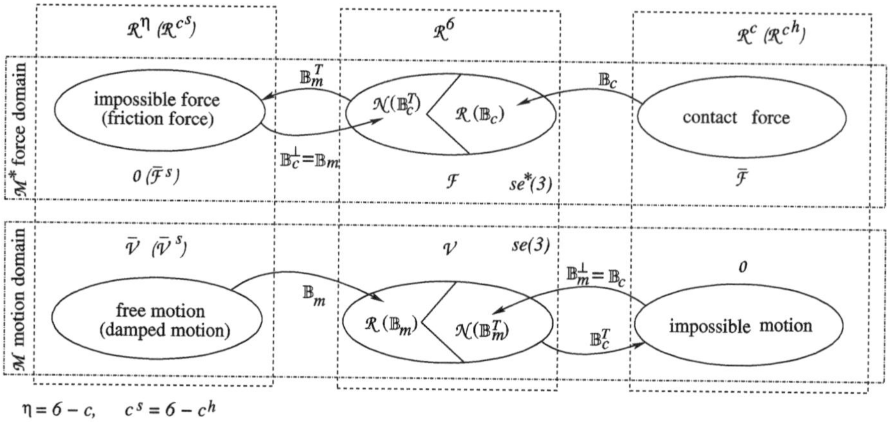

FIGURE 3.2 Motion/force duality relations across a contact joint. The 6D parameter space of motion/force domains $se(3)/se^*(3)$ is decomposed as $\Re^6 = \Re^c \oplus \Re^\eta$ (frictionless joint) or $\Re^6 = \Re^{c^h} \oplus \Re^{c^s}$ (joint with friction). The right angles in the $\Re^6$ subdomains represent orthogonal decomposition. Note that $\mathcal{R}(\mathbb{B}_m) = \mathcal{N}(\mathbb{B}_c^T)$ and vice versa. A frictionless joint comprises the *impossible-force* and the *free-motion* complementary subspaces. For a joint with friction, these subspaces become the *friction-force* and the friction-induced *damped-motion* subspaces, respectively.

Summary and Discussion

The *constraint-based* contact modeling method presented here refers to contact joints between rigid bodies. A contact joint may impose constraints characterized with or without friction. There are $c_k \leq 6$ motion constraints enforced by the contact joint. In the presence of friction, the contact joint model is determined by the pair $\{\mathbb{B}_c, FC_k\}$; $\mathbb{B}_c$ denotes the constraint basis for independent wrench components across the joint and FC_k stands for the friction cone that plays an important role in assessing the state of the frictional constraints. Most of the research in the field of humanoid robotics assumes that the *reaction* force at the contact is always positive or nonnegative. This is in contrast with the contact models used in the field of multifinger grasping [49] where the forces *impressed* by the fingers on the object are assumed positive.[1]

Constraint-based modeling is useful in motion/force analysis of constrained multibody systems like human figures in animation and humanoid robots. The implementation of the method in numerical environments for simulation and control, however, is not straightforward. For example, note that the friction cone (3.9) represents a system of nonlinear inequalities. When modeling frictional contacts numerically, the friction cone can be approximated by a pyramid [63] or by a convex PC for improved accuracy [31,62,3]. The problem can then be cast and solved as a linear complementarity problem (LCP) [5,62,2]. The respective standard solution methods lead to complexity, though, when applied to a multi-DoF system like a humanoid robot. Other problematic issues also exist. The related problems have been tackled via alternative modeling methods [39,51,17,70].

[1] These forces always point outward from the fingertips; see also [63].

3.4 KINETOSTATIC RELATIONS IN INDEPENDENT CLOSED-LOOP CHAINS

In Section 2.10 it was shown how important it is to determine instantaneous motion relations for kinematic chains comprising closed loops. The dual problem of determining *kinetostatic* forces within the closed loops is of equal significance. Such forces appear in multifinger grasping, dual-arm cooperating manipulation, and legged robots, including humanoids. The differential kinematic and kinetostatic force relations are complementary relations stemming from d'Alembert's principle of virtual work. The relations are obtained under the assumption of locked joints, s.t. $\dot{\boldsymbol{\theta}} = \mathbf{0}$ always.

Consider the case of two independent closed loops, as described in Section 2.10.1 (see also Fig. 3.6). Each closed loop is formed by the two contact joints at the respective end links. The force/moment relations within each closed loop are determined by the external wrenches acting at one or more of the links within the loop, the gravity field, and the joint torque vectors of the two limbs, $\boldsymbol{\tau}_k,\ k \in \{e_r,\ e_l\}$. The external wrench $-\mathcal{F}_\lambda$ acting on Link λ can be expressed as the following sum:

$$\mathcal{F}_\lambda = \sum_k \mathbb{T}^T_{\overleftarrow{\lambda k}}(\boldsymbol{q}_e)\mathcal{F}_k{}^\lambda. \tag{3.24}$$

Wrench $\mathcal{F}_k{}^\lambda$ is the contact wrench at contact joint k stemming from the external wrench $\mathcal{F}_\lambda$. By convention, the minus sign signifies that external wrenches are treated as being *impressed* by the robot on the environment. A positive wrench has then the meaning of a *reaction wrench*, in accordance with Newton's Third Law. Further on, the actions of all external wrenches, $\mathcal{F}_k{}^\lambda$ $(\lambda = 1, 2, ...)$, sum up at contact joint k as the *total contact wrench*, i.e. $\mathcal{F}_k = \sum_\lambda \mathcal{F}_k{}^\lambda$. On the other hand, the sum of all external wrenches, taken at the root link, determines the reaction to the net external wrench within the loop, i.e.

$$\mathcal{F}_R = \sum_\lambda \mathbb{T}^T_{\overleftarrow{R\lambda}}(\boldsymbol{q}_e)\mathcal{F}_\lambda = \sum_k \mathbb{T}^T_{\overleftarrow{Rk}}(\boldsymbol{q}_k)\mathcal{F}_k. \tag{3.25}$$

The term $\mathcal{F}_R$ will be henceforth referred to as the *root wrench*. A similar relation is used to map the sum of all external wrenches at the loop-closure link:

$$\mathcal{F}_e = \sum_\lambda \mathbb{T}^T_{\overleftarrow{e\lambda}}(\boldsymbol{q}_e)\mathcal{F}_\lambda = \sum_k \mathbb{T}^T_{\overleftarrow{ek}}(\boldsymbol{q}_k)\mathcal{F}_k. \tag{3.26}$$

The term $\mathcal{F}_e$ will be henceforth referred to as the *closure-link wrench*. Note that the above maps can be characterized as configuration-dependent *structural force* maps.

Finally, to ensure the closure of the loop, the total contact wrench $\mathcal{F}_k$ should be in balance with the contact wrench stemming from the joint torque of the respective limb. This implies that

$$\boldsymbol{\tau}_k = \boldsymbol{J}_R^T(\boldsymbol{q}_k)\,\mathcal{F}_k. \tag{3.27}$$

This relation is valid for a nonredundant limb. In the case of a kinematically redundant limb, a null-space term has to be added, as will be clarified shortly.

3.4.1 Orthogonal Decomposition of the Contact Wrench

The contact wrench $\mathcal{F}_k$ can be decomposed along the constrained (c) and unconstrained/mobility (m) directions in accordance with the relations discussed in Section 2.9. We have

$$\begin{aligned}\mathcal{F}_k &= \mathcal{F}_k^c + \mathcal{F}_k^m \\ &= \mathbb{B}_c(\boldsymbol{q}_k)\overline{\mathcal{F}}_k^c + \mathbb{B}_m(\boldsymbol{q}_k)\overline{\mathcal{F}}_k^m \\ &= \left[(\overline{\mathcal{F}}_k^c)^T \quad (\overline{\mathcal{F}}_k^m)^T\right]^T.\end{aligned} \tag{3.28}$$

The two wrenches in the sum are orthogonal, i.e. $\mathcal{F}_k^c \perp \mathcal{F}_k^m$. Component $\overline{\mathcal{F}}_k^c \in \Re^{c_k}$ denotes the *reactions* along the constrained directions at the contact joint; $\overline{\mathcal{F}}_k^m \in \Re^{\eta_k}$ stands for *inertial* force/moment components in the unconstrained directions. Thus, the above relations are of a *kinetostatic* nature.

The notation covers the two special cases of completely constrained or completely unconstrained motion, i.e. $c_k = 6$, $\eta_k = 0$, $\mathcal{F}_k = \mathcal{F}_k^c$ and $c_k = 0$, $\eta_k = 6$, $\mathcal{F}_k = \mathcal{F}_k^m$, respectively. The latter is trivial from the viewpoint of statics. The completely constrained case, on the other hand, deserves more attention in view of friction modeling and control. The friction models introduced in Section 3.3.2 can be employed, e.g. as defined in (3.9) or (3.13) for the cases of point or plane contacts, respectively. It is assumed that proper control is available (cf. Chapter 5) s.t. the contact wrench is always confined within the interior of the friction cone (FC_k: point contact) or the contact wrench cone (CWC_k: line or plane contact). This implies that the contact joint never *slips* and thus imposes a complete motion constraint. In this case, the reaction/contact wrench can be decomposed as

$$\begin{aligned}\mathcal{F}_k = \mathcal{F}_k^c &= \mathcal{F}_k^h + \mathcal{F}_k^s \\ &= \mathbb{B}_c(\boldsymbol{q}_k)\overline{\mathcal{F}}_k^h + \mathbb{B}_m(\boldsymbol{q}_k)\overline{\mathcal{F}}_k^s \\ &= \left[(\overline{\mathcal{F}}_k^h)^T \quad (\overline{\mathcal{F}}_k^s)^T\right]^T.\end{aligned} \tag{3.29}$$

The h and s superscripts refer to the normal and tangential directions at the contact joint. This decomposition is of the same form as (3.28). However, the meaning of the two components, $\overline{\mathcal{F}}_k^h$ and $\overline{\mathcal{F}}_k^s$, is different; they both are *reactions*. Thus, the above relations are characterized as pure static relations. An illustrative example will be presented in Section 3.6.5.

3.4.2 Orthogonal Decomposition of the Loop-Closure and Root Link Wrenches

Decomposition (3.28) of each contact wrench induces the following decomposition of the wrench at the loop-closure link:

$$\begin{aligned}\mathcal{F}_e &= \mathcal{F}_e^c + \mathcal{F}_e^m \\ &= \sum_k \mathbb{C}_{ce}(\boldsymbol{q}_k)\overline{\mathcal{F}}_k^c + \sum_k \mathbb{C}_{me}(\boldsymbol{q}_k)\overline{\mathcal{F}}_k^m \\ &= \mathbb{C}_{ce}(\boldsymbol{q}_e)\overline{\mathcal{F}}^c(\boldsymbol{q}_e) + \mathbb{C}_{me}(\boldsymbol{q}_e)\overline{\mathcal{F}}^m(\boldsymbol{q}_e),\end{aligned} \tag{3.30}$$

$\mathcal{F}^{(\circ)}(\boldsymbol{q}_e)$ denoting vectors of stacked $\overline{\mathcal{F}}_k^{(\circ)}$ components. Note that from the definition of the contact maps it follows that $\mathcal{F}_e^c \perp \mathcal{F}_e^m$. The importance of this loop-closure wrench decomposition will also be highlighted with an example in Section 3.6.5.

The root-link wrench is decomposed in the same way: in the above relations just replace contact map $\mathbb{C}_{(\circ)e}$ with $\mathbb{C}_{(\circ)R}$.

3.4.3 Decomposition of the Limb Joint Torque

The joint torque of Limb k, (3.27), will be decomposed into components stemming from the contact wrenches along the constrained and mobility directions, as follows. First, a relation dual to the first-order differential motion constraint (i.e. the upper part of (2.73)) will be derived via d'Alembert's principle of virtual work. Denote by $\delta\boldsymbol{\theta}_k$ and $\delta\mathcal{X}_k^c$ the virtual joint displacements and the relative virtual displacements at the contacts, respectively, the latter being associated with $\overline{\mathcal{V}}_k^c - \mathbb{C}_{cR}(\boldsymbol{q}_k)\mathcal{V}_R$. The virtual displacements can be related as $\mathcal{J}_{cR}(\boldsymbol{q}_k)\delta\boldsymbol{\theta}_k = \delta\mathcal{X}_k^c$. The instantaneous virtual work can then be expressed as

$$\delta\boldsymbol{\theta}_k^T\boldsymbol{\tau}_k^c = (\delta\mathcal{X}_k^c)^T\overline{\mathcal{F}}_k^c = \delta\boldsymbol{\theta}_k^T\mathcal{J}_{cR}^T(\boldsymbol{q}_k)\overline{\mathcal{F}}_k^c.$$

This relation is valid for any $\delta\boldsymbol{\theta}_k$. Thus, one obtains the joint torque component stemming from the contact wrench along the constrained-motion directions as

$$\boldsymbol{\tau}_k^c = \mathcal{J}_{cR}^T(\boldsymbol{q}_k)\overline{\mathcal{F}}_k^c. \tag{3.31}$$

The complementary quasistatic relationship can be obtained in a similar way from the lower part of (2.73) as

$$\boldsymbol{\tau}_k^m = \mathcal{J}_{mR}^T(\boldsymbol{q}_k)\overline{\mathcal{F}}_k^m. \tag{3.32}$$

The sum of the last two equations is

$$\boldsymbol{\tau}_k = \boldsymbol{\tau}_k^c + \boldsymbol{\tau}_k^m = \begin{bmatrix}\mathcal{J}_{cR}^T(\boldsymbol{q}_k) & \mathcal{J}_{mR}^T(\boldsymbol{q}_k)\end{bmatrix}\begin{bmatrix}\overline{\mathcal{F}}_k^c \\ \overline{\mathcal{F}}_k^m\end{bmatrix} = \mathcal{J}_R^T(\boldsymbol{q}_k)\mathcal{F}_k. \tag{3.33}$$

Jacobian $\mathcal{J}_R(\boldsymbol{q}_k)$ is the permuted Jacobian of the limb (cf. (2.77)). The above result is in agreement with the decomposition of the total wrench at the contact joint, as in (3.28), and the expression for the limb joint torque in (3.27). Note that the above decomposition is *not* orthogonal.

In the case of a kinematically redundant limb, the joint torque can be expressed as [38]

$$\boldsymbol{\tau}_k = \mathcal{J}_R^T(\boldsymbol{q}_k)\mathcal{F}_k + \left(\boldsymbol{E} - \mathcal{J}_R^T(\boldsymbol{q}_k)\mathcal{J}_R^{\#T}(\boldsymbol{q}_k)\right)\boldsymbol{\tau}_{k_u}. \tag{3.34}$$

The homogeneous component of the joint torque (the second component on the r.h.s.) induces self-motion via the vector parameter $\boldsymbol{\tau}_{k_u}$. A special subset of self-motion can be determined via an appropriated choice of the generalized inverse (cf. Section 4.5.2), s.t. the force balance at the end link, and hence the static contact conditions within the closed loop can be maintained; $(\boldsymbol{E} - \mathcal{J}_R^T(\boldsymbol{q}_k)\mathcal{J}_R^{\#T}(\boldsymbol{q}_k))$ is a projector onto the dual null space of the limb Jacobian,

$\mathcal{N}\left(\mathcal{J}_R^{\#T}(q_k)\right)$. This null space is identical with the intersection of the dual null spaces of the two sub-Jacobians in (3.33), i.e.

$$\mathcal{N}^*(\mathcal{J}_R(q_k)) = \mathcal{N}^*(\mathcal{J}_{cR}(q_k)) \cap \mathcal{N}^*(\mathcal{J}_{mR}(q_k)).$$

The asterisk superscript signifies that the dual null space is defined within the joint torque (force) domain, i.e.,

$$\mathcal{N}^*(A) \equiv \mathcal{N}(A^{\#T}). \tag{3.35}$$

Assuming a weighted generalized inverse, as in (2.41), it is straightforward to show that the null-space projector in the force domain is related to the null-space projector in the motion domain (as in (2.34) and (2.39)), via the following similarity transform:

$$W^{-1}\left(E - A^T A^{-WT}\right) W = \left(E - A^{-W} A\right). \tag{3.36}$$

The last equation implies that the two null spaces are isomorphic, i.e. $\mathcal{N}(A) \simeq \mathcal{N}(A^{\#T})$ and the respective (dual) vector bases are of the same dimension and of equal rank. Given a null-space projector, $N(\circ)$, the dual projector will be denoted as $N^*(\circ)$.

Further on, to honor the priority convention (2.86) of the velocity solution, the limb joint torque (3.34) can be rewritten as

$$\begin{aligned} \tau_k &= \mathcal{J}_{cR}{}^T(q_k)\overline{\mathcal{F}}_k^c + \overline{\mathcal{J}}_{mR}(q_k)\widetilde{\mathcal{F}}_k^m + \left(E - \mathcal{J}_R^T(q_k)\mathcal{J}_R^{\#T}(q_k)\right)\tau_{k_u}, \\ &= \tau_k^c + \tau_k^m + \tau_k^n, \text{ s.t. } \tau_k^c \succ \tau_k^m \succ \tau_k^n. \end{aligned} \tag{3.37}$$

Here $\overline{\mathcal{J}}_{mR}(q_k) = \mathcal{J}_{mR}(q_k)N^*(\mathcal{J}_{cR}(q_k))$ is the mobility Jacobian restricted by the dual null space of the joint-space constraint Jacobian, and $\widetilde{\mathcal{F}}_k^m = \overline{\mathcal{F}}_k^m - \mathcal{J}_{mR}^{\#T}(q_k)\mathcal{J}_{cR}^T(q_k)\overline{\mathcal{F}}_k{}^n$. The torque component τ_k^c ensures that the reaction forces at the contact are maintained with the highest priority; τ_k^m, on the other hand, is used as a control input w.r.t. the forces along the unconstrained motion directions at the contact (friction or inertial forces). Finally, τ_k^n can be used for self-motion control, as explained above.

Eq. (3.34) plays an important role in the inverse dynamic relations of redundant manipulators. Further details will be provided in Section 4.5.

3.5 THE WRENCH DISTRIBUTION PROBLEM

The kinetostatic relations derived in the last section state that the contact wrench components at the loop contact joints determine the root and closure-link wrenches, $\mathcal{F}_R$ and $\mathcal{F}_e$, respectively. The inverse problem, i.e. "Given the root (or closure-link) wrench, find appropriate reactions at the contact joints," plays an important role in balance, propulsion, and cooperative (dual-arm) manipulation control. Indeed, the only way to control a free-floating root link, such as the base link of a humanoid robot, is to generate appropriate reactions at the contact joints. The inverse problem is referred to as the *wrench distribution* (WD) problem.

In what follows, the WD problem will be explained with regard to the loop-closure link. The same relations can be used for the WD problem with regard to the root link, by replacement of the contact maps. First, it is important to note that the WD problem is a pure static problem. The total *reaction* wrench lumped at the loop-closure link is $\mathcal{F}_e = \mathcal{F}_e^c$. As already clarified, the other component of $\mathcal{F}_e$ appearing in (3.30), $\mathcal{F}_e^m$, stems from inertia forces along the mobility directions at frictionless contacts and, hence, should not appear in the static balance of forces. Thus, from (3.30), we obtain

$$\begin{aligned} \mathcal{F}_e &= \mathbb{C}_{ce}(\boldsymbol{q}_e)\overline{\mathcal{F}}^c(\boldsymbol{q}_e), \\ \overline{\mathcal{F}}^c(\boldsymbol{q}_e) &= \left[(\overline{\mathcal{F}}_{e_r}^c)^T \quad (\overline{\mathcal{F}}_{e_l}^c)^T\right]^T \in \Re^{c_e}. \end{aligned} \tag{3.38}$$

This notation also covers contact joints with friction, whereby $\overline{\mathcal{F}}_{e_j}^c = \mathcal{F}_{e_j}^c = \mathcal{F}_{e_j}$, for $j = r$ or $j = l$ or both. In the latter case, $\overline{\mathcal{F}}^c(\boldsymbol{q}_e) = \mathcal{F}^c(\boldsymbol{q}_e) = \mathcal{F}(\boldsymbol{q}_e) \in CWC_e$ should hold, where

$$CWC_e = CWC_{e_r} \times CWC_{e_l} \subset \Re^{12}$$

is the *loop CWC*.

3.5.1 General Solution to the Wrench Distribution Problem

The solution to the inverse problem may not be a trivial one since the conditioning of the linear system (3.38) depends on the type and number of constraints, their independence, and the configuration of the limbs. To satisfy the loop-closure conditions, the composite wrench $\overline{\mathcal{F}}^c(\boldsymbol{q}_e)$ should ensure contact stability and comply with the above CWC constraints. This implies the involvement of *inequality-type constraints*. In the general case, the loop-closure/root links are *unilaterally overconstrained* ($c_e > 6$), e.g. as in a double-stance posture with high friction. The inverse problem is then underdetermined, admitting an infinite number of solutions. This is also the case for an object held by the hands. Thus

$$\overline{\mathcal{F}}^c(\boldsymbol{q}_e) = \mathbb{C}_{ce}^{\#}(\boldsymbol{q}_e)\mathcal{F}_e + \overline{\mathcal{F}}^n(\boldsymbol{q}_e), \tag{3.39}$$
$$\overline{\mathcal{F}}^n(\boldsymbol{q}_e) = \boldsymbol{N}\left(\mathbb{C}_{ce}(\boldsymbol{q}_e)\right)\overline{\mathcal{F}}_a^c(\boldsymbol{q}_e) \tag{3.40}$$
$$= \boldsymbol{V}\left(\mathbb{C}_{ce}(\boldsymbol{q}_e)\right)\overline{\mathcal{F}}^{int}(\boldsymbol{q}_e). \tag{3.41}$$

The form of this solution is similar to the inverse kinematics solution of a kinematically redundant limb (cf. (2.34)), comprising particular and homogeneous solution components. The former is composed of contact wrench components that compensate the action of the closure-link wrench $\mathcal{F}_e$. The contact wrench component $\overline{\mathcal{F}}^n(\boldsymbol{q}_e) \in \mathcal{N}\left(\mathbb{C}_{ce}(\boldsymbol{q}_e)\right)$, on the other hand, being derived from the null space of the loop contact map, does not alter the wrench balance at the closure link; $\overline{\mathcal{F}}^n(\boldsymbol{q}_e)$ will be referred to as the *null-space contact wrench component*. Note that in (3.40) and (3.41), the null space is parametrized nonminimally/minimally by $\overline{\mathcal{F}}_a^c(\boldsymbol{q}_e)$ and $\overline{\mathcal{F}}^{int}(\boldsymbol{q}_e) \in \Re^{c_e-6}$, respectively.

3.5.2 Internal Force/Moments: the Virtual Linkage Model

The null-space contact wrench component $\overline{\mathcal{F}}^n$ is related to the *internal force/moments* within the closed loop. The internal force was first introduced in the field of multifinger grasping [58]. The components of the internal force do not contribute to the displacement of the grasped object; they can be used to squeeze it, and at the same time to ensure the friction and the joint-torque limit constraints [37,50]. In [43], the components of the internal force (referred to as the "interaction forces") were interpreted as pairs of equal and opposite forces acting along the lines joining the contact points. The internal force also plays an important role in the field of object manipulation by dual-arm and cooperating robots. In this case, bilateral contacts are quite often assumed [75,65,69]. Then the compressive and tension forces can both be controlled via the components of the internal force. In addition, *internal moments* can be controlled as well. The internal force/moments play an important role in humanoid robotics with regard to such subtasks as balance control under multiple contacts (e.g. with a double-stance posture), motion/force control, and object manipulation with a multifinger hand, a dual-arm robot, or cooperatively by multiple robots, as well as whole-body pushing [27,57,59,35,48].

The physical meaning of the internal force/moments will be clarified below with the help of the *Virtual Linkage* (VL) model [69].

Internal Forces

Assume that p *completely constrained* contact joints are formed at closure link $\{e\}$. Combining the contact joint pairs, one obtains the number ${}_pC_2 = p(p-1)/2$. In the case of point contacts, it is sufficient to consider only force components. The wrench impressed by the contact forces upon the loop closure link is defined as $\mathcal{F}_{e_f} = \begin{bmatrix} \boldsymbol{f}_e^T & \boldsymbol{m}_e^T \end{bmatrix}^T$. It can be expressed as

$$\mathcal{F}_{e_f} = \sum_k \mathbb{S}_L \mathbb{T}_{\overleftarrow{ek}}^T \boldsymbol{f}_k = \mathbb{T}_f^T \boldsymbol{f}^c;$$

$\mathbb{S}_L$ extracts the left column of $\mathbb{T}_{\overleftarrow{ek}}^T$, and $\boldsymbol{f}^c \in \Re^{3p}$ and $\mathbb{T}_f^T \in \Re^{6\times 3p}$ are composed of stacked $\boldsymbol{f}_k$ and $\mathbb{S}_L \mathbb{T}_{\overleftarrow{ek}}^T$ components, respectively. The solution to the WD problem can then be written as

$$\boldsymbol{f}^c = (\mathbb{T}_f^T)^{\#} \mathcal{F}_{e_f} + \boldsymbol{N}(\mathbb{T}_f^T) \boldsymbol{f}_a^c. \tag{3.42}$$

Note that the null space $\mathcal{N}(\mathbb{T}_f^T)$ is nonminimally parametrized by $\boldsymbol{f}_a^c \in \Re^{3p}$.

As an example, consider a closed loop with four bilateral point contacts, as shown in Fig. 3.3. According to the VL model, the pairs of forces between the contacts are represented by linearly actuated members that can apply compression/tension force components. With $p = 4$ the number of internal forces is ${}_pC_2 = 6$. The internal forces, denoted as f_{ij}^{int} $\left(i, j \in \{\overline{1,4}\}, i \neq j\right)$, act along the links of the VL model determined by unit vectors (cf. Fig. 3.3A)

$$\boldsymbol{e}_{ij} = \frac{\boldsymbol{r}_j - \boldsymbol{r}_i}{|\boldsymbol{r}_j - \boldsymbol{r}_i|}, \quad \boldsymbol{e}_{ij} = -\boldsymbol{e}_{ji},$$

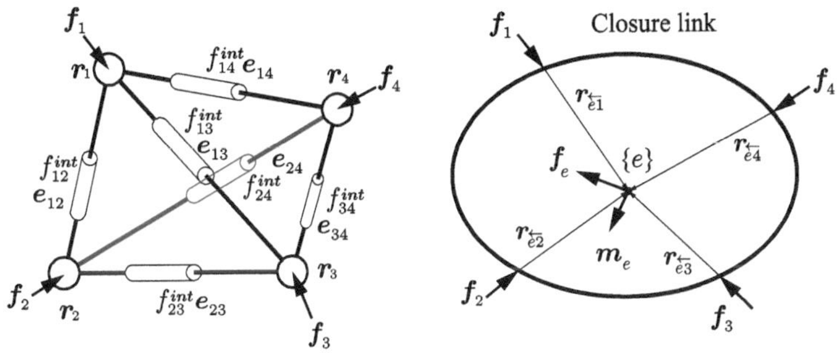

FIGURE 3.3 Virtual linkage model [69] for a closed loop with four bilateral point contacts. (A) Contact forces and tension/compression forces at the virtual links. (B) Resultant force/moment impressed at the center of frame $\{e\}$ on the loop-closure link.

$\boldsymbol{r}_{(\circ)}$ denoting the positions of the contact joints. The contact force components stemming from the internal forces can then be determined as

$$\begin{bmatrix} \boldsymbol{f}_1^n \\ \boldsymbol{f}_2^n \\ \boldsymbol{f}_3^n \\ \boldsymbol{f}_4^n \end{bmatrix} = \begin{bmatrix} \boldsymbol{e}_{12} & \boldsymbol{e}_{13} & \boldsymbol{e}_{14} & \mathbf{0} & \mathbf{0} & \mathbf{0} \\ \boldsymbol{e}_{21} & \mathbf{0} & \mathbf{0} & \boldsymbol{e}_{23} & \boldsymbol{e}_{24} & \mathbf{0} \\ \mathbf{0} & \boldsymbol{e}_{31} & \mathbf{0} & \boldsymbol{e}_{32} & \mathbf{0} & \boldsymbol{e}_{34} \\ \mathbf{0} & \mathbf{0} & \boldsymbol{e}_{41} & \mathbf{0} & \boldsymbol{e}_{42} & \boldsymbol{e}_{43} \end{bmatrix} \begin{bmatrix} f_{12}^{int} \\ f_{13}^{int} \\ f_{14}^{int} \\ f_{23}^{int} \\ f_{24}^{int} \\ f_{34}^{int} \end{bmatrix}. \tag{3.43}$$

The above equation is represented in compact form as

$$\boldsymbol{f}^n = \boldsymbol{V}_L \boldsymbol{f}^{int}, \tag{3.44}$$

where $\boldsymbol{f}^{int} \in \Re^{_pC_2}$ and $\boldsymbol{V}_L \in \Re^{3p \times {_pC_2}}$. The vector $\boldsymbol{f}^n \in \Re^{3p}$ is composed of the stacked contact force components $\boldsymbol{f}_k^n$. Note that this set of contact forces do not contribute to the closure-link wrench, i.e. this is a *null-space contact force component*. The solution to the WD problem (3.42) can then be rewritten as

$$\boldsymbol{f}^c = (\mathbb{T}_f^T)^{\#} \mathcal{F}_e + \boldsymbol{f}^n(\boldsymbol{f}^{int}).$$

Hereby, null space $\mathcal{N}^*(\mathbb{T}_f^T)$ is minimally parametrized by the internal force vector $\boldsymbol{f}^{int}$. It then becomes apparent that $\boldsymbol{V}_L$ is a mapping onto that null space, s.t. $\mathbb{T}_f^T \boldsymbol{V}_L \equiv \mathbf{0}_6$ holds.

Internal Moments

Consider now the case of plane contacts in 3D that can impress moment components upon the loop-closure link. These components represent in fact the internal moments within the closed loop. In the case of independent contact constraints, three internal moments will be generated at each contact joint [69]. The following relations can be established:

$$\begin{aligned} \boldsymbol{m}_k &= \boldsymbol{m}_k^{int}, \\ \boldsymbol{m}_e &= \sum_k \boldsymbol{m}_k, \\ \boldsymbol{m}^{int} &= \boldsymbol{m}^n \in \Re^{3p}, \end{aligned} \tag{3.45}$$

where $\boldsymbol{m}^{int}$ is composed of the stacked $\boldsymbol{m}_k^{int}$ components.

Internal Wrench

By definition, the sum of the contact wrench components stemming from the internal force/moment vectors do not alter the wrench balance at the loop-closure and root links, i.e.

$$\mathbb{T}_f^T \boldsymbol{f}^n + \mathbb{T}_m^T \boldsymbol{m}^n = \boldsymbol{0}. \tag{3.46}$$

Combining the force/moment relations derived so far, one obtains the system

$$\begin{bmatrix} \boldsymbol{0}_6 \\ \boldsymbol{f}^{int} \\ \boldsymbol{m}^{int} \end{bmatrix} = \begin{bmatrix} \mathbb{T}_f^T & \mathbb{T}_m^T \\ \boldsymbol{V}_L^+ & \boldsymbol{0}_{{}_pC_2 \times 3p} \\ \boldsymbol{0}_{3p} & \boldsymbol{E}_{3p} \end{bmatrix} \begin{bmatrix} \boldsymbol{f}^n \\ \boldsymbol{m}^n \end{bmatrix}. \tag{3.47}$$

The matrix $\mathbb{T}_m^T \in \Re^{6 \times 3p}$ is defined by the following relation:

$$\mathcal{F}_{e_m} = \begin{bmatrix} \boldsymbol{0} \\ \boldsymbol{m}_e \end{bmatrix} = \mathbb{T}_m^T \boldsymbol{m}^n.$$

Apparently, the first three rows of $\mathbb{T}_m^T \in \Re^{6 \times 3p}$ are zeros. The last three must be ones, in accordance with (3.45). Furthermore, note that the second equation in (3.47) stems from the approximate (least-squares) resolution of internal forces in (3.44), $\boldsymbol{V}_L^+$ denoting the *left* pseudoinverse. Note also that, according to that equation, contact moments $\boldsymbol{m}^n$ make no contribution to $\boldsymbol{f}^{int}$. It becomes then apparent that the above resolution of the internal forces is not an exact one. As noted in [69], such approximation is admissible under the assumption that the internal moments will be minimized, as is often the case in humanoid robotics. System (3.47) can be written in compact form as

$$\mathcal{F}^{int} = \boldsymbol{G} \mathcal{F}^n, \tag{3.48}$$

where $\boldsymbol{G} \in \Re^{(3p + {}_pC_2) \times 6p}$ is referred to as the *grasp description matrix* (GDM) [69].

For control purposes, the contact wrenches need to be determined. Assuming that the internal force/moments are known, the respective contact wrench components can be obtained from (3.48) and (3.47) as

$$\mathcal{F}^n = \boldsymbol{G}^{-1} \mathcal{F}^{int}, \tag{3.49}$$

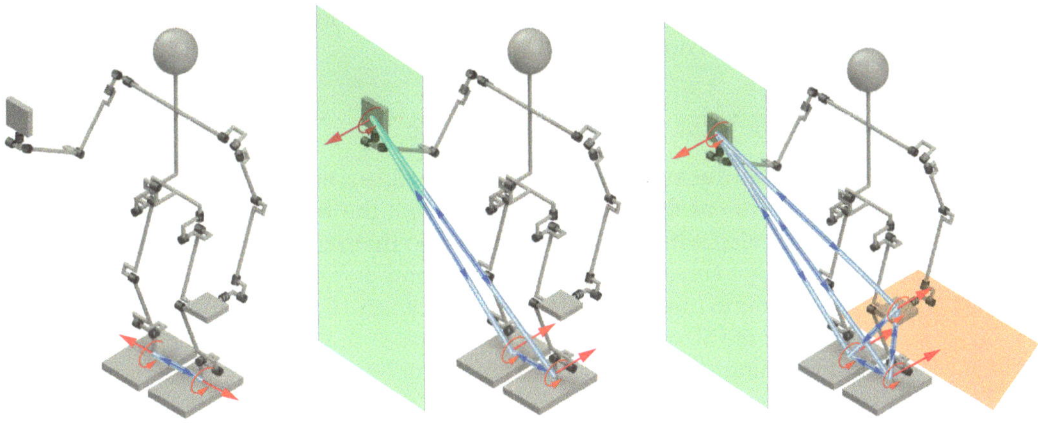

FIGURE 3.4 Virtual linkages for two (A), three (B), and four (C) plane contacts. The blue and the red arrows indicate the internal forces and the contact wrenches, respectively.

$$\begin{aligned}\begin{bmatrix}\boldsymbol{f}^n\\ \boldsymbol{m}^n\end{bmatrix} &= \begin{bmatrix}(\mathbb{T}_f^T)^+ & \boldsymbol{V}_L & -(\mathbb{T}_f^T)^+\mathbb{T}_m^T\\ (\mathbb{T}_m^T)^+ & \boldsymbol{0}_{3p\times {}_pC_2} & \boldsymbol{E}_{3p}\end{bmatrix}\begin{bmatrix}\boldsymbol{0}\\ \boldsymbol{f}^{int}\\ \boldsymbol{m}^{int}\end{bmatrix}\\ &= \begin{bmatrix}\boldsymbol{V}_L & -(\mathbb{T}_f^T)^+\mathbb{T}_m^T\\ \boldsymbol{0}_{3p\times {}_pC_2} & \boldsymbol{E}_{3p}\end{bmatrix}\begin{bmatrix}\boldsymbol{f}^{int}\\ \boldsymbol{m}^{int}\end{bmatrix}.\end{aligned} \tag{3.50}$$

Fig. 3.4 demonstrates the VL model applied to a humanoid robot in double stance ($p = 2$, Fig. 3.4A), double stance plus hand contact ($p = 3$, Fig. 3.4B), and double stance plus two-hand contact ($p = 4$, Fig. 3.4C).

The VL model is helpful in understanding the meaning of internal forces and moments. The model was introduced in humanoid robotics in [59]. There are some inherent problems, though, that hinder the implementation of the model. As already noted, the VL model is based on completely constrained contact joints; it cannot be directly adopted to account for frictionless contact joints. Frictionless contact joints are useful in such tasks as wiping a surface, for example. Clearly, the contact wrench components along the constrained motion directions contribute to the internal force/moments. Second, the VL model requires additional analysis to account for interdependent contact constraints. The contact constraints in the case of double stance, for example, are not independent. Referring to Fig. 3.4, it should be apparent that the two moments at the feet along the direction of the internal force are indeterminate: their sum contributes to the resultant moment at the loop-closure link, while their difference determines a component of the internal moment [59]. The dimension of $\mathcal{F}^{int}$ in (3.49) indicates seven internal force/moment components,[2] but in reality there are only six.

[2] Since $3p + {}_pC_2 = 6 + 1 = 7$.

Another problem with the VL model is that the parametrization of $\mathcal{R}(\boldsymbol{G}^T)$ via the internal force/moments, as in (3.49), is not straightforward.

3.5.3 Determining the Joint Torques in the Loop

The general solution to the WD problem, given in (3.39), is not plagued by the above-mentioned problems pertinent to the VL model. Indeed, the notation accounts for contact joints not only with, but also without friction. Note also that rank$\boldsymbol{N}\left(\mathbb{C}_{ce}(\boldsymbol{q}_e)\right)$ equals exactly the number of independent constraints in the closed loop. For example, for a double-stance posture with fully constrained feet contacts as above, rank$\boldsymbol{N}\left(\mathbb{C}_{ce}(\boldsymbol{q}_e)\right) = c_e - 6 = 12 - 6 = 6$. Furthermore, in the case of interdependent closed loops as those shown in Fig. 3.4B and C, there is a trivial but meaningful parametrization of the null-space contact wrench component, as will be shown in Section 3.6.2.

The net contact wrench components determined in (3.39) are realized with respective joint torques of the limbs constituting the closed loop. To clarify this, extract first the contact wrench components $\overline{\mathcal{F}}_k^c$ from the loop contact wrench (3.39) as follows:

$$\begin{aligned}\overline{\mathcal{F}}_k^c &= \mathbb{C}_{ck}^{\#}(\boldsymbol{q}_e)\mathcal{F}_e + \overline{\mathcal{F}}_k^n,\\ \overline{\mathcal{F}}_k^n &= \boldsymbol{N}_k\left(\mathbb{C}_{ce}(\boldsymbol{q}_e)\right)\overline{\mathcal{F}}_a^c(\boldsymbol{q}_e).\end{aligned} \tag{3.51}$$

Transforms $\mathbb{C}_{ck}^{\#}(\boldsymbol{q}_e) \in \Re^{c_k \times 6}$ and $\boldsymbol{N}_k(\mathbb{C}_{ce}(\boldsymbol{q}_e)) \in \Re^{c_k \times 12}$ are derived from the components of

$$\mathbb{C}_{ce}^{\#}(\boldsymbol{q}_e) \equiv \left[\frac{\mathbb{C}_{ce_r}^{\#}(\boldsymbol{q}_e)}{\mathbb{C}_{ce_l}^{\#}(\boldsymbol{q}_e)}\right], \quad \boldsymbol{N}(\mathbb{C}_{ce}(\boldsymbol{q}_e)) \equiv \left[\frac{\boldsymbol{N}_{e_r}(\mathbb{C}_{ce}(\boldsymbol{q}_e))}{\boldsymbol{N}_{e_l}(\mathbb{C}_{ce}(\boldsymbol{q}_e))}\right],$$

respectively. The two components of $\overline{\mathcal{F}}_k^c$ on the r.h.s. of (3.51) determine the respective joint torque components for the limb k, i.e.

$$\begin{aligned}\boldsymbol{\tau}_k^c &= \boldsymbol{\tau}_k^{ext} + \boldsymbol{\tau}_k^{int}\\ &= \mathcal{J}_{cR}{}^T(\boldsymbol{q}_k)\mathbb{C}_{ck}^{\#}(\boldsymbol{q}_e)\mathcal{F}_e + \mathcal{J}_{cR}{}^T(\boldsymbol{q}_k)\overline{\mathcal{F}}_k^n.\end{aligned} \tag{3.52}$$

The role of the first component is nominal compensation of the external wrench reaction $\mathcal{F}_e$. The second component can be used as internal wrench control input, to ensure that the reaction wrench $\mathcal{F}_k = \overline{\mathcal{F}}_k^c$ stays within CWC_k, and also as a force/moment control input. An illustrative example is presented in Section 3.6.5.

3.5.4 Which Generalized Inverse?

Determining the generalized inverse in the particular solution component in (3.39) is not a trivial problem. The Moore–Penrose pseudoinverse has been exploited in the field of multifinger grasping and multileg robots [43]. In analogy, the pseudoinverse was considered for double-stance balance of a humanoid robot [34,54]. It must be emphasized, though, that the use of the pseudoinverse for force distribution in humanoid robotics is inappropriate due to the following reasons. First, note that the dimensions of the reaction forces and moments are nonuniform.[3] This problem can be eventually alleviated by resorting to an appropriate weighted generalized inverse [22,21,76]. Second, even if uniform dimensions can be assumed, there is another problem: the magnitudes of the components of the pseudoinverse (minimum-norm) solution are of the same order. This implies that the gravity force will always be approximately evenly distributed between the feet, irrespectively of the location of the CoM. This type of distribution is *inconsistent with statics*: the robot will not be able to lift its foot to make a step, for example, since the static equilibrium would be immediately lost. To accomplish such task, an *asymmetric* contact wrench distribution is needed, such that the contact wrench at the foot before lift-off becomes (almost) zero. This cannot be achieved with the pseudoinverse.

Furthermore, note that with the pseudoinverse, the loop-closure (or the loop-root) wrench is distributed evenly not only among the contacts, but also along the normal and the tangential directions at each contact joint with friction. It would be desirable, however, to distribute this wrench in a way that accounts for the Coulomb friction model. This is not possible since the model is based on inequalities. It is possible though to employ a weighted generalized inverse instead of the pseudoinverse, whereby the weights are determined to be in proportion to the friction coefficients μ_k. With a weighted generalized inverse and without a null space component, the general solution (3.39) is rewritten as

$$\overline{\mathcal{F}}_e^c = \mathbb{C}_{ce}^{-W_\mu}(\boldsymbol{q}_e)\mathcal{F}_e. \tag{3.53}$$

The weight matrix is determined as $\boldsymbol{W}_\mu = \text{diag}\begin{bmatrix}\boldsymbol{E}_{c_e^h} & \mu_{e_r}\boldsymbol{E}_{c_{e_r}^s} & \mu_{e_l}\boldsymbol{E}_{c_{e_l}^s}\end{bmatrix}$, under the convention pertinent to the contact joints with friction, i.e. $c_k = 6 = c_k^h + c_k^s$, $\eta_k = 0$ (cf. Section 3.3.2). With this approach, the decomposition of the external wrench $\mathcal{F}_e$ into normal and tangential components at the contacts satisfies the equality component in the contact FC model, so that the reaction force is exactly at the boundary of the friction cone. This approach can thus prevent sliding to occur at the contacts. On the other hand, the distribution of the normal reactions is as with the pseudoinverse since $\boldsymbol{E}_{c_e^h}$ determines equal weights for those reactions. In Section 5.10.4, another weighted generalized inverse will be introduced that can alleviate this problem.

3.5.5 Priorities Among the Joint Torque Components

Notation (3.52) introduces one more priority level into priority scheme (3.37), so now we have

$$\boldsymbol{\tau}_k^{ext} \succ \boldsymbol{\tau}_k^{int} \succ \boldsymbol{\tau}_k^{m} \succ \boldsymbol{\tau}_k^{n}. \tag{3.54}$$

Apparently, the self-motion joint torque for a kinematically redundant limb, $\boldsymbol{\tau}_k^n$, is at the lowest priority level; neither the reaction forces nor the forces in the unconstrained motion directions at the contact joints will be affected. The latter forces, on the other hand, do not affect

[3] In the field of multifinger hands and multileg robots the pseudoinverse solution is admissible since point contacts are assumed. In this case, the reactions are pure forces of uniform dimensions.

the internal force distribution and external force compensation. It is also seen that the internal force control, via $\boldsymbol{\tau}_k^{int}$, is possible without affecting the external force compensation. The external force affects all the other components; they may need readjustment after a change in $\mathcal{F}_R$. Note also that to satisfy the friction cone conditions, both $\boldsymbol{\tau}_k^{ext}$ and $\boldsymbol{\tau}_k^{int}$ should be controlled, as already clarified.

3.6 KINETOSTATIC RELATIONS OF A HUMANOID ROBOT

The independent closed-loop model discussed so far can be used for loop-specific controller design. Such a control approach might be appropriate for the purpose of analysis, e.g. to focus on pure balance control whereby the hands do not contact the environment, or on pure cooperative hand force control whereby the legs are fixed. In real applications, though, the force relations within the closed loops are always interdependent. Consider as an example the case depicted in Fig. 3.6. As already clarified, in this case there are two loop-closure links. The gravity force impressed on the object, $-\mathcal{G}_O$, directly acts within the closed loop of the arms. But this force is also structurally mapped at the base link, which is the root link of the other (the leg) closed loop. Consequently, $\mathcal{G}_O$ will contribute to the contact reactions at the feet, in addition to those at the hands. Another example is multicontact balance, whereby one of the hands or both are used to stabilize the posture of the robot. The respective model should account for the presence of a single loop-closure link (i.e. the static environment) and multiple interdependent branches.

In what follows, the independent closed-loop kinetostatic force relations derived in the previous sections will be modified to reflect such dependencies. The ultimate goal is to determine the joint torque components that:

1. compensate for gravity and other external wrenches acting at specified links (the base link and a hand-held object);
2. ensure an appropriate internal force in each closed loop to maintain the friction cone conditions or to obtain a desired end-link wrench to be used in force control;
3. determine the motion along the end-link mobility directions, in the case of frictionless contact joints;
4. ensure joint load redistribution, e.g. to minimize loads at specific joints.

The notation introduced in Section 2.11 for the instantaneous motion relations will be used thereby.

3.6.1 The Composite Rigid Body (CRB) and the CRB Wrench

The kinetostatic relations of a humanoid robot are derived under the assumption of locked joints. The robot behaves then as a composite rigid body (CRB), i.e. a system of interconnected links characterized by the net CoM and the inertia tensor [67]. In statics, the inertia tensor is irrelevant but the CoM plays an important role. A coordinate frame, denoted as $\{C\}$, is attached to the CoM. The coordinate axes are parallel to those of the base frame $\{B\}$. In what follows, the coordinate frame $\{C\}$ will be used in motion/force relations to represent the CRB.

As clarified in Section 3.4, the sum of all external wrenches acting on the robot links (i.e. the net external wrench) can be mapped at a characteristic link of interest, e.g. the base link, the CRB virtual "link" or the common loop-closure link.

Assume first that the net external wrench is mapped at the base link as $\mathcal{F}_B$. According to (3.25), $\mathcal{F}_B$ can be represented by the sum of appropriately mapped contact (reaction) wrenches $\mathcal{F}_k^c = \mathcal{F}_k,\ k \in \{e_r, e_l\},\ e \in \{F,\ H\}$.[4] In the general case, i.e. when some of the contact joints are frictionless, $\mathcal{F}_B$ can be expressed according to (3.38) as the sum of reaction wrench components along the constrained motion directions, $\overline{\mathcal{F}}_k^c$, mapped by the contact map of the robot, i.e.

$$\mathcal{F}_B = \sum_k \mathbb{C}_{cB}(\boldsymbol{q}_k)\overline{\mathcal{F}}_k^c = \mathbb{C}_{cB}(\boldsymbol{q})\overline{\mathcal{F}}^c. \tag{3.55}$$

Here $\overline{\mathcal{F}}^c \in \Re^c$ is composed of the stacked $\overline{\mathcal{F}}_k^c$ contact wrenches, ordered according to the k, e subscript convention. The mapping of the net external wrench at the CRB frame can be written in a similar form, i.e.

$$\mathcal{F}_C = \sum_k \mathbb{C}_{cC}(\boldsymbol{q}_k)\overline{\mathcal{F}}_k^c = \mathbb{C}_{cC}(\boldsymbol{q})\overline{\mathcal{F}}^c. \tag{3.56}$$

As an example, consider a double-stance posture ($k \in \{F_r,\ F_l\}$) with frictional constraints. Such posture appears quite often in practice; $\mathcal{F}_{(\circ)}$ can then be written as

$$\mathcal{F}_{(\circ)} = \begin{bmatrix} \boldsymbol{f}_{(\circ)} \\ \boldsymbol{n}_{(\circ)} \end{bmatrix} = \sum_k \begin{bmatrix} \boldsymbol{E}_3 & \boldsymbol{0}_3 \\ -[\boldsymbol{r}^{\times}_{\overleftarrow{(\circ)k}}] & \boldsymbol{E}_3 \end{bmatrix} \begin{bmatrix} \boldsymbol{f}_k \\ \boldsymbol{m}_k \end{bmatrix}, \tag{3.57}$$

subscript $_{(\circ)}$ standing for either $_B$ or $_C$. $\boldsymbol{f}_k$ and $\boldsymbol{m}_k$ denote the ground reaction force (GRF) and the ground reaction moment (GRM) at the foot. They constitute the foot contact (reaction) wrench $\mathcal{F}_k^c$. Note that when the GRF-induced GRM is subtracted from the CRB wrench, one obtains the following net wrench:

$$\mathcal{F}_{net} \equiv \mathcal{F}_{(\circ)} - \sum_k \begin{bmatrix} \boldsymbol{0}_3 \\ [\boldsymbol{r}^{\times}_{\overleftarrow{(\circ)k}}]\boldsymbol{f}_k \end{bmatrix}, \tag{3.58}$$

s.t.

$$\mathcal{F}_{net} = \sum_k \mathcal{F}_k^c.$$

The net wrench $\mathcal{F}_{net}$ will be used in Section 5.10.4. Furthermore, note that the above relations can be reformulated for the case when the net external wrench is mapped at the common loop-closure link, i.e. the floor coordinate frame $\{F\}$. This is done straightforwardly by replacing the $_{(\circ)}$ subscripts with $_F$. A notation involving $\mathcal{F}_F$ will appear in Section 7.8.3. The net

[4] For clarity and without loss of generality, henceforth it is assumed that contact joints are established at the end links. In general, a contact joint can be established at anybody segment.

wrench acting on the CRB will be henceforth referred to as the *CRB wrench*, or simply as the *body wrench*.

The action of the gravity field on the CRB can be expressed either by a distributed model of gravity wrenches acting on each robot link, or by a lumped model, s.t. the net gravity *force* acts at the system CoM. These two representations are related to the two notations in the base and the mixed quasicoordinates, respectively. Consider first the distributed model. The reaction to the gravity wrench acting on the base link is written as

$$\mathcal{G}_B \equiv \begin{bmatrix} \boldsymbol{g}_f \\ \boldsymbol{g}_m \end{bmatrix} = M \begin{bmatrix} -\boldsymbol{E}_3 \\ [\boldsymbol{r}^{\times}_{\overleftarrow{BC}}] \end{bmatrix} \boldsymbol{a}_g. \tag{3.59}$$

Here $\boldsymbol{r}_{\overleftarrow{BC}}$ is the position of the base-link frame w.r.t. the CoM, $\boldsymbol{a}_g = \begin{bmatrix} 0 & 0 & -g \end{bmatrix}^T$, and g and M stand for the acceleration of the gravity and the total mass, respectively. The gravity wrenches acting on the rest of the links are compensated by the following gravity *joint torque*:

$$\boldsymbol{g}_\theta = M \boldsymbol{J}^T_{\overleftarrow{BC}} \boldsymbol{a}_g = -\boldsymbol{J}^T_{\overleftarrow{BC}} \boldsymbol{g}_f, \tag{3.60}$$

where $\boldsymbol{J}_{\overleftarrow{BC}}$ denotes the net-CoM Jacobian (cf. (2.117)).

The lumped-model representation yields a simpler relation. The reaction to the gravity wrench acting on the CRB is $\mathcal{G}_C \equiv \begin{bmatrix} \boldsymbol{g}_f^T & \boldsymbol{0}^T \end{bmatrix}^T$, i.e. there is no moment component, as expected.

3.6.2 Interdependent Closed Loops

The above notation refers to humanoid robot postures forming interdependent closed loops, in the general case of mixed friction/frictionless contact joints. For example, when the contact joints at the feet are with friction and those at the hands are frictionless, then

$$\overline{\mathcal{F}}^c = \begin{bmatrix} (\mathcal{F}^c_{F_r})^T & (\mathcal{F}^c_{F_l})^T & (\overline{\mathcal{F}}^c_{H_r})^T & (\overline{\mathcal{F}}^c_{H_l})^T \end{bmatrix}^T, \tag{3.61}$$

$\mathcal{F}^c_{F_j} = \mathcal{F}_{F_j}$ denoting the friction wrenches at the feet. When all contact joints are with friction ($c = 24$), then $\overline{\mathcal{F}}^c = \mathcal{F}^c = \mathcal{F} \in CWC$ and

$$CWC = CWC_{F_r} \times CWC_{F_l} \times CWC_{H_r} \times CWC_{H_r} \subset \Re^{24}.$$

In the case of multicontacts, e.g. double stance or double stance plus hand contact(s), as in Fig. 3.4, the base link will be unilaterally overconstrained since $c > 6$. Given the CRB wrench $\mathcal{F}_B$, there is an infinite set of contact wrenches (cf. (3.39)), i.e.

$$\overline{\mathcal{F}}^c = \overline{\mathcal{F}}^{ext} + \overline{\mathcal{F}}^n, \tag{3.62}$$

$$\overline{\mathcal{F}}^{ext} = \mathbb{C}^{\#}_{cB}(\boldsymbol{q})\mathcal{F}_B,$$

$$\overline{\mathcal{F}}^n = \boldsymbol{N}\left(\mathbb{C}_{cB}(\boldsymbol{q})\right)\overline{\mathcal{F}}^c_a. \tag{3.63}$$

The second term on the r.h.s. is a component from the null space $\mathcal{N}(\mathbb{C}_{cB})$, parametrized by an arbitrary composite contact wrench, $\overline{\mathcal{F}}_a^c \in \Re^c$. Recall that the number of internal force/moments within the closed loops equals the rank of the null-space projector, which is $c-6$. This is also apparent from the dimension of $\overline{\mathcal{F}}^{int}(\boldsymbol{q}_e)$ in (3.41), which is a minimal-parametrization vector. However, as noted at the end of Section 3.5.2, how to specify the internal force/moments is not always intuitively clear. The nonminimal parametrization, on the other hand, can be specified in a straightforward way since the meaning is that of a contact wrench. Indeed, note that a number of practical tasks require precise contact wrench control at the hands. In this case, with the nonminimal parametrization it is possible to specify the reference hand contact wrenches *directly* as null-space contact wrench components. To this end, (3.62) is partitioned first as

$$\begin{bmatrix} \overline{\mathcal{F}}_F^c \\ \overline{\mathcal{F}}_H^c \end{bmatrix} = \begin{bmatrix} \left(\mathbb{C}_{cB}^{\#}\right)_F \\ \hline \left(\mathbb{C}_{cB}^{\#}\right)_H \end{bmatrix} \mathcal{F}_B + \begin{bmatrix} \boldsymbol{N}_F(\mathbb{C}_{cB}) \\ \hline \boldsymbol{N}_H(\mathbb{C}_{cB}) \end{bmatrix} \overline{\mathcal{F}}_a^c, \tag{3.64}$$

the upper and lower blocks standing for contact wrench components for the feet and hands, respectively. Furthermore, it is assumed that the reference hand contact wrench $(\overline{\mathcal{F}}_H^c)^{ref}$ is obtained from a conventional force/moment controller. Then, the lower equation can be employed to obtain the parametrization vector as

$$\overline{\mathcal{F}}_a^c = \boldsymbol{N}_H^+(\mathbb{C}_{cB})\left((\overline{\mathcal{F}}_H^c)^{ref} - \left(\mathbb{C}_{cB}^{\#}\right)_H \mathcal{F}_B\right). \tag{3.65}$$

With this parametrization, the WD problem is solved as

$$\begin{bmatrix} \overline{\mathcal{F}}_F^c \\ \overline{\mathcal{F}}_H^c \end{bmatrix} = \begin{bmatrix} \left(\mathbb{C}_{cB}^{\#}\right)_F \mathcal{F}_B + \boldsymbol{N}_F(\mathbb{C}_{cB})\boldsymbol{N}_H^+(\mathbb{C}_{cB})\left(\overline{\mathcal{F}}_H^{ref} - \left(\mathbb{C}_{cB}^{\#}\right)_H \mathcal{F}_B\right) \\ (\overline{\mathcal{F}}_H^c)^{ref} \end{bmatrix}. \tag{3.66}$$

Hereby, $\boldsymbol{N}_H \boldsymbol{N}_H^+ = \boldsymbol{E}$ was used. Apparently, the reference hand contact wrenches appear *directly* as contact wrenches at the hands. The distribution of the CRB wrench at the feet depends on the type of generalized inverse (cf. Section 3.5.4), as well as on the reference hand contact wrenches.

3.6.3 Independent Closed Loops

In the case of two independent closed loops, i.e. as in the example in Fig. 3.6, the internal force/moments within each closed loop are noninterfering. Assume that the total external wrench at the base link is $-\mathcal{F}_B$. According to (3.25), this wrench is determined by the external wrenches acting at the links, e.g. the gravity wrenches at the base and the hand-held object, s.t. $-\mathcal{F}_B = -\mathcal{G}_B - \mathbb{T}_{\overleftarrow{BO}}^T \mathcal{G}_O$. Other external wrenches, e.g. resulting from a sudden push on an arbitrary link, could be added in a straightforward way. To account for the independent closed loops, rewrite (3.55) as

$$\mathcal{F}_B = \sum_k \mathbb{C}_{cB}(\boldsymbol{q}_k)\overline{\mathcal{F}}_k^c = \mathbb{C}_{cB}(\boldsymbol{q}_F)\overline{\mathcal{F}}_F^c + \mathbb{C}_{cB}(\boldsymbol{q}_H)\overline{\mathcal{F}}_H^c. \tag{3.67}$$

The term $\overline{\mathcal{F}}_e^c \in \Re^{c_e}$ is composed by stacking the two $\overline{\mathcal{F}}_{k=e_j}^c$ components. Further on, the external wrench acting on the loop-closure link (i.e. the hand-held object) is $-\mathcal{F}_H = -\mathcal{G}_O$. This wrench is represented as the following sum (cf. (3.30)):

$$\mathcal{F}_H = \sum_j \mathbb{C}_{cH_j}(\boldsymbol{q}_{H_j})\overline{\mathcal{F}}_{H_j}^c = \mathbb{C}_{cH}(\boldsymbol{q}_H)\overline{\mathcal{F}}_H^c. \tag{3.68}$$

Hand reaction wrenches can be obtained from this relation in an infinite number of ways. We have

$$\overline{\mathcal{F}}_H^c = \mathbb{C}_{cH}^{\#}(\boldsymbol{q}_H)\mathcal{F}_H + \boldsymbol{V}(\mathbb{C}_{cH}(\boldsymbol{q}_H))\overline{\mathcal{F}}_H^{int}. \tag{3.69}$$

Note that the second term on the r.h.s. determines the hand reaction wrenches from the null space $\mathcal{N}(\mathbb{C}_{cH}(\boldsymbol{q}_H))$ via the internal force/moments $\overline{\mathcal{F}}_H^{int} \in \Re^{c_H-6}$. Recall that $\boldsymbol{V}$ denotes the minimal representation of the null-space basis vectors. Then, substitute (3.69) into (3.67) and solve for the contact wrenches at the feet, i.e.

$$\begin{aligned}\overline{\mathcal{F}}_F^c &= \mathbb{C}_{cB}^{\#}(\boldsymbol{q}_F)\left(\mathcal{F}_B - \mathcal{F}_{B_H}\right) + \boldsymbol{V}(\mathbb{C}_{cB}(\boldsymbol{q}_F))\overline{\mathcal{F}}_F^{int}, \\ \mathcal{F}_{B_H} &\equiv \mathbb{C}_{cB}(\boldsymbol{q}_H)\left(\mathbb{C}_{cH}^{\#}(\boldsymbol{q}_H)\mathcal{F}_H + \boldsymbol{V}(\mathbb{C}_{cH}(\boldsymbol{q}_H))\overline{\mathcal{F}}_H^{int}\right).\end{aligned} \tag{3.70}$$

The net contact and internal wrenches are then assembled by stacking the respective components, i.e.

$$\begin{aligned}\overline{\mathcal{F}}^c &= \left[(\overline{\mathcal{F}}_F^c)^T \quad (\overline{\mathcal{F}}_H^c)^T\right]^T \in \Re^c, \\ \overline{\mathcal{F}}^{int} &= \left[(\overline{\mathcal{F}}_F^{int})^T \quad (\overline{\mathcal{F}}_H^{int})^T\right]^T \in \Re^{c-6}.\end{aligned}$$

3.6.4 Determining the Joint Torques

To determine the joint torque components, first derive the individual reaction wrench components, $\overline{\mathcal{F}}_k^c = \overline{\mathcal{F}}_k^{ext} + \overline{\mathcal{F}}_k^{n}$, from the total reaction wrench. This is done in a straightforward way by making use of (3.51). The internal force component, $\overline{\mathcal{F}}_k^{n} = \boldsymbol{N}_k(\mathbb{C}_{cR}(\boldsymbol{q}))\overline{\mathcal{F}}_a^c$, can then be used as a control input to ensure that the reaction wrench stays within the CWC_k at all times. This component can also be used in end-link force control. The limb joint torque components $\boldsymbol{\tau}_k^{ext}$ and $\boldsymbol{\tau}_k^{int}$ can then be calculated. These components are consequently summed up with components due to the motion in the mobility directions and the kinematic redundancy, $\boldsymbol{\tau}_k^{m}$ and $\boldsymbol{\tau}_k^{n}$, to obtain the limb joint torque $\boldsymbol{\tau}_k$. The joint torques of all limbs are stacked as follows:

$$\boldsymbol{\tau} = \left[\boldsymbol{\tau}_{F_r}^T \quad \boldsymbol{\tau}_{F_l}^T \quad \boldsymbol{\tau}_{H_r}^T \quad \boldsymbol{\tau}_{H_l}^T\right]^T \in \Re^n. \tag{3.71}$$

Note also that, similarly to (3.33),

$$\boldsymbol{\tau} = \left[\mathcal{J}_{cB}^T(\boldsymbol{q}) \quad \mathcal{J}_{mB}^T(\boldsymbol{q})\right]\begin{bmatrix}\overline{\mathcal{F}}^c \\ \mathcal{F}^m\end{bmatrix} = \boldsymbol{\tau}^c + \boldsymbol{\tau}^m. \tag{3.72}$$

Here $\overline{\mathcal{F}}^m \in \Re^{\eta}$ is composed of $\overline{\mathcal{F}}_k^m$ vectors stacked in the same order as the components of $\overline{\mathcal{F}}^c$. Further on, when kinematic redundancy is present, a null space component from $\mathcal{N}^*(\boldsymbol{J}_R(\boldsymbol{q}))$ can be added. As a result, the joint torque of the humanoid robot is expressed as the sum

$$\boldsymbol{\tau} = \boldsymbol{\tau}^{ext} + \boldsymbol{\tau}^{int} + \boldsymbol{\tau}^{m} + \boldsymbol{\tau}^{n}. \tag{3.73}$$

The components are arranged in the following priority scheme (cf. (3.54)):

$$\boldsymbol{\tau}^{ext} \succ \boldsymbol{\tau}^{int} \succ \boldsymbol{\tau}^{m} \succ \boldsymbol{\tau}^{n}.$$

Component $\boldsymbol{\tau}^n$ appears whenever there is kinematic redundancy. Component $\boldsymbol{\tau}^m$ is available when there are frictionless contact joints or completely unconstrained end links. The internal wrench component $\boldsymbol{\tau}^{int}$ stems from redundant actuation within a closed kinematic loop. Component $\boldsymbol{\tau}^{ext}$ is always present.

The joint torque that ensures the static balance can be expressed, either with the distributed or the lumped CRB model, as

$$\boldsymbol{\tau}_B = \boldsymbol{g}_\theta - \mathcal{J}_{cR}^T \mathcal{F}^c(\mathcal{G}_B), \tag{3.74}$$

$$\boldsymbol{\tau}_M = -\mathcal{J}_{cM}^T \overline{\mathcal{F}}^c(\mathcal{G}_C), \tag{3.75}$$

respectively. Using the expression for the Jacobian $\mathcal{J}_{cM}$ given in (2.126), it is straightforward to prove that $\boldsymbol{\tau}_B = \boldsymbol{\tau}_M$. This is the reason why the gravity joint torque $\boldsymbol{g}_\theta$ does not explicitly appear in the last equation. An important remark is due at this point with regard to the usage of the above expressions in a controller. In the case of balance control, for example, quite often a reference contact wrench $(\mathcal{F}^c)^{ref}$ is employed. In this case, the above two equations are not equivalent. Note that the gravity torque $\boldsymbol{g}_\theta$ in (3.74) can be used as an exact gravity compensation term. On the other hand, $\boldsymbol{\tau}_M$ does not provide for exact gravity compensation since the gravity joint torque becomes a function of $(\overline{\mathcal{F}}^c)^{ref}$.

3.6.5 Illustrative Examples

Double Stance on Flat Floor in 2D (Lateral Plane)

A lower-body biped model on the plane with four joints is shown in Fig. 3.5. It is assumed that the feet are fully constrained, s.t. $c_{F_j} = 3$. Thus, the contact wrenches $\mathcal{F}_{F_j}^c$ comprise one GRM and two GRF components. The static-balance condition for the closed loop is $\mathcal{G}_B = \mathbb{C}_{cB}(\boldsymbol{q}_F)\mathcal{F}^c(\boldsymbol{q}_F)$. The general solution to the WD problem (cf. (3.39)) can be written as

$$\begin{aligned}\mathcal{F}^c(\boldsymbol{q}_F) &= \mathbb{C}_{cB}^{\#}(\boldsymbol{q}_F)\mathcal{G}_B + \boldsymbol{N}\left(\mathbb{C}_{cB}(\boldsymbol{q}_F)\right)\mathcal{F}_a^c(\boldsymbol{q}_F)\\ &= \mathbb{C}_{cB}^{\#}(\boldsymbol{q}_F)\mathcal{G}_B + \boldsymbol{V}\left(\mathbb{C}_{cB}(\boldsymbol{q}_F)\right)\overline{\mathcal{F}}^{int}(\boldsymbol{q}_F).\end{aligned}$$

With the help of the VL model, this solution can be rewritten as

$$\mathcal{F}^c(\boldsymbol{q}_F) = \mathbb{C}_{cB}^{+}(\boldsymbol{q}_F)\mathcal{G}_B + \boldsymbol{V}_L f_{rl}^{int},$$

where

$$\boldsymbol{V}_L = \begin{bmatrix}\boldsymbol{e}_{rl}^T & \boldsymbol{e}_{lr}^T\end{bmatrix}^T = \begin{bmatrix}0 & 1 & 0 & 0 & -1 & 0\end{bmatrix}^T.$$

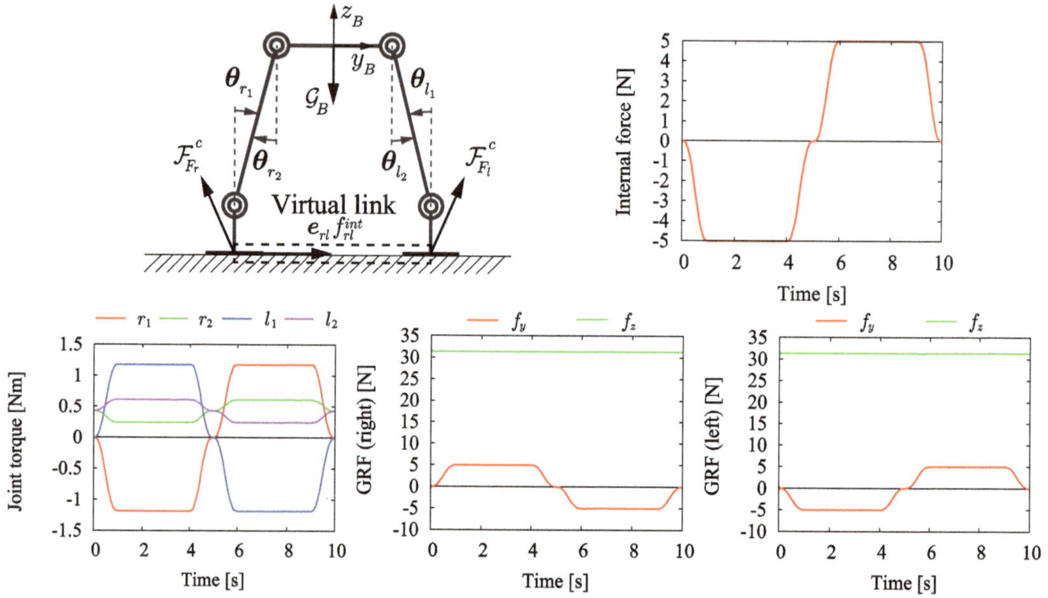

FIGURE 3.5 Double stance with a five-link biped on the plane. The magnitude of the internal force, f_{rl}^{int}, is shown in the upper-right figure. The internal force is changed smoothly, initially with compression (negative), then with extension (positive). The respective joint torque and GRF variations are shown in the lower row. The GRMs equal the ankle torques, denoted as r_1 and l_1 in the lower-left figure.

Three remarks are due. First, the number of the internal force/moment components is three: the internal force f_{rl}^{int} and the two GRMs m_{F_j}. Second, note that the internal moments are not considered in the above equation. Third, the pseudoinverse was adopted as a generalized inverse in the solution. With this choice, the vertical GRFs are distributed equally. No problem arises in this particular example with such distribution. This is not the case, however, in general, as already discussed. An example that highlights the problem stemming from the pseudoinverse-based WD will be introduced shortly.

The joint torque vector can be derived as

$$\begin{aligned}\boldsymbol{\tau} &= \boldsymbol{g}_\theta - (\boldsymbol{\tau}^{ext} + \boldsymbol{\tau}^{int}),\\ \boldsymbol{\tau}^{ext} &= \mathcal{J}_{cR}^T(\boldsymbol{q}_F)\mathbb{C}_{cB}^{+}(\boldsymbol{q}_F)\mathcal{G}_B,\\ \boldsymbol{\tau}^{int} &= \mathcal{J}_{cR}^T(\boldsymbol{q}_F)\boldsymbol{V}_L f_{rl}^{int}.\end{aligned}$$

The balance of force/moments in the closed loop is examined by varying the internal force within a symmetric posture, the ankle/hip joints being set at $5/-5$ degrees, respectively. The magnitude of the internal force is changed between ± 5 N smoothly, with the help of fifth-order splines. The respective variations of all force/moment and joint torque components are shown in the lower row of Fig. 3.5. The model parameters were derived from those of a miniature humanoid robot HOAP-2 (cf. Chapter A). Note that the ankle torques (the GRMs)

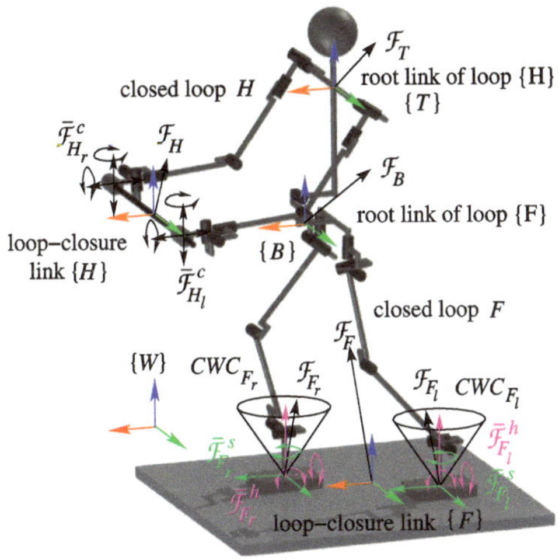

FIGURE 3.6 A humanoid robot with two independent closed loops, F and H, formed with contact joints at the feet (with friction) and hands (frictionless). The reaction wrenches acting at the loop-closure and loop-root links are denoted as $\mathcal{F}_e$, $e \in \{F, H\}$, and $\mathcal{F}_R$, $R \in \{B, T\}$, respectively. The reaction wrench components along the constrained and unconstrained motion directions at the hands are denoted as $\overline{\mathcal{F}}^c_{H_j}$, $j \in \{r, l\}$, and $\overline{\mathcal{F}}^m_{H_j}$, respectively (the latter are not shown). The reaction wrench components along the hard- and soft-constraint directions at the feet stem from friction. They are denoted as $\overline{\mathcal{F}}^h_{F_j}$ and $\overline{\mathcal{F}}^s_{F_j}$, respectively. The total friction/contact wrench $\mathcal{F}_{F_j}$ is confined within the contact wrench cone CWC_{F_j}.

determine the internal moment in the chain. Apparently, they are sufficiently small, s.t. the net CoP (not shown) stays well within the interior of the BoS. Therefore, no foot roll is observed.

Double Stance on Flat Floor With Friction

Next, consider the example shown in Fig. 3.6 of a humanoid robot in 3D on flat ground. The contact joints at the hands are assumed frictionless while those at the feet are with friction. The friction model is the polyhedral CWC (four-sided pyramid) one. Motion at the feet is then completely constrained, s.t. $c_{F_j} = 6$, $\eta_{F_j} = 0$. The motion constraints stem from the normal reactions $\overline{\mathcal{F}}^h_k \in \Re^{c^h_k}$ and the reactions to the tangential friction forces, $\overline{\mathcal{F}}^s_k \in \Re^{c^s_k}$. These reactions sum up as in (3.29), whereby contact wrenches $\mathcal{F}_{F_j} = \mathcal{F}^c_{F_j} \in CWC_{F_j}$. The feet do not slip therefore. It should be noted that the contact wrenches can be estimated (e.g. according to the algorithms explained in Chapter 4) and/or obtained from the sensors attached at the end links. The sensors could be either multiaxis force/torque sensors or arrays of single-axis pressure sensors, such as force sensing resistors (FSRs). Details about such sensors can be found in Section 3.2 of [36].

Further on, note that since the feet contacts are with friction, the base will be *unilaterally overconstrained*, i.e. $\mathcal{F}_B = \mathbb{T}^T_{\overleftarrow{BF_r}}(\boldsymbol{q}_{F_r})\mathcal{F}_{F_r} + \mathbb{T}^T_{\overleftarrow{BF_l}}(\boldsymbol{q}_{F_l})\mathcal{F}_{F_l}$. This also implies that the gravity wrench impressed on the base link can be compensated by the net reaction wrench at the base, so $\mathcal{F}_B = -\mathcal{G}_B = -(\mathcal{G}^h_B + \mathcal{G}^s_B)$.

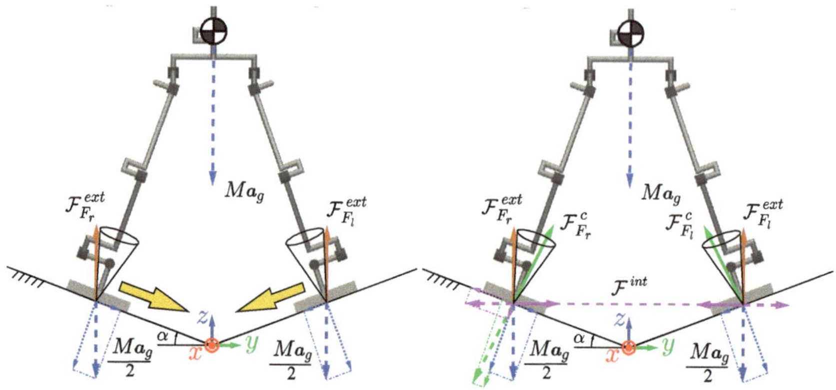

FIGURE 3.7 When supported on inclined surfaces, the force of gravity induces total reactions $\mathcal{F}_{F_j}^{ext}$ comprising components in the normal and tangential directions at each contact. Left (A): Zero internal force case. The tangential reaction components lead to slipping because of insufficient surface friction. Right (B): Nonzero internal force case. The slipping can be avoided with the help of a properly defined internal force or a weighted generalized inverse that modifies the magnitudes of the normal and tangential reaction components, s.t. the total reaction $\mathcal{F}_{F_j}^{c}$ is within the CWC.

Double Stance on Frictionless Flat Floor

Next, consider the case of frictionless contact joints at the feet. The net reaction wrench at the base then becomes $\mathcal{F}_B = \mathcal{F}_B^c = \mathbb{C}_{cR}(\boldsymbol{q}_{F_r})\overline{\mathcal{F}}_{F_r}^c + \mathbb{C}_{cR}(\boldsymbol{q}_{F_l})\overline{\mathcal{F}}_{F_l}^c$. Note that the quasistatic body wrench component $\mathcal{F}_B^m = -\mathcal{G}_B^m$ does not appear in the above relation.

Double Stance With Noncoplanar Contacts

The role of the component $-\mathcal{G}_B^m$ becomes apparent when the feet are placed on inclined support surfaces with relatively low friction, as in Fig. 3.7. This example demonstrates the case of an *underconstrained* base and thus of an uncontrollable robot. Since the feet are in contact with inclined surfaces, the gravity force generates components in both the normal and the tangential directions at the contact joints. Low friction implies that the reactions will be outside the CWC, as shown in Fig. 3.7A. The reactions in the tangential directions will then induce foot acceleration and the robot will become uncontrollable. This problem can be solved by applying appropriate joint torques that work against the reactions in the tangential directions, s.t. the CWC conditions will be satisfied. These joint torque components are computed from the internal force. In other words, $\boldsymbol{\tau}_{F_r}^{int}$ and $\boldsymbol{\tau}_{F_l}^{int}$ can be used to ensure the static equilibrium at the selected posture, despite the presence of a low friction (cf. Fig. 3.7B). This is evident from the simulation results shown in Fig. 3.8 and Video 3.6-1 [33]. In the simulation, a small-size robot with parameters similar to those of a HOAP-2 robot [25] was used (see Section A.1). Initially, a nonzero internal force is applied to ensure the static equilibrium. At $t = 1$ s, the internal force is gradually decreased which results in foot sliding (apparent from the foot position graphs). After keeping the internal force at zero for a while, it is then increased gradually to stop the sliding. It should be mentioned that this behavior cannot be achieved with pure static control because of the motion during the sliding phase. Notice that the robot is uncontrollable during this phase. This is why large errors appear when stopping

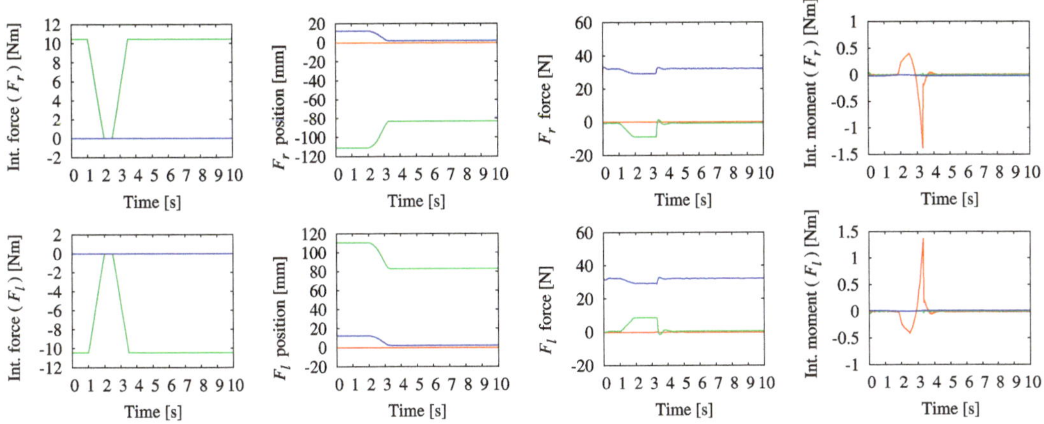

FIGURE 3.8 Simulation results of double stance on inclined surfaces (cf. Fig. 3.7). Right/left foot data are shown in the upper/lower rows, respectively. The RGB color convention is used for the xyz-components. The internal force is first used to initialize foot sliding, then to stop it. Details about the controller can be found in Section 5.10.4.

is initialized. These errors result in relatively large internal moments that need to be controlled to ensure that the feet CoPs stay within the interiors of the footprints. Further details about the controller used in this simulation can be found in Section 5.10.4.

Double Stance on Flat Floor With High/Zero Friction at the Right/Left Foot

Finally, for the sake of completeness, consider also the general case of contact joints with and without friction. For example, let the right foot be constrained with sufficient friction, while the left foot is placed on a frictionless surface. The net reaction wrench at the base link is $\mathcal{F}_B = \mathbb{T}^T_{\overleftarrow{BF_r}}(\boldsymbol{q}_{F_r})\mathcal{F}_{F_r} + \mathbb{C}_{cB}(\boldsymbol{q}_{F_l})\overline{\mathcal{F}}^c_{F_l}$. The base link is *unilaterally fully constrained* by the unilateral contact at the right foot. Note, however, that the wrench component from the left foot may introduce an additional independent constraint. Thus, the base link may become (conditionally) overconstrained.

3.6.6 Summary and Discussion

The kinetostatic relations in closed-loop chains are determined by the total external wrench acting at the loop and the (internal) joint torques of the limbs constituting the closed loop. The total external wrench is the sum of all external wrenches applied at one or more links within the loop, mapped at a link of preference, e.g. the base link. The mapping is structural, i.e. it depends on the configuration of the loop. The action of the joint torques, on the other hand, results in a wrench mapped via inverses of the transposed limb Jacobians at the base link. This wrench is in balance with the total external wrench.

Further on, the action of the external wrenches and the joint torques induces wrenches at the contact joints, called reaction wrenches. These wrenches should be controlled in appropriate ways to ensure the satisfaction of the friction cone conditions at the contact joints while compensating the action of the external wrenches. The joint torque of the limb that ensures a desired reaction wrench at the respective contact joint can be obtained in a straightforward manner. The internal force within the closed loop gives rise to another joint torque component to meet the friction cone conditions. The role of the internal force is to modify the contact wrenches without changing the force balance at the base link. For example, the maximum contact wrench along the unconstrained motion directions can be minimized via the internal force to avoid the initialization of slipping. This task can be performed independently from the postural balance task, which depends on the force balance at the base link. These relations play important roles in posture stability and torque optimization, as will be shown in the following section.

In motion/force control tasks, such as surface cleaning, it might be more appropriate to involve frictionless-contact joint models. Such models are also used in the case of completely unconstrained motion of the end links, e.g. of the hands and the foot of the swing leg during walking. The motion along the unconstrained (mobility) directions is determined by *task-based* constraints rather than by physical ones. When frictionless-joint models become involved, the relations derived in this section are better characterized as quasistatic rather than purely static. In fact, the motion along the mobility directions is determined by dynamic forces. This issue will be discussed in Chapter 4. Another example of a kinetostatic relation stems from a redundant actuation that is inherent to kinematically redundant limbs. A redundantly actuated limb can be subjected to an infinite number of joint torques that do not alter the contact wrench. An appropriate torque can then be selected, e.g. to minimize the load of a given actuator.

The static/kinetostatic force relations derived in this section are of a general nature. The results can be used in the case of varying contact conditions and topology changes, e.g. as a result of establishing/breaking contacts.

3.7 STATIC POSTURE STABILITY AND OPTIMIZATION

Static posture stability plays an important role in a motion generation approach based on a sequence of static postures [29,30,6,7,23]. A posture is said to be stable in the static sense when, given a static initial condition, the total external wrench acting on the CRB (e.g. the wrench of gravity plus the reaction wrenches) does not induce any motion. A statically stable posture is characterized with *robustness* [52,14]. Robustness implies that the stability of the posture will be ensured not only for the acting external wrench, but also for the nearby wrenches. Optimization of a static posture refers to the process of increasing its robustness, which can be achieved via the internal joint torque component. As already clarified, this torque component is related to friction. In the presence of high friction, for example, the optimization objective can be set as a minimization of the maximum component of the (reaction) contact wrench $\overline{\mathcal{F}}^c$. On the other hand, when the friction is not as high and there is a possibility of slip, the optimization is based on the objectives of an appropriate CRB posture and internal WD. As a result, the reaction wrenches can be eventually confined within the interior of the CWC. Referring to (3.62), a minimax optimization in the high-friction case can be achieved by employing the pseudoinverse in the particular solution. Such a solution is inappropriate, however, in the lower-friction case. The output of the optimization process yields the joint torque components $\boldsymbol{\tau}^{ext}$ and $\boldsymbol{\tau}^{int}$ (cf. (3.73)). Furthermore, the joint torque limits should also be taken in consideration during the optimization process, via the force component $\boldsymbol{\tau}^{n}$, when kinematic redundancy is present.

3.7.1 Static Posture Stability

In the field of multifinger grasping, the stability of a grasp has been related to the so-called *force closure* [49,10] (cf. Chapter 6). A force-closure grasp is defined by a set of reaction forces $\overline{\mathcal{F}}^c \in CWC$ such that $\mathbb{G}\overline{\mathcal{F}}^c = \mathcal{F}_O$, where $\mathcal{F}_O$ denotes the reaction to an external spatial force applied at the grasped object and $\mathbb{G}$ is the grasp map, which is an equivalent of the contact map $\mathbb{C}_{cR}$ (or $\mathbb{C}_{cC}$) introduced here. With a force-closure grasp, the instability problem due to the unilateral motion constraints at the fingertips can be effectively solved. From the viewpoint of a humanoid robot, force-closure grasps can be characterized as "strongly stable" since the grasp map and friction conditions admit *any* external force.

Humanoid robots, though, cannot attain force closure–like postures because of their human-like design. Therefore, the instability problem stemming from the unilateral contacts at the feet cannot be completely alleviated. A necessary condition for static stability on a flat ground is that the ground projection of the CoM (in what follows referred to as the *gCoM*) lies within the support polygon determined as the convex hull of all contact points [36]. Henceforth, the term "base of support" (BoS), borrowed from the biomechanics literature, will be used. Note that when the gCoM is at the boundary of the BoS, even a very small disturbance wrench (e.g. stemming from an air current) may induce uncontrollable CRB roll. In order to prevent this situation, usually a safety zone is introduced within the BoS. Constraining the gCoM within the safety zone increases the robustness of the posture.

Static postural stability has to be assessed, though, not only for coplanar contacts on flat ground but also for the general noncoplanar case [52,16]. As a theoretical base, a study on "imperfect" static stability conditions within a system of multiple rigid bodies in contact under Coulomb friction can be used [55]. Accordingly, the following four contact stability conditions can be distinguished:

- *weakly stable*: a valid contact wrench exists that induces zero accelerations;
- *strongly stable*: if every valid contact wrench induces zero accelerations;
- *weakly unstable*: if not strongly stable;
- *strongly unstable*: if not weakly stable.

A valid reaction/contact wrench is a wrench $\mathcal{F}_k$ that: (1) satisfies instantaneously the equation of motion under zero initial conditions, (2) is applied along the constraint basis ${}^k\mathbb{B}_c$ appearing in the contact map, and (3) lies within the contact wrench cone CWC_k. In the case of a humanoid robot with multiple contacts, valid wrenches can be obtained from the contact state determined by the contact map/CWC pair. Since the CoM position and its ground projection play an important role for stability, the CoM contact map $\mathbb{C}_{cC}$ (cf. (2.123)) is preferably used in lieu of $\mathbb{C}_{cB}$. Static stability w.r.t. the net CRB wrench (gravity force included), $\mathcal{F}_C$, can be thus assessed via $\{\mathbb{C}_{cC}, CWC\}$. Based on this information, the reaction/contact wrenches $\overline{\mathcal{F}}^c$ can be determined and examined for validity. As already noted, force closure–like, strongly stable postures may not be determined for a humanoid robot in the general case [32].

Simple Static Stability Test Based on Wrench Distribution

The simplest test for static stability of a given posture is to check whether the base link can be restrained to resist the CoM gravity wrench $-\mathcal{G}_C$. To this end, use can be made of the general solution to the WD problem, (3.62). The minimum-norm solution component

$$\overline{\mathcal{F}}^{ext} = \mathbb{C}_{cC}^{+} \mathcal{G}_C \tag{3.76}$$

usually yields valid contact wrenches for high-friction contacts. In the case of low-friction contacts and/or nonhorizontal contact surfaces, valid contact wrenches could be obtained by including the null-space solution component $\overline{\mathcal{F}}^n = N(\mathbb{C}_{cC})\overline{\mathcal{F}}_a^c$ into the above solution. A particular solution component with appropriate weighting, e.g. as in (3.53) (cf. Fig. 3.7), could also be useful. In some cases, it might be preferable to employ a particular solution component that is *consistent with statics*. The above pseudoinverse-based solution cannot be used then, as discussed in Section 3.5.4. A suitable weighted generalized inverse is introduced in Section 5.10.4.

CRB-Wrench Cone and Contact-Consistent CRB Wrench

The simple static stability test outlined above is not straightforward because of the existence of an infinite set of solutions to the WD problem. This problem can be solved by reformulating the test in terms of CRB wrenches instead of contact wrenches. A convex polyhedral wrench cone, henceforth referred to as the *CRB-wrench cone*, can be constructed from multiple point-contact PCs at the vertices of the end links, as suggested in [32]. An alternative, more straightforward approach is to obtain the CRB-wrench cone by intersecting all CWCs at the end links [14]. It is assumed that the CWCs have been calculated with the procedure described at the end of Section 3.3.2 for constructing a plane-contact CWC from point-contact convex polyhedral friction cones. The net external CRB wrench acting at the CoM is represented as in (3.56). Assume that the contact wrenches are confined within their CWCs, s.t. $\boldsymbol{U}_{CWC}\overline{\mathcal{F}}^c \preceq \mathbf{0}$. It follows then that the CRB wrench $\mathcal{F}_C \in \text{span}(\boldsymbol{V}_{BWC})$, where $\boldsymbol{V}_{BWC} = \mathbb{C}_{cC}\boldsymbol{U}_{CWC}^S$ denotes the span representation of the CRB-wrench cone. This implies that the net CRB wrench lies within the interior of the CRB-wrench cone, i.e. $\boldsymbol{U}_{BWC}\mathcal{F}_C \preceq \mathbf{0}$. The last relation provides a means to check the static stability of a posture with given contact states $\{\mathbb{C}_{cC}, CWC\}$, against the net external CRB wrench, $\mathcal{F}_C$, without having to consider the underdetermined WD problem. A CRB wrench that lies within the CRB-wrench cone, i.e.

$$\mathcal{F}_C \in \{\mathbb{C}_{cC}, BWC\}, \tag{3.77}$$

will be referred to as *contact-consistent*.

An Example

As an example, consider the pushing task shown in Fig. 3.9A. Depending on the friction intensity at the feet contacts, the posture can be characterized as either strongly or weekly stable. In the former case, high friction prevents the feet from sliding. In the latter case, sliding may occur due to low friction in the horizontal direction at one of the feet or both of them. To realize the task, then, it would be important to find another posture, if it exists, that can be characterized as strongly stable, even in the presence of such a low friction. Consider the

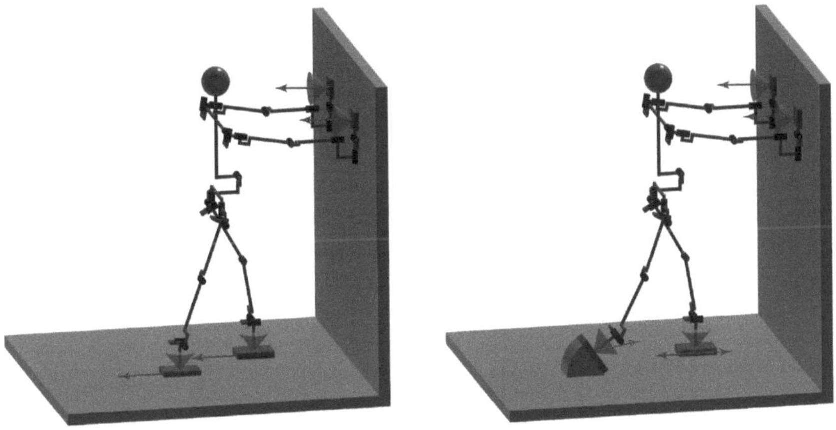

FIGURE 3.9 Pushing task. Left (A): Reactions at the hands induce slipping forces at the feet. Right (B): Postural change with pushing leg that can generate a force opposing slip at the support leg. The cones at each end link signify surface-contact friction conditions, obtained e.g. via the CWC (cf. Section 3.3.2).

posture shown in Fig. 3.9B as a candidate. With this posture, the robot attaches its pushing leg to an inclined surface in order to generate a reaction force that opposes sliding at the support foot.

Contact Planning

The last example highlights the importance of planning discrete sets of contact states [11] and respective static postures as part of the process of motion generation. The contact states may involve multiple contacts, not only at the end links but also at the intermittent links. The reason for such assumption is that the tasks of the humanoid robot should be made executable within various environments, e.g. typical for disaster scenes, involving an acyclic gait (climbing) on rough terrain [11], crawling under obstacles [71], climbing a ladder [66], and so on. The contact states are determined by a coarse model/map of the environment constructed in advance. The process is referred to as *contact planning* [23]. For each given contact state, a set of statically stable postures (or keyframes[5] [28]) is obtained. This process is referred to as the *posture generation* [13]. The posture generation is accomplished with the help of a constraint-based inverse kinematics and statics solver [6,8]. Finally, a dynamically feasible trajectory is generated to connect the keyframes by interpolation [41,28]. This approach is referred to as the "contact-before-motion" approach [29].

Joint Torque Limit Test

The joint torque component $\boldsymbol{\tau}^n$ is available in the case of a kinematically redundant robot. Then, the static-equilibrium posture can be tested for compliance with the joint torque limits [53,72,20,29]. In such case, the static posture test can be based on (3.73), without any motion along the unconstrained motion directions ($\boldsymbol{\tau}^m = \mathbf{0}$), i.e.

$$\boldsymbol{\tau} = \boldsymbol{\tau}^{ext} + \boldsymbol{\tau}^{int} + \boldsymbol{\tau}^n, \tag{3.78}$$

[5] A term borrowed from the field of animation.

$$\boldsymbol{\tau}^{n} = \boldsymbol{N}(\mathcal{J}_{cM})\boldsymbol{\tau}_a, \ |\tau_i| \le \tau_i^{max},$$
$$\boldsymbol{\tau}^{ext} + \boldsymbol{\tau}^{int} = \mathcal{J}_{cM}^T(\overline{\mathcal{F}}^{ext} + \overline{\mathcal{F}}^n),$$
$$\overline{\mathcal{F}}^{ext} + \overline{\mathcal{F}}^n = \overline{\mathcal{F}}^c \in CWC.$$

The Jacobian $\mathcal{J}_{cM}$ is defined in (2.126); $\overline{\mathcal{F}}^{ext}$ is given in (3.76). The internal joint torque component $\boldsymbol{\tau}^{int} = \mathcal{J}_{cM}^T \overline{\mathcal{F}}^n$ can be used in the contact wrench optimization; τ_i^{max} denotes the joint torque limit for joint i, $i \in \overline{1,n}$. Apparently, the joint torque component $\boldsymbol{\tau}^n(\boldsymbol{\tau}_a)$ can be used in the joint torque optimization.

3.7.2 Static Posture Optimization

The above formulas for static-posture stability tests provide means for optimization of: (1) the reaction forces at the contact joints (the force optimization problem [37,10]) and (2) the joint torque (redundant actuation–based optimization). The problem of force distribution optimization/control in locomotion (with a multilegged robot) and manipulation with closed kinematic chains was addressed in the pioneering work [53]. To deal with the underdetermined problem, the authors applied a linear programming with an objective function that combines energy consumption (via local power minimization) and maximum reaction (contact) force minimization.

More recent approaches to the force optimization problem have been based on a convex optimization approach [9] that involves matrix inequalities or second-order cone inequalities [10,56,12]. Standard solution algorithms for semidefinite programming problems (SDPs) or second-order cone programs (SOCPs) can then be applied. The SOCP problem can be defined as [45]

$$\min_x \boldsymbol{f}^T\boldsymbol{x}, \tag{3.79}$$

subject to

$$\|\boldsymbol{A}_i\boldsymbol{x} + \boldsymbol{b}_i\| \le \boldsymbol{c}_i^T\boldsymbol{x} + d_i, \ \ i = 1, ..., N, \tag{3.80}$$

where $\boldsymbol{x} \in \Re^n$ is the optimization vector (variable), while $\boldsymbol{f} \in \Re^n$, $\boldsymbol{A}_i \in \Re^{(n_i-1)\times n}$, $\boldsymbol{b}_i \in \Re^{n_i-1}$, $\boldsymbol{c}_i \in \Re^n$, and $d_i \in \Re$ are the problem parameters (constants); $\|\circ\|$ is the standard Euclidean norm. Constraint (3.80) is called a *second-order cone constraint of dimension* n_i. Apparently, the friction cone constraints introduced in Section 3.3.2 can be readily represented in the above form.

In [12], for example, the so-called *support region* has been introduced to guarantee the static stability of a multipoint contact posture. The support region is defined as the 2D projection of a nonlinear convex set that is determined by the properties of each contact joint. Static posture stability is guaranteed when the CoM projection is within the support region. An iterative algorithm for convex minimization (SOCP) under inequality (friction) constraints is employed to determine the boundaries of the support region, with properly bounded error. Standard solvers, however, can be computationally demanding. This drawback has been addressed in [10], where a fast method for feasibility check was proposed based on a dual problem formulation.

It is possible to formulate the contact wrench optimization problem as a QP task (cf. (2.57)). Further details will be presented in Section 5.14.

3.8 POSTURE CHARACTERIZATION AND DUALITY RELATIONS

The differential-motion and kinetostatics relations, treated in Chapter 2 and the present one, respectively, can be characterized as dual relations. These relations were derived in coordinate form, referring either to the base-link frame $\{B\}$ or the CRB one, $\{C\}$. Five transforms were involved in each case. In the notation w.r.t. $\{B\}$, for example, these are the three Jacobian matrices $\mathcal{J}_{cB}(\boldsymbol{q})$, $\mathcal{J}_{mB}(\boldsymbol{q})$, $\boldsymbol{J}_B(\boldsymbol{q})$ and the two contact maps $\mathbb{C}_{cB}(\boldsymbol{q})$ and $\mathbb{C}_{mB}(\boldsymbol{q})$. The three Jacobians are characterized by a complementarity condition: any two will specify the third one uniquely. Contact maps $\mathbb{C}_{cB}(\boldsymbol{q})$ and $\mathbb{C}_{mB}(\boldsymbol{q})$ are also complementary. In the following analysis, $\mathcal{J}_{cB}(\boldsymbol{q})$, $\boldsymbol{J}_B(\boldsymbol{q})$, and $\mathbb{C}_{cB}(\boldsymbol{q})$ will be employed as *fundamental* transforms that provide complete insight into the differential kinematics and kinetostatics relations. They also play an important role in dynamic relations.

The above transforms are functions of the generalized coordinates that determine the robot's posture in a unique way. Typical postures are listed in Table 3.1. A *regular posture* is characterized by the base link being mobile in any direction. This implies that the row space of contact map $\mathbb{C}_{cB}$ should be a subspace of the range (column) space of the constraint Jacobian $\mathcal{J}_{cB}$. At a multicontact posture, the base-link motion can be generated only with constraint-consistent joint velocities. When the regular-posture condition is not satisfied, the posture is characterized as *singular*. For example, when pushing a heavy object, (near) singular arm/leg configurations are used. Then, one or more of the limb Jacobians, $\boldsymbol{J}_B(\boldsymbol{q}_{e_j})$, (and hence, the constraint Jacobian $\mathcal{J}_{cB}$) will be rank-deficient. The mobility of the base link will be then constrained: a solution to the inverse kinematics problem might not exist when the base-link velocity is assigned in an arbitrary way. Further on, at a singular posture, the force impressed on the base link from the respective end links becomes a *structural force*, according to the structural-force posture condition shown in Table 3.1.[6] For instance, the quiet-stance upright posture with straightened legs induces a structural-force pattern w.r.t. the force of gravity.

Next, an *internal-wrench* posture is a multicontact posture, s.t. the contact wrenches sum up to yield a nonzero net wrench on the base link (CRB wrench). In this case, the contact wrenches can be used to ensure that the friction cone conditions are met. At an internal joint torque posture, on the other hand, motor-load redistribution can be achieved, e.g. to avoid the joint torque limits and/or to minimize the joint torques (cf. (3.73)). Note that sometimes the term "overactuation" appears in texts referring to redundant actuation (a local, posture-dependent characteristic). This may lead to confusion, though, since a humanoid robot is characterized as an underactuated system (a global, posture-independent characteristic).

The next three postures in Table 3.1 refer to the types of motion constraints imposed on the base via the contact joints. When the robot is in a static single-stance posture with planar contact, the foot can be assumed to be fixed to the ground as long as the gCoM is within

[6] This term is borrowed from grasp analysis [49].

TABLE 3.1 Types of humanoid robot postures

Posture	Condition	Characterization
regular	$\mathcal{R}(\mathbb{C}_{cB}^T) \subset \mathcal{R}(\boldsymbol{\mathcal{J}}_{cB})$	base motion possible in any direction
singular	$\mathcal{R}(\mathbb{C}_{cB}^T) \not\subset \mathcal{R}(\boldsymbol{\mathcal{J}}_{cB})$	impossible base motion in some directions
structural force	$\mathbb{C}_{cB}^{+}\mathcal{F}_B \in \mathcal{N}^*(\boldsymbol{\mathcal{J}}_{cB}{}^T)$	external forces are resisted without motor load
internal wrench	$\overline{\mathcal{F}}^n \notin \mathcal{N}^*(\mathbb{C}_{cB}) \cap CWC$	a nonzero CRB wrench is generated
internal joint torque	$\boldsymbol{\tau}^n \in \mathcal{N}^*(\boldsymbol{J}_B)$	does not alter the end-link wrench
fully constrained (unilaterally)	$c = 6$ (single stance)	any joint velocity admissible
overconstrained (unilaterally)	$c > 6$ (double stance)	only constraint-consistent joint vel. admissible
underconstrained	$c < 6,\ \mathcal{V}_B \in \mathcal{N}(\mathbb{C}_{cB}^T)$	base motion resulting from lack of constraints
stationary base	$\dot{\boldsymbol{\theta}} \in \mathcal{N}(\boldsymbol{\mathcal{J}}_{cB})$	end-link and internal-link motion
stationary base and end links	$\dot{\boldsymbol{\theta}} \in \mathcal{N}(\boldsymbol{J}_B)$	internal-link motion only
stationary end links	$\dot{\boldsymbol{\theta}} \in \mathcal{N}(\boldsymbol{J}_B) \cap \mathcal{R}(\boldsymbol{\mathcal{J}}_{cB}^T)$	internal-link motion with mobile base

the BoS. This posture is characterized as a *(unilaterally) fully constrained* posture: there are no closed loops and, therefore, any joint velocity is admissible. When in double stance, on the other hand, the posture is characterized as *unilaterally overconstrained*: there is a closed loop and, therefore, only constraint-consistent joint velocities are admissible. Furthermore, for a posture, s.t. the gCoM is at the BoS boundary, an arbitrary small disturbance may induce a change in the contact joint type (e.g. from planar to line or point contact). The posture then becomes *underconstrained*: there is at least one base mobility direction that cannot be constrained via the contacts. Such a posture may eventually result in a fall. Another example of an underconstrained posture is when the robot is in mid-air, i.e. while running or jumping. Then, all six mobility directions of the base link are unconstrained. Note that the term "underconstrained" should not be confused with "underactuated." The former, being posture-dependent, provides a local characterization while the latter refers to a global feature: there are no actuators attached to the base.

Next, a *stationary-base* posture is characterized by motion in the links whereby the base remains stationary. When the motion is obtained with joint velocities from the null space of the joint-space constraint Jacobian, the end links will move. The associated motion patterns are "motion with stationary base" (e.g. limb self-motion) and "end-link motion with stationary base." Further restricting the joint velocities to the null space of the complete Jacobian immobilizes the end links; the only associated motion pattern is "limb self-motion." At such a posture, an internal-link motion is available for obstacle and/or singularity avoidance and other secondary subtasks in the *motion* domain.

Fig. 3.10 displays a graphic representation of the conditions along with representative postures from the subspaces of the three fundamental transforms. The transforms decompose the joint, the end-link, and the base-link spaces as follows. The joint space is decomposed in two ways, by $\boldsymbol{\mathcal{J}}_{cB}$ and $\boldsymbol{J}_B$. The end-link space is also decomposed in two ways, by $\boldsymbol{\mathcal{J}}_{cB}$ and $\mathbb{C}_{cB}$. The base-link space is decomposed by $\mathbb{C}_{cB}$ only. The right angles signify the orthogonality relations pertinent to the fundamental subspaces of the transforms.

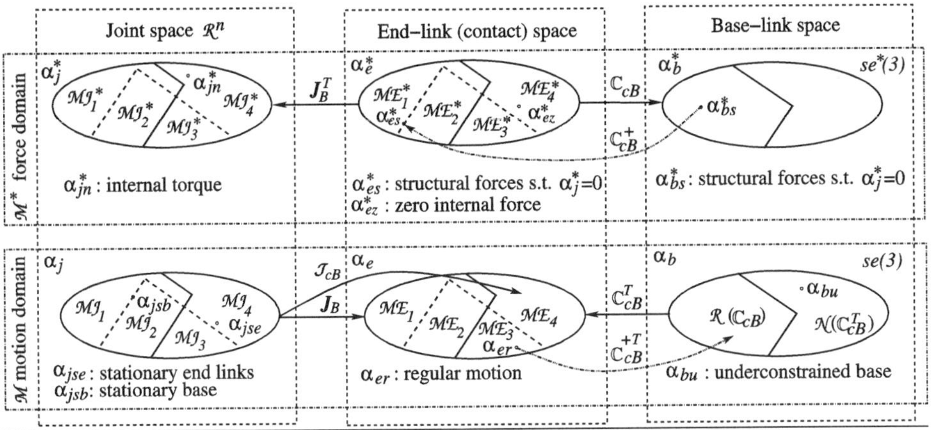

Joint space ($\in \mathcal{M}$)		End-link (contact) space ($\in \mathcal{M}$)	
$\mathcal{MJ}_1 = \mathcal{N}(J_B) \cap \mathcal{N}(\mathcal{J}_{cB})$	$\mathcal{N}(J_B) = \mathcal{MJ}_1 \cup \mathcal{MJ}_4$	$\mathcal{ME}_1 = \mathcal{N}(\mathbb{C}_{cB}) \cap \mathcal{N}(\mathcal{J}_{cB}^T)$	$\mathcal{N}(\mathbb{C}_{cB}) = \mathcal{ME}_1 \cup \mathcal{ME}_4$
$\mathcal{MJ}_2 = \mathcal{R}(J_B^T) \cap \mathcal{N}(\mathcal{J}_{cB})$	$\mathcal{R}(J_B^T) = \mathcal{MJ}_2 \cup \mathcal{MJ}_3$	$\mathcal{ME}_2 = \mathcal{R}(\mathbb{C}_{cB}^T) \cap \mathcal{N}(\mathcal{J}_{cB}^T)$	$\mathcal{R}(\mathbb{C}_{cB}^T) = \mathcal{ME}_2 \cup \mathcal{ME}_3$
$\mathcal{MJ}_3 = \mathcal{R}(J_B^T) \cap \mathcal{R}(\mathcal{J}_{cB}^T)$	$\mathcal{N}(\mathcal{J}_{cB}) = \mathcal{MJ}_1 \cup \mathcal{MJ}_2$	$\mathcal{ME}_3 = \mathcal{R}(\mathbb{C}_{cB}^T) \cap \mathcal{R}(\mathcal{J}_{cB})$	$\mathcal{N}(\mathcal{J}_{cB}^T) = \mathcal{ME}_1 \cup \mathcal{ME}_2$
$\mathcal{MJ}_4 = \mathcal{N}(J_B) \cap \mathcal{R}(\mathcal{J}_{cB}^T)$	$\mathcal{R}(\mathcal{J}_{cB}^T) = \mathcal{MJ}_3 \cup \mathcal{MJ}_4$	$\mathcal{ME}_4 = \mathcal{N}(\mathbb{C}_{cB}) \cap \mathcal{R}(\mathcal{J}_{cB})$	$\mathcal{R}(\mathcal{J}_{cB}) = \mathcal{ME}_3 \cup \mathcal{ME}_4$

FIGURE 3.10 Joint space, end-link (contact) space, and base-link space with their dual motion/force subspaces. Right angles signify orthogonality decompositions stemming from the three fundamental transforms, $\mathcal{J}_{cB}$, J_B, and $\mathbb{C}_{cB}$. The table explains the meaning of the subspace notation in the motion subdomain. The respective meaning in the force subdomain is analogous. The αs denote motion elements (velocities), the α^*s are their duals (forces). Representative postures are specified according to the relations in Table 3.1.

References

[1] Y. Abe, M. Da Silva, J. Popović, Multiobjective control with frictional contacts, in: ACM SIGGRAPH/Eurographics Symposium on Computer Animation, 2007, pp. 249–258.

[2] M. Anitescu, F.A. Potra, Formulating dynamic multi-rigid-body contact problems with friction as solvable linear complementarity problems, Nonlinear Dynamics 14 (1997) 231–247.

[3] D.J. Balkcom, J. Trinkle, Computing wrench cones for planar rigid body contact tasks, The International Journal of Robotics Research 21 (2002) 1053–1066.

[4] R. Ball, The Theory of Screws: A Study in the Dynamics of a Rigid Body, Hodges, Foster, and Company, 1876.

[5] D. Baraff, Fast contact force computation for nonpenetrating rigid bodies, in: 21st Annual Conference on Computer Graphics and Interactive Techniques – SIGGRAPH '94, ACM Press, New York, New York, USA, 1994, pp. 23–34.

[6] K. Bouyarmane, A. Kheddar, Static multi-contact inverse problem for multiple humanoid robots and manipulated objects, in: IEEE-RAS International Conference on Humanoid Robots, Nashville, TN, USA, 2010, pp. 8–13.

[7] K. Bouyarmane, A. Kheddar, Using a multi-objective controller to synthesize simulated humanoid robot motion with changing contact configurations, in: IEEE/RSJ International Conference on Intelligent Robots and Systems, 2011, pp. 4414–4419.

[8] K. Bouyarmane, A. Kheddar, Humanoid robot locomotion and manipulation step planning, Advanced Robotics 26 (2012) 1099–1126.

[9] S.P. Boyd, L. Vandenberghe, Convex Optimization, Cambridge University Press, 2004.

[10] S.P. Boyd, B. Wegbreit, Fast computation of optimal contact forces, IEEE Transactions on Robotics 23 (2007) 1117–1132.

[11] T. Bretl, Motion planning of multi-limbed robots subject to equilibrium constraints: the free-climbing robot problem, The International Journal of Robotics Research 25 (2006) 317–342.

[12] T. Bretl, S. Lall, Testing static equilibrium for legged robots, IEEE Transactions on Robotics 24 (2008) 794–807.

[13] S. Brossette, A. Escande, G. Duchemin, B. Chrétien, A. Kheddar, Humanoid posture generation on non-Euclidean manifolds, in: IEEE-RAS International Conference on Humanoid Robots, 2015, pp. 352–358.
[14] S. Caron, Q. Cuong Pham, Y. Nakamura, Leveraging cone double description for multi-contact stability of humanoids with applications to statics and dynamics, in: Robotics: Science and Systems XI, Rome, Italy, 2015, pp. 28–36.
[15] S. Caron, Q.-C. Pham, Y. Nakamura, Stability of surface contacts for humanoid robots: closed-form formulae of the contact wrench cone for rectangular support areas, in: IEEE International Conference on Robotics and Automation, Seattle, Washington, USA, 2015, pp. 5107–5112.
[16] S. Caron, Q.-C. Pham, Y. Nakamura, ZMP support areas for multicontact mobility under frictional constraints, IEEE Transactions on Robotics 33 (2017) 67–80.
[17] J.-R. Chardonnet, S. Miossec, A. Kheddar, H. Arisumi, H. Hirukawa, F. Pierrot, K. Yokoi, Dynamic simulator for humanoids using constraint-based method with static friction, in: IEEE International Conference on Robotics and Biomimetics, 2006, pp. 1366–1371.
[18] C. Collette, A. Micaelli, C. Andriot, P. Lemerle, Dynamic balance control of humanoids for multiple grasps and non coplanar frictional contacts, in: IEEE-RAS International Conference on Humanoid Robots, 2007, pp. 81–88.
[19] J.K. Davidson, K.H. Hunt, Robots and Screw Theory: Applications of Kinematics and Statics to Robotics, Oxford University Press, 2004.
[20] P.G. De Santos, J. Estremera, E. Garcia, M. Armada, Including joint torques and power consumption in the stability margin of walking robots, Autonomous Robots 18 (2005) 43–57.
[21] K.L. Doty, C. Melchiorri, C. Bonivento, A theory of generalized inverses applied to robotics, The International Journal of Robotics Research 12 (1993) 1–19.
[22] J. Duffy, The fallacy of modern hybrid control theory that is based on "orthogonal complements" of twist and wrench spaces, Journal of Robotic Systems 7 (1990) 139–144.
[23] A. Escande, A. Kheddar, S. Miossec, Planning contact points for humanoid robots, Robotics and Autonomous Systems 61 (2013) 428–442.
[24] R. Featherstone, Rigid Body Dynamics Algorithms, Springer Science+Business Media, LLC, Boston, MA, 2008.
[25] Fujitsu, Miniature Humanoid Robot HOAP-2 Manual, 1st edition, Fujitsu Automation Co., Ltd, 2004 (in Japanese).
[26] cdd and cddplus homepage, https://www.inf.ethz.ch/personal/fukudak/cdd_home/ [online].
[27] K. Harada, M. Kaneko, Analysis of internal force in whole body manipulation by a human type robotic mechanisms, Journal of the Robotics Society of Japan 21 (2003) 647–655 (in Japanese).
[28] K. Hauser, Fast interpolation and time-optimization with contact, The International Journal of Robotics Research 33 (2014) 1231–1250.
[29] K. Hauser, T. Bretl, J.C. Latombe, Non-gaited humanoid locomotion planning, in: IEEE-RAS International Conference on Humanoid Robots, 2005, pp. 7–12.
[30] K. Hauser, T. Bretl, J.-C. Latombe, K. Harada, B. Wilcox, Motion planning for legged robots on varied terrain, The International Journal of Robotics Research 27 (2008) 1325–1349.
[31] S. Hirai, Analysis and Planning of Manipulation Using the Theory of Polyhedral Convex Cones, Ph.D. thesis, Kyoto University, 1991.
[32] H. Hirukawa, S. Hattori, K. Harada, S. Kajita, K. Kaneko, F. Kanehiro, K. Fujiwara, M. Morisawa, A universal stability criterion of the foot contact of legged robots – adios ZMP, in: IEEE International Conference on Robotics and Automation, 2006, pp. 1976–1983.
[33] M. Hosokawa, D. Nenchev, Sliding control with a double-stance posture on V-shaped support, Robotic Life Support Laboratory, Tokyo City University, 2017 (Video clip), https://doi.org/10.1016/B978-0-12-804560-2.00010-9.
[34] S.-H. Hyon, J. Hale, G. Cheng, Full-body compliant human-humanoid interaction: balancing in the presence of unknown external forces, IEEE Transactions on Robotics 23 (2007) 884–898.
[35] Y. Jun, A. Alspach, P. Oh, Controlling and maximizing humanoid robot pushing force through posture, in: 9th International Conference on Ubiquitous Robots and Ambient Intelligence, URAI, 2012, pp. 158–162.
[36] S. Kajita, H. Hirukawa, K. Harada, K. Yokoi, Introduction to Humanoid Robotics, Springer Verlag, Berlin, Heidelberg, 2014.
[37] J. Kerr, B. Roth, Analysis of multifingered hands, The International Journal of Robotics Research 4 (1986) 3–17.
[38] O. Khatib, A unified approach for motion and force control of robot manipulators: the operational space formulation, IEEE Journal on Robotics and Automation 3 (1987) 43–53.

[39] E. Kokkevis, Practical physics for articulated characters, in: Game Developers Conference, 2004, 2004, pp. 1–16.

[40] T. Koolen, S. Bertrand, G. Thomas, T. de Boer, T. Wu, J. Smith, J. Englsberger, J. Pratt, Design of a momentum-based control framework and application to the humanoid robot Atlas, International Journal of Humanoid Robotics 13 (2016) 1650007.

[41] J.J. Kuffner, S. Kagami, K. Nishiwaki, M. Inaba, H. Inoue, Dynamically-stable motion planning for humanoid robots, Autonomous Robots 12 (2002) 105–118.

[42] S. Kuindersma, R. Deits, M. Fallon, A. Valenzuela, H. Dai, F. Permenter, T. Koolen, P. Marion, R. Tedrake, Optimization-based locomotion planning, estimation, and control design for the atlas humanoid robot, Autonomous Robots 40 (2016) 429–455.

[43] V. Kumar, K. Waldron, Force distribution in closed kinematic chains, IEEE Journal on Robotics and Automation 4 (1988) 657–664.

[44] S.-H.H. Lee, A. Goswami, A momentum-based balance controller for humanoid robots on non-level and non-stationary ground, Autonomous Robots 33 (2012) 399–414.

[45] M.S. Lobo, L. Vandenberghe, S. Boyd, H. Lebret, Applications of second-order cone programming, Linear Algebra and Its Applications 284 (1998) 193–228.

[46] N. Mansard, O. Khatib, Continuous control law from unilateral constraints, in: IEEE International Conference on Robotics and Automation, 2008, pp. 3359–3364.

[47] D. Mansour, A. Micaelli, P. Lemerle, A computational approach for push recovery in case of multiple noncoplanar contacts, in: IEEE/RSJ International Conference on Intelligent Robots and Systems, IEEE, San Francisco, CA, USA, 2011, pp. 3213–3220.

[48] M. Murooka, S. Nozawa, Y. Kakiuchi, K. Okada, M. Inaba, Whole-body pushing manipulation with contact posture planning of large and heavy object for humanoid robot, in: IEEE International Conference on Robotics and Automation, 2015, pp. 5682–5689.

[49] R.M. Murray, Z. Li, S.S. Sastry, A Mathematical Introduction to Robotic Manipulation, CRC Press, 1994.

[50] Y. Nakamura, Minimizing object strain energy for coordination of multiple robotic mechanisms, in: American Control Conference, Atlanta, GA, USA, 1988, pp. 499–509.

[51] S. Nakaoka, S. Hattori, F. Kanehiro, S. Kajita, H. Hirukawa, Constraint-based dynamics simulator for humanoid robots with shock absorbing mechanisms, in: IEEE/RSJ International Conference on Intelligent Robots and Systems, San Diego, CA, USA, 2007, pp. 3641–3647.

[52] Y. Or, E. Rimon, Computation and graphical characterization of robust multiple-contact postures in two-dimensional gravitational environments, The International Journal of Robotics Research 25 (2006) 1071–1086.

[53] D.E. Orin, S.Y. Oh, Control of force distribution in robotic mechanisms containing closed kinematic chains, Journal of Dynamic Systems, Measurement, and Control 103 (1981) 134–141.

[54] C. Ott, M.A. Roa, G. Hirzinger, Posture and balance control for biped robots based on contact force optimization, in: IEEE-RAS International Conference on Humanoid Robots, Bled, Slovenia, 2011, pp. 26–33.

[55] J.-S. Pang, J. Trinkle, Stability characterizations of rigid body contact problems with Coulomb friction, ZAMM – Zeitschrift für Angewandte Mathematik und Mechanik (Journal of Applied Mathematics and Mechanics) 80 (2000) 643–663.

[56] I.-W. Park, J.-Y. Kim, J. Lee, J.-H. Oh, Mechanical design of the humanoid robot platform HUBO, Advanced Robotics 21 (2007) 1305–1322.

[57] J. Park, Control Strategies for Robots in Contact, Ph.D. thesis, Stanford University, USA, 2006.

[58] J.K. Salisbury, J.J. Craig, Articulated hands: force control and kinematic issues, The International Journal of Robotics Research 1 (1982) 4–17.

[59] L. Sentis, O. Khatib, Compliant control of multicontact and center-of-mass behaviors in humanoid robots, IEEE Transactions on Robotics 26 (2010) 483–501.

[60] B. Siciliano, O. Khatib (Eds.), Handbook of Robotics, Springer Verlag, Berlin, Heidelberg, 2008.

[61] W. Son, K. Kim, N.M. Amato, J.C. Trinkle, A generalized framework for interactive dynamic simulation for multirigid bodies, IEEE Transactions on Systems, Man and Cybernetics. Part B. Cybernetics 34 (2004) 912–924.

[62] D.E. Stewart, J.C. Trinkle, An implicit time-stepping scheme for rigid body dynamics with inelastic collisions and coulomb friction, International Journal for Numerical Methods in Engineering 39 (1996) 2673–2691.

[63] J.C. Trinkle, J.-S. Pang, S. Sudarsky, G. Lo, On dynamic multi-rigid-body contact problems with Coulomb friction, ZAMM – Zeitschrift für Angewandte Mathematik und Mechanik (Journal of Applied Mathematics and Mechanics) 77 (1997) 267–279.

[64] J.C. Trinkle, J.A. Tzitzouris, J.S. Pang, Dynamic multi-rigid-body systems with concurrent distributed contacts, Philosophical Transactions of the Royal Society of London A: Mathematical, Physical and Engineering Sciences 359 (2001) 2575–2593.

[65] M. Uchiyama, P. Dauchez, Symmetric kinematic formulation and non-master/slave coordinated control of two-arm robots, Advanced Robotics 7 (1992) 361–383.

[66] J. Vaillant, A. Kheddar, H. Audren, F. Keith, S. Brossette, A. Escande, K. Bouyarmane, K. Kaneko, M. Morisawa, P. Gergondet, E. Yoshida, S. Kajita, F. Kanehiro, Multi-contact vertical ladder climbing with an HRP-2 humanoid, Autonomous Robots 40 (2016) 561–580.

[67] M.W. Walker, D.E. Orin, Efficient dynamic computer simulation of robotic mechanisms, Journal of Dynamic Systems, Measurement, and Control 104 (1982) 205.

[68] P.M. Wensing, G. Bin Hammam, B. Dariush, D.E. Orin, Optimizing foot centers of pressure through force distribution in a humanoid robot, International Journal of Humanoid Robotics 10 (2013) 1350027.

[69] D. Williams, O. Khatib, The virtual linkage: a model for internal forces in multi-grasp manipulation, in: IEEE International Conference on Robotics and Automation, Atlanta, GA, USA, 1993, pp. 1025–1030.

[70] K. Yamane, Y. Nakamura, A numerically robust LCP solver for simulating articulated rigid bodies in contact, in: Robotics: Science and Systems IV, MIT Press, Zurich, Switzerland, 2008, pp. 89–104.

[71] K. Yokoi, E. Yoshida, H. Sanada, Unified motion planning of passing under obstacles with humanoid robots, in: IEEE International Conference on Robotics and Automation, 2009, pp. 1185–1190.

[72] Y. Yokokohji, S. Nomoto, T. Yoshikawa, Static evaluation of humanoid robot postures constrained to the surrounding environment through their limbs, in: IEEE International Conference on Robotics and Automation, 2002, pp. 1856–1863.

[73] Yu Zheng, K. Yamane, Y. Zheng, K. Yamane, Human motion tracking control with strict contact force constraints for floating-base humanoid robots, in: IEEE-RAS International Conference on Humanoid Robots, Atlanta, GA, USA, 2013, pp. 34–41.

[74] Y. Zheng, C.-M. Chew, Fast equilibrium test and force distribution for multicontact robotic systems, Journal of Mechanisms and Robotics 2 (2010) 021001.

[75] Y. Zheng, J. Luh, Joint torques for control of two coordinated moving robots, in: IEEE International Conference on Robotics and Automation, 1986, pp. 1375–1380.

[76] D. Zlatanov, D.N. Nenchev, On the use of metric-dependent methods in robotics, in: ASME Proceedings, 5th International Conference on Multibody Systems, Nonlinear Dynamics and Control, vol. 6, ASME, 2005, pp. 703–710.

CHAPTER

4

Dynamics

4.1 INTRODUCTION

Biped locomotion is a complex dynamical motion/force phenomenon. Healthy humans can keep balance while walking due to skills acquired by learning. Under such skills, proper force/moment components can be impressed on the stance foot during the single-leg support phase, or at both feet, during the double-leg support phase. The respective reaction force/moments are used then to propel the body in a desired way. The impressed force/moment components stem from the force of gravity and the inertial force/moments generated by the acceleration of the total CoM and the angular accelerations of the body segments. Friction forces at the foot contacts and impact forces at foot landing have to be considered as well. During walking, the stability of the gait is the main concern. Fortunate enough, though, walking on regular ground is a periodic phenomenon. Thus, despite its complexity, biped locomotion can be modeled with simplified dynamical models such as the linear inverted pendulum (LIP) or the table-cart models [54].

The situation changes dramatically, however, when the cyclic motion process is subjected to irregular inputs. Recently, research has focused on the role of external disturbances during walking or while standing upright. Avoiding a fall, e.g. by modifying the length of the step or by making a side step in the direction of the disturbance, is highly desirable. The robot should also be able to walk over unknown and varying irregular terrain, e.g. as in disaster-related environments. This obviously cannot be done with a perfect periodic gait. Research results suggest that in these cases, the simple dynamical models mentioned above are insufficient. Enhanced simplified models may still be used for motion generation in the presence of such external perturbations [146], but for stabilization purposes a complete dynamical model might be needed [149]. Such a model is also advantageous in multifinger and dual-arm object manipulation, as well as in whole-body motions like jumping, kicking a ball, and so on.

This chapter discusses dynamical models of biped humanoid robots. Such models are used in the analysis and model-based control. The chapter is organized in fourteen sections. In the following Section 4.2, the focus is on the characterization of humanoid robots as underactuated systems with a "floating" base. Simple underactuated models on the plane and in 3D are discussed in Section 4.3 and Section 4.4, respectively. The dynamic modeling of fixed-base manipulators is highlighted in Section 4.5. In Section 4.6, the first-order differential motion relations of a free-floating manipulator in zero gravity are analyzed from the viewpoint of spatial momentum. In Section 4.7, the concept of the Reaction Null Space (RNS) is introduced. The equation of motion of a free-floating manipulator is discussed in Section 4.8. A solution

Humanoid Robots
https://doi.org/10.1016/B978-0-12-804560-2.00011-0

to the inverse dynamics is derived with the help of the RNS in Section 4.9. The spatial momentum of a humanoid robot is discussed in Section 4.10. In Section 4.11, the equation of motion of a humanoid robot is represented in terms of different types of quasivelocities. Constraint elimination methods for constrained multibody systems are introduced in Section 4.12. Reduced-form representations of the equation of motion of a humanoid robot are discussed in Section 4.13. Section 4.14 explores inverse dynamics solutions.

It is assumed that the reader is familiar with the basic concepts of rigid-body mechanics in 3D space, such as mass and inertia characteristics of a rigid body, the CoM and rotational dynamics, the concepts of momentum and angular momentum, and the equation of motion. Also, knowledge about fixed-base robot dynamical relations in 3D space, including analytical and recursive methods of forward and inverse robot dynamics, is a prerequisite.

4.2 UNDERACTUATED ROBOT DYNAMICS

Mechatronic systems having fewer actuators than the number of generalized coordinates are called *underactuated systems*. There is a variety of such systems, including humanoid robots. Underactuated systems can be classified into two main groups [167]. The first one is characterized by passive DoFs *distributed* throughout the kinematic chain. Flexible-link manipulators composed of slender, light-weight links are one representative example. Their joints are actuated, the passive DoFs stemming from elastic deflections in the links. These manipulators were the first underactuated manipulators to be studied extensively, since the mid-1970s [13,63]. Another representative example are manipulators with flexible joints, comprising the so-called series-elastic actuators. Each joint is actuated but there is an elastic element (e.g. a torsional spring) that yields the passive DoF [75]. Flexible-joint manipulators have drawn a lot of attention due to their potential for ensuring safety in physical human/robot interaction tasks. Successful implementations have been reported in [48,2]. More recently, implementations in humanoid robots have appeared as well [34,150,30]. Furthermore, there are underactuated manipulators with distributed passive DoFs that comprise active as well as fully passive joints. These type of underactuated manipulators were studied in the beginning of the 1990s, motivated by "minimalism" [4,88]. Finally, there is a class of underactuated manipulators comprising passive joints exclusively. Motion is induced by reaction wheels (the so-called torque-unit manipulator) [111,110], or by control momentum gyros [17]. These type of underactuated manipulators have been considered for use in microgravity environments as space robots. Similarly, motion of the so-called "Gyrobot" is induced by a gyrating link [38]. The desired configuration is attained with the help of brakes at the passive joints.

The second major group of underactuated systems is characterized with passive DoFs *concentrated* at the root (base) link of the kinematic chain. Such systems can be modeled by connecting the base link via a virtual rigid-body (6-DoF) joint to inertial ground. Hence, there is a "floating" base, as in the case of a humanoid robot. Other representative examples within this subclass include spacecraft and free-flying space robots, aerial (vertical take-off and landing [VTOL] aircraft), and marine vehicles (surface and underwater), as well as manipulators mounted on a flexible base. The "floating" property of the base depends on the type of environment and the respective interaction forces. Apparently, the environment conditions are quite different. This complicates the development of a uniform modeling and control approach.

Assuming a total of $n = n_p + n_a$ DoFs, where the p and a subscripts denote quantities related to the passive and active joints, respectively, the equation of motion representing a "generic" underactuated system can be written as

$$\begin{bmatrix} \boldsymbol{M}_p & \boldsymbol{M}_{pa} \\ \boldsymbol{M}_{pa}^T & \boldsymbol{M}_a \end{bmatrix} \begin{bmatrix} \ddot{\boldsymbol{q}}_p \\ \ddot{\boldsymbol{q}}_a \end{bmatrix} + \begin{bmatrix} \boldsymbol{c}_p \\ \boldsymbol{c}_a \end{bmatrix} + \begin{bmatrix} \boldsymbol{g}_p \\ \boldsymbol{g}_a \end{bmatrix} = \begin{bmatrix} \boldsymbol{0} \\ \mathcal{F}_a \end{bmatrix}, \tag{4.1}$$

where $\boldsymbol{q}_p \in \Re^{n_p}$ and $\boldsymbol{q}_a \in \Re^{n_a}$ denote the generalized coordinates, $\mathcal{F}_a \in \Re^{n_a}$ stands for the generalized force, $\boldsymbol{M}_p \in \Re^{n_p \times n_p}$, $\boldsymbol{M}_a \in \Re^{n_a \times n_a}$, and $\boldsymbol{M}_{pa} \in \Re^{n_p \times n_a}$ are submatrices of the system inertia matrix, $\boldsymbol{c}_p \in \Re^{n_p}$ and $\boldsymbol{c}_a \in \Re^{n_a}$ denote the nonlinear velocity-dependent forces, and $\boldsymbol{g}_p \in \Re^{n_p}$ and $\boldsymbol{g}_a \in \Re^{n_a}$ are the gravity-dependent forces. Note that no other external forces than gravity are taken in consideration at this point.

The matrix-vector form of the above equation of motion clearly displays the passive/active DoF structure of an underactuated manipulator with concentrated passive-joint disposition, e.g. a free-floating space robot or a humanoid robot (without elasticity in the joints). Especially, the explicit appearance of the inertial coupling between the passive and active DoFs via submatrix $\boldsymbol{M}_{pa}$ is important. This matrix is referred to as the *coupling inertia matrix* [96]. The inertial coupling property plays an important role in control law design as shown in [140]. Input/output *partial feedback linearization* (PFL) is possible, either w.r.t. the actuated coordinates (collocated PFL) or w.r.t. the passive coordinates (noncollocated PFL). The latter is especially interesting since then the internal system dynamics become explicit. These dynamics are quite useful, e.g. for generating *reactionless motion*. The method was first developed for reactionless manipulator control of a free-floating space robot [97,94] and later applied in the balance control of a humanoid robot in response to external disturbances [101]. Further details will be discussed in Section 4.6 and Section 7.6. The above matrix-vector form representation of the equation of motion will be preferred throughout this text in lieu of an often used notation involving a "selection" matrix for the active coordinates (cf. (2.102)). The latter is more compact, but the passive/active structure is hidden.

It is important to emphasize that the behavior of the above-mentioned underactuated systems, stemming from their unforced dynamics, is quite distinctive. Take as an example the subclass of underactuated manipulators with distributed active and fully passive joints. A simple planar 2R manipulator with a passive joint at the root and an active joint at the elbow (the so-called "Acrobot" [46]) behaves quite differently from that of a similar 2R manipulator obtained by exchanging the passive/active roles of the joints (known as the "Pendubot" [137,31]). Furthermore, these two types of passive-joint robots operate within the vertical plane and hence under gravity. If they were to operate on the horizontal plane, and hence under zero gravity, completely different types of behavior would be obtained [53,4,88].

Fewer actuators than the number of generalized coordinates means also fewer control inputs. This is the reason why classic methods of linear control theory, such as continuous time-invariant feedback control for example, cannot be applied to underactuated systems [15]. Efforts made since the 1990s have led to the development of a variety of new control methods. The PFL approach mentioned above is one of them; others are based on nonlinear state feedback, passivity, and energy control, as well as hybrid and switching control

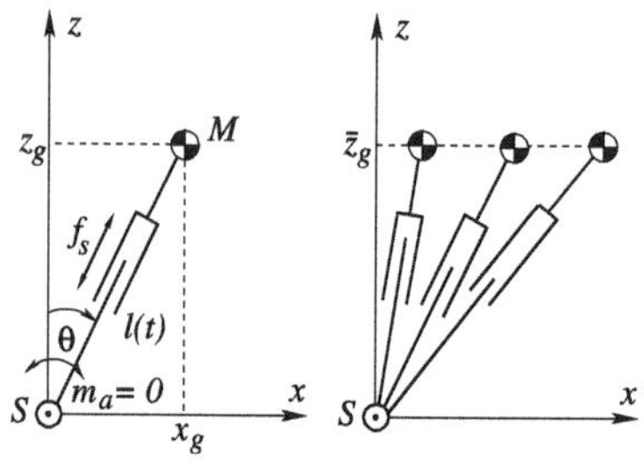

FIGURE 4.1 The motion of the support leg is modeled in the plane (sagittal or frontal). (A) Left: Inverted pendulum with massless leg and varying length. (B) Right: Linear inverted pendulum. The CoM is constrained to move along a line (in this case, a horizontal one): $z_g = \bar{z}_g = \text{const}$ or $z_g = kz_x$, $k = \text{const}$. The ankle joint S is assumed passive. The stroke force f_s is actively controlled.

[53,136,123,104]. The reason for such diversity in control methods is that control law design for underactuated manipulators relies on a deeper understanding of their distinct dynamics, as explained above. Humanoid robots are no exception.

4.3 SIMPLE UNDERACTUATED MODELS ON THE PLANE

The dynamics of humanoid robots are essential for their control. Multi-DoF dynamic models are quite complex though. Therefore, control law design has always relied on simplified models. For instance, for gait pattern generation and control within a restricted environment, e.g. walking on a level ground, very simple models on the plane have proven useful. These will be introduced below.

4.3.1 The Linear Inverted Pendulum Model

The linear inverted pendulum (LIP) and cart-on-table are simple models used to design balance controllers for a number of humanoid robots [54]. The availability of a closed-form solution for these models facilitates analysis and cyclic motion generation for gaits on level ground, as well as control law design for gait stabilization. Real-time implementation can also be ensured. The underlying model is an inverted pendulum (IP) with point mass M at the end of a massless leg of variable length $l(t)$, as shown in Fig. 4.1A. The IP is used as a model of the support leg during gait in the sagittal plane. The varying leg length, actively controlled with stroke force f_s, accounts for the motion in the knee joint. The mass point represents the total mass of the robot concentrated at the hip joint. The pendulum is pivoted at the ankle joint which is assumed to be passive (nonactuated). Thus, the model represents a simple underactuated system.

To derive the equation of motion, it is convenient to use polar coordinates as generalized coordinates; such coordinates are in full harmony with the physical system. Denote by $\theta(t)$ the angular displacement from the vertical. The angular coordinate is passive by assumption; hence, the respective generalized force component (moment m_a at support point S) is zero.

The other generalized force component is the leg stroke force f_s. The equation of motion is

$$\begin{aligned} Ml^2\ddot{\theta} + 2Ml\dot{l}\dot{\theta} - Mlg\sin\theta &= 0, \\ M\ddot{l} + Ml\dot{\theta}^2 + Mg\cos\theta &= f_s, \end{aligned} \tag{4.2}$$

where g denotes the gravity acceleration. The vector matrix form of this equation is:

$$\begin{bmatrix} M_p & 0 \\ 0 & M_a \end{bmatrix} \begin{bmatrix} \ddot{\theta} \\ \ddot{l} \end{bmatrix} + \begin{bmatrix} c_p \\ c_a \end{bmatrix} + \begin{bmatrix} g_p \\ g_a \end{bmatrix} = \begin{bmatrix} 0 \\ f_s \end{bmatrix}, \tag{4.3}$$

where $M_p = Ml^2$ and $M_a = M$ stand for the inertia/mass properties, $c_p = 2Ml\dot{l}\dot{\theta}$ and $c_a = Ml\dot{\theta}^2$ are the nonlinear velocity-dependent forces, and $g_p = -Mlg\sin\theta$ and $g_a = Mg\cos\theta$ are the gravity force terms, g denoting the acceleration of the gravity force. The passive/active structure, revealed with (4.1), is immediately apparent.

Note that in the above equation, there is no *inertial coupling* between the generalized coordinates, i.e. $M_{pa} = M_{ap} = 0$. Hence, under linearization, two decoupled differential equations would be obtained. The *linear* IP model, however, does not rely upon linearization. Instead, the following *task-based motion constraint* is applied [146]: keep the vertical position of the CoM constant, so $z_g \equiv \bar{z}_g =$ const. This constraint yields the important advantage of having a closed-form solution to the nonlinear system.

To impose the $\bar{z}_g =$ const constraint, it is convenient to rewrite the equation of motion in Cartesian coordinates, so we have

$$\begin{aligned} M\ddot{x}_g &= |f_s|\sin\theta, \\ M(\ddot{z}_g + g) &= |f_s|\cos\theta. \end{aligned} \tag{4.4}$$

Here x_g is the projection of the CoM on the ground, henceforth referred to as the *gCoM*. The two terms on the r.h.s. represent the components of a vector, denoted as $\boldsymbol{f}_r$. This vector is referred to as the Ground Reaction Force (GRF) vector.

Furthermore, with $\ddot{z}_g = 0$, the stroke force f_s can be eliminated from the above equations to obtain the following linear ordinary differential equation (ODE):

$$\ddot{x}_g = \bar{\omega}^2 x_g, \tag{4.5}$$

where $\bar{\omega} = \sqrt{g/\bar{z}_g}$ is the (constant) natural angular frequency of the LIP. The closed-form solution to this equation is [54]

$$x_g(t) = x_{g0}\cosh(\bar{\omega}t) + \frac{v_{g0}}{\bar{\omega}}\sinh(\bar{\omega}t), \tag{4.6}$$

$x_{g0} = x_g(0)$, $v_{g0} = \dot{x}_g(0)$ denoting the initial values. The above equation of motion facilitates gait generation on level ground with the appropriate timing [58,54]. It should be noted, though, that the resulting gait is with bent knees. Such gait lacks efficiency in terms of energy [25]. Also, the appearance of the gait is "ape-like," i.e. it is quite different from the erect human gait.

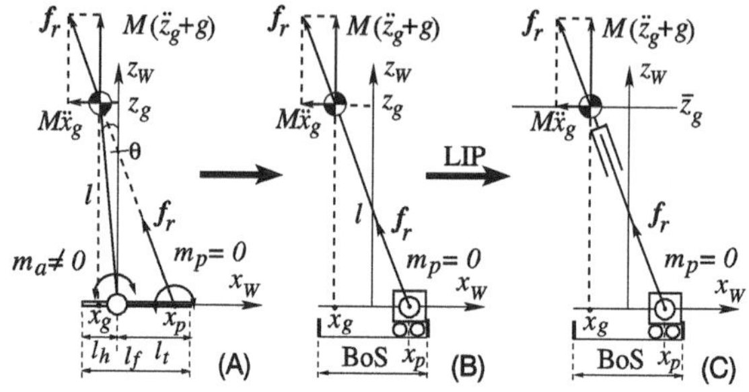

FIGURE 4.2 IP type models with BoS constraint $l_f = l_h + l_t$ (the foot length). (A) Nonlinear (constant-length) IP-on-foot model. The ankle torque ($m_a \neq 0$) alters the GRM, thereby displacing the CoP and accelerating/decelerating the CoM. (B) Nonlinear (constant-length) IP-on-cart model. The CoM horizontal motion is determined by the displacement of the cart, associated with that of the CoP. (C) Variable-length, constant-height LIP-on-cart model. In all models, the GRF acts along the line connecting the CoP and the CoM.

4.3.2 Foot Modeling: CoM Dynamics Driven by the Center of Pressure

The LIP model can be improved by adding a model of the foot. It is assumed that the foot is in multipoint/line contact with the supporting ground. The foot serves as an interface for the impressed/reaction forces that act via the contact points. Note that the resultant force is the GRF, $\boldsymbol{f}_r$. This force is applied at a specific point called Center of Pressure (CoP). This situation is depicted in Fig. 4.2A, where x_p denotes the CoP. In the case of multiple links in contact with the environment, CoPs are defined for each link. These CoPs exist only within the Base-of-Support (BoS) boundaries of the links in contact.

A number of studies in the field of biomechanics have clarified the important role of the CoP in human balance control [85,10,162]. The importance stems from the fact that the tangential components of the Ground Reaction Moment (GRM) are zero at this point. Thus, the location of the CoP within the BoS can be used as an indicator of contact stability. Assuming an unknown external disturbance, the contact will be more stable when the CoP is in the vicinity of the BoS center rather than in that of the BoS boundaries. Further insight into stability issues are presented in Chapter 5.

In the special case of feet support on flat ground, i.e. such as with a double-stance posture, a *net CoP* is defined [162]. The net CoP is not necessarily located within the BoS of a single foot; it may be located between the feet. Nevertheless, such location still falls within the double-stance BoS, which is defined as the convex hull of all contact points on the flat ground [39,120,54]. It is worth noting that in the early studies on humanoid robots, the special case of flat ground was exclusively considered and the term "zero-moment point" (ZMP) was coined [158]. On flat ground, the ZMP is identical to the CoP/net CoP (in the cases of single/double stance). On nonflat ground, though, these points do not coincide [120]. In the following discussion, whenever ambiguity is avoidable, the terms CoP and ZMP will be used alternatively.

Linearized IP-on-Foot Model

A constant-length (nonlinear) IP model attached to a foot is shown in Fig. 4.2A. The GRF $\boldsymbol{f}_r$ acts at the CoP, x_p, along the line determined by the CoP and the CoM. The two components of this force result from gravity and the CoM acceleration. Note that the moment at the CoP is zero. Note also that the ankle joint is not anymore passive, as it was with the LIP model. The ankle torque $m_a \neq 0$ alters both, the CoM acceleration and the CoP. The equation of motion of the simple IP is

$$Ml^2\ddot{\theta} = Mgl\sin\theta + m_a. \tag{4.7}$$

The CoP is determined from the GRM balance equation as

$$x_p = -\frac{m_a}{M(g+\ddot{z}_g)} \underset{(4.7)}{=} \frac{gl\sin\theta - l^2\ddot{\theta}}{g - l\dot{\theta}^2\cos\theta - l\ddot{\theta}\sin\theta}. \tag{4.8}$$

Furthermore, since the foot length, l_f, (the BoS) is limited, the pendulum cannot deviate significantly from the vertical. Otherwise, the foot will begin to rotate and the pendulum will eventually tip over. In the following discussion it is assumed that the deviation is limited, s.t. the foot remains always in full contact with the ground. The equation of motion can then be linearized around the vertical as

$$\ddot{\theta} = \omega^2\theta + \frac{m_a}{Ml^2}. \tag{4.9}$$

Here $\omega \equiv \omega_{IP} = \sqrt{g/l}$ denotes the natural angular frequency of the pendulum.

The linearized IP-on-foot model has been employed for human balance stability analysis in the field of biomechanics [162,113]. The equation of motion (4.9) has also been used in the field of humanoid robotics [6]. Most works make use, however, of a representation in Cartesian coordinates. Since the aim is to gain insight into the role of the CoP, the focus is on the component in the horizontal direction, i.e.

$$\ddot{x}_g = \omega^2(x_g - x_p), \tag{4.10}$$

or

$$\ddot{x}_g - \omega^2 x_g = -\omega^2 x_p. \tag{4.11}$$

Apparently, the CoP plays the role of a forcing term for the gCoM dynamics.

In the special case of a fixed CoP, denoted as $\bar{x}_p$, the solution to the above equation can be obtained in closed-form. We have

$$x_g(t) = \left(x_{g0} - \bar{x}_p\right)\cosh(\omega t) + \frac{v_{g0}}{\omega}\sinh(\omega t) + \bar{x}_p \tag{4.12}$$

$$= \frac{1}{2}\left(x_{g0} - \bar{x}_p + \frac{v_{g0}}{\omega}\right)e^{\omega t} + \frac{1}{2}\left(x_{g0} - \bar{x}_p + \frac{v_{g0}}{\omega}\right)e^{-\omega t} + \bar{x}_p. \tag{4.13}$$

This equation plays an important role in balance stability analysis (cf. Chapter 5).

IP-on-Cart Model

The ODE (4.11) clarifies that the CoM motion trajectory can be controlled via the forcing term, i.e. by appropriate position control of the CoP. This approach resembles the way of balancing a stick on the palm. The respective model is shown in Fig. 4.2B. This model is widely used in the field of control under the name "IP-on-cart model." Note that the distance between the CoM and the cart (the CoP) is assumed constant. The equation of motion in the horizontal direction is of the same form as (4.11). Note, however, that the coefficient ω^2 is not anymore constant; now we have

$$\omega(t) = \sqrt{\frac{\ddot{z}_g + g}{z_g - z_p}}. \tag{4.14}$$

The vertical CoP coordinate z_p is usually assumed constant (e.g. zero on a level ground). Solving (4.11) for the CoP with the above expression for ω, one obtains

$$x_p = x_g - \frac{\ddot{x}_g}{\ddot{z}_g + g} z_g = x_g - \frac{f_{rx}}{f_{rz}} z_g. \tag{4.15}$$

Here f_{rx} and f_{rz} denote the two components of the GRF. This equation is used for balance control under the name "ZMP manipulation" or "indirect ZMP" control [80,144], as explained in Section 5.4.

LIP-on-Cart Model

A variable ω, as in (4.14), complicates the analysis since the solution cannot be obtained in such a simple form as that for the linearized IP model. To alleviate this problem, the task-based (LIP) constraint

$$z_g = \bar{z}_g = \text{const} \Rightarrow \omega \equiv \bar{\omega} = \sqrt{g/\bar{z}_g} = \text{const} \tag{4.16}$$

is imposed. The resulting model is shown in Fig. 4.2C. Note that the form of the equation of motion, (4.11), will be preserved. Thus, any displacement of the cart (or equivalently, the CoP) will alter the horizontal CoM acceleration via the term $\bar{\omega}^2 x_p$. This also implies respective changes in the direction of the GRF. We have

$$\frac{f_{rx}}{Mg} = \frac{x_g - x_p}{\bar{z}_g} \tag{4.17}$$

and hence, of the GRM, i.e. $(x_g - x_p)Mg = \bar{z}_g f_{rx}$. Note that with this model, the vertical component of the GRF equals the gravity force; $f_{rz} = Mg$. Furthermore, using (4.8) with $\ddot{z}_g = 0$, the equation of motion can be represented as

$$\ddot{x}_g = \omega^2 x_g + \frac{m_a}{M\bar{z}_g}. \tag{4.18}$$

The LIP-on-cart model is used in numerous studies on balance control. Further details will be presented in Chapter 5.

The LIP constraint is also used in the so-called "cart-on-table" model [55,54], shown in Fig. 4.3. The equation of motion of the cart-on-table model is identical to that of the IP models, (4.11). The difference lies only in the interpretation, when used in controller design. In

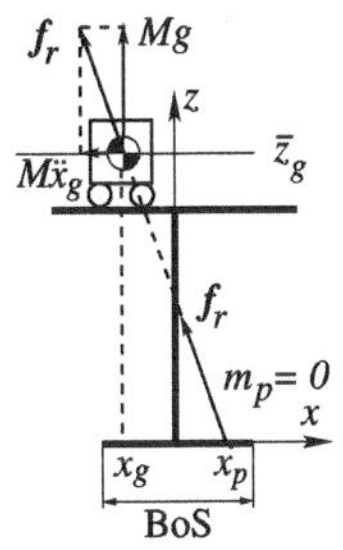

FIGURE 4.3 Cart-on-table model: the CoP location is determined by the horizontal acceleration component of the CoM.

fact, all these models demonstrate that *balance control can be based on the relative CoP-CoM motion*. In the table-cart model, the CoM motion trajectory is used as the input while the CoP position is subject to control. This is in contrast to the IP models, where the CoP is used as control input, as clarified above.

CoP manipulation via CoM acceleration has also been exploited under the so-called "angular momentum inducing IP model" (AMPM) [62]. Through the CoP relation, an additional moment (a controllable feedforward component) stemming from the rate of change of the *moment of momentum* of the pendulum is imposed on the foot. It is important to note that with this enhancement, a closed-form solution is still available to facilitate gait planning and control with the capability of dealing with external disturbances. Note also that the term "moment of momentum" is deliberately introduced here to distinguish this type of angular momentum from the "intrinsic" (cf. [32], p. 31) or "centroidal" angular momentum. The latter can be used to control the imposed moment on the foot as well. This will be addressed in Section 4.3.3.

4.3.3 Linear Reaction Wheel Pendulum Model and Centroidal Moment Pivot

With the LIP model, a feedforward control component to be used in CoP-based gait generation can be obtained. Then, an appropriate balance of the reactive force components at the foot that result from the gravity and inertia forces imposed by the vertical and horizontal acceleration of the CoM, respectively, can be ensured. The inclusion of the CoP into the model leads to the desirable property of a controllable moment at the foot, as already explained with the cart-on-table model. It should also be noted that this property can be obtained in an alternative way, via a controlled change of *centroidal* angular momentum. The simplest way to accomplish this is with the *reaction wheel pendulum* (RWP) model. The RWP is a simple pendulum with a massless leg of constant length, comprising a reaction wheel (RW) at the CoM. Controllable angular acceleration/deceleration of the RW induces a variation in the joint rate of the passive pivot joint, via inertial coupling. In this way, the stability of the system can be improved.

The RWP was introduced in the beginning of the century [138] and studied extensively as a landmark underactuated example in nonlinear control [12]. The RWP has been adopted in humanoid robotics as well, e.g. as a means to determine appropriate foot locations for a

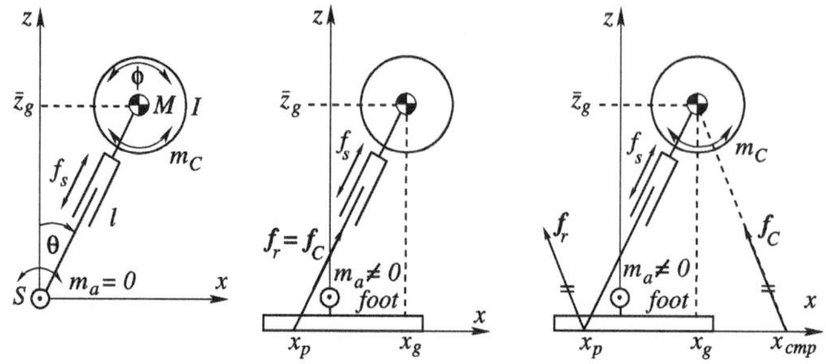

FIGURE 4.4 Linear RWP models. (A) Left: Ankle joint as passive joint. (B) Middle: The linear RWP-foot model is identical to the LIP when the RW torque is zero. (C) Right: A nonzero RW moment m_C alters the direction of the GRF $\boldsymbol{f}_r$ so that its line of action deviates from the CoM. The centroidal (reaction) force $\boldsymbol{f}_C$ is in the same direction as the GRF but its line of action passes through the CoM. The application point of $\boldsymbol{f}_C$ defines the centroidal moment pivot (CMP) [40,120].

reactive-step strategy, the so-called "Capture point" [121]. As with the LIP model, the leg length of the RWP is made variable (cf. Fig. 4.4A). The general form of the equation of motion for such system is

$$\begin{bmatrix} M_p & \boldsymbol{M}_{pa} \\ \boldsymbol{M}_{pa}^T & \boldsymbol{M}_a \end{bmatrix} \begin{bmatrix} \ddot{q}_p \\ \ddot{\boldsymbol{q}}_a \end{bmatrix} + \begin{bmatrix} c_p \\ \boldsymbol{c}_a \end{bmatrix} + \begin{bmatrix} g_p \\ \boldsymbol{g}_a \end{bmatrix} = \begin{bmatrix} 0 \\ \mathcal{F}^a \end{bmatrix}. \tag{4.19}$$

Here, $q_p = \theta$, $\boldsymbol{q}_a = \begin{bmatrix} l & \phi \end{bmatrix}^T$ denote the generalized coordinates, $\mathcal{F}^a = \begin{bmatrix} f_s & m_c \end{bmatrix}^T$ is the generalized force, $M_p = Ml^2$, $\boldsymbol{M}_{pa} = \begin{bmatrix} 0 & I \end{bmatrix}$, $\boldsymbol{M}_a = \text{diag}\begin{bmatrix} m & I \end{bmatrix}$ are the inertia matrix components, I denotes the RW moment of inertia, $c_p = 2Ml\dot{l}\dot{\theta}$, $\boldsymbol{c}_a = \begin{bmatrix} Ml\dot{\theta}^2 & 0 \end{bmatrix}^T$ are the nonlinear velocity-dependent forces, and $g_p = -Mgl\sin\theta$, $\boldsymbol{g}_a = \begin{bmatrix} Mg\cos\theta & 0 \end{bmatrix}^T$ are the gravity-dependent forces. Denote by l_y the angular momentum w.r.t. the CoM, i.e. the *centroidal angular momentum*. The RW torque, m_c, alters the centroidal angular momentum as $\frac{d}{dt}l_{cy} = \frac{d}{dt}I\dot{\phi} = m_c$. Note that the RW rotation angle ϕ represents a *cyclic* or *ignorable* generalized coordinate. It is known that the Lagrangian is independent of such coordinates and that the respective momentum will be conserved [65]. This is the essence of the inertia coupling. As already explained, according to the partial-feedback linearization approach, the (linear) inertia coupling for the passive coordinate (pendulum rotation θ) is essential for stability. As apparent from the coupling inertia matrix (row matrix $\boldsymbol{M}_{pa}$), the coupling is realized via the RW moment $I\ddot{\phi}$.

To impose the task-based LIP constraint, rewrite the equation of motion in Cartesian coordinates, i.e.

$$\begin{aligned} M\ddot{x}_g &= |f_s|\sin\theta - \frac{m_c}{l}\cos\theta = f_{cx}, \\ M(\ddot{z}_g + g) &= |f_s|\cos\theta + \frac{m_c}{l}\sin\theta = f_{cz}, \end{aligned} \tag{4.20}$$

$$I\ddot{\phi} = m_C. \tag{4.21}$$

The vector $\boldsymbol{f}_C = \begin{bmatrix} f_{cx} & f_{cz} \end{bmatrix}^T$, thus defined, acts in the same direction as the GRF vector $\boldsymbol{f}_r$ (cf. Fig. 4.4C). The application points however are different. This follows from the fact that the line of action of the GRF does not pass anymore through the CoM; the direction of $\boldsymbol{f}_r$ depends on the centroidal moment, m_C. The line of action of $\boldsymbol{f}_C$, on the other hand, always passes through the CoM. The point of application of this vector will be clarified in short.

Next, apply constraint $\dot{z}_g = \ddot{z}_g = 0$ to reduce the dimension of the system, i.e.

$$\begin{aligned} \ddot{x}_g &= \omega^2 x_g - \frac{1}{M\bar{z}_g} m_C, \\ \ddot{\phi} &= \frac{1}{I} m_C, \end{aligned} \tag{4.22}$$

where $\omega = \bar{\omega} = \sqrt{g/\bar{z}_g}$. The two equations obtained are the ODEs coupled via m_C. The coupling term on the r.h.s. of (4.22) alters the centroidal angular momentum via $m_C = I\ddot{\phi}$, as explained. When compared to (4.18), it becomes apparent that the role of the centroidal moment m_C is the same as that of the ankle torque m_a: to induce an additional CoM acceleration in the horizontal direction.

The above model facilitates the design of control laws for increased balance stability of the LIP-on-foot model since the CoP can be manipulated with the rate of change of the centroidal angular momentum $\dot{l}_{cy} = I\ddot{\phi}$, in addition to the CoM acceleration. This becomes apparent from the CoP equation. We have

$$\begin{aligned} x_p &= x_g - \frac{\ddot{x}_g}{\omega^2} - \frac{\dot{l}_{cy}}{Mg} \\ &= x_g - \frac{f_{rx}}{f_{rz}}\bar{z}_g - \frac{m_C}{f_{rz}}. \end{aligned} \tag{4.23}$$

When the RW is nonaccelerating, i.e. stationary or spinning with constant angular velocity, then $\ddot{\phi} = 0$ and the relations are the same as with the LIP-on-cart model (cf. Fig. 4.4B). An accelerating RW, on the other hand, alters the direction of the reaction force $\boldsymbol{f}_r$ (cf. Fig. 4.4C). This means that the point on the ground where the moment of the reaction force is zero may be outside the BoS. This point cannot be referred to as the CoP (or ZMP) anymore since there will be a contradiction with the definition of the CoP or ZMP [158]. To alleviate the problem, new terms were introduced as follows. In [40], the point was referred to as the *zero rate of change of angular momentum* point, while in [118] the *zero spin center of pressure* was used. Another term has become commonly accepted, though: *centroidal moment pivot* (CMP) [120]. The CMP is defined by the following equation:

$$x_{cmp} = x_g - \frac{f_{rx}}{f_{rz}}\bar{z}_g. \tag{4.24}$$

Its relation to the CoP is obtained in a straightforward manner from the last two equations as

$$x_{cmp} = x_p + \frac{m_C}{f_{rz}}. \tag{4.25}$$

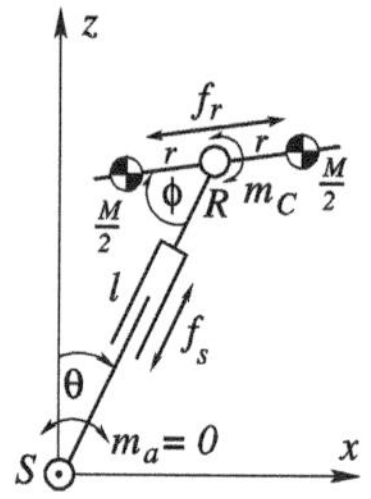

FIGURE 4.5 Model of a planar reaction mass pendulum. The two mass points slide under the action of force f_r, keeping thereby equal distance r from the center point R. Thus, the moment of inertia Mr^2 is controllable via f_r. Note that f_r and m_C are redundant control inputs with regard to the inertia moment $mr^2\ddot{\phi}$ that ensures the inertial coupling with passive coordinate θ.

Apparently, the CoP and the CMP will coincide whenever the centroidal angular momentum is conserved (this condition implies $m_C = 0$).

It would be appropriate to refer to the RWP model as the *linear RWP*. This model has been successfully implemented for gait generation and control of ASIMO [147].

Concluding this subsection, it is worth noting that RWs have been used for a long time as actuators in satellite attitude control systems. An RW can induce a desired variation in attitude via a controlled rate of change in the angular momentum. An RW should not be confused though with a "momentum wheel." The latter is used to store a constant bias momentum that ensures the satellite attitude stabilization against external disturbances and, hence, rotates only in one direction. In this sense, the terminology found in the robotics literature appears inconsistent; terms like "angular momentum pendulum," "flywheel pendulum," or "inertial wheel pendulum" have been used to denote the same model. Based on the above discussion, the term "reaction wheel pendulum" seems to be more appropriate.

4.3.4 Reaction Mass Pendulum Model

The linear RWP model can be further enhanced by resorting to a variable inertia moment $I = I(t)$. The variable centroidal inertia model matches the variable, posture-dependent centroidal dynamics of a humanoid robot to a greater extent. This can be demonstrated with the so-called *reaction mass pendulum* (RMP) model [67,68,130]. The RMP is an underactuated system comprising a pendulum with a massless telescopic leg and a "reaction mass" assembly at the tip of the pendulum. The mass assembly represents the total (constant) mass and the (varying) aggregate inertia of the humanoid in a concentrated form at the CoM.

A planar version of the RMP model is shown in Fig. 4.5. The mass assembly consists of two constant mass points with variable relative position r, symmetric w.r.t. the rotation joint R at the tip of the leg (the "hip" joint). Note that this joint is actuated via torque m_C. The inertia moment around R is adjustable by varying the mass distance r. The general form of the equation of motion is the same as (4.19). We have

$$\begin{bmatrix} \boldsymbol{M}_p & \boldsymbol{M}_{pa} \\ \boldsymbol{M}_{pa}^T & \boldsymbol{M}_a \end{bmatrix} \begin{bmatrix} \ddot{\boldsymbol{q}}_p \\ \ddot{\boldsymbol{q}}_a \end{bmatrix} + \begin{bmatrix} \boldsymbol{c}_p \\ \boldsymbol{c}_a \end{bmatrix} + \begin{bmatrix} \boldsymbol{g}_p \\ \boldsymbol{g}_a \end{bmatrix} = \begin{bmatrix} \boldsymbol{0} \\ \mathcal{F}^a \end{bmatrix}, \tag{4.26}$$

where $q_p = \theta$, $\boldsymbol{q}_a = \begin{bmatrix} l & r & \phi \end{bmatrix}^T$ denote the generalized coordinates, $\mathcal{F}^a = \begin{bmatrix} f_s & f_r & m_C \end{bmatrix}^T$ stands for the generalized force, $M_p = M(l^2 + r^2)$, $\boldsymbol{M}_{pa} = \begin{bmatrix} 0 & 0 & Mr^2 \end{bmatrix}$, $\boldsymbol{M}_a = \text{diag}\begin{bmatrix} M & M & Mr^2 \end{bmatrix}$ are the inertia matrix components, $c_p = 2M((l\dot{l} + r\dot{r})\dot{\theta} + r\dot{r}\dot{\phi})$, $\boldsymbol{c}_a = \begin{bmatrix} Ml\dot{\theta}^2 & Mr(\dot{\theta}+\dot{\phi})^2 & 2Mr\dot{r}(\dot{\theta}+\dot{\phi}) \end{bmatrix}$ are the nonlinear velocity-dependent forces, and $g_p = -Mgl\cos\theta$, $\boldsymbol{g}_a = \begin{bmatrix} Mg\sin\theta & 0 & 0 \end{bmatrix}^T$ are the gravity-dependent forces.

As apparent from the coupling inertia matrix (row matrix $\boldsymbol{M}_{pa}$), there is one inertia coupling term that produces the coupling torque $mr^2\ddot{\phi}$. For a desired inertia coupling moment, a larger r would imply less effort, i.e. less RW torque m_C, or equivalently, less variation in angular acceleration $\ddot{\phi}$. This is the idea behind the "inertia shaping" approach discussed in [67,68,130].

4.3.5 Multilink Models on the Plane

Besides the simple IP-on-foot models discussed in Section 4.3.2, quite often multilink models on the plane (either sagittal or lateral) or in 3D are found to be useful. Such models are used to evaluate balance stability, as in the pioneering work [159,157]. They are also used in the field of biomechanics for evaluating balance stability of humans via single-leg models, e.g. the triple-IP-on-foot sagittal-plane model comprising foot, thigh (femur), shank (tibia), and trunk, as described in [52].[1]

In this work, examples can be found in Section 3.6.5 (lateral plane) and Section 7.6.3 (sagittal plane).

4.4 SIMPLE UNDERACTUATED MODELS IN 3D

When applied to a real robot, an extension of the above planar models to the third dimension is needed. With a 3D model, it becomes possible to account for the lateral acceleration of the CoM in the y-direction. Basically, there are two simple models, the 3D IP and the spherical IP that are supported on 3- and 2-DoF pivots, respectively. It is assumed that the mass is concentrated at the tip of the pendulum that determines the location of the CoM. Note also that these basic models are constant-length pendulums, s.t. the CoM motion is restricted on a spherical surface with radius equal to the pendulum length. The equation of motion of the 3D IP model, for example, can be represented in the form of Euler's equation of motion plus gravity term, i.e. as a set of three coupled nonlinear equations [134]. The equation of motion of the spherical IP is also nonlinear and coupled, in the x- and y-directions. Under the assumption of small deviations from the vertical, linearization and decoupling can be achieved [166].

4.4.1 3D Inverted Pendulum With Variable Length

The main goal is to make use of the 3D or spherical IP models to approximate the complex dynamics of a biped. In this case, a model with varying pendulum length would be appro-

[1] The dynamic equations are included in this work.

priate to account for arbitrary vertical displacements of the CoM. Note, however, that when the variations in length are left unconstrained, the equation of motion will be further complicated. The main problem in this regard is the abovementioned nonlinear coupling between the three components of the equation. As already noted, decoupling can be achieved under the assumption of small deviations from the vertical. In the case of a humanoid robot, though, this assumption is unrealistic. Fortunately, more suitable approaches do exist. One possibility is to restrict the CoM motion on piecewise linear surfaces that locally approximate the varying terrain height [171]. To this end, derive first the equation of motion of the 3D IP without the constraint. It is assumed that the pendulum forms a point contact at $\boldsymbol{r}_P$. The dynamic balance of moments can then be written as

$$\boldsymbol{r}_P \times \boldsymbol{f}_r = \boldsymbol{r}_C \times (\boldsymbol{f}_C + M\boldsymbol{a}_g) + \boldsymbol{m}_C. \tag{4.27}$$

Furthermore, since the balance of forces yields

$$\boldsymbol{f}_r = \boldsymbol{f}_C + M\boldsymbol{a}_g, \tag{4.28}$$

(4.27) can be rewritten as

$$(\boldsymbol{f}_C + M\boldsymbol{a}_g) \times (\boldsymbol{r}_C - \boldsymbol{r}_P) = \boldsymbol{m}_C. \tag{4.29}$$

Since point contact is assumed, the GRM and centroidal moment $\boldsymbol{m}_C$ are both zero.[2] Note also that $\boldsymbol{f}_C = M\ddot{\boldsymbol{r}}_C$. With these relations, the CoM acceleration components become

$$\begin{aligned}
\ddot{r}_{Cx} &= \frac{(r_{Cx} - r_{Px})(\ddot{r}_{Cz} + g)}{r_{Cz} - r_{Pz}}, \\
\ddot{r}_{Cy} &= \frac{(r_{Cy} - r_{Py})\ddot{r}_{Cx}}{r_{Cx} - r_{Px}}, \\
\ddot{r}_{Cx} &= \frac{(r_{Cz} - r_{Pz})\ddot{r}_{Cy}}{r_{Cy} - r_{Py}} - g.
\end{aligned} \tag{4.30}$$

It can be seen that these equations are coupled, nonlinear equations.

Next, constrain the CoM motion on the plane $r_{Cz} = a_p r_{Cx} + b_p$, $\forall r_{Cy}$. The CoM acceleration will then be constrained by $\ddot{r}_{Cz} = a_p \ddot{r}_{Cx}$. Insert these constraints into (4.30) to obtain the CoM acceleration in the sagittal and lateral planes as

$$\begin{aligned}
\ddot{r}_{Cx} - \omega^2 r_{Cx} &= -\omega^2 r_{Px}, \\
\ddot{r}_{Cy} - \omega^2 r_{Cy} &= -\omega^2 r_{Py},
\end{aligned} \tag{4.31}$$

respectively, where $\omega^2 \equiv g/(a_p r_{Px} + b_p - r_{Pz})$. Note that the form of these two equations is the same as that of the linearized IP-on-foot model in (4.11). Apparently, the above two equations are decoupled ODEs. Under the assumption of a constant CoP, explicit solutions in the form of (4.12) will be available.

[2] The case of plane contact that yields nonzero GRM and centroidal moment will be discussed shortly.

4.4.2 Spherical IP-on-Foot and Sphere-on-Plane Models

Consider now a spherical IP of variable length attached to a foot, i.e. a spherical IP-on-foot model. It should be noted at this point that since humanoid robots usually comprise 2-DoF ankle joints, the spherical IP model is more appropriate than the 3D IP one.

Furthermore, flat ground is assumed, s.t. the feet contacts are coplanar. Then, the projection of the CoM on the ground, i.e. the gCoM, plays an important role. This projection will be denoted as $\boldsymbol{r}_g = \begin{bmatrix} x_g & y_g \end{bmatrix}^T$. The net CoP, on the other hand, will be denoted as[3] $\boldsymbol{r}_p = \begin{bmatrix} x_p & y_p \end{bmatrix}^T$. As already noted, the CoP is constrained to lie within the convex hull of all contact points. It is convenient to approximate the convex hull by a convex polygon. The constraint, henceforth referred to as the *CoP-in-BoS* constraint, can then be formally expressed as

$$\boldsymbol{P}_s \boldsymbol{r}_p \preceq \boldsymbol{c}, \tag{4.32}$$

where $\boldsymbol{c} \in \Re^p$ is a set of constants, p denoting the number of polygon sides. The equation of the ith side[4] is given by $\boldsymbol{p}_{si} \boldsymbol{r}_p = c_i, i \in \{\overline{1, p}\}$, $\boldsymbol{p}_{si}$ denoting the ith row in $\boldsymbol{P}_s$.

The coupling problem pertinent to the equation of motion will be addressed as in Section 4.4.1, by restricting the CoM to move within a plane. In this case, the spherical IP model is referred to as the *3D LIP* [54]. The equation of motion is then determined by two ODEs that are decoupled in x and y, as in (4.31). The 3D LIP model can thus be regarded as an extension of the LIP-on-cart model (cf. Fig. 4.2C) in 3D.

Next, consider the extension of the planar IP-on-foot and the IP-on-cart models (shown in Fig. 4.2A and B, respectively) to 3D. The universal joint at the foot of the respective spherical IP (the ankle joint) is assumed actuated. The equation of motion of each of these models has the form of (4.10). Thus

$$\ddot{\boldsymbol{r}}_g = \omega^2(\boldsymbol{r}_g - \boldsymbol{r}_p). \tag{4.33}$$

Here, $\omega = \omega_{IP} = \sqrt{g/l}$ for the constant-length ($l = \text{const}$) IP model, while $\omega = \bar{\omega} = \sqrt{g/\bar{z}_g}$ for the constant-height ($\bar{z}_g = \text{const}$) IP model. In the case of a variable-stroke model without the CoM vertical-motion constraint, use $\omega = \omega(t)$ as in (4.14). The equation of motion, expressed in terms of the CoP coordinates, assumes then the form of (4.15), i.e.

$$\boldsymbol{r}_p = \boldsymbol{r}_g - \frac{z_g}{\ddot{z}_g + g} \ddot{\boldsymbol{r}}_g. \tag{4.34}$$

Furthermore, the LIP-on-cart model in Fig. 4.2C can also be extended to 3D by replacing the cart with a sphere rolling on a horizontal plane of constant height $\bar{z}_g$ [23,22]. The tangential components of the GRM, $\boldsymbol{m}_t = \begin{bmatrix} m_x & m_y \end{bmatrix}^T$, are relevant. They can be obtained from the equation of motion as

$$\boldsymbol{m}_t = Mg\mathbb{S}_2^{\times} \left(\frac{1}{\omega^2} \ddot{\boldsymbol{r}}_g - \boldsymbol{r}_g \right), \tag{4.35}$$

3 A lower-case subscript is used to distinguish a 2D vector from a 3D one.

4 Recall that the curly inequality sign denotes componentwise operation.

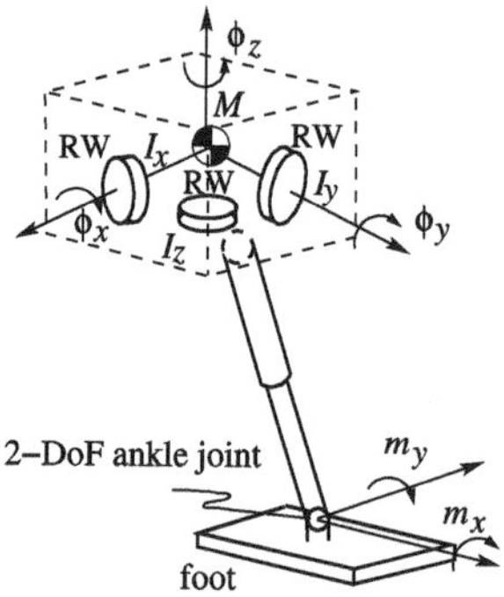

FIGURE 4.6 Model of a 3D reaction wheel pendulum with a telescopic massless leg. The spherical IP is connected to the foot via a 2-DoF ankle joint. The three RWs are mounted on three orthogonal axes. RW accelerations/torques produce a controlled variation of centroidal angular momentum. This results in a variation of the three reaction moments at the stance foot. The tangential reaction moments m_x and m_y induce displacements of the CoP. The vertical reaction moment m_z is used to compensate the gyroscopic moment [64].

where

$$\mathbb{S}_2^{\times} \equiv \begin{bmatrix} 0 & 1 \\ -1 & 0 \end{bmatrix}.$$

The subscript t signifies that the respective vector quantity is composed of tangential (i.e. x- and y-) components. The CoP is then determined as

$$\boldsymbol{r}_p = -\frac{1}{Mg}\mathbb{S}_2^{\times}\boldsymbol{m}_t. \tag{4.36}$$

4.4.3 The 3D Reaction Wheel Pendulum Model

An interesting question is whether the linear RWP model can be extended to a 3D one, in a similar way as the LIP was extended to the 3D LIP. It can be expected that a 3D linear RWP model will match humanoid gait dynamics to a greater extent and contribute to a further increase in balance stability w.r.t. the sagittal (pitch), frontal (roll), and transverse plane (yaw) dynamic components.

3D pendulum models with actuation have been introduced near the end of the 2000s as a means of teaching the foundations of nonlinear dynamics and control [134]. As shown in [20], with the help of the nonlinear control theory, it is possible to stabilize the IP to its equilibrium manifold. These models have been embodied recently in the form of self-balancing cubes [36,83,76]. The cubes comprise an assembly with three RWs mounted on mutually orthogonal axes, thus resembling the arrangement used in the three-axis attitude stabilization of spacecraft. In its essence, this system represents a 3D RWP.

The idea can be applied in humanoid robotics according to the conceptual representation in Fig. 4.6. The figure depicts the model of a 3D RWP supported by a telescopic massless leg. The spherical IP is connected to the foot via an actuated universal joint. The three RWs are mounted on three orthogonal axes. The RW accelerations/torques produce a controlled variation in the centroidal angular momentum, in accordance with the Euler equation for rigid-body rotation, i.e.

$$\boldsymbol{I}\dot{\boldsymbol{\omega}} + \boldsymbol{\omega} \times \boldsymbol{I}\boldsymbol{\omega} = \boldsymbol{m}_{RW}. \tag{4.37}$$

Here $\boldsymbol{\omega}$ is the angular velocity in the inertia frame, $\boldsymbol{m}_{RW} = \dot{\boldsymbol{l}}_C = \boldsymbol{I}\ddot{\boldsymbol{\phi}}$, and $\boldsymbol{I}$ denotes the diagonal inertia matrix of the RW assembly. The RW moments are controlled to induce desirable GRMs at the stance foot. The moment about the vertical is undesirable; it leads to torsional slip at the foot. Thus, it is assumed that $\omega_z = 0$ and that the vertical RW moment is controlled to eliminate the gyroscopic moment, so $m_{RWz} = -(I_x - I_y)\omega_x\omega_y$. This also implies that $\dot{\omega}_z = 0$ [64].

The CoP equations can be obtained from (4.34) by adding a component for the rate of change of the angular momentum [71,120]. We have

$$\begin{aligned} \boldsymbol{r}_p &= \boldsymbol{r}_g - \frac{1}{\ddot{z}_g + g}\left(z_g \ddot{\boldsymbol{r}}_g - \frac{1}{M}\mathbb{S}_2^{\times}\dot{\boldsymbol{l}}_t\right) \\ &= \boldsymbol{r}_g - \frac{1}{f_{rz}}\left(z_g \boldsymbol{f}_t - \mathbb{S}_2^{\times}\boldsymbol{m}_t\right), \end{aligned} \tag{4.38}$$

where M denotes the total mass of the RW assembly. The equations have the same structure as the CoP equation of the planar RWP (4.23). Furthermore, the CMP can be obtained in analogy with (4.24) and (4.25) as (see also [120])

$$\boldsymbol{r}_{cmp} = \boldsymbol{r}_g - \frac{z_g}{f_{rz}}\boldsymbol{f}_t \tag{4.39}$$

and

$$\boldsymbol{r}_{cmp} = \boldsymbol{r}_p + \frac{1}{f_{rz}}\mathbb{S}_2^{\times}\boldsymbol{m}_t, \tag{4.40}$$

respectively, where $\boldsymbol{r}_{cmp} = \begin{bmatrix} x_{cmp} & y_{cmp} \end{bmatrix}^T$. On flat ground, z_{cmp} is assumed zero, similarly to the assumption that $z_p = 0$ (cf. (4.34)).

The above 3D RWP model can be further enhanced to match the humanoid robot dynamics to a higher degree. For example, a variable, posture-dependent inertia tensor can be introduced to mimic the variable inertia of the humanoid, e.g. as with the "Eulerian ZMP resolution" method [152].

4.4.4 The 3D Reaction Mass Pendulum Model

It is straightforward to extend the planar RMP model discussed in Section 4.3.4 to three dimensions with the help of a 3D reaction mass assembly. The assembly comprises three pairs of point masses sliding along three nonplanar linear tracks that intersect at the CoM. The distances between each pair of point masses are controlled independently. Thereby, the centroidal inertia can be "shaped" in a desirable way. This results in improved balance capabilities w.r.t. a desired angular momentum equilibrium, as shown in [67,68,130]. An implementation of the 3D RMP model for compass-type biped walking is discussed in [141].

4.4.5 Multilink Models in 3D

Multilink models in 3D have been introduced at an early stage in biped robot modeling. The dynamic model of such a robot was derived in [145] from the time derivatives of the linear and angular momenta of the robot. This approach will also be employed in this chapter. At this point, it is instructive to introduce the ZMP formula for the 3D multilink model suggested in [145] (see also [54], p. 97). To avoid complexity, each link of the robot is represented as a point mass. Then, the two components of the ZMP are

$$r_{Pt} = \frac{\sum_{i=1}^{n} M_i \left[(\ddot{r}_{Czi} + g) r_{Cti} - (r_{Czi} - r_{Pz}) \ddot{r}_{Cti} \right]}{\sum_{i=1}^{n} M_i (\ddot{r}_{Czi} + g)}, \quad t \in \{x, y\}, \tag{4.41}$$

where $\boldsymbol{r}_P = \begin{bmatrix} r_{Px} & r_{Py} & r_{Pz} \end{bmatrix}^T$ denotes the ZMP, M_i stands for the mass of the ith link, and

$$\boldsymbol{r}_{Ci} = \begin{bmatrix} r_{Cxi} & r_{Cyi} & r_{Czi} \end{bmatrix}^T$$

denotes its CoM position. The total number of links is n.

4.5 DYNAMIC MODELS OF A FIXED-BASE MANIPULATOR

Dynamic models of a fixed-base serial-link manipulator are helpful in understanding the limb dynamics of a humanoid robot. The focus here will be on modeling kinematically redundant manipulators and redundancy resolution with dynamic models (inverse dynamics). The models will be formulated in terms of joint-space and end-link (spatial) coordinates. Thereby, it is important to distinguish the cases of completely free (unconstrained) and constrained motion dynamics.

4.5.1 Dynamic Model in Joint-Space Coordinates

Consider first the completely free motion of a fixed-base manipulator. The kinetic energy, expressed in joint-space coordinates, is written as

$$T(\boldsymbol{\theta}) = \frac{1}{2} \dot{\boldsymbol{\theta}}^T \boldsymbol{M}_\theta(\boldsymbol{\theta}) \dot{\boldsymbol{\theta}}, \tag{4.42}$$

$\boldsymbol{M}_\theta(\boldsymbol{\theta}) \in \Re^{n \times n}$ denoting the joint-space manipulator inertia matrix. As known from introductory texts, this matrix can be expressed as

$$\boldsymbol{M}_\theta = \sum_{i=1}^{n} \left\{ M_i \boldsymbol{J}_{vi}^T \boldsymbol{J}_{vi} + \boldsymbol{J}_{\omega i}^T \boldsymbol{I}_i \boldsymbol{J}_{\omega i} \right\}. \tag{4.43}$$

Here M_i, $\boldsymbol{I}_i \in \Re^{3\times3}$ are the ith link's mass and inertia tensor and $\boldsymbol{J}_{vi}(\boldsymbol{\theta}) \in \Re^{3 \times n}$ (defined in (2.118)) and $\boldsymbol{J}_{\omega i}$ are the Jacobians for the ith link's CoM velocity and angular velocity, respectively. The latter is written as

$$J_{\omega i}(\theta) = \begin{bmatrix} e_1 & e_2 & \cdots & e_i & 0 & \cdots & 0 \end{bmatrix} \in \Re^{3\times n}, \tag{4.44}$$

$e_j = R_j{}^j e_j$, ${}^j e_j = [0\ 0\ 1]^T$. The link inertia tensor M_θ is positive definite and, hence, its inverse can always be calculated. The respective Lagrangian form of manipulator dynamics is

$$M_\theta(\theta)\ddot{\theta} + c_\theta(\theta, \dot{\theta}) + g_\theta(\theta) = \tau. \tag{4.45}$$

Here $\tau \in \Re^n$ is the joint torque, $g_\theta(\theta) \in \Re^n$ is the gravity torque, and $c_\theta(\theta, \dot{\theta})$ denotes the nonlinear velocity-dependent torque. The latter can be expressed as

$$c_\theta(\theta, \dot{\theta}) = C_\theta(\theta, \dot{\theta})\dot{\theta} = \dot{M}_\theta(\theta)\dot{\theta} - \frac{1}{2}\left(\frac{\partial}{\partial \theta}\dot{\theta}^T M_\theta(\theta)\dot{\theta}\right)^T, \tag{4.46}$$

where

$$C_\theta(\theta, \dot{\theta}) = \frac{1}{2}\dot{M}_\theta(\theta) + S^\times(\theta, \dot{\theta})$$

and

$$S^\times(\theta, \dot{\theta})\dot{\theta} = \frac{1}{2}\left[\dot{M}_\theta(\theta)\dot{\theta} - \left(\frac{\partial}{\partial \theta}\dot{\theta}^T M_\theta(\theta)\dot{\theta}\right)^T\right];$$

$S^\times(\theta, \dot{\theta})$ is a skew-symmetric matrix that determines the passivity property of the equation of motion. This property plays an important role in control design [5,139]. Furthermore, note that since the end link does not contact any object, the only acting external force is the gravity. The equation of motion is an ODE. With an ODE solver, it is straightforward to find the solution to the *forward dynamics* problem: given the joint torque, find the joint acceleration, velocity, and positions. Given the current state, $(\theta, \dot{\theta})$, the joint acceleration is obtained as

$$\ddot{\theta} = M_\theta^{-1}(\theta)\left(\tau - c_\theta(\theta, \dot{\theta}) - g_\theta(\theta)\right) \tag{4.47}$$

and is integrated twice to obtain the new state. The forward dynamics problem is used in simulation.

Furthermore, the following *inverse dynamics* problem needs to be solved in the case of dynamic control: given the state of the manipulator, $(\theta, \dot{\theta})$, and the desired acceleration, $\ddot{\theta}$, find the joint torque. Note that in many cases, the motion task is specified in terms of end-link spatial coordinates. The inverse dynamics solution can then be obtained via the second-order inverse kinematics solution, (2.18) for a nonredundant manipulator, or (2.39) for a redundant one. The latter case is more interesting since there is an infinite number of solutions. Inserting (2.39) into (4.45) yields

$$\begin{aligned} \tau &= \tau_{lin} + \tau_n + \tau_{nl}, \\ \tau_{lin} &= M_\theta J^{\#}\dot{\mathcal{V}}, \\ \tau_n &= M_\theta\left(E - J^{\#}J\right)\ddot{\theta}_a, \\ \tau_{nl} &= c_\theta + g_\theta - M_\theta J^{\#}\dot{J}\dot{\theta}. \end{aligned} \tag{4.48}$$

The linear component $\boldsymbol{\tau}_{lin}$ results from the desired end-link acceleration, $\dot{\mathcal{V}}$; the null space component $\boldsymbol{\tau}_n$ represents an infinite number of joint torque vectors resulting from the self-motion acceleration vector parameter $\ddot{\boldsymbol{\theta}}_a$. Finally, $\boldsymbol{\tau}_{nl}$ is a nonlinear, state-dependent torque component. The behavior of the manipulator depends very much on the choice of the generalized inverse and the self-motion acceleration. To determine a specific joint torque, an additional task should be imposed, as discussed in Section 2.7. In [51], an attempt was made to use the null-space torque to locally minimize the deviation from the mid-range torque, i.e. $\|\boldsymbol{\tau} - \boldsymbol{\tau}^{mid}\|^2$, where $\boldsymbol{\tau}^{mid} = 0.5(\boldsymbol{\tau}^{max} + \boldsymbol{\tau}^{min})$. Substituting $\boldsymbol{\tau}$ from (4.48), the minimization yields the following self-motion acceleration:

$$\ddot{\boldsymbol{\theta}}_a = \left(\boldsymbol{W}^{1/2}\boldsymbol{M}_\theta(\boldsymbol{E} - \boldsymbol{J}^+\boldsymbol{J})\right)^+ \boldsymbol{W}^{1/2}\hat{\boldsymbol{\tau}}^{mid}. \tag{4.49}$$

The "hat" notation accounts for the linear and nonlinear torque components. The weight matrix $\boldsymbol{W} = \boldsymbol{W}^{1/2}\boldsymbol{W}^{1/2}$ scales the torque range of the individual joints.

Unfortunately, the above local torque minimization approach is plagued by a major problem: the solution lacks integrability. As a consequence, there is no control over the self-motion velocity. This quite often results in undesirable velocity build-up. Respective analysis has been provided by several authors [74,73,105]; see also the analysis in Section 2.11.3. Nevertheless, the problem is still actual, as apparent from more recent results in [112,14,127,18,170].

4.5.2 Dynamic Model in Spatial Coordinates

Humanoid robots interact with the environment mainly via their end links (feet and hands). Therefore, end-link motion/force control tasks are specified in spatial coordinates. When designing a controller, it would be reasonable to involve dynamic models expressed in terms of (end-link) spatial coordinates, rather than generalized (joint) coordinates. Indeed, a number of such controllers for fixed-base manipulators have been developed throughout the years. They are known as the "task space" [8], workspace/Cartesian space [50], or "operational space" [60] controllers. More recently, it has been shown that this type of controllers can be redesigned for use with floating-base humanoid robots. This section introduces basic relations pertinent to dynamic models in spatial coordinates for fixed-base manipulators. The application of such models to humanoid robots will be discussed in Section 4.13.3.

Consider first the completely free motion case. The derivation of a dynamic model in spatial coordinates is straightforward when the manipulator is a nonredundant one. Then, under the assumption that the inverse kinematic map exists, spatial coordinates can play the role of generalized coordinates. Making use of the inverse kinematics relation for the velocities, $\dot{\boldsymbol{\theta}} = \boldsymbol{J}(\boldsymbol{\theta})^{-1}\mathcal{V}$, kinetic energy (4.42) is expressed in spatial coordinates as [8,50]

$$T(\boldsymbol{\theta}) = \frac{1}{2}\mathcal{V}^T\mathbb{M}_e(\boldsymbol{\theta})\mathcal{V}. \tag{4.50}$$

Here $\mathbb{M}_e(\boldsymbol{\theta})$ is the manipulator inertia matrix represented in the end-link spatial coordinates. Its inverse equals the *end-effector mobility tensor* [49,50]

$$\mathbb{M}_e^{-1}(\boldsymbol{\theta}) = \boldsymbol{J}(\boldsymbol{\theta})\boldsymbol{M}_\theta^{-1}(\boldsymbol{\theta})\boldsymbol{J}^T(\boldsymbol{\theta}).$$

The term $\mathbb{M}_e(\boldsymbol{\theta})$ is also known as the "operational space inertia matrix" [19].

Furthermore, (4.45) is transformed into spatial coordinates through the following four steps. First premultiply by $\boldsymbol{J}\boldsymbol{M}_\theta^{-1}$ and then make use of the second-order kinematic relation $\boldsymbol{J}(\boldsymbol{\theta})\ddot{\boldsymbol{\theta}} = \dot{\mathcal{V}} - \dot{\boldsymbol{J}}(\boldsymbol{\theta})\dot{\boldsymbol{\theta}}$. The resultant equation is

$$\dot{\mathcal{V}} - \dot{\boldsymbol{J}}\dot{\boldsymbol{\theta}} = \boldsymbol{J}\boldsymbol{M}_\theta^{-1}\boldsymbol{\tau} - \boldsymbol{J}\boldsymbol{M}_\theta^{-1}(\boldsymbol{c}_\theta + \boldsymbol{g}_\theta). \tag{4.51}$$

Then, substitute the static force relation $\boldsymbol{\tau} = \boldsymbol{J}^T\mathcal{F}^m$ and solve the resulting equation for spatial force $\mathcal{F}^m$. The final result is

$$\begin{aligned}\mathcal{F}^m &= \mathbb{M}_e(\dot{\mathcal{V}} - \dot{\boldsymbol{J}}\dot{\boldsymbol{\theta}}) + \mathbb{M}_e\boldsymbol{J}\boldsymbol{M}_\theta^{-1}(\boldsymbol{c}_\theta + \boldsymbol{g}_\theta)\\ &= \mathbb{M}_e(\boldsymbol{\theta})\dot{\mathcal{V}} + \mathcal{C}_e(\boldsymbol{\theta}, \dot{\boldsymbol{\theta}}) + \mathcal{G}_e(\boldsymbol{\theta}),\end{aligned} \tag{4.52}$$

where $\mathcal{C}_e(\boldsymbol{\theta}, \dot{\boldsymbol{\theta}}) = \boldsymbol{J}^{-T}\boldsymbol{c} - \mathbb{M}_e\dot{\boldsymbol{J}}\dot{\boldsymbol{\theta}}$ denotes the nonlinear velocity-dependent force and $\mathcal{G}_e(\boldsymbol{\theta}) = \boldsymbol{J}^{-T}\boldsymbol{g}_\theta$ is the gravity force; these quantities are expressed as end-link wrenches. Furthermore, note that $\mathcal{F}^m = \boldsymbol{J}^{-T}\boldsymbol{\tau}$ is a quasistatic wrench stemming from the motion/force duality principle; it is not an external force. The m superscript is used to emphasize that this wrench should be related to mobility rather than to an external contact wrench.

Operational Space Method [60]

The dynamical model of a *kinematically redundant manipulator* could be transformed into end-link spatial coordinates with the same procedure as above. But the resultant equation would be an incomplete representation of the dynamics since the null-space joint torque does not appear in the equation. Indeed, for a redundant manipulator, the underlying coordinate transform should involve a generalized inverse of the Jacobian matrix. Since there is an infinite number of such inverses, the joint space dynamics cannot be uniquely represented in terms of spatial end-link coordinates. It is possible to choose a particular generalized inverse, but then subsequent analysis is needed to clarify the properties of the specific map. This approach was used in the operational space formulation [60], where the joint-to-spatial coordinates transform involves the inertia-weighted pseudoinverse of the Jacobian, i.e.

$$\begin{aligned}\boldsymbol{J}^{-M}(\boldsymbol{\theta}) &= \boldsymbol{M}_\theta^{-1}\boldsymbol{J}^T\left(\boldsymbol{J}\boldsymbol{M}_\theta^{-1}\boldsymbol{J}^T\right)^{-1}\\ &= \boldsymbol{M}_\theta^{-1}\boldsymbol{J}^T\mathbb{M}_e.\end{aligned} \tag{4.53}$$

Matrix $\boldsymbol{J}^{-MT}$ replaces $\boldsymbol{J}^{-T}$ in the nonlinear, gravity, and quasistatic wrench terms in (4.52). This particular inverse yields the so-called "dynamically consistent relationship" [61,33]: the end-link and the null-space dynamics become completely decoupled. Such decoupling of the dynamics is highly desirable from the viewpoint of controller design. A number of such controllers have been developed throughout the years, including the compliance [112] and impedance controllers [90,91,3]. The method has been applied to humanoid robots as well. Details will be given in Section 4.13.3.

To understand the dynamical decoupling property, assume a stationary manipulator configuration and zero gravity. Dynamical models (4.45) and (4.52) express then only linear motion/force relationships $\boldsymbol{M}_\theta\ddot{\boldsymbol{\theta}} = \boldsymbol{\tau}$ and $\mathbb{M}_e\dot{\mathcal{V}} = \mathcal{F}^m$, respectively. The kinetostatic relations, on the other hand, are $J\ddot{\theta} = \dot{\mathcal{V}}$ and $\tau = J^T \mathcal{F}^m$. In the presence of kinematic redundancy, there is an infinite set of control joint torque vectors. The last relation is then rewritten as [60,61]

$$\tau_f = J^T \mathcal{F}^m + \left(E - J^T J^{-WT}\right) \tau_a. \tag{4.54}$$

Here τ_a is an arbitrary joint torque that parametrizes (dual) null space $\mathcal{N}^*(J) \equiv \mathcal{N}(J^{-WT})$ (recall from Section 3.4.3 that $\left(E - J^T J^{-WT}\right)$ denotes a projector onto this null space). The joint torque component obtained from the projection has no contribution to the resultant forces at the end link.

On the other hand, when the second-order inverse kinematic solution of a redundant manipulator, (2.39), is inserted into (4.45) (under the assumption of using a weighted pseudoinverse, a stationary configuration, and zero gravity), another infinite set of control joint torques can be derived [33], i.e.

$$\tau_m = M_\theta J^{-W} \mathbb{M}_e^{-1} \mathcal{F}^m + M_\theta \left(E - J^{-W} J\right) \ddot{\theta}_a. \tag{4.55}$$

Here $\ddot{\theta}_a$ is an arbitrary joint acceleration that parametrizes the null space $\mathcal{N}(J)$. Its projection yields a joint acceleration that could be applied without affecting the resultant spatial acceleration of the end link. All $\ddot{\theta}_a$-induced joint torques $\{\tau_m\}$ produce the same end-link acceleration as the end-link force $\mathcal{F}^m$.

It is straightforward to show that the two joint torque sets (4.54) and (4.55) are compatible (or *dynamically consistent*), only when the inertia-weighted pseudoinverse of the Jacobian is used in the equations, i.e. require $W = M_\theta$. This leads to a *complete dynamical decoupling* between the particular components responsible for the motion/force control task and the null space components. It was shown in [33] that the following four conditions will then be satisfied:

$$\mathcal{F}_n = M_\theta \mathcal{M}_n, \quad \mathcal{F}_r = M_\theta \mathcal{M}_r \tag{4.56}$$

and

$$\mathcal{F}_n \perp \mathcal{M}_r, \quad \mathcal{F}_r \perp \mathcal{M}_n, \tag{4.57}$$

where $\mathcal{M}$ and $\mathcal{F} \equiv \mathcal{M}^*$ denote the joint motion (generalized velocity/acceleration) and dual (generalized momentum/joint torque) domains, respectively. Subscripts "r" and "n" stand for components from the row and null subspaces of the underlying transform, i.e. the Jacobian matrix.

The property of complete dynamical decoupling plays an important role in motion/force and impedance control design, since the task and null space control components can be designed independently.

Constrained Dynamics in Spatial Coordinates

Consider now a motion/force control scenario wherein the end link is in contact with the environment and moves along the contact surface. A closed loop is formed via the contact joint, giving rise to the kinetostatic conditions introduced in Chapter 2. The first-order differential motion relation is of the same form as (2.95). Accordingly, the manipulator Jacobian decomposes into sub-Jacobians for the constrained and unconstrained directions: $\boldsymbol{J}_c(\boldsymbol{\theta}) \in \Re^{c\times n}$ and $\boldsymbol{J}_m(\boldsymbol{\theta}) \in \Re^{\eta\times n}$, respectively, c and η ($c+\eta=6$) denoting the number of constrained and unconstrained directions at the contact joint. The equation of motion is written as

$$\boldsymbol{M}_\theta(\boldsymbol{\theta})\ddot{\boldsymbol{\theta}} + \boldsymbol{c}_\theta(\boldsymbol{\theta},\dot{\boldsymbol{\theta}}) + \boldsymbol{g}_\theta(\boldsymbol{\theta}) = \boldsymbol{\tau} + \boldsymbol{J}_c^T(\boldsymbol{\theta})\overline{\mathcal{F}}^c, \tag{4.58}$$

$\overline{\mathcal{F}}^c$ denoting the contact wrench components (reactions). Furthermore, the Jacobian decomposition specifically structures the dynamical terms that are represented in spatial coordinates. Consequently, the operational space inertia matrix and the nonlinear velocity and gravity terms assume the form [28]

$$\mathbb{M}_e(\boldsymbol{\theta}) = \begin{bmatrix} \boldsymbol{J}_c\boldsymbol{M}_\theta^{-1}\boldsymbol{J}_c^T & \boldsymbol{J}_c\boldsymbol{M}_\theta^{-1}\boldsymbol{J}_m^T \\ \boldsymbol{J}_m\boldsymbol{M}_\theta^{-1}\boldsymbol{J}_c^T & \boldsymbol{J}_m\boldsymbol{M}_\theta^{-1}\boldsymbol{J}_m^T \end{bmatrix}^{-1} = \begin{bmatrix} \boldsymbol{M}_c & \boldsymbol{M}_{cm} \\ \boldsymbol{M}_{cm}^T & \boldsymbol{M}_m \end{bmatrix}, \tag{4.59}$$

$$\mathcal{C}_e(\boldsymbol{\theta},\dot{\boldsymbol{\theta}}) = \mathbb{M}_e \begin{bmatrix} \boldsymbol{J}_c\boldsymbol{M}_\theta^{-1}\boldsymbol{c}_\theta - \dot{\boldsymbol{J}}_c\dot{\boldsymbol{\theta}} \\ \boldsymbol{J}_m\boldsymbol{M}_\theta^{-1}\boldsymbol{c}_\theta - \dot{\boldsymbol{J}}_m\dot{\boldsymbol{\theta}} \end{bmatrix} = \begin{bmatrix} \mathcal{C}_c \\ \mathcal{C}_m \end{bmatrix}, \tag{4.60}$$

$$\mathcal{G}_e(\boldsymbol{\theta}) = \mathbb{M}_e \begin{bmatrix} \boldsymbol{J}_c \\ \boldsymbol{J}_m \end{bmatrix} \boldsymbol{M}_\theta^{-1}\boldsymbol{g}_\theta = \begin{bmatrix} \mathcal{G}_c \\ \mathcal{G}_m \end{bmatrix},$$

respectively. With the above notations, the equation of motion can be represented in spatial coordinates as

$$\begin{bmatrix} \boldsymbol{M}_c & \boldsymbol{M}_{cm} \\ \boldsymbol{M}_{cm}^T & \boldsymbol{M}_m \end{bmatrix} \begin{bmatrix} \boldsymbol{0} \\ \dot{\mathcal{V}}^m \end{bmatrix} + \begin{bmatrix} \mathcal{C}_c \\ \mathcal{C}_m \end{bmatrix} + \begin{bmatrix} \mathcal{G}_c \\ \mathcal{G}_m \end{bmatrix} = \begin{bmatrix} \overline{\mathcal{F}}^c \\ \overline{\mathcal{F}}^m \end{bmatrix}, \tag{4.61}$$

where $\mathcal{V}^m = \boldsymbol{J}_m\dot{\boldsymbol{\theta}} \in \Re^\eta$ is the velocity along the unconstrained motion directions at the contact joint. The dual quantity $\overline{\mathcal{F}}^m$ denotes the quasistatic force (along the unconstrained directions). Note that the acceleration $\dot{\mathcal{V}}^m$ does not produce any reaction force. Likewise, the reaction force $\overline{\mathcal{F}}^c$ does not generate any acceleration along the unconstrained motion directions. This results in dynamic decoupling. Consequently, the end-effector motion and force are controllable in an independent manner, via $\overline{\mathcal{F}}^m$ and $\overline{\mathcal{F}}^c$, respectively.

Complete Dynamic Decoupling via the KD-JSD Method [116]

The kinematically decoupled joint-space decomposition (KD-JSD) method (cf. Section 2.7.5) can be applied to achieve complete dynamic decoupling, as with the operational space formulation. The minimal parametrization property of the method yields an important advantage when the dynamics are expressed in end-link coordinates, though.

The infinite set of static joint torques (4.54) can be represented with the help of the KD-JSD notation as

$$\boldsymbol{\tau}_f = \boldsymbol{J}^T\mathcal{F} + \boldsymbol{W}\boldsymbol{Z}^T\overline{\boldsymbol{\tau}}_a, \tag{4.62}$$

where $\overline{\boldsymbol{\tau}}_a$ is an arbitrary r-vector that parametrizes the dual null space $\mathcal{N}^*(\boldsymbol{J}) \equiv \mathcal{F}_n$ in a minimal way. The weighting matrix $\boldsymbol{W}(\boldsymbol{\theta}) \in \Re^{n\times n}$, the full row-rank matrix $\boldsymbol{Z}(\boldsymbol{\theta}) \in \Re^{r\times n}$, and the "asymmetrically" weighted generalized inverse $\boldsymbol{Z}_W^{\#}(\boldsymbol{\theta})$ have been defined in Section 2.7.5. It can be shown that for a given nonminimal-parametrization vector $\boldsymbol{\tau}_a \in \Re^n$, there is a $\bar{\boldsymbol{\tau}}_a \in \Re^r$ s.t. the null space components (the second terms on the r.h.s.) in (4.54) and (4.62) are equal.

On the other hand, there is another infinite set of joint torques that result from the set of possible joint accelerations. The latter are obtained from the KD-JSD second-order inverse kinematic solution (derived via differentiation of (2.48) w.r.t. time) and inserted into (4.45). Under the assumption of a stationary configuration and zero gravity, the acceleration-induced joint torques are expressed as

$$\boldsymbol{\tau}_m = \boldsymbol{M}_\theta \begin{bmatrix} \boldsymbol{J}^{-W}(\boldsymbol{\theta}) & \boldsymbol{Z}_W^{\#}(\boldsymbol{\theta}) \end{bmatrix} \begin{bmatrix} \dot{\mathcal{V}} \\ \dot{\boldsymbol{b}} \end{bmatrix}. \tag{4.63}$$

Arbitrary r-vector $\dot{\boldsymbol{b}}$ parametrizes in a minimal way the null space $\mathcal{N}(\boldsymbol{J}) \equiv \mathcal{M}_n$. This null space comprises joint accelerations that could be applied without affecting the resultant spatial accelerations of the end link, as already discussed.

Under the joint-space decomposition scheme, the linear motion/force relations in joint space can be expressed as

$$\begin{bmatrix} \mathcal{F} \\ \bar{\boldsymbol{\tau}}_a \end{bmatrix} = \begin{bmatrix} \mathbb{M}_e(\boldsymbol{\theta}) & \boldsymbol{H}_{er}(\boldsymbol{\theta}) \\ \boldsymbol{H}_{er}^T(\boldsymbol{\theta}) & \boldsymbol{M}_r(\boldsymbol{\theta}) \end{bmatrix} \begin{bmatrix} \dot{\mathcal{V}} \\ \dot{\boldsymbol{b}} \end{bmatrix}, \tag{4.64}$$

where

$$\mathbb{M}_e = \boldsymbol{J}^{-WT} \boldsymbol{M}_\theta \boldsymbol{J}^{-W},$$

$$\boldsymbol{H}_{er} = \boldsymbol{J}^{-WT} \boldsymbol{M}_\theta \boldsymbol{Z}_W^{\#},$$

$$\boldsymbol{M}_r = \boldsymbol{Z}_W^{\#T} \boldsymbol{M}_\theta \boldsymbol{Z}_W^{\#}.$$

A dynamically consistent relationship for joint torques, (4.62) and (4.63) (i.e. complete dynamical decoupling), can be ensured only when the weighting matrix equals the joint-space inertia, i.e. $\boldsymbol{W}(\boldsymbol{\theta}) = \boldsymbol{M}_\theta(\boldsymbol{\theta})$. Then, $\boldsymbol{H}_{er} = \boldsymbol{0}$ since $\boldsymbol{J}\boldsymbol{Z}_W^{\#} = \boldsymbol{0}$. Hence, the inertia matrix in (4.64) assumes the block-diagonal form diag$\begin{bmatrix} \mathbb{M}_e & \boldsymbol{M}_r \end{bmatrix}$. This feature facilitates the design of control laws, as will be explained in Section 4.5.3.

4.5.3 Null-Space Dynamics With Dynamically Decoupled Hierarchical Structure

The complete dynamic decoupling approach discussed in Section 4.5.2 provides a basis to develop asymptotically stable null-space dynamic control. To this end, the approach is first reformulated for the general case of r prioritized tasks, in terms of differential kinematics [29], i.e.

$$\mathcal{V}_k = \bar{\boldsymbol{J}}_k \dot{\boldsymbol{\theta}}, \tag{4.65}$$

where $\mathcal{V}_k \in \Re^{m_k}$ and $\bar{\boldsymbol{J}}_k \in \Re^{m_k \times n}$, $k \in \{\overline{2,r}\}$ are the additional-task velocities and (null-space) restricted Jacobians, respectively. Recall that the Jacobians have to be determined in a way s.t.

the kinematic decoupling for all additional tasks is ensured (cf. Section 2.7.5). The kinematic decoupling also yields dynamic decoupling, as already discussed in Section 4.5.2.

The dynamic decoupling can be ensured when the restricted Jacobians are determined in the following form [29]:

$$\bar{J}_k = \left(Z_k M_\theta Z_k^T\right)^{-1} Z_k M_\theta. \tag{4.66}$$

Here Z_k represents a full row-rank null-space basis derived from matrix $J_{C_{k-1}}$ appearing in the recursive scheme (2.52), s.t. $J_{C_{k-1}} Z_k^T = \mathbf{0}$. Apparently, the link inertia matrix, M_θ, plays the role of a weighting matrix in the above expression. With this notation, the inverse kinematics solution is expressed as

$$\dot{\theta} = J^{-M_\theta}\mathcal{V} + \sum_{k=2}^{r} Z_k^T \mathcal{V}_k. \tag{4.67}$$

The Z_k null-space bases can be determined via singular-value decomposition (SVD) of the $J_{C_{k-1}}$ matrices. Note also that the following relations hold:

$$\begin{aligned} \bar{J}_k^T Z_k J_k^T &= N^*(J_{C_{k-1}}) J_k^T, \\ N^*(J_{C_{k-1}}) &\equiv E - J_{C_{k-1}}^T J_{C_{k-1}}^{-M\,T}. \end{aligned} \tag{4.68}$$

Furthermore, in accordance with (2.32), the null-space projector can be decomposed as $N^*(J_{C_{k-1}}) = V_{k-1} V_{k-1}^T$. Then,

$$\begin{aligned} Z_k &= J_k V_{k-1}\left(V_{k-1}^T M_\theta V_{k-1}\right)^{-1} V_{k-1}^T, \ k = 2, 3, \ldots, (r-1), \\ Z_r &= V_{r-1}. \end{aligned} \tag{4.69}$$

As discussed in [29], when implemented appropriately in a controller, the above decomposition can ensure asymptotic stability without involving external force measurement, guaranteeing thereby successive convergence in accordance with the order of priority.

4.6 SPATIAL MOMENTUM OF A MANIPULATOR FLOATING FREELY IN ZERO GRAVITY

The complete dynamic model of a humanoid robot accounts for the presence of a floating base and external forces, such as gravity and contact (reaction) forces. Floating-base dynamic models comprise a specific component that distinguishes them from the fixed-base dynamic models discussed in Section 4.5. For the in-depth understanding of floating-base dynamic models, it will prove useful to focus first on the dynamics of a free-floating robot in zero gravity. In the absence of external forces, these dynamics refer solely to the motion of the robot. The properties of motion dynamics play an important role as an inherent part of the complete dynamics. Note also that the study of pure free-floating, i.e. contactless, underactuated system models, is justifiable from a practical viewpoint as well. Indeed, there are humanoid robots that are able to attain a flight phase wherein the feet loose contact with the ground, e.g. during running or jumping [21,148,149,147,26]. To arrive at an appropriate posture for landing, the motion of the limbs needs to be controlled accordingly in mid-air. With an appropriate landing posture control, excessive reactions upon touchdown can be avoided. Balance stability in the postimpact phase is then also easier to achieve. Free-floating robot models are especially helpful in understanding the important role of angular momentum.

4.6.1 Brief Historical Background

Free-floating system models appeared first in the field of space robotics during the 1980s. In the pioneering work [154], the nonholonomic nature of a free-floating space robot was revealed. Attention was drawn to the disturbance of the spatial position of the base that is induced by the motion of the manipulator. Also, a related difficulty for the manipulator motion planning problem was highlighted. The use of kinematic redundancy for momentum compensation has been first investigated in [97,122]. In [97,94], velocity-based equations for simultaneous control of the end-link and base motions were established and a novel redundancy resolution approach with prioritization of tasks was developed to minimize or maximize the reactions at the base. Also, the so-called "manipulator inversion task" was introduced wherein the end link remains fixed in inertial space while the orientation of the floating base varies in a desirable way. This behavior is achieved via a specific null space: that of the coupling inertia matrix. Later it was shown that the same null space can also play an important role for balance control of a humanoid robot subjected to an external disturbance [102,101,168].

4.6.2 Spatial Momentum

Consider a free-floating serial-link chain in zero gravity comprising an n-joint manipulator arm mounted on a rigid-body satellite, as shown in Fig. 4.7A. The satellite represents the floating base of the system. It is assumed that the base is not actuated (there are no thrusters or RWs) while the manipulator joints are. Thus, the system is underactuated.

The linear momentum of the system is expressed uniquely as

$$p = \sum_{i=0}^{n} M_i \dot{r}_i = M \dot{r}_C, \tag{4.70}$$

M_i and r_i denoting the mass and CoM position of Link i; M stands for the total mass and r_C is the position of the system CoM. The expression for the angular momentum depends on the point of reference. In the field of space robotics, expressions w.r.t. the origins of the inertial and the base-link frames as well as the system CoM are used. In the field of humanoid robotics, the latter two are exclusively employed. The simplest expression is obtained when

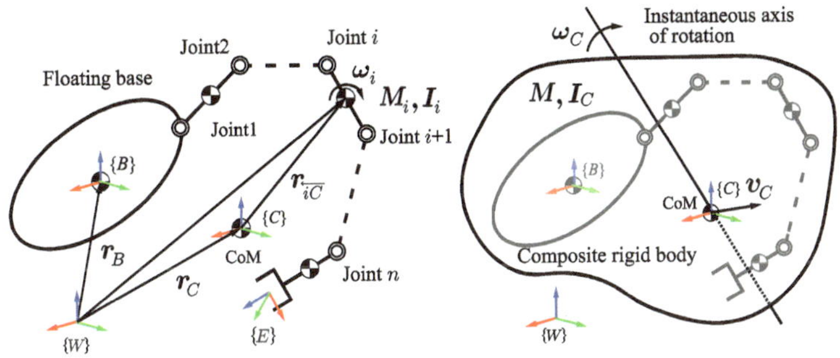

FIGURE 4.7 Model of a serial-link manipulator on a free-floating base. (A) Left: main coordinate frames and system parameters. (B) Right: representation as a composite rigid body (CRB) (all joints locked).

the angular momentum is represented w.r.t. the system CoM, i.e.

$$l_C = \sum_{i=0}^{n} I_i \omega_i. \tag{4.71}$$

Here I_i and ω_i denote the inertia tensor and angular velocity of Link i. Note that the base is designated as Link 0; related quantities, such as the base position, velocity, and inertia, will be denoted by subscript $(\circ)_B$. All quantities are expressed in the inertia frame, $\{W\}$. In the notation for angular momentum, the subscript signifies the reference point.

It is convenient to combine the above two momentum vectors into a single 6D vector that will represent the *spatial momentum* of the floating-base system w.r.t. its CoM, i.e.

$$\mathcal{L}_C \equiv \begin{bmatrix} p \\ l_C \end{bmatrix}. \tag{4.72}$$

The spatial momentum is an element of $se^*(3)$, as is the spatial force. Note that in the absence of external forces (e.g. gravity), the spatial momentum is conserved. In the presence of gravity, on the other hand, only the angular momentum is conserved. Note also that the angular momentum is nonintegrable and, hence, its conservation imposes a nonholonomic motion constraint.

4.6.3 Locked Joints: the Composite Rigid Body

When the manipulator joints are locked, the free-floating system behaves as a composite rigid body (CRB). Recall that the notation of a CRB as a composite system of links characterized by its CoM was introduced in Section 3.6. The CRB is characterized also by an inertia tensor defined as

$$I_C(q) \equiv \sum_{i=0}^{n} \left(I_i + M_i [r^{\times}_{\overleftarrow{iC}}][r^{\times}_{\overleftarrow{Ci}}] \right) \in \Re^{3\times 3}. \tag{4.73}$$

Hereby, $r_{\overleftarrow{iC}} = r_i - r_C$ and $q = (\mathcal{X}_B, \theta)$ denotes the generalized coordinates of the floating-base system, $\mathcal{X}_B$ standing for the 6D position of the floating base. At this point it is instructive to note that the states of the manipulator joints and the floating base are derived from the joint angle encoder and the inertial measurement sensor (IMU) readings, respectively. Sensor fusion techniques may increase the accuracy of the latter [126].

Furthermore, as shown in Fig. 4.7B, the CRB coordinate frame $\{C\}$ is conveniently attached to the system CoM. Since the CRB is not a real body, the coordinate axes are chosen to be parallel to those of the base-link frame $\{B\}$. With this choice, the spatial velocity of the CRB can be expressed by $\mathcal{V}_M$ (defined in Section 2.11.4), with components CoM velocity $\boldsymbol{v}_C = \dot{\boldsymbol{r}}_C$ and the angular velocity of the base link $\boldsymbol{\omega}_B$. As an alternative expression, the so-called *system spatial velocity* $\mathcal{V}_C = \begin{bmatrix} \boldsymbol{v}_C^T & \boldsymbol{\omega}_C^T \end{bmatrix}^T$ will be introduced here. Angular velocity $\boldsymbol{\omega}_C$ is referred to as the *system angular velocity*[5]; its meaning will be clarified in Section 4.6.4.

The relation between the spatial momentum and the system spatial velocity can be expressed as

$$\mathcal{L}_C = \mathbb{M}_C \mathcal{V}_C. \tag{4.74}$$

Matrix

$$\mathbb{M}_C(\boldsymbol{q}) \equiv \begin{bmatrix} M\boldsymbol{E} & \boldsymbol{0} \\ \boldsymbol{0} & \boldsymbol{I}_C(\boldsymbol{q}) \end{bmatrix} \in \Re^{6\times 6} \tag{4.75}$$

has the structure of a rigid-body *spatial inertia tensor* [32] (or a *locked inertia tensor* [86,109]). It plays an important role in floating-base system modeling, including humanoid robots. Note that the subscript indicates dependence on the reference point. In this case, the reference point is the system CoM (or centroid), thus $\mathbb{M}_C(\boldsymbol{q})$ is referred to as the *centroidal CRB inertia* [67]; $\mathcal{L}_C$, on the other hand, has been called a *centroidal spatial momentum* [107]. Other representations of the spatial inertia tensor, e.g. w.r.t. the origin of the floating-base or the inertia frame, will be given in what follows. The angular momentum derived from (4.74) is

$$\boldsymbol{l}_C = \boldsymbol{I}_C(\boldsymbol{q})\boldsymbol{\omega}_c. \tag{4.76}$$

The quantities $\boldsymbol{I}_C(\boldsymbol{q})$ and $\boldsymbol{l}_C$ are referred to as the *centroidal* inertia tensor and the *centroidal* angular momentum, respectively; $\boldsymbol{l}_C$ is also called a *spin* [119,120] or an *intrinsic* [32,152] angular momentum.

The preferable reference point for the spatial momentum and the respective spatial inertia tensor may differ in each application field. In the field of humanoid robots, the CoM may be the preferred choice (e.g. as in [55]). In the field of space robotics, on the other hand, the origin of the floating base frame is usually used. In this case, the spatial momentum is expressed as

$$\mathcal{L}_B = \mathbb{T}_{\overleftarrow{BC}}^T \mathcal{L}_C \tag{4.77}$$

or

$$\mathcal{L}_B \equiv \begin{bmatrix} \boldsymbol{p} \\ \boldsymbol{l}_B \end{bmatrix} = \begin{bmatrix} \boldsymbol{E} & \boldsymbol{0} \\ -[\boldsymbol{r}_{\overleftarrow{BC}}^{\times}] & \boldsymbol{E} \end{bmatrix} \begin{bmatrix} \boldsymbol{p} \\ \boldsymbol{l}_C \end{bmatrix} = \begin{bmatrix} \boldsymbol{p} \\ \boldsymbol{l}_C \end{bmatrix} - \begin{bmatrix} \boldsymbol{0} \\ [\boldsymbol{r}_{\overleftarrow{BC}}^{\times}]\boldsymbol{p} \end{bmatrix}.$$

[5] The system angular/spatial velocity $\boldsymbol{\omega}_C/\mathcal{V}_C$ have been named the *average* angular/spatial velocity in [107].

Note that since spatial momentum is an element of $se^*(3)$, in (4.77) the transformation rule for wrenches (cf. (3.4)) was used. Note also that the transformation is valid in general, i.e. not only for the locked-joint case. Furthermore, by inserting (4.74) into (4.77) and using $\mathcal{V}_C = \mathbb{T}_{\overleftarrow{CB}}\mathcal{V}_B$ (cf. (2.4)), one obtains

$$\widetilde{\mathcal{L}}_B = \mathbb{M}_B \mathcal{V}_B.$$

Here $\mathcal{V}_B = \begin{bmatrix} \boldsymbol{v}_B^T & \boldsymbol{\omega}_B^T \end{bmatrix}^T$ denotes the spatial velocity of the base. The overtilde notation signifies that the above expression for the spatial momentum in the locked-joint case may differ from that in the unlocked-joint one. The matrix

$$\mathbb{M}_B \equiv \mathbb{T}_{\overleftarrow{BC}}^T \mathbb{M}_C \mathbb{T}_{\overleftarrow{CB}} \tag{4.78}$$

represents the spatial inertia tensor. Relation (4.78) provides a useful formula for transforming spatial inertia tensors. The above tensor can be expanded as

$$\mathbb{M}_B(\boldsymbol{q}) = \begin{bmatrix} M\boldsymbol{E} & -M[\boldsymbol{r}_{\overleftarrow{CB}}^{\times}] \\ -M[\boldsymbol{r}_{\overleftarrow{CB}}^{\times}]^T & \boldsymbol{I}_B \end{bmatrix}, \tag{4.79}$$

where

$$\begin{aligned} \boldsymbol{I}_B(\boldsymbol{q}) &= \boldsymbol{I}_C(\boldsymbol{q}) + M[\boldsymbol{r}_{\overleftarrow{CB}}^{\times}][\boldsymbol{r}_{\overleftarrow{BC}}^{\times}] \\ &= \boldsymbol{I}_0 + \sum_{i=1}^{n}\left(\boldsymbol{I}_i + M_i[\boldsymbol{r}_{\overleftarrow{iB}}^{\times}][\boldsymbol{r}_{\overleftarrow{Bi}}^{\times}]\right). \end{aligned} \tag{4.80}$$

Here $\boldsymbol{I}_0$ denotes the inertia tensor of the base link.

Furthermore, the spatial momentum can be expressed w.r.t. the inertial frame, if needed, as

$$\mathcal{L}_W = \mathbb{T}_{\overleftarrow{WC}}^T \mathcal{L}_C. \tag{4.81}$$

The expanded form is given as

$$\mathcal{L}_W \equiv \begin{bmatrix} \boldsymbol{p} \\ \boldsymbol{l}_W \end{bmatrix} = \begin{bmatrix} \boldsymbol{E} & \boldsymbol{0} \\ -[\boldsymbol{r}_{\overleftarrow{WC}}^{\times}] & \boldsymbol{E} \end{bmatrix} \begin{bmatrix} \boldsymbol{p} \\ \boldsymbol{l}_C \end{bmatrix} = \begin{bmatrix} \boldsymbol{p} \\ \boldsymbol{l}_C \end{bmatrix} - \begin{bmatrix} \boldsymbol{0} \\ [\boldsymbol{r}_{\overleftarrow{WC}}^{\times}]\boldsymbol{p} \end{bmatrix}.$$

The CRB angular momentum can then be written as

$$\boldsymbol{l}_W = \boldsymbol{I}_C\boldsymbol{\omega}_c + [\boldsymbol{r}_{\overleftarrow{CW}}^{\times}]\boldsymbol{p} = \sum_{i=0}^{n}\left(\boldsymbol{I}_i\boldsymbol{\omega}_i + M_i[\boldsymbol{r}_{\overleftarrow{iW}}^{\times}]\dot{\boldsymbol{r}}_i\right). \tag{4.82}$$

4.6.4 Joints Unlocked: Multibody Notation

It will be shown in what follows that, when the joints are unlocked, the representation of spatial momentum depends on the particular choice of the quasivelocity. Recall that, so far, two kinds of quasivelocity were used: the base quasivelocity $\dot{q}_B = (\mathcal{V}_B, \dot{\theta})$ and the mixed quasivelocity $\dot{q}_M = (\mathcal{V}_M, \dot{\theta})$. The system spatial velocity $\mathcal{V}_C$ gives rise of a third type of quasivelocity: the *centroidal quasivelocity* $\dot{q}_C = (\mathcal{V}_C, \dot{\theta})$.

To clarify the dependence on the quasivelocity formulation, note first that the (linear) momentum can be represented as

$$\begin{aligned} p(q, \dot{q}_B) &= M v_B + \sum_{i=1}^{n} M_i \left(-[r^{\times}_{\overleftarrow{iB}}]\omega_B + J_{vi}(\theta)\dot{\theta}\right) \\ &= M\left(v_B - [r^{\times}_{\overleftarrow{CB}}]\omega_B + J_{\overleftarrow{CB}}(\theta)\dot{\theta}\right), \end{aligned} \tag{4.83}$$

$J_{\overleftarrow{CB}}(\theta)$ denoting the CoM Jacobian of the robot.[6] Since $p = p(q, \dot{q}_B) = M v_C$, it follows that

$$v_C = v_B - [r^{\times}_{\overleftarrow{CB}}]\omega_B + J_{\overleftarrow{CB}}\dot{\theta}. \tag{4.84}$$

On the other hand, the *system angular momentum* (SAM) w.r.t. the CoM can be represented as a sum, i.e.

$$l_C(q, \dot{q}_\omega) = I_C(q)\omega_B + H_C(q)\dot{\theta}, \tag{4.85}$$

where $\dot{q}_\omega = (\omega_B, \dot{\theta})$ and

$$H_C(q) = \sum_{i=1}^{n} \left(I_i J_{\omega i}(\theta) + M_i [r^{\times}_{\overleftarrow{iC}}] J_{vi}(\theta)\right), \tag{4.86}$$

$J_{\omega i}(\theta)$ being defined in (4.44). Here and henceforth, $H_{(\circ)}$ will be used to denote a map that is composed of mass/inertia-Jacobian products. This type of map represents the *mechanical connection* [86,109] in coordinate form; $H_{(\circ)}$ is referred to as the *coupling inertia* matrix [96].

Furthermore, since the CRB (locked) inertia tensor $I_C(q)$ is p.d., its inverse exists. Premultiply (4.85) by I_C^{-1} to obtain

$$I_C^{-1} l_C(q, \dot{q}_\omega) = \omega_B + J_\omega(\theta)\dot{\theta}, \tag{4.87}$$

where $J_\omega \equiv I_C^{-1} H_C$. The compact form representation can be written as

$$\omega_C \equiv \hat{J}_\omega(\theta)\dot{q}_\omega, \tag{4.88}$$

where $\hat{J}_\omega \equiv \begin{bmatrix} E & J_\omega \end{bmatrix}$. The angular velocity ω_C is the sum of the angular velocities of all bodies that constitute the system; it is the *system angular velocity* that was introduced in Section 4.6.3. Note that when the joints are locked, $\omega_C = \omega_B = \omega_i$, $i \in \{\overline{1, n}\}$.

The simplest expression of the *system spatial momentum* (SSM) w.r.t. the CoM (i.e. the centroidal SSM) is obtained by employing the centroidal quasivelocity. Then,

$$\mathcal{L}_C(q, \dot{q}_C) = \mathcal{L}_C(q, \mathcal{V}_C) = \mathbb{M}_C \mathcal{V}_C. \tag{4.89}$$

[6] Jacobians $J_{\overleftarrow{CB}}(\theta)$ and $J_{vi}(\theta)$ have been defined in (2.118).

Apparently, there is no dependency on the joint rates whatsoever. In other words, in this representation, the SSM equals the *CRB spatial momentum* (CRB-SM); we can write: $\mathcal{L}_C(\boldsymbol{q}, \mathcal{V}_C) = \widetilde{\mathcal{L}}_C(\boldsymbol{q}, \mathcal{V}_C)$.

When expressed in terms of the mixed quasivelocity, the SSM assumes the form

$$\begin{aligned}\mathcal{L}_C(\boldsymbol{q}, \dot{\boldsymbol{q}}_M) &= \widetilde{\mathcal{L}}_C(\boldsymbol{q}, \mathcal{V}_M) + \mathcal{L}_{CM}(\boldsymbol{q}, \dot{\boldsymbol{\theta}}) \\ &= \mathbb{M}_C \mathcal{V}_M + \boldsymbol{H}_{CM} \dot{\boldsymbol{\theta}} \\ &= \begin{bmatrix} M\boldsymbol{E} & \boldsymbol{0} \\ \boldsymbol{0} & \boldsymbol{I}_C \end{bmatrix} \begin{bmatrix} \boldsymbol{v}_C \\ \boldsymbol{\omega}_B \end{bmatrix} + \begin{bmatrix} \boldsymbol{0} \\ \boldsymbol{I}_C \boldsymbol{J}_\omega \end{bmatrix} \dot{\boldsymbol{\theta}} \\ &= \boldsymbol{\mathcal{A}}_C(\boldsymbol{q}) \dot{\boldsymbol{q}}_M. \end{aligned} \tag{4.90}$$

There are two components: a CRB-SM one, $\widetilde{\mathcal{L}}_C(\boldsymbol{q}, \mathcal{V}_M)$, and a joint velocity-dependent one, $\mathcal{L}_{CM}$. The latter is referred to as the *coupling spatial momentum (CSM)* [96].

The SSM can also be expressed with the base quasivelocity, as follows:

$$\begin{aligned}\mathcal{L}_C(\boldsymbol{q}, \dot{\boldsymbol{q}}_B) &= \widetilde{\mathcal{L}}_C(\boldsymbol{q}, \mathcal{V}_B) + \mathcal{L}_{CB}(\boldsymbol{q}, \dot{\boldsymbol{\theta}}) \\ &= \mathbb{M}_{CB} \mathcal{V}_B + \boldsymbol{H}_{CB} \dot{\boldsymbol{\theta}} \\ &= \begin{bmatrix} M\boldsymbol{E} & -M[\boldsymbol{r}^{\times}_{\overleftarrow{CB}}] \\ \boldsymbol{0} & \boldsymbol{I}_C \end{bmatrix} \begin{bmatrix} \boldsymbol{v}_B \\ \boldsymbol{\omega}_B \end{bmatrix} + \begin{bmatrix} M\boldsymbol{J}_{\overleftarrow{CB}} \\ \boldsymbol{I}_C \boldsymbol{J}_\omega \end{bmatrix} \dot{\boldsymbol{\theta}} \\ &= \boldsymbol{\mathcal{A}}_{\overleftarrow{CB}}(\boldsymbol{q}) \dot{\boldsymbol{q}}_B. \end{aligned} \tag{4.91}$$

In this representation, there are also two components: CRB-SM $\widetilde{\mathcal{L}}_C(\boldsymbol{q}, \mathcal{V}_B)$ and CSM $\mathcal{L}_{CB}$. The map from the base quasivelocity to the centroidal SSM, $\boldsymbol{\mathcal{A}}_{\overleftarrow{CB}}$, has been named "centroidal momentum matrix" [106,107].

The last equation can be transformed with (4.77) to represent the SSM w.r.t. the base centroid. One arrives then at

$$\begin{aligned}\mathcal{L}_B(\boldsymbol{q}, \dot{\boldsymbol{q}}_B) &= \widetilde{\mathcal{L}}_B(\boldsymbol{q}, \mathcal{V}_B) + \mathcal{L}_{BB}(\boldsymbol{q}, \dot{\boldsymbol{\theta}}) \\ &= \mathbb{M}_B \mathcal{V}_B + \boldsymbol{H}_{BB} \dot{\boldsymbol{\theta}} \\ &= \begin{bmatrix} M\boldsymbol{E} & -M[\boldsymbol{r}^{\times}_{\overleftarrow{CB}}] \\ -M[\boldsymbol{r}^{\times}_{\overleftarrow{CB}}]^T & \boldsymbol{I}_B \end{bmatrix} \begin{bmatrix} \boldsymbol{v}_B \\ \boldsymbol{\omega}_B \end{bmatrix} + \mathbb{T}^T_{\overleftarrow{BC}} \begin{bmatrix} M\boldsymbol{J}_{\overleftarrow{CB}} \\ \boldsymbol{I}_C \boldsymbol{J}_\omega \end{bmatrix} \dot{\boldsymbol{\theta}} \\ &= \boldsymbol{\mathcal{A}}_B \dot{\boldsymbol{q}}_B. \end{aligned} \tag{4.92}$$

Hereby, the parallel-axis inertia transform (4.80) was used. The two components of the SSM in (4.92) are CRB-SM $\widetilde{\mathcal{L}}_B(\boldsymbol{q}, \mathcal{V}_B)$ and CSM $\mathcal{L}_{BB}$.

The following remarks are due. First, note that the locked inertia tensors $\mathbb{M}_C$ and $\mathbb{M}_B$, appearing in (4.89)/(4.90) and (4.92), respectively, are symmetric and positive definite. Thus, they comprise the property of a rigid-body inertia tensor. On the other hand, note that $\mathbb{M}_{CB}$ in (4.91) is p.d. (since $[\boldsymbol{r}^{\times}_{\overleftarrow{CB}}]\boldsymbol{\omega}_B = -\boldsymbol{v}_B$) but not symmetric. Furthermore, the representation in terms of the mixed quasivelocity, (4.90), provides for inertial decoupling between the linear and angular CRB momentum components and the independence of the former from the joint rates. This representation yields important advantages from the viewpoint of humanoid robot balance control design (cf. Chapter 5):

1. the block-diagonal form of the spatial (locked) inertia tensor is quite desirable for control decoupling,
2. the CoM velocity plays an essential role in balance control, and
3. the angular velocity of the base relates directly to orientation control of the base link (and, hence, of the upper body).

4.6.5 Instantaneous Motion of a Free-Floating Manipulator

From the definitions of the centroidal and mixed quasivelocities and the system angular velocity in (4.87), the following relation can be established:

$$\dot{\boldsymbol{q}}_C = \boldsymbol{T}_{\overleftarrow{CM}}(\boldsymbol{q})\dot{\boldsymbol{q}}_M. \tag{4.93}$$

The matrix

$$\boldsymbol{T}_{\overleftarrow{CM}}(\boldsymbol{q}) = \begin{bmatrix} \boldsymbol{E} & \boldsymbol{0} & \boldsymbol{0} \\ \boldsymbol{0} & \boldsymbol{E} & \boldsymbol{J}_\omega(\boldsymbol{q}) \\ \boldsymbol{0} & \boldsymbol{0} & \boldsymbol{E} \end{bmatrix} \tag{4.94}$$

plays the role of a coordinate transform for the quasivelocities. Note that the common rule for coordinate transforms, $\boldsymbol{T}^{-1}_{\overleftarrow{CM}} = \boldsymbol{T}_{\overleftarrow{MC}}$, applies. From (4.93), the following relation for the spatial velocities can be obtained:

$$\mathcal{V}_C = \mathcal{V}_M + \begin{bmatrix} \boldsymbol{0} \\ \boldsymbol{J}_\omega \end{bmatrix} \dot{\boldsymbol{\theta}}. \tag{4.95}$$

The following is an important observation: given the system spatial velocity $\mathcal{V}_C$,[7] *the joint rates will determine the angular velocity of the base link in a unique way*. This is the formulation of the instantaneous-motion forward kinematic problem w.r.t. the floating base of a free-floating system. On the other hand, the determination of the joint rates that ensure the desired rotational state of the floating base can be recognized as the respective inverse problem. The solution to the inverse problem will be presented in Section 4.7. Apparently, the forward and inverse kinematic problems of the free-floating system are expressed in the same form as the instantaneous-motion kinematic problems discussed in Chapter 2.

Furthermore, from the SSM expressions (4.89) and (4.90) one obtains

$$\mathcal{V}_C = \mathbb{M}_C^{-1}\mathcal{L}_C(\boldsymbol{q}, \dot{\boldsymbol{q}}_M) = \mathcal{V}_M + \mathbb{M}_C^{-1}\boldsymbol{H}_{CM}\dot{\boldsymbol{\theta}}. \tag{4.96}$$

Comparing this result with (4.95), we arrive at the identity

$$\mathbb{M}_C^{-1}\boldsymbol{H}_{CM} \equiv \begin{bmatrix} \boldsymbol{0}^T & \boldsymbol{J}_\omega^T \end{bmatrix}^T, \tag{4.97}$$

which holds for any $\mathcal{V}_C$, $\mathcal{V}_M$ and $\dot{\boldsymbol{\theta}}$.

[7] Note that in zero gravity and absence of other external forces, the system spatial velocity $\mathcal{V}_C$ is constant (as is $\boldsymbol{\omega}_C$).

Next, consider the relation between the centroidal and base quasivelocities. Combine (4.84) with (4.87) to obtain

$$\begin{aligned}\mathcal{V}_C &= \mathbb{T}_{\overleftarrow{CB}} \mathcal{V}_B + \boldsymbol{J}_\theta(\boldsymbol{\theta})\dot{\boldsymbol{\theta}} && (4.98)\\ &= \boldsymbol{J}_q(\boldsymbol{q})\dot{\boldsymbol{q}}_B, && (4.99)\end{aligned}$$

where

$$\boldsymbol{J}_q(\boldsymbol{q}) \equiv \begin{bmatrix} \mathbb{T}_{\overleftarrow{CB}} & \boldsymbol{J}_\theta(\boldsymbol{\theta}) \end{bmatrix} \in \Re^{6\times(n+6)}$$

and

$$\boldsymbol{J}_\theta(\boldsymbol{\theta}) \equiv \begin{bmatrix} \boldsymbol{J}^T_{\overleftarrow{CB}}(\boldsymbol{\theta}) & \boldsymbol{J}^T_\omega(\boldsymbol{\theta}) \end{bmatrix}^T \in \Re^{6\times n}.$$

Spatial velocity relation (4.98) can be rewritten as

$$\mathcal{V}_B + \mathbb{T}_{\overleftarrow{BC}} \boldsymbol{J}_\theta \dot{\boldsymbol{\theta}} = \mathbb{T}_{\overleftarrow{BC}} \mathcal{V}_C. \tag{4.100}$$

Then, the following relation between quasivelocities $\dot{\boldsymbol{q}}_B$ and $\dot{\boldsymbol{q}}_C$ can be obtained:

$$\dot{\boldsymbol{q}}_C = \boldsymbol{T}_{\overleftarrow{CB}}(\boldsymbol{q})\dot{\boldsymbol{q}}_B. \tag{4.101}$$

The coordinate transform is

$$\boldsymbol{T}_{\overleftarrow{CB}} = \begin{bmatrix} \boldsymbol{E} & -[\boldsymbol{r}^\times_{\overleftarrow{CB}}] & \boldsymbol{J}_{\overleftarrow{CB}}(\boldsymbol{\theta}) \\ \boldsymbol{0} & \boldsymbol{E} & \boldsymbol{J}_\omega(\boldsymbol{q}) \\ \boldsymbol{0} & \boldsymbol{0} & \boldsymbol{E} \end{bmatrix}. \tag{4.102}$$

On the other hand, from (4.91) we have

$$\mathcal{V}_B + \mathbb{M}^{-1}_{CB} \boldsymbol{H}_{CB} \dot{\boldsymbol{\theta}} = \mathbb{M}^{-1}_{CB} \mathcal{L}_C(\boldsymbol{q}, \dot{\boldsymbol{q}}_B). \tag{4.103}$$

Comparing the last equation with (4.100), the following identity can be established:

$$\mathbb{M}^{-1}_{CB} \boldsymbol{H}_{CB} \equiv \mathbb{T}_{\overleftarrow{BC}} \boldsymbol{J}_\theta, \tag{4.104}$$

and thus

$$\mathbb{M}^{-1}_{CB} \mathcal{L}_C(\boldsymbol{q}, \dot{\boldsymbol{q}}_B) \equiv \mathbb{T}_{\overleftarrow{BC}} \mathcal{V}_C. \tag{4.105}$$

These identities hold for any $\mathcal{V}_C$, $\mathcal{V}_B$ and $\dot{\boldsymbol{\theta}}$.

Finally, a rule for transforming quasivelocity $\dot{\boldsymbol{q}}_B$ to quasivelocity $\dot{\boldsymbol{q}}_M$ will be obtained. To this end, make use of (2.121):

$$\dot{\boldsymbol{q}}_M = \boldsymbol{T}_{\overleftarrow{MB}}(\boldsymbol{q})\dot{\boldsymbol{q}}_B. \tag{4.106}$$

The coordinate transform is written as (see also [47])

$$T_{\overleftarrow{MB}}(q) = \begin{bmatrix} E & -[r^{\times}_{\overleftarrow{CB}}] & J_{\overleftarrow{CB}}(\theta) \\ 0 & E & 0 \\ 0 & 0 & E \end{bmatrix}. \tag{4.107}$$

The following relation for the spatial velocities can then be obtained:

$$\mathcal{V}_B + \mathbb{T}_{\overleftarrow{BM}} \begin{bmatrix} J_{\overleftarrow{CB}} \\ 0 \end{bmatrix} \dot{\theta} = \mathbb{T}_{\overleftarrow{BM}} \mathcal{V}_M. \tag{4.108}$$

4.7 MOMENTUM-BASED REDUNDANCY RESOLUTION

4.7.1 The Momentum Equilibrium Principle

It is assumed that a completely unconstrained space robot comprising a kinematically redundant arm ($n > 6$) is floating freely in zero gravity. The SSM can be represented as a sum, i.e.

$$\text{SSM} = \text{CRB-SM} + \text{CSM}.$$

This relation is valid for all SSM expressions in the form of a sum, i.e. $\mathcal{L}_C(q, \dot{q}_M)$, $\mathcal{L}_C(q, \dot{q}_B)$, and $\mathcal{L}_B(q, \dot{q}_B)$. The above relation, henceforth referred to as the *momentum equilibrium principle*, plays a fundamental role in momentum-based redundancy resolution. It is worth noting that the momentum equilibrium principle results from the dynamic equilibrium of the wrenches in the spatial dynamics component of the equation of motion, that is observed when no external forces act on the floating-base system (cf. Section 4.8).

Furthermore, note that the instantaneous motion in the joints, expressed by the joint velocity vector $\dot{\theta}$, is explicitly present in the CSM term of the above relation. The joint velocity can be used as a control input in motion and balance control tasks. It should be noted that although any instantaneous motion in the joints induces a variation in the CSM, it does alter the SSM. Indeed, in the case of a free-floating space robot, the SSM is conserved during the unconstrained manipulator motion since the environmental forces (i.e. the solar pressure, the air drag when operating in low orbit, etc.) acting on the system can be ignored; they only matter in a relatively long term. It should be also noted that, since the system is underdetermined, there is a set of joint velocities that do not alter the CRB-SM at all. This type of instantaneous motion plays an important role, as will be shown in what follows.

Furthermore, the angular momentum component of spatial momentum deserves special attention. This component plays a more important role than the linear one, as revealed in a number of studies in the field of free-floating space robotics [97,98,94,93].[8] In the field of humanoid robotics, the CRB rotational motion plays likewise an important role (e.g. in balance control). When redundancy resolution is based on angular momentum only, the dimension of the system can be decreased. This approach helps to avoid overconstrained states and to

[8] The floating-base mounted communication antenna should always be pointing with high accuracy in the direction of the remote control center.

decrease the computational cost. It should also be noted that the special case of *angular momentum conservation* is quite essential in relation to such humanoid robot tasks as jumping or running. The representation of the spatial momentum in terms of mixed quasivelocity is quite useful in angular momentum–based analysis, motion generation, and control.

4.7.2 Spatial Momentum–Based Redundancy Resolution

Without loss of generality, the following derivations will be based on the base link–centered SSM representation in terms of the base quasivelocity, $\mathcal{L}_B(\boldsymbol{q}, \dot{\boldsymbol{q}}_B)$. The momentum equation (4.92) can be solved for the manipulator joint rates that can then be used as input variables in velocity–based motion control schemes. Since the equation is linear in the velocities, its solution type depends on the number of manipulator joints n. The system is assumed underdetermined and thus, there is an infinite number of solutions for the joint velocities. The general solution can be represented as a sum of two orthogonal components, very much like in the case of a kinematically redundant limb (cf. Section 2.7, (2.34)), i.e.

$$\dot{\boldsymbol{\theta}} = \boldsymbol{H}_{BB}^{+}(\mathcal{L}_B - \mathbb{M}_B \mathcal{V}_B) + \boldsymbol{N}(\boldsymbol{H}_{BB})\dot{\boldsymbol{\theta}}_a. \tag{4.109}$$

Here $\boldsymbol{H}_{BB}^{+}$ and $\boldsymbol{N}(\boldsymbol{H}_{BB})$ denote the (right) pseudoinverse of the coupling inertia matrix and a projector onto its null space, respectively.

Coupling Spatial Momentum Conservation: the Reaction Null Space

The CSM is conserved at zero ($\mathcal{L}_{BB} \equiv \boldsymbol{H}_{BB}(\boldsymbol{q})\dot{\boldsymbol{\theta}} = \boldsymbol{0}$) when the SSM equals the CRB-SM, i.e. $\mathcal{L}_B = \widetilde{\mathcal{L}}_B = \mathbb{M}_B \mathcal{V}_B$. Vector $\dot{\boldsymbol{\theta}}_a$ in (4.109) is a joint velocity vector that parametrizes the following infinite set of joint velocities:

$$\{\dot{\boldsymbol{\theta}}_{csm}(\boldsymbol{q}, \dot{\boldsymbol{\theta}}_a) \in \mathcal{N}(\boldsymbol{H}_{BB}(\boldsymbol{q})) : \dot{\boldsymbol{\theta}} = \boldsymbol{N}(\boldsymbol{H}_{BB}(\boldsymbol{q})\dot{\boldsymbol{\theta}}_a, \forall \dot{\boldsymbol{\theta}}_a\}. \tag{4.110}$$

Joint velocities derived from the set $\{\dot{\boldsymbol{\theta}}_{csm}(\boldsymbol{q}, \dot{\boldsymbol{\theta}}_a)\}$ are solutions to the homogeneous equation $\boldsymbol{H}_{BB}(\boldsymbol{q})\dot{\boldsymbol{\theta}} = \boldsymbol{0}$. Thus, they can be characterized as *CSM conserving*.

The above set is orthogonal to the particular solution component (the pseudoinverse term in (4.109)). This implies that whatever the CSM conserving joint velocity, it will not affect the state of the CRB. Notice that the set is defined in a similar way as the set of self-motion velocities (2.29) of a fixed-base kinematically redundant manipulator; $\mathcal{N}(\boldsymbol{H}_{BB}(\boldsymbol{q}))$ determines a subspace at $\boldsymbol{q}$ that is tangent to a manifold in the joint space (similar to the self-motion manifold mentioned in Section 2.7.1). Locally, the dimension of this manifold, henceforth referred to as the *CSM conserving manifold*, equals the rank of the null-space projector. At a nonsingular manipulator configuration s.t. the coupling inertia matrix is full row-rank, $\text{rank}\boldsymbol{N}(\boldsymbol{H}_{BB}) = n - 6$. With a 7-DOF manipulator, for example, the manifold will be just 1D. In this case, the set of CSM conserving joint velocities can be represented in the form of an autonomous dynamical system, i.e.

$$\dot{\boldsymbol{\theta}} = b\boldsymbol{n}_{BB}(\boldsymbol{q}). \tag{4.111}$$

Scalar b is an arbitrary parameter having the dimension of joint rate. The null space vector $\boldsymbol{n}_{BB}(\boldsymbol{q}) \in \mathcal{N}(\boldsymbol{H}_{BB}(\boldsymbol{q}))$ determines the *CSM conserving vector field* in the joint space. Its integral curves constitute the CSM conserving manifold. The projection of an integral curve onto the workspace of the manipulator (via forward kinematics) will be referred to as a *CSM conserving path*.

From the above discussion it becomes apparent that the role of the coupling inertia matrix of a free-floating system is similar to that of the Jacobian of a fixed-base manipulator, in the sense that $\boldsymbol{H}_{BB}$ induces a local, orthogonal decomposition formalism on the CSM conserving manifold. Thereby, the null space of the coupling inertia matrix, $\mathcal{N}(\boldsymbol{H}_{BB})$, represents the tangent subspace at $\boldsymbol{q}$. The respective map is defined via the null-space projector $\boldsymbol{N}(\boldsymbol{H}_{BB}) = \boldsymbol{E} - \boldsymbol{H}_{BB}^{+}\boldsymbol{H}_{BB}$. The orthogonal complement of this map, $\boldsymbol{H}_{BB}^{+}\boldsymbol{H}_{BB}$, determines the normal component at the manifold. It will be shown below (cf. Section 4.8.1) that when an instantaneous motion originates from within the CSM conserving set, no force/moment will be impressed upon the base link. Thus, the term "*reactionless motion*" can be used as an alternative to the "CSM conserving motion." The set (4.110) (i.e. the null space of the coupling inertia matrix) is referred to as the (spatial momentum) *Reaction Null Space* (RNS) [96,95]. It follows then that the RNS is the tangent subspace to the CSM conserving manifold at $\boldsymbol{q}$.

The decomposition formalism outlined above is the essence of the *RNS method*. This method plays an important role in motion analysis, generation, and control of floating-base systems, including humanoid robots [100]. Readers interested in the theoretical aspect of the decomposition formalism for generic underactuated systems are referred to [86,109].

System Spatial-Momentum Conservation

According to the spatial velocity relation (4.99), there is an infinite number of base-link and coupling spatial velocity combinations for a given (constant) system spatial velocity, i.e.

$$\dot{\boldsymbol{q}}_B(\dot{\boldsymbol{q}}_a) = \boldsymbol{J}_q^{+}(\boldsymbol{q})\mathcal{V}_C + \boldsymbol{N}(\boldsymbol{J}_q)\dot{\boldsymbol{q}}_a. \tag{4.112}$$

Here $\dot{\boldsymbol{q}}_a$ parametrizes the null space $\mathcal{N}(\boldsymbol{J}_q)$ in a nonminimal way. This null space comprises instantaneous motion components that preserve the system spatial velocity and thus the spatial momentum of the system, i.e.

$$\{\dot{\boldsymbol{q}}_{sm} \in \mathcal{N}(\boldsymbol{J}_q) : \dot{\boldsymbol{q}}_B = \boldsymbol{N}(\boldsymbol{J}_q)\dot{\boldsymbol{q}}_a,\ \forall \dot{\boldsymbol{q}}_a\};$$

$\mathcal{N}(\boldsymbol{J}_q)$ will be henceforth referred to as the *SSM conserving null space*.

When the SSM is conserved, as usually assumed in the field of free-floating space robotics, the system spatial velocity will be constant: $\mathcal{V}_C = \mathcal{V}_{\text{const}}$. In the general case of instantaneous joint motion that does not originate from within the (spatial momentum) RNS, the base twist will be altered via the CSM component, as apparent from (4.103), i.e.

$$\mathcal{V}_B = \mathcal{V}_{\text{const}} - \mathbb{M}_B^{-1}\boldsymbol{H}_{BB}\dot{\boldsymbol{\theta}}. \tag{4.113}$$

In the special case of reactionless motion, $\boldsymbol{H}_{BB}\dot{\boldsymbol{\theta}} = \boldsymbol{0},\ \forall\dot{\boldsymbol{\theta}} \Rightarrow \mathcal{V}_B = \mathcal{V}_{\text{const}}$.

4.7.3 Angular Momentum–Based Redundancy Resolution

The redundancy resolution approach outlined above can be recast in terms of angular momentum only, thus decreasing the dimension of the linear system from six to three. The representation of the centroidal SSM in terms of the mixed quasivelocity is quite suitable for this purpose, as already clarified. The SAM relation, given in (4.85), is an underdetermined linear system of three equations in n unknowns. As already noted, this relation can be rewritten in terms of angular velocity (cf. (4.87)). The general solution of (4.87) can be written as

$$\dot{\boldsymbol{\theta}} = \boldsymbol{J}_{\omega}^{+}(\boldsymbol{\theta})\Delta\boldsymbol{\omega} + \boldsymbol{N}(\boldsymbol{J}_{\omega}(\boldsymbol{\theta}))\dot{\boldsymbol{\theta}}_{a}. \tag{4.114}$$

The particular solution component (the pseudoinverse term) can be used to control the *relative angular velocity* $\Delta\boldsymbol{\omega} = \boldsymbol{\omega}_C - \boldsymbol{\omega}_B$, i.e. the system angular velocity, $\boldsymbol{\omega}_C$, w.r.t. a given base angular velocity $\boldsymbol{\omega}_B$. The homogeneous solution component, on the other hand, is useful to reconfigure the limbs in a desirable way, by employing $\dot{\boldsymbol{\theta}}_a$ as a control input. Thereby, the relative angular velocity will be unaffected. Since the two components are orthogonal, they provide the base for decoupled control of these two control objectives.

Coupling Angular Momentum Conservation

The homogeneous component in (4.114) is obtained by setting the relative angular velocity at zero, i.e. $\boldsymbol{\omega}_C = \boldsymbol{\omega}_B$. This condition implies that the coupling angular momentum (CAM) will be conserved, e.g. at zero, so $\boldsymbol{H}_C\dot{\boldsymbol{\theta}} = \boldsymbol{0}$. The set of joint velocities satisfying this condition is obtained as

$$\{\dot{\boldsymbol{\theta}}_{cam}(\boldsymbol{\theta}, \dot{\boldsymbol{\theta}}_a) \in \mathcal{N}(\boldsymbol{J}_{\omega}(\boldsymbol{\theta})) : \dot{\boldsymbol{\theta}} = \boldsymbol{N}(\boldsymbol{J}_{\omega}(\boldsymbol{\theta}))\dot{\boldsymbol{\theta}}_a, \forall\dot{\boldsymbol{\theta}}_a\}.$$

Any joint velocity from within the set $\{\dot{\boldsymbol{\theta}}_{cam}\}$ will be referred to as a *CAM conserving* joint velocity; $\{\dot{\boldsymbol{\theta}}_{cam}(\boldsymbol{\theta}, \dot{\boldsymbol{\theta}}_a)\}$ determines the tangent subspace of a manifold (the CAM conserving manifold) at $\boldsymbol{\theta}$. Since $\mathcal{N}(\boldsymbol{J}_{\omega}) = \mathcal{N}(\boldsymbol{H}_C)$ and $\text{rank}\boldsymbol{N}(\boldsymbol{H}_C) = n - 3 > \text{rank}\boldsymbol{N}(\boldsymbol{H}_{BB}) = n - 6$, it can be concluded that the tangent bundle of the CAM conserving manifold is a *superset* of that of the CSM conserving manifold. In analogy to the spatial momentum RNS, the tangent subspace of the CAM conserving manifold at $\boldsymbol{\theta}$ can be referred to as the *angular momentum RNS*. Noting that angular momentum plays an important role in motion generation and control of free-floating space robots and humanoid robots alike (e.g. in satellite-base orientation and balance control, respectively), the angular momentum RNS will be henceforth simply referred to as the RNS.

System Angular Momentum Conservation

According to the angular velocity relation (4.88), there is an infinite number of base-link and coupling angular velocity combinations for a given system angular velocity, i.e.

$$\dot{\boldsymbol{q}}_{\omega} = \hat{\boldsymbol{J}}_{\omega}^{+}(\boldsymbol{\theta})\boldsymbol{\omega}_C + \boldsymbol{N}(\hat{\boldsymbol{J}}_{\omega})\dot{\boldsymbol{q}}_{\omega a}; \tag{4.115}$$

$\dot{\boldsymbol{q}}_{\omega a} = \begin{bmatrix}\boldsymbol{\omega}_{Ba}^T & \dot{\boldsymbol{\theta}}_a^T\end{bmatrix}^T$ parametrizes the null space $\mathcal{N}(\hat{\boldsymbol{J}}_{\omega})$ in a nonminimal way. This null space comprises instantaneous motion components that do not alter the system angular velocity and thus conserve the SAM ($\boldsymbol{l}_C = \boldsymbol{I}_C\boldsymbol{\omega}_C = \text{const.}$), so we have

$$\{\dot{\boldsymbol{q}}_{\bar{\omega}} \in \mathcal{N}(\hat{\boldsymbol{J}}_{\omega}) : \dot{\boldsymbol{q}}_{\omega} = \boldsymbol{N}(\hat{\boldsymbol{J}}_{\omega}(\boldsymbol{\theta}))\dot{\boldsymbol{q}}_{\omega a}, \forall\dot{\boldsymbol{q}}_{\omega a}\}. \tag{4.116}$$

The null space $\mathcal{N}(\hat{\boldsymbol{J}}_{\omega})$ will be henceforth referred to as the *SAM conserving null space*. Base-link/joint velocity combinations from within this null space are used when a humanoid robot is airborne.

Example: a Dual-Task Scenario With a Free-Floating Space Manipulator

It is straightforward to apply the RNS method to a dual-task scenario wherein the instantaneous motion of each end link (base and end effector) of the free-floating manipulator is to be controlled [94]. Denote by $\boldsymbol{J}_E(\boldsymbol{q}) \in \Re^{6\times n}$ the *fixed-base* manipulator end-effector Jacobian expressed w.r.t. the inertial frame. The spatial velocity of the end effector results from the instantaneous motion of the base and the manipulator joints; hence

$$\mathcal{V}_E = \mathbb{T}_{\overleftarrow{EB}}\mathcal{V}_B + \boldsymbol{J}_E\dot{\boldsymbol{\theta}}. \tag{4.117}$$

Furthermore, assume zero initial momentum, s.t. $\mathcal{V}_{\text{const}} = \mathbf{0}$. Then, the momentum conservation (4.113) can be used as a constraint to eliminate $\mathcal{V}_B$ from the last equation. One arrives at

$$\mathcal{V}_E = \hat{\boldsymbol{J}}_E(\boldsymbol{q})\dot{\boldsymbol{\theta}}. \tag{4.118}$$

The matrix $\hat{\boldsymbol{J}}_E = \boldsymbol{J}_E - \mathbb{T}_{\overleftarrow{EB}}\mathbb{M}_B^{-1}\boldsymbol{H}_{BB}$ is called the "generalized Jacobian" [153]. The terminology stems from the fact that the last equation can be used not only for floating-base systems in zero gravity, but also for fixed-base systems. Indeed, when the CRB inertia approaches infinity, e.g. as a result of using a very heavy or fixed base, the $\mathbb{M}_B^{-1}$ term becomes infinitesimally small. Hence, the generalized Jacobian approaches that of the fixed-base manipulator, i.e. $\hat{\boldsymbol{J}}_E \to \boldsymbol{J}_E$. Note also that, using (4.104), the generalized Jacobian can be represented as

$$\hat{\boldsymbol{J}}_E = \boldsymbol{J}_E - \mathbb{T}_{\overleftarrow{EC}}\boldsymbol{J}_\theta. \tag{4.119}$$

When the manipulator is kinematically redundant, the generalized Jacobian comprises a nontrivial null space, so $\mathcal{N}(\hat{\boldsymbol{J}}_E) \neq \emptyset$. The redundancy can be resolved in accordance with the approaches introduced in Chapter 2. For instance, a task prioritization can be introduced. The end-effector task can be solved within the RNS, i.e. as a lower-priority task, or vice versa. Thereby, the self-motion obtained from the null space of the generalized Jacobian yields the so-called "manipulator inversion task," wherein the end effector is fixed in inertial space but the base is not [94,93]. Furthermore, when the degree of redundancy is sufficiently large ($n > 12$), it becomes possible to immobilize both end links s.t. only the intermittent links will be in motion. The motion of the free-floating manipulator resembles then the self-motion of a kinematically redundant fixed-base manipulator. To achieve this, the joint velocity should be derived from within the intersection of the two null spaces, i.e. $\mathcal{N}(\hat{\boldsymbol{J}}_E) \cap \mathcal{N}(\boldsymbol{H}_{BB})$.

To avoid the frequent appearance of an overconstrained state, it would be useful to introduce task priorities with regard to the translational and rotational motion subtasks of the satellite base. From a practical viewpoint, the latter subtask should have a higher priority, as already mentioned. The translational motion constraint can be, in fact, fully relaxed. In other words, the angular momentum RNS should be employed instead of the spatial momentum RNS. This is the so-called "selective RNS" approach [99].

4.7.4 Motion of a Free-Floating Humanoid Robot in Zero Gravity

Motion analysis of a free-floating humanoid robot in zero gravity is helpful for better understanding the role of the spatial momentum components introduced in the previous sections. It is also helpful for designing control methods for humanoid robots [132]. As an example, consider the implementation of the RNS-based redundancy resolution method in the following multitask scenario. The robot is required to track the motion of a translating and rotating box with its hands (first motion task), aiming at a precontact posture with zero relative twists. To ensure sufficient time for the tracking, the rotation of the base link of the robot has to be synchronized with that of the box, otherwise the box could leave the workspace before the robot attains the correct posture for grasping. Base reorientation control is achieved via cyclic (circular) motion in the feet, without a phase difference (this is the second motion task). Note, however, that if the motion of the box is arbitrary, the tracking could result in an overconstrained system and/or self-collision. To avoid this, the motion of the box is constrained within the sagittal plane. It is then sufficient to reorient the base link in the pitch direction only (this is the third motion control task). The result of the simulation is shown in Video 4.7-1 [45].

Note that if a phase difference is introduced in the feet motion, significant shaking in the yaw direction will be observed. The problem can be alleviated with the help of a yaw compensation control. This can be achieved with inertial coupling induced by the arm motion, as shown in Video 4.7-2 [44]. Apparently, to compensate the deviation of the base in the yaw direction, the arms have to move significantly since their inertia is much lower than that of the CRB yaw inertia. Such motion is practically infeasible, though, because of the high probability of self-collision. A possible solution to the problem is shown in Video 4.7-3 [43]. The goal is achieved in two phases. During the first phase, only base pitch control is invoked whereby the hand position/orientation control is relaxed. In this way, the ill-conditioning leading to a significant arm motion can be avoided. After achieving the desired base pitch orientation, the hand control is invoked to attain a suitable grasp posture (this is the second phase). Thereby, the feet move slightly to compensate the base-link deviation induced by the approaching motion of the hands. This example demonstrates how important motion generation is in preventing the frequent occurrence of an overconstrained system.

It is worth noting that the method of floating-base reorientation control of a space-borne manipulator via end-link cyclic motion was introduced in the pioneering work [154]. Note also that this approach is used by an astronaut when reorienting his/her body via limb motion [143,142]. A simulation for body-roll reorientation via cyclic arm motion is shown in Video 4.7-4 [42]. Such reorientation can also be achieved with cyclic motion in the feet, as shown in Video 4.7-5 [103]. In this relation, it is interesting to observe that while walking in zero gravity, the orientation of the body in the pitch direction changes gradually, as shown in Video 4.7-6 [82]. This effect stems from the inertial coupling, as in the rest of the above examples. The following important conclusion can then be made: angular momentum control in the pitch direction is essential during normal gait on flat terrain (see also [161]).

4.8 EQUATION OF MOTION OF A FREE-FLOATING MANIPULATOR IN ZERO GRAVITY

The equation of motion of a free-floating manipulator in zero gravity will be derived in terms of the base, the mixed, and the centroidal quasivelocity.

4.8.1 Representation in Terms of Base Quasivelocity

The kinetic energy of the system can be written as

$$T = \frac{1}{2}\dot{\boldsymbol{q}}_B^T \boldsymbol{M}_B(\boldsymbol{q})\dot{\boldsymbol{q}}_B, \tag{4.120}$$

where

$$\boldsymbol{M}_B(\boldsymbol{q}) = \begin{bmatrix} \mathbb{M}_B & \boldsymbol{H}_{BB} \\ \boldsymbol{H}_{BB}^T & \boldsymbol{M}_{\theta B} \end{bmatrix}, \tag{4.121}$$

$\boldsymbol{M}_{\theta B}(\boldsymbol{\theta})$ denoting the joint-space inertia matrix of the *fixed-base* manipulator (cf. (4.43)). The equation of motion, obtained from the Lagrangian

$$\frac{\partial}{\partial t}\left(\frac{\partial T}{\partial \dot{\boldsymbol{q}}_B}\right) - \frac{\partial T}{\partial \boldsymbol{q}} = \mathcal{Q},$$

can be written as

$$\boldsymbol{M}_B(\boldsymbol{q})\ddot{\boldsymbol{q}}_B + \boldsymbol{c}_B(\boldsymbol{q}, \dot{\boldsymbol{q}}_B) = \mathcal{Q}. \tag{4.122}$$

The nonlinear velocity-dependent force term $\boldsymbol{c}_B(\boldsymbol{q}, \dot{\boldsymbol{q}}_B)$ can be expressed in the form of (4.46), $\mathcal{Q} = \boldsymbol{S}^T\boldsymbol{\tau} = \begin{bmatrix}\boldsymbol{0}^T & \boldsymbol{\tau}^T\end{bmatrix}^T \in \Re^{6+n}$ denotes the *generalized force*, $\boldsymbol{S} = \begin{bmatrix}\boldsymbol{0}_{n\times 6} & \boldsymbol{E}_n\end{bmatrix}$ is the underactuation filtering matrix (cf. (2.102)), and $\boldsymbol{\tau} \in \Re^n$ denotes the joint torque of the manipulator. Note that since the system is underactuated, the generalized force does not depend on the specific type of quasivelocity. Thus, there is no subscript attached to $\mathcal{Q}$.

Furthermore, for the purpose of analysis and control, the following expanded-form representations of the equation of motion are found to be useful:

$$\begin{bmatrix} \mathbb{M}_B & \boldsymbol{H}_{BB} \\ \boldsymbol{H}_{BB}^T & \boldsymbol{M}_{\theta B} \end{bmatrix}\begin{bmatrix} \dot{\mathcal{V}}_B \\ \ddot{\boldsymbol{\theta}} \end{bmatrix} + \begin{bmatrix} \mathcal{C}_B \\ \boldsymbol{c}_{\theta B} \end{bmatrix} = \begin{bmatrix} \boldsymbol{0} \\ \boldsymbol{\tau} \end{bmatrix} \tag{4.123}$$

and

$$\begin{bmatrix} M\boldsymbol{E} & -M[\boldsymbol{r}^{\times}_{\overleftarrow{CB}}] & M\boldsymbol{J}_{\overleftarrow{CB}} \\ -M[\boldsymbol{r}^{\times}_{\overleftarrow{CB}}]^T & \boldsymbol{I}_B & \boldsymbol{H}_B \\ M\boldsymbol{J}^T_{\overleftarrow{CB}} & \boldsymbol{H}_B^T & \boldsymbol{M}_{\theta B} \end{bmatrix}\begin{bmatrix} \dot{\boldsymbol{v}}_B \\ \dot{\boldsymbol{\omega}}_B \\ \ddot{\boldsymbol{\theta}} \end{bmatrix} + \begin{bmatrix} \boldsymbol{c}_{fB} \\ \boldsymbol{c}_{mB} \\ \boldsymbol{c}_{\theta B} \end{bmatrix} = \begin{bmatrix} \boldsymbol{0} \\ \boldsymbol{0} \\ \boldsymbol{\tau} \end{bmatrix}; \tag{4.124}$$

$\mathcal{C}_B = \begin{bmatrix}\boldsymbol{c}_{fB}^T & \boldsymbol{c}_{mB}^T\end{bmatrix}^T$ is a nonlinear velocity-dependent wrench stemming from the spatial dynamics component. We have

$$\begin{aligned} \mathcal{C}_B = {} & \overbrace{\dot{\mathbb{M}}_B\mathcal{V}_B + \dot{\boldsymbol{H}}_{BB}\dot{\boldsymbol{\theta}}} && + \overbrace{\mathcal{C}'_B} \\ = {} & \begin{bmatrix} M\left(-[\dot{\boldsymbol{r}}^{\times}_{\overleftarrow{CB}}]^T\boldsymbol{\omega}_B + \dot{\boldsymbol{J}}_{\overleftarrow{BC}}\dot{\boldsymbol{\theta}}\right) \\ \dot{\boldsymbol{I}}_B\boldsymbol{\omega}_B + \dot{\boldsymbol{H}}_{BB}\dot{\boldsymbol{\theta}} - M[\dot{\boldsymbol{r}}^{\times}_{\overleftarrow{CB}}]^T\boldsymbol{v}_B \end{bmatrix} && + \begin{bmatrix} \boldsymbol{0} \\ M[\dot{\boldsymbol{r}}^{\times}_{\overleftarrow{CB}}]^T\boldsymbol{v}_C \end{bmatrix}. \end{aligned} \tag{4.125}$$

Note that $\mathcal{C}'_B$ is a derivative of the moment of momentum. The $\boldsymbol{c}_{\theta B}$-term, on the other hand, denotes the nonlinear velocity-dependent joint torque of the fixed-base manipulator (cf. (4.46)).

In the Case of Momentum Conservation (no External Forces)

In the above derivation of the equation of motion it was assumed that no external forces act on the system. In this case, spatial momentum is conserved with $\mathcal{V}_C = \text{const}$. In the special case of momentum conservation at zero (i.e. $\boldsymbol{v}_C = \boldsymbol{0}$), $\mathcal{C}'_B$ in (4.125) becomes zero.

Furthermore, from analytical mechanics it is known that conserved quantities in the equation of motion are related to *ignorable or cyclic coordinates* [65]. This implies that the dynamics of the system can be represented in *reduced* form. Indeed, by inserting velocity relation (4.113) into the expression for the kinetic energy, (4.120), one obtains

$$T = \frac{1}{2}\left(\dot{\boldsymbol{\theta}}^T \hat{\boldsymbol{M}}_{\theta B}\dot{\boldsymbol{\theta}} + \mathcal{V}_{\text{const}}^T \mathbb{M}_B \mathcal{V}_{\text{const}}\right). \tag{4.126}$$

The matrix

$$\hat{\boldsymbol{M}}_{\theta B} = \boldsymbol{M}_{\theta B} - \boldsymbol{H}_{BB}^T \mathbb{M}_B^{-1} \boldsymbol{H}_{BB} \tag{4.127}$$

is referred to as the *generalized inertia tensor* [163]. In this case, the equation of motion obtained from the Lagrangian

$$\frac{\partial}{\partial t}\left(\frac{\partial T}{\partial \dot{\boldsymbol{\theta}}}\right) - \frac{\partial T}{\partial \boldsymbol{\theta}} = \boldsymbol{\tau}$$

assumes the form

$$\hat{\boldsymbol{M}}_{\theta B}\ddot{\boldsymbol{\theta}} + \hat{\boldsymbol{c}}_{\theta B} = \boldsymbol{\tau}. \tag{4.128}$$

The term

$$\hat{\boldsymbol{c}}_{\theta B} = \boldsymbol{c}_{\theta B} - \boldsymbol{H}_{BB}^T \mathbb{M}_B^{-1} \mathcal{C}_B$$

denotes the nonlinear velocity-dependent joint torque in the reduced-form representation. The above equation stands for the system dynamics mapped onto joint space. The dimension of the system has thus been reduced from $n+6$ to n. This form of the equation of motion of the floating-base system is similar to that of a fixed-base manipulator. An important observation is that *the joint torque does not alter the spatial momentum of the system whatsoever.*

The reduced-form representation (4.128) exhibits explicitly how the fixed-base manipulator dynamics are generalized to that of a free-floating base system; with a heavy base s.t. $\mathbb{M}_B \to \infty$, the terms containing $\mathbb{M}_B^{-1}$ become infinitesimally small and, hence, $\hat{\boldsymbol{M}}_{\theta B} \to \boldsymbol{M}_{\theta B}$, $\hat{\boldsymbol{c}}_{\theta B} \to \boldsymbol{c}_{\theta B}$. This was also the case with the generalized Jacobian appearing in (4.118). Readers interested in a rigorous theoretical foundation with regard to the reduced-form dynamics in relation to conserved quantities are referred to [86,109].

In the Presence of External Forces

As clarified in Chapter 3, the sum of all external wrenches can be mapped at a specific link of interest. For clarity and without loss of generality, in what follows it will be assumed that the net external wrench acts at Link $n \equiv E$, i.e. at the end effector. Then, the equation of motion (4.123) can be written as

$$\begin{bmatrix} \mathbb{M}_B & \boldsymbol{H}_{BB} \\ \boldsymbol{H}_{BB}^T & \boldsymbol{M}_{\theta B} \end{bmatrix} \begin{bmatrix} \dot{\mathcal{V}}_B \\ \ddot{\boldsymbol{\theta}} \end{bmatrix} + \begin{bmatrix} \mathcal{C}_B \\ \boldsymbol{c}_{\theta B} \end{bmatrix} = \begin{bmatrix} \boldsymbol{0} \\ \boldsymbol{\tau} \end{bmatrix} + \begin{bmatrix} \mathbb{T}_{\overleftarrow{BE}}^T \\ \boldsymbol{J}_E^T \end{bmatrix} \mathcal{F}_E. \tag{4.129}$$

From the upper part of the above equation and the expression for SSM $\mathcal{L}_B$ in (4.92), the following relation is derived:

$$\frac{d}{dt}\mathcal{L}_B(\boldsymbol{q}, \dot{\boldsymbol{q}}_B) = \mathbb{M}_B \dot{\mathcal{V}}_B + \dot{\mathbb{M}}_B \mathcal{V}_B + \mathcal{C}'_B + \boldsymbol{H}_{BB}\ddot{\boldsymbol{\theta}} + \dot{\boldsymbol{H}}_{BB}\dot{\boldsymbol{\theta}} = \mathcal{F}_B, \tag{4.130}$$

where $\mathcal{C}'_B$ is defined in (4.125) and $\mathcal{F}_B = \mathbb{T}_{\overleftarrow{BE}}^T \mathcal{F}_E$. This equation reveals the following important points:

- the spatial dynamics are derived as the time differential of the SSM;
- the rate of change of the SSM depends on the external wrench only.

The manipulator joint motion/torque does not alter the rate of change of the SSM whatsoever; joint motion induces a zero dynamic force component, i.e.

$$\frac{d}{dt}\mathcal{L}_{BB} = \boldsymbol{H}_{BB}\ddot{\boldsymbol{\theta}} + \dot{\boldsymbol{H}}_{BB}\dot{\boldsymbol{\theta}} = \boldsymbol{0}. \tag{4.131}$$

Thus, when subjected to an *nonimpulsive* external force,[9] the free-floating manipulator will behave as a rigid body, as if the joints were locked, so we have

$$\frac{d}{dt}\widetilde{\mathcal{L}}_B = \mathbb{M}_B \dot{\mathcal{V}}_B + \dot{\mathbb{M}}_B \mathcal{V}_B + \mathcal{C}'_B = \mathcal{F}_B. \tag{4.132}$$

This relation also implies that $\mathcal{F}_B = \mathcal{F}_C$ holds, $\mathcal{F}_C = \begin{bmatrix} \boldsymbol{f}_C^T & \boldsymbol{m}_C^T \end{bmatrix}^T = \mathbb{T}_{\overleftarrow{CE}}^T \mathcal{F}_E$ denoting the external wrench mapped at the system CoM. It can then be reconfirmed that instantaneous motion derived from the tangent bundle of the CSM conserving manifold (cf. (4.110)) does not disturb the dynamic force balance at the CRB. This is the reason why the tangent bundle was named the (spatial momentum) RNS in Section 4.7 (cf. also [96,95,100]).

Finally, note that in the presence of an external force, the reduced form of the equation of motion, (4.128), becomes

$$\hat{\boldsymbol{M}}_{\theta B}\ddot{\boldsymbol{\theta}} + \hat{\boldsymbol{c}}_{\theta B} = \boldsymbol{\tau} + \hat{\boldsymbol{J}}_E^T \mathcal{F}_E, \tag{4.133}$$

where $\hat{\boldsymbol{J}}_E$ stands for the generalized Jacobian (cf. (4.119)). Apparently, the equation of motion of a floating-base underactuated system can be expressed as that of a fixed-base manipulator (cf. (4.58)).

[9] The case of an impulsive external force will be discussed in Section 7.8.

4.8.2 Representation in Terms of Mixed Quasivelocity

The representation of spatial momentum in terms of the mixed quasivelocity is expected to yield simpler expressions in the equation of motion because of the decoupling of the linear CoM motion component. The equation of motion, expressed in terms of the base quasivelocity (4.122), can be transformed with the help of the quasivelocity relation (4.106) and the transform (4.107) as

$$M_M(q)\ddot{q}_M + c_M(q,\dot{q}_M) = \mathcal{Q}, \tag{4.134}$$

where

$$M_M(q) \equiv T^T_{\overleftarrow{MB}} M_B T_{\overleftarrow{BM}}, \tag{4.135}$$

$$c_M(q,\dot{q}_M) \equiv T^T_{\overleftarrow{MB}} \left(c_B(q,\dot{q}_B) + M_B \dot{T}_{\overleftarrow{BM}} \dot{q}_B \right). \tag{4.136}$$

The expanded-form representations of (4.134) are

$$\begin{bmatrix} \mathbb{M}_C & H_{CM} \\ H^T_{CM} & M_{\theta M} \end{bmatrix} \begin{bmatrix} \dot{\mathcal{V}}_M \\ \ddot{\theta} \end{bmatrix} + \begin{bmatrix} \mathcal{C}_M \\ c_{\theta M} \end{bmatrix} = \begin{bmatrix} 0 \\ \tau \end{bmatrix} \tag{4.137}$$

and

$$\begin{bmatrix} ME & 0 & 0 \\ 0 & I_C & H_C \\ 0 & H^T_C & M_{\theta M} \end{bmatrix} \begin{bmatrix} \dot{v}_C \\ \dot{\omega}_B \\ \ddot{\theta} \end{bmatrix} + \begin{bmatrix} 0 \\ c_{mM} \\ c_{\theta M} \end{bmatrix} = \begin{bmatrix} 0 \\ 0 \\ \tau \end{bmatrix}, \tag{4.138}$$

where

$$M_{\theta M}(q) \equiv M_{\theta B}(q) - M J^T_{\overleftarrow{CB}}(\theta) J_{\overleftarrow{CB}}(\theta). \tag{4.139}$$

Apparently, $\mathcal{C}_M = \begin{bmatrix} 0^T & c^T_{mM} \end{bmatrix}^T$, with the nonlinear velocity-dependent moment

$$c_{mM}(q,\dot{q}_B) = \dot{I}_C \omega_B + \dot{H}_C \dot{\theta}, \tag{4.140}$$

whereby

$$\dot{I}_C(q) = \sum_{i=1}^{n} \left([\omega_i^\times] I_i - I_i [\omega_i^\times] - M_i \left([\dot{r}^\times_{\overleftarrow{Ci}}][r^\times_{\overleftarrow{iC}}] + [r^\times_{\overleftarrow{Ci}}][\dot{r}^\times_{\overleftarrow{iC}}] \right) \right), \tag{4.141}$$

$$\dot{H}_C(q) = \sum_{i=1}^{n} \left(\left([\omega_i^\times] I_i - I_i [\omega_i^\times] \right) J_{\omega_i} + I_i \dot{J}_{\omega_i} + M_i \left([\dot{r}^\times_{\overleftarrow{Ci}}] J_{vi} + [r^\times_{\overleftarrow{Ci}}] \dot{J}_{vi} \right) \right), \tag{4.142}$$

and

$$\dot{J}_{vi} = \begin{bmatrix} [\dot{e}_1^\times] r_{\overleftarrow{i1}} + [e_1^\times] \dot{r}_{\overleftarrow{i1}} & [\dot{e}_2^\times] r_{\overleftarrow{i2}} + [e_2^\times] \dot{r}_{\overleftarrow{i2}} & \cdots & [\dot{e}_j^\times] r_{\overleftarrow{ij}} + [e_j^\times] \dot{r}_{\overleftarrow{ij}} & 0 & \cdots & 0 \end{bmatrix} \in \Re^{3\times n},$$

$$\dot{J}_{\omega i} = \begin{bmatrix} \dot{e}_1 & \dot{e}_2 & \cdots & \dot{e}_i & 0 & \cdots & 0 \end{bmatrix} \in \Re^{3\times n},$$
$$\dot{e}_i = [\omega_i^{\times}] R_i{}^i e_i.$$

Here $e_i = R_i{}^i e_i$, ${}^i e_i = [0\ 0\ 1]^T$, and $r_{\overleftarrow{ij}}$ is the distance vector from the jth joint axis to the ith link CoM ($1 \le j \le i$).

With this notation, the nonlinear velocity-dependent wrench in the spatial dynamics component can be represented in the following form:

$$\mathcal{C}_M = \dot{\mathbb{M}}_C \mathcal{V}_M + \dot{H}_{CM}\dot{\theta}. \tag{4.143}$$

Note also that the nonlinear velocity-dependent wrench (4.125) and the joint torque can be written as

$$\mathcal{C}_B = \begin{bmatrix} a \\ c_{mM} + [r^{\times}_{\overleftarrow{CB}}]^T a \end{bmatrix} \tag{4.144}$$

and

$$c_{\theta M}(q, \dot{q}_M) = c_{\theta B}(\theta, \dot{\theta}) - \dot{J}^T_{\overleftarrow{CB}} a, \tag{4.145}$$

respectively, where $a \equiv M\left(-[\dot{r}^{\times}_{\overleftarrow{CB}}]^T \omega_B + \dot{J}_{\overleftarrow{BC}}\dot{\theta}\right)$.

External Wrench Applied

An external wrench alters the rate of change of the SSM, as follows:

$$\frac{d}{dt}\mathcal{L}_C(q, \dot{q}_M) = \mathbb{M}_C\dot{\mathcal{V}}_M + \dot{\mathbb{M}}_C\mathcal{V}_M + H_{CM}\ddot{\theta} + \dot{H}_{CM}\dot{\theta} = \mathcal{F}_M, \tag{4.146}$$

where $\mathcal{F}_M = \begin{bmatrix} f_C^T & m_B^T \end{bmatrix}^T = \mathbb{T}^T_{\overleftarrow{ME}}\mathcal{F}_E$ is the external wrench mapped in accordance with the mixed quasivelocity notation. Relations similar to those with the base quasivelocity notation in Section 4.8.1 exist. That is, the CSM rate of change is zero, i.e.

$$\frac{d}{dt}\mathcal{L}_{CM} = H_{CM}\ddot{\theta} + \dot{H}_{CM}\dot{\theta} = 0. \tag{4.147}$$

This implies that the system response w.r.t. an applied nonimpulsive wrench is that of a rigid body, i.e.

$$\frac{d}{dt}\tilde{\mathcal{L}}_C(q, \dot{q}_M) = \mathbb{M}_C\dot{\mathcal{V}}_M + \dot{\mathbb{M}}_C\mathcal{V}_M = \mathcal{F}_M. \tag{4.148}$$

Thus, $\mathcal{F}_M = \mathcal{F}_C$ holds.

4.8.3 Representation in Terms of Centroidal Quasivelocity

An expression for the spatial dynamics in terms of the centroidal spatial velocity will be derived first. In analogy to (4.132), the rate of change of the centroidal SSM is determined by the external wrench acting on the system. Using the time differential of (4.89), the spatial dynamics can be expressed as

$$\frac{d}{dt}\mathcal{L}_C(\boldsymbol{q}, \mathcal{V}_C) = \mathbb{M}_C \dot{\mathcal{V}}_C + \dot{\mathbb{M}}_C \mathcal{V}_C = \mathcal{F}_C. \tag{4.149}$$

Note that since the centroidal quasivelocity is employed, the SSM equals the CRB-SM, as in the locked-joints case; there is no contribution from the joint motion at all. It is then straightforward to conclude that the above expression is equivalent to that in (4.132) and (4.148).

The expanded form of the above equation can be written as

$$\begin{bmatrix} ME & \boldsymbol{0} \\ \boldsymbol{0} & \boldsymbol{I}_C \end{bmatrix} \begin{bmatrix} \dot{\boldsymbol{v}}_C \\ \dot{\boldsymbol{\omega}}_C \end{bmatrix} + \begin{bmatrix} \boldsymbol{0} \\ \boldsymbol{c}_\omega \end{bmatrix} = \begin{bmatrix} \boldsymbol{f}_C \\ \boldsymbol{m}_C \end{bmatrix}. \tag{4.150}$$

Recall that when the joints are locked, $\boldsymbol{\omega}_C = \boldsymbol{\omega}_B$ and $\mathcal{V}_C = \mathcal{V}_M$. The structure of (4.150) resembles the Newton–Euler equation of motion of a free-floating rigid body. It should be noted, though, that the nonlinear moment $\boldsymbol{c}_\omega = \dot{\boldsymbol{I}}_C(\boldsymbol{q})\boldsymbol{\omega}_C$ comprises terms containing time and partial derivatives of the posture-dependent inertia matrix $\boldsymbol{I}_C(\boldsymbol{q})$ (cf. (4.141)). Such terms do not appear in the nonlinear-velocity term of the rigid-body Euler equation (i.e. in the gyroscopic torque $\boldsymbol{\omega}_C \times \bar{\boldsymbol{I}}_C \boldsymbol{\omega}_C$, $\bar{\boldsymbol{I}}_C = \text{const.}$).

The CRB motion induced by the impressed external wrench depends on the line of action of the force component. A special case deserves to be mentioned. Let the external wrench, acting at point E, represent a pure force, i.e. $\mathcal{F}_E = \begin{bmatrix} \boldsymbol{f}_E^T & \boldsymbol{0}^T \end{bmatrix}^T$. When the line of action of this force goes through the CRB centroid, the angular momentum will be conserved since the moment component of $\mathcal{F}_C = \mathbb{T}_{\overleftarrow{CE}}^T \mathcal{F}_E$ is identically zero; $\boldsymbol{m}_C = \boldsymbol{0}$. On the other hand, when the line of action is apart from the centroid, a change in the angular momentum will be induced (cf. Fig. 4.8), so

$$\frac{d}{dt}\boldsymbol{l}_C \equiv \boldsymbol{m}_C = -[\boldsymbol{r}_{\overleftarrow{CE}}^{\times}]\boldsymbol{f}_E \neq \boldsymbol{0}.$$

Assuming an external force of a unit magnitude, the magnitude of the rate of change of the angular momentum will be in proportion to the distance $\|\boldsymbol{r}_{\overleftarrow{CE}}\|$. This simple relation plays an important role in balance stability of humanoid robots (cf. Chapter 5).

Finally, to obtain the system dynamics expressed in terms of the centroidal quasivelocity, adjoin the decoupled CRB dynamics (4.149) to the reduced-form dynamics (4.133). We have

$$\begin{bmatrix} \mathbb{M}_C & \boldsymbol{0} \\ \boldsymbol{0} & \hat{\boldsymbol{M}}_{\theta B} \end{bmatrix} \begin{bmatrix} \dot{\mathcal{V}}_C \\ \ddot{\boldsymbol{\theta}} \end{bmatrix} + \begin{bmatrix} \mathcal{C}_C \\ \hat{\boldsymbol{c}}_{\theta B} \end{bmatrix} = \begin{bmatrix} \boldsymbol{0} \\ \boldsymbol{\tau} \end{bmatrix} + \begin{bmatrix} \mathbb{T}_{\overleftarrow{CE}}^T \\ \hat{\boldsymbol{J}}_E^T \end{bmatrix} \mathcal{F}_E, \tag{4.151}$$

$\mathcal{C}_C \equiv \dot{\mathbb{M}}_C \mathcal{V}_C$. This form of the equation of motion of a floating-base system was first revealed in [164]. With this notation, the three distinctive partial dynamic components (i.e. the linear and angular CRB and the joint-space dynamics) have been *completely* dynamically decoupled.

It should be noted that the above result can be confirmed by transforming the equation of motion expressed in terms of the base quasivelocity, (4.129), with a similar procedure as in Section 4.8.2, by making use of the quasivelocity relation and transform given in (4.101) and (4.102), respectively.

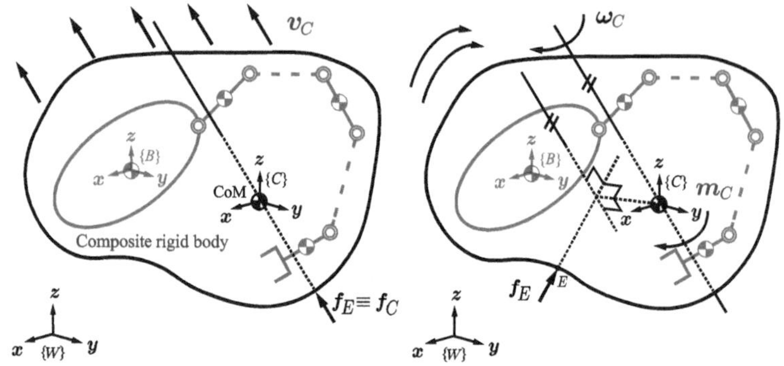

FIGURE 4.8 External wrench $\boldsymbol{f}_E \neq \boldsymbol{0}$, $\boldsymbol{m}_E = \boldsymbol{0}$ acting on the CRB. Left (A): The angular momentum is conserved when the line of action of $\boldsymbol{f}_E$ passes through the CRB centroid. Right (B): When the line of action of the external force does not pass through the CoM, the force induces an angular momentum with rate of change in proportion to the distance between C and E.

4.9 REACTION NULL SPACE–BASED INVERSE DYNAMICS

The spatial dynamics equation, (e.g. (4.130)), can be resolved for the manipulator joint acceleration. The solution can be used to derive control inputs in dynamic control schemes, e.g. resolved acceleration control [72] or computed torque control [24]. Since the case of a nonredundant manipulator has a unique solution, it will be more interesting to focus on the redundant manipulator case with an infinite number of solutions. The joint acceleration can be obtained as

$$\ddot{\boldsymbol{\theta}} = \boldsymbol{H}_{BB}^{+}\left(\mathcal{F}_B - \mathbb{M}_B\dot{\mathcal{V}}_B - \mathcal{C}_B\right) + \boldsymbol{N}(\boldsymbol{H}_{BB})\ddot{\boldsymbol{\theta}}_a. \tag{4.152}$$

The homogeneous solution component $\boldsymbol{N}(\boldsymbol{H}_{BB})\ddot{\boldsymbol{\theta}}_a$ determines the following infinite set of joint accelerations:

$$\{\ddot{\boldsymbol{\theta}}_{rl} \in \mathcal{N}(\boldsymbol{H}_{BB}) : \ddot{\boldsymbol{\theta}} = \boldsymbol{N}(\boldsymbol{H}_{BB})\ddot{\boldsymbol{\theta}}_a,\ \forall \ddot{\boldsymbol{\theta}}_a\}.$$

Apparently, joint accelerations $\{\ddot{\boldsymbol{\theta}}_{rl}\}$ belong to the (spatial momentum) RNS. They can be characterized as *reactionless* since they do not alter the force balance at the base whatsoever. On the other hand, the particular solution component (the pseudoinverse term) yields joint acceleration that ensures optimal inertial coupling in terms of the coupling kinetic energy minimization (cf. (4.120)). Such minimization is a highly desirable feature from a control viewpoint. For instance, a reactionless motion control objective [94] would be based on two control components: (i) reactionless-motion generation, via the null-space term (a feedforward control component) and (ii) base deviation[10] compensation via the pseudoinverse term (a feedback control component). Coupling energy minimization will thereby ensure better performance with regard to the error compensation dynamics.

[10] Stemming e.g. from modeling errors.

The inverse dynamics solution can be derived in a straightforward manner by inserting the joint acceleration (4.152) into the lower part of the equation of motion, (4.129). We have

$$\begin{aligned}\boldsymbol{\tau} &= \boldsymbol{H}_{BB}^T \dot{\mathcal{V}}_B + \boldsymbol{M}_{\theta B}\ddot{\boldsymbol{\theta}} + \boldsymbol{c}_{\theta B} \\ &= \hat{\boldsymbol{M}}_{\theta B}\dot{\mathcal{V}}_B + \overline{\boldsymbol{M}}_{\theta B}\ddot{\boldsymbol{\theta}}_a + \hat{\boldsymbol{c}}_{\theta B}.\end{aligned} \tag{4.153}$$

The "hat" notation stems from the reduced-form representation of the equation of motion introduced in Section 4.8.1. This notation is typical for underactuated system inverse dynamics; it will also appear in the humanoid robot inverse dynamics notation, to be introduced in what follows. The overbar notation, on the other hand, denotes a *restricted matrix* [92] stemming from the constrained least-squares redundancy resolution (cf. (2.49)); $\overline{\boldsymbol{M}}_{\theta B} \equiv \boldsymbol{M}_{\theta B}\boldsymbol{N}(\boldsymbol{H}_{BB})$ will be referred to as the *RNS-restricted joint-space inertia matrix*. According to the abovementioned control scenario, $\ddot{\boldsymbol{\theta}}_a$ and $\dot{\mathcal{V}}_B$ could be used in the feedforward and error compensation (feedback) control components, respectively.

4.10 SPATIAL MOMENTUM OF A HUMANOID ROBOT

From the discussion on simple dynamic models in Section 4.3 it became apparent that the two components of the spatial momentum, determined by the CoM motion and a centroidal moment, play an important role in balance control. Furthermore, in Section 4.8 it was shown that the spatial dynamics of a manipulator floating freely in zero gravity are determined by the rate of change of the spatial momentum resulting from the external forces (cf. (4.130)); the joint-space partial dynamics have no contribution at all. This is true for floating-base systems in general (i.e. under gravity), humanoid robots included.

Most of the early studies on gait generation and balance control of humanoid robots were based on IP models; they focused almost exclusively on the linear component of the spatial momentum determined by the CoM motion (cf. (4.70)). The role of the angular momentum component was also discussed at an early stage [128,129]. Nevertheless, follow-up research began appearing much later [89,41,119,81]. It has been clarified that the impressed moment on the foot can be controlled in two ways: (1) via the rate of change of the *moment of momentum*, by accelerating the CoM horizontally, as in the so-called "angular momentum pendulum model" (AMPM) approach [62], and (2) via the centroidal angular momentum, as in the RWP approach.

Besides in gait generation and balance control, spatial momentum also plays an important role in whole-body motion control. The focus in this area has been especially on the spatial momentum represented w.r.t. the CRB centroid in terms of the base quasivelocity, (4.91), as in the so-called "resolved momentum control" method [56,135,57]. This representation also appears in other studies on balance control [40,69,107,160].

The following properties of the spatial momentum and its two components, linear and angular momentum, and the respective rates of change are derived from the free-floating space robot models in the previous sections; these properties play an important role in humanoid robotics as well.

- The reference point for angular momentum is essential; the base and CRB CoMs are of particular interest as points of reference.
- The spatial momentum of the system (the SSM) is expressed as the sum of CRB (the CRB-SM) and coupling spatial momentum (the CSM), i.e.

$$\text{SSM} = \text{CRB-SM} + \text{CSM}.$$

  This relation is referred to as the *momentum equilibrium principle*. Momentum equilibrium stems from the dynamic equilibrium of wrenches in the spatial dynamics component of the equation of motion, in the absence of external forces (e.g. as in (4.123)).
- The expression of spatial momentum depends on the chosen type of quasivelocity:
  - The representation in terms of the centroidal quasivelocity, $\mathcal{V}_C$, is the simplest one; it is independent from the motion in the joints and yields inertial decoupling between all dynamic components, i.e. *complete* inertial decoupling.
  - The representation with the base quasivelocity, $\mathcal{V}_B$, yields inertial coupling between all dynamic components. This coupling determines the *coupling spatial momentum conserving* manifold in joint space. The tangent subspace to this manifold at $\boldsymbol{q}$ is called the *spatial momentum Reaction Null Space.*
  - The representation with the mixed quasivelocity, $\mathcal{V}_M$, yields inertial decoupling between the CoM dynamics on one side and the angular momentum and joint-space partial dynamics on the other. The latter two components are inertially coupled; the coupling determines the *coupling angular momentum conserving* manifold in joint space. The tangent subspace to this manifold at $\boldsymbol{q}$ is called the *angular momentum Reaction Null Space.*
- The angular component of the spatial momentum is conserved:
  - in the absence of external forces;
  - when the CRB centroid lies on the line of action of the external force (as with the IP models in Fig. 4.2).

Using the appropriate representation of the angular momentum is important in balance control design, as will be shown in Chapter 5.

Furthermore, note that with the exception of the flight phase during running and jumping, a humanoid robot is always in contact with the environment. The posture is then typically characterized as single/double stance, or multicontact. In the latter case, multiple interdependent closed loops are formed via the hand contacts, in addition to those at the feet. In this situation, the momentum control objectives have to be determined as constraint-consistent. This can be done in a straightforward manner, by employing the constraint-consistent generalized velocity/accelerations, as described in Section 2.11. To this end, note that it would be sufficient to restrict the spatial momentum matrix, $\boldsymbol{\mathcal{A}}_{(\circ)}$, by the null space of the appropriate constraint Jacobian. Two expressions of the *constrained-consistent centroidal spatial momentum* can then be obtained, as follows:

$$\overline{\mathcal{L}}_C(\boldsymbol{q}, \dot{\boldsymbol{q}}_{(\circ)}) = \overline{\boldsymbol{\mathcal{A}}}_{(\circ)}(\boldsymbol{q})\dot{\boldsymbol{q}}_u. \tag{4.154}$$

This notation covers expressions in terms of the base ($\dot{\boldsymbol{q}}_{(\circ)} = \dot{\boldsymbol{q}}_B$) and the mixed ($\dot{\boldsymbol{q}}_{(\circ)} = \dot{\boldsymbol{q}}_M$) quasivelocity, whereby $\overline{\boldsymbol{\mathcal{A}}}_{\overleftarrow{CB}}(\boldsymbol{q}) = \boldsymbol{\mathcal{A}}_{\overleftarrow{CB}}\boldsymbol{N}(\boldsymbol{J}_{cB})$ and $\overline{\boldsymbol{\mathcal{A}}}_C(\boldsymbol{q}) = \boldsymbol{\mathcal{A}}_C\boldsymbol{N}(\boldsymbol{J}_{cM})$, respectively; $\overline{\boldsymbol{\mathcal{A}}}_{(\circ)}$ is referred to as a *restricted centroidal momentum matrix*, $\dot{\boldsymbol{q}}_u$ is an arbitrary quasivelocity vector of the unconstrained system. Note that the spatial momentum $\mathcal{L}_B(\boldsymbol{q}, \dot{\boldsymbol{q}}_B)$ can be restricted in a similar way. Note also that the expression of centroidal spatial momentum in terms of the centroidal quasivelocity, $\mathcal{L}_C(\boldsymbol{q}, \mathcal{V}_C)$, does not have to be restricted because of the complete dynamic decoupling property.

4.11 EQUATION OF MOTION OF A HUMANOID ROBOT

A humanoid robot is modeled as an underactuated multibody system on a floating base. The Lagrangian form of the equation of motion of the robot is of the same form as that of a free-floating space robot, as described in Section 4.8. The main difference is the presence of the gravity field. Other external wrenches include determinate reactions at the contact joints as well as indeterminate disturbance wrenches. Both continuous and impact-type reaction/disturbance wrenches have to be taken in consideration. The Lagrangian form of the equation of motion will be used as a base for the dynamic modeling.

From the discussion in Section 4.8 it is apparent that the representations of the equation of motion depend upon the particular choice for the quasivelocity. The classical representation is in terms of the base quasivelocity [35]. From the viewpoint of a whole-body balance control, more convenient is the mixed quasivelocity, though. The simplest expression is obtained with the centroidal quasivelocity. These notations will be introduced below. First, consider the classical representation

$$\begin{bmatrix} \mathbb{M}_B & \boldsymbol{H}_{BB} \\ \boldsymbol{H}_{BB}^T & \boldsymbol{M}_{\theta B} \end{bmatrix} \begin{bmatrix} \dot{\mathcal{V}}_B \\ \ddot{\theta} \end{bmatrix} + \begin{bmatrix} \mathcal{C}_B \\ \boldsymbol{c}_{\theta B} \end{bmatrix} + \begin{bmatrix} \mathcal{G}_B \\ \boldsymbol{g}_\theta \end{bmatrix} = \begin{bmatrix} \boldsymbol{0} \\ \boldsymbol{\tau} \end{bmatrix} + \begin{bmatrix} \mathbb{C}_{cB} \\ \mathcal{J}_{cB}^T \end{bmatrix} \overline{\mathcal{F}}^c. \tag{4.155}$$

The upper and lower parts express the spatial and joint-space partial dynamics, respectively; $\overline{\mathcal{F}}^c \in CWC$, and $\mathcal{G}_B$ and $\boldsymbol{g}_\theta$ stand for the reactions stemming from the contact joints and the gravity wrench acting on the base link and the gravity joint torque vector. Their expressions are given in (3.55), (3.59), and (3.60), respectively. In the following derivations it will be assumed that the contact joints are formed by one or more end links (feet/hands). In practice, other links, such as fingers and intermittent links, may establish contacts as well. Such contacts can be easily incorporated without altering the qualitative properties of the model described herein. The reactions are mapped, first at the base link via the contact map of the robot (cf. (2.74)) and second to the joint torque via the transpose of the joint-space constraint Jacobian, $\mathcal{J}_{cB}^T \in \Re^{n \times c}$ (cf. (2.75)).

Equation of motion (4.155) together with the motion constraint (2.104)–(2.105) and the contact wrench cones will be referred to as the *complete model*. With this model, the spatial and joint-space partial dynamics are made apparent. The two parts play important and distinctive roles in analysis and controller design.

Some models make use of a more detailed representation of the equation of motion. The following expanded-form representation renders the spatial rotational and translational dynamic motion components explicitly visible (cf. also (4.124)):

$$\begin{bmatrix} ME & -M[r^{\times}_{\overleftarrow{CB}}] & MJ_{\overleftarrow{CB}} \\ -M[r^{\times}_{\overleftarrow{CB}}]^T & I_B & H_B \\ MJ^T_{\overleftarrow{CB}} & H^T_B & M_{\theta B} \end{bmatrix} \begin{bmatrix} \dot{v}_B \\ \dot{\omega}_B \\ \ddot{\theta} \end{bmatrix} + \begin{bmatrix} c_{fB} \\ c_{mB} \\ c_{\theta B} \end{bmatrix} + \begin{bmatrix} g_f \\ g_m \\ g_\theta \end{bmatrix} = \begin{bmatrix} 0 \\ 0 \\ \tau \end{bmatrix} + \begin{bmatrix} \mathbb{C}_{cB_f} \\ \mathbb{C}_{cB_m} \\ \mathcal{J}_{cB}{}^T \end{bmatrix} \mathcal{F}^c. \tag{4.156}$$

This equation appeared first in [35]. The meaning of the nonlinear velocity-dependent and gravity terms is apparent from (4.125) and (3.59), respectively. The force and moment component maps, $\mathbb{C}_{cB_f}$, $\mathbb{C}_{cB_m} \in \Re^{3\times c}$, are defined in (2.120).

The above representations of the equation of motion can be rewritten in terms of the *mixed quasivelocity* in accordance with the procedure described in Section 4.8.2. The compact and expanded-form representations are

$$\begin{bmatrix} \mathbb{M}_C & H_{CM} \\ H^T_{CM} & M_{\theta M} \end{bmatrix} \begin{bmatrix} \dot{\mathcal{V}}_M \\ \ddot{\theta} \end{bmatrix} + \begin{bmatrix} \mathcal{C}_M \\ c_{\theta M} \end{bmatrix} + \begin{bmatrix} \mathcal{G}_C \\ 0 \end{bmatrix} = \begin{bmatrix} 0 \\ \tau \end{bmatrix} + \begin{bmatrix} \mathbb{C}_{cC} \\ \mathcal{J}^T_{cM} \end{bmatrix} \mathcal{F}^c \tag{4.157}$$

and

$$\begin{bmatrix} ME & 0 & 0 \\ 0 & I_C & H_C \\ 0 & H^T_C & M_{\theta M} \end{bmatrix} \begin{bmatrix} \dot{v}_C \\ \dot{\omega}_B \\ \ddot{\theta} \end{bmatrix} + \begin{bmatrix} 0 \\ c_{mM} \\ c_{\theta M} \end{bmatrix} + \begin{bmatrix} g_f \\ 0 \\ 0 \end{bmatrix} = \begin{bmatrix} 0 \\ 0 \\ \tau \end{bmatrix} + \begin{bmatrix} \mathbb{C}_{cC_f} \\ \mathbb{C}_{cC_m} \\ \mathcal{J}^T_{cM}(q) \end{bmatrix} \mathcal{F}^c, \tag{4.158}$$

respectively. The link inertia tensor, $M_{\theta M}$, the nonlinear velocity-dependent moment, c_{mM}, and the nonlinear velocity-dependent joint torque, $c_{\theta M}$, are defined in (4.139), (4.140), and (4.145), respectively. The constraint Jacobian and contact map components are defined in (2.126) and (2.127), respectively. The decoupling of the linear spatial dynamics component (i.e. the CoM partial dynamics in the upper row of (4.158)) from the rest is apparent. The decoupling yields an advantage in balance controller design. This topic will be discussed in Chapter 5.

The simplest expression of the equation of motion is obtained with the *centroidal quasivelocity*, i.e.

$$\begin{bmatrix} \mathbb{M}_C & 0 \\ 0 & \hat{M}_{\theta B} \end{bmatrix} \begin{bmatrix} \dot{\mathcal{V}}_C \\ \ddot{\theta} \end{bmatrix} + \begin{bmatrix} \mathcal{C}_C \\ \hat{c}_{\theta B} \end{bmatrix} + \begin{bmatrix} \mathcal{G}_C \\ 0 \end{bmatrix} = \begin{bmatrix} 0 \\ \tau \end{bmatrix} + \begin{bmatrix} \mathbb{C}_{cC} \\ \mathcal{J}^T_{cC} \end{bmatrix} \mathcal{F}^c. \tag{4.159}$$

The expanded-form representation is

$$\begin{bmatrix} ME & 0 & 0 \\ 0 & I_C & 0 \\ 0 & 0 & \hat{M}_{\theta B} \end{bmatrix} \begin{bmatrix} \dot{v}_C \\ \dot{\omega}_C \\ \ddot{\theta} \end{bmatrix} + \begin{bmatrix} 0 \\ c_\omega \\ \hat{c}_{\theta B} \end{bmatrix} + \begin{bmatrix} g_f \\ 0 \\ 0 \end{bmatrix} = \begin{bmatrix} 0 \\ 0 \\ \tau \end{bmatrix} + \begin{bmatrix} \mathbb{C}_{cC_f} \\ \mathbb{C}_{cC_m} \\ \mathcal{J}^T_{cC} \end{bmatrix} \mathcal{F}^c. \tag{4.160}$$

The link inertia matrix, $\hat{M}_{\theta B}$, and the nonlinear velocity-dependent joint torque, $\hat{c}_{\theta B}$, are defined in (4.127) and (4.128), respectively, whereas $c_\omega = \dot{I}_C \omega$. The Jacobian

$$\mathcal{J}_{cC} = \mathcal{J}_{cB} - \mathbb{C}_{cC} J_\theta \tag{4.161}$$

is referred to as the *generalized joint-space constraint Jacobian*. It is defined in analogy to the generalized Jacobian (4.119). Apparently, all three partial dynamic components in the above equation are inertially decoupled.

When using a standard solver (e.g. a quadratic programming solver) for the inverse dynamics, the following compact-form representation of the equation of motion is quite often preferred:

$$\boldsymbol{M}_{(\circ)}(\boldsymbol{q})\ddot{\boldsymbol{q}}_{(\circ)} + \boldsymbol{c}_{(\circ)}(\boldsymbol{q}, \dot{\boldsymbol{q}}_{(\circ)}) + \boldsymbol{g}_{(\circ)}(\boldsymbol{q}) = \mathcal{Q} - \mathcal{Q}_{c(\circ)}(\boldsymbol{q}). \tag{4.162}$$

The $(\circ)$ subscript is replaced with B, M, and C for the notations in terms of the base, mixed, and centroidal quasivelocities, respectively. The inertia, nonlinear velocity-dependent and the gravity force components are apparent from the respective expanded-form representations. The generalized force $\mathcal{Q}$ is defined in (4.122). The terms $\mathcal{Q}_{c(\circ)}(\boldsymbol{q}) \equiv \boldsymbol{J}_{c(\circ)}^T(\boldsymbol{q})\boldsymbol{\lambda}$ stand for the *generalized constraint force*, $\boldsymbol{\lambda} = -\overline{\mathcal{F}}^c$ denoting Lagrange's multiplier vector. Jacobians $\boldsymbol{J}_{cB}$ and $\boldsymbol{J}_{cM}$ were defined in (2.96) and (2.129), respectively. The Jacobian

$$\boldsymbol{J}_{cC} = \begin{bmatrix} \mathbb{C}_{cC}^T & \mathcal{J}_{cC} \end{bmatrix}.$$

Apparently, the expression of the generalized constraint force depends on the respective quasivelocity. From the principle of virtual work it follows that one type of a generalized constraint force can be transformed into another, via the transpose of the quasivelocity transform, e.g. $\mathcal{Q}_{cC} = \boldsymbol{T}_{\overleftarrow{CB}}^T \mathcal{Q}_{cB}$, $\boldsymbol{T}_{\overleftarrow{CB}}$ being defined in (4.102).

It should be noted that Lagrange's multiplier vector is frequently used in constrained optimization problems. In optimization frameworks that focus on extrema, the sign of the vector would be irrelevant. When used in the field of robotics, though, the vector is quite often endowed with a physical meaning. The sign is then of utmost importance, e.g. to honor the requirement that the normal reaction force at a unilateral contact must be positive (cf. Section 2.9.3). This can be ensured by taking the sign to be opposite to that of the constraint force, as above.

4.12 CONSTRAINT-FORCE ELIMINATION METHODS

Constraint-force elimination is a well-known approach within the field of constrained multibody system studies. The generalized constraint force is considered unknown and eliminated from the equation of motion [9,151,32,66,84]. The method can be directly applied in the field of humanoid robotics. The following derivations are valid for any type of quasivelocity. For simplicity, but with some abuse in the notation, the quasivelocity-related subscripts will be omitted. In what follows, holonomic (cf. Section 2.11.1), hard constraints will be assumed, s.t. $\boldsymbol{J}_c(\boldsymbol{q})\dot{\boldsymbol{q}} = \boldsymbol{0}$ (cf. (2.96)) holds. Also, the initial state is assumed to be consistent with the constraints $\boldsymbol{q}_0 \in \boldsymbol{\gamma}(\boldsymbol{q}_0)$ and $\dot{\boldsymbol{q}}_0 \in \mathcal{N}\left(\boldsymbol{J}_c(\boldsymbol{q}_0)\right)$. Combining the compact-form representation of the equation of motion, (4.162), with the time differential of the constraint equation yields

$$\begin{bmatrix} \boldsymbol{M} & \boldsymbol{J}_c^T \\ \boldsymbol{J}_c & \boldsymbol{0} \end{bmatrix} \begin{bmatrix} \ddot{\boldsymbol{q}} \\ \boldsymbol{\lambda} \end{bmatrix} = \begin{bmatrix} \mathcal{Q}_u \\ -\dot{\boldsymbol{J}}_c\dot{\boldsymbol{q}} \end{bmatrix}. \tag{4.163}$$

The generalized force appearing on the r.h.s. is defined as $\mathcal{Q}_u \equiv \mathcal{Q} - (\boldsymbol{c} + \boldsymbol{g})$. In the analysis, all terms of this force are assumed known; the generalized input force is determined by the controller, while the other two terms are state-dependent. The subscript "u" is used to denote the dynamics of the *unconstrained system* ($\boldsymbol{\lambda} = \mathbf{0}$). We have

$$\boldsymbol{M}\ddot{\boldsymbol{q}}_u = \mathcal{Q}_u. \tag{4.164}$$

Note that any initial state is admissible for the unconstrained system. The generalized acceleration $\ddot{\boldsymbol{q}}_u$ is obtained in a straightforward way from the last equation since the inverse of the inertia matrix exists at all system postures. The quantities $\ddot{\boldsymbol{q}}_u$ and $\mathcal{Q}_u$ play an important role in the following analysis. They will be referred to in short as the *unconstrained* generalized acceleration and generalized force, respectively.

The compact form of the complete model, (4.163), represents a set of differential-algebraic equations (DAEs). The forces of constraint, collected in $\boldsymbol{\lambda}$, are algebraic variables, i.e. no time derivatives of these variables appear in the equation. As already mentioned, it is customary to eliminate these forces from the equation of motion. As a result, one arrives at a set of ODEs which are easier to solve with conventional solvers. Since (4.163) was constructed by combining the equation of motion with the second-order motion constraint relation (2.105), it can be anticipated that the latter will be useful to eliminate the unknown Lagrange multiplier, $\boldsymbol{\lambda}$. There are different ways to eliminate the forces of constraint in a constraint multibody system formulation [151,66]. The following methods have been adopted in the field of humanoid robotics.

4.12.1 Gauss' Principle of Least Constraint

Gauss [37] was surprised to find out that ... *"The motion of a system of mass particles, whatever their relative displacements and the constraints they obey, takes place at every instant in maximum accordance with the free motion of the system, or under least constraint, whereby the measure of constraint (that the whole system is subjected to) is considered as the sum of products of mass and the square of the deviation of each mass particle from its free motion."*.[11]

The principle is formulated as a quadratic minimization problem: Minimize

$$\mathcal{Z} = \frac{1}{2}\sum_j M_j \left(\ddot{\boldsymbol{r}}_j - \frac{\boldsymbol{f}_j}{M_j}\right)^T \left(\ddot{\boldsymbol{r}}_j - \frac{\boldsymbol{f}_j}{M_j}\right), \tag{4.165}$$

where $\mathcal{Z}$ is the *measure of the constraint* (Zwang in German), M, $\boldsymbol{r}$ denote the particle mass and position, and $\boldsymbol{f}$ is the force acting on the particle.

The principle has been applied to constrained multibody systems [70,59,151]. It was introduced in the field of robotics in the pioneering work [117]; see also [16,28]. In the case of a multibody system, the above measure of constraint assumes the form [151]

$$\mathcal{Z} = \frac{1}{2}\left(\ddot{\boldsymbol{q}}_u - \ddot{\boldsymbol{q}}\right)^T \boldsymbol{M}\left(\ddot{\boldsymbol{q}}_u - \ddot{\boldsymbol{q}}\right) = \frac{1}{2}\ddot{\boldsymbol{q}}_c^T \boldsymbol{M}\ddot{\boldsymbol{q}}_c. \tag{4.166}$$

[11] Translation by the author of this chapter.

Here $\ddot{\boldsymbol{q}}_u$ denotes the generalized acceleration of the unconstrained system, $\ddot{\boldsymbol{q}}$ stands for the constraint-consistent generalized acceleration, and the difference $\ddot{\boldsymbol{q}}_c = \ddot{\boldsymbol{q}}_u - \ddot{\boldsymbol{q}}$ is the constraint enforcing generalized acceleration (cf. Section 2.11.3). It is important to note that, at each time instant of the motion, the state $(\boldsymbol{q}, \dot{\boldsymbol{q}})$ and the generalized input force $\boldsymbol{\tau}$ are assumed to be known. Therefore, the generalized acceleration of the unconstrained system can be determined uniquely via (4.164). With this notation, the equation of motion can be rewritten as

$$\boldsymbol{M}\left(\ddot{\boldsymbol{q}}_u - \ddot{\boldsymbol{q}}\right) = \boldsymbol{J}_c^T \boldsymbol{\lambda} \tag{4.167}$$

or

$$\boldsymbol{M}\ddot{\boldsymbol{q}}_c = \mathcal{Q}_c. \tag{4.168}$$

From the last relation it is seen that $\mathcal{Q}_c$ (resp. $\boldsymbol{\lambda}$) determines $\boldsymbol{q}_c$ in a unique way:

$$\ddot{\boldsymbol{q}}_c = \boldsymbol{M}^{-1}\mathcal{Q}_c = \boldsymbol{M}^{-1}\boldsymbol{J}_c^T\boldsymbol{\lambda}. \tag{4.169}$$

To determine the unknown Lagrange multiplier (and hence the constraint enforcing acceleration), the (second-order) motion constraint has to be imposed on the equation of motion. To this end, solve (4.167) for the constraint-consistent generalized acceleration and substitute into (2.105) to obtain

$$\boldsymbol{\lambda} = \boldsymbol{M}_c(\dot{\boldsymbol{J}}_c\dot{\boldsymbol{q}} + \boldsymbol{J}_c\ddot{\boldsymbol{q}}_u), \tag{4.170}$$

where $\boldsymbol{M}_c \equiv (\boldsymbol{J}_c\boldsymbol{M}^{-1}\boldsymbol{J}_c^T)^{-1}$ is the system inertia matrix mapped along the constrained-motion directions. Then, inserting (4.170) into (4.169), the constraint enforcing generalized acceleration is obtained as

$$\ddot{\boldsymbol{q}}_c = \boldsymbol{J}_c^{-M}(\dot{\boldsymbol{J}}_c\dot{\boldsymbol{q}} + \boldsymbol{J}_c\ddot{\boldsymbol{q}}_u), \tag{4.171}$$

where

$$\boldsymbol{J}_c^{-M} = \boldsymbol{M}^{-1}\boldsymbol{J}_c^T\left(\boldsymbol{J}_c\boldsymbol{M}^{-1}\boldsymbol{J}_c^T\right)^{-1} = \boldsymbol{M}^{-1}\boldsymbol{J}_c^T\boldsymbol{M}_c \tag{4.172}$$

is the inertia-weighted (right) pseudoinverse. The generalized constraint force can also be derived, if needed, via (4.168). We have

$$\mathcal{Q}_c = \boldsymbol{J}_c^T\boldsymbol{T}_c^{-1}(\dot{\boldsymbol{J}}_c\dot{\boldsymbol{q}} + \boldsymbol{J}_c\ddot{\boldsymbol{q}}_u). \tag{4.173}$$

Note that $\boldsymbol{\lambda}$ (and hence $\mathcal{Q}_c$ and $\ddot{\boldsymbol{q}}_c$) comprises a linear and a nonlinear component. The former results from the generalized acceleration of the unconstrained system. The latter is state-dependent, whereby the velocity state is constraint-consistent, i.e. it is in accordance with (2.100).

Next, the constraint-consistent generalized acceleration is derived by substituting (4.170) into (4.167). We have

$$\ddot{\boldsymbol{q}} = -\boldsymbol{J}_c^{-M}\dot{\boldsymbol{J}}_c\dot{\boldsymbol{q}} + \ddot{\boldsymbol{q}}_m, \tag{4.174}$$

$$\ddot{\boldsymbol{q}}_m \equiv \left(\boldsymbol{E} - \boldsymbol{J}_c^{-M}\boldsymbol{J}_c\right)\ddot{\boldsymbol{q}}_u. \tag{4.175}$$

This is recognized as the general solution to the underdetermined second-order differential constraint, (2.105), comprising a particular and a homogeneous solution component (the first and second terms on the r.h.s., respectively). The particular solution component stands for the nonlinear, state-dependent acceleration. The homogeneous component, on the other hand, is determined uniquely via the projection of the unconstrained generalized acceleration $\ddot{\boldsymbol{q}}_u$ onto null space $\mathcal{N}(\boldsymbol{J}_c)$. The resultant unique acceleration, $\ddot{\boldsymbol{q}}_m$, will be called *constraint maintaining*: this acceleration does not impress any inertial force in the constrained directions. Note also that the unconstrained generalized acceleration $\ddot{\boldsymbol{q}}_u$ is decomposed respectively via (4.171) and (4.175).

With the above notation, the equation of motion (4.167) can be rewritten as

$$\begin{aligned} \boldsymbol{M}\ddot{\boldsymbol{q}} &= -\boldsymbol{J}_c^T \boldsymbol{M}_c \dot{\boldsymbol{J}}_c \dot{\boldsymbol{q}} + \boldsymbol{M}\left(\boldsymbol{E} - \boldsymbol{J}_c^{-M} \boldsymbol{J}_c\right)\ddot{\boldsymbol{q}}_u \\ &= -\boldsymbol{J}_c^T \boldsymbol{M}_c \dot{\boldsymbol{J}}_c \dot{\boldsymbol{q}} + \boldsymbol{M}\ddot{\boldsymbol{q}}_m. \end{aligned} \tag{4.176}$$

4.12.2 Direct Elimination

It is interesting to note that the results obtained with Gauss' principle of least constraint can be confirmed via direct elimination, as follows. First, solve (4.163) for $\boldsymbol{\lambda}$, i.e.

$$\boldsymbol{\lambda} = \boldsymbol{M}_c\left(\dot{\boldsymbol{J}}_c \dot{\boldsymbol{q}} + \boldsymbol{J}_c \boldsymbol{M}^{-1} \mathcal{Q}_u\right). \tag{4.177}$$

Then, substitute (4.177) back into the upper part of (4.163) to obtain

$$\boldsymbol{M}\ddot{\boldsymbol{q}} = \mathcal{Q}_u - \mathcal{Q}_c, \tag{4.178}$$

$$\mathcal{Q}_c = \boldsymbol{J}_c^T \boldsymbol{M}_c\left(\dot{\boldsymbol{J}}_c \dot{\boldsymbol{q}} + \boldsymbol{J}_c \boldsymbol{M}^{-1} \mathcal{Q}_u\right) \tag{4.179}$$

or

$$\boldsymbol{M}\ddot{\boldsymbol{q}} = -\boldsymbol{J}_c^T \boldsymbol{M}_c \dot{\boldsymbol{J}}_c \dot{\boldsymbol{q}} + \mathcal{Q}_m, \tag{4.180}$$

$$\mathcal{Q}_m = \left(\boldsymbol{E} - \boldsymbol{J}_c^T \boldsymbol{J}_c^{-MT}\right)\mathcal{Q}_u. \tag{4.181}$$

To show that this result is identical to the result obtained with Gauss' principle of least constraint, note first that since $\boldsymbol{M}^{-1}\mathcal{Q}_u = \ddot{\boldsymbol{q}}_u$, relations (4.177) and (4.179) are identical to (4.170) and (4.173), respectively. Note also that (4.178) can be directly obtained from the definition of the constraint enforcing generalized acceleration, i.e. $\ddot{\boldsymbol{q}}_c = \ddot{\boldsymbol{q}}_u - \ddot{\boldsymbol{q}}$. In addition, solving (4.180) for the generalized acceleration yields

$$\ddot{\boldsymbol{q}} = -\boldsymbol{J}_c^{-M} \dot{\boldsymbol{J}}_c \dot{\boldsymbol{q}} + \boldsymbol{M}^{-1} \mathcal{Q}_m. \tag{4.182}$$

The second term on the r.h.s. can be represented as

$$\boldsymbol{M}^{-1}\mathcal{Q}_m = \boldsymbol{M}^{-1}\left(\boldsymbol{E} - \boldsymbol{J}_c^T \boldsymbol{J}_c^{-MT}\right)\mathcal{Q}_u = \left(\boldsymbol{E} - \boldsymbol{J}_c^{-M}\boldsymbol{J}_c\right)\boldsymbol{M}^{-1}\mathcal{Q}_u = \left(\boldsymbol{E} - \boldsymbol{J}_c^{-M}\boldsymbol{J}_c\right)\ddot{\boldsymbol{q}}_u = \ddot{\boldsymbol{q}}_m. \tag{4.183}$$

The above relations were obtained with the help of the similarity transform (3.36), whereby the inertia matrix $\boldsymbol{M}$ was used as the weighing matrix. These relations imply that (4.182) and (4.174) are identical. Note also that the dynamical system (4.180) is driven by the force $\mathcal{Q}_m$ which is a projection of the unconstrained generalized force $\mathcal{Q}_u$ onto the (dual) null space $\mathcal{N}^*(\boldsymbol{J}_c)$; $\mathcal{Q}_m$ will be called a *constraint maintaining* generalized force, in analogy to $\ddot{q}_m$. It is straightforward to confirm that, with the help of the above identity relations, the equation of motion derived under Gauss' principle, (4.168), is identical to (4.180).

The resultant equations in both approaches, Gauss' approach and the direct elimination one, are obtained in a specific form, expressed via the inertia-weighted pseudoinverse $\boldsymbol{J}_c^{-M}$ and its transpose, as in (4.174) and (4.182), respectively. Indeed, note that (4.182) and (4.174) likewise are generalized solutions to the motion constraint (2.105). Recall that (2.106) also represents a general solution to the same motion constraint. As already clarified, the role of its particular solution component is to account for the nonlinearity in the constraints, within the normal subspace at the constraint manifold (i.e. for centrifugal and centripetal forces). This is also the role of the particular solution components in (4.182) and (4.174). Recall further that the homogeneous component in (2.106) was specified in a particular way to ensure integrability.

The direct elimination approach was used in [115,133] to derive the equation of motion of a humanoid robot.

4.12.3 Maggi's Equations (Null-Space Projection Method)

Lagrange's multiplier vector can be eliminated by projecting the equation of motion onto the null space of the system constraint matrix, $\mathcal{N}(\boldsymbol{J}_c)$. In constrained multibody system analysis, the method is known as *Maggi's equations* [114,11,151,66]. The first step of the procedure is to complement the differential constraints (2.96) by a set of $r = n + 6 - c$ independent kinematic relations, i.e.

$$\begin{bmatrix} \boldsymbol{J}_c \\ \boldsymbol{A}_r \end{bmatrix} \dot{q} = \begin{bmatrix} \boldsymbol{0} \\ \mathcal{V}_r \end{bmatrix}. \tag{4.184}$$

The newly introduced relations do not necessarily have to be integrable; $\mathcal{V}_r$ is therefore sometimes referred to as the *quasivelocity* or *virtual speeds* or *independent kinematic parameters* [11] (see also Section 2.11.1). Matrix $\boldsymbol{A}_r \in \Re^{r\times(n+6)}$ is composed in such a way that the *extended* matrix $\boldsymbol{A}_e \equiv \begin{bmatrix} \boldsymbol{J}_c^T & \boldsymbol{A}_r^T \end{bmatrix}^T \in \Re^{(n+6)\times(n+6)}$ is always invertible. The inverse is represented as

$$\boldsymbol{A}_e^{-1} = \begin{bmatrix} \boldsymbol{V}_c & \boldsymbol{V}_r \end{bmatrix} \tag{4.185}$$

so that

$$\begin{bmatrix} \boldsymbol{J}_c^T & \boldsymbol{A}_r^T \end{bmatrix}^T \begin{bmatrix} \boldsymbol{V}_c & \boldsymbol{V}_r \end{bmatrix} = \text{diag}\begin{bmatrix} \boldsymbol{E}_c & \boldsymbol{E}_r \end{bmatrix}.$$

The constraint-consistent generalized velocity can then be expressed as

$$\dot{q} = \boldsymbol{V}_r \mathcal{V}_r. \tag{4.186}$$

Since $\boldsymbol{J}_c \boldsymbol{V}_r = \boldsymbol{0}$, it follows that $\boldsymbol{V}_r \in \Re^{(n+6)\times r}$ spans the null space of $\boldsymbol{J}_c$.

It is interesting to observe that (4.184) has the same form as (2.46) and that matrix $\boldsymbol{A}_e$ plays the same role as the extended Jacobian used in kinematic redundancy resolution (cf. Section 2.7.5). Moreover, it can be concluded that if $\boldsymbol{V}_r$ is composed in analogy to matrix $\boldsymbol{Z}_W^{\#}$ of the KD-JSD method discussed in the same section, then the invertability of the extended matrix, $\boldsymbol{A}_e$, will be ensured. Furthermore, note that $\boldsymbol{V}_r\boldsymbol{V}_r^T$ is a null-space projector onto $\mathcal{N}^*(\boldsymbol{J}_c)$. This implies that $\boldsymbol{V}_r$ can be derived from the SVD of the system constraint Jacobian $\boldsymbol{J}_c$, in analogy to the null-space projector of a kinematically redundant manipulator (cf. Section 2.7.1).

The next step is to differentiate the generalized velocity (4.186) w.r.t. time. We have

$$\ddot{q} = \boldsymbol{V}_r\dot{\mathcal{V}}_r + \dot{\boldsymbol{V}}_r\mathcal{V}_r. \tag{4.187}$$

Finally, substitute into the equation of motion and premultiply the result by $\boldsymbol{V}_r^T$. Then we have

$$\boldsymbol{V}_r^T\boldsymbol{M}\left(\boldsymbol{V}_r\dot{\mathcal{V}}_r + \dot{\boldsymbol{V}}_r\mathcal{V}_r\right) = \boldsymbol{V}_r^T\mathcal{Q}_u. \tag{4.188}$$

This equation represents the dynamics projected onto the null space of the system constraint matrix and expressed in terms of the quasivelocity. Apparently, the generalized constraint force $\mathcal{Q}_c = \boldsymbol{J}_c^T\boldsymbol{\lambda}$ has been annihilated (by $\boldsymbol{V}_r^T$). Note also that the dimension has been reduced from $n+6$ to r, to obtain an undetermined system with $n+6$ unknowns, as in (4.187). The general solution can be obtained via the pseudoinverse $\boldsymbol{V}_r^{+T} = \boldsymbol{V}_r$, i.e.

$$\boldsymbol{M}\left(\boldsymbol{V}_r\dot{\mathcal{V}}_r + \dot{\boldsymbol{V}}_r\mathcal{V}_r\right) = \boldsymbol{V}_r\boldsymbol{V}_r^T\mathcal{Q}_u + (\boldsymbol{E} - \boldsymbol{V}_r\boldsymbol{V}_r^T)\mathcal{Q}_a. \tag{4.189}$$

The generalized force $\mathcal{Q}_a$ parametrizes the null space $\mathcal{N}(\boldsymbol{V}_r^T)$. An appropriate way for the determination of this force will be introduced in short. Note that in the above derivation, the columns of $\boldsymbol{V}_r$ were assumed orthonormal (i.e. as derived from the SVD of $\boldsymbol{J}_c$). The following identities were used thereby:

$$\boldsymbol{V}_r^T\boldsymbol{V}_r = \boldsymbol{E}_r, \tag{4.190}$$

$$\boldsymbol{V}_r^{+T}\boldsymbol{V}_r^T = \boldsymbol{V}_r\boldsymbol{V}_r^T. \tag{4.191}$$

From (4.189) it can be confirmed that the null-space projector $\boldsymbol{V}_r\boldsymbol{V}_r^T$ maps the generalized forces. Thus, it is characterized as a projector onto the *dual* null space of $\boldsymbol{J}_c$, i.e. $\boldsymbol{N}^*(\boldsymbol{J}_c) \equiv \boldsymbol{V}_r\boldsymbol{V}_r^T$. There is an infinite number of representations of this projector, when parametrized by a generalized inverse. Consider the particular choice

$$\boldsymbol{N}^*(\boldsymbol{J}_c) = (\boldsymbol{E} - \boldsymbol{J}_c^T\boldsymbol{J}_c^{-MT}), \tag{4.192}$$

which also appears in the direct elimination/Gauss approach. Replacing $\boldsymbol{V}_r\boldsymbol{V}_r^T$ with $(\boldsymbol{E} - \boldsymbol{J}_c^T\boldsymbol{J}_c^{-MT})$ in (4.189) and using (4.187), the constraint-consistent generalized acceleration can be derived as

$$\begin{aligned}\ddot{q} &= M^{-1} J_c^T J_c^{-MT} \mathcal{Q}_a + M^{-1}(E - J_c^T J_c^{-MT})\mathcal{Q}_u \\ &= J_c^{-M} J_c M^{-1} \mathcal{Q}_a + M^{-1} \mathcal{Q}_m.\end{aligned} \tag{4.193}$$

When the arbitrary force $\mathcal{Q}_a$ is set to satisfy the second-order differential constraint (2.105),

$$\begin{aligned}J_c M^{-1} \mathcal{Q}_a &= -\dot{J}_c \dot{q}, \\ \mathcal{Q}_a &= M\ddot{q},\end{aligned} \tag{4.194}$$

the constraint-consistent generalized acceleration (4.193) becomes identical to (4.182) from the direct elimination/Gauss approach. Under these conditions and from (4.193), the equation of motion can be written as

$$M\ddot{q} = -J_c^T M_c \dot{J}_c \dot{q} + \mathcal{Q}_m. \tag{4.195}$$

This is identical to (4.180) from the direct elimination/Gauss approach. Thus, the equivalence between the direct elimination/Gauss and the Maggi method has been confirmed.

The constraint force can be recovered, if needed. To this end, premultiply the equation of motion by V_c^T. Then we have

$$\lambda = V_c^T (M\ddot{q} - \mathcal{Q}_u) = V_c^T M \ddot{q}_c.$$

Hereby, the relations $J_c V_c = E_c$, $\mathcal{Q}_u = M\ddot{q}_u$, and $\ddot{q}_c = \ddot{q}_u - \ddot{q}$ were used. The generalized constraint force can also be confirmed, via $\mathcal{Q}_c = J_c^T \lambda = J_c^T V_c^T M \ddot{q}_c = M\ddot{q}_c$.

The projection of the dynamics onto the null space of the Jacobian of constrained fixed-base manipulators has been introduced in the field of robotics in [1]. The method has been later adopted for use with underactuated manipulators [78], humanoid robots [77,125,124], and a dual-arm robot [108]. In the case of humanoid robots, the motivation was to alleviate problems related to contact force sensing, e.g. the lack of sensors at certain contacts or the presence of noisy signals and/or time delays resulting from filtering [77]. Note also that when the constraints are eliminated, there is no need to include the contact joints into the model. In this way, the difficulty in handling the inequality constraints pertinent to the friction cone can be alleviated [124].

According to [1], the dynamics are projected onto $\mathcal{N}(J_c)$ with null-space projector $N(J_c) = (E - J_c^+ J_c)$, i.e.

$$NM\ddot{q} = N\mathcal{Q}_u. \tag{4.196}$$

If a solution to the generalized acceleration could be found, it would be compatible with the constraints. However, since the projector is singular ($\text{rank}N = r < n + 6$), the solution cannot be obtained directly. As shown in [1], the problem can be tackled by adding the motion constraint (2.115) to the above projected equation of motion. This is possible because the two equations are orthogonal. Therefore, the respective parts can be summed as follows:

$$\tilde{M}\ddot{q} = N\mathcal{Q}_u + \dot{N}\dot{q}, \tag{4.197}$$

where

$$\tilde{M} \equiv \{NM + M(E - N)\}. \tag{4.198}$$

The matrix $\tilde{M}$ is called the *constrained joint-space inertia matrix* [1]. It can be represented as $\tilde{M} = M + \tilde{S}$, $\tilde{S} \equiv NM - (NM)^T$. Since M is positive definite and $\tilde{S}$ is skew-symmetric ($\tilde{S} = -\tilde{S}^T$), it follows that $\tilde{M}$ is also positive definite. Hence, this matrix will be always invertible, irrespective of the particular set of constraints.

It is important to note that the constrained joint-space inertia matrix does not retain the properties of an inertia matrix; $\tilde{M}$ is *asymmetric* and *nonunique*. This leads to problems pertinent to the specific choice of this matrix. Fortunately, there is a straightforward approach to avoid the appearance of $\tilde{M}$ into the projected dynamics. Since the null-space projector N and its dual, N^*, are isomorphic, they can be decomposed in the same way. Thus,

$$N = V_r V_r^T \Rightarrow V_r^T N = V_r^T. \tag{4.199}$$

Then, premultiply projected dynamics (4.196) by V_r^T to obtain

$$V_r^T M\ddot{q} = V_r^T \mathcal{Q}_u. \tag{4.200}$$

This equation is identical to (4.188) from Maggi's method. By following the rest of the steps of this method, one arrives at the equation of motion represented as in (4.195). Thus, the important conclusion can be made that the N-based projection of the dynamics yields the same result as Gauss'/direct elimination/Maggi's approaches. In this way, the problem associated with the asymmetric and nonunique inertia-like matrix, $\tilde{M}$, as in (4.197), can be avoided.

4.12.4 Range-Space Projection Method

As clarified in Section 4.12.3, the essence of Maggi's method is projecting the system dynamics onto the null space of the system constraint matrix. As an alternative, the dynamics can be projected onto the range space of that matrix [115,131,133]. The elimination of the constraint forces is accomplished in accordance with the four steps used to transform the fixed-base manipulator dynamics in Section 4.5.2. First, the system dynamics in the upper part of (4.163) are projected onto $\mathcal{R}(J_c)$ by premultiplication with $J_c M^{-1}$. Second, making use of the second-order motion constraint from the lower part of the same equation, one arrives at

$$M_c^{-1}\lambda + J_c M^{-1}(c + g) - \dot{J}_c\dot{q} = J_c M^{-1}\mathcal{Q}. \tag{4.201}$$

Here $M_c^{-1} = (J_c M^{-1} J_c^T)$ stands for the *system mobility tensor* mapped along the constrained motion directions. This matrix exists at nonsingular postures where J_c has a full row rank. The third step is to obtain Lagrange's multiplier vector (the constraint force), i.e.

$$\lambda = J_c^{-MT}\mathcal{Q} + M_c\dot{J}_c\dot{q} - J_c^{-MT}(c + g). \tag{4.202}$$

Here $J_c^{-M} = M^{-1} J_c M_c$ denotes the inertia-weighted system constraint matrix. In the final step, substitute the constraint force $\boldsymbol{\lambda}$ back into the upper part of (4.163), i.e.

$$M\ddot{q} + J_c^T M_c \dot{J}_c \dot{q} + N^*(J_c)(c + g) = N^*(J_c)\mathcal{Q}. \tag{4.203}$$

The null-space projector $N^*(J_c)$ is given in (4.192).[12] This form of the equation of motion appeared in [133]. The equation can be further simplified as

$$M\ddot{q} = -J_c^T M_c \dot{J}_c \dot{q} + \mathcal{Q}_m. \tag{4.204}$$

This is exactly the same equation as (4.180) and (4.195), obtained with the direct elimination/Gauss and Maggi methods, respectively. It becomes then clear that the range-space projection method is identical to all the other methods discussed so far.

4.12.5 Summary and Conclusions

It has been confirmed that the direct elimination, Gauss' principle of least constraint, Maggi's null-space projection, and the range-space projection methods yield identical equations: (4.168), (4.180), (4.195), and (4.204), respectively. Their equivalence was confirmed under the following conditions:

- the dynamics are projected onto $\mathcal{N}^*(J_c)$,
- the respective null-space projector is parametrized as in (4.192),
- the arbitrary generalized force $\mathcal{Q}_a$ is set as in (4.194).

These conditions are relevant in the case of *ideal* constraints, whereby motion is always in harmony with Gauss' principle of least constraint. With a given generalized input force, the motion is determined in a unique way. This also implies uniqueness of all the components: the Lagrange's multiplier (4.170), the generalized constraint force (4.169), the unconstrained generalized acceleration $\ddot{q}_u$, and the constraint-consistent generalized acceleration with its two components shown in (4.174) (the linear, tangential $\ddot{q}_m$ and the nonlinear, state-dependent component in the normal direction).

In a real system, however, the constraints are never ideal; there is always some friction and compliance at the contact joints. This renders the constraint equations inhomogeneous and calls for active control of the constraint forces. Especially, in multilimb/finger/leg robots, the control of these forces is essential for reaction-based propulsion and manipulation. Such control is achievable in an infinite number of ways since the generalized forces $\mathcal{Q}_u$ and $\mathcal{Q}_a$, appearing in (4.181) and (4.189), respectively, can be chosen freely. The former is relevant to the tangential-force control, the latter ensures the controllability in the normal direction and, thus, of the internal force. Note also that, through the choice of $\mathcal{Q}_u$, Lagrange's multiplier vector (4.177), the constraint force $\mathcal{Q}_c$ (4.179), and the constraint maintaining force $\mathcal{Q}_m$ (4.181) will be determined in a unique way. Furthermore, recall that nonideal constraints will render the motion of the system nonintegrable at the acceleration level. As a consequence, a joint velocity drift may appear, which will necessitate an internal motion (self-motion) control.

[12] This projector is sometimes referred to as the "dynamically consistent" null-space projector [115,131,133].

4.13 REDUCED-FORM REPRESENTATIONS OF THE EQUATION OF MOTION

Reduced-form representations of the equation of motion lead to a system of reduced dimensions. An example of such representation for a space robot model floating in zero gravity, without and with the presence of an external force, was given in (4.127) and (4.133), respectively. In the example, the system dynamics were expressed in terms of the joint-space partial dynamics. Another possibility is to express the system dynamics in terms of the spatial partial dynamics. As will be shown in what follows, these reduced-form representations also result in the elimination of the constraint forces. Another type of reduced-form representation can be obtained via projections onto the end-link (sub)space, yielding dynamic relations in the end-link spatial coordinates.

For clarity but without loss of generality, the derivations will be based on the equation of motion expressed in terms of base quasivelocity, (4.155).

4.13.1 Joint-Space Dynamics–Based Representation

A reduced-form representation of the equation of motion can be obtained by mapping the spatial dynamics component onto the joint space. The joint acceleration is derived from the lower part of (4.155) as

$$\ddot{\boldsymbol{\theta}} = \boldsymbol{M}_{\theta B}^{-1}(\boldsymbol{\tau}_u - \boldsymbol{H}_{BB}^T \dot{\mathcal{V}}_B) - \boldsymbol{M}_{\theta B}^{-1} \mathcal{J}_{cB}^T \boldsymbol{\lambda}. \tag{4.205}$$

Hereby, definitions $\boldsymbol{\lambda} \equiv -\overline{\mathcal{F}}^c$ and $\boldsymbol{\tau}_u \equiv \boldsymbol{\tau} - \boldsymbol{c}_{\theta B} - \boldsymbol{g}_\theta$ were used. To enforce the motion constraints, insert (4.205) into (2.104), under the assumption of hard constraints ($\dot{\mathcal{V}}^c = \mathbf{0}$). The joint acceleration will then be eliminated from the latter equation. The resultant equation is solved for Lagrange's multiplier vector, i.e.

$$\boldsymbol{\lambda} = \mathcal{J}_{cB}^{-M_{\theta B}T}(\boldsymbol{\tau}_u - \boldsymbol{H}_{BB}^T \dot{\mathcal{V}}_B) + \boldsymbol{M}_{cB}\left(\boldsymbol{h}_{\theta B} + \mathbb{C}_{cB}^T \dot{\mathcal{V}}_B\right), \tag{4.206}$$

where $\boldsymbol{h}_{\theta B} \equiv \dot{\mathcal{J}}_{cB}\dot{\boldsymbol{\theta}} + \dot{\mathbb{C}}_{cB}^T \mathcal{V}_B$. The matrix

$$\boldsymbol{M}_{cB} \equiv \left(\mathcal{J}_{cB}\boldsymbol{M}_{\theta B}^{-1}\mathcal{J}_{cB}^T\right)^{-1} \in \Re^{c\times c} \tag{4.207}$$

stands for the *joint-space inertia tensor* mapped along the constrained motion directions. Substitute this result back into (4.205) to eliminate the constraint forces, so we have

$$\ddot{\boldsymbol{\theta}} = \overline{\boldsymbol{T}}_{\theta B}(\boldsymbol{\tau}_u - \boldsymbol{H}_{BB}^T \dot{\mathcal{V}}_B) - \mathcal{J}_{cB}^{-M_{\theta B}}\left(\boldsymbol{h}_{\theta B} + \mathbb{C}_{cB}^T \dot{\mathcal{V}}_B\right). \tag{4.208}$$

Here $\mathcal{J}_{cB}^{-M_{\theta B}}$ is the inertia-weighted generalized inverse of the joint-space constraint Jacobian (cf. also (4.53)). The matrix

$$\begin{aligned}\overline{\boldsymbol{T}}_{\theta B} &\equiv \boldsymbol{M}_{\theta B}^{-1}\left(\boldsymbol{E} - \mathcal{J}_{cB}^T \mathcal{J}_{cB}^{-M_{\theta B}T}\right) \\ &= \boldsymbol{M}_{\theta B}^{-1} - \boldsymbol{M}_{\theta B}^{-1}\mathcal{J}_{cB}^T \boldsymbol{M}_{cB}\mathcal{J}_{cB}\boldsymbol{M}_{\theta B}^{-1}\end{aligned} \tag{4.209}$$

is symmetric and positive definite, under the assumption of independent constraints. This is the *joint-space mobility tensor* restricted by the null space of the joint-space constraint Jacobian; $\overline{\boldsymbol{T}}_{\theta B}$ will be referred to as the *constraint-consistent joint-space mobility tensor*. Note that, for a given set of independent constraints, $\overline{\boldsymbol{T}}_{\theta B}$ is unique. Premultiply (4.208) by $\boldsymbol{M}_{\theta B}$ to obtain the equation of motion as

$$\boldsymbol{M}_{\theta B}\ddot{\boldsymbol{\theta}} = \left(\boldsymbol{E} - \mathcal{J}_{cB}^T \mathcal{J}_{cB}^{-M_{\theta B}T}\right)(\boldsymbol{\tau}_u - \boldsymbol{H}_{BB}^T \dot{\mathcal{V}}_B) - \mathcal{J}_{cB}^T \left(\mathcal{J}_{cB} \boldsymbol{M}_{\theta B}^{-1} \mathcal{J}_{cB}^T\right)^{-1} \left(\boldsymbol{h}_{\theta B} + \mathbb{C}_{cB}^T \dot{\mathcal{V}}_B\right). \quad (4.210)$$

In this representation, the dimension of the equation of motion has been reduced from $n+6$ to n. The arbitrary joint torque $\boldsymbol{\tau}_u$ can play the role of a control input component. It is filtered by the null-space projector $\mathcal{N}^*(\mathcal{J}_{cB})$ to produce a *constraint maintaining* joint torque, $\boldsymbol{\tau}_m = \left(\boldsymbol{E} - \mathcal{J}_{cB}^T \mathcal{J}_{cB}^{-M_{\theta B}T}\right)\boldsymbol{\tau}_u$, that is useful for motion control along the unconstrained motion directions. This joint torque does not produce any acceleration along the unconstrained motion directions, i.e. it is dynamically consistent.

4.13.2 Spatial Dynamics–Based Representation (Lagrange–d'Alembert Formulation)

The representation of the equation of motion in terms of spatial dynamics is known as Lagrange–d'Alembert's formulation. This formulation has been employed in the field of multifinger grasp modeling ([87], p. 280). Indeed, note that the grasped object is not directly actuated. It is then interesting to observe that the equation of motion of the system (i.e. multiple actuated fingers plus nonactuated object) comprises the same structure as that of a humanoid robot (multiple actuated limbs plus nonactuated base link).

The formulation can be derived from d'Alembert's principle of virtual work. The virtual work of the generalized input and constraint forces along virtual displacement $\delta \boldsymbol{q}_B = [\delta \mathcal{X}_B^T \ \ \delta\boldsymbol{\theta}^T]^T$ is expressed as $\delta W = \delta \boldsymbol{q}_B^T(\mathcal{Q} - \mathcal{Q}_{cB})$. According to the principle, the generalized constraint force does no virtual work; $\delta \boldsymbol{q}_B^T \mathcal{Q}_{cB} = \boldsymbol{0}$. Recalling that $\mathcal{Q}_{cB} = \boldsymbol{M}\ddot{\boldsymbol{q}}_B - \mathcal{Q}_u$, $\mathcal{Q}_u$ standing for the generalized force of the unconstrained system (cf. (4.163)), the principle of virtual work can be rewritten as

$$\delta \boldsymbol{q}_B^T\left(\mathcal{Q}_u - \boldsymbol{M}\ddot{\boldsymbol{q}}_B\right) = 0, \quad (4.211)$$

or

$$[\delta \mathcal{X}_B^T \ \ \delta\boldsymbol{\theta}^T]\left(\begin{bmatrix} \boldsymbol{0} \\ \mathcal{Q} \end{bmatrix} - \begin{bmatrix} \mathbb{M}_B & \boldsymbol{H}_{BB} \\ \boldsymbol{H}_{BB}^T & \boldsymbol{M}_{\theta B} \end{bmatrix}\begin{bmatrix} \dot{\mathcal{V}}_B \\ \ddot{\boldsymbol{\theta}} \end{bmatrix} - \begin{bmatrix} \mathcal{C}_B \\ \boldsymbol{c}_{\theta B} \end{bmatrix} - \begin{bmatrix} \mathcal{G}_B \\ \boldsymbol{g}_\theta \end{bmatrix}\right) = 0. \quad (4.212)$$

Thus

$$\begin{aligned}
0 &= \delta\boldsymbol{\theta}^T \left(\mathcal{Q} - \boldsymbol{M}_{\theta B}\ddot{\boldsymbol{\theta}} - \boldsymbol{H}_{BB}^T \dot{\mathcal{V}}_B - \boldsymbol{c}_{\theta B} - \boldsymbol{g}_\theta\right) - \delta \mathcal{X}_B^T \left(\boldsymbol{H}_{BB}\ddot{\boldsymbol{\theta}} + \mathbb{M}_B \dot{\mathcal{V}}_B + \mathcal{C}_B + \mathcal{G}_B\right) \\
&= \left(\mathcal{J}_{cB}{}^{\#}\mathbb{C}_{cB}^T \delta \mathcal{X}_B\right)^T \left(\mathcal{Q} - \boldsymbol{M}_{\theta B}\ddot{\boldsymbol{\theta}} - \boldsymbol{H}_{BB}^T \dot{\mathcal{V}}_B - \boldsymbol{c}_{\theta B} - \boldsymbol{g}_\theta\right) - \delta \mathcal{X}_B^T \left(\boldsymbol{H}_{BB}\ddot{\boldsymbol{\theta}} + \mathbb{M}_B \dot{\mathcal{V}}_B + \mathcal{C}_B + \mathcal{G}_B\right) \\
&= \left(\mathcal{J}_{cB}{}^{\#}\mathbb{C}_{cB}^T\right)^T \left(\mathcal{Q} - \boldsymbol{M}_{\theta B}\ddot{\boldsymbol{\theta}} - \boldsymbol{H}_{BB}^T \dot{\mathcal{V}}_B - \boldsymbol{c}_{\theta B} - \boldsymbol{g}_\theta\right) - \left(\boldsymbol{H}_{BB}\ddot{\boldsymbol{\theta}} + \mathbb{M}_B \dot{\mathcal{V}}_B + \mathcal{C}_B + \mathcal{G}_B\right).
\end{aligned}$$

Hereby, the virtual displacements in the joints were expressed using (2.91) as

$$\delta\boldsymbol{\theta} = \mathcal{J}_{cB}^{\#}\mathbb{C}_{cB}^{T}\,\delta\mathcal{X}_B \tag{4.213}$$

for any virtual displacement of the base link $\delta\mathcal{X}_B$. Finally, substitute the constraint-consistent joint velocity and acceleration, i.e.

$$\dot{\boldsymbol{\theta}} = \mathcal{J}_{cB}^{\#}\mathbb{C}_{cB}^{T}\mathcal{V}_B \tag{4.214}$$

and

$$\ddot{\boldsymbol{\theta}} = \mathcal{J}_{cB}^{\#}\left(\mathbb{C}_{cB}^{T}\dot{\mathcal{V}}_B + \dot{\mathbb{C}}_{c}^{T}\mathcal{V}_B - \dot{\mathcal{J}}_{cB}\dot{\boldsymbol{\theta}}\right) \tag{4.215}$$

(derived from (2.91) and (2.104), respectively), into the above relations to arrive at

$$\widetilde{\mathbb{M}}_B\dot{\mathcal{V}}_B + \widetilde{\mathcal{C}}_B + \widetilde{\mathcal{G}}_B = \mathcal{F}_B. \tag{4.216}$$

Hereby,

$$\begin{aligned}
\widetilde{\mathbb{M}}_B &= \mathbb{M}_B + \mathbb{C}_{cB}\mathcal{J}_{cB}^{\#T}\boldsymbol{M}_{\theta B}\mathcal{J}_{cB}^{\#}\mathbb{C}_{cB}^{T},\\
\widetilde{\mathcal{C}}_B &= \mathcal{C}_B + \mathbb{C}_{cB}\mathcal{J}_{cB}^{\#T}\left\{\boldsymbol{c}_{\theta B} + \boldsymbol{M}_{\theta B}\mathcal{J}_{cB}^{\#}\left(\dot{\mathbb{C}}_{cB}^{T} - \dot{\mathcal{J}}_{cB}\mathcal{J}_{cB}^{\#}\mathbb{C}_{cB}^{T}\right)\mathcal{V}_B\right\},\\
\widetilde{\mathcal{G}}_B &= \mathcal{G}_B + \mathbb{C}_{cB}\mathcal{J}_{cB}^{\#T}\boldsymbol{g}_{\theta},\\
\mathcal{F}_B &= \mathbb{C}_{cB}\mathcal{J}_{cB}^{\#T}\boldsymbol{\tau}.
\end{aligned}$$

With (4.216), the equation of motion of the humanoid robot is expressed in the form of spatial dynamics. The dimension of the system has been significantly reduced, from $n+6$ to six. The forces of constraint appear only implicitly (the $\mathcal{J}_{cB}^{\#T}\boldsymbol{\tau}$-term in the body wrench $\mathcal{F}_B$). Note, however, that with this representation, the partial dynamic relations stemming from the presence of kinematic redundancy cannot be accounted for.

Adjoining the Object Dynamics

Lagrange–d'Alembert's formulation is quite suitable to represent the combined dynamics of the humanoid robot and an object grasped by the robot. The object and the arms form an independent closed loop, the object playing thereby the role of the loop-closure link. The general case of unilateral hand/object contact joints is assumed, $\mathbb{C}_{cH}$ denoting the contact map. The coordinate frame of the object, $\{H\}$, is fixed at its CoM. The object's mass and inertia tensors are denoted as M_H and $\boldsymbol{I}_H$, respectively. Newton–Euler's equation of motion of the object is written as

$$\begin{bmatrix} M_H\boldsymbol{E}_3 & \boldsymbol{0}\\ \boldsymbol{0} & \boldsymbol{I}_H\end{bmatrix}\begin{bmatrix}\dot{\boldsymbol{v}}_H\\ \dot{\boldsymbol{\omega}}_H\end{bmatrix} + \begin{bmatrix}\boldsymbol{0}\\ \boldsymbol{\omega}_H\times\boldsymbol{I}_H\boldsymbol{\omega}_H\end{bmatrix} + \begin{bmatrix}M_H\boldsymbol{g}\\ \boldsymbol{0}\end{bmatrix} = \begin{bmatrix}\boldsymbol{f}_H\\ \boldsymbol{m}_H\end{bmatrix}. \tag{4.217}$$

The compact form representation of this equation is

$$\mathbb{M}_H\dot{\mathcal{V}}_H + \mathcal{C}_H + \mathcal{G}_H = \mathcal{F}_H, \tag{4.218}$$

where $\mathbb{M}_H$ is the constant spatial inertia of the object, $\mathcal{C}_H$ contains the gyroscopic torque term, and $\mathcal{G}_H$ is the gravity wrench. The net wrench, $\mathcal{F}_H$, is in a quasistatic balance with the contact wrenches *impressed* (by the hands) upon the object, i.e.

$$\mathcal{F}_H = -\mathbb{C}_{cH}(\boldsymbol{q}_H)\overline{\mathcal{F}}_H^c.$$

Furthermore, adjoin the object dynamics to those of the humanoid robot. For clarity and without loss of generality, the compact form representations (4.162) and (4.218) will be used, the former being expressed in terms of the base quasivelocity. We have

$$\begin{bmatrix} \mathbb{M}_H & \boldsymbol{H}_{HB} \\ \boldsymbol{H}_{HB}^T & \boldsymbol{M}_B \end{bmatrix}\begin{bmatrix} \dot{\mathcal{V}}_H \\ \ddot{\boldsymbol{q}}_B \end{bmatrix} + \begin{bmatrix} \mathcal{C}_H \\ \boldsymbol{c}_B \end{bmatrix} + \begin{bmatrix} \mathcal{G}_H \\ \boldsymbol{g}_B \end{bmatrix} = \begin{bmatrix} \boldsymbol{0} \\ \mathcal{Q} \end{bmatrix} + \begin{bmatrix} -\mathbb{C}_{cH}(\boldsymbol{q}) \\ \boldsymbol{\mathcal{J}}_{cB}^T \end{bmatrix}\overline{\mathcal{F}}^c. \tag{4.219}$$

Here $\boldsymbol{H}_{HB} \in \Re^{6\times(n+6)}$ is a map accounting for the inertial coupling between the object and the hands. Note that $\boldsymbol{H}_{HB}$ and $\mathbb{C}_{cH}(\boldsymbol{q})$ map the joint accelerations of the legs and the contact wrenches at the feet, respectively, at zero since these quantities do not contribute to the object dynamics. Lagrange–d'Alembert's formulation can be derived from the above equation of motion by employing the principle of virtual work with regard to the virtual displacements of the object and the robot. The final result can be written as

$$\widetilde{\mathbb{M}}_H\dot{\mathcal{V}}_H + \widetilde{\mathcal{C}}_H + \widetilde{\mathcal{G}}_H = \mathcal{F}_H, \tag{4.220}$$

where

$$\begin{aligned}
\widetilde{\mathbb{M}}_H &= \mathbb{M}_H + \mathbb{C}_{cH}\boldsymbol{\mathcal{J}}_{cB}^{\#T}\boldsymbol{M}_{\theta B}\boldsymbol{\mathcal{J}}_{cB}^{\#}\mathbb{C}_{cH}^T + \mathbb{C}_{cH}\boldsymbol{\mathcal{J}}_{cB}^{\#T}\boldsymbol{H}_{HB} + \left(\mathbb{C}_{cH}\boldsymbol{\mathcal{J}}_{cB}^{\#T}\boldsymbol{H}_{HB}\right)^T,\\
\widetilde{\mathcal{C}}_H &= \mathcal{C}_H + \mathbb{C}_{cH}\boldsymbol{\mathcal{J}}_{cB}^{\#T}\left\{\boldsymbol{c}_{\theta B} + \boldsymbol{M}_{\theta B}\boldsymbol{\mathcal{J}}_{cB}^{\#}\left(\dot{\mathbb{C}}_{cH}^T - \dot{\boldsymbol{\mathcal{J}}}_{cB}\boldsymbol{\mathcal{J}}_{cB}^{\#}\mathbb{C}_{cH}\right)\mathcal{V}_H\right\}\\
&\quad + \boldsymbol{H}_{HB}^T\boldsymbol{\mathcal{J}}_{cB}^{\#}\left(\dot{\mathbb{C}}_{cH}^T - \dot{\boldsymbol{\mathcal{J}}}_{cB}\boldsymbol{\mathcal{J}}_{cB}^{\#}\mathbb{C}_{cH}\right)\mathcal{V}_H,\\
\widetilde{\mathcal{G}}_H &= \mathcal{G}_H + \mathbb{C}_{cH}\boldsymbol{\mathcal{J}}_{cB}^{\#T}\boldsymbol{g}_\theta,\\
\mathcal{F}_H &= \mathbb{C}_{cH}\boldsymbol{\mathcal{J}}_{cB}^{\#T}\boldsymbol{\tau}.
\end{aligned}$$

4.13.3 Equation of Motion in End-Link Spatial Coordinates

The joint acceleration can be eliminated from the equation of motion to arrive at a dynamic model represented in terms of the end-link spatial coordinates. The motivation behind such an approach for the case of a fixed-base manipulator was clarified in Section 4.5.2. The same motivation is valid in the case of a humanoid robot, as shown in [115] (see also [131, 133]). In these works, the system dynamics were first projected with a dynamically consistent null-space projector along the constrained motion directions at the contacts. With such a model, the *motion control tasks* for the unconstrained motion directions and balance control are achieved within the null space of the projected dynamics. In this way, the *contact control task* can be performed in a dynamically decoupled manner, with the highest task priority. To ensure the performance of the lower-priority (motion) tasks, the system dynamics have to be projected along the unconstrained motion directions, in a subsequent step.

In the remainder of this section, the end-link spatial coordinate representations of the constrained dynamic models derived so far will be obtained by projecting their system dynamics along:

- the constrained motion directions (force-based projection);
- the unconstrained motion directions (motion-based projection);
- both the constrained and the unconstrained motion directions.

Constrained Dynamics Projection Along the Constrained Motion Directions

The projection of the constrained dynamics onto the subspace of constraints in end-link coordinates is achieved in accordance with the four steps used to transform the fixed-base manipulator dynamics in Section 4.5.2 (see also [115]). At the first step, project the constrained equation of motion (e.g. (4.204)) along the constrained motion directions by premultiplication with $\boldsymbol{J}_c\boldsymbol{M}^{-1}$. The second step is to employ the second-order differential kinematics along the constrained motion directions, i.e. $\boldsymbol{J}_c\ddot{\boldsymbol{q}} = \dot{\mathcal{V}}^c - \dot{\boldsymbol{J}}_c\dot{\boldsymbol{q}}$. Note that the general case of soft constraints is assumed with a nonzero $\dot{\mathcal{V}}^c$. The resultant equation is simple:

$$\begin{aligned}\dot{\mathcal{V}}^c &= \boldsymbol{J}_c\overline{\boldsymbol{T}}\mathcal{Q}_m \\ &= \boldsymbol{J}_c\overline{\boldsymbol{T}}(\mathcal{Q} - \boldsymbol{c} - \boldsymbol{g}).\end{aligned} \tag{4.221}$$

Here $\overline{\boldsymbol{T}} = \boldsymbol{M}^{-1}\boldsymbol{N}^*(\boldsymbol{J}_c)$ is the system mobility tensor restricted by the (dual) null space of the system constraint Jacobian; $\overline{\boldsymbol{T}}$ is henceforth referred to as the *constraint-consistent system mobility tensor*. The third step is to employ the quasistatic force relation along the constrained motion directions, i.e.

$$\mathcal{Q} = \boldsymbol{J}_c^T\overline{\mathcal{F}}^c + \boldsymbol{N}^*(\boldsymbol{J}_c)\mathcal{Q}_a, \tag{4.222}$$

$\mathcal{Q}_a$ denoting an arbitrary generalized force. Substituting (4.222) into (4.221) results in

$$\dot{\mathcal{V}}^c = \boldsymbol{J}_c\overline{\boldsymbol{T}}\boldsymbol{J}_c^T\overline{\mathcal{F}}^c + \boldsymbol{J}_c\overline{\boldsymbol{T}}(\mathcal{Q}_a - \boldsymbol{c} - \boldsymbol{g}). \tag{4.223}$$

Hereby, the idempotence property of the null-space projector was employed. The final step is to solve for $\overline{\mathcal{F}}^c$. We have

$$\overline{\mathcal{F}}^c = \overline{\boldsymbol{M}}_c\dot{\mathcal{V}}^c - (\boldsymbol{J}_c^{-\overline{T}})^T(\mathcal{Q}_a - \boldsymbol{c} - \boldsymbol{g}). \tag{4.224}$$

The matrix $\overline{\boldsymbol{M}}_c \equiv \left(\boldsymbol{J}_c\overline{\boldsymbol{T}}\boldsymbol{J}_c{}^T\right)^{-1}$ is the *constraint-consistent system inertia mapped along the constrained directions* and $\boldsymbol{J}_c^{-\overline{T}}$ is the (right) pseudoinverse of $\boldsymbol{J}_c$ weighted by the constraint-consistent mobility tensor $\overline{\boldsymbol{T}}$.

The last equation represents the constrained dynamics mapped along the constrained motion directions. It is easy to confirm that the same result can be derived from the null space–projected equation of motion (4.196) [78].

System Dynamics Projection Along the Unconstrained Motion Directions

The same procedure as above is applied to project the system dynamics along the unconstrained motion directions. The projection procedure is initialized with premultiplication of the constrained equation of motion by $\boldsymbol{J}_m\boldsymbol{M}^{-1}$. The second step is to employ the second-order differential kinematics along the unconstrained motion directions, i.e. $\boldsymbol{J}_m\ddot{\boldsymbol{q}} = \dot{\bar{\mathcal{V}}}^m - \dot{\boldsymbol{J}}_m\dot{\boldsymbol{q}}$. After these steps, the equation of motion assumes the form

$$\dot{\bar{\mathcal{V}}}^m - \dot{\boldsymbol{J}}_m\dot{\boldsymbol{q}} + \boldsymbol{J}_m\boldsymbol{J}_c^{-M}\dot{\boldsymbol{J}}_c\dot{\boldsymbol{q}} + \boldsymbol{J}_m\overline{\boldsymbol{T}}(\boldsymbol{c}+\boldsymbol{g}) = \boldsymbol{J}_m\overline{\boldsymbol{T}}\mathcal{Q}. \tag{4.225}$$

In the third step, substitute the quasistatic force relation along the unconstrained motion directions, i.e.

$$\mathcal{Q} = \boldsymbol{J}_m^T\bar{\mathcal{F}}^m + \boldsymbol{N}^*(\boldsymbol{J}_m)\mathcal{Q}_a. \tag{4.226}$$

The term on the r.h.s. of (4.225) then assumes the form

$$\boldsymbol{J}_m\overline{\boldsymbol{T}}\mathcal{Q} = \overline{\boldsymbol{M}}_m\bar{\mathcal{F}}^m + \boldsymbol{J}_m\overline{\boldsymbol{T}}\boldsymbol{N}^*(\boldsymbol{J}_m)\mathcal{Q}_a, \tag{4.227}$$

$\overline{\boldsymbol{M}}_m \equiv \left(\boldsymbol{J}_m\overline{\boldsymbol{T}}\boldsymbol{J}_m{}^T\right)^{-1}$ denoting the *constraint-consistent system inertia mapped along the unconstrained motion directions*. Note that in the second term, the arbitrary generalized force $\mathcal{Q}_a$ is projected onto the null space $\mathcal{N}^*(\boldsymbol{J}) = \mathcal{N}^*(\boldsymbol{J}_c) \cap \mathcal{N}^*(\boldsymbol{J}_m)$. This term will result in zero in the case of a robot with *nonredundant* limbs.

The final step is to solve for the quasistatic force $\bar{\mathcal{F}}^m$, i.e.

$$\bar{\mathcal{F}}^m = \overline{\boldsymbol{M}}_m\dot{\bar{\mathcal{V}}}^m + \mathcal{C}_m + \mathcal{G}_m - \overline{(\boldsymbol{J}_m^{-\overline{T}})^T}\mathcal{Q}_a. \tag{4.228}$$

Here $\boldsymbol{J}_m^{-\overline{T}}$ is the (right) pseudoinverse of $\boldsymbol{J}_m$ weighted by the constraint-consistent mobility tensor $\overline{\boldsymbol{T}}$ and $\overline{(\boldsymbol{J}_m^{-\overline{T}})^T}$ stands for the restriction of its transpose by the (dual) null space of the mobility Jacobian:

$$\overline{(\boldsymbol{J}_m^{-\overline{T}})^T} = (\boldsymbol{J}_m^{-\overline{T}})^T\boldsymbol{N}^*(\boldsymbol{J}_m).$$

The nonlinear velocity-dependent and gravity terms are

$$\mathcal{C}_m = (\boldsymbol{J}_m^{-\overline{T}})^T\boldsymbol{c} + \overline{\boldsymbol{M}}_m(\boldsymbol{J}_m\boldsymbol{J}_c^{-M}\dot{\boldsymbol{J}}_c - \dot{\boldsymbol{J}}_m)\dot{\boldsymbol{q}},$$
$$\mathcal{G}_m = (\boldsymbol{J}_m^{-\overline{T}})^T\boldsymbol{g}.$$

Projection Along the Constrained and Unconstrained Motion Directions

Consider the system dynamics written in terms of the base quasivelocity. The first-order instantaneous motion relation of the humanoid robot, (2.93), can be written as

$$\begin{bmatrix} \mathcal{V}_B \\ \mathcal{V} \end{bmatrix} = \begin{bmatrix} \boldsymbol{E} & \boldsymbol{0} \\ \mathbb{C}_B^T & \mathcal{J}_B \end{bmatrix}\begin{bmatrix} \mathcal{V}_B \\ \dot{\boldsymbol{\theta}} \end{bmatrix}. \tag{4.229}$$

Denote by $\boldsymbol{T}_B$ the 4×4 block matrix on the r.h.s. Assuming a robot with *nonredundant* limbs in a nonsingular configuration, the inverse transformation can be written as

$$\begin{bmatrix} \mathcal{V}_B \\ \dot{\boldsymbol{\theta}} \end{bmatrix} = \begin{bmatrix} \boldsymbol{E} & \boldsymbol{0} \\ -\mathcal{J}_B^{-1}\mathbb{C}_B^T & \mathcal{J}_B^{-1} \end{bmatrix}\begin{bmatrix} \mathcal{V}_B \\ \mathcal{V} \end{bmatrix}, \tag{4.230}$$

the 4 × 4 block matrix on the r.h.s. standing for the inverse $\boldsymbol{T}_B^{-1}$. The compact form of the equation of motion, (4.162), is projected with the help of $\boldsymbol{T}_B$ as

$$\boldsymbol{M}_B(\boldsymbol{q})\begin{bmatrix}\dot{\mathcal{V}}_B\\ \dot{\mathcal{V}}\end{bmatrix}+\boldsymbol{C}_B(\boldsymbol{q},\dot{\boldsymbol{q}})\begin{bmatrix}\mathcal{V}_B\\ \mathcal{V}\end{bmatrix}+\boldsymbol{g}_B(\boldsymbol{q})=\begin{bmatrix}-\mathbb{C}_B\\ \boldsymbol{E}\end{bmatrix}\mathcal{J}_B^{-T}\boldsymbol{\tau}+\boldsymbol{T}_B^{-T}\mathcal{Q}_{cB}, \tag{4.231}$$

where $\boldsymbol{M}_B \equiv \boldsymbol{T}_B^{-T}\boldsymbol{M}\boldsymbol{T}_B^{-1}$, $\boldsymbol{C}_B \equiv \boldsymbol{T}_B^{-T}\boldsymbol{c}+\boldsymbol{T}_B^{-T}\boldsymbol{M}\frac{d}{dt}(\boldsymbol{T}_B^{-1})$, and $\boldsymbol{g}_B \equiv \boldsymbol{T}_B^{-T}\boldsymbol{g}$. Since the contact (reaction) wrenches are assumed to act at the contact frames, the last term on the r.h.s. of (4.231) simplifies s.t. the equation of motion becomes

$$\boldsymbol{M}_B(\boldsymbol{q})\begin{bmatrix}\dot{\mathcal{V}}_B\\ \dot{\mathcal{V}}\end{bmatrix}+\boldsymbol{C}_B(\boldsymbol{q},\dot{\boldsymbol{q}})\begin{bmatrix}\mathcal{V}_B\\ \mathcal{V}\end{bmatrix}+\boldsymbol{g}_B(\boldsymbol{q})=\begin{bmatrix}-\mathbb{C}_B\\ \boldsymbol{E}\end{bmatrix}\mathcal{J}_B^{-T}\boldsymbol{\tau}+\begin{bmatrix}\boldsymbol{0}\\ \boldsymbol{E}\end{bmatrix}\mathcal{F}^c. \tag{4.232}$$

Note that there is a duality between this representation and (4.155) [47]. In the above representation, the joint torque appears in all equations while the contact wrenches appear only in the lower part. In (4.155) the situation is just the opposite: the contact wrenches appear in all equations while the joint torque appears only in the lower part. This duality can be useful in balance controller design, as will be clarified in Chapter 5 (see e.g. Section 5.10.6). Note also that the upper part of (4.232) is in the form of Lagrange–d'Alembert's formulation, (4.216). A drawback in the above notation is its limitation to nonredundant systems (because of the appearance of the inverse Jacobian). A balance controller based on the above transformation is presented in Section 5.13.2.

4.13.4 Summary and Discussion

There is a variety of reduced-form representations of the equation of motion. This variety stems from the subspaces and the respective transforms pertinent to both the kinetostatic and the dynamic interrelations. The understanding of these interrelations is quite important in view of a control algorithm design. The interrelations can be represented graphically via a characteristic map for the states of the system. An example of a characteristic map for the states of a dynamic model expressed in terms of the base quasivelocity is shown in Fig. 4.9. This map is actually an extension of Fig. 3.10, appearing at the end of Chapter 3. Recall that three transforms, the Jacobians $\boldsymbol{J}_B, \mathcal{J}_{cB}$ and the contact map $\mathbb{C}_{cB}$, were designated as the fundamental kinetostatic transforms. The latter two transforms appear in both figures (shown with horizontal arrows).

Furthermore, in Fig. 4.9 also dynamic transforms are included (displayed with vertical and diagonal arrows). The vertical arrows reveal three *inertia-type* dynamic transforms, $\boldsymbol{M}_{\theta B}, \mathbb{M}_B$, and $\boldsymbol{M}_{cB}^{-1}$. These transforms act between the dual subspaces of the joint, end-link, and base-link spaces, in the direction from the motion ($\mathcal{M}$) to the force ($\mathcal{F} \equiv \mathcal{M}^*$) domain. They do not induce decompositions of the relevant subspaces. Obviously, $\boldsymbol{M}_{\theta B}$ and $\mathbb{M}_B$ play a fundamental role in the dynamic relations; $\boldsymbol{M}_{cB}$ is a derivative (cf. (4.207)). The diagonal arrows denote coupling inertia–type dynamic transforms (that express *mechanical connection maps* [86,109]). The transform from the joint-motion subspace to the base-link force subspace is represented by the coupling inertia matrix $\boldsymbol{H}_{BB}$. The dual transform, from the base-link motion subspace

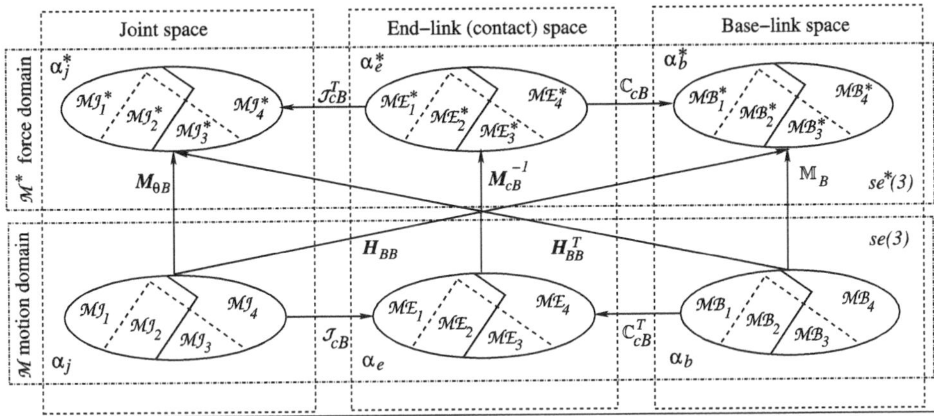

Joint space ($\mathcal{M}$)		Base-link space ($\mathcal{M}$)	
$\mathcal{MJ}_1 = \mathcal{N}(H_{BB}) \cap \mathcal{N}(\mathcal{J}_{cB})$	$\mathcal{N}(H_{BB}) = \mathcal{MJ}_1 \cup \mathcal{MJ}_4$	$\mathcal{MB}_1 = \mathcal{N}(\mathbb{C}_{cB}) \cap \mathcal{N}(H_{BB}^T)$	$\mathcal{N}(\mathbb{C}_{cB}) = \mathcal{MB}_1 \cup \mathcal{MB}_4$
$\mathcal{MJ}_2 = \mathcal{R}(H_{BB}^T) \cap \mathcal{N}(\mathcal{J}_{cB})$	$\mathcal{R}(H_{BB}^T) = \mathcal{MJ}_2 \cup \mathcal{MJ}_3$	$\mathcal{MB}_2 = \mathcal{R}(\mathbb{C}_{cB}^T) \cap \mathcal{N}(H_{BB}^T)$	$\mathcal{R}(\mathbb{C}_{cB}^T) = \mathcal{MB}_2 \cup \mathcal{MB}_3$
$\mathcal{MJ}_3 = \mathcal{R}(H_{BB}^T) \cap \mathcal{R}(\mathcal{J}_{cB}^T)$	$\mathcal{N}(\mathcal{J}_{cB}) = \mathcal{MJ}_1 \cup \mathcal{MJ}_2$	$\mathcal{MB}_3 = \mathcal{R}(\mathbb{C}_{cB}^T) \cap \mathcal{R}(H_{BB})$	$\mathcal{N}(H_{BB}^T) = \mathcal{MB}_1 \cup \mathcal{MB}_2$
$\mathcal{MJ}_4 = \mathcal{N}(H_{BB}) \cap \mathcal{R}(\mathcal{J}_{cB}^T)$	$\mathcal{R}(\mathcal{J}_{cB}^T) = \mathcal{MJ}_3 \cup \mathcal{MJ}_4$	$\mathcal{MB}_4 = \mathcal{N}(\mathbb{C}_{cB}) \cap \mathcal{R}(H_{BB})$	$\mathcal{R}(H_{BB}) = \mathcal{MB}_3 \cup \mathcal{MB}_4$

FIGURE 4.9 Characteristic map for the states of a dynamic model expressed in terms of the base quasivelocity. Right angles signify orthogonality decompositions within the motion and force subdomains, stemming from the fundamental transforms $\mathcal{J}_{cB}$, J, and $\mathbb{C}_{cB}$ (kinetostatics) and $M_{\theta B}$, $\mathbb{M}_B$, and H_{BB} (dynamics). The αs denote motion elements (velocities/accelerations), the α^*s are their duals (momenta/forces). The horizontal arrows signify the kinetostatic transforms. The vertical arrows denote the inertia-type motion-to-force dynamic transforms across the dual subspaces. The diagonal arrows stand for the coupling inertia–type motion-to-force transforms across the nondual subspaces (i.e. the mechanical connection [86,109]).

to the joint torque subspace, is represented by the transposed coupling inertia, H_{BB}^T. The coupling-inertia transform does induce a decomposition of the relevant subspaces, as do the kinetostatic transforms.

To keep the presentation tractable, not all of the range and null spaces of the kinematic, kinetostatic, inertia, and coupling-inertia transforms are shown in Fig. 4.9. In the lower part of the figure, subdomains determined by the range and null spaces of the underlying transforms are shown for the motion subdomains within the joint and base-link subspaces. Similar relations can be obtained for the respective force subdomains. Representative dynamic states within the $\mathcal{MJ}$ domain are:

- $\alpha_{j1} \in \mathcal{MJ}_1$: motion component along the unconstrained motion directions induced by reactionless motion;
- $\alpha_{j2} \in \mathcal{MJ}_2$: motion component along the unconstrained motion directions induced by a motion with maximum inertial coupling;
- $\alpha_{j3} \in \mathcal{MJ}_3$: motion component along the constrained motion directions induced by a motion with maximum inertial coupling;
- $\alpha_{j4} \in \mathcal{MJ}_4$: motion component along the constrained motion directions induced by reactionless motion.

4.14 INVERSE DYNAMICS

A model-based control design makes use of inverse dynamic relations to determine the joint torque as control input. Sought are expressions for the joint torque that explicitly comprise components to be used for the realization of:

- end-link motion/force control,
- internal motion/force control,
- spatial dynamics control (i.e. CoM and angular momentum control).

In Section 4.13, a few ways of representing the equation of motion have been introduced. Depending on the notation, one or more of the above joint torque components may not be exhibited explicitly in the inverse dynamics solution. The following discussion is meant to provide an understanding about the choice of an appropriate notation in controller design.

4.14.1 Based on the Direct Elimination/Gauss/Maggi/Projection Methods

To derive the inverse dynamics solution, recall first that the generalized input force $\mathcal{Q} = \boldsymbol{S}^T \boldsymbol{\tau}$ is one of the components of the unconstrained generalized force $\mathcal{Q}_u$. The latter appears as a projection in the equations (i.e. as constraint maintaining force $\mathcal{Q}_m$ in (4.180) and (4.193)). In other words, the system is singular and underdetermined. To remove the c extra equations, make use of Maggi's approach (the null-space projection method). Applying transformations (4.199) to (4.180) yields

$$\boldsymbol{V}_r^T \mathcal{Q}_u = \boldsymbol{V}_r^T \left(\boldsymbol{M}\ddot{\boldsymbol{q}} + \boldsymbol{J}_c^T \boldsymbol{M}_c \dot{\boldsymbol{J}}_c \dot{\boldsymbol{q}} \right). \tag{4.233}$$

This is an undetermined linear system of $r = n + 6 - c$ equations in $n + 6$ unknowns. The general solution can be written as

$$\mathcal{Q}_u = \boldsymbol{V}_r \boldsymbol{V}_r^T \left(\boldsymbol{M}\ddot{\boldsymbol{q}} + \boldsymbol{J}_c^T \boldsymbol{M}_c \dot{\boldsymbol{J}}_c \dot{\boldsymbol{q}} \right) + \left(\boldsymbol{E} - \boldsymbol{V}_r \boldsymbol{V}_r^T \right) \mathcal{Q}_a, \tag{4.234}$$

whereby identities (4.190) were used. Since $\boldsymbol{N}^*$ is parametrized as in (4.192), the last equation can be rewritten as

$$\mathcal{Q} = \left(\boldsymbol{E} - \boldsymbol{J}_c^T \boldsymbol{J}_c^{-MT} \right) \boldsymbol{M}\ddot{\boldsymbol{q}} + \boldsymbol{J}_c^T \boldsymbol{J}_c^{-MT} \mathcal{Q}_a + \boldsymbol{c} + \boldsymbol{g}, \tag{4.235}$$

whereby the expression $\mathcal{Q}_u = \mathcal{Q} - \boldsymbol{c} - \boldsymbol{g}$ and the orthogonality relation

$$\left(\boldsymbol{E} - \boldsymbol{J}_c^T \boldsymbol{J}_c^{-MT} \right) \boldsymbol{J}_c^T \boldsymbol{M}_c = \boldsymbol{0}$$

were used. The joint torque can be obtained from the last n rows of solution (4.235), via the relation $\mathcal{Q} = \boldsymbol{S}^T \boldsymbol{\tau}$. This solution is characterized by simplicity, dynamic consistency, and instantaneous kinetic energy minimization. The quantities suitable as control inputs are the constraint-consistent generalized acceleration $\ddot{\boldsymbol{q}}$ and the arbitrary generalized force $\mathcal{Q}_a$. The former can be used for internal and/or end-link motion control, via the inverse kinematics.

It can also be used to control the tangential force. The latter is useful for internal and/or end-link force control (e.g. for reactive propulsion/manipulation). In such case, relations (4.194) will be violated, though, which implies a case of imperfect constraints. Note also that there is no explicit control input for the spatial dynamics. Nevertheless, the base link acceleration $\dot{\mathcal{V}}_B$, a component of constraint-consistent generalized acceleration $\ddot{\boldsymbol{q}}$, can be used to control these dynamics indirectly, via (4.99).

If the task of the robot does not require internal force control capability, $\mathcal{Q}_a$ in (4.235) can be set in accordance with (4.194). We have

$$\mathcal{Q} = \left(\boldsymbol{E} - \boldsymbol{J}_c^T \boldsymbol{J}_c^{-MT}\right) \boldsymbol{M}\ddot{\boldsymbol{q}} - \boldsymbol{J}_c^T \boldsymbol{M}_c \dot{\boldsymbol{J}}_c \dot{\boldsymbol{q}} + \boldsymbol{c} + \boldsymbol{g}. \tag{4.236}$$

This is the inverse dynamics solution in the case of ideal constraints. Only $\ddot{\boldsymbol{q}}$ is available as control input, the rest of the terms are nonlinear and state-dependent.

The inverse dynamics solution of a humanoid robot can also be obtained through the null-space projection method [124]. To this end, rewrite (4.200) as

$$\boldsymbol{V}_r^T(\boldsymbol{M}\ddot{\boldsymbol{q}} + \boldsymbol{c} + \boldsymbol{g}) = \boldsymbol{V}_r^T \boldsymbol{S}^T \boldsymbol{W}^{-\frac{1}{2}} \boldsymbol{W}^{\frac{1}{2}} \boldsymbol{\tau}, \tag{4.237}$$

where $\boldsymbol{W} \in \Re^{n\times n}$ is a p.d. weighting matrix. Solving for the joint torque, one arrives at

$$\boldsymbol{\tau} = \boldsymbol{V}^\dagger \boldsymbol{V}_r^T(\boldsymbol{M}\ddot{\boldsymbol{q}} + \boldsymbol{c} + \boldsymbol{g}) + \left(\boldsymbol{E} - \boldsymbol{V}^\dagger \boldsymbol{V}_r^T \boldsymbol{S}^T\right)\boldsymbol{\tau}_a, \tag{4.238}$$

where the generalized inverse $\boldsymbol{V}^\dagger = \boldsymbol{W}^{-\frac{1}{2}}\left(\boldsymbol{V}_r^T \boldsymbol{S}^T \boldsymbol{W}^{-\frac{1}{2}}\right)^+ \in \Re^{n\times r}$ and the relations $\mathcal{Q}_u = \mathcal{Q} - \boldsymbol{c} - \boldsymbol{g}$, $\mathcal{Q} = \boldsymbol{S}^T\boldsymbol{\tau} = \begin{bmatrix}0 & \boldsymbol{\tau}^T\end{bmatrix}^T$ were used. The two orthogonal components on the r.h.s. are parametrized by $\boldsymbol{W}$. The normal component is also parametrized by the arbitrary joint torque vector $\boldsymbol{\tau}_a$. The parametrization can be used to minimize the joint torque $\boldsymbol{\tau}$ and, indirectly, the force of the constraint (the reactions). With the latter minimization, the direct handling of the inequality constraints pertinent to the friction cone can be avoided. This can be done either via minimization of the tangential contact forces, or by tracking a suitably designed reaction in the normal direction. No contact joint models thereby need to appear in the equation [124].

In the previous discussion, it was mentioned that the minimum null-space basis operator, $\boldsymbol{V}_r$, can be derived from the SVD of the system constraint matrix $\boldsymbol{J}_c$. It is possible to make use of other known decomposition methods, e.g. the QR factorization. The transpose of the system constraint Jacobian can be factorized as $\boldsymbol{J}_c^T = \boldsymbol{Q}\begin{bmatrix}\boldsymbol{R}^T & 0\end{bmatrix}^T$, whereby $\boldsymbol{R} \in \Re^{c\times c}$ and $\boldsymbol{Q} \in \Re^{(n+6)\times(n+6)}$ denote full-rank upper triangular and orthogonal matrices, respectively [77]. The latter can be decomposed as $\boldsymbol{Q} = \begin{bmatrix}\boldsymbol{Q}_c & \boldsymbol{Q}_r\end{bmatrix}$, whereby $\boldsymbol{Q}_r$ provides the minimum null-space basis. With this factorization/decomposition, the inverse dynamics solution attains exactly the same form as (4.238), wherein $\boldsymbol{V}_r$ is just replaced by $\boldsymbol{Q}_r$. Note also that Lagrange's multiplier vector (i.e. the reaction force) can be obtained in a straightforward manner, if needed, as follows:

$$\boldsymbol{\lambda} = \boldsymbol{R}^{-1}\boldsymbol{Q}_c^T(\boldsymbol{M}\ddot{\boldsymbol{q}} + \boldsymbol{c} + \boldsymbol{g} - \boldsymbol{S}^T\boldsymbol{\tau}). \tag{4.239}$$

Both results, (4.235) and (4.238), comprise two orthogonal components; they provide the base for a dynamic controller design. The latter solution has an advantage though: there is no need at all to calculate the inverse M^{-1} that appears in the J_c^{-MT}-term in (4.235). On the other hand, solution (4.238) does not guarantee the dynamic decoupling and kinetic energy minimization, as does (4.235). Also, solution (4.238) includes explicitly the V_r (or Q_r) decomposition component. The decomposition imposes an additional computational burden.

4.14.2 Based on Lagrange–d'Alembert's Formulation

According to Lagrange–d'Alembert's representation of the equation of motion, (4.216), the joint torque can be derived from the following expression:

$$\mathcal{F}_B = \mathbb{C}_{cB}\mathcal{J}_{cB}^{\#T}\boldsymbol{\tau}. \tag{4.240}$$

Solving for the joint torque, one arrives at

$$\begin{aligned}\boldsymbol{\tau} &= \mathcal{J}_{cB}^T\mathbb{C}_{cB}^{\#}\mathcal{F}_B + \mathcal{J}_{cB}^T\overline{\mathcal{F}}^n,\\ \overline{\mathcal{F}}^n &= (\boldsymbol{E} - \mathbb{C}_{cB}^{\#}\mathbb{C}_{cB})\overline{\mathcal{F}}_a^c.\end{aligned} \tag{4.241}$$

The net wrench acting at the base link, $\mathcal{F}_B$, is obtained from Lagrange–d'Alembert's formulation, as in (4.216); $\overline{\mathcal{F}}_a^c$ parametrizes the null space $\mathcal{N}(\mathbb{C}_{cB})$, i.e. the internal-wrench null space (cf. (3.62)). Evaluating this result from the prospective control viewpoint, it becomes apparent that two control components are explicitly present. The spatial dynamics and the internal wrench can be controlled with appropriate choices of $\mathcal{F}_B/\dot{\mathcal{V}}_B$ and $\overline{\mathcal{F}}_a^c$. On the other hand, note that there is no explicit component for the joint/end-link motion control.

4.14.3 Based on the Joint-Space Dynamics Elimination Approach

First, recall that equation of motion (4.210) represents a system of n equations. It is assumed that the number of joints n exceeds the number of constraints c. The difference $r = n - c$ denotes the degree of redundancy. The linear subsystem, determined by the (dual) null-space projection operator $\boldsymbol{N}^*(\mathcal{J}_{cB}) = (\boldsymbol{E} - \mathcal{J}_{cB}^T\mathcal{J}_{cB}^{-M_{\theta B}T})$, is singular (rank$\boldsymbol{N}^*(\mathcal{J}_{cB}) = r$) and, hence, underdetermined. To obtain the joint torque, the procedure used with the direct elimination approach will be applied. Accordingly, assume that $\boldsymbol{N}^*(\mathcal{J}_{cB})$ is represented by $\boldsymbol{V}_{rB}\boldsymbol{V}_{rB}^T$, with an appropriate modification in the dimensions, i.e.

$$\boldsymbol{V}_{rB}\boldsymbol{V}_{rB}^T(\boldsymbol{\tau}_u - \boldsymbol{H}_{BB}^T\dot{\mathcal{V}}_B) = \boldsymbol{M}_{\theta B}\ddot{\boldsymbol{\theta}} + \mathcal{J}_{cB}^T\left(\mathcal{J}_{cB}\boldsymbol{M}_{\theta B}^{-1}\mathcal{J}_{cB}^T\right)^{-1}\left(\boldsymbol{h}_{\theta B} + \mathbb{C}_{cB}^T\dot{\mathcal{V}}_B\right). \tag{4.242}$$

Then, reduce the dimension of the above equation to r by annihilating the extra c equations. This is accomplished by premultiplication with $\boldsymbol{V}_{rB}^T$. The resultant equation is solved for the joint torque of the unconstrained system, $\boldsymbol{\tau}_u$. We have

$$\begin{aligned}\boldsymbol{\tau}_u - \boldsymbol{H}_{BB}^T\dot{\mathcal{V}}_B &= \boldsymbol{V}_{rB}\boldsymbol{V}_{rB}^T\boldsymbol{M}_{\theta B}\ddot{\boldsymbol{\theta}}\\ &+ \boldsymbol{V}_{rB}\boldsymbol{V}_{rB}^T\mathcal{J}_{cB}^T\left(\mathcal{J}_{cB}\boldsymbol{M}_{\theta B}^{-1}\mathcal{J}_{cB}^T\right)^{-1}\left(\boldsymbol{h}_{\theta B} + \mathbb{C}_{cB}^T\dot{\mathcal{V}}_B\right) + \mathcal{J}_{cB}^T\mathcal{J}_{cB}^{-M_{\theta B}T}\boldsymbol{\tau}_a,\end{aligned} \tag{4.243}$$

where $\boldsymbol{V}_{rB}^{+T} = \boldsymbol{V}_{rB}$ and $\boldsymbol{E} - \boldsymbol{V}_{rB}\boldsymbol{V}_{rB}^{T} = \mathcal{J}_{cB}^{T}\mathcal{J}_{cB}^{-M_{\theta B}T}$ were used. The input joint torque is then finally obtained as

$$\boldsymbol{\tau} = (\boldsymbol{E} - \mathcal{J}_{cB}^{T}\mathcal{J}_{cB}^{-M_{\theta B}T})\boldsymbol{M}_{\theta B}\ddot{\boldsymbol{\theta}} + \boldsymbol{H}_{BB}^{T}\dot{\mathcal{V}}_{B} + \mathcal{J}_{cB}^{T}\overline{\mathcal{F}}_{a}^{c} + \boldsymbol{c}_{\theta B} + \boldsymbol{g}_{\theta}. \tag{4.244}$$

Hereby, the orthogonality relation $(\boldsymbol{E} - \mathcal{J}_{cB}^{T}\mathcal{J}_{cB}^{-M_{\theta B}T})\mathcal{J}_{cB}^{T}(\mathcal{J}_{cB}\boldsymbol{M}_{\theta B}^{-1}\mathcal{J}_{cB}^{T})^{-1} = \boldsymbol{0}$ was used and the term $\mathcal{J}_{cB}^{-M_{\theta B}T}\boldsymbol{\tau}_{a}$ was replaced with an arbitrary contact wrench $\overline{\mathcal{F}}_{a}^{c}$. The normal joint torque component $\mathcal{J}_{cB}^{T}\overline{\mathcal{F}}_{a}^{c}$ can be used for reaction/internal and/or end-link force control. The constraint-consistent joint acceleration $\ddot{\boldsymbol{\theta}}$ is available for tangential force and/or internal motion control. It can also be used for the end-link motion control, via the inverse kinematic relations. Furthermore, it is apparent that $\boldsymbol{H}_{BB}^{T}\dot{\mathcal{V}}_{B}$, $\boldsymbol{c}_{\theta B}$, and $\boldsymbol{g}_{\theta}$ comprise both normal and tangential components. The latter two are usually compensated in the controller. The former term, on the other hand, represents a prospective control input for the base-link motion/spatial dynamics.

4.14.4 Summary and Conclusions

The inverse dynamics solutions are determined under the assumption that the state $(\boldsymbol{q}, \dot{\boldsymbol{q}}_{(\circ)})$, the generalized acceleration $\ddot{\boldsymbol{q}}_{(\circ)} = \begin{bmatrix} \dot{\mathcal{V}}_{(\circ)}^{T} & \ddot{\boldsymbol{\theta}}^{T} \end{bmatrix}^{T}$, and the contact forces $\overline{\mathcal{F}}^{c}$ are known. The state can be obtained either from the calculations during the previous time step in a simulator, or from sensor signals in the real robot. The generalized acceleration and the contact forces are usually determined as reference values that include feedforward and feedback terms in the controller. There are two major issues that need to be tackled. First, since humanoid robots are usually subjected to multiple motion/force task constraints, the system quite frequently may arrive at an overconstrained state (cf. the discussion in Section 2.8). At such a state, no solution to the inverse dynamics problem can be found. Second, the inequality constraints pertinent to the friction cone cannot be handled directly. There are a few ways to deal with these two problems. Two of them, the assignment of priorities (fixed or variable) among the tasks and the use of general solvers to handle inequality constraints, have been discussed in Chapter 2 from the kinetostatic point of view. Others, that are based on dynamic relations, will be introduced in Section 5.14.

Concluding this section, it should be noted that due to the significant computational cost, the complete model of a humanoid robot is most often used in off-line algorithms, mainly in simulators. Complete models also appear in both off-line and on-line motion generators. In the former case, for example, the so-called "dynamics filter" has been introduced to transform the captured (complex) whole-body movements of an actor into motions of a real robot [165,169,172]. In the latter case, the complete model can be used to confirm the correctness of the motion [155] or to compensate the errors generated when using a simple dynamic model [149]. Furthermore, complete models have also been used in the field of parameter identification [156,79,7]. Readers interested in an on-line implementation of a whole-body control approach with a resolved acceleration type controller are referred to [27].

References

[1] F. Aghili, A unified approach for inverse and direct dynamics of constrained multibody systems based on linear projection operator: applications to control and simulation, IEEE Transactions on Robotics 21 (2005) 834–849.

[2] A. Albu-Schaffer, O. Eiberger, M. Grebenstein, S. Haddadin, C. Ott, T. Wimbock, S. Wolf, G. Hirzinger, Soft robotics, IEEE Robotics & Automation Magazine 15 (2008) 20–30.

[3] A. Albu-Schaffer, C. Ott, U. Frese, G. Hirzinger, Cartesian impedance control of redundant robots: recent results with the DLR-light-weight-arms, in: 2003 IEEE International Conference on Robotics and Automation, IEEE, 2003, pp. 3704–3709 (Cat. No. 03CH37422).

[4] H. Arai, S. Tachi, Position control of manipulator with passive joints using dynamic coupling, IEEE Transactions on Robotics and Automation 7 (1991) 528–534.

[5] S. Arimoto, Control Theory of Non-Linear Mechanical Systems: A Passivity-Based and Circuit-Theoretic Approach, Clarendon Press, 1996.

[6] D. Asmar, B. Jalgha, A. Fakih, Humanoid fall avoidance using a mixture of strategies, International Journal of Humanoid Robotics 09 (2012) 1250002.

[7] K. Ayusawa, G. Venture, Y. Nakamura, Identifiability and identification of inertial parameters using the underactuated base-link dynamics for legged multibody systems, The International Journal of Robotics Research 33 (2014) 446–468.

[8] A. Balestrino, G. De Maria, L. Sciavicco, Adaptive control of manipulators in the task oriented space, in: 13th Int. Symp. Ind. Robots, 1983, pp. 131–146.

[9] E. Bayo, R. Ledesma, Augmented Lagrangian and mass-orthogonal projection methods for constrained multibody dynamics, Nonlinear Analysis 9 (1996) 113–130.

[10] B.J. Benda, P.O. Riley, D.E. Krebs, Biomechanical relationship between center of gravity and center of pressure during standing, IEEE Transactions on Rehabilitation Engineering 2 (1994) 3–10.

[11] W. Blajer, A geometric unification of constrained system dynamics, Multibody System Dynamics 1 (1997) 3–21.

[12] D.J. Block, K.J. Åström, M.W. Spong, The Reaction Wheel Pendulum, Morgan & Claypool Publishers, 2007.

[13] W. Book, O. Maizza-Neto, D. Whitney, Feedback control of two beam, two joint systems with distributed flexibility, Journal of Dynamic Systems, Measurement, and Control 97 (1975) 424.

[14] A. Bowling, S. Harmeyer, Repeatable redundant manipulator control using nullspace quasivelocities, Journal of Dynamic Systems, Measurement, and Control 132 (2010) 031007.

[15] R. Brockett, Asymptotic stability and feedback stabilization, in: R.W. Brockett, R.S. Millmann, H.J. Sussmann (Eds.), Differential Geometric Control Theory, Birkhäuser, Boston, Basel, Stuttgart, 1983, pp. 181–191.

[16] H. Bruyninckx, O. Khatib, Gauss' principle and the dynamics of redundant and constrained manipulators, in: IEEE International Conference on Robotics and Automation, 2000, pp. 2563–2568.

[17] M. Carpenter, M.A. Peck, Reducing base reactions with gyroscopic actuation of space-robotic systems, IEEE Transactions on Robotics 25 (2009) 1262–1270.

[18] M. Cefalo, G. Oriolo, M. Vendittelli, Planning safe cyclic motions under repetitive task constraints, in: IEEE International Conference on Robotics and Automation, 2013, pp. 3807–3812.

[19] K.-S. Chang, O. Khatib, Efficient algorithm for extended operational space inertia matrix, in: IEEE/RSJ International Conference on Intelligent Robots and Systems, 1999, pp. 350–355.

[20] N.a. Chaturvedi, N.H. McClamroch, D.S. Bernstein, Asymptotic smooth stabilization of the inverted 3-D pendulum, IEEE Transactions on Automatic Control 54 (2009) 1204–1215.

[21] C. Chevallereau, E. Westervelt, J.W. Grizzle, Asymptotically stable running for a five-link, four-actuator, planar bipedal robot, The International Journal of Robotics Research 24 (2005) 431–464.

[22] Y. Choi, D. Kim, Y. Oh, B.-j. You, Posture/walking control for humanoid robot based on kinematic resolution of CoM Jacobian with embedded motion, IEEE Transactions on Robotics 23 (2007) 1285–1293.

[23] Y. Choi, B.-j. You, S.-r. Oh, On the stability of indirect ZMP controller for biped robot systems, in: IEEE/RSJ International Conference on Intelligent Robots and Systems (IROS), 2004, pp. 1966–1971.

[24] J. Craig, Introduction to Robotics: Mechanics & Control, 3rd edition, Prentice Hall, 2004.

[25] R.H. Crompton, L. Yu, W. Weijie, M. Günther, R. Savage, The mechanical effectiveness of erect and "bent-hip, bent-knee" bipedal walking in Australopithecus afarensis, Journal of Human Evolution 35 (1998) 55–74.

[26] H. Dai, A. Valenzuela, R. Tedrake, Whole-body motion planning with centroidal dynamics and full kinematics, in: IEEE-RAS International Conference on Humanoid Robots, Madrid, Spain, 2014, pp. 295–302.

[27] B. Dariush, G.B. Hammam, D. Orin, Constrained resolved acceleration control for humanoids, in: IEEE/RSJ International Conference on Intelligent Robots and Systems, 2010, pp. 710–717.

[28] V. De Sapio, O. Khatib, Operational space control of multibody systems with explicit holonomic constraints, in: IEEE International Conference on Robotics and Automation, 2005, pp. 2950–2956.

[29] A. Dietrich, C. Ott, A. Albu-Schaffer, Multi-objective compliance control of redundant manipulators: hierarchy, control, and stability, in: IEEE/RSJ International Conference on Intelligent Robots and Systems, IEEE, 2013, pp. 3043–3050.

[30] J. Englsberger, A. Werner, C. Ott, B. Henze, M.A. Roa, G. Garofalo, R. Burger, A. Beyer, O. Eiberger, K. Schmid, A. Albu-Schaffer, Overview of the torque-controlled humanoid robot TORO, in: IEEE-RAS International Conference on Humanoid Robots, IEEE, Madrid, Spain, 2014, pp. 916–923.

[31] I. Fantoni, R. Lozano, M.W. Spong, Energy based control of the pendubot, IEEE Transactions on Automatic Control 45 (2000) 725–729.

[32] R. Featherstone, Rigid Body Dynamics Algorithms, Springer Science+Business Media, LLC, Boston, MA, 2008.

[33] R. Featherstone, O. Khatib, Load independence of the dynamically consistent inverse of the Jacobian matrix, The International Journal of Robotics Research 16 (1997) 168–170.

[34] C. Fitzgerald, Developing baxter, in: IEEE Conference on Technologies for Practical Robot Applications, 2013, pp. 1–6.

[35] Y. Fujimoto, A. Kawamura, Simulation of an autonomous biped walking robot including environmental force interaction, IEEE Robotics & Automation Magazine 5 (1998) 33–42.

[36] M. Gajamohan, M. Merz, I. Thommen, R. D'Andrea, The Cubli: a cube that can jump up and balance, in: IEEE/RSJ International Conference on Intelligent Robots and Systems, 2012, pp. 3722–3727.

[37] C.F. Gauß, Über ein neues allgemeines Grundgesetz der Mechanik, Journal für die Reine und Angewandte Mathematik 4 (1829).

[38] J.M. Gilbert, Gyrobot: control of multiple degree of freedom underactuated mechanisms using a gyrating link and cyclic braking, IEEE Transactions on Robotics 23 (2007) 822–827.

[39] A. Goswami, Postural stability of biped robots and the foot-rotation indicator (FRI) point, The International Journal of Robotics Research 18 (1999) 523–533.

[40] A. Goswami, V. Kallem, Rate of change of angular momentum and balance maintenance of biped robots, in: IEEE International Conference on Robotics and Automation, New Orleans, LA, USA, 2004, pp. 3785–3790.

[41] J. Grizzle, Planar bipedal robot with impulsive foot action, in: IEEE Conf. on Decision and Control, Atlantis, Paradise Island, Bahamas, 2004, pp. 296–302.

[42] M. Hara, D. Nenchev, Body roll reorientation via cyclic arm motion in zero gravity, Robotic Life Support Laboratory, Tokyo City University, 2017 (Video clip), https://doi.org/10.1016/B978-0-12-804560-2.00011-0.

[43] M. Hara, D. Nenchev, Object tracking with a humanoid robot in zero gravity (asynchronous leg motion in two phases), Robotic Life Support Laboratory, Tokyo City University, 2017 (Video clip), https://doi.org/10.1016/B978-0-12-804560-2.00011-0.

[44] M. Hara, D. Nenchev, Object tracking with a humanoid robot in zero gravity (asynchronous leg motion), Robotic Life Support Laboratory, Tokyo City University, 2017 (Video clip), https://doi.org/10.1016/B978-0-12-804560-2.00011-0.

[45] M. Hara, D. Nenchev, Object tracking with a humanoid robot in zero gravity (synchronous leg motion), Robotic Life Support Laboratory, Tokyo City University, 2017 (Video clip), https://doi.org/10.1016/B978-0-12-804560-2.00011-0.

[46] J. Hauser, R.M. Murray, Nonlinear controllers for non-integrable systems: the Acrobot example, in: 1990 American Control Conference, 1990, pp. 669–671.

[47] B. Henze, M.A. Roa, C. Ott, Passivity-based whole-body balancing for torque-controlled humanoid robots in multi-contact scenarios, The International Journal of Robotics Research 35 (2016) 1522–1543.

[48] G. Hirzinger, N. Sporer, A. Albu-Schaffer, M. Hahnle, R. Krenn, A. Pascucci, M. Schedl, DLR's torque-controlled light weight robot III-are we reaching the technological limits now?, in: IEEE International Conference on Robotics and Automation, 2002, pp. 1710–1716.

[49] N. Hogan, Impedance control: an approach to manipulation, in: American Control Conference, 1984, pp. 304–313.

[50] N. Hogan, Impedance control: an approach to manipulation: Part I–theory, Journal of Dynamic Systems, Measurement, and Control 107 (1985) 1.

[51] J.M. Hollerbach, Ki C. Suh, Redundancy resolution of manipulators through torque optimization, IEEE Journal on Robotics and Automation 3 (1987) 308–316.

[52] K. Iqbal, Y.-c. Pai, Predicted region of stability for balance recovery: motion at the knee joint can improve termination of forward movement, Journal of Biomechanics 33 (2000) 1619–1627.
[53] A. Jain, G. Rodriguez, An analysis of the kinematics and dynamics of underactuated manipulators, IEEE Transactions on Robotics and Automation 9 (1993) 411–422.
[54] S. Kajita, H. Hirukawa, K. Harada, K. Yokoi, Introduction to Humanoid Robotics, Springer Verlag, Berlin, Heidelberg, 2014.
[55] S. Kajita, F. Kanehiro, K. Kaneko, K. Fujiwara, K. Harada, K. Yokoi, H. Hirukawa, Biped walking pattern generation by using preview control of zero-moment point, in: IEEE International Conference on Robotics and Automation, 2003, pp. 1620–1626.
[56] S. Kajita, F. Kanehiro, K. Kaneko, K. Fujiwara, K. Harada, K. Yokoi, H. Hirukawa, Resolved momentum control: humanoid motion planning based on the linear and angular momentum, in: IEEE/RSJ International Conference on Intelligent Robots and Systems, Las Vegas, Nevada, 2003, pp. 1644–1650.
[57] S. Kajita, F. Kanehiro, K. Kaneko, K. Fujiwara, K. Harada, K. Yokoi, H. Hirukawa, Resolved momentum control: motion generation of a humanoid robot based on the linear and angular momenta, Journal of the Robotics Society of Japan 22 (2004) 772–779.
[58] S. Kajita, T. Yamaura, A. Kobayashi, Dynamic walking control of a biped robot along a potential energy conserving orbit, IEEE Transactions on Robotics and Automation 8 (1992) 431–438.
[59] R.E. Kalaba, F.E. Udwadia, Equations of motion for nonholonomic, constrained dynamical systems via Gauss's principle, ASME Journal of Applied Mechanics 60 (1993) 662–668.
[60] O. Khatib, A unified approach for motion and force control of robot manipulators: the operational space formulation, IEEE Journal on Robotics and Automation 3 (1987) 43–53.
[61] O. Khatib, Inertial properties in robotic manipulation: an object-level framework, The International Journal of Robotics Research 14 (1995) 19–36.
[62] T. Komura, H. Leung, J. Kuffner, S. Kudoh, J. Kuffner, A feedback controller for biped humanoids that can counteract large perturbations during gait, in: IEEE International Conference on Robotics and Automation, 2005, pp. 1989–1995.
[63] A. Konno, M. Uchiyama, M. Murakami, Configuration-dependent vibration controllability of flexible-link manipulators, The International Journal of Robotics Research 16 (1997) 567–576.
[64] T. Koolen, T. de Boer, J. Rebula, A. Goswami, J. Pratt, Capturability-based analysis and control of legged locomotion, Part 1: theory and application to three simple gait models, The International Journal of Robotics Research 31 (2012) 1094–1113.
[65] C. Lanczos, The Variational Principles of Mechanics, Dover Publications, 1970, 4th, 1986 edition.
[66] A. Laulusa, O.A. Bauchau, Review of classical approaches for constraint enforcement in multibody systems, Journal of Computational and Nonlinear Dynamics 3 (2008) 011004.
[67] S.-H. Lee, A. Goswami, Reaction Mass Pendulum (RMP): an explicit model for centroidal angular momentum of humanoid robots, in: IEEE International Conference on Robotics and Automation, 2007, pp. 4667–4672.
[68] S.-H. Lee, A. Goswami, The reaction mass pendulum (RMP) model for humanoid robot gait and balance control, in: B. Choi (Ed.), Humanoid Robots, InTech, 2009, p. 396, chapter 9.
[69] S.-H.H. Lee, A. Goswami, A momentum-based balance controller for humanoid robots on non-level and non-stationary ground, Autonomous Robots 33 (2012) 399–414.
[70] L. Lilov, M. Lorer, Dynamic analysis of multirigid-body system based on the gauss principle, ZAMM – Zeitschrift für Angewandte Mathematik und Mechanik 62 (1982) 539–545.
[71] H.-o. Lim, Y. Kaneshima, A. Takanishi, Online walking pattern generation for biped humanoid robot with trunk, in: IEEE International Conference on Robotics and Automation, Washington, DC, USA, 2002, pp. 3111–3116.
[72] J. Luh, M. Walker, R. Paul, Resolved-acceleration control of mechanical manipulators, IEEE Transactions on Automatic Control 25 (1980) 468–474.
[73] S. Ma, D.N. Nenchev, Local torque minimization for redundant manipulators: a correct formulation, Robotica 14 (1996) 235–239.
[74] A. Maciejewski, Kinetic limitations on the use of redundancy in robotic manipulators, IEEE Transactions on Robotics and Automation 7 (1991) 205–210.
[75] R. Marino, M.W. Spong, Nonlinear control techniques for flexible joint manipulators: a single link case study, in: 1986 IEEE International Conference on Robotics and Automation, 1986, pp. 1030–1036.

[76] J. Mayr, F. Spanlang, H. Gattringer, Mechatronic design of a self-balancing three-dimensional inertia wheel pendulum, Mechatronics 30 (2015) 1–10.
[77] M. Mistry, J. Buchli, S. Schaal, Inverse dynamics control of floating base systems using orthogonal decomposition, in: IEEE International Conference on Robotics and Automation, IEEE, Anchorage, AK, USA, 2010, pp. 3406–3412.
[78] M. Mistry, L. Righetti, Operational space control of constrained and underactuated systems, in: H. Durrant-Whyte, N. Roy, P. Abbeel (Eds.), Robotics: Science and Systems VII, MIT Press, 2012, pp. 225–232.
[79] M. Mistry, S. Schaal, K. Yamane, Inertial parameter estimation of floating base humanoid systems using partial force sensing, in: IEEE-RAS International Conference on Humanoid Robots, Paris, France, 2009, pp. 492–497.
[80] K. Mitobe, G. Capi, Y. Nasu, Control of walking robots based on manipulation of the zero moment point, Robotica 18 (2000) 651–657.
[81] K. Mitobe, G. Capi, Y. Nasu, A new control method for walking robots based on angular momentum, Mechatronics 14 (2004) 163–174.
[82] S. Miyahara, D. Nenchev, Walking in zero gravity, Robotic Life Support Laboratory, Tokyo City University, 2017 (Video clip), https://doi.org/10.1016/B978-0-12-804560-2.00011-0.
[83] M. Muehlebach, G. Mohanarajah, R. D'Andrea, Nonlinear analysis and control of a reaction wheel-based 3D inverted pendulum, in: 52nd IEEE Conference on Decision and Control, Florence, Italy, 2013, pp. 1283–1288.
[84] A. Müller, Motion equations in redundant coordinates with application to inverse dynamics of constrained mechanical systems, Nonlinear Dynamics 67 (2011) 2527–2541.
[85] M.P. Murray, A. Seireg, R.C. Scholz, Center of gravity, center of pressure, and supportive forces during human activities, Journal of Applied Physiology 23 (1967) 831–838.
[86] R.M. Murray, Nonlinear control of mechanical systems: a Lagrangian perspective, Annual Reviews in Control 21 (1997) 31–42.
[87] R.M. Murray, Z. Li, S.S. Sastry, A Mathematical Introduction to Robotic Manipulation, CRC Press, 1994.
[88] Y. Nakamura, T. Suzuki, M. Koinuma, Nonlinear behavior and control of a nonholonomic free-joint manipulator, IEEE Transactions on Robotics and Automation 13 (1997) 853–862.
[89] N. Naksuk, Y. Mei, C. Lee, Humanoid trajectory generation: an iterative approach based on movement and angular momentum criteria, in: IEEE/RAS International Conference on Humanoid Robots, Los Angeles, CA, USA, 2004, pp. 576–591.
[90] C. Natale, B. Siciliano, L. Villani, Spatial impedance control of redundant manipulators, in: IEEE International Conference on Robotics and Automation, 1999, pp. 1788–1793.
[91] B. Nemec, L. Zlajpah, Null space velocity control with dynamically consistent pseudo-inverse, Robotica 18 (2000) 513–518.
[92] D. Nenchev, Restricted Jacobian matrices of redundant manipulators in constrained motion tasks, The International Journal of Robotics Research 11 (1992) 584–597.
[93] D. Nenchev, A controller for a redundant free-flying space robot with spacecraft attitude/manipulator motion coordination, in: IEEE/RSJ International Conference on Intelligent Robots and Systems, 1993, pp. 2108–2114.
[94] D. Nenchev, Y. Umetani, K. Yoshida, Analysis of a redundant free-flying spacecraft/manipulator system, IEEE Transactions on Robotics and Automation 8 (1992) 1–6.
[95] D. Nenchev, K. Yoshida, Impact analysis and post-impact motion control issues of a free-floating space robot subject to a force impulse, IEEE Transactions on Robotics and Automation 15 (1999) 548–557.
[96] D. Nenchev, K. Yoshida, M. Uchiyama, Reaction null-space based control of flexible structure mounted manipulator systems, in: 5th IEEE Conference on Decision and Control, Kobe, Japan, 1996, pp. 4118–4123.
[97] D. Nenchev, K. Yoshida, Y. Umetani, Introduction of redundant arms for manipulation in space, in: IEEE International Workshop on Intelligent Robots, 1988, pp. 679–684.
[98] D. Nenchev, K. Yoshida, Y. Umetani, Analysis, design and control of free-flying space robots using fixed-attitude-restricted Jacobian matrix, in: The Fifth International Symposium on Robotics Research, MIT Press, Cambridge, MA, USA, 1990, pp. 251–258.
[99] D. Nenchev, K. Yoshida, P. Vichitkulsawat, M. Uchiyama, Reaction null-space control of flexible structure mounted manipulator systems, IEEE Transactions on Robotics and Automation 15 (1999) 1011–1023.
[100] D.N. Nenchev, Reaction null space of a multibody system with applications in robotics, Mechanical Sciences 4 (2013) 97–112.
[101] D.N. Nenchev, A. Nishio, Ankle and hip strategies for balance recovery of a biped subjected to an impact, Robotica 26 (2008) 643–653.

[102] A. Nishio, K. Takahashi, D. Nenchev, Balance control of a humanoid robot based on the reaction null space method, in: IEEE/RSJ Int. Conf. on Intelligent Robots and Systems, Beijing, China, 2006, pp. 1996–2001.

[103] K. Nishizawa, D. Nenchev, Body roll reorientation via cyclic foot motion in zero gravity, Robotic Life Support Laboratory, Tokyo City University, 2017 (Video clip), https://doi.org/10.1016/B978-0-12-804560-2.00011-0.

[104] R. Olfati-Saber, Nonlinear Control of Underactuated Mechanical Systems with Application to Robotics and Aerospace Vehicles, Ph.D. thesis, MIT, 2001.

[105] K. O'Neil, Divergence of linear acceleration-based redundancy resolution schemes, IEEE Transactions on Robotics and Automation 18 (2002) 625–631.

[106] D.E. Orin, A. Goswami, Centroidal momentum matrix of a humanoid robot: structure and properties, in: IEEE/RSJ International Conference on Intelligent Robots and Systems, IROS, Nice, France, 2008, pp. 653–659.

[107] D.E. Orin, A. Goswami, S.H. Lee, Centroidal dynamics of a humanoid robot, Autonomous Robots 35 (2013) 161–176.

[108] V. Ortenzi, M. Adjigble, J.A. Kuo, R. Stolkin, M. Mistry, An experimental study of robot control during environmental contacts based on projected operational space dynamics, in: IEEE-RAS International Conference on Humanoid Robots, Madrid, Spain, 2014, pp. 407–412.

[109] J.P. Ostrowski, Computing reduced equations for robotic systems with constraints and symmetries, IEEE Transactions on Robotics and Automation 15 (1999) 111–123.

[110] K. Osuka, T. Nohara, Attitude control of torque unit manipulator via trajectory planning – treatment as a nonholonomic system, Journal of the Robotics Society of Japan 18 (2000) 612–615 (in Japanese).

[111] K. Osuka, K. Yoshida, T. Ono, New design concept of space manipulator: a proposal of torque-unit manipulator, in: 33rd IEEE Conference on Decision and Control, 1994, pp. 1823–1825.

[112] C. Ott, A. Kugi, Y. Nakamura, Resolving the problem of non-integrability of nullspace velocities for compliance control of redundant manipulators by using semi-definite Lyapunov functions, in: IEEE International Conference on Robotics and Automation, 2008, pp. 1999–2004.

[113] Y.-c. Pai, J. Patton, Center of mass velocity-position predictions for balance control, Journal of Biomechanics 30 (1997) 347–354.

[114] J.G. Papastavridis, Maggi's equations of motion and the determination of constraint reactions, Journal of Guidance, Control, and Dynamics 13 (1990) 213–220.

[115] J. Park, Control Strategies for Robots in Contact, Ph.D. thesis, Stanford University, USA, 2006.

[116] J. Park, W. Chung, Y. Youm, On dynamical decoupling of kinematically redundant manipulators, in: IEEE/RSJ International Conference on Intelligent Robots and Systems, 1999, pp. 1495–1500.

[117] E. Popov, A. Vereschagin, S. Zenkevich, Manipulating Robots. Dynamics and Algorithms, Nauka, Moscow, USSR, 1978 (in Russian).

[118] M. Popovic, A. Hofmann, H. Herr, Angular momentum regulation during human walking: biomechanics and control, in: IEEE International Conference on Robotics and Automation, 2004, pp. 2405–2411.

[119] M. Popovic, A. Hofmann, H. Herr, Zero spin angular momentum control: definition and applicability, in: IEEE/RAS International Conference on Humanoid Robots, 2004, pp. 478–493.

[120] M.B. Popovic, A. Goswami, H. Herr, Ground reference points in legged locomotion: definitions, biological trajectories and control implications, The International Journal of Robotics Research 24 (2005) 1013–1032.

[121] J. Pratt, J. Carff, S. Drakunov, A. Goswami, Capture point: a step toward humanoid push recovery, in: IEEE-RAS International Conference on Humanoid Robots, Genoa, Italy, 2006, pp. 200–207.

[122] R.D. Quinn, J. Chen, C. Lawrence, Redundant manipulators for momentum compensation in a micro-gravity environment, in: AIAA Guidance, Navigation and Control Conference, 1988, AIAA PAPER 88-4121.

[123] M. Reyhanoglu, A. van der Schaft, N.H. McClamroch, I. Kolmanovsky, Dynamics and control of a class of underactuated mechanical systems, IEEE Transactions on Automatic Control 44 (1999) 1663–1671.

[124] L. Righetti, J. Buchli, M. Mistry, M. Kalakrishnan, S. Schaal, Optimal distribution of contact forces with inverse-dynamics control, The International Journal of Robotics Research 32 (2013) 280–298.

[125] L. Righetti, J. Buchli, M. Mistry, S. Schaal, Inverse dynamics control of floating-base robots with external constraints: a unified view, in: IEEE International Conference on Robotics and Automation, IEEE, 2011, pp. 1085–1090.

[126] N. Rotella, M. Bloesch, L. Righetti, S. Schaal, State estimation for a humanoid robot, in: IEEE/RSJ International Conference on Intelligent Robots and Systems, 2014, pp. 952–958.

[127] H. Sadeghian, L. Villani, M. Keshmiri, B. Siciliano, Dynamic multi-priority control in redundant robotic systems, Robotica 31 (2013) 1–13.

[128] A. Sano, J. Furusho, Realization of natural dynamic walking using the angular momentum information, in: IEEE International Conference on Robotics and Automation, Tsukuba, Japan, 1990, pp. 1476–1481.

[129] A. Sano, J. Furusho, Control of torque distribution for the BLR-G2 biped robot, in: Fifth International Conference on Advanced Robotics, vol. 1, IEEE, 1991, pp. 729–734.

[130] A.K. Sanyal, A. Goswami, Dynamics and balance control of the reaction mass pendulum: a three-dimensional multibody pendulum with variable body inertia, Journal of Dynamic Systems, Measurement, and Control 136 (2014) 021002.

[131] L. Sentis, Synthesis and Control of Whole-Body Behaviors in Humanoid Systems, Ph.D. thesis, Standford University, 2007.

[132] L. Sentis, O. Khatib, Control of free-floating humanoid robots through task prioritization, in: IEEE International Conference on Robotics and Automation, IEEE, 2005, pp. 1718–1723.

[133] L. Sentis, O. Khatib, Compliant control of multicontact and center-of-mass behaviors in humanoid robots, IEEE Transactions on Robotics 26 (2010) 483–501.

[134] J.S.J. Shen, A. Sanyal, N. Chaturvedi, D. Bernstein, H. McClamroch, Dynamics and control of a 3D pendulum, in: 43rd IEEE Conference on Decision and Control, Nassau, Bahamas, 2004, pp. 323–328.

[135] N.E. Sian, K. Yokoi, S. Kajita, F. Kanehiro, K. Tanie, Whole body teleoperation of a humanoid robot – a method of integrating operator's intention and robot's autonomy, in: IEEE International Conference on Robotics and Automation, 2003, pp. 1613–1619.

[136] M.W. Spong, Underactuated mechanical systems, in: Control Problems in Robotics and Automation, Springer-Verlag, London, 1998, pp. 135–150.

[137] M.W. Spong, D.J. Block, The Pendubot: a mechatronic system for control research and education, in: 1995 34th IEEE Conference on Decision and Control, New Orleans, LA, USA, 1995, pp. 555–556.

[138] M.W. Spong, P. Corke, R. Lozano, Nonlinear control of the reaction wheel pendulum, Automatica 37 (2001) 1845–1851.

[139] M.W. Spong, S. Hutchinson, M. Vidyasagar, Robot Modeling and Control, Wiley, 2006.

[140] M.W.M. Spong, Partial feedback linearization of underactuated mechanical systems, in: IEEE/RSJ International Conference on Intelligent Robots and Systems, IEEE, 1994, pp. 314–321.

[141] K. Sreenath, A.K. Sanyal, The reaction mass biped: equations of motion, hybrid model for walking and trajectory tracking control, in: IEEE International Conference on Robotics and Automation, 2015, pp. 5741–5746.

[142] L. Stirling, K. Willcox, D. Newman, Development of a computational model for astronaut reorientation, Journal of Biomechanics 43 (2010) 2309–2314.

[143] L.A. Stirling, Development of Astronaut Reorientation Methods, Ph.D. thesis, MIT, 2008.

[144] T. Sugihara, Y. Nakamura, H. Inoue, Real-time humanoid motion generation through ZMP manipulation based on inverted pendulum control, in: IEEE Int. Conf. on Robotics and Automation, Washington, DC, 2002, pp. 1404–1409.

[145] A. Takanishi, M. Ishida, Y. Yamazaki, I. Kato, The realization of dynamic walking by the biped walking robot WL-10RD, Journal of the Robotics Society of Japan 3 (1985) 325–336.

[146] A. Takanishi, H. o. Lim, M. Tsuda, I. Kato, Realization of dynamic biped walking stabilized by trunk motion on a sagittally uneven surface, in: IEEE International Workshop on Intelligent Robots and Systems, 1990, pp. 323–330.

[147] T. Takenaka, T. Matsumoto, T. Yoshiike, Real time motion generation and control for biped robot – 3rd report: dynamics error compensation, in: IEEE/RSJ International Conference on Intelligent Robots and Systems, St. Louis, USA, 2009, pp. 1594–1600.

[148] T. Takenaka, T. Matsumoto, T. Yoshiike, T. Hasegawa, S. Shirokura, H. Kaneko, A. Orita, Real time motion generation and control for biped robot – 4th report: integrated balance control, in: IEEE/RSJ Int. Conf. on Intelligent Robots and System, 2009, pp. 1601–1608.

[149] T. Takenaka, T. Matsumoto, T. Yoshiike, S. Shirokura, Real time motion generation and control for biped robot – 2nd report: running gait pattern generation, in: IEEE/RSJ International Conference on Intelligent Robots and Systems, 2009, pp. 1092–1099.

[150] N.G. Tsagarakis, S. Morfey, G. Medrano Cerda, Z. Li, D.G. Caldwell, COMpliant huMANoid COMAN: optimal joint stiffness tuning for modal frequency control, in: 2013 IEEE International Conference on Robotics and Automation, 2013, pp. 673–678.

[151] F.-C. Tseng, Z.-D. Ma, G.M. Hulbert, Efficient numerical solution of constrained multibody dynamics systems, Computer Methods in Applied Mechanics and Engineering 192 (2003) 439–472.

[152] B. Ugurlu, A. Kawamura, Bipedal trajectory generation based on combining inertial forces and intrinsic angular momentum rate changes: Eulerian ZMP resolution, IEEE Transactions on Robotics 28 (2012) 1406–1415.
[153] Y. Umetani, K. Yoshida, Resolved motion rate control of space manipulators with generalized Jacobian matrix, IEEE Transactions on Robotics and Automation 5 (1989) 303–314.
[154] Z. Vafa, S. Dubowsky, On the dynamics of manipulators in space using the virtual manipulator approach, in: IEEE International Conference on Robotics and Automation, 1987, pp. 579–585.
[155] B. Vanderborght, B. Verrelst, R. Van Ham, M. Van Damme, D. Lefeber, Objective locomotion parameters based inverted pendulum trajectory generator, Robotics and Autonomous Systems 56 (2008) 738–750.
[156] G. Venture, K. Ayusawa, Y. Nakamura, Dynamics identification of humanoid systems, in: CISM-IFToMM Symp. on Robot Design, Dynamics, and Control, ROMANSY, 2008, pp. 301–308.
[157] M. Vukobratović, Contribution to the study of anthropomorphic systems, Kybernetika 8 (1972) 404–418.
[158] M. Vukobratovic, B. Borovac, Zero-moment point – thirty five years of its life, International Journal of Humanoid Robotics 01 (2004) 157–173.
[159] M. Vukobratović, A.A. Frank, D. Juricić, On the stability of biped locomotion, IEEE Transactions on Bio-Medical Engineering 17 (1970) 25–36.
[160] P.M. Wensing, L.R. Palmer, D.E. Orin, Efficient recursive dynamics algorithms for operational-space control with application to legged locomotion, Autonomous Robots 38 (2015) 363–381.
[161] P.-B. Wieber, Holonomy and nonholonomy in the dynamics of articulated motion, in: M. Diehl, K. Mombaur (Eds.), Fast Motions in Biomechanics and Robotics, Springer-Verlag, Berlin, Heidelberg, 2006, pp. 411–425, Lecture no edition.
[162] D. Winter, Human balance and posture control during standing and walking, Gait & Posture 3 (1995) 193–214.
[163] Y. Xu, T. Kanade (Eds.), Space Robotics: Dynamics and Control, The Kluwer International Series in Engineering and Computer Science, vol. 188, Springer US, Boston, MA, 1993.
[164] K. Yamada, K. Tsuchiya, Force control of a space manipulator, in: i-SAIRAS, Kobe, Japan, 1990, pp. 255–258.
[165] K. Yamane, Y. Nakamura, Dynamics filter – concept and implementation of online motion generator for human figures, IEEE Transactions on Robotics and Automation 19 (2003) 421–432.
[166] R. Yang, Y.-Y. Kuen, Z. Li, Stabilization of a 2-DOF spherical pendulum on X-Y table, in: IEEE International Conference on Control Applications, 2000, pp. 724–729.
[167] K. Yoshida, D. Nenchev, A general formulation of under-actuated manipulator systems, in: S. Hirose, Y. Shirai (Eds.), Robotics Research: The 8th International Symposium, Springer, London, New York, 1997, pp. 72–79.
[168] Y. Yoshida, K. Takeuchi, Y. Miyamoto, D. Sato, D.N. Nenchev, Postural balance strategies in response to disturbances in the frontal plane and their implementation with a humanoid robot, IEEE Transactions on Systems, Man, and Cybernetics: Systems 44 (2014) 692–704.
[169] Y. Zheng, K. Yamane, Human motion tracking control with strict contact force constraints for floating-base humanoid robots, in: IEEE-RAS International Conference on Humanoid Robots, IEEE, Atlanta, GA, USA, 2013, pp. 34–41.
[170] Y. Zhang, D. Guo, S. Ma, Different-level simultaneous minimization of joint-velocity and joint-torque for redundant robot manipulators, Journal of Intelligent & Robotic Systems 72 (2013) 301–323.
[171] Y. Zhao, L. Sentis, A three dimensional foot placement planner for locomotion in very rough terrains, in: IEEE-RAS International Conference on Humanoid Robots, Humanoids 2012, IEEE, 2012, pp. 726–733.
[172] Y. Zheng, K. Yamane, Adapting human motions to humanoid robots through time warping based on a general motion feasibility index, in: IEEE International Conference on Robotics and Automation, IEEE, 2015, pp. 6281–6288.

CHAPTER

5

Balance Control

5.1 OVERVIEW

Balance control refers to the ability of a human to attain appropriate postures for a given task (e.g. standing, walking, running) within a given environment and to maintain the (dynamic) equilibrium of the body segments during task execution [61]. Numerous studies in the field of biomechanics and physical therapy have contributed to the understanding of the mechanisms used by humans during balance control [105,108,162,86,94,56,116]. Balance control of a humanoid robot bears many similarities to that of a human. Proper balance control is vital for the daily activities of humans. Likewise, balance control plays a central role within the overall control architecture of a humanoid robot. Improper balance control usually results in a fall that hinders the execution of the preplanned tasks. In addition, damage to the robot hardware may be incurred. Similar to balance control in humans [56], the control objectives for humanoid robot balance control can be divided into two large groups:

- balance control during *proactive* (preplanned) tasks;
- *reactive* balance control in response to unexpected external disturbances.

A preplanned task is usually determined by a set of motion/force trajectories, $\{T_{tg}(t)\}$, $t \in \{\overline{t_0, t_f}\}$. The motion trajectory of the robot quite often is expressed in terms of generalized coordinates and their time derivatives. Note that this information encodes the desired motion of the center of mass (CoM), provided one of the feet or both of them are attached to the ground. Furthermore, the force trajectory specifies a set of contact reaction wrenches that constrain the floating base in an appropriate manner, e.g. to support the robot against the gravity force or to ensure the desired way of propulsion. This information, together with that from the motion trajectory, encodes the desired motion of the centers of pressure (CoPs), which also play an important role in balance. Trajectory $T_{tg}(t)$ can be obtained in various ways, either off-line and/or on-line. The balance controller is based on an appropriate dynamic model of the robot, the task, and the environment (see Fig. 5.1). The task-space controller design depends on the low-level controller that is based either on position/velocity or torque control. The task-space controller provides feedback to the on-line trajectory modifier whose role is to ensure the *feasibility* of the input task, $T^{des}(t) = T_{tg}(t) + T_{tm}(t)$, via the $T_{tm}(t)$ component.

Early research on balance control of humanoid robots focused exclusively on preplanned tasks, mainly on cyclic, dynamic gaits on a flat ground [68,145]. As pointed out in [67], besides

Humanoid Robots
https://doi.org/10.1016/B978-0-12-804560-2.00012-2

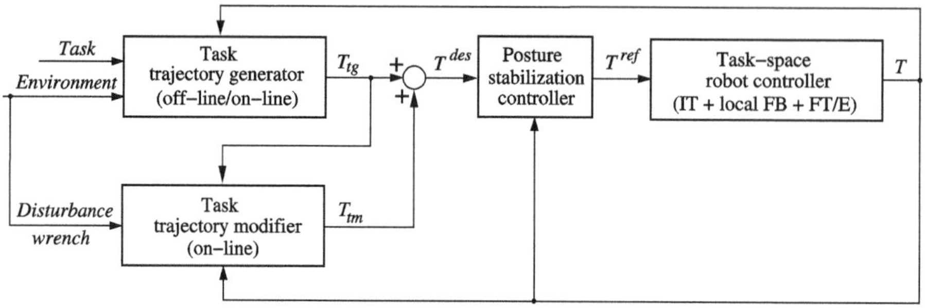

FIGURE 5.1 Generic balance controller structure. The task motion/force trajectory T_{tg} can be generated either off-line and then modified on-line [163,84], or directly on-line. The task trajectory modifier alters the generated trajectory via T_{tm}, s.t. the input to the posture stabilization controller (stabilizer), T^{des}, is *feasible*. The task-space controller: (1) uses inverse kinematics/dynamics transforms (ITs) to generate reference joint angle/torque for position/torque-controlled robots; (2) performs joint-level local feedback (FB) control; (3) uses forward transforms and estimation (FT/E) to generate the current states in terms of motion/force task coordinates (e.g. foot and CoM positions, CoP, reaction wrenches, trunk inclination).

the input trajectory determined by the output of a walking pattern generator, a "stabilizer" needs to be involved. In the ideal situation, the robot would follow a properly generated, dynamically consistent walking pattern. In reality, however, without a "stabilizer," i.e. appropriate feedback control, it would be impossible to ensure steady gait; even small imperfections in the model and/or the environment would lead sooner than later to a fall. Note also that the performance of the controller depends very much on the input trajectory, i.e. the feedforward component. For example, to improve the dynamic stability of a gait, it is recommended to generate input joint angle trajectories not only for the legs but also for the torso [148,164,123] and arm motion [4]. Another example is walking on a slope. In this case, the trunk should be slightly bent forward s.t. the CoM ground projection remains always within the vicinity of the BoS center. In this way, a sufficiently large stability margin can be ensured [169]. Besides gait, there are other proactive tasks that require appropriate balance control. Examples are pushing [35], lifting [36], kicking [151], hitting [76], and retrieving [84] objects. The respective trajectory generation and balance control methods will be discussed in this chapter and in Chapter 7.

Reactive balance control, on the other hand, is used to accommodate *unexpected* external disturbances in order to avoid the loss of balance. This is done via the on-line trajectory modifier. A typical response to a disturbance is composed of two phases: reaction (reflexive type) and recovery. The reaction/recovery trajectory to be generated depends very much on the direction, magnitude, and point of application of the disturbance wrench. Note that the abovementioned two-phase reaction/recovery response pattern is known from studies in the field of human biomechanics and physical therapy, where reactive balance control methods have been used to assess balance disorders [56]. In the field of humanoid robotics, reactive balance control is likewise of utmost importance, endowing the robot with the ability to withstand the unexpected disturbances inherent to human-centered environments. There are various situations whereby such disturbances may be impressed on the robot. The disturbances may stem from:

- the ground (e.g. unstable [1] or uneven [72,32,70,62]),
- an external push (e.g. due to a physical human-humanoid or humanoid-humanoid interaction),
- the sudden appearance of an obstacle in front of the robot [101,75,152].

From a historical perspective, initially research on balance control on flat ground was solely based on the CoM/zero-moment point (ZMP) dynamical relations [157,156]. Later on, attention was drawn to the role of the angular momentum [129,130]. To ensure a real-time response capability, many studies on balance control resort to simple models, such as those described in Section 4.2. The focus with such models is on the spatial dynamics component of the equation of motion (cf. (4.155)) that includes the CoM states plus the angular momentum and its rate of change. Impedance-based control methods can be designed with such models to ensure a desirable whole-body compliance/admittance type of behavior in response to relatively small external disturbances. With regard to larger disturbances, research in the field of biomechanics on human balance control [162] has revealed that they are better accommodated by involving more complex responses, characterized by specific multijoint movements and muscle activation patterns (synergies). This implies the involvement of the joint-space partial dynamics, in addition to the spatial dynamics ones. Synergy-based balance control methods will be discussed in Chapter 7.

This chapter is organized in 14 sections. The following section defines the dynamic postural stability. In Section 5.3, stability analysis based on a simple inverted-pendulum (IP)-on-foot model is introduced. In Section 5.4, the ZMP manipulation-type stabilization on flat ground is discussed. Capture point (CP)-based analysis and stabilization are addressed in Section 5.5. Stabilization based on the maximum output admissible set method is explained in Section 5.7. The important role of the angular momentum in stability analysis and control is highlighted in Section 5.6. Methods of balance control based on the spatial momentum and its rate of change are introduced in Section 5.8. In Section 5.9, the role of the task-space controller in balance control is clarified. The following four sections are devoted to noniterative optimization methods for balance control. The optimization approaches to the body wrench distribution (WD) problem are discussed in Section 5.10. Spatial dynamics–based motion optimization is addressed in Section 5.11. Noniterative whole-body motion/force optimization methods are introduced in Section 5.12. In Section 5.13, balance control approaches that ensure the compliant whole-body behaviors in response to weak external disturbances are highlighted. In the final section, a number of methods for iterative optimization in balance control are introduced.

5.2 DYNAMIC POSTURAL STABILITY

Methods of balance stability analysis make predominant use of the simple IP models discussed in Section 4.3.2. Such methods were developed in the field of biomechanics to assess postural stability in humans, using the (linearized) constant-length IP-cart model. Thereby, the important role of the CoM/CoP dynamic relation has been revealed. In the field of robotics, balance stability analysis was based on the ZMP concept, originally introduced to assess dynamic postural stability during biped gait [154,155]. Recall that on flat ground, the

ZMP coincides with the CoP. The preferable model in the field of robotics is the LIP-cart model. As will be shown below, the linearized IP and LIP models yield identical results from the viewpoint of stability. One should bear in mind, though, that the LIP model restricts the set of postures by excluding a subset comprising the straightened leg (a kinematically singular posture) and postures in its vicinity. In this sense, the LIP-cart model–based analysis lacks completeness.

The LIP-cart balance stability model can be straightforwardly implemented with a real robot. In 3D, the spherical IP models described in Section 4.4 are relevant. In the general case, the equation of motion of a spherical IP is coupled in the sagittal and frontal planes. When linearized around the vertical, though, decoupling can be ensured. Thus, balance stability analysis and balance control design can be simplified by making use of two identical IP models on each of these planes. In the following discussions, both scalar and vectorial notations pertinent to IP models on the plane and spherical IP models in 3D, respectively, will be used.

The ZMP concept was helpful in establishing a number of balance control methods, leading to reliable results obtained with real robots. Nevertheless, gradually it became apparent that ZMP-based stability assessment is insufficient for the complete characterization of postural stability [127,50,7,143]. The inherent limitations of ZMP methods can be summarized as follows. They are:

1. valid only with planar contact models on flat ground;
2. cannot deal with:
   a. multicontact (feet and hand contact) postures,
   b. the presence of friction,
   c. foot rotation around an edge of the BoS polygon;
3. do not incorporate important information about:
   a. the full state (i.e. the CoM velocity),
   b. the stability margin that results from a limited BoS area.

Efforts have been made to alleviate the abovementioned ZMP-related problems. A number of researchers tackled for example the flat-ground and multicontact limitations of the ZMP concept by involving projections of the ZMP onto a virtual plane [147,145,66,131,136, 132,9]. The problem of foot rotation around an edge of the BoS polygon stemming from an unbalanced moment was assessed via the so-called *foot rotation indicator* (FRI) [29] (called "fictitious ZMP" in [155]). These efforts have contributed to new algorithms and improvements in ZMP-based balance control algorithms. On the other hand, regarding implementation with a real robot, the most successful example are Honda's P2 [48] and ASIMO [150] robots. The so-called "model-ZMP" method was introduced to deal with an undesirable upper-body motion that appears during a cyclic gait because of model inaccuracies in both the robot and the environment. The balance controller makes use of a number of balance stability mechanisms, such as (cf. [67], p. 150):

- ankle torque control,
- foot placement modification (e.g. in a cyclic gait),
- desired (LIP-cart model) and actual CoM/ZMP error dynamics,
- the so-called "divergent component of motion" [149] (see Section 5.6.2),
- angular momentum control via the RWP model,

- impact absorption in the foot landing phase (passive and active type),
- full-body control.

Recently, a growing interest can be observed toward environments with relatively large terrain irregularities, e.g. as in a disaster zone. Generally, the same balance stability mechanisms as above can be applied. There are some differences though. Note that the vertical motion of the CoM cannot be ignored anymore, as with the LIP-cart model. Also, the role of angular momentum control increases. Since the rate of change of angular momentum can only be controlled via the contact wrenches, their appropriate distribution at the contacts has to be tackled as well.

In the remainder of this chapter, stability analysis methods for balance control used in the fields of biomechanics and humanoid robotics will be presented. It should be noted that rigorous balance stability analysis is available only for some of the balance mechanism components mentioned above, but not for the system as a whole. The reason is the complexity of the model (human or robot), characterized as underactuated, on a floating base, comprising a multi-DoF varying structure, and subjected to various types of external disturbances, including impacts. It should also be noted that the term "stability," as used in the two fields, lacks formal rigor due to the same reason. Attempts have been made to devise a more rigorous treatment e.g. based on the *viability theory* [5]. This theory was successfully applied to mobile robots on wheels. Note, however, that the success can be attributed to the simple point mass model. An adaptation to humanoid robots has been proposed via the so-called *viability kernel* [159]. The method is general but lacks constructiveness from the viewpoint of controller design. A combination with model-predictive control has been considered to alleviate the problem [160]. Nevertheless, the rigorous treatment of stability for complex models still remains an open issue. With simplified models, such as the IP-on-foot–type ones, rigorous results can be obtained, as will be shown in the following discussion.

5.3 INVERTED PENDULUM-ON-FOOT STABILITY ANALYSIS

The simple, nonlinear or linearized IP-on-foot models introduced in Section 4.3 (cf. Fig. 4.2A) have been employed in studies on balance control in the fields of biomechanics and physical therapy [116,64,52,51,54].

5.3.1 The Extrapolated CoM and the Dynamic Stability Margin

A dynamic stability region can be identified within the gCoM position/velocity phase plane with the help of an optimization procedure that accounts for gravity, friction, CoP,[1] and ankle torque constraints [116]. The static (BoS) and dynamic stability regions are shown in Fig. 5.2A. Apparently, the dynamic stability region is much larger than the static one. Consider, for example, a representative state outside the static stability region, State A. At this state, the body is strongly leaning forward. Nevertheless, forward fall can be prevented with

[1] In the biomechanics literature, the term CoP is exclusively used. As discussed in Section 4.3.2 this term has a broader meaning than the ZMP.

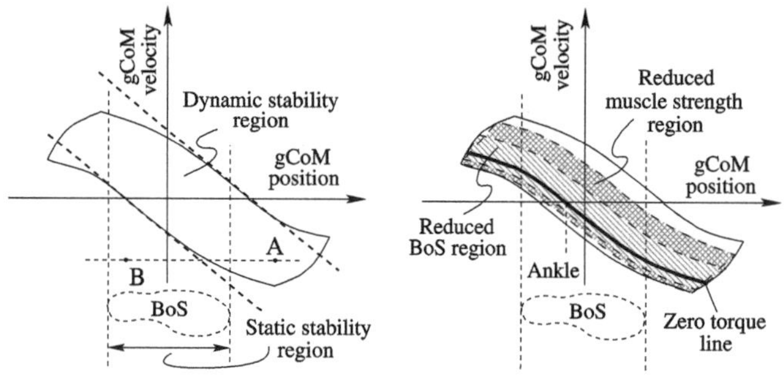

FIGURE 5.2 Left (A): Phase plane of the gCoM with static and dynamic stability regions [116]. The dynamic stability region can be approximated with a linear relation (the inclined dotted lines) to facilitate real-time balance controller design [52]. *A* and *B* are two representative states. Right (B): The dynamic stability region area shrinks for weaker (elderly) persons. This also happens with a smaller BoS area.

a backward (negative) CoM velocity. The same velocity is applied at State B, located within the static stability region. Note, however, that the system is dynamically unstable in this state: a backward fall will occur because the body is not sufficiently leaning forward. Furthermore, as apparent from Fig. 5.2B, the dynamic stability region shrinks significantly with a smaller or a slippery BoS. The region also shrinks considerably for (elderly) persons with weaker muscles. This finding can be directly transferred to the field of humanoid robots, thus emphasizing the important role of the ankle torque constraint in balance control.

The optimization procedure used to identify the dynamic stability region may not be suitable for on-line implementation due to computational cost limitations. The problem can be alleviated by approximating the region with a linear relation derived from a closed-form solution (4.12) [52]. This solution expresses the relative motion of the gCoM, $x_g(t)$, w.r.t. to a given CoP, say $\bar{x}_p$. As apparent from (4.12), the balance control mechanism depends on the initial state, $(x_{g0},\ v_{g0})$. On the other hand, note that the CoP is constrained by the size of the BoS, i.e. $-l_f \leq \bar{x}_p \leq l_f$, l_f denoting the half foot length. As an example, consider a state with positive initial gCoM velocity ($v_{g0} > 0$). It may then happen that the backward (negative) acceleration, $-m\ddot{x}_g$, is insufficient to prevent the gCoM x_g to pass CoP $\bar{x}_p$. When the gCoM passes this value, the CoM will accelerate further forward. As a consequence, CoP $x_p(t)$ will arrive at the BoS boundary l_f and the foot will begin to rotate in the clockwise direction. This is a potentially dangerous state that may result in posture instability and a fall. In order to prevent this from happening, the CoM should never pass the given CoP, i.e. $x_g(t) < \bar{x}_p,\ \forall t$. This condition can be rewritten with the help of (4.12) as

$$x_{g0} + \frac{v_{g0}}{\omega} \leq \bar{x}_p, \tag{5.1}$$

where $\omega = \omega_{IP} = \sqrt{g/l}$. In the above derivation $|\tanh(\omega t)|_{t\to\infty} \to 1$ was used. One can conclude then that to avoid destabilization, the CoP $x_p(t)$ should move sufficiently fast toward the positive BoS boundary, l_f. In other words, the gCoM should always follow the CoP motion and come to a stop exactly when it reaches the CoP. Thus, when the CoP eventually arrives at the boundary, followed by the gCoM, the movement will end there with a stationary state. The same line of thought also holds for motion in the opposite direction, i.e. with a negative initial gCoM speed. By replacing $\bar{x}_p$ with the BoS boundaries, (5.1) can be rewritten in the general form, i.e.

$$-l_f \leq x_{ex}(t) \leq l_f, \tag{5.2}$$

where

$$x_{ex}(t) \equiv x_g(t) + \frac{\dot{x}_g(t)}{\omega}. \tag{5.3}$$

Here x_{ex} is named the "extrapolated CoM" (henceforth abbreviated as xCoM) [52]. The xCoM is useful in approximating the dynamic stability region in Fig. 5.2A. The approximation is indicated by the inclined dashed lines. Furthermore, the xCoM can be used to define the *dynamic stability margin*, i.e.

$$s(t) = \left|\pm l_f - x_{ex}(t)\right|. \tag{5.4}$$

The dynamic stability margin plays an important role in choosing an appropriate controller action, as will be shown in what follows.

5.3.2 Extrapolated CoM Dynamics

By investigating the dynamics of the xCoM, important conclusions about balance control stability can be obtained [51]. To this end, first recall the CoM/CoP dynamic relation for the IP-on-foot model (cf. (4.10)). We have

$$\ddot{x}_g = \omega^2(x_g - x_p). \tag{5.5}$$

Next, solve (5.3) for the gCoM velocity, i.e.

$$\dot{x}_g = -\omega(x_g - x_{ex}). \tag{5.6}$$

On the other hand, the time differential of (5.3) is

$$\dot{x}_{ex} = \dot{x}_g + \frac{\ddot{x}_g}{\omega}. \tag{5.7}$$

Insert (5.5) and (5.6) into the above equation to obtain the xCoM dynamics as

$$\dot{x}_{ex} = \omega(x_{ex} - x_p). \tag{5.8}$$

The gCoM and xCoM dynamics ((5.6) and (5.8), respectively) comprise two autonomous dynamic system components: a stable one, $\dot{x}_g = -\omega x_g$, and an unstable one, $\dot{x}_{ex} = \omega x_{ex}$. It can then be concluded that to obtain a stable CoM motion, it would suffice to stabilize the motion of the xCoM [51]. In other words, with a stable xCoM trajectory there is no need to care about the CoM trajectory at all; the CoM will follow the xCoM motion faithfully. This conclusion plays an important role in balance controller design, as will be shown in what follows.

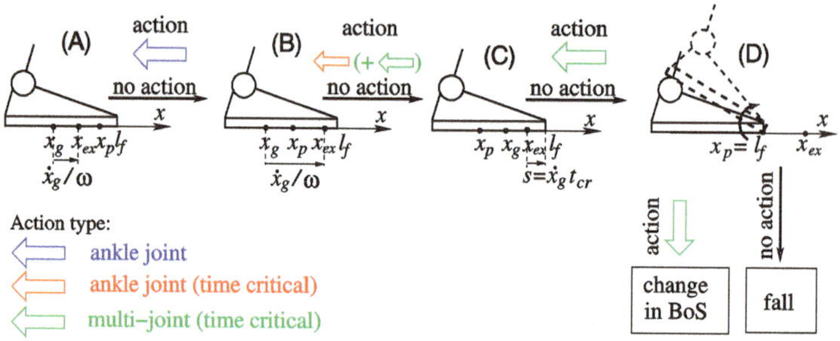

FIGURE 5.3 The balance stability mechanism depends on transitions between states of gCoM x_g, CoP x_p, and the extrapolated CoM (xCoM), x_{ex}, as well as on the BoS (foot length l_f) [52]. The figure shows some discrete states with $\dot{x}_g > 0$. In order to maintain balance, a controller action is needed. Without such action, the posture may become unstable as apparent from the rolling foot State (D). The ankle-joint action at State (B) is not time-critical. At State (C), a time-critical action is needed (depending on the dynamic stability margin $s = v_g t_{cr}$). The action may involve the ankle joint only, or alternatively, a multijoint motion could be invoked that yields a change in the angular momentum. At State (D), there are two possible time-critical actions invoked through a multijoint motion: either reverse foot rotation via a change in the partial angular momentum, or change the BoS (e.g. via a reactive step).

The xCoM dynamics provide an additional insight into the stability of the system. Using (5.6) and (5.8), the IP-on-foot dynamics can be represented in the following state-space form [23]:

$$\frac{d}{dt}\begin{bmatrix} x_g \\ x_{ex} \end{bmatrix} = \begin{bmatrix} -\omega & \omega \\ 0 & \omega \end{bmatrix} \begin{bmatrix} x_g \\ x_{ex} \end{bmatrix} + \begin{bmatrix} 0 \\ -\omega \end{bmatrix} x_p. \tag{5.9}$$

This representation exhibits the stable (upper row) and unstable (lower row) autonomous dynamical system components mentioned above. As already noted, it would be sufficient to design a balance controller for the unstable part only. Such a design is presented in Section 5.6.3.

5.3.3 Discrete States With Transitions

The above analysis has clarified that three characteristic points, gCoM, xCoM, and CoP, denoted as x_g, x_{ex}, and x_p, respectively, play an important role in dynamic stability along with the BoS. In what follows, it will be shown that the relative positions of these points within the BoS can provide deeper insight into the balance control mechanism. Consider the following four states [52] (cf. Fig. 5.3):

(A) $x_g < x_{ex} < x_p < l_f$,
(B) $x_g < x_p < x_{ex} < l_f$,
(C) $x_p < x_g < x_{ex} < l_f$,
(D) $x_p = l_f < x_{ex}$.

Note that x_g, x_{ex}, and x_p provide information for the CoM position, velocity, and acceleration, respectively. Assume a positive initial gCoM speed and consider the possible transitions between the states for a given xCoM. First, note that with State (a), the gCoM will never reach the CoP. At some time instant, though, the gCoM motion will reverse direction. Then, State (b) will be invoked. With this state, the gCoM will pass the CoP and arrive at State (c), accelerating thereby further in the positive direction, as explained above. A time-critical action is then needed. The xCoM will reach the BoS boundary at the critical time instant t_{cr} that can be calculated from the closed-form solution (4.12). By assuming a constant $\bar{x}_p$, t_{cr} can be approximated as

$$t_{cr} \approx \frac{l_f - x_{ex}(t)}{\dot{x}_g(t)} = \frac{s(t)}{\dot{x}_g(t)}. \tag{5.10}$$

The time-critical action depends on the posture and the initial conditions. The ankle joint torque could be sufficient to ensure the transition to State (b). When the required fast response cannot be achieved with an ankle torque alone, a multijoint motion should be invoked to generate a change in the angular momentum.

State (d) is characterized by a change in the contact condition, from a line contact to a point contact in the 2D example, or from a plane contact to a line contact in 3D. This change stems form an unbalanced moment at the foot that may lead to a foot rolling around the BoS boundary. The magnitude of this moment can be assessed via the so-called *foot rotation indicator* (FRI) [29] (called "fictitious ZMP" in [155]). Note that the unbalanced moment implies a change in the *partial angular momentum*. The CoM equations (4.38) reveal that the state of the robot is determined by the CRB partial dynamics, including the rate of change of the *CRB angular momentum*. When the joints are locked, the unbalanced moment at the foot applies to the whole CRB, yielding eventually postural instability and a fall. When the joints are unlocked, the change in the partial angular momentum at the foot could be compensated via a change in the partial angular momentum of the rest of the links, e.g. trunk, arms and head. Thus, a possibility to avoid postural destabilization despite foot rotation exists. In [30], the so-called *zero rate of change of angular momentum* (ZRAM) point[2] was introduced to assess postural instability in relation to the angular momentum. Further details with regard to the role of the angular momentum in balance control will be provided in Section 5.6. When the robot is incapable of generating the required change of the angular momentum, another multijoint action can be invoked. This type of action is adopted from the so-called "change-in-support" strategy used by humans to maintain their upright stance [94,93]. The strategy refers to establishing a hand contact with the environment and/or initializing reactive stepping (step strategy). Further details will be given in Section 7.7.5.

5.3.4 Dynamic Stability Region in 2D

The BoS of a humanoid robot on flat ground is quite often represented by a convex 2D polygon. The respective CoP, gCoM, and xCoM then become 2D vector quantities. The CoM/CoP dynamics (5.5) are written in vector form as

$$\ddot{\boldsymbol{r}}_g = \omega^2(\boldsymbol{r}_g - \boldsymbol{r}_p). \tag{5.11}$$

[2] Referred to as "centroidal moment pivot (CMP)" in [121] (defined in (4.24)).

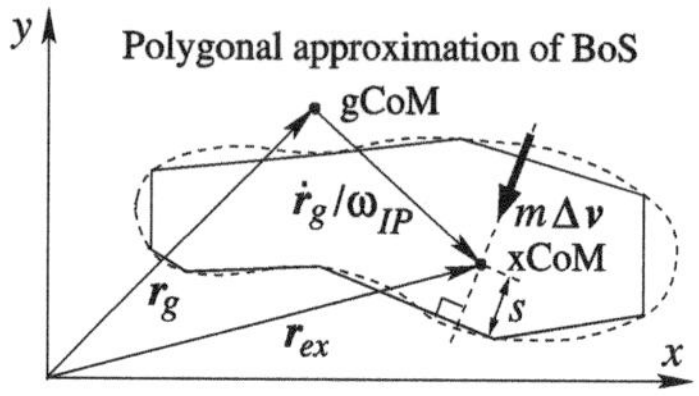

FIGURE 5.4 Dynamic stability margin s is defined in 2D as the shortest distance from the xCoM vector, $\boldsymbol{r}_{ex}$, to the BoS polygon [52]. The dynamic stability margin can also be interpreted as the minimum force impulse $M\Delta\boldsymbol{v}$ that, when acting along the line determined by the shortest distance, would deteriorate the balance.

Here $\boldsymbol{r}_g(t)$ and $\boldsymbol{r}_p(t)$ stand for the gCoM and CoP position vectors, respectively. Also, referring to (5.3), the xCoM can be written as

$$\boldsymbol{r}_{ex}(t) = \boldsymbol{r}_g(t) + \frac{\dot{\boldsymbol{r}}_g(t)}{\omega}. \tag{5.12}$$

In 2D, the dynamic stability margin (5.4) is interpreted as the shortest distance between the xCoM and the BoS polygon (cf. Fig. 5.4). It also can be interpreted as the minimum change of momentum, $M\Delta\boldsymbol{v}$, that would compromise the balance when applied in the direction determined by the shortest distance between the xCoM and the BoS polygon [52]. It then becomes apparent that the dynamic stability margin can vary with the direction of the applied disturbance.

To formally constrain the xCoM within the BoS polygon, use notation (4.32) introduced for the CoP constraint. Then we have

$$\boldsymbol{B}_s\boldsymbol{r}_{ex} \preceq \boldsymbol{c}. \tag{5.13}$$

It should be noted that a 2D stability margin definition appeared also in the robotics literature as the shortest distance between the CoP and the BoS polygon [59]. This definition, however, does not include full information about the current state since the gCoM velocity does not appear in the CoP definition.

5.4 ZMP MANIPULATION–TYPE STABILIZATION ON FLAT GROUND

Early research on biped gait control was mainly based on the LIP-cart model on flat ground. In Section 4.3.2 it was clarified that on flat ground, the (net) CoP coincides with the ZMP (see also [121]). The dynamic relations pertinent to the LIP-cart model were also introduced there. Note that the equation of motion, (4.5), is of the same form as that of the linearized IP-on-foot model used in Section 5.3. There is an explicit solution, (4.6), which facilitates balance stability analysis and control. Referring to the LIP-cart dynamics (4.18), the following simple ankle torque controller ($u = m_y$) can be employed [99,146,27,73]:

$$u = -k_d\dot{x}_g - k_p x_g. \tag{5.14}$$

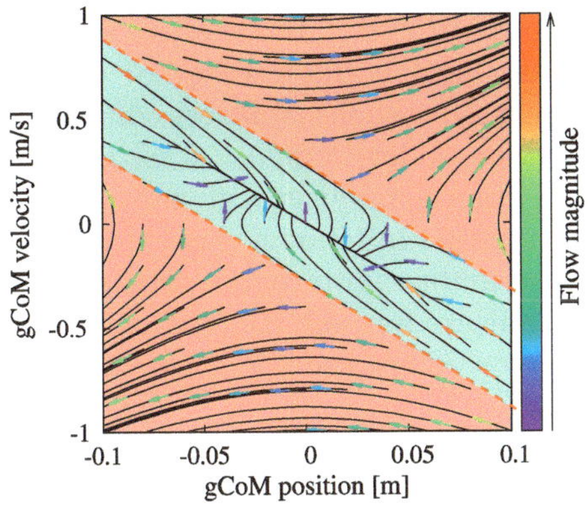

FIGURE 5.5 gCoM phase portrait for the IP/LIP-on-cart models. The approximation of the dynamic stability region in accordance with (5.2) is shown in light blue.

Although feedback gains k_p and k_d can be readily selected to obtain a desirable closed-loop response, the implementation of this controller is prone to a number of difficulties [67]. To tackle this problem, consider the following state-space representation of the gCoM dynamics (5.5):

$$\frac{d}{dt}\begin{bmatrix} x_g \\ \dot{x}_g \end{bmatrix} = \begin{bmatrix} 0 & 1 \\ \omega^2 & 0 \end{bmatrix}\begin{bmatrix} x_g \\ \dot{x}_g \end{bmatrix} + \begin{bmatrix} 0 \\ -\omega^2 \end{bmatrix} x_p, \tag{5.15}$$

where $\omega = \overline{\omega} = \sqrt{g/\bar{z}_g}$ and $\bar{z}_g$ denotes the constant height of the LIP-cart model. Apparently, the CoM state is used as the state variable and the CoP as the input ($u = x_p$). The solution to the above equation can be obtained in explicit form as [23]

$$\begin{bmatrix} x_g \\ \dot{x}_g \end{bmatrix} = \begin{bmatrix} \cosh(\omega t) & \frac{1}{\omega}\sinh(\omega t) \\ \omega\sinh(\omega t) & \cosh(\omega t) \end{bmatrix}\begin{bmatrix} x_{g0} \\ v_{g0} \end{bmatrix} + \begin{bmatrix} 1-\cosh(\omega t) \\ -\omega\sinh(\omega t) \end{bmatrix} x_p. \tag{5.16}$$

The state-space representation (5.15) can be used for quantitative analysis of the IP/LIP-on-foot models, via their phase portraits. The unforced system generates the vector field and the respective flow. To obtain the dynamic stability region approximation, (5.2), make use of the forced system with input x_p in (5.15) set at the two BoS boundaries. As an example, consider the phase portrait in Fig. 5.5, generated with the LIP-cart model.[3] We have

$$\begin{aligned} \bar{z}_g &= 0.279\,\text{m} \to \omega = 5.94\ 1/\text{s}, \\ l_f &= 0.096\,\text{m}. \end{aligned} \tag{5.17}$$

The dynamic stability region in the figure is shown in light blue. As already clarified, the boundaries of this region play an important role in making decisions about proper controller actions for balance control.

[3] The data are derived from a small-size HOAP-2 robot [26] (see Section A.1).

The main result from the above brief stability analysis is that, to alleviate the problem of controller (5.14), the balance stabilizer should use the CoP/ZMP as a control input in lieu of the ankle torque. In what follows, a few such stabilizers will be introduced. The stabilizers can be formulated either in a scalar or vector form, $u = x_p$ or $\boldsymbol{u} = \boldsymbol{r}_p$, respectively. Note that the latter is needed when implementing a stabilizer with a real robot. In this case, the two vector components are obtained from the 3D LIP model. Since this model comprises two equations of motion that can be decoupled in the sagittal and frontal planes, it is possible to design two independent and identical stabilizers for each coordinate. In the following discussion, both the scalar and the vector notations will be used.

5.4.1 The ZMP Manipulation–Type Stabilizer

A stabilizer that uses the ZMP as a control input was first introduced in [97] under the name "ZMP manipulation" controller. Referring to the IP equation of motion in Cartesian coordinates, (4.4), the following feedback control laws are designed:

$$f_z = |f| \cos\theta = z_g^{des} - z_g - \dot{z}_g + Mg, \tag{5.18}$$

$$\begin{aligned} m\ddot{x}_g &= \frac{\sin\theta}{\cos\theta}(z_g^{des} - z_g - \dot{z}_g + Mg) \\ &= \frac{x_g - x_p}{z_g}(z_g^{des} - z_g - \dot{z}_g + Mg). \end{aligned} \tag{5.19}$$

The quantities appearing in these equations have been defined in Section 4.3.2; see also Fig. 4.2. First, it is easy to confirm that the closed-loop dynamics in the vertical direction yield exponential convergence of state z_g to z_g^{des}. Next, consider the following equation for the horizontal coordinate:

$$\beta\ddot{x}_g = x_g - x_p. \tag{5.20}$$

Observe that (5.20) can be rendered in the form (4.10) with $\beta^{des}(t) = 1/\omega^2(t)$, $\omega(t)$ being defined in (4.14). Having in mind (5.19), set

$$\beta(t) = \frac{Mz_g}{z_g^{des} - z_g - \dot{z}_g + Mg}. \tag{5.21}$$

As a result of the exponential convergence in the vertical direction, $\beta(t)$ in (5.21) will converge to $\beta^{des}(t)$, also exponentially.

Furthermore, by setting the ZMP as the control input in (5.20), $u = x_p$, two feedback control laws can be designed, as follows:

$$u = x_g + \dot{x}_g - v_g^{des} \tag{5.22}$$

and

$$u = 2x_g + \dot{x}_g - x_g^{des}. \tag{5.23}$$

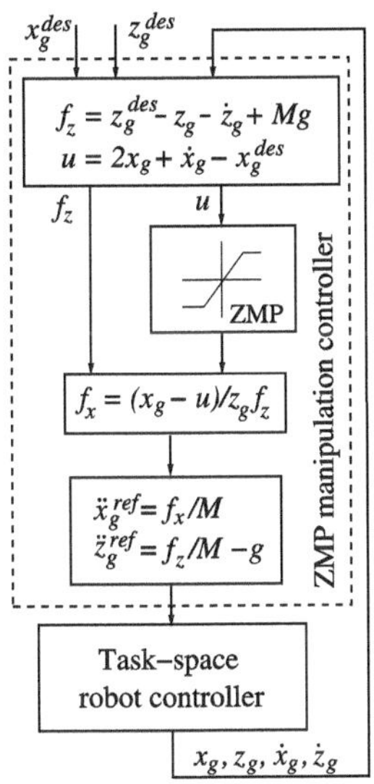

FIGURE 5.6 Block diagram for ZMP manipulation–type balance regulator based on the constant-length IP-on-cart model in Fig. 4.2B [97]. With reference to Fig. 5.1, it is apparent that: (1) the task trajectory T is defined in terms of the CoM coordinates and (2) the ZMP manipulation controller plays the role of the stabilizer.

Control law (5.22) is useful to track the reference gCoM speed v_g^{des}, e.g. as required in the case of gait control. Control law (5.23), on the other hand, is useful as a regulator w.r.t. the desired CoM position x_g^{des}, e.g. as required in upright balance control. The controller block diagram for the latter case is shown in Fig. 5.6. Note that the control input u passes through a limiter that saturates at the BoS boundaries. The task-space controller was implemented as a dynamic torque controller in [97].

5.4.2 Velocity-Based ZMP Manipulation–Type Stabilization in 3D

The ZMP manipulation–type stabilizer can be formulated in terms of velocity relations [145]. The design is based on the nonlinear spherical IP model introduced in Section 4.4.2. Thus, the ZMP and CoM coordinates are 2D and 3D quantities, respectively. Recall that the 3D CoM coordinate vector is denoted as $\boldsymbol{r}_C = \begin{bmatrix} \boldsymbol{r}_g^T & z_g \end{bmatrix}^T$. The control input is the ZMP, as in the original ZMP manipulation stabilizer discussed above: $\boldsymbol{u} = \boldsymbol{r}_p^{ref} \in \Re^2$. The block scheme of the controller is shown in Fig. 5.7. First, the *ZMP planner* calculates the control input from the desired CoM velocity vector, $\dot{\boldsymbol{r}}_C^{des} \in \Re^3$. Recall that the spherical IP dynamics can be considered decoupled in x and y, under the assumption of small deviation from the vertical. Hence, it is possible to design the spherical IP controller as two independent linearized IP-on-cart (cf. Fig. 4.2B) controllers, using either the conventional (e.g. as in [20]) or any other approach (e.g. as in [166]). Note that control input $\boldsymbol{u}$ has to be limited to comply with the ZMP constraint. Furthermore, the *ZMP manipulator* determines the reference CoM velocity vector $\dot{\boldsymbol{r}}_C^{ref}$

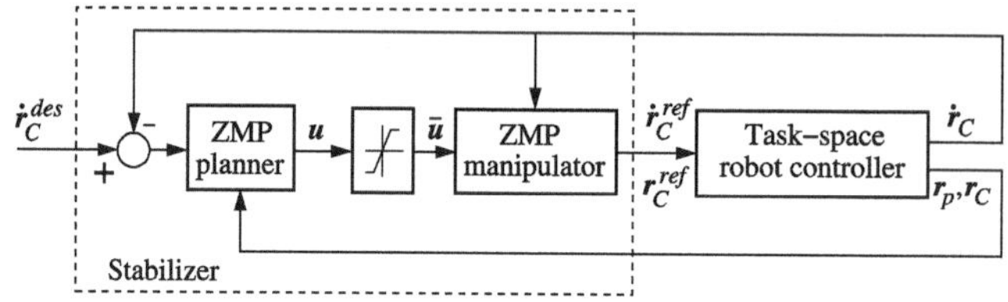

FIGURE 5.7 Block diagram of a velocity-based ZMP manipulation–type balance controller [145]. With reference to Fig. 5.1, it is apparent that: (1) the task trajectory T is defined in terms of the CoM velocity and (2) the stabilizer comprises two parts. The ZMP planner is in fact an IP-on-cart controller. The ZMP manipulator determines the (stabilized) reference CoM trajectory. The task-space controller makes use of the CoM inverse kinematics in combination with other velocity-level constraints and includes a local joint-angle feedback controller.

by integrating the respective accelerations. Then we obtain

$$\begin{aligned} \ddot{z}_g^{ref} &= k_z(\dot{z}_g^{ref} - \dot{z}_g), \\ \beta^{ref}\ddot{\boldsymbol{r}}_g^{ref} &= (\boldsymbol{r}_g^{ref} - \boldsymbol{u}), \end{aligned} \tag{5.24}$$

where k_z is a feedback gain and $\beta^{ref} = (z_g - z_p^{ref})/(\ddot{z}_g^{ref} + g)$ (cf. (4.14)). Note that the vertical position of the ZMP is not constant and that $\dot{\boldsymbol{r}}_C^{ref}$ is fed into the task-space robot controller. This controller is designed for a position-controlled robot and comprises an inverse kinematics solver that transforms the input into reference joint velocities. The solver also takes care of the joint-space and other constraints expressed in terms of velocities. The reference joint velocities drive a joint-space local feedback controller. Finally, current joint and ZMP data are obtained from the robot sensors and transformed into the current CoM velocity and ZMP vectors, $\dot{\boldsymbol{r}}_C$ and $\boldsymbol{r}_p$, respectively.

The velocity-based ZMP manipulation–type balance controller is efficient from a computational viewpoint: real-time implementation with position-controlled humanoid robots is possible. Also, the CoM and CoP are allowed to deviate in the vertical direction. Thus, the controller can be used for proactive tasks such as walking/stepping on an irregular terrain.

One notable drawback of the above controller is the lack of stability analysis. This problem can be alleviated by introducing the constant-height CoM (LIP) constraint, as shown in [10] (see also [12,11]). The LIP constraint is enforced in 3D via the sphere-on-plane model introduced in Section 4.4.2. With this constraint, there is no need to account for the vertical CoM coordinate. Thus, the controller can be designed based on the gCoM only. The block scheme is shown in Fig. 5.8. The desired gCoM position, $\boldsymbol{r}_g^{des}$, and its time derivatives are determined by a task motion planner. The ZMP planner calculates just the desired ZMP, $\boldsymbol{r}_p^{des}$, from the CoM/ZMP dynamics (5.11). Thereby, note that $\omega = \bar{\omega}$, according to the LIP constraint. The same equation is also used to obtain the actual ZMP, $\boldsymbol{r}_p$, whereby the actual gCoM velocity is given as

$$\dot{\boldsymbol{r}}_g = \boldsymbol{u} + \boldsymbol{\epsilon}. \tag{5.25}$$

Here $\boldsymbol{\epsilon}$ stands for the control error. The control input, on the other hand, is defined as

$$\boldsymbol{u} = \dot{\boldsymbol{r}}_g^{des} + \boldsymbol{K}_g\boldsymbol{e}_g - \boldsymbol{K}_p\boldsymbol{e}_p, \tag{5.26}$$

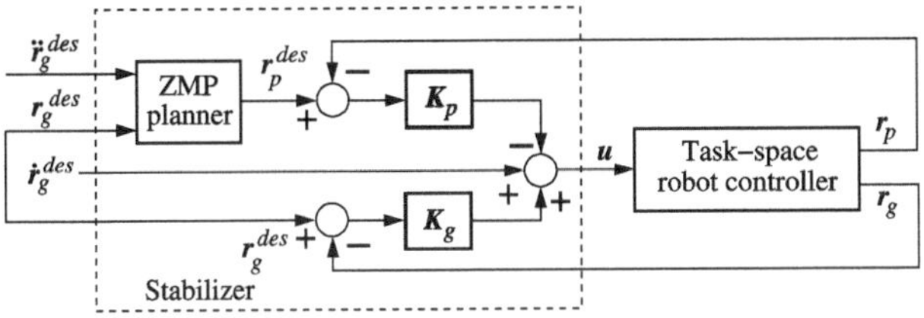

FIGURE 5.8 Block diagram of another velocity-based ZMP manipulation–type balance controller [10]. With reference to Fig. 5.1, it is apparent that: (1) the task trajectory T is defined in terms of the CoM motion and (2) the stabilizer comprises a ZMP planner and two loops for the CoM and ZMP motion proportional feedback control. The task-space controller makes use of a CoM inverse kinematics solution with a whole-body motion capability and includes a local joint-angle feedback controller.

where $\boldsymbol{e}_g \equiv \boldsymbol{r}_g^{des} - \boldsymbol{r}_g$ and $\boldsymbol{e}_p \equiv \boldsymbol{r}_p^{des} - \boldsymbol{r}_p$ are the gCoM and CoP errors, respectively; $\boldsymbol{K}_g = \text{diag}(k_g^x, k_g^y)$, $\boldsymbol{K}_p = \text{diag}(k_p^x, k_p^y)$ are control gain matrices. When the control gains are selected to satisfy the following conditions:

$$k_g^i > \varpi > 1, \quad 0 < k_p^i < \varpi - (\alpha^2/\varpi) - \gamma^2, \quad i \in \{x, y\},$$

for any positive constants $\alpha < \varpi$ and $\gamma < \sqrt{\varpi - (\alpha^2/\varpi)}$, then the controller is said to be input-to-state–stable, $(\boldsymbol{\epsilon}, \dot{\boldsymbol{\epsilon}})$ and $(\boldsymbol{e}_p, \dot{\boldsymbol{e}}_p)$ standing for the input and state, respectively. Readers interested in the proof are referred to [11].

The task-space robot controller in Fig. 5.8 comprises similar components as the one in Fig. 5.7. A notable difference can be found in the resolution method of the CoM inverse kinematics. The 3D CoM velocity vector is expressed as the sum of two components. One of the components results from the motion of the support leg(s). The other component is determined via the inverse kinematics solution w.r.t. the desired spatial velocities of the other end links. In this way, the motion in the limbs will not disturb the balance determined from the simple CoM/ZMP dynamics model. This method has been named "Kinematic resolution of CoM Jacobian with embedded motion" [10]. Similar to the previous velocity-based ZMP manipulation controller, there is an advantage of computational efficiency and possibility for real-time implementation with position-controlled humanoid robots. Note, though, that in this controller no provisions have been made to constrain the desired ZMP (the output of the ZMP planner) to stay within the BoS. Thus, there is no guarantee that the foot will always be in full contact with the flat ground.

5.4.3 Regulator-Type ZMP Stabilizer

Making use of the state-space representation (5.15), a regulator-type ZMP stabilizer can be designed [143]. Again, the ZMP is used as the control input. First, the gCoM, ZMP, and BoS limit coordinates are redefined w.r.t. the desired gCoM position, x_g^{des}, as $x_g^* = x_g - x_g^{des}$, $x_p^* = x_p - x_g^{des}$, and $l_f^* = l_f - x_g^{des}$, respectively. Then, the reference input ZMP, $u = x_p^{ref}$, is determined as

$$u = -k_1\dot{x}_g^* - k_2 x_g^*, \tag{5.27}$$

k_1, k_2 denoting feedback gains. The state-space representation of the stabilizer can then be written as

$$\frac{d}{dt}\begin{bmatrix} x_g^* \\ \dot{x}_g^* \end{bmatrix} = \begin{bmatrix} 0 & 1 \\ \omega^2(k_1+1) & \omega^2 k_2 \end{bmatrix}\begin{bmatrix} x_g^* \\ \dot{x}_g^* \end{bmatrix}. \tag{5.28}$$

Since there is no guarantee that u will honor the BoS constraints, $-l_f^* \leq u \leq l_f^*$, a limiter has to be involved, as was the case with the ZMP manipulation–type stabilizers discussed above. Under saturation at the boundaries, the state evolves according to

$$\frac{d}{dt}\begin{bmatrix} x_g^* \\ \dot{x}_g^* \end{bmatrix} = \begin{bmatrix} 0 & 1 \\ \omega^2 & 0 \end{bmatrix}\begin{bmatrix} x_g^* \\ \dot{x}_g^* \end{bmatrix} + \begin{bmatrix} 0 \\ -\omega^2 \end{bmatrix}(\pm l_f^*). \tag{5.29}$$

This equation is in the same form as (5.15). It is important to note at this point that the BoS boundaries constitute a set of equilibrium points. We have

$$\{(x_g^*,\ \dot{x}_g^*) : (\pm l_f^*,\ \dot{x}_g^*),\ \forall \dot{x}_g^*\}.$$

When the state arrives at the boundary, it will be trapped there and, hence, will not converge to the desired (isolated) equilibrium state $(x_g^{des}, 0)$. This means that the boundaries should be excluded as admissible states. The dynamic stability region can then be determined in the same form as (5.2), so we have

$$-l_f^* < x_{ex}^* < l_f^*. \tag{5.30}$$

Here $x_{ex}^* \equiv x_g^* + \dot{x}_g^*/\omega$ is recognized as the xCoM represented in modified coordinates. The strict inequalities imply that the undesirable equilibria at the boundaries are to be avoided.

Furthermore, assume that the desired poles are given as ωq_1 and ωq_2. The feedback gains can then be determined as $k_1 = -(q_1 q_2 + 1)$ and $k_2 = (q_1 + q_2)/\omega$. Poles with negative real values will steer the state toward the isolated equilibrium $(x_g^*,\ \dot{x}_g^*) = (0,\ 0)$. Consequently, the CoM will approach the desired state, $(x_g^{des},\ 0)$.

Furthermore, a "constrained-control" region can be defined by restricting the action of control law (5.27) within the BoS boundaries. Then we have

$$-k_1\dot{x}_g^* - k_2 x_g^* = \pm l_f^*. \tag{5.31}$$

Apparently, the constrained-control region is parametrized by the feedback gains. Whenever control u satisfies the constraint (5.31) for some given feedback gains and the respective xCoM (i.e. the CoM state) lies within the dynamic stability region (5.30), the posture of the robot will be stable.

It is interesting to observe that the pole assignment also determines whether the dynamic stability region and the constrained-control one intersect. Intersection implies a smaller operational domain for the regulator. To obtain a regulator with the largest possible operational domain requires the two regions to overlap. From the expressions for the four boundary conditions, (5.30) and (5.31), it is straightforward to determine that $k_2 = k_1/\omega$ should then hold. This condition implies that the poles are correlated via $(q_1+1)(q_2+1) = 0$. It will then suffice to assign the poles as $(-1,\ q)$, $q < 0$. The value of q can be chosen to satisfy any other per-

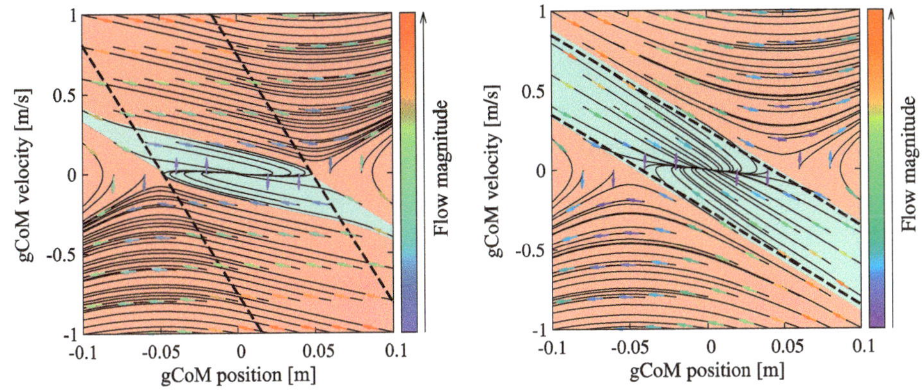

FIGURE 5.9 gCoM phase portraits with constrained-control and dynamic stability regions. The former are enclosed between the dashed lines. The dynamic stability regions are shown in light blue. Left (A): Intersecting (nonoptimal) case. Right (B): Overlapping case (named the "best CoM-ZMP regulator" in [143]).

formance constraints, e.g. the maximum torque limits. A regulator based on this type of pole assignment has been named the "best CoM-ZMP regulator" [143].

Two sample phase portraits of the regulator are shown in Fig. 5.9. They were generated with $x_g^{des} = 0$ and data set (5.17). With pole assignment $(q_1, q_2) = (-0.2, -0.2)$ and $(q_1, q_2) = (-1, -0.2)$, it can be confirmed that the constrained-control and dynamic stability regions will be intersecting and overlapping, as in the plot on the left and right side, respectively. When the xCoM is confined to the interior of the dynamic stability region, stable motion of the gCoM will be guaranteed. When the xCoM arrives at the boundary or leaves the region, reactive stepping should be initialized. This problem will be discussed in Section 7.7.5.

5.4.4 ZMP Stabilization in the Presence of GRF Estimation Time Lag

Most existing humanoid robots are position-controlled. The balance controllers discussed so far were designed for such robots. As already noted, their tasks are specified in terms of position/velocity. In order to implement a ZMP manipulation–type stabilizer, one has to account for the inherent GRF estimation lag in the actual GRF/ZMP readings. This lag results from the shock absorbing passive elements (rubber bushing) embedded into the robot soles, as well as from the GRF/ZMP sensor controller. For practical purposes, the following simple dynamic model can be used [71,98]:

$$x_p(s) = \frac{1}{T_p s + 1} x_p^{des}(s) \tag{5.32}$$

or

$$\dot{x}_p = -F_p(x_p - x_p^{des}), \tag{5.33}$$

where $F_p \equiv 1/T_p$. The time constant T_p can be determined experimentally. Combining this relation with CoM/ZMP dynamics (5.5), one arrives at the following linear state-space equation:

$$\dot{\boldsymbol{x}} = \boldsymbol{A}\boldsymbol{x} + \boldsymbol{B}u, \tag{5.34}$$

where $\boldsymbol{x} \equiv \begin{bmatrix} x_g & \dot{x}_g & x_p \end{bmatrix}^T$ is the state vector and

$$\boldsymbol{A} \equiv \begin{bmatrix} 0 & 1 & 0 \\ \omega^2 & 0 & -\omega^2 \\ 0 & 0 & -f_p \end{bmatrix}, \quad \boldsymbol{B} \equiv \begin{bmatrix} 0 \\ 0 \\ f_p \end{bmatrix}.$$

A tracking stabilizer can then be designed in a straightforward manner with the following control input:

$$u = x_p^{des} - \boldsymbol{K}\boldsymbol{e}_x,$$

where $\boldsymbol{e}_x \equiv (\boldsymbol{x}^{des} - \boldsymbol{x})$ is the tracking error. Feedback gains $\boldsymbol{K} \equiv \begin{bmatrix} k_1 & k_2 & k_3 \end{bmatrix}$ can be determined e.g. via pole placement, from the closed-loop dynamics

$$\frac{d}{dt}\boldsymbol{e}_x = (\boldsymbol{A} - \boldsymbol{B}\boldsymbol{K})\,\boldsymbol{e}_x.$$

This stabilizer was implemented under the LIP mode, whereby the stable pole of the pendulum ($\omega = \bar{\omega}$) was used to obtain the performance of the "best CoM/ZMP regulator" introduced in Section 5.4.3. The respective balance controller has been successfully tested by realizing a walking task with the HRP-4C humanoid robot [71].

5.4.5 Torso Position Compliance Control (TPCC)

The torso position compliance control (TPCC) [106] is a stabilization approach that uses the ZMP as a control input. The TPCC approach is similar to the ZMP manipulation–type approaches discussed in the previous subsections. The TPCC method will be derived below by representing the ZMP equation in terms of finite differences. TPCC comprises two components: (a) the IP control and (b) the ZMP compensation control.

(a) IP control: The 3D LIP model (cf. Section 4.4.2) illustrated in Fig. 5.10 will be employed in the design of the stabilizer; θ_y and θ_x (not shown in the figure) denote the rotation angles about the y- and x-axes, respectively. Assuming relatively small deviations of the 3D LIP from the vertical, the relationship between (θ_y, θ_x) and $\boldsymbol{r}_g - \boldsymbol{r}_p$ can be approximated as

$$\boldsymbol{\theta} \triangleq \begin{bmatrix} \theta_y \\ -\theta_x \end{bmatrix} \simeq \frac{1}{z_g}\left(\boldsymbol{r}_g - \boldsymbol{r}_p\right). \tag{5.35}$$

The linearized dynamics of the 3D LIP model around the vertical can be written as

$$z_g\ddot{\boldsymbol{\theta}} + \Delta\ddot{\boldsymbol{r}}_p = g\boldsymbol{\theta}; \tag{5.36}$$

$\Delta\ddot{\boldsymbol{r}}_p$, denoting a desired deviation of the ZMP, is considered as a control input.

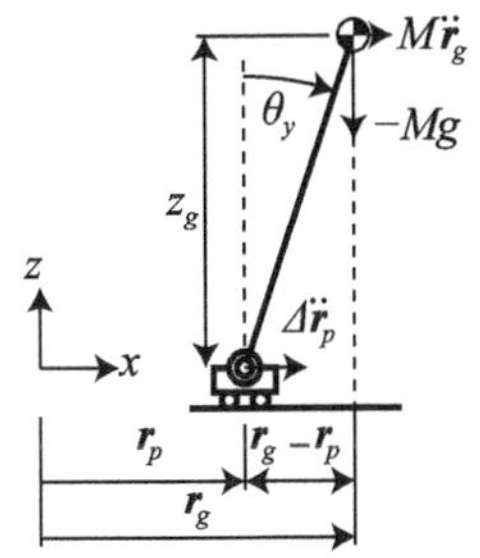

FIGURE 5.10 3D LIP-on-cart model.

A simple PD controller can be designed as follows:

$$\Delta\ddot{\boldsymbol{r}}_p = K\left(\boldsymbol{\theta} + T_D\dot{\boldsymbol{\theta}}\right). \tag{5.37}$$

Substituting (5.37) into (5.36), the following second-order system is obtained:

$$\ddot{\boldsymbol{\theta}} + \frac{KT_D}{z_g}\dot{\boldsymbol{\theta}} + \frac{K-g}{z_g}\boldsymbol{\theta}. = \boldsymbol{0}. \tag{5.38}$$

This system can be stabilized with appropriate gains K ($> g$) and T_D (> 0). Using the centered difference formula, the PD controller (5.37) is discretized as follows:

$$\Delta\boldsymbol{r}_p(t_{k+1}) = \Delta t^2 K\boldsymbol{\theta}(t_k) + \Delta t K T_D\left(\boldsymbol{\theta}(t_k) - \boldsymbol{\theta}(t_{k-1})\right) + \Delta\boldsymbol{r}_p(t_k) + \left(\Delta\boldsymbol{r}_p(t_k) - \Delta\boldsymbol{r}_p(t_{k-1})\right). \tag{5.39}$$

(b) ZMP compensation control: The IP controller determines the desired deviation of the ZMP. In order to track the desired deviation, a ZMP compensation controller is needed.

Recall the gCoM dynamics in 2D given in (5.11). Suppose that the ZMP $\boldsymbol{r}_p$ tracks the desired ZMP $\boldsymbol{r}_p^{des}$ by an increment in the CoM acceleration denoted as $\Delta\ddot{\boldsymbol{r}}_g$. The gCoM dynamics then assume the following form:

$$\ddot{\boldsymbol{r}}_g + \Delta\ddot{\boldsymbol{r}}_g = \omega^2\left(\boldsymbol{r}_g - \boldsymbol{r}_p^{des}\right). \tag{5.40}$$

Subtracting (5.11) from (5.40), the following equation is obtained:

$$\Delta\ddot{\boldsymbol{r}}_g = -\omega^2\Delta\boldsymbol{r}_p, \tag{5.41}$$

where $\Delta\boldsymbol{r}_p \triangleq \boldsymbol{r}_p^{des} - \boldsymbol{r}_p$. The above equation suggests that if $\Delta\boldsymbol{r}_p$ is positive, $\Delta\ddot{\boldsymbol{r}}_g$ must be negative (see Fig. 5.11). Using the centered difference formula, (5.41) is discretized as follows:

$$\Delta\boldsymbol{r}_g(t_{k+1}) = -\Delta t^2\omega^2\Delta\boldsymbol{r}_p(t_k) + \Delta\boldsymbol{r}_g(t_k) + \left(\Delta\boldsymbol{r}_g(t_k) - \Delta\boldsymbol{r}_g(t_{k-1})\right). \tag{5.42}$$

(c) Simplified TPCC: According to the two steps (a) and (b) above, first the desired deviation of the ZMP is calculated using (5.39) to stabilize a humanoid robot body; then the deviation

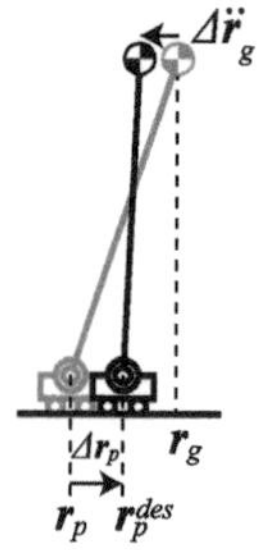

FIGURE 5.11 ZMP compensation control.

of the CoM is calculated using (5.42) to realize the desired deviation of the ZMP. It was experimentally confirmed in [106] that the simplified versions of the IP and the ZMP compensation controllers, written as

$$\Delta \boldsymbol{r}_p(t_{k+1}) = \Delta t^2 K \boldsymbol{\theta}(t_k) + \Delta t K T_D (\boldsymbol{\theta}(t_k) - \boldsymbol{\theta}(t_{k-1})) \tag{5.43}$$

and

$$\Delta \boldsymbol{r}_g(t_{k+1}) = -\Delta t^2 \omega^2 \Delta \boldsymbol{r}_p(t_k) + \Delta \boldsymbol{r}_g(t_k), \tag{5.44}$$

respectively, are also effective.

5.5 CAPTURE POINT–BASED ANALYSIS AND STABILIZATION

Recall the important result in the field of human balance control that the dynamic stability margin for such control is defined w.r.t. the xCoM (cf. Section 5.3). A similar result has been reported in the field of humanoid robotics, though from a different perspective. The goal was to determine the most appropriate foot placement when the change-of-BoS action (cf. Section 5.3.3) is to be undertaken to stabilize the posture after the application of an unknown disturbance. Such type of action is also known from studies on human postural stability, under the name "stepping strategy" or "stumbling strategy" [139,91,93]. This strategy will be discussed in Section 7.7.5.

5.5.1 Capture Point (CP) and Instantaneous Capture Point (ICP)

The special foot placement mentioned above was referred to as the "capture point" (CP) in [122] (see also [125]). The CP was obtained from the orbital energy pertinent to the LIP-on-cart model introduced in Section 4.3.1.

The CP is the only point *outside the BoS* where the LIP-on-cart model will become stationary after the step. In other words, at the CP, the gCoM coincides with a given CoP, $\bar{x}_p$, and also, the gCoM speed is zero; we have

$$x_g(t)|_{t\to\infty} = \bar{x}_p, \tag{5.45}$$

$$\dot{x}_g(t)|_{t\to\infty} = 0. \tag{5.46}$$

This result can also be derived [124,119] from the explicit solution of the gCoM dynamics (cf. (4.13)):

$$x_g(t) = \frac{1}{2}\left(x_{g0} - \bar{x}_p + \frac{v_{g0}}{\omega}\right)e^{\omega t} + \frac{1}{2}\left(x_{g0} - \bar{x}_p + \frac{v_{g0}}{\omega}\right)e^{-\omega t} + \bar{x}_p. \tag{5.47}$$

Hereby, the zero in the subscript notations signifies an initial value. The term that comprises the positive exponent is unstable; it has a destabilizing effect on the gCoM motion, as $t \to \infty$. This problem is tackled by letting the coefficient of the positive exponent become identically zero, so we have

$$x_{g0} - \bar{x}_p + \frac{v_{g0}}{\omega} = 0 \Rightarrow \bar{x}_p = x_{g0} + \frac{v_{g0}}{\omega}. \tag{5.48}$$

It is also worth noting that the above result can be confirmed via the explicit solution for the state-space representation of the gCoM dynamics, (5.16) [23]. Inserting (5.45) into the upper row of (5.16) and using the condition $|\tanh(\omega t)|_{t\to\infty} \to 1$, as in (5.1), it becomes apparent that $\bar{x}_p = x_{g0} + v_{g0}/\omega$. Then, inserting $\bar{x}_p$ into the lower row of (5.16), it is easy to confirm that condition (5.46) will also be satisfied.

The above result also implies that, with an arbitrary gCoM state $(x_g(t), \dot{x}_g(t))$, an expression that is *identical to the xCoM* (5.3) will be obtained. The only difference is in the natural angular frequency $\omega = \omega_{IP}$ for the linearized IP-on-foot model and $\omega = \bar{\omega}$ for the LIP-on-cart one.

Furthermore, note that in the above derivation, it was assumed that the step is taken *instantaneously*. This implies that an infinitely large ankle torque has to be employed. As will be shown in what follows, the above CP theoretical result can be implemented in motion generation and balance control schemes whereby the assumption will be relaxed. In other words, the step will be taken as fast as possible, but in finite time. The stepping time will depend on the maximum velocity and the torque limits of the robot actuators. When taking a step in finite time, the xCoM trajectory evolves in time as in (5.3). In [78] it was suggested to use the term "instantaneous capture point" (ICP); at the end of the step the ICP will coincide with the CP.

5.5.2 ICP-Based Stabilization

Recall the state-space representation of the gCoM and xCoM (or ICP) dynamics (cf. (5.9)):

$$\frac{d}{dt}\begin{bmatrix} x_g \\ x_{ex} \end{bmatrix} = \begin{bmatrix} -\omega & \omega \\ 0 & \omega \end{bmatrix}\begin{bmatrix} x_g \\ x_{ex} \end{bmatrix} + \begin{bmatrix} 0 \\ -\omega \end{bmatrix} x_p. \tag{5.49}$$

The goal is to design a controller that can stabilize the unstable component, i.e. the xCoM/ICP dynamics in the lower row, via an appropriate control input $u = x_p$. This can be done via the explicit solution for the unstable component. Given a CP $\bar{x}_p$, the solution is

$$x_{ex}(t) = (x_{ex0} - \bar{x}_p)e^{\omega t} + \bar{x}_p. \tag{5.50}$$

The control input can be derived from this equation by replacing $x_{ex}(t) \leftarrow x_{ex}^{des}$, x_{ex}^{des} denoting the desired CP, $x_{ex0} \leftarrow x_{ex}(t)$, and $\bar{x}_p \leftarrow x_p(t)$. One then obtains [23]

$$u = \frac{x_{ex}^{des} - e^{\omega dT} x_{ex}}{1 - e^{\omega dT}}, \tag{5.51}$$

where dT is the (desired) time span until the arrival at the desired CP. In the following analysis, the convenient shorthand notation $b \equiv e^{\omega dT}$ will be introduced. The above control law can then be rewritten as

$$u = \frac{1}{1-b} x_{ex}^{des} - \frac{b}{1-b} x_{ex}. \tag{5.52}$$

It is straightforward to confirm that the unstable component of the xCoM/ICP dynamics, $\dot{x}_{ex} = \omega x_{ex}$, will be stabilized by the above control input via a negative feedback, with gain $b/(1-b) > 0$. A positive gain is obtained when dT is chosen to be positive. By inserting u into the state-space representation (5.9), one obtains the closed-loop dynamics as

$$\frac{d}{dt}\begin{bmatrix} x_g \\ x_{ex} \end{bmatrix} = \begin{bmatrix} -\omega & \omega \\ 0 & \frac{\omega}{1-b} \end{bmatrix} \begin{bmatrix} x_g \\ x_{ex} \end{bmatrix} + \begin{bmatrix} 0 \\ -\frac{\omega}{1-b} \end{bmatrix} x_{ex}^{des}, \tag{5.53}$$

with the eigenvalues $-\omega$ and $\omega/(1-b)$. The first eigenvalue is always stable, while the second one is stable for any $dT > 0$. This condition also determines the global stability of the system.

5.5.3 ICP Stabilization in the Presence of GRF Estimation Time Lag

Stabilizer (5.52) can be modified to deal with the GRF/ZMP lag discussed in Section 5.4.4. It will suffice to adjoin the first-order lag dynamics (5.33) to the system dynamics (5.49), by redefining the state-space vector $\boldsymbol{x}$ in (5.34) as $\boldsymbol{x} \equiv \begin{bmatrix} x_g & x_{ex} & x_p \end{bmatrix}^T$. The system matrices assume the form

$$\boldsymbol{A} \equiv \begin{bmatrix} -\omega & \omega & 0 \\ 0 & \omega & -\omega \\ 0 & 0 & -f_p \end{bmatrix}, \quad \boldsymbol{B} \equiv \begin{bmatrix} 0 \\ 0 \\ f_p \end{bmatrix}.$$

Inserting u from (5.52) into the system dynamics, one arrives at the closed-loop dynamics

$$\frac{d}{dt}\begin{bmatrix} x_g \\ x_{ex} \\ x_p \end{bmatrix} = \begin{bmatrix} -\omega & \omega & 0 \\ 0 & \omega & -\omega \\ 0 & -\frac{b}{1-b} f_p & -f_p \end{bmatrix} \begin{bmatrix} x_g \\ x_{ex} \\ x_p \end{bmatrix} + \begin{bmatrix} 0 \\ 0 \\ \frac{1}{1-b} f_p \end{bmatrix} x_{ex}^{des}. \tag{5.54}$$

The eigenvalues of the system matrix $\boldsymbol{A}$ are

$$\lambda_{1,2} = \frac{\omega T_p - 1 \pm \sqrt{r}}{2T_p}, \quad \lambda_3 = -\omega,$$

where $r = (1 + \omega T_p)^2 + 4\omega T_p\, b/(1-b)$. Since λ_3 is always stable, the analysis focuses on the other two eigenvalues. Critical damping is achieved with $r = 0$. This condition determines

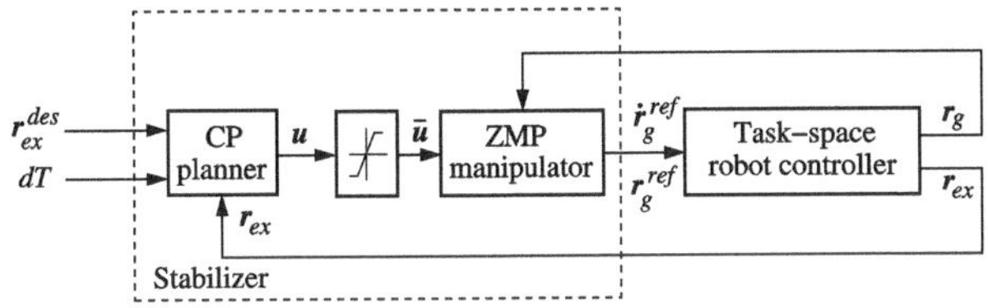

FIGURE 5.12 Block diagram of a CP type balance controller. With reference to Fig. 5.1, it is apparent that: (1) the task trajectory T is defined in terms of the CP motion and (2) the stabilizer comprises a CP planner and the "conventional" ZMP manipulator. The task-space controller makes use of the CoM inverse kinematics solution and includes a local joint-angle feedback controller. The actual CP is calculated from the actual gCoM and its velocity. The limiter restricts the control input to the BoS via a projection [23].

the following critical value for the time duration of the step:

$$dT_{cr} = \frac{2}{\omega} \ln \frac{1 + \omega T_p}{1 - \omega T_p}.$$

Furthermore, an oscillatory response stemming from the nonzero imaginary eigenvalues is undesirable. This case can be avoided as long as $dT > dT_{cr}$. On the other hand, note that dT_{cr} is a function of the time lag constant T_p. The influence of this constant becomes apparent from the eigenvalue behavior as dT grows to infinity, as we have $\lambda_1|_{dT\to\infty} = \omega - 1/T_p$, $\lambda_2|_{dT\to\infty} = 0$. From this result it can be concluded that the system will be stable with time lag constants $T_p < 1/\omega$.

It is worth noting that the above ICP stabilizer is robust w.r.t. a constant gCoM error Δx_g: the error results only in the constant control input offset $b\Delta x_g$. Since the system matrix remains unchanged, the global stability can be preserved [23].

5.5.4 ICP Dynamics and Stabilization in 2D

The implementation of the CP stabilizer in a real robot requires a 2D formulation. Such formulation is obtained in a straightforward manner from the decoupled dynamics of the 3D LIP model (cf. Section 4.4.2). The vector quantities are expressed in accordance with the notation introduced in Section 5.3.4. The xCoM/CP (5.3), the CP dynamics (5.8), and the CP based controller (5.52) are represented in 2D space as

$$\boldsymbol{r}_{ex}(t) = \boldsymbol{r}_g(t) + \frac{\dot{\boldsymbol{r}}_g(t)}{\omega}, \tag{5.55}$$

$$\dot{\boldsymbol{r}}_{ex} = \omega\left(\boldsymbol{r}_{ex} - \boldsymbol{r}_p\right), \tag{5.56}$$

$$\boldsymbol{r}_p^{ref} = \frac{1}{1 - e^{\omega dT}}\boldsymbol{r}_{ex}^{des} - \frac{e^{\omega dT}}{1 - e^{\omega dT}}\boldsymbol{r}_{ex}. \tag{5.57}$$

From the CP dynamics (5.56) it is apparent that the CoP $\boldsymbol{r}_p$ "pushes" the ICP $\boldsymbol{r}_{ex}(t)$ away, in the direction of the CoM velocity. Thus, geometrically the CoP and the ICP lie on the line determined by the CoM velocity.

The block scheme of the controller is shown in Fig. 5.12. An implementation of the method for gait generation and walking control is given in Section 7.2.1.

5.6 STABILITY ANALYSIS AND STABILIZATION WITH ANGULAR MOMENTUM COMPONENT

From the LIP-cart model–based stability analysis in Section 5.4 it became apparent that the underlying constant-height CoM constraint plays an important role in the proof of the stability. This constraint, however, limits the application of the model to flat-ground environments, theoretically. On the other hand, there always has been considerable interest in prospective applications within environments with irregular terrain. Basically, there are three possible approaches to deal with this problem. The first approach is based on the ZMP concept: find a properly defined virtual plane and project the ZMP on that plane [147,145,66,131,136,132,9]. The second approach also uses the ZMP, but in combination with the linear reaction wheel pendulum (LRWP)-on-foot model described in Section 4.3.3. This approach was suggested as an improvement to the original Model-ZMP method [48] used in Honda's robots [150]. The third approach is based on the extension of the ICP concept to three dimensions.

From a historical perspective, past studies on biped gait on flat ground focused almost exclusively on ZMP-based balance control, completely ignoring thereby the angular momentum component. A few exceptions deserve to be mentioned. In [148], variations in angular momentum generated indirectly via the optimized trunk motion were used to stabilize the gait. In [128,129], a simple angular momentum feedback controller was implemented to determine the ankle joint torque of the support leg during the gait. The importance of angular momentum control as a component of a more complex, optimization-based balance controller was revealed in [151,65]. The ankle torque–based angular momentum control was also experimentally tested with other proactive tasks, such as single-leg balance, squats, and kicking motion [74]. This work evolved into the "resolved momentum control" method for a whole-body motion control based on the CRB momentum [69]. It was shown that a stable gait pattern can be realized when the reference CRB angular momentum is set at zero. Later it was revealed that humans, indeed, actively regulate their CRB angular momentum (around zero) during walking on flat ground [120,39].

5.6.1 Stability Analysis Based on the LRWP Model

As shown in Chapter 4 with the LRWP models on the plane and in 3D, the rate of change of the CRB (centroidal) angular momentum alters the CoP by an additional component. The sum determines the "centroidal moment pivot" (CMP) defined in (4.25), (4.40) (see also (5.59) below). In this way, the centroidal moment (and hence the rate of change of the centroidal angular momentum) can be directly related to balance control. The LRWP model thus provides an additional control input that can be used to control the foot rotation occurring when the xCoM leaves the BoS, as depicted in Fig. 5.3D. Note that in this case, the CoP control input used in the ZMP methods is unavailable. The real robot can generate the desired centroidal moment via acceleration in its upper body joints, e.g. with trunk and arm "windmilling" rotation–type maneuvers, as observed in [82,92].

The gCoM dynamics can be obtained from (4.22) as

$$\ddot{x}_g - \omega^2 x_g = -\omega^2 x_{cmp}, \tag{5.58}$$

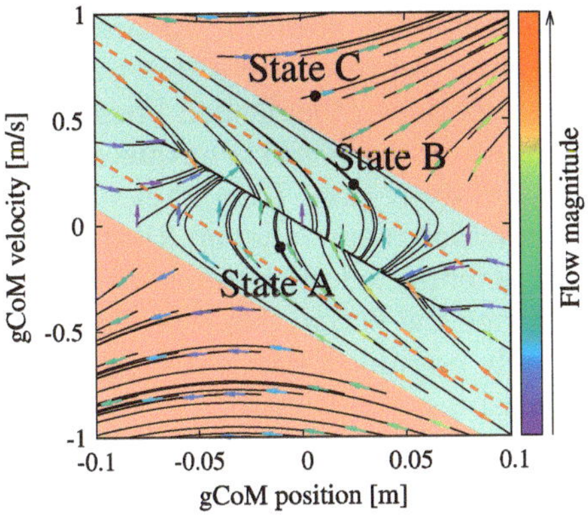

FIGURE 5.13 gCoM phase portrait for the LRWP-on-foot model. The RW torque effectively enlarges the dynamic stability region. The dynamic stability region with zero RW torque (dashed lines) is identical with that of the LIP-on-cart model in Fig. 5.5. State A is inside the LIP dynamic stability region (stable). State B, on the other hand, is outside this region (unstable). However, since State B is within the LRWP dynamic stability region, it would be possible to use a centroidal RW torque (e.g. arm "windmilling") to steer the state toward the LIP dynamic stability region. Finally, State C is outside the LRWP dynamic stability region (unstable). A change-of-BoS (e.g. a reactive step) is needed at this state to avoid a fall.

$$x_{cmp} \equiv x_p + x_{RW}, \tag{5.59}$$

$$x_{RW} \equiv \frac{m_c}{Mg} = \frac{I}{Mg}\ddot{\phi}_c. \tag{5.60}$$

Here x_{cmp} stands for the CMP; x_{RW} defines the additional component. Note that the definition assumes that the vertical CoM acceleration is negligible, s.t. $\ddot{z}_g \ll g$ and hence, the LIP constraint (4.16) holds. It turns out that this assumption can be validated with practical controllers for robots, not only on flat ground but also while negotiating irregular terrain [22,79, 161]. When compared to the gCoM dynamics of the LIP-on-cart model, (5.5), it is apparent that the reaction wheel contributes an additional forcing term. The results from the stability analysis in the previous sections apply directly; just replace x_p with x_{cmp} everywhere. For example, the xCoM dynamics, (5.8), become

$$\dot{x}_{ex} = \omega(x_{ex} - x_{cmp}). \tag{5.61}$$

Furthermore, note that with the LRWP model, the dynamic stability region (5.2) will be effectively enlarged as follows:

$$-x_{RW}^{max} - l_f \leq x_{ex}(t) \leq l_f + x_{RW}^{max}. \tag{5.62}$$

Here $x_{RW}^{max} \equiv m_c^{max}/(Mg)$, where m_c^{max} denotes the maximum RW torque. This is also true for the dynamic stability margin (5.4) and the critical time (5.10): just replace $l_f \rightarrow l_f + x_{RW}^{max}$ in the respective equations. Note also that by varying the RW torque from zero to m_c^{max}, the CP grows to a capture region with maximum area determined by (5.62). This is shown with the phase portrait in Fig. 5.13. It was generated with $m_c^{max} = 2$ Nm, in addition to the data set (5.17). The increased dynamic stability region is apparent, when compared to that of the LIP-on-cart model in Fig. 5.5. Three representative states are shown. State A, being within the LIP dynamic stability region (between the dashed lines), is dynamically stable and thus, balance control can be ensured only with the ankle torque. States B and C, on the other hand, are outside the LIP dynamic stability region. Thus, they are unstable. State B is within the LRWP dynamic stability region which means that an RW-type torque can be used to restore the dynamic balance. As already mentioned, such torque can be generated via the upper-body motion. There is no need to change the BoS in this case. Lastly, State C is outside the LRWP dynamic stability region. The only possible balance strategy in this case is the change-of-BoS one (e.g. reactive stepping or grasping a handrail).

To summarize, the LRWP model contribution to stability is threefold. First, it allows to deal with a rolling foot when the CoP control input is not available. Second, the dynamic stability region and margin, as well as the critical time limit, are enlarged when compared to those obtained with the LIP model. Third, when the foot/feet is/are in flat planar contact, the additional control input provided by the LRWP model increases the robustness of the balance controller w.r.t. disturbances stemming from external wrenches and/or unmodeled dynamics.

5.6.2 Stability Analysis in 3D: the Divergent Component of Motion

The stability analysis based on the linearized IP-on-foot model in Section 5.3.2 and the CP in Section 5.5 has shown that the gCoM comprises a stable and an unstable motion component. An important result was obtained: to achieve an overall stability it would be sufficient to stabilize the unstable component only. The question then is: to facilitate the implementation with a real robot, could this result be extended from 2D to 3D? In order to find an answer, the complete 3-DoF motion of the CoM has to be considered, instead of only the 2-DoF ground projection.

To this end, consider the following direct extension of the xCoM expression (5.3) to 3D:

$$\boldsymbol{r}_X \equiv \boldsymbol{r}_C + \frac{1}{\bar{\omega}_X}\dot{\boldsymbol{r}}_C, \tag{5.63}$$

where $\boldsymbol{r}_C = \begin{bmatrix}\boldsymbol{r}_g^T & z_g\end{bmatrix}^T$ is the CoM position (recall that $\boldsymbol{r}_g \in \Re^2$ stands for the gCoM position). The above expression was introduced in [21] under the name *divergent component of motion* (DCM), a term borrowed from [149]. The DCM is not bounded to the BoS; it floats in 3D space as the CoM does. Note also that in (5.63), the CoM motion is not subjected to any task-induced constraint, neither by a constant-length pendulum, as with the IP-on-foot model, nor by a constant-height one, as with the LIP model. Nevertheless, in the following derivations $\bar{\omega}_X$, denoting the natural angular frequency of the DCM dynamic system, is assumed constant. This condition can be relaxed, though, as shown in [55].

The CoM dynamics are obtained from (5.63) as

$$\dot{\boldsymbol{r}}_C = -\bar{\omega}_X(\boldsymbol{r}_C - \boldsymbol{r}_X). \tag{5.64}$$

The two components on the r.h.s. determine two vector field components: $-\bar{\omega}_X \boldsymbol{r}_C$ is stable (convergent); $\bar{\omega}_X \boldsymbol{r}_X$, on the other hand, is unstable (divergent); this is the DCM. Next, differentiate (5.63) w.r.t. time to obtain the DCM dynamics as

$$\dot{\boldsymbol{r}}_X = \dot{\boldsymbol{r}}_C + \frac{1}{\bar{\omega}_X}\ddot{\boldsymbol{r}}_C. \tag{5.65}$$

The CoM acceleration, appearing in the above formula, is determined by Newton's Third Law: $\ddot{\boldsymbol{r}}_C = \boldsymbol{f}_C/M$, $\boldsymbol{f}_C$ denoting the total *reaction* acting at the CoM; $\boldsymbol{f}_C$ results from the parallel shift of GRF $\boldsymbol{f}_r$ that acts at the CoP. The shifted GRF can be decomposed as follows:

$$\boldsymbol{f}_C = \boldsymbol{f}_I + \boldsymbol{f}_G, \tag{5.66}$$

$\boldsymbol{f}_G = Mg\boldsymbol{e}_z$ and $\boldsymbol{f}_I$ denoting the gravity and inertia reactions, respectively; $\boldsymbol{e}_z$, g, and M stand for the unit vector along the vertical, gravity acceleration, and total mass of the robot, respectively. The inertia reaction comprises three components stemming from the horizontal and the vertical CoM acceleration, as well as from a centroidal moment. The line of action of $\boldsymbol{f}_C$ intersects the (irregular) ground surface at the CMP, $\boldsymbol{r}_{cmp}$. As explained in Section 5.6.1 with the LRWP model (cf. (5.59)), a CMP that does not coincide with the CoP, i.e. $\boldsymbol{r}_{cmp} - \boldsymbol{r}_p \equiv \boldsymbol{r}_{RW} \neq \boldsymbol{0}$, is an indicator of the presence of a nonzero centroidal moment. In the 3D case, this moment is "generated" by a fictitious 3D RW assembly (cf. Section 4.4.3) around the normal to the plane determined by the CoP, the CMP, and the CoM. Note that $\boldsymbol{r}_{RW}$ is the modifier of the CMP obtained from the roll-pitch-yaw components of the centroidal moment.

Combining (5.64), (5.65), and (5.66), one obtains

$$\dot{\boldsymbol{r}}_X - \bar{\omega}_X\boldsymbol{r}_X - \bar{\omega}_X\boldsymbol{r}_C + \frac{g}{\bar{\omega}_X}\boldsymbol{e}_z + \frac{1}{M\bar{\omega}_X}\boldsymbol{f}_I. \tag{5.67}$$

The DCM, $\bar{\omega}_X\boldsymbol{r}_X$, is explicitly visible in the above equation. To deal with the divergence, ensure first the decoupling of the DCM from the CoM motion with an appropriate design of inertial reaction $\boldsymbol{f}_I$. To this end, in [22] the so-called "enhanced CMP" (eCMP) was defined. The eCMP, $\boldsymbol{r}_{ecmp}$, is located on the line of action of $\boldsymbol{f}_C$, s.t. $\boldsymbol{f}_C = k_a(\boldsymbol{r}_C - \boldsymbol{r}_{ecmp})$ holds; k_a is a positive constant to be determined as follows. First, the inertial reaction is expressed as

$$\boldsymbol{f}_I = \boldsymbol{f}_C - \boldsymbol{f}_G = k_a(\boldsymbol{r}_C - \boldsymbol{r}_{ecmp}) - Mg\boldsymbol{e}_z. \tag{5.68}$$

Then, insert the above $\boldsymbol{f}_I$ into (5.67) to obtain

$$\dot{\boldsymbol{r}}_X = \bar{\omega}_X\boldsymbol{r}_X + \frac{k_a - M\bar{\omega}_X^2}{M\bar{\omega}_X}\boldsymbol{r}_C - \frac{k_a}{M\bar{\omega}_X}\boldsymbol{r}_{ecmp} - \frac{g}{\bar{\omega}_X}\boldsymbol{e}_z. \tag{5.69}$$

The dependence of the DCM dynamics on the CoM motion can be nullified by choosing $k_a = M\bar{\omega}_X^2$. Then, (5.69) becomes

$$\dot{\boldsymbol{r}}_X = \bar{\omega}_X\left(\boldsymbol{r}_X - \boldsymbol{r}_{ecmp} - \bar{z}_{vrp}\boldsymbol{e}_z\right), \tag{5.70}$$

where $\bar{z}_{vrp} \equiv g/\bar{\omega}_X^2$. Constant $\bar{z}_{vrp}$ can be interpreted as the average CoM height obtained while the robot walks across irregular terrain [22]. Note that the definition of $\bar{z}_{vrp}$ implies that $\bar{\omega}_X = \sqrt{g/\bar{z}_{vrp}}$. This reveals that $\bar{\omega}_X$ has the meaning of a natural angular frequency for the DCM dynamics.

The above equation represents an important result clarifying that the DCM can be stabilized via the eCMP. The eCMP component, $-\bar{\omega}_X \boldsymbol{r}_{ecmp}$, can be characterized as a "DCM damper." There is also a constant damping component in the vertical direction stemming from the acceleration of gravity. For convenience, the sum of these two damping terms will be denoted as the so-called "virtual repellent point" (VRP) [22]. We have

$$\boldsymbol{r}_{vrp} \equiv \boldsymbol{r}_{ecmp} + \bar{z}_{vrp}\boldsymbol{e}_z. \tag{5.71}$$

The DCM dynamics and the shifted GRF can then be rewritten as

$$\dot{\boldsymbol{r}}_X = \bar{\omega}_X(\boldsymbol{r}_X - \boldsymbol{r}_{vrp}) \tag{5.72}$$

and

$$\begin{aligned}\boldsymbol{f}_C &\underset{(5.68)}{=} M\bar{\omega}_X^2(\boldsymbol{r}_C - \boldsymbol{r}_{ecmp})\\ &\underset{(5.70)}{=} M\bar{\omega}_X^2\left(\boldsymbol{r}_C - \boldsymbol{r}_{vrp} + \frac{g}{\bar{\omega}_X^2}\boldsymbol{e}_z\right)\\ &= M\bar{\omega}_X^2(\boldsymbol{r}_C - \boldsymbol{r}_{vrp}) + Mg\boldsymbol{e}_z,\end{aligned} \tag{5.73}$$

respectively. Note that since $\boldsymbol{f}_G = Mg\boldsymbol{e}_z$, from (5.66) it follows that $\boldsymbol{f}_I = M\bar{\omega}_X^2(\boldsymbol{r}_C - \boldsymbol{r}_{vrp})$ must hold.

The above model implies the existence of a centroidal moment. We have

$$\boldsymbol{m}_C = [\boldsymbol{r}^{\times}_{\overleftarrow{PC}}]\boldsymbol{f}_C.$$

Here $\boldsymbol{m}_C$ comprises two components stemming from the inertia and gravity force components (cf. (5.66)):

$$\begin{aligned}\boldsymbol{m}_C &= \boldsymbol{m}_I + \boldsymbol{m}_G,\\ \boldsymbol{m}_I &= [\boldsymbol{r}^{\times}_{\overleftarrow{PC}}]\boldsymbol{f}_I,\\ \boldsymbol{m}_G &= Mg[\boldsymbol{r}^{\times}_{\overleftarrow{PC}}]\boldsymbol{e}_z.\end{aligned} \tag{5.74}$$

A geometrical interpretation of these results in 2D and 3D is shown in Fig. 5.14.

Furthermore, by introducing state-space coordinates $(\boldsymbol{r}_C,\ \boldsymbol{r}_X)$, the DCM dynamics (5.72) can be represented in state-space form as

$$\frac{d}{dt}\begin{bmatrix}\boldsymbol{r}_C\\ \boldsymbol{r}_X\end{bmatrix} = \begin{bmatrix}-\bar{\omega}_X & \bar{\omega}_X\\ 0 & \bar{\omega}_X\end{bmatrix}\begin{bmatrix}\boldsymbol{r}_C\\ \boldsymbol{r}_X\end{bmatrix} + \begin{bmatrix}0\\ -\bar{\omega}_X\end{bmatrix}\boldsymbol{r}_{vrp}. \tag{5.75}$$

The form of this equation is the same as that in (5.9) for the planar case. Note that in (5.75), the VRP determines the forcing term, in place of the CoP in (5.9).

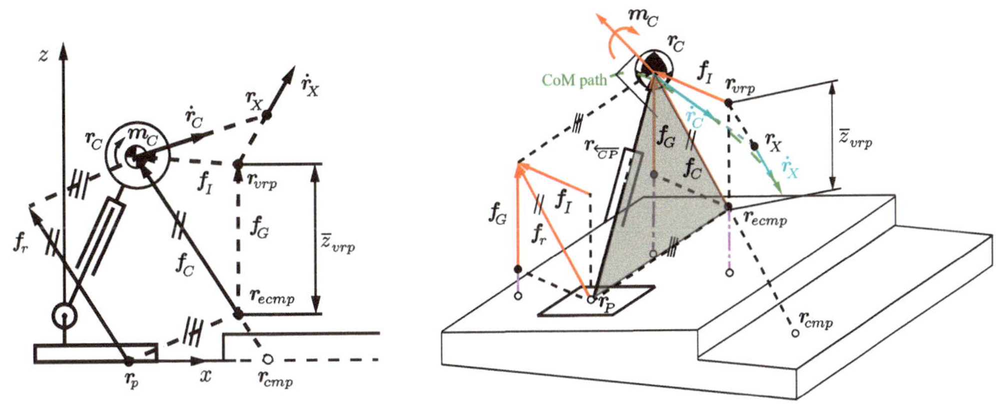

FIGURE 5.14 Geometrical interpretation of DCM relations with specific points: the CoP ($\boldsymbol{r}_P$), CoM ($\boldsymbol{r}_C$), CMP ($\boldsymbol{r}_{cmp}$), enhanced CMP (eCMP) ($\boldsymbol{r}_{ecmp}$), and VRP ($\boldsymbol{r}_{vrp}$) (based on [22]). Left (A): 2D representation. Right (B): 3D representation. All forces denote reactions (by convention reactions are positive). The system is stabilized via inertia reaction $\boldsymbol{f}_I$ that determines an appropriate VRP. The parallel shift of total GRF $\boldsymbol{f}_r = \boldsymbol{f}_C = \boldsymbol{f}_G + \boldsymbol{f}_I$ via the eCMP implies that the system is stabilized in the presence of a centroidal moment, $\boldsymbol{m}_C$, in the direction normal to the sagittal plane (2D case) or to the CoP-CoM-eCMP plane (the gray plane in the 3D figure).

5.6.3 DCM Stabilizer

The above stability analysis has led to the important result that the unstable DCM dynamics in 3D can be stabilized with an appropriate control law defined via the VRP. The envisioned balance controller should track the desired DCM $\boldsymbol{u} \equiv \begin{bmatrix} (\boldsymbol{r}_X^{des})^T & (\dot{\boldsymbol{r}}_X^{des})^T \end{bmatrix}^T$ faithfully. To this end, consider the following error dynamics [22]:

$$\dot{\boldsymbol{r}}_X^{ref} = \dot{\boldsymbol{r}}_X^{des} - k_X(\boldsymbol{r}_X - \boldsymbol{r}_X^{des}). \tag{5.76}$$

This system will be asymptotically stable with $k_x > 0$. The reference VRP can be expressed as (cf. (5.72))

$$\boldsymbol{r}_{vrp}^{ref} = \boldsymbol{r}_X - \frac{1}{\overline{\omega}_X}\dot{\boldsymbol{r}}_X^{ref}. \tag{5.77}$$

The closed-loop dynamics are derived in the conventional form

$$\dot{\boldsymbol{x}} = \boldsymbol{A}\boldsymbol{x} + \boldsymbol{B}\boldsymbol{u}, \tag{5.78}$$

$\boldsymbol{x} \equiv \begin{bmatrix} \boldsymbol{r}_C^T & \boldsymbol{r}_X^T \end{bmatrix}^T$ denoting the state vector. By inserting the control law (5.77) into the state-space equation (5.75) ($\boldsymbol{u} \equiv \boldsymbol{r}_{vrp}^{ref}$), one obtains

$$\boldsymbol{A} \equiv \begin{bmatrix} -\overline{\omega}_X \boldsymbol{E} & \overline{\omega}_X \boldsymbol{E} \\ 0 & -k_x \boldsymbol{E} \end{bmatrix}, \quad \boldsymbol{B} \equiv \begin{bmatrix} 0 & 0 \\ -k_x \boldsymbol{E} & \boldsymbol{E} \end{bmatrix}.$$

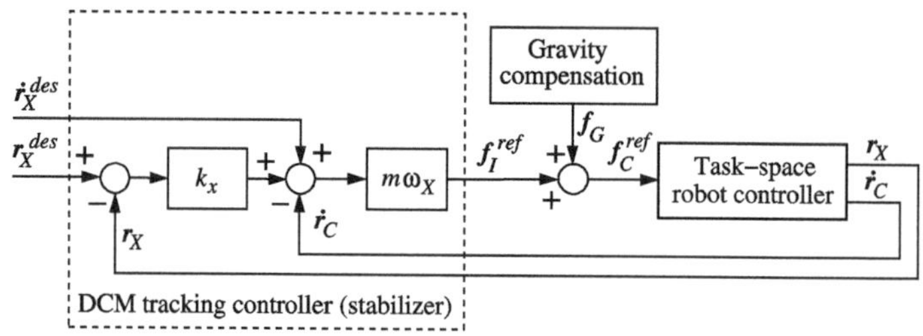

FIGURE 5.15 Block diagram of a DCM-type balance controller. With reference to Fig. 5.1, it is apparent that the task trajectory T is defined in terms of the DCM motion.

Since the system matrix $\boldsymbol{A}$ is stable for positive k_x and the control input $\boldsymbol{u}$ is assumed bounded, the above linear system can be characterized as bounded-input, bounded-output (BIBO)-stable.

The above result can be used to design a balance controller that incorporates the following reference (shifted) GRF:

$$\begin{aligned} f_C^{ref} &\underset{(5.73)}{=} M\bar{\omega}_X^2\left(\boldsymbol{r}_C - \boldsymbol{r}_{vrp}^{ref} + \frac{g}{\bar{\omega}_X^2}\boldsymbol{e}_z\right) \\ &\underset{(5.77)(5.64)}{=} M\bar{\omega}_X\left(\dot{\boldsymbol{r}}_X^{ref} - \dot{\boldsymbol{r}}_C\right) + Mg\boldsymbol{e}_z. \end{aligned} \tag{5.79}$$

From (5.66) it follows that the first control term on the r.h.s. can be interpreted as a reference inertial force, $(f_I)^{ref}$. This force ensures faithful tracking of the desired DCM, provided the BoS constraints are met. The block diagram of the DCM stabilizer is shown in Fig. 5.15. The design parameters are k_x and $\bar{\omega}_X = \sqrt{g/\bar{z}_{vrp}}$. The stabilizer is characterized with robustness against[4]: (1) a time-varying CoM motion tracking error, (2) a GRF/CoP estimation time-lag, (3) a mass estimation error, and (4) a constant external disturbance.

5.6.4 Summary and Conclusions

The DCM-based stabilization approach introduced above has been implemented with a number of real robots and has shown superior performance when compared to other balance controllers. The method has become indispensable in all areas of balance control such as walking gait generation and control, not only on flat ground but also on irregular terrain, and reactive stepping in response to an external perturbation. Implementation examples will be given in Chapter 7.

The method can be further improved by resorting to a time-varying reformulation, $\omega_X = \omega_X(t)$, as suggested in [55]. With such improvement, the somewhat fuzzy "average" CoM height notation $\bar{z}_{vrp}$ can be avoided. Any feasible desired vertical CoM trajectories can then be tracked under the same stability conditions. In a recent work, it has been shown how to use this improvement for a stretched-leg negotiation of irregular terrain [31].

[4] The robustness proof can be found in [22].

5.7 MAXIMUM OUTPUT ADMISSIBLE SET BASED STABILIZATION

The *maximum output admissible set* concept has been introduced as a linear control system approach whereby the state and/or the input are constrained [28]. The initial state of an unforced linear system is said to be output admissible w.r.t. a constraint set Y if the resulting output function satisfies the pointwise-in-time condition $y(t) \in Y,\ t \geq 0$. The set of all such possible initial conditions is called the maximum output admissible (MOA) set. It can then be guaranteed that all future states will comply with the constraints. Note that, as in the ICP/DCM methods, the initial state plays an important role.

The MOA set method is especially suitable for linear time-discrete systems. In this case, the MOA set[5] can be constructed iteratively, without a significant computational cost. Furthermore, the method can be used for systems with so-called exogenous inputs such as a large disturbance input [49]. From this brief introduction it becomes apparent that the MOA set method could be quite suitable for balance control, e.g. to enforce the BoS constraints during proactive tasks and, also, to ensure an appropriate reactive behavior in response to an external disturbance.

An implementation of the MOA set method for humanoid robot balance control has been pioneered in [165]. The implementation is based on the spherical IP model. Differently from the ICP-based stabilizer, the CoM is allowed to move in the vertical direction. A tracking stabilizer for the desired motion in this direction can be formulated as follows:

$$f_{rz}^{ref} = m(\ddot{z}_g + g) + k_{pv}(z_g^{des} - z_g) + k_{dv}(\dot{z}_g^{des} - \dot{z}_g), \tag{5.80}$$

k_{pv} and k_{dv} denoting the feedback gains. The control laws for the CoM motion in the horizontal directions, on the other hand, can be obtained via the gCoM/CoP dynamic relation (5.5). Since the spherical IP is linearized, the equation of motion in the x-direction is decoupled from that in y. It is important to note, though, that since relatively large vertical CoM accelerations are admissible (e.g. when running or jumping), the common coefficient in the gCoM/CoP dynamic relations is not constant (cf. (4.14)); we have

$$\omega(t) = \sqrt{\frac{\ddot{z}_g + g}{z_g}}.$$

Note also that under the assumption of zero disturbance in the vertical direction, it will be possible to obtain data sets $z_g(t)$ and $\omega(t)$ for the entire motion *in advance*.

Consider the state-space form of the gCoM/ZMP dynamic relation as given in (5.15) for the x-direction. It is straightforward to append a second equation of the same form for the y-direction. Furthermore, the recursive calculation of the MOA requires system discretization. The state-space equation for the two horizontal directions can then be written as

$$\boldsymbol{x}_{k+1} = \boldsymbol{A}_k \boldsymbol{x}_k + \boldsymbol{B}_k \boldsymbol{u}_k,$$

where $\boldsymbol{x}_k \equiv \begin{bmatrix} x_g(t_k) & \dot{x}_g(t_k) & y_g(t_k) & \dot{y}_g(t_k) \end{bmatrix}^T$ stands for the state-space vector at time t_k. We have matrices $\boldsymbol{A}_k = \Delta t \boldsymbol{A}(t_k) + \boldsymbol{E}$ and $\boldsymbol{B}_k = \Delta t \boldsymbol{B}(t_k)$, Δt is the sampling time, and $\boldsymbol{A}(t_k)$ and

[5] Also known as the constrained positively invariant (CPI) set [49].

$\boldsymbol{B}(t_k)$ are block-diagonal matrices with two identical block components apparent from (5.15). These matrices are functions of time since ω is a time variable. The control input is $\boldsymbol{u}_k$.

As a proactive task, consider a CoM motion regulator

$$\boldsymbol{u}_k = -\boldsymbol{K}\boldsymbol{x}_k,$$

where $\boldsymbol{K} \equiv \text{diag}(k_{px},\ k_{dx},\ k_{py},\ k_{dy})$ is a feedback gain matrix. The closed-loop dynamics are obtained as

$$\boldsymbol{x}_{k+1} = \Delta\boldsymbol{A}_k\boldsymbol{x}_k, \tag{5.81}$$

where $\Delta\boldsymbol{A}_k \equiv \boldsymbol{A}_k - \boldsymbol{B}_k\boldsymbol{K}$. It is straightforward to determine $\boldsymbol{K}$ by well-known methods such as the pole placement or the LQ. Thus, the closed-loop dynamics will be asymptotically stable.

The MOA set can be constructed for an asymptotically stable system like the one in (5.81) iteratively. First, the control input is constrained within the BoS. Referring to (5.13),

$$\boldsymbol{B}_{sk}\boldsymbol{u}_k \preceq \boldsymbol{c}_k,$$

where $\boldsymbol{B}_{sk},\ \boldsymbol{c}_k$ determine the BoS at time instant t_k. Next, the state space is constrained within the MOA set defined as

$$O_\infty = \{\boldsymbol{x} : \boldsymbol{S}\boldsymbol{x} \preceq \boldsymbol{a}\}.$$

The MOA set can be represented as a convex polyhedron in state space. The iterative construction is summarized as

$$\boldsymbol{S} = \begin{bmatrix} \boldsymbol{B}_s & 0 & \cdots & 0 \\ 0 & \boldsymbol{B}_s & \cdots & 0 \\ \vdots & \vdots & \ddots & \vdots \\ 0 & 0 & \cdots & \boldsymbol{B}_s \end{bmatrix} \begin{bmatrix} -\boldsymbol{K} \\ -\boldsymbol{K}\Delta\boldsymbol{A} \\ \vdots \\ -\boldsymbol{K}\Delta\boldsymbol{A}^l \end{bmatrix}, \quad \boldsymbol{a} = \begin{bmatrix} \boldsymbol{c} \\ \boldsymbol{c} \\ \vdots \\ \boldsymbol{c} \end{bmatrix},$$

where l denotes the iteration number. Given initial state $\boldsymbol{x}_0$, at the kth iteration $\boldsymbol{x}_k = \Delta\boldsymbol{A}^k\boldsymbol{x}_0$ and, hence, $\boldsymbol{u}_k = -\boldsymbol{K}\Delta\boldsymbol{A}^k\boldsymbol{x}_0$. Thus, $\boldsymbol{S}_k\boldsymbol{x}_0 \preceq \boldsymbol{a}$, where $\boldsymbol{S}_k$ is calculated from the above equation with $l = k$. The set $O_k = \{\boldsymbol{x} : \boldsymbol{S}_k\boldsymbol{x} \preceq \boldsymbol{a}\}$ is the kth output admissible set. If the initial state at t_0 belongs to this set, it will be guaranteed that all future states at $t_1,\ t_2, \ldots, t_k$ will satisfy the BoS constraint. Moreover, when the closed-loop system is asymptotically stable, as already confirmed, a finite number of iterations will be required, whereby l is determined by the condition $O_l = O_{l+1}$.

In addition to the regulator-type task, the performance of the MOA set stabilizer was also confirmed with a time-varying input task and BoS constraint, including a jumping/hopping task [165].

5.8 BALANCE CONTROL BASED ON SPATIAL MOMENTUM AND ITS RATE OF CHANGE

The spatial momentum–based balance control can ensure balance while performing a broad scope of tasks (both proactive and reactive) within various types of environment with coplanar, noncoplanar, and nonplanar, as well as with time-varying contact models. The method also plays an important role in whole-body control. Because of these advantages, the spatial momentum–based balance control has become a mainstream control approach in humanoid robotics and physically based computer animation [74,68,30,1,109,80,92,53,18,87, 88,112,16,85,77,161,41].

Spatial momentum–based controllers comprise two main components: a stabilizer and a task-space controller (cf. Fig. 5.1). The stabilizer generates the input for the task-space controller in terms of spatial momentum or its rate of change. Stabilizers developed at an early stage used mainly the ZMP manipulation technique and were implemented in position-controlled humanoid robots (e.g. Honda's P2 [48]). Note that the ZMP manipulation–type stabilizer in Fig. 5.7 can be regarded as a linear momentum stabilizer since the produced control input is in terms of CoM velocity. Velocity-based spatial momentum balance controllers are simpler than dynamic balance controllers. Dynamic balance controllers make use of the rate of change of spatial momentum; they provide means for direct control of the contact/reaction wrenches, the overactuation in the closed chains and the CoPs. Velocity-based controllers do not have such capabilities. Nevertheless, they can control the net CoP, though only indirectly, as in the ZMP manipulation approach. It should be noted that the majority of the existing humanoid robots so far are position-controlled. Thus, velocity-based balance control based on spatial momentum plays an important role, as does dynamic balance control based on the rate of change of spatial momentum.

5.8.1 Fundamental Functional Dependencies in Balance Control

When designing a momentum-based balance controller it is important to have an in-depth understanding of the functional dependencies between the contact wrenches, the rates of change of linear/angular momenta, and the CoPs at the contact surfaces. Consider first the case of coplanar contacts at the feet. In this case, the tangential components of the rates of change of linear and angular momenta are relevant. These quantities, denoted as $\dot{\boldsymbol{p}}_t$ and $\dot{\boldsymbol{l}}_{C_t}$, respectively, appear in the ZMP/CoP equation, (4.38) (note that $\dot{\boldsymbol{p}}_t = \boldsymbol{f}_t = M\ddot{\boldsymbol{r}}_g$). The following functional dependencies exist:

$$\begin{aligned} \dot{\boldsymbol{p}} &= \dot{\boldsymbol{p}}(\boldsymbol{f}_C), \\ \dot{\boldsymbol{l}}_{C_t} &= \dot{\boldsymbol{l}}_{C_t}(\dot{\boldsymbol{p}}_t, \boldsymbol{r}_P), \\ \boldsymbol{r}_P &= \boldsymbol{r}_P(\dot{\boldsymbol{p}}_t, \dot{\boldsymbol{l}}_{C_t}). \end{aligned} \tag{5.82}$$

The first two dependencies are apparent from (4.38), the last one from the CRB dynamics (4.150). An important consequence of the above dependencies is that the rates of change of the two spatial momentum components are coupled. As apparent from (4.39) and (4.40), the coupling is via the CMP. We have

$$\begin{bmatrix} \dot{\boldsymbol{p}}_t \\ \dot{\boldsymbol{l}}_{C_t} \end{bmatrix} = \begin{bmatrix} \boldsymbol{f}_t \\ \boldsymbol{m}_t \end{bmatrix} = M\omega_{IP}^2 \begin{bmatrix} (\boldsymbol{r}_g - \boldsymbol{r}_{cmp}) \\ -z_g \mathbb{S}^{\times}(\boldsymbol{r}_p - \boldsymbol{r}_{cmp}) \end{bmatrix} \in \Re^4. \tag{5.83}$$

This type of coupling exists not only for coplanar contacts, but also for noncoplanar ones. Indeed, using (5.73) and (5.74), it is straightforward to express the CRB momentum rates as

$$\dot{\mathcal{L}}_C = \begin{bmatrix} \dot{\boldsymbol{p}} \\ \dot{\boldsymbol{l}}_C \end{bmatrix} = \begin{bmatrix} \boldsymbol{f}_I \\ \boldsymbol{m}_I \end{bmatrix} = M\omega_X^2 \begin{bmatrix} \boldsymbol{E} \\ [\boldsymbol{r}^{\times}_{\overleftarrow{PC}}] \end{bmatrix} (\boldsymbol{r}_C - \boldsymbol{r}_{vrp}) \in \Re^6. \tag{5.84}$$

From this equation, functional dependencies of the type (5.82) can be derived. The coupling between the two momentum rate components is apparently via $\boldsymbol{r}_{vrp}$.

The coupling problem plays an important role in balance controller design as well as in motion/force input generation for balance control. Consider as an example a humanoid robot in a double-stance posture. *Independent* control of the six spatial momentum components can only be achieved if all 12 leg actuators (6-DoF legs are assumed) are engaged. Thus, there is no actuator left to also control the net CoP, in an *independent* way. Attempting to design a balance controller with all control inputs being independent will result in an overconstrained system. The situation is similar when the robot is in a single-stance posture, when the 6D position of the upper body is to be controlled, in addition to that of the swing foot. It should be noted, though, that a rigorous assessment of the coupling problem from the viewpoint of input generation and control of balance is still missing. Various approaches have been suggested to deal with this problem, such as ignoring some of the angular momentum components (pitch and roll) [69], introducing weights [92] or prioritization [88,89]. These approaches have been shown to work in certain cases, but nevertheless, the ambiguity w.r.t. parameter settings (weights, priorities) remains.

A practical approach to the coupling problem is to design the momentum stabilizer in terms of the linear component only, based on a suitable reference CMP/VRP. The angular component will then be stabilized as a "byproduct" of the coupling, without the need to specify an explicit reference. This approach was used, in fact, in the DCM stabilizer described in Section 5.6.3. It was also used in [77,161]. The expression for $\dot{\boldsymbol{l}}_{C_t}$ in (5.83) shows that, as long as the difference between the CoP and the reference CMP is small, the angular momentum component will not contribute significantly to balance control (e.g. as with States (A)–(C) in Fig. 5.3). This is the usual situation while keeping balance at a standing posture or walking on regular terrain. In this case, the rate of change of angular momentum is quite small [120, 39]. This also implies that any proactive postural variations that *minimize (or conserve) the angular momentum* will not deteriorate the balancing capability. One example of such postural variation is the so-called "hip strategy," characterized by antiphase rotations in the leg and trunk body segments (cf. Section 7.6.3). On the other hand, note that when the xCoM leaves the BoS, as with State (D) in Fig. 5.3, a change of angular momentum via an appropriate upper-body maneuver will be needed to bring back the xCoM within the BoS limits [161]. Further details will be given in Section 5.8.6.

As a final remark it should be noted that the above dependencies are valid in the case of irregular terrain as well.

5.8.2 Resolved Momentum Control

The idea to design a balance controller based on the spatial momentum relation with the base quasivelocity, (4.91), was introduced in [69] under the name "resolved momentum control." The spatial momentum of the system is decomposed into three components stemming from the linear and angular motion of the base link ($\mathcal{V}_B$), the feet ($\mathcal{V}_{F_j}$), and the remaining joints ($\dot{\boldsymbol{\theta}}_{free}$). We have

$$\mathcal{L}_C(\boldsymbol{q}, \dot{\boldsymbol{q}}_B) = \mathcal{T}_B \begin{bmatrix} \mathcal{V}_B \\ \dot{\boldsymbol{\theta}}_{free} \end{bmatrix} + \sum_{j\in\{r,l\}} \mathcal{T}_j \mathcal{V}_{F_j}, \tag{5.85}$$

$\mathcal{T}_{(\circ)}$ denoting appropriate transforms that also account for the contact constraints. The following velocity control law was derived from the solution of the above equation:

$$\begin{bmatrix} \mathcal{V}_B \\ \dot{\boldsymbol{\theta}}_{free} \end{bmatrix} = \overline{\mathcal{T}}_B^{+} \left(\tilde{\mathcal{L}}_C^{ref} - \sum_{j\in\{r,l\}} \mathcal{T}_j \mathcal{V}_{F_j}^{ref} \right) + (\boldsymbol{E} - \overline{\mathcal{T}}_B^{+} \overline{\mathcal{T}}_B) \begin{bmatrix} \mathcal{V}_B^{ref} \\ \dot{\boldsymbol{\theta}}_{free}^{ref} \end{bmatrix}. \tag{5.86}$$

Here $\tilde{\mathcal{L}}_C^{ref}$ and $\mathcal{V}_{F_j}^{ref}$ are reference values for the CRB spatial momentum and feet motion. These quantities constitute the high-priority task; $\mathcal{V}_B^{ref}$ and $\dot{\boldsymbol{\theta}}_{free}^{ref}$ are reference values for the base twist and the motion in the remaining joints (i.e. arm motion); they constitute the lower-priority task. The base twist $\mathcal{V}_B$, obtained from the above control law, is used to calculate the joint velocity in each leg. We have

$$\dot{\boldsymbol{\theta}}_{F_j} = \boldsymbol{J}^{-1}(\boldsymbol{q}_{F_j})\mathcal{V}_{F_j} - \boldsymbol{J}^{-1}(\boldsymbol{q}_{F_j})\mathbb{T}_{\overleftarrow{F_j B}}\mathcal{V}_B. \tag{5.87}$$

The overbar notation in (5.86) denotes a selective approach s.t. some of the components of spatial momentum are relaxed (ignored). This approach was introduced in [69] to avoid undesirable upper-body rotation observed when the above control law is applied to walking and kicking motion tasks. The problem with the original control law (without the relaxation) is that $\mathcal{V}_B^{ref}$ is constrained within the null space $\mathcal{N}(\mathcal{T}_B)$. This means that there is no guarantee that the desired CRB trajectories, specified by the reference base-link twist, will be tracked faithfully.

5.8.3 Whole-Body Balance Control With Relative Angular Momentum/Velocity

The resolved momentum framework was the result of a pioneering effort toward a velocity-based *whole-body balance control*. The control law formulation was flawed, however, since the coupling between the linear and angular momentum components was not addressed appropriately. As a result, not all of the components of the desired base-link twist could be controlled. Also, a rigorous stability assessment regarding the numerical instabilities reported in [69] was lacking. So far, stability assessment of velocity-based balance control has been done for simple IP models only, e.g. those used in the ZMP manipulation approach (cf. Section 5.4).

In this section, a *velocity-based whole-body balance controller* will be derived that can track the desired CRB trajectories with asymptotic stability. The design of the controller is based on the *momentum equilibrium principle* described in Section 4.7. In the case when the centroidal spatial momentum is expressed in terms of the mixed quasivelocity, the equilibrium relation can be written as (cf. (4.90))

$$\boldsymbol{H}_{CM}\dot{\boldsymbol{\theta}} = \mathbb{M}_C\mathcal{V}_C - \mathbb{M}_C\mathcal{V}_M. \tag{5.88}$$

The term on the l.h.s. is the coupling momentum; $\mathbb{M}_C\mathcal{V}_C$ and $\mathbb{M}_C\mathcal{V}_M$ denote the system and the CRB spatial momentum, respectively.[6] Since the rate of change of the SSM is determined by the total external force acting on the system (cf. (4.148)), it can be concluded that $\mathbb{M}_C\mathcal{V}_C$ stems from the reaction (contact) wrenches. Moreover, noting that the CRB (locked) inertia tensor is positive definite, the above momentum equilibrium can be recast as a *spatial velocity equilibrium* as follows:

$$\mathbb{M}_C^{-1}\boldsymbol{H}_{CM}\dot{\boldsymbol{\theta}} = \mathcal{V}_C - \mathcal{V}_M. \tag{5.89}$$

Here $\mathcal{V}_C \equiv \mathbb{M}_C^{-1}\mathcal{L}_C$ is the *centroidal twist*. The term on the l.h.s., referred to as the *coupling spatial velocity*, is in balance with the *relative spatial velocity* $\Delta\mathcal{V} = \mathcal{V}_C - \mathcal{V}_M$ on the r.h.s.

Furthermore, recall that $\mathbb{M}_C^{-1}\boldsymbol{H}_{CM} = \begin{bmatrix}\boldsymbol{0}^T & \boldsymbol{J}_\omega^T\end{bmatrix}^T$ (cf. (4.97)). Thus, the above spatial velocity equilibrium relation can be represented componentwise as

$$\boldsymbol{v}_{C_I} = \boldsymbol{v}_{C_R}, \tag{5.90}$$

$$\boldsymbol{J}_\omega\dot{\boldsymbol{\theta}} = \boldsymbol{\omega}_C - \boldsymbol{\omega}_B \equiv \Delta\boldsymbol{\omega}, \tag{5.91}$$

where $\boldsymbol{J}_\omega(\boldsymbol{\theta}) = \boldsymbol{I}_C^{-1}(\boldsymbol{q})\boldsymbol{H}_C(\boldsymbol{q})$ (cf. (4.86)) and $\boldsymbol{\omega}_C = \boldsymbol{I}_C^{-1}\boldsymbol{l}_C$ is the *system angular velocity* (cf. (4.87)). The notation in the upper equation clarifies that the CoM velocity can be interpreted twofold: $\boldsymbol{v}_{C_I}$ is of *inertial* origin, while $\boldsymbol{v}_{C_R}$ stems from the *net* system twist that is of *reactive* origin. The origins can be distinguished only when the CoM motion is expressed in terms of acceleration, though (cf. Section 5.11.2). The velocities are indistinguishable, thus $\boldsymbol{v}_{C_I} = \boldsymbol{v}_{C_R} = \boldsymbol{v}_C$, $\Delta\boldsymbol{v}_C = \boldsymbol{0}$. Next, note that the lower equation expresses an *angular velocity equilibrium*: the *coupling angular velocity* $\boldsymbol{J}_\omega\dot{\boldsymbol{\theta}}$ is in balance with the *relative angular velocity* $\Delta\boldsymbol{\omega} \neq \boldsymbol{0}$. An important conclusion that can be made is that *the angular velocities of the system and the base link can be assigned in an independent way*. This conclusion clarifies why the flaw in the resolved momentum–type controller is avoidable.

It is assumed that the reference values for the CoM velocity and the angular velocity of the base link are known from the task assignment. The desired trajectories of the CRB twist, $\mathcal{V}_M^{des} = \begin{bmatrix}(\boldsymbol{v}_C^{des})^T & (\boldsymbol{\omega}_B^{des})^T\end{bmatrix}^T$, can be tracked with a conventional velocity controller derived from the constraint-consistent joint velocity solution (2.132). Then we have

$$\dot{\boldsymbol{\theta}}_1 = -\mathcal{J}_{cM}^{+}\mathbb{C}_{cC}^T\mathcal{V}_M^{ref} + \boldsymbol{N}(\mathcal{J}_{cM})\,\dot{\boldsymbol{\theta}}_u. \tag{5.92}$$

Note that $\overline{\mathcal{V}}^c = \boldsymbol{0}$ was used in the derivation, under the assumption of hard feet constraints on stationary ground; $\mathcal{V}_M^{ref}$ comprises the *independent* feedforward/feedback control compo-

[6] The respective abbreviations, SSM and CRB-SM, were introduced in Section 4.6.4.

nents

$$\boldsymbol{v}_C^{ref} = \boldsymbol{v}_C^{des} + K_{PC}\boldsymbol{e}_{PC}, \tag{5.93}$$

$$\boldsymbol{\omega}_B^{ref} = \boldsymbol{\omega}_B^{des} + K_{OB}\boldsymbol{e}_{OB}, \tag{5.94}$$

$\boldsymbol{e}_{PC} = \boldsymbol{r}_C^{des} - \boldsymbol{r}_C$ and $\boldsymbol{e}_{OB}$ denoting the CoM position error and the orientation error of the base link, respectively. The latter is expressed in any convenient form, e.g. with quaternions or using Euler's axis/angle formula (see e.g. [140], p. 139); $K_{(\circ)}$ are feedback gains. The control law (5.92) guarantees that $\mathcal{V}_M(t) = \mathcal{V}_M^{ref}(t)$ asymptotically, provided the contact states are maintained and the joint-space constraint Jacobian $\mathcal{J}_{cM}$ is full (row) rank.

The control input $\dot{\boldsymbol{\theta}}_1$ is useful in case of a double stance; the feet are fully constrained, s.t. $\mathbb{B}_c(\boldsymbol{q}_F) = \boldsymbol{E}$, $\mathbb{B}_m(\boldsymbol{q}_F) = \boldsymbol{0}$, while the hands are completely unconstrained, s.t. $\mathbb{B}_c(\boldsymbol{q}_H) = \mathbb{B}_m(\boldsymbol{q}_H) = \boldsymbol{0}$. In a single stance, the swing-foot motion control task could be embedded as a lower-priority task. To this end, determine the arbitrary joint velocity vector $\dot{\boldsymbol{\theta}}_u$ in (5.92) using the instantaneous-motion equation (2.129). The control joint velocity then assumes the form of (2.86), so we have

$$\dot{\boldsymbol{\theta}}_2 = -\mathcal{J}_{cM}^{+}\mathbb{C}_{cC}^{T}\mathcal{V}_M^{ref} + \overline{\mathcal{J}}_{mM}^{+}(\widetilde{\mathcal{V}}^{m})^{ref} + \boldsymbol{N}(\mathcal{J}_{cM})\boldsymbol{N}(\overline{\mathcal{J}}_{mM})\dot{\boldsymbol{\theta}}_u, \tag{5.95}$$

where $\overline{\mathcal{J}}_{mM} = \mathcal{J}_{mM}\boldsymbol{N}(\mathcal{J}_{cM})$ is the restricted end-link mobility Jacobian and

$$(\widetilde{\mathcal{V}}^{m})^{ref} = (\overline{\mathcal{V}}^{m})^{ref} + \left(\mathcal{J}_{mM}\mathcal{J}_{cM}^{+}\mathbb{C}_{cC}^{T} - \mathbb{C}_{mC}^{T}\right)\mathcal{V}_M^{ref}.$$

Here $(\overline{\mathcal{V}}^{m})^{ref}$ comprises a nonzero component for the swing leg, i.e.

$$\mathcal{V}_{F_{SW}}^{ref} = \mathcal{V}_{F_{SW}}^{des} + \boldsymbol{K}_{F_{SW}}\boldsymbol{e}_{F_{SW}}, \tag{5.96}$$

and the rest of the components of $(\overline{\mathcal{V}}^{m})^{ref}$ are zeros. The subscript SW stands for the swing leg, $\boldsymbol{e}_{F_{SW}}$ is the error twist, and $\boldsymbol{K}_{F_{SW}}$ is a feedback gain. Note that control input $\dot{\boldsymbol{\theta}}_2$ can also be used in the case of double stance, by adjusting the constraint conditions via the $\mathbb{B}_{(\circ)}$ matrices.

Relative Angular Momentum/Velocity (RAM/V) Balance Control

To obtain a controller with an enhanced balance control capability, it is recommendable to add a control term for the angular momentum. To this end, make use of the system angular velocity $\boldsymbol{\omega}_C$. Insert (5.95) into (5.91) and solve for the arbitrary $\dot{\boldsymbol{\theta}}_u$. Then, insert back into (5.92) to finally obtain the enhanced control law as

$$\begin{aligned}\dot{\boldsymbol{\theta}} &= -\mathcal{J}_{cM}^{+}\mathbb{C}_{cC}^{T}\mathcal{V}_M^{ref} + \overline{\mathcal{J}}_{mM}^{+}(\widetilde{\mathcal{V}}^{m})^{ref} + \overline{\boldsymbol{J}}_{\omega}^{+}\left(\Delta\boldsymbol{\omega}^{ref} - \tilde{\boldsymbol{\omega}}\right) + \boldsymbol{N}(\mathcal{J}_{cM})\boldsymbol{N}(\overline{\mathcal{J}}_{mM})\boldsymbol{N}(\overline{\boldsymbol{J}}_{\omega})\dot{\boldsymbol{\theta}}_u^{ref} \\ &= \dot{\boldsymbol{\theta}}^{c} + \dot{\boldsymbol{\theta}}^{m} + \dot{\boldsymbol{\theta}}^{am} + \dot{\boldsymbol{\theta}}^{n}, \quad \text{s.t. } \dot{\boldsymbol{\theta}}^{c} \succ \dot{\boldsymbol{\theta}}^{m} \succ \dot{\boldsymbol{\theta}}^{am} \succ \dot{\boldsymbol{\theta}}^{n}.\end{aligned} \tag{5.97}$$

Hereby, $\Delta\boldsymbol{\omega}^{ref} = \boldsymbol{\omega}_C^{ref} - \boldsymbol{\omega}_B^{ref}$,

$$\overline{\boldsymbol{J}}_{\omega} = \boldsymbol{J}_{\omega}\boldsymbol{N}(\mathcal{J}_{cM})\boldsymbol{N}(\overline{\mathcal{J}}_{mM}) \quad \text{and} \quad \tilde{\boldsymbol{\omega}} = \boldsymbol{J}_{\omega}\left(-\mathcal{J}_{cM}^{+}\mathbb{C}_{cC}^{T}\mathcal{V}_M^{ref} + \overline{\mathcal{J}}_{mM}^{+}(\widetilde{\mathcal{V}}^{m})^{ref}\right).$$

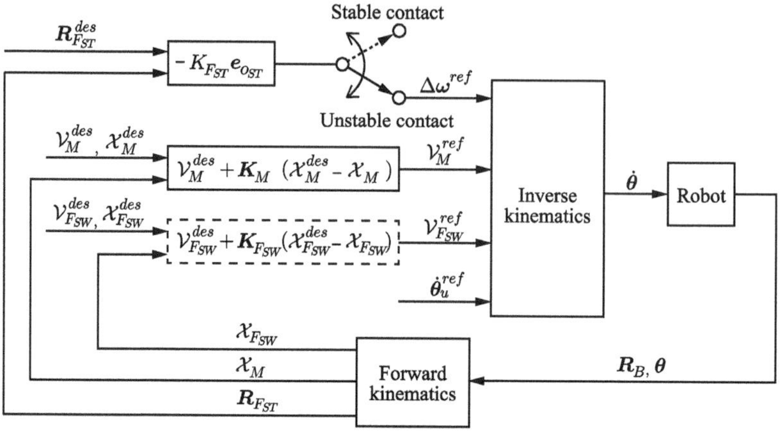

FIGURE 5.16 Block diagram of the RAM/V controller. The Inverse kinematics block calculates the control joint velocity in accordance with (5.97). The swing leg (the quantities denoted by the $_{F_{FSW}}$ subscripts) control is optional. The stance leg(s) (the quantities denoted by the $_{F_{ST}}$ subscripts) control is activated only when the contact(s) are unstable.

The control input $\dot{\theta}$ in the last equation is composed of four components arranged in hierarchical order. The highest-priority component, $\dot{\theta}^c$, is used to control the instantaneous motion of the CRB, via the contact constraints. The desired CRB translational (i.e. of the CoM) and rotational (of the base link) displacements are achieved via the movements in the leg(s). The rest of the control components are derived from within the null space $\mathcal{N}(\mathcal{J}_{cM})$. These components will not disturb the main (the CRB) control task. The role of the second term, $\dot{\theta}^m$, is to control the desired motion of the swing leg, when the robot is in a single stance. The role of the third term, $\dot{\theta}^{am}$, is to control the system angular velocity in a way to ensure an appropriate inertial coupling w.r.t. the desired CRB rotational motion. Such coupling can only be achieved via motion in the arms since the legs and the upper body are controlled by the first two components. The last, fourth component, $\dot{\theta}^n$, is used to enforce the joint velocity/angular constraints. To this end, the additional control input $\dot{\theta}_u^{ref}$ can be determined via the gradient projection approach with the joint-limit avoidance potential function introduced in Section 2.7.4.

Note that when the robot is in a single stance and there is no desired motion task for the swing leg, the second component, $\dot{\theta}^m$, becomes irrelevant. The motion of the swing leg will then be determined by the angular momentum component $\dot{\theta}^{am}$. This means that the motion of the swing leg will contribute to postural stabilization, as does the motion in the arms. This contribution plays an important role when a large external disturbance is applied to the robot, as will be shown in Section 7.7.6.

The above controller will be referred to as the *relative angular momentum/velocity (RAM/V) controller*. The block diagram of the controller is shown in Fig. 5.16. The desired values for the CoM motion, the base-link rotation, the swing-leg motion, and the system angular momentum/velocity can be specified in an *independent* way. The controller does not account for the constraints (e.g. the CoP-in-BoS and friction cone constraints). But it provides means to avoid destabilization, via an appropriate arm (and possibly swing leg) motion generated by the relative angular velocity (RAV) control input $\Delta\boldsymbol{\omega}^{ref}$. Moreover, the controller can be endowed with a *self-stabilization* property. With such a property, the stability can be recovered even when the state is destabilized (e.g. a state with rolling foot/feet) by an inappropriate (desired) input or by an external disturbance of large magnitude, as will be explained below.

Special Cases: Balance Control That Conserves the System or the Coupling Angular Momentum

A straightforward approach to deal with the coupling problem in balance control is to constrain the motion of the robot within the angular momentum–conserving subset of motion (cf. (4.116)). This is achieved by setting the reference system angular velocity in (5.97) at zero throughout the motion. Then we have

$$\boldsymbol{\omega}_C^{ref}(t) = \mathbf{0} \Rightarrow \Delta\boldsymbol{\omega}^{ref} = -\boldsymbol{\omega}_B^{ref} = -(\boldsymbol{\omega}_B^{des} + K_{oB}\boldsymbol{e}_{oB}). \tag{5.98}$$

This means that the CoP (or ZMP) will depend only on the DCM stabilization task; the CoP will not be disturbed by the angular momentum task (i.e. the desired base-link rotation).

As an example, consider a humanoid robot in a double stance and performing a body twist motion task. During this task, the CoM is regulated to the initial position via control law (5.93). The torso twists around the vertical, first counterclockwise and then clockwise (twist angle ±10 degrees). The control of this motion is achieved via control law (5.94). The result is shown in animated form in Video 5.8-1 [43]. Note that the torso and arm rotations are antiphase; the arms are used to cancel out the moment component at the feet that is induced by twisting the torso. Thus, the CoP variation is almost zero (a slight deviation occurs because of the error terms).

Another possible approach to deal with the coupling problem is to make use of the instantaneous motion component $\dot{\boldsymbol{\theta}}_{cam}$ from within the Reaction Null Space (RNS) $\mathcal{N}(\boldsymbol{J}_\omega)$. In this case, the *coupling angular momentum* will be conserved at zero (cf. Section 4.7.3). This is achieved by setting the reference system angular velocity to equal the base angular velocity throughout the motion. Then we have

$$\boldsymbol{\omega}_C^{ref} = \boldsymbol{\omega}_B^{ref} \Rightarrow \Delta\boldsymbol{\omega}^{ref} = \mathbf{0}. \tag{5.99}$$

The resulting motion endows the RAM/V controller with the important property of *self-stabilization through angular momentum damping*. This property becomes apparent when the RAM/V controller is rewritten in terms of acceleration, as shown in Section 5.11.2. The property will be demonstrated below with an example.

Consider the same motion task as in the above example, performed with the RNS control constraint (5.99). The result is shown in an animated form in Video 5.8-2 [44]. Observe that, in this case, the torso and arm rotation is in-phase. That is, the arms are "actively" used to support the desired swift twisting motion of the torso.

For comparison, the same task was performed under the resolved momentum control law. The result is shown in an animated form in Video 5.8-3 [45]. Apparently, the desired swift torso twist is impossible to be achieved due to the reasons already explained.

A demonstration of a ball kicking task is shown in Video 5.8-4 [42]. There are three phases. First, the support foot is loaded by shifting the gCoM within the foot BoS. Thereby, the hands

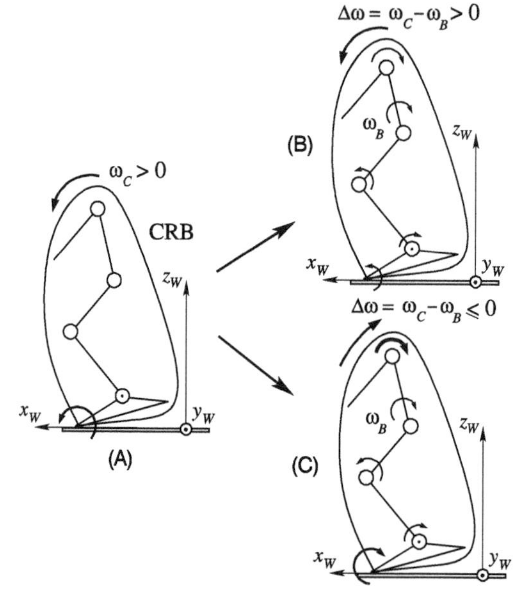

FIGURE 5.17 Stabilization with the relative angular velocity (RAV) $\Delta\omega$. (A) An unstable state of the CRB (joints locked) with positive system angular velocity ($\omega_C > 0$). (B) The system cannot be stabilized when the RAV is in the same direction as the system angular velocity. (C) A necessary condition for system stabilization is to generate an RAV in the direction opposite to the system angular velocity.

do not deviate significantly from their initial positions. Second, a prekick posture is attained whereby the kicking leg is extended to the side and the back. During this phase, the RNS control constraint is employed. This results in the arm motion being involved in the balancing. Then, during the final, kicking phase, the RAV constraint is switched from coupling to system angular momentum conservation. In this way robust balance control with minimum CoP deviation is ensured. As seen from the video, the motion looks quite "realistic." Note that the motion of the arms is determined solely by the two reference angular velocities, $\boldsymbol{\omega}_C^{ref}$ and $\boldsymbol{\omega}_B^{ref}$. This example demonstrates that a human-like performance of various dynamic motion tasks can be achieved by appropriately defining the relative angular velocity. This issue is an ongoing research effort.

5.8.4 RNS-Based Stabilization of Unstable Postures

Assume the robot has been destabilized, either proactively or by an external force. This means that the foot (when in a single stance) or the feet (when in a double stance) have begun to roll. If no swift action is taken, a fall will be unavoidable, as already clarified in Section 5.3.3 (cf. State D in Fig. 5.3).

The required swift action for contact stabilization can be generated in a straightforward way with the RAM/V controller introduced in Section 5.8.3. This will be explained with the help of the simple sagittal-plane model shown in Fig. 5.17. Assume the foot rolls around the toe tip counterclockwise, s.t. the system angular speed is positive in the chosen coordinate frame. Recall that when the joints are locked, $\omega_C = \omega_B = \omega_i$, $i \in \{\overline{1,n}\}$ is the angular speed of the CRB and also, the relative angular speed is zero, i.e. $\Delta\omega = \omega_C - \omega_B = 0$. When the robot links are allowed to rotate, in general the system angular speed will be different from that of the base link, and thus, the relative angular speed will be nonzero. For this particular example, when $\Delta\omega > 0$, the foot roll will persist and result in a fall. On the other hand, when $\Delta\omega \leq 0$, the foot will start rotating in the opposite (clockwise) direction, resulting in the recovery of the line contact and, eventually, of a stable posture.

Stabilization of Postures With Rolling Feet

Note that with the RNS input $\Delta\omega^{ref} = 0$ an appropriate arm motion will be generated to counter the foot rotation. This motion, however, does not yield contact stabilization but only a rocking motion, as shown in Video 5.8-5 [47]. To generate an arm motion that leads to contact stabilization, make use of the following control law:

$$\Delta\boldsymbol{\omega}^{ref} = -K_{F_{ST}}\boldsymbol{e}_{o_{ST}}, \tag{5.100}$$

where $\boldsymbol{e}_{o_{ST}}$ is the stance foot (the rolling foot) orientation error and $K_{F_{ST}}$ is a feedback gain. Motion control for contact stabilization with this control input is shown in Video 5.8-6 [46]. As seen in the video, the orientation of the feet converged to the desired state and the contacts could be stabilized. After the stabilization, the control input $\Delta\boldsymbol{\omega}^{ref}$ should be switched off to ensure that the arms do not move anymore.

Posture Stabilization on a Balance Board

As another example, consider a robot placed on a slightly damped balance board. High-friction contacts are employed to avoid a slip at the feet. The RAM/V controller with the RNS constraint is applied to stabilize the unstable states attained whenever the CoM of the robot is displaced from the (unstable) set of equilibrium points, i.e. from the vertical passing through the center of rotation of the balance board. The CoM motion control task is a regulation in the horizontal direction toward the equilibrium line. The base-link rotation task is also a regulation one: keeping the upper body upright. It is assumed that the actual base-link angular velocity is obtained from the IMU of the robot. Thus, there is no need to account for the rotational displacement of the balance board.

Snapshots from the simulation are displayed in Fig. 5.18. The result is shown in an animated form in Video 5.8-7 [107] and in graphic form in Fig. 5.19. Note that the initial posture is unstable. It takes about five seconds to stabilize the posture. Immediately after the stabilization, the robot destabilizes itself with a swift forward bend in the hips. Apparently, the RAM/V controller can handle such type of destabilization as well.

Summary and Conclusions

The above two examples, as well as those introduced in Section 5.8.3, demonstrate that the RAM/V controller can play a significant role in balance stability and control, especially when the angular momentum damping component from the RNS is used. The formulation is simple and yet quite efficient—there is no need to modify the contact model to account for the transitions between the stable and unstable states. When the contacts are stable, the RNS-based control input (5.100) should be switched off to avoid unnecessary arm movements. When the

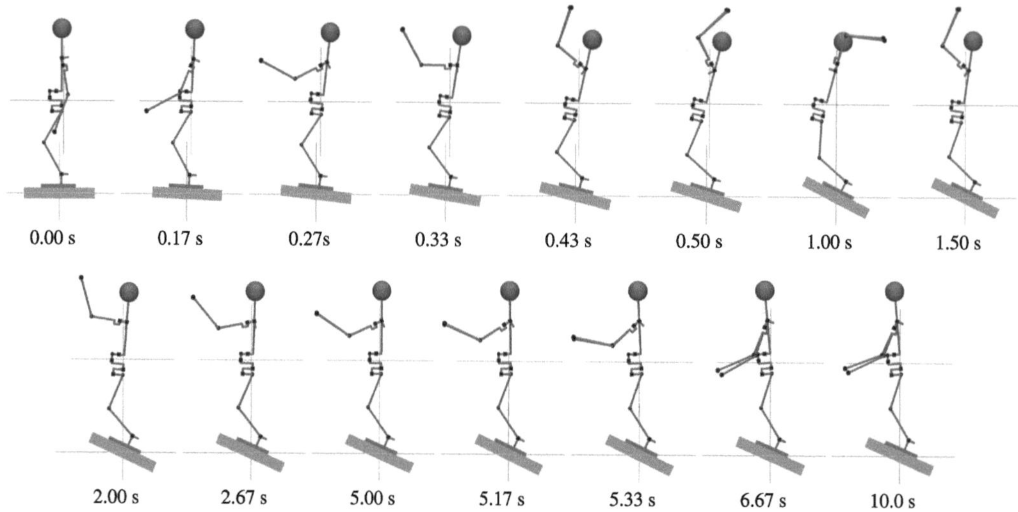

FIGURE 5.18 Human-like posture stabilization on a balance board under RAM/V plus RNS control. The initial posture is unstable since the CoM is displaced from the vertical passing through the center of rotation of the balance board. The posture is stabilized swiftly via the upper-limb motion (the trunk is commanded to keep its upright orientation). At $t = 5$ s the robot self-destabilizes by a fast forward bent. The final posture at $t = 10$ s represents a stable, static state.

contacts are unstable (the feet rotate), the control input can be used to ensure the recovery of the stability via the induced arm motion. Thereby, the system is forced to behave as a CRB, for any desired base-link state. The recovery is swift, with a subtle arm motion, for relatively small deviation of the feet from their equilibrium states. For larger deviations, the recovery can be prolonged; this results in a "windmilling" arm motion pattern for stabilization that is also sometimes used by humans to stabilize their posture in a critical situation.

The performance of the RAM/V controller can be further improved by a reformulation in terms of acceleration (to be presented in Section 5.11.2). With such a reformulation it becomes possible to control the angular momentum damping injected into the system. With such damping, the convergence to a stable contact state can be ensured, e.g. without the direct involvement of the foot rotation error, as in the rolling feet example.

5.8.5 An Approach to Contact Stabilization Within the Resolved Momentum Framework

The stabilization of the contacts can be ensured by introducing the friction cone constraints into the resolved momentum framework [50]. An appropriate CRB (reaction) wrench can then be generated that relates to the spatial momentum. The CRB wrench also relates to the reference CoP via the tangential GRM components. In the special case of a flat ground, the method yields the ZMP equation (4.38). Furthermore, the method provides a solution to the coupling problem as well. The solution comprises two steps, as apparent from Fig. 5.20. The first step involves the time derivatives of (5.85) and (5.86). Given the desired motion of the base link,

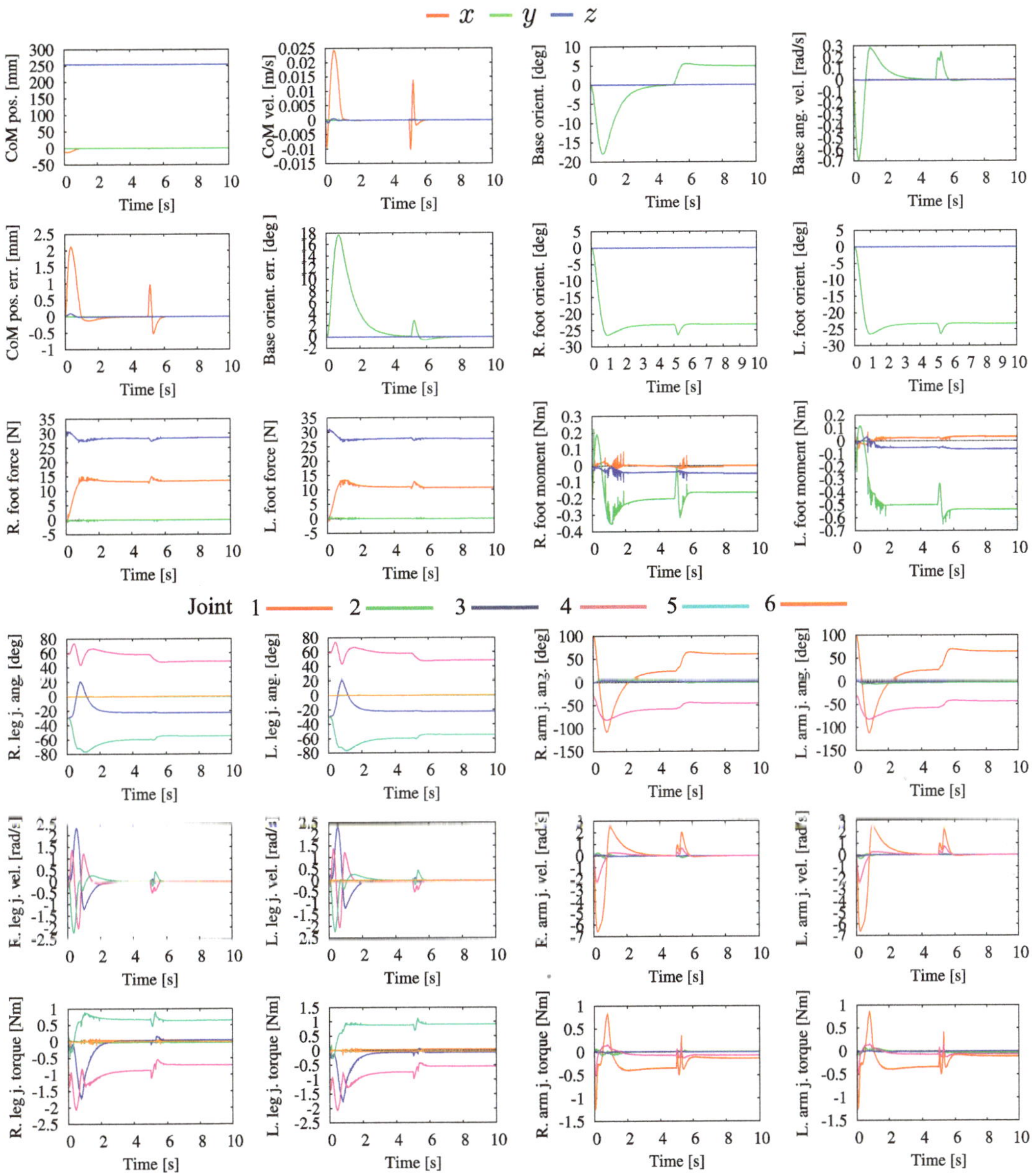

FIGURE 5.19 RNS-based stabilization on a balance board. Initially, the CoM is slightly displaced from the vertical equilibrium line passing through the balance board center, yielding an unstable posture. Stabilization of the base link requires about 5 s. After that, the robot self-destabilizes by a fast forward bent. The stabilization property of the controller can be reconfirmed for such type of destabilization. The joint numbering is as in the model described in Section A.1.

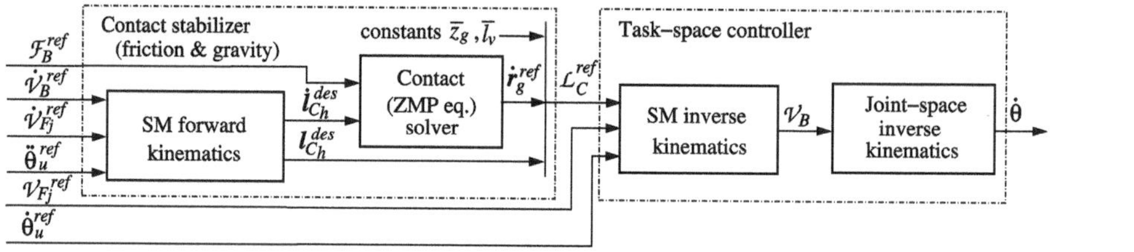

FIGURE 5.20 Block diagram of a Resolved-momentum (RM) trajectory generator [69], modified to account for friction at the contacts, via CRB contact wrench $\mathcal{F}_B^{ref}$ [50]. The spatial momentum (SM) forward kinematics block uses (5.85). The SM and joint-space inverse kinematics blocks refer to (5.86) and (5.87), respectively.

the end links (along the unconstrained motion directions), and additional posture variations in terms of acceleration, the desired *angular momentum rate* components in the horizontal directions $\dot{l}_{C_h}^{des}$ are determined. Since the time derivative of (5.85) does not include information about the contact constraints, they will be ignored at the first step. At the second step, $\dot{l}_{C_h}^{des}$ is inserted into the ZMP equation. This equation is then solved (under the LIP mode constraint) for the reference CoM trajectory that will account for the contact constraints and the desired momenta. The solution is obtained with the help of the preview control approach [68].

The task-space controller for resolving the CRB motion thus obtained is designed at the velocity level, based on the momentum equation (5.85) (cf. Fig. 5.20). Given the reference angular and linear momentum trajectories obtained at the first and second steps of the motion generation algorithm with contact constraints, the equation is resolved for the base quasivelocity $\mathcal{V}_B$. Finally, the control input for the joints rates, $\dot{\theta}$, is obtained from (5.87) with $\mathcal{V}_B$. From the block diagram in Fig. 5.20 it should be apparent that the reference spatial acceleration of the base, $\dot{\mathcal{V}}_B^{ref}$, is modified by the algorithm in order to account for the contact conditions. In this way, the inherent inconsistency of the independently specified spatial momentum components can be resolved.

The performance of the resolved momentum controller was experimentally verified while tele-operating the end-link movements of a humanoid robot. Thereby, the base-link motion was generated autonomously to enlarge the workspace of the end link. The contacts were stabilized, also autonomously, via the ZMP equations [109]. The method has also been tested in a multicontact scenario with gait stabilization while holding a handrail [80].

5.8.6 Spatial Momentum Rate Stabilization Parametrized by the CMP/VRP

Because of the existing coupling between the rates of change of the linear and angular momenta, as mentioned in Section 5.8.1, it would be sufficient to formulate the stabilizer in terms of the linear component only. First, assume that the environment model includes a flat ground. In this case, the LIP mode constraint $\bar{\omega} = \sqrt{g/\bar{z}_g} = \text{const}$ can be invoked and consequently, the stability results from the preceding sections can be applied. The reference momentum rate is obtained from the upper row of (5.83) as

$$\dot{p}_t^{ref} = M\bar{\omega}^2(r_g - r_{cmp}^{ref}). \tag{5.101}$$

The reference CMP can be derived from the xCoM/ICP dynamics of the LRWP, (5.61), as

$$\boldsymbol{r}_{cmp}^{ref} = \boldsymbol{r}_{ex} - \frac{1}{\overline{\omega}}\dot{\boldsymbol{r}}_{ex}^{des} + k_x(\boldsymbol{r}_{ex} - \boldsymbol{r}_{ex}^{des}), \tag{5.102}$$

k_x denoting a feedback gain. The angular momentum rate component will also be parametrized by $\boldsymbol{r}_{cmp}^{ref}$ because of the coupling, as already noted. The output of this CMP stabilizer is determined by inserting the CMP control law (5.102) into (5.101). The rate of change of the angular momentum does not have to be computed explicitly; it will be stabilized as a "byproduct" of the coupling, as noted in Section 5.8.1.

Furthermore, it is interesting to note that the above control law has been shown to perform well not only on flat ground but also on irregular terrain, using e.g. the edges of concrete blocks as footholds [161]. In critical situations, i.e. whenever the xCoM leaves the BoS, the angular momentum component, stabilized via the coupling with (5.102), will ensure a suitable upper-body lunge maneuver to drive the xCoM back toward the BoS. This result may not be too surprising, though, when noticing that the form of control law (5.102) is the same as that of the VRP control law (5.77). Recall that the latter was developed for irregular terrain without imposing the LIP mode constraint.

5.8.7 CRB Motion Trajectory Tracking With Asymptotic Stability

The CMP/VRP stabilizer has shown superior performance in locomotion on irregular terrain [22], including stepping with minimalistic footprints derived from approximated line- and point-type contacts [161]. This stabilizer, however, does not control the rate of change of the angular momentum independently from that of the linear momentum. Thus, there is no possibility to invoke proactive upper-body maneuvers with a desired angle, such as flexion/extension (i.e. forward/backward bending), inversion/eversion (i.e. side bending), and rotations around the vertical. These types of maneuvers or any combinations thereof are needed when performing such proactive tasks as picking up an object from the floor or reaching out for a target. Such maneuvers are also used in synergy-based accommodation of external perturbations during reactive balance tasks, as will be discussed in Section 7.6. The CRB momentum controllers discussed so far in the literature lack the capability of asymptotic tracking of the arbitrary angular momentum trajectories. In [88,112], for example, there is no angular orientation feedback at all. Also a proof of stability of the proposed controller is missing. This is also the case in [41]. In what follows, a CRB spatial momentum controller will be designed that ensures asymptotic tracking of any *constraint-consistent CRB motion trajectories.*

The CRB motion trajectories are specified via the rate of change of the spatial momentum. This quantity can be derived from the spatial dynamics component of the equation of motion. For simplicity and without loss of generality, the derivation will be based on the notation with the mixed quasivelocity (cf. Section 4.8.2). Implementations in terms of the base quasivelocity are discussed in Sections 5.10.4 and 5.12.1. The spatial dynamics, given in the upper part of (4.157), can be written as

$$\dot{\mathcal{L}}_C + \mathcal{G}_C = \mathcal{F}_C, \tag{5.103}$$

where $\dot{\mathcal{L}}_C \equiv \frac{d}{dt}\mathcal{L}_C(\boldsymbol{q}, \dot{\boldsymbol{q}}_M)$. The wrench $\mathcal{F}_C \equiv \mathbb{C}_{cM}\overline{\mathcal{F}}^c$ is the net body wrench. This wrench, henceforth referred to as the *system wrench*,[7] drives the spatial dynamics. The reference rate of change of the spatial momentum can then be determined as (cf. (4.146)) as follows:

$$\dot{\tilde{\mathcal{L}}}_C^{ref} = \mathbb{M}_C \dot{\mathcal{V}}_M^{ref} + \dot{\mathbb{M}}_C \mathcal{V}_M. \tag{5.104}$$

Recall that the overtilde notation stands for the *CRB spatial momentum* (CRB-SM), that is, the system spatial momentum (SSM) component in the locked-joint case. Also recall that throughout this chapter it is assumed that the desired trajectories for the stabilizer are feasible linear and angular CRB motion trajectories. Such trajectories can be tracked under the conventional feedforward plus PD feedback control. We have

$$\dot{\boldsymbol{v}}_C^{ref} = \dot{\boldsymbol{v}}_C^{des} + K_{vC}\dot{\boldsymbol{e}}_{pC} + K_{pC}\boldsymbol{e}_{pC}, \tag{5.105}$$

$$\dot{\boldsymbol{\omega}}_B^{ref} = \dot{\boldsymbol{\omega}}_B^{des} + K_{\omega B}\boldsymbol{e}_{\omega B} + K_{oB}\boldsymbol{e}_{oB}. \tag{5.106}$$

Here $\boldsymbol{e}_{\omega B} = \boldsymbol{\omega}_B^{des} - \boldsymbol{\omega}_B$ denotes the error in the angular velocity of the base link, its orientation error $\boldsymbol{e}_{oB}$ and the CoM position error $\boldsymbol{e}_{pC}$ being defined in Section 5.8.3. The $K_{(\circ)}$ quantities stand for the PD feedback gains. The above two reference components constitute the reference twist, $\dot{\mathcal{V}}_M^{ref} = \left[(\dot{\boldsymbol{v}}_C^{ref})^T \quad (\dot{\boldsymbol{\omega}}_B^{ref})^T\right]^T$; they are determined in a form widely used in end-link trajectory tracking with fixed-base robots. In the case of balance control of a floating-base humanoid robot, however, it is much more preferable to use the DCM stabilization result derived in Section 5.6.3. This result can be expressed in terms of the CoM acceleration as

$$\dot{\boldsymbol{v}}_C^{ref} = \omega_X\left(\dot{\boldsymbol{r}}_X^{des} + k_x(\boldsymbol{r}_X^{des} - \boldsymbol{r}_X) - \dot{\boldsymbol{r}}_C\right). \tag{5.107}$$

In the derivation, relations (5.76) and (5.79) were used.

The reference body wrench can be written as

$$\mathcal{F}_C^{ref} = \dot{\tilde{\mathcal{L}}}_C^{ref} + \mathcal{G}_C. \tag{5.108}$$

The role of this wrench is to compensate the effect of the gravity force and to linearize the CRB trajectory error dynamics. This is a necessary condition to achieve an asymptotic CRB motion trajectory tracking. Sufficiency depends on the feasibility of the desired CRB motion trajectories provided by the motion generator and the on-line motion modifier that uses feedback from the task-space controller (cf. Fig. 5.1). The latter imposes various motion/force constraints to ensure appropriate balance control. Details will be presented in Section 5.9.

5.9 TASK-SPACE CONTROLLER DESIGN FOR BALANCE CONTROL

The momentum rate–based balance controllers comprise task-space controllers that admit the rate of change of the spatial momentum (or equivalently, a body wrench) as the reference

[7] In [8], the term "gravito-inertial wrench" was used.

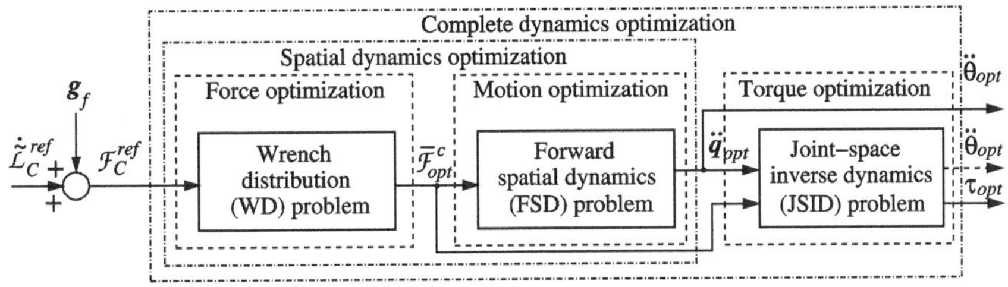

FIGURE 5.21 Generic task-space controller structure for optimization-based balance control. The momentum rate input $\dot{\mathcal{L}}_C^{ref}$ (or alternatively, $\mathcal{F}_C^{ref}$) is supplied by a stabilizer. The output of the controller can be specified either in terms of the joint torque or the joint acceleration, suitable for torque- or position-controlled robots, respectively.

input. A task-space controller design should account for the fact that the spatial dynamics do not *directly* depend on the joint torque. The reference input should be resolved to obtain joint acceleration as control input. In the case of a position-controlled humanoid robot, this control input is integrated twice to obtain joint angles to be fed to the control system. In the case of a torque-controlled robot, on the other hand, the joint torques for control can be derived from the joint-space partial dynamics component (the lower part of the equation of motion, e.g. as in (4.158)). This means that, in addition to the joint acceleration, the contact wrenches should be made available, i.e. either measured or calculated.

5.9.1 Generic Task-Space Controller Structure

The above considerations determine the structure of the task-space controller shown in Fig. 5.21. The input, supplied by the stabilizer, is the reference spatial momentum rate $\dot{\mathcal{L}}_C^{ref}$ or, alternatively, the reference CRB wrench $\mathcal{F}_C^{ref}$. The controller first generates the contact wrenches by solving the quasistatic *wrench distribution (WD) problem*. As discussed in Section 3.5.2, this is an underdetermined problem that is formulated as a constrained optimization task in the force domain. The solution is found either noniteratively (i.e. via generalized inverses) or iteratively (via a general solver). The constraints stem from the friction cone relations and the unilateral contacts at the feet. The friction cones can be approximated with polyhedral convex cones (cf. Section 3.3.2) to reduce the computational burden. Since these constraints are inequality-type, a general solver (i.e. a quadratic program or QP solver) might be preferred in lieu of noniterative optimization. Furthermore, the wrench optimization task is formulated to include other subtasks as well, such as the CoP localization (via the GRM) and/or the contact wrench minimization. As a result of the wrench optimization task, the difference $\|\mathcal{F}_C^{ref} - \mathbb{C}_c\overline{\mathcal{F}}_{opt}^c\|_W$ will be minimized. It is important to note that the outcome of the optimization task depends very much on the reference CRB wrench, $\mathcal{F}_C^{ref}$. The ideal case is when $\mathbb{C}_c\overline{\mathcal{F}}_{opt}^c = \mathcal{F}_C^{ref}$. This implies that the reference CRB wrench lies within the CRB-wrench cone, i.e. $\mathcal{F}_C^{ref}$ is a *contact-consistent* wrench (cf. Section 3.76); then we have

$$\dot{\mathcal{L}}_C^{ref} + \mathcal{G}_C = \mathcal{F}_C^{ref} \in \mathfrak{F}_{BWC} = \{\mathbb{C}_{cC}, BWC\}.$$

Once the optimal contact wrenches $\overline{\mathcal{F}}^c_{opt}$ are found, the *forward spatial dynamics (FSD)* problem can be solved. The generalized acceleration $\ddot{q}_{opt} = (\dot{\mathcal{V}}^{opt}_M, \ddot{\theta}^{opt})$ is obtained from the spatial dynamics equation, i.e.

$$\mathbb{M}_C \dot{\mathcal{V}}^{opt}_M + \boldsymbol{H}_{CM}\ddot{\boldsymbol{\theta}}^{opt} + \mathcal{C}_M + \mathcal{G}_C = \mathbb{C}_{cC}\overline{\mathcal{F}}^c_{opt}. \tag{5.109}$$

Note that the FSD problem is also underdetermined. Thus, (differential) *motion optimization* has to be involved. The conventional approach is to introduce a subtask prioritization for the motion control tasks along the constrained and unconstrained motion directions, as well as for the limb and whole-body self-motion control (cf. Section 2.8.2). As a result of the optimization, a *constraint-consistent* $\ddot{q}_{opt}$ is obtained.

The two-step optimization process just described represents a *sequential*-type optimization. This approach has been used in [87,88,167]. It is also possible to combine the generalized acceleration and force optimization steps into a single optimization task. This combination introduces, however, some computational overhead, as will be explained in Section 5.11.1. The formulation of two smaller optimization problems is preferable in lieu of a single but large one [40].

As already noted, in the case of a position-controlled humanoid robot, there is no need to calculate the joint torque; $\ddot{\theta}^{opt}$ and its time integrals can be applied as a control input. In the case of a torque-controlled robot, on the other hand, the *joint-space inverse dynamics (JSID) problem* needs to be solved. Note that once the optimal solutions for the WD and FSD problems have been obtained, the joint torque can be calculated in a *unique* way from the lower part of (4.157). We have

$$\boldsymbol{\tau} = \boldsymbol{M}_{\theta M}\ddot{\boldsymbol{\theta}}^{opt} + \boldsymbol{H}^T_{CM}\dot{\mathcal{V}}^{opt}_M + \boldsymbol{c}_{\theta M} - \mathcal{J}^T_{cM}\mathcal{F}^c_{opt}. \tag{5.110}$$

Note that the last term on the r.h.s. implicitely compensates the action of the gravity force. As noted in Section 3.6.4, this gravity compensation approach introduces errors; to avoid them the equivalent expression (3.74) should be preferably used. The last equation assumes then the form:

$$\boldsymbol{\tau} = \boldsymbol{M}_{\theta M}\ddot{\boldsymbol{\theta}}^{opt} + \boldsymbol{H}^T_{CM}\dot{\mathcal{V}}^{opt}_M + \boldsymbol{c}_{\theta M} + \boldsymbol{g}_\theta - \mathcal{J}^T_{cB}\overline{\mathcal{F}}^c_{opt}. \tag{5.111}$$

The optimization tasks can be structured as outlined in Fig. 5.22. There are two basic structures based on the FSD (upper two schemes) and JSID (lower two schemes) optimization. Each of these structures can be realized in a sequential/nonsequential form. Nonsequential realizations usually involve computational overhead. Therefore, when aiming at real-time control, the sequential schemes should be preferably used, provided contact-consistent inputs are available.

5.9.2 Optimization Task Formulation and Constraints

From the generic task-space controller structure introduced in Section 5.9.1 it became apparent that constrained optimization tasks have to be solved over the motion $\mathcal{M}$ and force $\mathcal{F}$ domains. When organizing the optimization process, three possibilities should be considered:

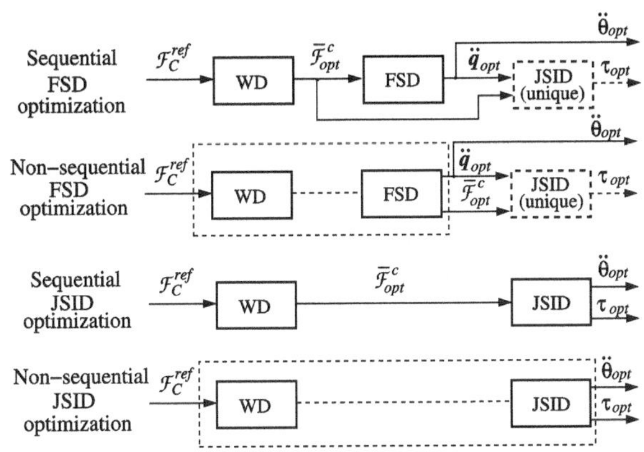

FIGURE 5.22 Optimization task structures for task-space control.

- a noniterative approach based on generalized inverses,
- an iterative approach based on a general solver, and
- a mixed, iterative/noniterative approach.

These possibilities will be explored in the following sections.

Furthermore, the optimization process is based on subsets of equality/inequality-type of constraints defined within the motion and force domains. The most frequently used sets of equality- and inequality-type constraints are presented in Table 5.1 and Table 5.2, respectively. Inequality-type constraints can be embedded into iterative optimization schemes in a straightforward manner. It is also possible to include such type of constraints in noniterative optimization task formulations [118].

One of the main concerns when formulating an optimization task is the lack of solution due to multiple contradicting constraints. This problem can be alleviated by relaxing one or more of the equality-type constraints, $\boldsymbol{A}\boldsymbol{x} - \boldsymbol{a} = \boldsymbol{0}$. This is done by reformulating the constraint as a quadratic objective function of the type

$$c_l = \|\boldsymbol{A}\boldsymbol{x} - \boldsymbol{a}\|^2_{W_l}.$$

Here $\boldsymbol{W}_l$ denotes a weight matrix. This approach will be henceforth referred to as the *constraint relaxation*. The subscripts appearing in Table 5.1 will be used to denote the penalty terms derived via constraint relaxation. For example, relaxing the joint acceleration tracking constraint, $\mathcal{M}_{JA}$, yields the penalty term

$$c_{JA} = \|\ddot{\boldsymbol{\theta}} - \ddot{\boldsymbol{\theta}}^{ref}\|^2_{W_{JA}}$$

in the objective function. It should be noted that inequality-type constraints can also be relaxed, via slack variables (cf. Section 2.60). Further details will be given in Section 5.14. Note that strict, equality-type constraints are sometimes referred to as the *hard* constraints. Penalty-type constraints $c_{(\circ)}$, on the other hand, are known as *soft* constraints [25]. The optimization

TABLE 5.1 Equality-type constraints

Name	Notation	Equation
Equation of motion	$\mathcal{F}_{EoM}$	(4.162)
Closed-chain constraint	$\mathcal{M}_{CC}$	(2.105)
Spatial dynamics tracking	$\mathcal{F}_{SD}$	$\mathbb{C}_c\bar{\mathcal{F}}^c = \mathcal{F}_C^{ref}$
Spatial momentum rate tracking	$\mathcal{F}_{SM}$	$\dot{\mathcal{L}}_C = \dot{\mathcal{L}}_C^{ref}$
Angular momentum minimization	$\mathcal{F}_{AM}$	$\boldsymbol{l}_C = \boldsymbol{0}$
CoP tracking	$\mathcal{F}_{CoP}$	(5.179)
End-link motion tracking	$\mathcal{M}_{EL}$	$\mathcal{J}_m\ddot{\boldsymbol{\theta}} = (\dot{\mathcal{V}}^m)^{ref} - \dot{\mathcal{J}}_m\dot{\boldsymbol{\theta}}$
Contact wrench tracking	$\mathcal{F}_{CW}$	$\bar{\mathcal{F}}^c = (\bar{\mathcal{F}}^c)^{ref}$
Joint torque tracking	$\mathcal{F}_{JT}$	$\boldsymbol{\tau} = \boldsymbol{\tau}^{ref}$
Joint acceleration tracking	$\mathcal{M}_{JA}$	$\ddot{\boldsymbol{\theta}} = \ddot{\boldsymbol{\theta}}^{ref}$
Posture tracking	$\mathcal{M}_{PT}$	$\boldsymbol{\theta} = \boldsymbol{\theta}^{ref}$

TABLE 5.2 Inequality-type constraints

Name	Notation	Relation
Point-contact friction cone	$\mathcal{F}_{FC}$	(3.9)
Plane-contact torsional (yaw) friction	$\mathcal{F}_{FT}$	as in (3.11)
CRB wrench cone	$\mathcal{F}_{BWC}$	(3.77)
CoP-in-BoS	$\mathcal{F}_{BoS}$	(4.32), rewritten for each contact separately
Torque limits	$\mathcal{F}_{TL}$	$\tau_i^{min} \le \tau_i \le \tau_i^{max}$
Joint acceleration limits	$\mathcal{M}_{AL}$	$\ddot{\theta}_i^{min} \le \ddot{\theta}_i \le \ddot{\theta}_i^{max}$
Joint velocity limits	$\mathcal{M}_{VL}$	$\dot{\theta}_i^{min} \le \dot{\theta}_i \le \dot{\theta}_i^{max}$
Joint angle limits	$\mathcal{M}_{\theta L}$	$\theta_i^{min} \le \theta_i \le \theta_i^{max}$
Collision avoidance	$\mathcal{M}_{Col}$	

task can be written as

$$\min_{\boldsymbol{x}\in S} \sum_l c_l, \tag{5.112}$$

where S is a set of (usually convex) hard constraints. Note that the relaxation process is of relative character. When multiple constraints are relaxed, it might be difficult to adjust the numerous weights while aiming at a broad spectrum of tasks [18]. To obtain a realistic, "naturally looking" whole-body motion that avoids excessive postural variations, the weights should be determined in a way consistent with the given tasks [34]. This problem is still unsolved, though.

The form of the solution vector $\boldsymbol{x}$ depends on the optimization subtask; we have

$$\boldsymbol{x} = \begin{cases} \left[(\bar{\mathcal{F}}^c)^T\right]^T & \text{contact wrench optimization,} \\ \left[(\bar{\mathcal{F}}^c)^T \quad \ddot{\boldsymbol{q}}^T\right]^T & \text{spatial dynamics–based optimization,} \\ \left[(\bar{\mathcal{F}}^c)^T \quad \ddot{\boldsymbol{q}}^T \quad \boldsymbol{\tau}^T\right]^T & \text{complete dynamics optimization.} \end{cases} \tag{5.113}$$

The optimization task formulations outlined here can be used in noniterative as well as in multistep (iterative) optimization schemes. From the viewpoint of computational cost, a noniterative, generalized inverse–based optimization is preferable. The problem is that inequality-type constraints cannot be handled directly in this case. Nevertheless, as already noted, it is possible to employ an indirect approach based on appropriately defined quadratic objectives (i.e. a penalty-type approach) [118].

5.10 NONITERATIVE BODY WRENCH DISTRIBUTION METHODS

To obtain the noniterative solution to the wrench distribution problem, solve (5.109) for the contact wrenches as follows:

$$\overline{\mathcal{F}}_{opt}^{c} = \mathbb{C}_{cC}^{-W}(\dot{\mathcal{L}}_{C}^{ref} + \mathcal{G}_{C}) + N(\mathbb{C}_{cC})\overline{\mathcal{F}}_{a}^{c}. \tag{5.114}$$

This solution is of the same form as (3.62). From the discussion in Section 3.5.2, recall that the null-space term (the second term on the r.h.s.) can be parametrized by the internal wrench. The role of this term is to generate *quasistatic* tangential contact wrench components. These components can be used to enforce the friction constraints. Furthermore, the body wrench ($\mathcal{F}_{C}^{ref} = \dot{\mathcal{L}}_{C}^{ref} + \mathcal{G}_{C}$) is distributed via the particular solution term (the first term on the r.h.s.). Note that a weighted generalized inverse of the contact map is employed. With the trivial choice, $W = E$, the body wrench will be distributed via the pseudoinverse. Such distribution, however, is problematic (cf. Section 3.5.4). The problem becomes apparent if one considers the simple proactive task of maintaining a static posture on even floor. Since there is no motion in the joints, $\dot{\mathcal{L}}_{C}^{ref}$ can be assumed zero.[8] Thus, only the gravity term $\mathcal{G}_{C}$ will be distributed. With a symmetric posture, the net CoP will be at the center of the *net BoS*; $\mathcal{G}_{C}$ is distributed equally over the feet with the pseudoinverse. This implies consistency with the statics. Next, consider an asymmetric posture s.t. the net CoP is located within the left or right *foot BoS*. The respective foot will be referred to as "loaded." The normal reaction force at the loaded foot compensates the impressed gravity force. The normal reaction at the unloaded foot, on the other hand, is almost zero. This implies a highly asymmetric distribution. Note that in this case, the pseudoinverse solution will be *inconsistent* with the static state. From this observation it becomes apparent that the pseudoinverse-based distribution is inappropriate for tasks that involve foot loading/unloading, e.g. while walking or stepping in response to a disturbance.

5.10.1 Pseudoinverse-Based Body-Wrench Distribution

On the other hand, a pseudoinverse-based distribution can be quite useful in adaptation to unknown terrain. Consider a general contact model with multiple (user-specified) point

[8] In reality, there will always be small errors stemming from flexibilities in the joints, the environment, etc., which means that $\dot{\mathcal{L}}_{C}^{ref}$ cannot be completely ignored.

contacts at each end link [63]. Note that with such a model, the contacts do not have to be coplanar. The *net CoP* can be written as

$$x_P = \frac{\sum_{i=1}^{\kappa} x_j f_{zi}}{\sum_{i=1}^{\kappa} f_{zi}}, \quad y_P = \frac{\sum_{i=1}^{\kappa} y_j f_{zi}}{\sum_{i=1}^{\kappa} f_{zi}}, \quad \boldsymbol{f}_C = \sum_{i=1}^{\kappa} \boldsymbol{f}_i,$$

where $\boldsymbol{r}_i,\ \boldsymbol{f}_i,\ i \in \{\overline{1,\kappa}\}$ denote the position and reaction force at the contact points. These relations can be rewritten in the following form:

$$\begin{bmatrix} x_P \\ y_P \\ 1 \end{bmatrix} f_{Cz} = \begin{bmatrix} x_1 & x_2 & \dots & x_\kappa \\ y_1 & y_2 & \dots & y_\kappa \\ 1 & 1 & \dots & 1 \end{bmatrix} \begin{bmatrix} f_{z1} \\ f_{z2} \\ \dots \\ f_{z\kappa} \end{bmatrix},$$

$$\begin{bmatrix} x_P \\ 1 \end{bmatrix} f_{Cx} = \begin{bmatrix} x_1 & x_2 & \dots & x_\kappa \\ 1 & 1 & \dots & 1 \end{bmatrix} \begin{bmatrix} f_{x1} \\ f_{x2} \\ \dots \\ f_{x\kappa} \end{bmatrix},$$

$$\begin{bmatrix} y_P \\ 1 \end{bmatrix} f_{Cy} = \begin{bmatrix} y_1 & y_2 & \dots & y_\kappa \\ 1 & 1 & \dots & 1 \end{bmatrix} \begin{bmatrix} f_{y1} \\ f_{y2} \\ \dots \\ f_{y\kappa} \end{bmatrix}.$$

Given x_P^{des}, y_P^{des}, and $\boldsymbol{f}_C^{des}$, it is straightforward to solve the above matrix-vector equations for $\boldsymbol{f}_i^{des}$, by employing the pseudoinverse. Because of the minimum-norm solution, the CRB wrench will be distributed evenly over the contact points. If contacts are lost, e.g. by stepping over an edge, the even distribution yields foot rotation until another contact is made. In this way, the system can adapt to an unknown terrain, without force measurements.

Furthermore, the minimum-norm solutions in the tangential (x, y) directions can be modified by adding a null-space term for internal force control. It then becomes possible to address the friction problem and to avoid a slip. The internal force can be derived, for example, from an impedance control law, as suggested in [62].

This method has been used in reactive balance control with a torque-controlled robot and shown a satisfactory response to a relatively weak unknown disturbance [63,62]. On the other hand, this method is not suitable for reactive step-type balance control under a strong disturbance, since the feet cannot be loaded/unloaded with the pseudoinverse-based distribution, as already noted. The method cannot be used in gait control either, due to the same reason.

5.10.2 The ZMP Distributor

Aiming at walking, the ZMP distributor approach attempts to distribute the body wrench in a way consistent with the static states [71]. This implies asymmetric distribution. When the net CoP is within the BoS of Foot $j,\ j \in \{r, l\}$, that foot is fully loaded (which implies that the other foot will be fully unloaded). This means that the gravity force term f_G will be distributed in a *discrete manner*, i.e. without accounting for the distance between the CoP and the foot BoS boundary. The contact force components thus obtained are then used to calculate the desired ankle torques (or contact moment components), from

$$\bar{\boldsymbol{m}}_j^c = (\boldsymbol{r}_{A_j} - \boldsymbol{r}_{P_j}^{des}) \times \bar{\boldsymbol{f}}_j^c. \tag{5.115}$$

Here $\bar{\boldsymbol{f}}_j^c$ and $\bar{\boldsymbol{m}}_j^c$ are the components of the contact wrench and $\boldsymbol{r}_{P_j}^{des}$ and $\boldsymbol{r}_{A_j}$ are the positions of the CoP and the ankle, respectively. On the other hand, when the net CoP is in between the feet, a *force distribution ratio* is determined *heuristically* and the net force is distributed accordingly. The desired contact force components are then used to obtain the desired net ankle torque. This torque is also distributed via the heuristics.

It should be noted, though, that because of the discretization and the heuristics, the solution cannot be fully consistent with the static states. Note also that the desired CoP is needed for the distribution. Although the method has shown satisfactory performance during experiments with a real robot (HRP-4C) [71], better solutions that avoid the abovementioned shortcomings do exist, as will be shown in Sections 5.10.3 and 5.10.4.

5.10.3 Proportional Distribution Approach

A proportional body wrench distribution approach can be devised based on the observation that the vertical GRF appears in the friction constraints (friction cone and torsional friction) as well as in the CoP-in-BoS constraint [89]. A force distribution ratio can be determined from the vertical GRF component as

$$\eta = f_{rz}/f_{Gz},$$

where $f_{Gz} = f_{rz} + f_{lz}$ is the reaction stemming from the gravity force. Note that $0 \leq \eta \leq 1$. This ratio is then used to calculate all GRF components from

$$f_{ri} = \eta f_{Gi}, \quad f_{li} = (1 - \eta) f_{Gi}, \tag{5.116}$$

where $i \in \{x, y, z\}$. The total reaction acting at the CoM (the CRB force) can then be distributed in a *continuous way*, by varying η appropriately. For example, with $\eta = 0.5$, a symmetric distribution is obtained. With $\eta = 0$ or $\eta = 1$, on the other hand, a highly asymmetric distribution is obtained. In this way, consistency with the static states can be ensured. Note also that with the above ratio, the friction cone constraints will always be satisfied.

Furthermore, η could also be used to distribute the CRB moment in a similar fashion, i.e.

$$m_{ri} = \eta(m_{ri} + m_{li}), \quad m_{li} = (1 - \eta)(m_{ri} + m_{li}).$$

There is no guarantee, however, that the CoP-in-BoS and torsional friction constraints will be satisfied, since the three GRM components depend on the GRF. In [89], analytically derived bounds for η were suggested to alleviate the problem.

The proportional distribution approach avoids the disadvantage of the discrete wrench distribution of the ZMP distributor approach. An implementation example, to be introduced in Section 5.12.2, will provide further insight.

5.10.4 The DCM Generalized Inverse

A noniterative wrench distribution approach that can distribute the body wrench dynamically at multicontact postures is based on the so-called DCM generalized inverse (DCM-GI) [58]. The main idea is to design the weight matrix in (5.114) as a function of the DCM. The design should also avoid the discrete distribution disadvantage of the ZMP distributor, allowing for complete loading/unloading of the links in contact. In this way, undesired foot/hand roll during contact transitions can be avoided. The distribution approach should also ensure *consistency with statics*, taking into account the friction cone, torsional friction and CoP-in-BoS constraints, in a similar fashion as the proportional distribution approach from Section 5.10.3, but with the added capability of *multicontact transitions*.

Assuming κ robot links are in contact with the environment, the body wrench can be expressed as (cf. (3.57))

$$\mathcal{F}_C = \begin{bmatrix} \boldsymbol{f}_C \\ \boldsymbol{m}_C \end{bmatrix} = \sum_{k=1}^{\kappa} \left(\begin{bmatrix} \boldsymbol{E}_3 & \boldsymbol{0}_3 \\ -[\boldsymbol{r}^{\times}_{\overleftarrow{Ck}}] & \boldsymbol{E}_3 \end{bmatrix} \begin{bmatrix} \boldsymbol{f}_k \\ \boldsymbol{m}_k \end{bmatrix} \right). \tag{5.117}$$

Here $\boldsymbol{r}_{\overleftarrow{Ck}}$ is the position vector pointing from the kth local coordinate frame to the centroidal coordinate frame. The local coordinate frame is attached to the projection of the ankle/wrist joint on the contact surface. It is apparent that the body-GRM $\boldsymbol{m}_C$ depends on the foot/hand reaction wrenches. This dependence should be eliminated before deciding the weight coefficients for the distribution. This can be done by subtracting the GRF-induced GRM from the body wrench (cf. (3.58)), as follows:

$$\mathcal{F}_{net} \equiv \mathcal{F}_C - \sum_{k=1}^{\kappa} \begin{bmatrix} \boldsymbol{0}_3 \\ [\boldsymbol{r}^{\times}_{\overleftarrow{Ck}}]\boldsymbol{f}_k \end{bmatrix}. \tag{5.118}$$

From Section 3.6.1 recall that this operation ensures that $\mathcal{F}_{net}$ can be represented as a sum of contact wrenches, i.e. $\mathcal{F}_{net} = \sum_{k=1}^{\kappa} \mathcal{F}_k$. Here $\mathcal{F}_{net}$ is to be distributed with the help of weights appearing on the diagonal of the following weighting matrix:

$$\boldsymbol{W} = \text{diag}(\boldsymbol{W}_1 \ \ \boldsymbol{W}_2 \ \ \dots \ \ \boldsymbol{W}_\kappa) \in \Re^{c \times c},$$

where $\boldsymbol{W}_k \in \Re^{c_k \times c_k}$, c_k denoting the number of contact constraints at Contact k. When $c_k = 6$, we have

$$\boldsymbol{W}_k \equiv \text{diag}\left(\boldsymbol{w}^f_k \ \ \boldsymbol{w}^m_k\right). \tag{5.119}$$

Here $\boldsymbol{w}^f_k = \begin{bmatrix} w^f_{k_x} & w^f_{k_y} & w^f_{k_z} \end{bmatrix}^T$ and $\boldsymbol{w}^m_k = \begin{bmatrix} w^m_{k_x} & w^m_{k_y} & w^m_{k_z} \end{bmatrix}^T$ are vectors composed of positive weights for the distribution of the force and moment components of the net wrench. Since the weighing matrix is diagonal, the weights will determine the distribution ratios for the force and moment components as

$$w^f_{k_i}/w^f_{\bar{k}_i} = f_{\bar{k}_i}/f_{k_i}, \tag{5.120}$$

$$w^m_{k_i}/w^m_{\bar{k}_i} = m_{\bar{k}_i}/m_{k_i},$$

respectively, $i \in \{x, y, z\}$. Index $\bar{k}$ stands for "not k," that is, if $k \in \{\overline{1,\kappa}\}$ then $\bar{k} \in \{\overline{1,\kappa}\}/k$. From the above relation, it is apparent that the net wrench is distributed componentwise, i.e.

$$f_{k_i} = \frac{\tilde{w}^f_i}{w^f_{k_i}} f_{net_i}, \; m_{k_i} = \frac{\tilde{w}^m_i}{w^m_{k_i}} m_{net_i}, \tag{5.121}$$

f_{net_i} and m_{net_i} denoting the scalar force and moment components, respectively, of the net wrench. The tilde notation over the weigh coefficients are introduced to account for the case of multiple contacts, i.e.

$$\tilde{w}^f_i = \frac{P^f_i}{\sum_{k=1}^{\kappa} \frac{P^f_i}{w^f_k}}, \; P^f_i = \prod_{k=1}^{\kappa} w^f_{k_i}, \; \tilde{w}^m_i = \frac{P^m_i}{\sum_{k=1}^{\kappa} \frac{P^m_i}{w^m_k}}, \; P^m_i = \prod_{k=1}^{\kappa} w^m_{k_i}.$$

In what follows, simple examples will be presented for better understanding of formula (5.121). For clarity but without loss of generality, the DCM-GI-based approach will be applied first to a double-stance (DS) posture, s.t. $c = 12$. The envisioned DS to single-stance (SS) as well as SS-to-DS transitions can be achieved with the design based on the following three policies.

Vertical GRF Force Distribution Policy

Denote the weighting matrix for double stance as

$$\boldsymbol{W}_X = \mathrm{diag}(\boldsymbol{W}_R \;\; \boldsymbol{W}_L) \in \Re^{12\times 12}.$$

The net contact wrench is $\mathcal{F}_F = \mathcal{F}_R + \mathcal{F}_L$,[9] in accordance with (5.118). Referring to (5.120), the following ratio will be used to realize the approach proposed herein:

$$f_{R_z}/f_{L_z} = \frac{||\boldsymbol{r}_P - (\boldsymbol{r}_L + \boldsymbol{r}_{\overleftarrow{P_L L}})||}{||\boldsymbol{r}_P - (\boldsymbol{r}_R + \boldsymbol{r}_{\overleftarrow{P_R R}})||}. \tag{5.122}$$

Here $\boldsymbol{r}_P \in \Re^3$ denotes the net CoP, $\boldsymbol{r}_k$ is the position of the ankle center k, and $\boldsymbol{r}_{\overleftarrow{P_k k}}$ is the position vector pointing from $\boldsymbol{r}_k$ to the local CoP $\boldsymbol{r}_{P_k}$. The vertical reaction component at each foot contact is determined by the following inner product:

$$w^f_{k_z} = s_k \begin{bmatrix}\Delta r_x & \Delta r_y & 0\end{bmatrix} \left(\boldsymbol{r}_X - \boldsymbol{r}^{ds}_k\right), \tag{5.123}$$

where $\boldsymbol{r}^{ds}_k = \boldsymbol{r}_k + \Delta\boldsymbol{r}_k$. As shown in Fig. 5.23, the horizontal components of the $\Delta\boldsymbol{r}_k$ vectors specify the DS area relative to the local contact frames, fixed at the ground projections $\boldsymbol{r}_k$ of

[9] This is also true for the force/moment component vectors, i.e. $\boldsymbol{f}_F = \boldsymbol{f}_R + \boldsymbol{f}_L, \boldsymbol{m}_F = \boldsymbol{m}_R + \boldsymbol{m}_L$.

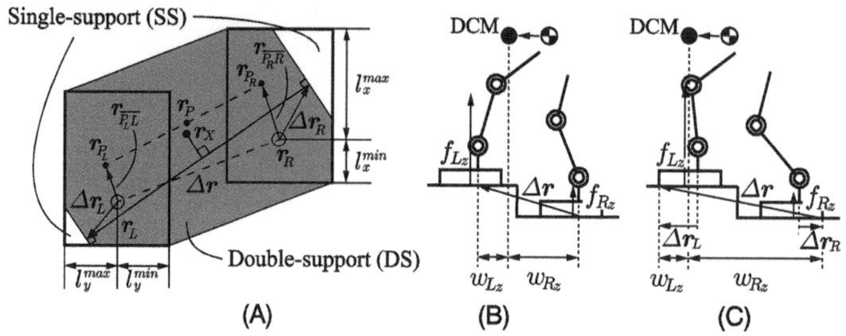

FIGURE 5.23 Geometrical relations pertinent to the wrench distribution policy. (A) The double-support area is specified either by the user or the system, via $\Delta \boldsymbol{r}_k$. (B) With $\Delta \boldsymbol{r}_k$ set at zero, the vertical reactions pass through the ankles. (C) With $\Delta \boldsymbol{r}_k$ set at the outer vertices, the entire BoS can be used as a double-support area.

the ankle centers. The $\Delta \boldsymbol{r}_k$ vectors are geometrically constrained within the foot BoS; we have

$$l_h^{min} \leq \Delta r_{k_h} \leq l_h^{max}, \ h \in \{x, y\}. \tag{5.124}$$

As apparent from (5.123), the vertical component of vector $\Delta \boldsymbol{r} = \boldsymbol{r}_L^{ds} - \boldsymbol{r}_R^{ds}$ is set at zero. This renders the wrench distribution problem a planar one. The terms $s_R = 1$, $s_L = -1$ ensure the positive definiteness of the vertical reactions, in agreement with the unilateral contact condition at the feet. Note that the distribution weight (5.123) depends on the distance between the DCM and the dynamic stability area. Note also that when the difference $\boldsymbol{r}_X - \boldsymbol{r}_k^{ds} = \boldsymbol{0}$, the weighting matrix becomes singular; this case should be avoided.

Consider first the static case, $\dot{\boldsymbol{r}}_C = \boldsymbol{0}$ and $\boldsymbol{r}_X = \boldsymbol{r}_C$. When the CoM is within the DS area, $w_{k_z}^f < 1$. At DS-SS boundary k ($\boldsymbol{r}_C = \boldsymbol{r}_k^{ds}$), $w_{k_z}^f = 1$ and $w_{\bar{k}_z}^f = 0$. The vertical force distribution policy can then be characterized as continuous and consistent with the statics. For example, when $\Delta \boldsymbol{r}_k = \boldsymbol{0}$, as in Fig. 5.23B, minimal load in the ankle joints will be obtained with a static stance s.t. the CoM is over ankle k. The reaction force at that ankle becomes then a *structural force* (i.e. the reaction does not generate a moment). At ankle $\bar{k}$ there will be no reaction at all. Humans sometimes use such a load/unload static posture to relax. In another example, shown in Fig. 5.23C, $\Delta \boldsymbol{r}_k$ is set at the outer vertices of the BoS. In this way, the entire BoS constitutes the area of dynamic stability. This also implies that no transition is meant to occur, i.e. only a swaying task is envisioned.

In addition to statics, the above vertical GRF distribution policy is also consistent with the dynamic states, thanks to the use of the DCM. This will be demonstrated with dynamic contact transitions in an example below.

Friction Policy

The net wrench distribution relation (5.121) can be used to express the friction cone constraint (3.9) as follows:

$$\frac{\sqrt{\left((\tilde{w}_x^f / w_{k_x}^f) f_{net_x}\right)^2 + \left((\tilde{w}_y^f / w_{k_y}^f) f_{net_y}\right)^2}}{(\tilde{w}_z^f / w_{k_z}^f) f_{net_z}} \leq \mu_k.$$

This is an inequality constraint that cannot be adopted directly. To minimize the effect of friction, make use of the following equality-based friction cone policy. From the above relation,

$$\frac{\sqrt{\left((\tilde{w}_x^f/w_{k_x}^f)f_{net_x}\right)^2+\left((\tilde{w}_y^f/w_{k_y}^f)f_{net_y}\right)^2}}{(\tilde{w}_z^f/w_{k_z}^f)f_{net_z}}=\frac{\sqrt{f_{net_x}^2+f_{net_y}^2}}{f_{net_z}}$$

or

$$\left((\tilde{w}_x^f/w_{k_x}^f)f_{net_x}\right)^2+\left((\tilde{w}_y^f/w_{k_y}^f)f_{net_y}\right)^2=(\tilde{w}_z^f/w_{k_z}^f)^2(f_{net_x}^2+f_{net_y}^2).$$

Now, assume that $w_{k_x}^f=w_{k_y}^f\equiv w_{k_h}^f$, $h\in\{x,y\}$. Then

$$(\tilde{w}_h^f/w_{k_h}^f)^2(f_{net_x}^2+f_{net_y}^2)=(\tilde{w}_z^f/w_{k_z}^f)^2(f_{net_x}^2+f_{net_y})^2, \tag{5.125}$$

where $\tilde{w}_h^f/w_{k_h}^f=\tilde{w}_z^f/w_{k_z}^f$ and $w_{k_h}^f=w_{k_z}^f$.

Next, consider the torsional friction constraint $\mathcal{F}_{FT}$ (cf. Table 5.2 and (3.11)), rewritten with the help of (5.121) as

$$\frac{(\tilde{w}_z^m/w_{k_z}^m)|m_{net_z}|}{(\tilde{w}_z^f/w_{k_z}^f)f_{net_z}}\le\gamma_k.$$

Based on this inequality, the following torsional friction distribution policy will be adopted:

$$\frac{(\tilde{w}_z^m/w_{k_z}^m)|m_{net_z}|}{(\tilde{w}_z^f/w_{k_z}^f)f_{net_z}}=\frac{|m_{net_z}|}{f_{net_z}},$$

and hence,

$$\tilde{w}_z^n/w_{k_z}^n=\tilde{w}_z^f/w_{k_z}^f.$$

CoP Allocation Policy

The net wrench distribution relation (5.121) can be used to express the foot CoP in the local contact frame as follows:

$$\boldsymbol{r}_{P_k}=\begin{bmatrix}-\left((\tilde{w}_y^m/w_{k_y}^m)m_{net_y}\right)/\left((\tilde{w}_z^f/w_{k_z}^f)f_{net_z}\right)\\ \left((\tilde{w}_x^m/w_{k_x}^m)m_{net_x}\right)/\left((\tilde{w}_z^f/w_{k_z}^f)f_{net_z}\right)\end{bmatrix}.$$

Apparently, the foot CoP location depends on the vertical force distribution. This implies that when the foot is going to be unloaded, the foot CoP will be shifted toward the BoS boundary, eventually resulting in a foot roll. To avoid this problem, the effect of the asymmetric vertical reaction distribution should be compensated for in the CoP equation. This can be done when

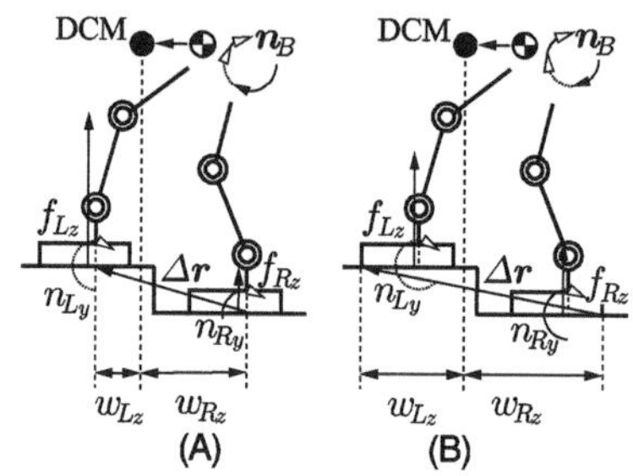

FIGURE 5.24 Relations between vertical foot reactions, tangential reaction moments and the net base moment. (A) A smaller DS area setting yields smaller reaction moments at the feet. (B) A larger DS area setting yields larger reaction moments and increases the possibility of a foot roll.

the following relation holds:

$$\begin{bmatrix} -\left((\tilde{w}_y^m/w_{k_y}^m)m_{net_y}\right)/\left((\tilde{w}_z^f/w_{k_z}^f)f_{net_z}\right) \\ \left((\tilde{w}_x^m/w_{k_x}^m)m_{net_x}\right)/\left((\tilde{w}_z^f/w_{k_z}^f)f_{net_z}\right) \end{bmatrix} = \begin{bmatrix} -m_{net_y}/f_{net_z} \\ m_{net_x}/f_{net_z} \end{bmatrix}.$$

Hence,

$$\tilde{w}_x^m/w_{k_x}^m = \tilde{w}_y^m/w_{k_y}^m = \tilde{w}_z^f/w_{k_z}^f.$$

The CoP allocation problem can be related to the setting of the dynamic stability area, as follows. First, note that from (5.117) (see also Fig. 5.24), it is apparent that when the contribution of the net vertical GRF $\boldsymbol{f}_{net}$ to the body moment $\boldsymbol{m}_B$ is relatively large (filled arrow heads in Fig. 5.24), that of the reaction moments (transparent arrow heads in Fig. 5.24) will be relatively small. Thus, larger force weights will yield a smaller net moment to be distributed, i.e. a smaller net foot GRM $\boldsymbol{m}_{net}$. On the other hand, according to (5.123), the force weights are determined from the dynamic stability area. With a smaller dynamic stability area, larger force weights will be obtained. In this way, the reaction moments will be minimized and the CoPs will be allocated closer to the local frames at the ankle/wrist projections. With a larger dynamic stability area it is more likely that a foot roll will occur since the CoPs will allocate apart from the local frames due to the larger ground reaction moments (see Fig. 5.24B).

Final Result

As a result of the friction cone, torsional friction, and the CoP allocation policies, the weight matrix in the DCM-GI assumes the simple form

$$\boldsymbol{W}_k = (w_{k_z}^f \boldsymbol{E}_6). \tag{5.126}$$

This result emphasizes the important role of an appropriate vertical reaction distribution. It should be noted that the DCM-GI distribution algorithm introduces an error since the net wrench (5.118) that is distributed is different from the real body wrench. But the error is small and only in the force component.

The DCM-GI wrench distribution approach can be extended in a straightforward manner to cope with noncoplanar contact postures [58]. In such a case, internal wrench control is needed. This can be achieved via the null space component in (5.114). The method can also be applied at multicontact (e.g. triple-support) postures. An implementation example for this case will be presented below. Furthermore, the method is quite suitable for reactive balance control in response to unknown external disturbances. An implementation example will be presented in Section 7.7.

Simple Balance Controller With DCM-GI-Based Body Wrench Distribution

The reference contact wrenches obtained under the DCM-GI based body wrench distribution approach can be written as (cf. (5.114))

$$(\mathcal{F}^c)^{ref} = \mathbb{C}_{cC}^{-W_X}(\dot{\mathcal{L}}_C^{ref} + \mathcal{G}_C) + N(\mathbb{C}_{cC})\overline{\mathcal{F}}_a^c. \tag{5.127}$$

As already noted, the internal wrench control component (i.e. the null-space term) is needed only in the case of noncoplanar contacts. The linear and angular components of $\dot{\mathcal{L}}_C^{ref}$ can be determined from the asymptotically stable CRB motion tracking controller described in Section 5.8.7.

A relatively simple balance controller for torque-controlled robots can be derived from the JSID solution, (4.244). We have

$$\boldsymbol{\tau} = \boldsymbol{g}_\theta - \mathcal{J}_{cB}^T(\overline{\mathcal{F}}^c)^{ref} - N^*(\mathcal{J}_{cM})\boldsymbol{M}_{\theta_M}\boldsymbol{K}_{\theta_D}\dot{\boldsymbol{\theta}}. \tag{5.128}$$

The last term stands for joint damping that is used to suppress the motion in the arms. This term is derived from the null space of the mixed quasivelocity constraint Jacobian to ensure that there will be no interference with the other two terms. As clarified at the end of Section 3.6.4, the distributed gravity compensation with $\boldsymbol{g}_\theta$, in combination with the base quasivelocity constraint Jacobian map $\mathcal{J}_{cB}^T$, yields better results than the lumped gravity compensation with $\mathcal{J}_{cM}^T\overline{\mathcal{F}}^c(\mathcal{G}_C)$. The interference between the gravity compensation and the $(\overline{\mathcal{F}}^c)^{ref}$ term can then be avoided. The control block diagram is shown in Fig. 5.25. This controller can be used for balance control in single or double stance and during the respective transitions. An implementation with a proactive balance control task will be given in Section 5.12.2. Another implementation with a reactive task can be found in Section 7.7.

Implementation Example

The following example demonstrates the DCM-GI body wrench distribution control capability in a proactive task with multicontact transitions. The reference rates of change of the linear and angular momentum can be determined from (5.107) and (5.106), respectively. In this particular example, a virtual viscous damper for the angular motion is used, i.e. $\dot{l}_B^{ref} = -D_\omega\omega_B$, D_ω denoting the damping gain.

The performance of the DCM-GI balance controller is examined via simulations in OpenHRP3 (cf. Chapter 8). A small-size humanoid robot model with parameters similar to those of the HOAP-2 robot [26] is used. For the numbering of the joints and other relevant data, see Section A.1. The footprint (BoS) is $(l_x^{max}, l_x^{min}, l_y^{max}, l_y^{min}) = (58, -40, 31.5, -31.5)$ mm w.r.t. the ankle joint center. The gains in the spatial momentum rate stabilizer are set at

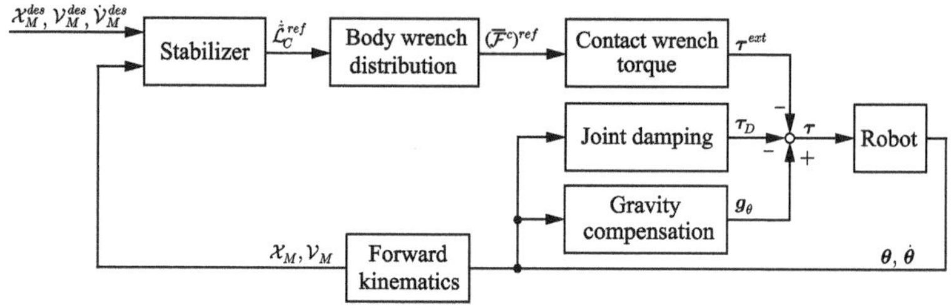

FIGURE 5.25 Control block-diagram for balance control of a torque-controlled robot in single/double stance. The controller ensures asymptotic trajectory tracking of feasible desired CRB trajectories and a DCM-GI based body-wrench distribution approach.

$k_X = 300$ and $D_\omega = 50$. The torque-based whole-body controller described in Section 5.12.1 is used. The motion of the feet and hand end links in the unconstrained motion directions is controlled as a lower-priority subtask, via a joint torque component obtained from a conventional kinematic PD feedback plus feedforward control law (cf. Fig. 5.30).

Initially, the robot is in a triple-support (TS) posture: double-support (DS) stance with plane contacts at the feet plus right hand in point contact. Since the foot contacts are coplanar, no internal wrench control is needed. Thus, the null-space term in (5.114) is set at zero. The boundaries of the dynamic stability area are determined by the locations of the two ankle joints and the hand contact, i.e. set $\Delta \boldsymbol{r}_k = \boldsymbol{0}$, $k = 1, 2, 3$. The proactive balance control task is defined as a DCM tracking task to obtain two consecutive contact transitions: first lifting the left foot (TS-to-DS) and then lifting the hand (DS-to-SS[10]). The desired DCM values are set at $\boldsymbol{r}_X^{des}(t_0) = (0, 0, 0)$, $\boldsymbol{r}_X^{des}(t_1) = (30, -75, 0)$ mm, $\boldsymbol{r}_X^{des}(t_2) = (25, -40, 0)$ mm, respectively, whereby $t_0 = 3$ s, $t_1 = 7$ s, and $t_2 = 9$ s. These positions are connected via a fifth-order spline.

Snapshots and graphs from the result of the simulation are displayed in Fig. 5.26 and Fig. 5.27, respectively. The result is shown in animated form in Video 5.10-1 [57]. The trajectory of the CoM is apparent from Fig. 5.27A. First, the CoM is shifted toward the right foot and slightly forward, toward the hand contact. The contact force at the left foot decreases thereby gradually (cf. Fig. 5.27B). At $t_1 = 7$ s, the left foot contact force becomes zero. A continuous contact transition is then ensured from TS to DS, resulting in right foot and hand contact only. Such transition is possible because the body wrench distribution method ensures the same distribution ratio. After the transition, the left foot lifts off vertically. At the same time, the CoM is moved above the right-foot ankle joint, whereby the hand contact force decreases gradually (cf. Fig. 5.27C). At $t_2 = 9$ s, the hand contact force becomes zero. A contact transition is then ensured from DS to SS without generating excessive reactions. The contact force at the right foot is shown in Fig. 5.27D.

5.10.5 The VRP Generalized Inverse

As shown in Section 5.10.4, the DCM-GI wrench distribution approach can be used in multicontact transitions. It should be noted, though, that for tasks that require swift changes

[10] SS stands for "single support."

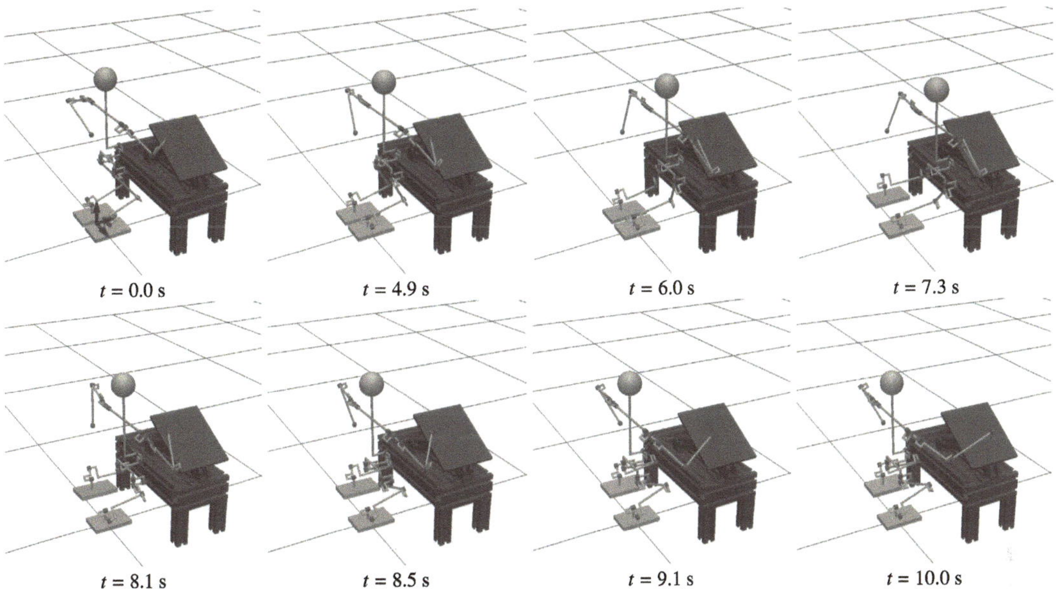

FIGURE 5.26 Multicontact transition task. Snapshots: initially, the robot is in double stance on flat ground, the right hand being in (point) contact with an inclined surface (a triple-support (TS) posture). The body wrench is redistributed via the DCM-GI, as follows. First, the left foot contact is gradually unloaded s.t. the foot can be lifted off ground (at $t = 7.3$ s, a double-support (DS) posture). Then, the hand contact is gradually unloaded s.t. the hand can be lifted off the table (at $t = 10.0$ s, a single-support (SS) posture).

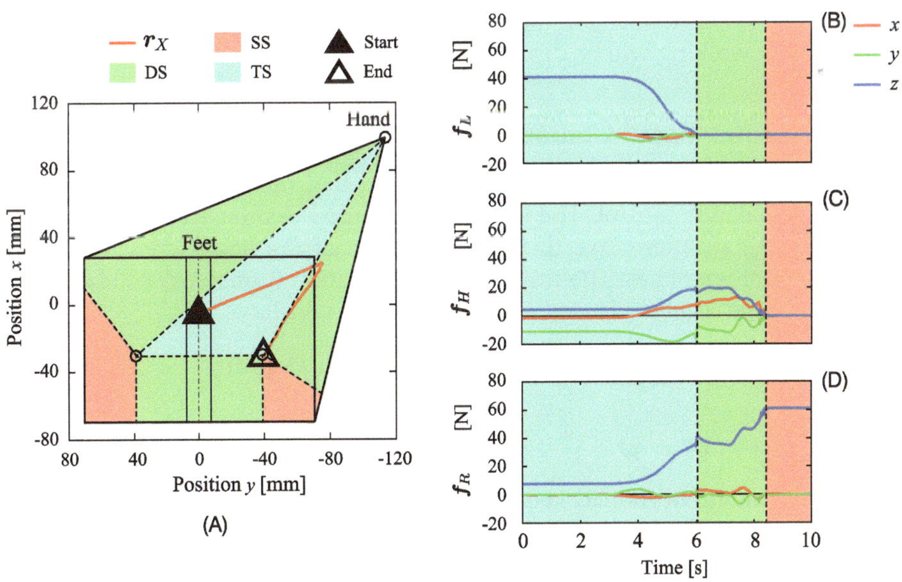

FIGURE 5.27 Multicontact transition task. Graph (A): TS/DS/SS dynamic stability area setting and trajectory of the DCM. The circles within the feet denote the ankle joints. Graphs (B)–(D): GRFs at the contacts.

in the contact states, such as dynamic walking or reactive stepping, the distribution should be based on a second-order formulation rather that on a first-order one, as is the case with the DCM $\boldsymbol{r}_X$. The VRP $\boldsymbol{r}_{vrp}$ represents a second-order formulation that turns out to be quite suitable for such type of distribution.

A wrench distribution based on the VRP can be achieved in a straightforward way within the DCM-GI framework: in (5.123) just substitute $\boldsymbol{r}_X$ with $\boldsymbol{r}_{vrp}$. All other relations derived in Section 5.10.4 remain valid. The weighting matrix obtained in this way will be denoted as $\boldsymbol{W}_{vrp}$. The respective weighted generalized inverse will be referred to as the *VRP generalized inverse* or VRP-GI for short. Note that, since rapid contact transitions are assumed under VRP-GI-based wrench distribution, there is no need to specify a stability area that is different from the minimal one, i.e. $\Delta \boldsymbol{r}_k = \boldsymbol{0}$.

To understand the implications of using the VRP instead of the DCM, consider a *dynamic state* s.t. the robot is in a double stance, the rate of change of the angular momentum is zero, and thus, the net CoP is the vertical projection of the VRP. Furthermore, assume that the local CoP $\boldsymbol{r}_{\overleftarrow{P}_k k}$ is at the ground projection of the respective ankle joint (cf. Fig. 5.23A). This means that the net CoP lies on the line connecting the projections of the ankle joints (i.e. on the line determined by $\Delta \boldsymbol{r}$). The dynamic stability area is set at the minimum, as already noted. At such a state, it is desirable to obtain a wrench distribution that does not generate GRMs.

It can be confirmed that the abovementioned wrench distribution can be achieved with the VRP-GI. On the contrary, this cannot be done with the DCM-GI since the DCM does not lie on the above line; at a dynamic state the ground projection of the DCM (i.e. the xCoM) is always offset from the line (cf. Fig. 5.23A). In general, at states that differ from the above special case, the moments at the ankle joints induced by the VRP-GI distribution will be smaller than those induced by the DCM-GI distribution. This also supports faster contact transitions—a highly desirable property in a dynamic gait task, as already mentioned. An implementation of the VRP-GI in dynamic walking will be discussed in Section 7.4.

5.10.6 Joint Torque–Based Contact Wrench Optimization

From the preceding discussion it became apparent that the body WD problem plays an important role in balance control. The solutions provided demonstrate that the problem is not a trivial one. As an alternative to the distribution of the body wrench one could just minimize the contact wrenches. The respective optimization framework can be based on the inverse solution to the *complete dynamics* (cf. (4.238)); then we have

$$\boldsymbol{\tau} = \boldsymbol{V}^{\dagger} \boldsymbol{V}_r^T (\boldsymbol{M}\ddot{\boldsymbol{q}} + \boldsymbol{h}) + \left(\boldsymbol{E} - \boldsymbol{V}^{\dagger} \boldsymbol{V}_r^T \boldsymbol{S}^T\right) \boldsymbol{W}^{-1} \boldsymbol{\tau}_a. \tag{5.129}$$

Recall that $\boldsymbol{V}^{\dagger} = \boldsymbol{W}^{-\frac{1}{2}} \left(\boldsymbol{V}_r^T \boldsymbol{S}^T \boldsymbol{W}^{-\frac{1}{2}}\right)^+ \in \Re^{n \times r}$, $\boldsymbol{V}_r$ providing the minimal null-space basis associated with the underdetermined task; $\boldsymbol{S} \in \Re^{n \times (n+6)}$ is the underactuation filtering matrix s.t. $\mathcal{Q} = \boldsymbol{S}^T \boldsymbol{\tau}$ represents the generalized force vector (cf. (2.102)). Note that since the above solution is derived with the constraint force elimination approach (cf. Section 4.12), the contact constraints are absent. Note also that the spatial dynamics do not appear explicitly. This means that the WD problem cannot be directly addressed. It is possible, though, to optimize the contact wrenches indirectly. Indeed, notice that the solution is parametrized by a symmetric positive definite weight matrix, $\boldsymbol{W}$, and an arbitrary joint torque vector, $\boldsymbol{\tau}_a$. These parameters can be determined via optimization. An appropriate objective function that combines quadratic and linear components can be formulated as follows [126]:

$$\min_{\boldsymbol{x}} \left(c_{JT} + c_{CW} + \boldsymbol{b}_{JT}^T \boldsymbol{\tau} + \boldsymbol{b}_{CW}^T \boldsymbol{\tau}^c \right), \tag{5.130}$$

where $\boldsymbol{x} = \begin{bmatrix} \boldsymbol{\tau}^T & (\overline{\mathcal{F}}^c)^T \end{bmatrix}^T$. The terms c_{JT} and c_{CW} are penalty-type, quadratic form relaxations of the equality constraints $\mathcal{F}_{JT}|_{\boldsymbol{\tau}^{ref}=0}$ and $\mathcal{F}_{CW}|_{(\overline{\mathcal{F}}^c)^{ref}=0}$, respectively; $\boldsymbol{b}_{(\circ)}$ denote arbitrary vectors that parametrize the linear components. Note that the quadratic components $c_{(\circ)}$ are also parametrized, via appropriate weight matrices $\boldsymbol{W}_{(\circ)}$. The terms c_{JT} and c_{CW} minimize the total joint torque and the contact wrenches, respectively. The solution to the above minimization task is obtained as [126]:

$$\boldsymbol{W} = \boldsymbol{W}_{JT} + \boldsymbol{S}\boldsymbol{W}_{CW}\boldsymbol{S}^T, \tag{5.131}$$

$$\boldsymbol{\tau}_a = -\boldsymbol{b}_{JT} + \boldsymbol{S}\boldsymbol{W}_{CW}(\boldsymbol{M}\ddot{\boldsymbol{q}} + \boldsymbol{h}) + \boldsymbol{S}\boldsymbol{b}_{CW}. \tag{5.132}$$

Although the contact wrenches do not appear explicitly in (5.129), it is possible to minimize them via the c_{CW} term. Note that with

$$\boldsymbol{W}_{CW} = \boldsymbol{J}_c \boldsymbol{W}_{\kappa} \boldsymbol{J}_c^T, \tag{5.133}$$

the joint torque stemming from the constraints will be minimized, i.e. $c_{CW} = \|\boldsymbol{\tau}^c\|^2_{\boldsymbol{W}_\kappa}$. This means that *the contact wrench minimization problem can be reformulated as a joint torque minimization problem.* In the above formula, $\boldsymbol{J}_c$ stands for the system constraint Jacobian.

There are a few possibilities to tackle the (implicit) contact wrench optimization problem. The simplest approach is just to minimize the quadratic cost for the contact wrenches, with a suitably defined $\boldsymbol{W}_{CW}$. To this end, set $\boldsymbol{W}_{JT} = \boldsymbol{0}$ and $\boldsymbol{b}_{JT} = \boldsymbol{0} = \boldsymbol{b}_{CW}$ in (5.131) and (5.132). Note that with $\boldsymbol{W}_{CW}$ as in (5.133) and $\boldsymbol{W}_\kappa = \boldsymbol{M}^{-1}$, the c_{CW} minimization subtask will yield a dynamically consistent (in the sense of Gauss' principle of least constraint) solution. Next, to tackle the friction constraints, one could attempt to minimize the tangent reaction force and the reaction moment components. This can be done with the help of the following weight matrix [126]:

$$\boldsymbol{W}_{CW} = \begin{bmatrix} \mathbb{R}_{F_r}^T \boldsymbol{W}_{F_r} \mathbb{R}_{F_r} & \boldsymbol{0} \\ \boldsymbol{0} & \mathbb{R}_{F_l}^T \boldsymbol{W}_{F_l} \mathbb{R}_{F_l} \end{bmatrix}. \tag{5.134}$$

The submatrices $\boldsymbol{W}_{F_j} = \text{diag}(\alpha_{f_jx}, \alpha_{f_jy}, 1, \alpha_{m_jx}, \alpha_{m_jy}, \alpha_{m_jz})$ contain the weights for the components. This type of minimization yields a relatively conservative solution.

The above two contact wrench minimization approaches are of limited use, though, since they yield equal vertical force distribution that was shown to be inconsistent with the static states. A better approach would be a force control–based tracking of dynamically consistent reference contact wrench components. The objective function can then be formulated as

$$\min_{\boldsymbol{x}} c_{CW},$$

where $\boldsymbol{x} = \overline{\mathcal{F}}^c$, $c_{CW} = \|\overline{\mathcal{F}}^c - (\overline{\mathcal{F}}^c)^{ref}\|_{W_t}$, $(\overline{\mathcal{F}}^c)^{ref} \neq \mathbf{0}$. the minimization is achieved with the choice of $\boldsymbol{W}_{CW} = \boldsymbol{W}_t$ and $\boldsymbol{b}_{CW}$ defined from $\boldsymbol{J}_c\, \boldsymbol{b}_{CW} = -(\overline{\mathcal{F}}^c)^{ref} \boldsymbol{W}_t$ [126]. However, since this is a penalty-type approach; full consistency with the static states and/or exact contact wrench tracking is not guaranteed. Also, the determination of dynamically consistent reference contact wrench components is a nontrivial problem.

5.11 NONITERATIVE SPATIAL DYNAMICS–BASED MOTION OPTIMIZATION

Motion can be optimized as a subtask of a spatial dynamics–based sequential optimization (cf. Fig. 5.21). As a result of the optimization, a solution to the FSD problem will be obtained. This type of optimization is suitable for position-controlled robots. In torque-controlled robots, motion optimization is used as a control component. Implementation examples with torque-based control will be presented in Section 5.12.

5.11.1 Independent Motion Optimization With CRB Wrench-Consistent Input

To obtain a solution to the FSD problem, insert the solution to the WD problem, (5.114), into the spatial dynamics equation, (5.109). Then we have

$$\mathbb{M}_C \dot{\mathcal{V}}_M^{opt} + \boldsymbol{H}_{CM} \ddot{\boldsymbol{\theta}}^{opt} + \mathcal{C}_M + \mathcal{G}_C = \mathbb{C}_{cC} \mathbb{C}_{cC}^{-W} (\dot{\mathcal{L}}_C^{ref} + \mathcal{G}_C) + \mathbb{C}_{cC} \boldsymbol{N}(\mathbb{C}_{cC}) \overline{\mathcal{F}}_a^c. \tag{5.135}$$

This equation simplifies to

$$\mathbb{M}_C \dot{\mathcal{V}}_M^{opt} + \boldsymbol{H}_{CM} \ddot{\boldsymbol{\theta}}^{opt} = \dot{\mathcal{L}}_C^{ref} - \mathcal{C}_M. \tag{5.136}$$

The gravity and null-space terms were annihilated by using $\mathbb{C}_{cC}\mathbb{C}_{cC}^{-W} = \boldsymbol{E}$ and $\mathbb{C}_{cC}\boldsymbol{N}(\mathbb{C}_{cC}) = \mathbf{0}$, respectively. Also, the nonlinear velocity-dependent term $\mathcal{C}_M$ will be partially annihilated under the assumption of perfect model matching.

The above result shows that motion optimization could be performed *independently* from the WD problem. The reference rate of change of the CRB momentum is assumed *contact-consistent*, s.t. $\dot{\mathcal{L}}_C^{ref} + \mathcal{G}_C = \mathcal{F}_C^{ref} \in \mathcal{F}_{BWC}$ holds. This is a necessary condition for using (5.136) in an independent motion optimization. If this condition is not satisfied, an on-line modification of the input trajectories will be required (cf. Fig. 5.1). In this case, the output from the WD problem, $\mathbb{C}_{cC}\overline{\mathcal{F}}_{opt}^c$, is used in place of $\dot{\mathcal{L}}_C^{ref}$ in (5.136) to obtain an "admissible" rate of change of the CRB momentum [88]. This implies that the WD and motion optimization problems cannot be solved independently. On the other hand, when the initially generated reference input $\dot{\mathcal{L}}_C^{ref}$ is feasible, there is no necessity to involve a *combined* motion/force optimization task. Thus, the computational overhead stemming from the annihilation of the null-space (internal force), gravity, and nonlinear velocity force components can be avoided. Moreover, the WD problem and motion optimization tasks can be invoked in parallel threads, leading to a further decrease in computational time. In conclusion, it is highly desirable to rely upon a contact and motion planning system capable of generating feasible inputs. Asymptotically stable whole-body balance control can then be achieved, as clarified in Section 5.8.7, in addition to the decreased computational overhead.

5.11.2 Stabilization With Angular Momentum Damping

A motion optimization with unconstrained hand motion implies that the motion in the arms, obtained as a result of the optimization, supports the main balance stabilization task. Some results in this direction have already been reported in the literature. In an early work [81], an arm "windmilling" motion was generated as the outcome of an iterative, QP optimization. In [112], a centroidal momentum balance controller generated an "unintentional" arm motion. Note, however, that the mechanism of such type of arm motion generation and the interference with the other balance control subtasks has not been clarified yet.

From Section 5.8.3 recall that the RAM/V plus RNS balance controller generated motion in the unconstrained limbs that injected angular momentum damping into the system. In this way, the balance controller was endowed with a highly desirable self-stabilization property. The controller was able to stabilize the unstable (foot roll) contact states without any special provisions, such as contact transition modeling. In this subsection, the RAM/V balance control approach will be reformulated using the *rate of change of the RAM/V*. Analysis at the acceleration level can then be performed, e.g. to exactly reveal the angular momentum damping mechanism.

The spatial dynamics of the robot can be represented in terms of spatial acceleration by premultiplying (5.109) by the inverse of the CRB inertia tensor; then we have

$$\mathbb{M}_C^{-1}\boldsymbol{H}_{CM}\ddot{\boldsymbol{\theta}}+\mathbb{M}_C^{-1}(\mathcal{C}_M+\mathcal{G}_C)=\dot{\mathcal{V}}_C-\dot{\mathcal{V}}_M. \tag{5.137}$$

Here $\mathbb{M}_C^{-1}\boldsymbol{H}_{CM}=\begin{bmatrix}\boldsymbol{0}^T & \boldsymbol{J}_\omega^T\end{bmatrix}$ (cf. (4.97)). Spatial velocity $\dot{\mathcal{V}}_C\equiv\mathbb{M}_C^{-1}\mathbb{C}_{cC}\overline{\mathcal{F}}^c$ is referred to as the *system spatial acceleration*. This quantity should be generated in compliance with the *force-task constraints* (i.e. friction cone and CoP-in-BoS constraints). The *CRB spatial acceleration* $\dot{\mathcal{V}}_M$, on the other hand, should be generated in agreement with the *motion task constraints*. The control joint acceleration can then be determined by solving a *noniterative* least-squares optimization task. Following the procedure in Section 5.8.3 for the velocities, first rewrite (5.137) componentwise, i.e.

$$\boldsymbol{a}_g=\dot{\boldsymbol{v}}_{C_R}-\dot{\boldsymbol{v}}_{C_I}, \tag{5.138}$$

$$\boldsymbol{J}_\omega\ddot{\boldsymbol{\theta}}=\dot{\boldsymbol{\omega}}_C-\dot{\boldsymbol{\omega}}_B-\boldsymbol{I}_C^{-1}\boldsymbol{c}_{mM}. \tag{5.139}$$

The nonlinear velocity-dependent moment $\boldsymbol{c}_{mM}$ is defined in (4.140). From the upper equation it is apparent that the acceleration of the gravity force, $\boldsymbol{a}_g$, is expressed as a *relative CoM acceleration*, $\Delta\dot{\boldsymbol{v}}_C=\dot{\boldsymbol{v}}_{C_R}-\dot{\boldsymbol{v}}_{C_I}$. The CoM acceleration component, $\dot{\boldsymbol{v}}_{C_R}\equiv\dot{\boldsymbol{v}}_C(\overline{\mathcal{F}}^c)$, that stems from the reaction (contact) wrenches is clearly distinguished from the *inertial* CoM acceleration component, $\dot{\boldsymbol{v}}_{C_I}\equiv\dot{\boldsymbol{v}}_C$. The CoM acceleration $\dot{\boldsymbol{v}}_{C_R}$ compensates the gravity CoM acceleration $\boldsymbol{a}_g$ at all postures, for any inertial acceleration of the CoM. In other words, *the effect of the gravity field on the spatial dynamics can be completely ignored*. It also becomes apparent that it will suffice to *formulate the optimization task for the joint acceleration in terms of the relative angular acceleration*, i.e.

$$\Delta\dot{\omega} = \dot{\omega}_C - \dot{\omega}_B. \tag{5.140}$$

The desired CRB trajectories, $\dot{\mathcal{V}}_M^{des} = \begin{bmatrix}(\dot{v}_{C_I}^{des})^T & (\dot{\omega}_B^{des})^T\end{bmatrix}^T$, can be tracked with asymptotic stability in accordance with the results in Section 5.8.7, under the following kinematic control law:

$$\ddot{\theta} = \mathcal{J}_{cM}^{+}\left(\dot{\bar{\mathcal{V}}}^c - \mathbb{C}_{cC}^T\dot{\mathcal{V}}_M^{ref} - \dot{\mathcal{J}}_{cM}\dot{\theta} - \dot{\mathbb{C}}_{cC}^T\mathcal{V}_M\right) + N(\mathcal{J}_{cM})\,\ddot{\theta}_u. \tag{5.141}$$

This control law was derived from the time differential of the instantaneous motion constraint (2.125).

Usually, it is assumed that the desired CRB trajectories are feasible (contact-consistent) and that $\mathcal{J}_{cM}(q)$ is full (row) rank throughout the motion. When there is no guarantee that the desired CRB trajectories are contact-consistent or when modeling and other errors yield an unbounded reference input, the contact conditions will be violated, resulting e.g. in a foot roll that may lead to loss of balance. This problem can be addressed by adding a *relative angular acceleration control component*. To this end, determine the unconstrained joint acceleration $\ddot{\theta}_u$ in (5.141) with the help of the angular acceleration relation (5.139). As a result, one arrives at

$$\ddot{\theta}^{ref} = (E - \bar{J}_\omega^{+}J_\omega)\,\mathcal{J}_{cM}^{+}\left(\dot{\bar{\mathcal{V}}}^c - \mathbb{C}_{cC}^T\dot{\mathcal{V}}_M^{ref}\right) + \bar{J}_\omega^{+}\Delta\dot{\omega}^{ref} + N(\mathcal{J}_{cM})N(\bar{J}_\omega)\ddot{\theta}_u^{ref} + \dot{\theta}_{nl}, \tag{5.142}$$

where $\bar{J}_\omega = J_\omega N(\mathcal{J}_{cM})$ and

$$\dot{\theta}_{nl} = -(E - \bar{J}_\omega^{+}J_\omega)\,\mathcal{J}_{cM}^{+}\left(\dot{\mathcal{J}}_{cM}\dot{\theta} + \dot{\mathbb{C}}_{cC}^T\mathcal{V}_M\right) - \bar{J}_\omega^{+}I_C^{-1}c_{mM}.$$

Here $\ddot{\theta}_u^{ref}$ is an arbitrary joint acceleration that comprises several additional joint-level control inputs. In the case of a single stance, for example, the motion of the swing foot can be constrained by the time derivative of (2.130), i.e.

$$J_{mM}\ddot{q}_M + \dot{J}_{mM}\dot{q}_M = \dot{\bar{\mathcal{V}}}^m. \tag{5.143}$$

This constraint is treated as a lower-priority task w.r.t. the balance control task. In other words, the arbitrary $\ddot{\theta}_u^{ref}$ in (5.142) is to be resolved via (5.143). The form of the final solution is quite complicated and will be omitted here. The other components of $\ddot{\theta}_u^{ref}$ should be designed to ensure joint velocity damping and joint-limit avoidance.

The control law (5.142) will be referred to as the *relative angular acceleration (RAA) balance controller*. The RAA controller can function as a DCM stabilizer. This becomes apparent by setting the reference inertial CoM trajectories $\dot{v}_{C_I}^{ref}(t)$ in accordance with (5.107). Note that the DCM stabilization is achieved as a high-priority task. In addition to this important function, the RAA controller provides an input for controlling the rate of change of the angular momentum in an *independent way*. Being set at a lower-priority level, this subtask will not disturb the main DCM stabilization task.

In order to understand the role of the *independent* control of the rate of change of the angular momentum, consider first the two constraints for the (centroidal) system and coupling angular momentum conservation, (5.98) and (5.99), respectively. At the acceleration level, these conditions assume the following form:

$$\boldsymbol{\omega}_C^{ref} = \mathbf{0} \Rightarrow \dot{\boldsymbol{\omega}}_C^{ref} = -D_\omega \boldsymbol{\omega}_C, \tag{5.144}$$

$$\Delta\boldsymbol{\omega}^{ref} = \mathbf{0} \Rightarrow \Delta\dot{\boldsymbol{\omega}}^{ref} = -D_\omega \Delta\boldsymbol{\omega}, \tag{5.145}$$

respectively, D_ω denoting a PD damping gain. With these formulations, the integrability of the control joint acceleration (5.142) will be ensured. In other words, the RAA balance controller will behave as a RAM/V balance controller at the velocity level. Recall that the latter was capable of handling critical situations with unstable contacts (rolling feet), without the need to use any contact models. The above control inputs provide for a system angular damping that cannot be achieved with the high-priority DCM stabilization task.

First, consider the role of the system angular velocity damper (5.144). The optimal joint acceleration can be obtained from the rotational dynamics in the equation of motion with mixed quasivelocity, (4.158). We have

$$\boldsymbol{I}_C\dot{\boldsymbol{\omega}}_B^{ref} + \boldsymbol{H}_C\ddot{\boldsymbol{\theta}} + \boldsymbol{c}_{mM} = -D_\omega \boldsymbol{\omega}_C. \tag{5.146}$$

Note that the driving term $\mathbb{C}_{cC_m}\overline{\mathcal{F}}^c$ in (4.158) has been replaced with the damping term to ensure the angular momentum conservation constraint. Note also that the constraint-consistent solution to (5.146) yields exactly the RAA control law (5.142), wherein the reference system angular acceleration is replaced with the damping term.

Next, consider the role of the RAV damper (5.145). With this constraint, the RAA control law can be used to generate reactive synergies in response to large impact–type disturbances. Note that even when the impact induces a rolling feet motion, the RAA controller is capable of restoring a stable posture, unless the impact is very large, of course.

The RAA controller can be embedded into a general torque controller with a wrench distribution component. The block diagram of such a controller is shown in Fig. 5.28. The output of the RAA controller is the quasiacceleration $\ddot{\boldsymbol{q}}_M^{ref}$. The reference rate of change of the system spatial momentum is obtained in a straightforward way from this output and the reference CRB trajectories $\dot{\mathcal{V}}_M^{ref}$ as

$$\dot{\mathcal{L}}_C^{ref} = \mathcal{A}_C\ddot{\boldsymbol{q}}_M^{ref} + \dot{\mathcal{A}}_C\dot{\boldsymbol{q}}_M, \tag{5.147}$$

where

$$\mathcal{A}_C = \begin{bmatrix} \mathbb{M}_C & \boldsymbol{H}_{CM} \end{bmatrix} \in \Re^{6\times(6+n)}.$$

Note that this matrix is similar to the centroidal momentum matrix appearing in [111,112]. The sum of $\dot{\mathcal{L}}_C^{ref}$ and the gravity wrench are then distributed at the contacts, e.g. with the VRP-GI (cf. Section 5.10.5). Finally, the control joint torque is obtained via the JSID solution.

The above controller provides for enhanced balance control. It is very efficient from a computational point of view since it does not rely on recursive optimization. Implementations of this controller will be presented in Section 7.7.4.

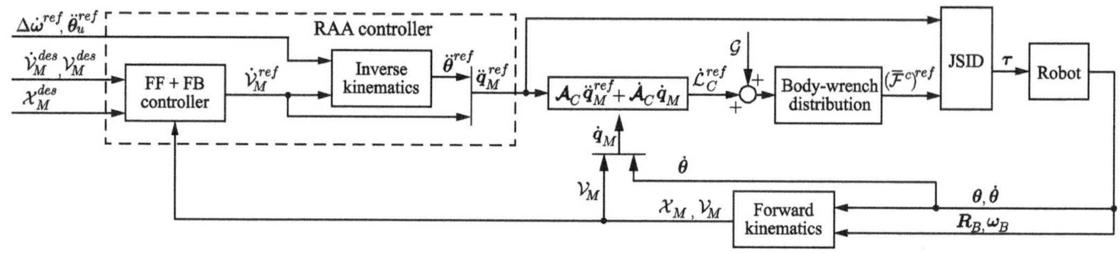

FIGURE 5.28 Block diagram of a torque controller with a RAA control component. The FF+FB controller includes feedforward and feedback control terms for CRB trajectory tracking. In particular, DCM stabilization can be embedded within this component along with the desirable base-link orientation.

5.11.3 Motion Optimization With Task-Based Hand Motion Constraints

The balance controller described in Section 5.11.2 is quite suitable for dynamic motion generation and control whereby the upper-body motion plays an important role and the arms are free to move. In tasks where the hand(s) (or other parts) come in contact with the environment or are required to track a desired trajectory, a slightly different approach to motion optimization is needed. The following derivations will be based on the formulation of spatial momentum w.r.t. the base link, $\mathcal{L}_B(\boldsymbol{q}, \dot{\boldsymbol{q}}_B) = \boldsymbol{\mathcal{A}}_B \dot{\boldsymbol{q}}_B$, as in (4.92). With this formulation, there is a slight expense to pay, though, since the transform (2.121) and its time derivative have to be involved in the balance control component.

The motion optimization task is defined by combining the rate of change of the spatial momentum with the second-order contact constraints. We have

$$\begin{bmatrix} \boldsymbol{\mathcal{A}}_B \\ \boldsymbol{J}_{cB} \end{bmatrix} \ddot{\boldsymbol{q}}_B + \begin{bmatrix} \dot{\boldsymbol{\mathcal{A}}}_B \\ \dot{\boldsymbol{J}}_{cB} \end{bmatrix} \dot{\boldsymbol{q}}_B = \begin{bmatrix} \dot{\mathcal{L}}_B^{ref} \\ \boldsymbol{0} \end{bmatrix}. \tag{5.148}$$

The following compact-form representation of the above equation will be used in the derivations:

$$\overline{\boldsymbol{\mathcal{A}}}_B \ddot{\boldsymbol{q}}_B + \dot{\overline{\boldsymbol{\mathcal{A}}}}_B \dot{\boldsymbol{q}}_B = \dot{\overline{\mathcal{L}}}_B^{ref}, \tag{5.149}$$

where $\overline{\boldsymbol{\mathcal{A}}}_B$ and $\dot{\overline{\mathcal{L}}}_B^{ref}$ denote the coefficient matrix for the acceleration and the term on the r.h.s. in (5.148), respectively. The above equation is solved for the joint acceleration. Then we have

$$\ddot{\boldsymbol{q}}_B = \overline{\boldsymbol{\mathcal{A}}}_B^+ \left(\dot{\overline{\mathcal{L}}}_B^{ref} - \dot{\overline{\boldsymbol{\mathcal{A}}}}_B \dot{\boldsymbol{q}}_B \right) + \boldsymbol{N}(\overline{\boldsymbol{\mathcal{A}}}_B) \ddot{\boldsymbol{q}}_{Ba}. \tag{5.150}$$

Note that the solution is constraint-consistent and that the overbar notation is in agreement with the *constraint-consistent spatial momentum* introduced in Section 4.10 (cf. (4.154)). Note also that the two subtasks of controlling the rate of change of the spatial momentum and the motion constraints are arranged as a single control task, without any priority. This is reasonable since both subtasks play a crucial role in balance control.

Furthermore, the motion of the hand(s) along the unconstrained motion directions will be constrained by the time derivative of (2.95), i.e.

$$J_{mB}\ddot{q}_B + \dot{J}_{mB}\dot{q}_B = \dot{\bar{\mathcal{V}}}^m. \tag{5.151}$$

This constraint is treated as a lower-priority task w.r.t. the balance control task. In other words, the arbitrary $\ddot{q}_{Ba}$ in (5.150) is to be resolved via (5.151). The final solution is then to be obtained as

$$\ddot{q}_B = \ddot{q}_1 + \bar{J}_{mB}^{+}\left((\dot{\bar{\mathcal{V}}}^m)^{ref} - J_{mB}\ddot{q}_1\right) + N(\bar{\mathcal{A}}_B)N\left(\bar{J}_{mB}\right)\ddot{q}_{Ba} + h. \tag{5.152}$$

Hereby, $\ddot{q}_1 = \bar{\mathcal{A}}_B^{+}\dot{\bar{\mathcal{L}}}_B^{ref}$ is the solution to the highest-priority subtask (the balance control task), $\bar{J}_{mB} = J_{mB}N(\bar{\mathcal{A}}_B)$, and the nonlinear term

$$h = -\left(\left(E - \bar{J}_{mB}^{+}J_{mB}\right)\bar{\mathcal{A}}_B^{+}\dot{\bar{\mathcal{A}}}_B + \bar{J}_{mB}^{+}\dot{J}_{mB}\right)\dot{q}_B.$$

The reference end-link twist is determined in the conventional way (feedforward plus PD feedback control) as follows:

$$(\dot{\bar{\mathcal{V}}}^m)^{ref} = (\dot{\bar{\mathcal{V}}}^m)^{des} + K_{v_m}\left((\bar{\mathcal{V}}^m)^{des} - \bar{\mathcal{V}}^m\right) + K_{p_m}\left((\bar{\mathcal{X}}^m)^{des} - \bar{\mathcal{X}}^m\right), \tag{5.153}$$

$K_{\mathcal{V}_m} = \text{diag}(K_{v_m}E,\ K_{\omega_m}E)$ and $K_{p_m} = \text{diag}(K_{p_m}E,\ K_{o_m}E)$ denoting PD feedback gains. The arbitrary $\ddot{q}_{Ba}$ in (5.152) can be reused to impose further constraints, e.g. joint limits and damping.

5.12 NONITERATIVE WHOLE-BODY MOTION/FORCE OPTIMIZATION

The WD (force) and FSD (motion) optimization approaches discussed in the last two sections can be combined in the design of a whole-body motion/force optimization–based balance control methods.

5.12.1 Multicontact Motion/Force Controller Based on the Closed-Chain Model

Balance control at multicontact postures, i.e. when the end links form interdependent closed loops via a common closure link (the environment), is envisioned. The controller should also ensure the end-link trajectory tracking along the unconstrained motion directions, as well as contact transitions. The structure of such controller is presented in Fig. 5.29. Feasible (contact-consistent) desired task trajectories are assumed, s.t. the stabilizer generates a feasible reference rate of change of the spatial momentum $\dot{\mathcal{L}}_B^{ref}$. As already noted, asymptotic stability can then be ensured. In addition, the force and motion optimization tasks can be performed in parallel, as apparent from the figure. The unique solution to the JSID problem can then be obtained in a straightforward way, as explained in Section 5.9.1 (cf. (5.111)).

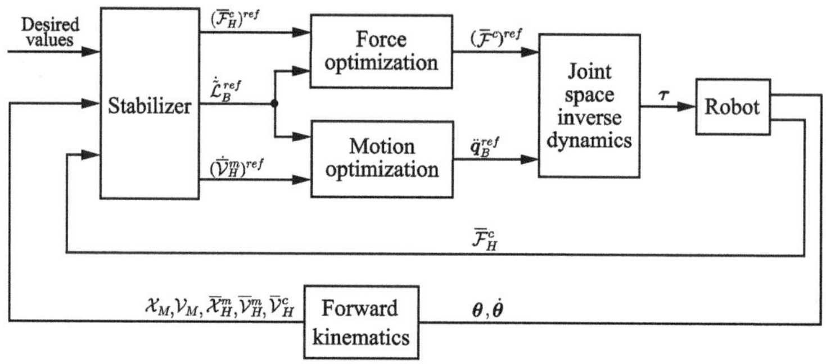

FIGURE 5.29 Outline of a motion/force controller based on the closed-chain formulation. The controller ensures balance control (also in the presence of unexpected external disturbances) while performing motion/force control tasks with the hands.

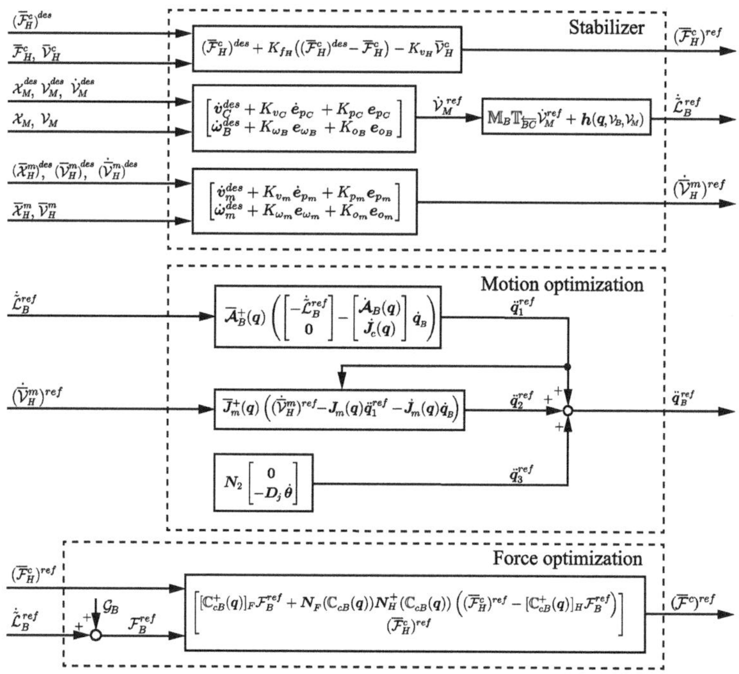

FIGURE 5.30 The equations used in the Stabilizer, Force and Motion optimization blocks in Fig. 5.29.

The equations used in the Stabilizer, Force, and Motion optimization blocks are shown in Fig. 5.30. Most of these equations have already been introduced in various parts of this text. Apparently, the Stabilizer comprises three control components. From the top to the bottom they are: (1) conventional force control in the constrained motion directions of the hands, (2) the asymptotic stabilization approach for balance control that was introduced in Section 5.8.7, and (3) conventional motion control in the unconstrained motion directions of the hands, as in (5.153). Note that in this particular implementation of the balance stabilizer, spatial momentum is represented in terms of the base coordinates, while the asymptotic stabilization controller is based on a mixed quasivelocity. A transformation is needed in this case. This transformation appears in the block on the right side of the Stabilizer.[11] The term $\boldsymbol{h}(\boldsymbol{q}, \mathcal{V}_B, \mathcal{V}_M) \approx \dot{\mathbb{T}}_{\overleftarrow{BC}} \mathcal{V}_M + \dot{\mathbb{M}}_B \mathcal{V}_B$ collects the nonlinear terms in the transformation.

Furthermore, the Motion optimization block in Fig. 5.30 also comprises three control components. From the top to the bottom they are: (1) motion resolution from the rate of change of the CRB momentum, (2) motion resolution along the unconstrained motion directions, and (3) a joint damping term that is used to suppress excessive motion stemming from kinematic redundancy. The term $\boldsymbol{N}_2$ denotes a projector onto the common null space of the two higher-priority subtasks (balance and end-link motion control). The equations used are those derived in Section 5.11.3. The Force optimization block solves the wrench distribution problem in parallel with motion optimization, via the quasistatic relation derived in Section 3.6.2 (cf. (3.66)).

The above controller is highly efficient from computational viewpoint and robust w.r.t. disturbances. The performance can be confirmed with the following example. The robot's task is to clean an inclined flat surface with the right hand. The hand moves in the sagittal plane along a straight-line trajectory in a cyclic manner, applying a desired force of magnitude 2 Nm in the normal (constrained motion) direction. The model of a small-size robot[12] is used in the simulation. The result is shown in Video 5.12-1 [138]. The data graphs are displayed in Fig. 5.31. As apparent from the right-arm position/orientation error and the hand-force graphs (the measured hand force is displayed in the local sensor frame), the controller is capable of performing the cleaning task accurately while keeping balance. The right-hand position/orientation errors can be further decreased by shifting the hand motion control task to the upper level in the hierarchy. In doing so, a controller comprising only two levels of hierarchy will be obtained: a higher-priority level for the motion control in the constrained and unconstrained motion directions and a lower-priority level for the joint damping control. It is also worth noting that this controller is capable of controlling the reaction to an external disturbance, in a compliant way via base-link motion. Thereby, the hand motion/force task performance degrades only slightly. This will be confirmed in Section 5.13.3.

5.12.2 Motion/Force Optimization Based on the Operational-Space Formulation

A hierarchical framework for a noniterative motion/force optimization based on the Operational-space formulation was developed in [134,133,135]. The motion/force tasks are embedded into the following quasistatic notation:

$$\boldsymbol{\tau} = \sum_i \boldsymbol{J}_i^T \mathcal{F}_i^{ref} + \boldsymbol{N}^*(\boldsymbol{\tau}^n)^{ref} + (\boldsymbol{\tau}^{int})^{ref}. \tag{5.154}$$

[11] An approximation is used in the implementation under the assumption that the CoM does not accelerate significantly in the case of a multicontact posture.

[12] The model parameters are similar to those of the HOAP-2 robot [26]. For the numbering of the joints and other relevant data, see Section A.2.

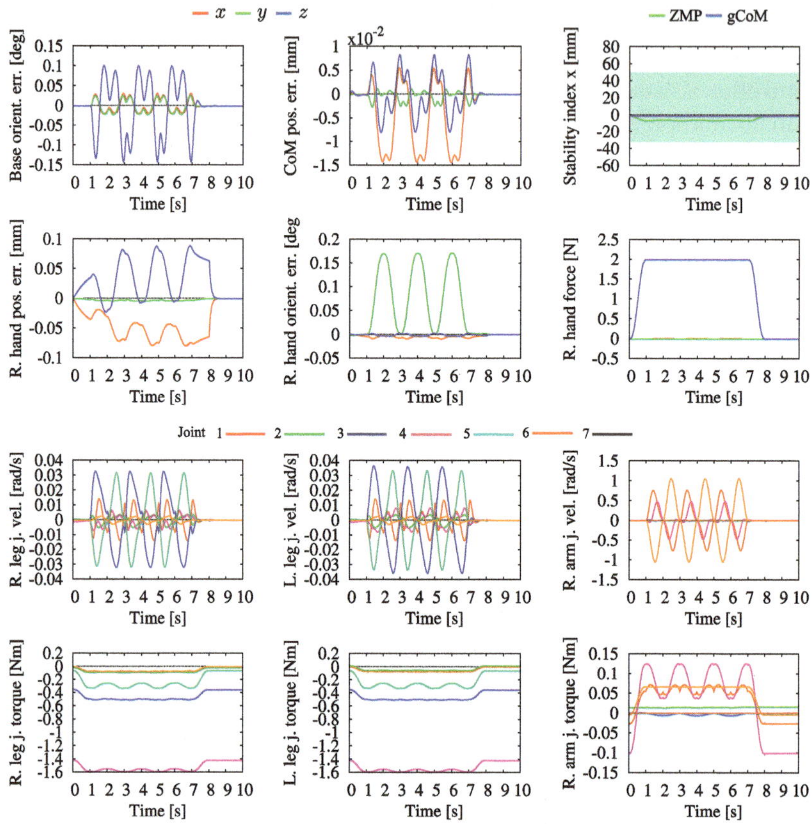

FIGURE 5.31 Data graphs from the simulation experiment with the multicontact motion/force controller, whereby the right hand cleans an inclined flat surface along a cyclic straight-line path by applying a desired force of 2 N. The desired CoM position and base orientation are constant (the initial values). R and L stand for "right" and "left," respectively. No external disturbance is present.

Here $\bar{\boldsymbol{J}}_i$ is the Jacobian for task i restricted by the null space of the contact constraints as well as by the null spaces of higher-priority tasks, $\boldsymbol{N}^*$ is a projector onto the intersection of all null spaces, and $\mathcal{F}_i^{ref}$, $(\boldsymbol{\tau}^n)^{ref}$, and $(\boldsymbol{\tau}^{int})^{ref}$ are reference inputs for motion/force control task i, postural variation control performed with the remaining DoFs, and internal force control, respectively.

The method provides a general platform for multicontact, whole-body balance control. The solution has been derived with the help of the dynamically consistent generalized inverse (cf. Section 4.52). Hence, the internal force controller can be designed in an independent way, without influencing the rest of the tasks. The controller adjusts the internal force to satisfy the friction cone and the CoP allocation constraints; $(\boldsymbol{\tau}^{int})^{ref}$ can also ensure adaptation to a dynamically varying environment, similarly to the pseudoinverse-based wrench distribution approach described in Section 5.10.1.

The method, however, has some shortcomings. First, since an off-line approach[13] is used to verify the feasibility of the solution, the method is unsuitable for real-time control. Second, since the solution to the WD problem is based on static relations only, the acceleration of the CoM is excessively restricted. The end-link CoPs are also allocated in a conservative manner, i.e. within the vicinity of the centers of the respective BoS.

Real-Time Implementation for Balance Control

A real-time implementation of the above method for balance control has been suggested in [89]. The proportional wrench distribution policy described in Section 5.10.3 is employed to ensure continuous double/single stance transitions, consistent with statics. A special task-priority assignment policy is adopted to handle overconstraints pertinent to the legs. Recall that such overconstraints stem from the spatial momentum/CoP interdependencies outlined in Section 5.8.1. The overconstraint problem can be tackled by introducing a two-level task hierarchy, T_{high} and T_{low}, denoting the high- and low-priority tasks, respectively. Eq. (5.154) is rewritten as

$$\boldsymbol{\tau} = \overline{\boldsymbol{J}}^T_{low}\mathcal{F}^{ref}_{low} + \boldsymbol{N}^*_{high}\boldsymbol{\tau}^{ref}_a + (\boldsymbol{\tau}^{int})^{ref}. \tag{5.155}$$

In [89], the high-priority task was set as the CoM motion in the vertical direction. Recall that CoM acceleration in this direction generates the vertical GRF that appears as denominator in the GRM equations; $\boldsymbol{\tau}^{ref}_a$ is determined in an appropriate way from the desired horizontal GRM components (i.e. the ankle torques). In this way, the CoP control that works even when the vertical GRF becomes zero[14] can be ensured. Note that the null-space CoP control term is not decoupled from the low-priority task term (the first term on the r.h.s.) since the null space of the high-priority task is different from $\mathcal{N}(\overline{\boldsymbol{J}}_{low})$. All low-priority control tasks, i.e. the CoM motion and base orientation control, will be disturbed by the null-space term. But this disturbance will be only a temporary one, under the assumption that the CoP control does not have to be active at all times. This becomes apparent from the flowchart of the balance controller shown in Fig. 5.32.

Example

The performance of the above balance controller will be demonstrated with a simple example. The robot is in a double stance, its feet being aligned. High friction contacts are assumed, s.t. $c_F = 12$. The goal is to shift the xCoM as much as possible forward and then to stabilize the posture without losing balance. This task can be accomplished with the simple controller in Fig. 5.25 that ensures asymptotic trajectory tracking of feasible desired CRB trajectories and the DCM-GI-based body wrench distribution approach. In this example, the null-space solution component in (5.127) does not need to be included since double stance on a flat floor is assumed and, thus, there is no need of internal wrench control.

[13] "Stability cloud" scanning.

[14] Recall that a zero vertical GRF is a necessary condition for continuous double/single-stance transitions, consistent with statics.

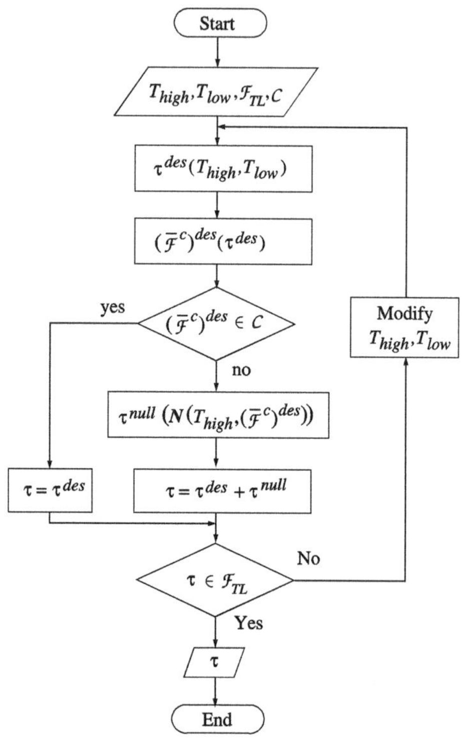

FIGURE 5.32 Flowchart of a real-time balance controller based on the Operational-space formulation [89]. The null-space of the high-priority task, T_{high}, is used to enforce the contact wrench constraints, s.t. $(\overline{\mathcal{F}}^c)^{des} \in CWC$. The tasks may have to be modified to avoid torque limit violation (the branch on the r.h.s.).

The controller in Fig. 5.25 is augmented with the high-priority control component $\boldsymbol{N}^*_{high}\boldsymbol{\tau}^{ref}_a$, whereby $\boldsymbol{N}^*_{high} = \boldsymbol{N}(\boldsymbol{\mathcal{J}}^T_{cM_Z})$, $\boldsymbol{\mathcal{J}}_{cM_Z} \in \Re^{1\times n}$ denoting the constraint Jacobian component in the vertical z-direction. This component is invoked in critical situations only, e.g. when the ZMP enters a predetermined safety margin within the BoS. The high-priority control component is capable of inducing slight torso rotations and CoM vertical displacements. As a result, the output of the "conventional stabilizer" will be modified to constrain any excessive torque in the ankles. The block diagram of the augmented controller is shown in Fig. 5.33.

The details of the implementation are as follows. The reference joint torque for the high-priority task is

$$\boldsymbol{\tau}^{ref}_a = -\overline{\boldsymbol{\mathcal{J}}}^T \boldsymbol{m}^{ref} \in \Re^n, \tag{5.156}$$

$$\boldsymbol{m}^{ref} = \mathbb{S}^{\times} K_p(\boldsymbol{r}^{SM}_p - \boldsymbol{r}_p) \in \Re^2. \tag{5.157}$$

Here $\boldsymbol{m}^{ref}$ denotes the reference pitch moments in the two ankles. These are obtained via tangential forces $K_p(\boldsymbol{r}^{SM}_p - \boldsymbol{r}_p)$, where $\boldsymbol{r}^{SM}_p$ is the predetermined BoS safety margin, K_p denoting

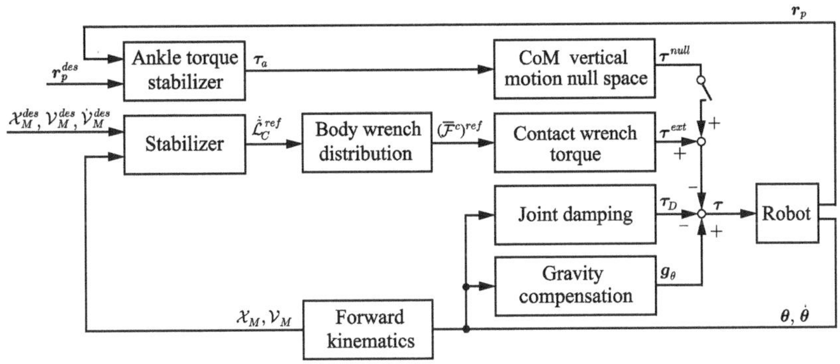

FIGURE 5.33 The control block-diagram of the simple balance control in Fig. 5.25 is augmented with the τ^{null} control component. This component is invoked in critical situations only, to ensure stabilization via slight upper-body postural variations, e.g. trunk rotation and CoM motion in the vertical direction. These variations result from the control algorithm; there is no need to preprogram them.

a p.d. feedback gain. The Jacobian

$$\overline{\mathcal{J}} = S_p {}^{M}\mathbb{X}_{F_j \overleftarrow{FM}} \mathcal{J}_{cM} \in \Re^{2\times n},$$

where

$$\begin{bmatrix} {}^{M}\mathbb{X}_{F_R \overleftarrow{FM}} & {}^{M}\mathbb{X}_{F_L \overleftarrow{FM}} \end{bmatrix} \in \Re^{6\times 12}.$$

Here $S_p \in \Re^{2\times 6}$ selects the pitch angular velocity components in the ankles. The graphs from the simulation are displayed in Fig. 5.34. The stability margin is set at 10 mm from the "toe" line of the BoS, as apparent from the CoP graphs (the pink area). The CoP/ZMP enters the margin at $t = 20$ s. As a result, the high-priority control component is invoked, resulting in a slight displacement of the CoM downward and a slight base rotation, as apparent from the respective graphs. The joint velocities change drastically but in a stable way. The pitch ankle torques are constrained and the posture can be stabilized without a fall. The plots in the bottom line display the pitch ankle torques and zoomed-in CoP motion for the right and left foot. The red/blue line graphs are for the cases when the high-priority control component is activated/deactivated. Note that the CoP trajectories are identical during most of the time upon entering the stability margin. At the end, the high-priority component stabilizes the posture at the "toe" boundary of the BoS, which is quite amazing. In contrast, without the activation, it is impossible to stabilize the posture. The simulation result is presented in animated form in Video 5.12-2 [33].

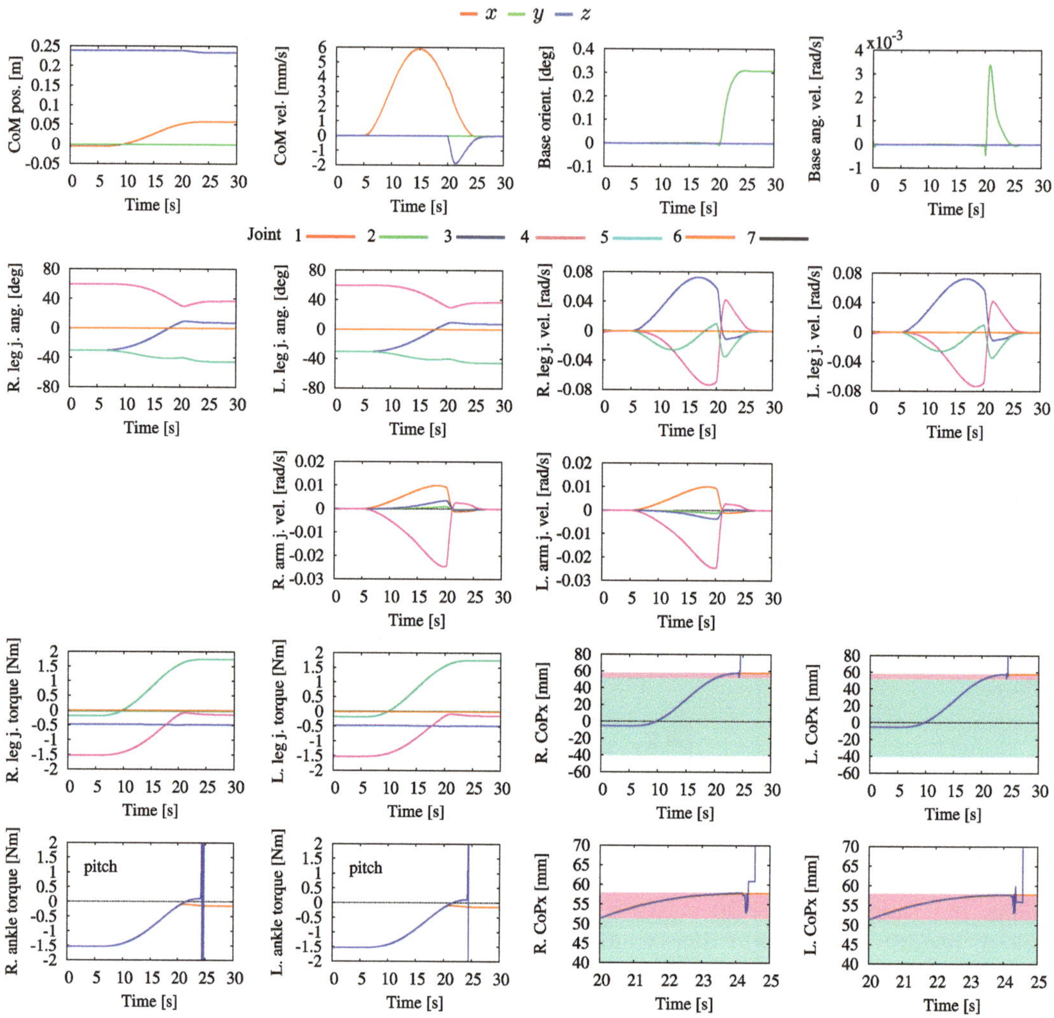

FIGURE 5.34 The high-priority τ^{null} component is invoked in a critical situation, i.e. when the CoPs enter the BoS safety margin (the pink area). The robot is then able to stabilize the posture at the edge of the BoS (apparent from the red lines in the bottom-row graphs) via a slight upper-body rotation and the CoM downward motion, as seen in the upper-row graphs. Such a stabilization results in limiting the ankle pitch torques, which is impossible to achieve without activating τ^{null} (see the blue lines in the bottom-row graphs).

5.13 REACTIVE BALANCE CONTROL IN RESPONSE TO WEAK EXTERNAL DISTURBANCES

An important feature of the balance controller is its robustness w.r.t. external disturbances. This issue was left out of the discussion so far. In Section 5.5 it was only mentioned that the LRWP model provides a possibility to increase the robustness via the angular momentum component. In Section 5.6.3 it was also mentioned that the DCM stabilizer is robust against a constant external disturbance. In this section, compliance-based control approaches that can ensure balance in the presence of relatively small external disturbances will be introduced. Balance control methods for accommodating larger disturbances will be introduced in Section 7.7.

5.13.1 Gravity Compensation–Based Whole-Body Compliance With Passivity

A whole-body compliant–type behavior is appropriate to accommodate relatively weak unknown external disturbances. Such type of behavior can be obtained with a humanoid robot controlled by a simple gravity compensating torque controller [63]. The control model is derived from the equation of motion in mixed quasicoordinates, (4.158). Using the notation $\boldsymbol{f}_C = \mathbb{C}_{cC_f}\mathcal{F}^c$, $\mathbb{C}_{cC_f}$ defined in (2.127), the CoM partial dynamics (given in the upper row of (4.158)) can be written as

$$M\dot{\boldsymbol{v}}_C = \boldsymbol{f}_C - \boldsymbol{g}_f, \tag{5.158}$$

$\boldsymbol{g}_f = M\boldsymbol{a}_g$. Hereby, $\mathbb{C}_{cC_f}\mathbb{C}^{\#}_{cC_f} = \boldsymbol{E}$ was used. To derive the control joint torque, consider the lower part of (4.158), rewritten as

$$\begin{bmatrix} \boldsymbol{I}_C & \boldsymbol{H}_C \\ \boldsymbol{H}_C^T & \boldsymbol{M}_{\theta_M} \end{bmatrix} \begin{bmatrix} \dot{\boldsymbol{\omega}}_B \\ \ddot{\boldsymbol{\theta}} \end{bmatrix} + \begin{bmatrix} \boldsymbol{c}_{mM} \\ \boldsymbol{c}_{\theta_M} \end{bmatrix} = \begin{bmatrix} \boldsymbol{0} \\ \boldsymbol{\tau} \end{bmatrix} + \begin{bmatrix} \mathbb{C}_{cC_m}\mathbb{C}^{\#}_{cC_f} - [\boldsymbol{r}^{\times}_{\overleftarrow{C}B}] \\ \mathcal{J}^T_{cM}\mathbb{C}^{\#}_{cC_f} - \boldsymbol{J}^T_{\overleftarrow{C}B} \end{bmatrix} \boldsymbol{f}_C. \tag{5.159}$$

The coefficient block matrices on the l.h.s. and r.h.s. will be denoted as $\boldsymbol{I}$ and $\boldsymbol{J}^T$, respectively. The control joint torque that compensates the gravity force can be derived from

$$\begin{bmatrix} \boldsymbol{0} \\ \boldsymbol{\tau} \end{bmatrix} = \boldsymbol{J}^T(\boldsymbol{f}_C^{des} - \boldsymbol{g}_f). \tag{5.160}$$

Note that, by assuming a quasistatic state, the appearance of nonlinear velocity terms has been avoided.[15] Furthermore, it is straightforward to show [63] that with

$$\boldsymbol{f}_C^{des} = \left(\boldsymbol{E} + (M\boldsymbol{J}\boldsymbol{I}^{-1}\boldsymbol{J}^T)^{-1}\right)\boldsymbol{f}_C^{ref} \tag{5.161}$$

the closed-loop system yields $\boldsymbol{f}_C \approx \boldsymbol{f}_C^{ref} + \boldsymbol{g}_f$ and, ultimately, $M\dot{\boldsymbol{v}}_C \approx \boldsymbol{f}_C^{ref}$. It is possible to avoid the inverse inertia term in (5.161) by using $\boldsymbol{f}_C^{ref}$ in place of $\boldsymbol{f}_C^{des}$ in (5.160). Then, with $\boldsymbol{f}_C^{ref} = \boldsymbol{0}$, the joint torque merely compensates the gravity force, conserving thereby the linear momentum.

On the other hand, noting that Jacobian $\boldsymbol{J}$ is nonsquare, a null-space term needs to be added in (5.160). This term will induce internal motion (or self-motion), though, that needs to be suppressed. The simplest way to do this is to add joint damping. The closed-loop system dynamics then assume the following form:

$$M\dot{\boldsymbol{v}}_C = \boldsymbol{f}_C^{ref} + \boldsymbol{f}_D + \boldsymbol{f}_{ext}, \tag{5.162}$$

[15] Such terms can cause problems stemming from noisy signal readings.

$$I\ddot{q}_\omega + C = J^T f_C^{ref} - D\dot{q}_\omega, \tag{5.163}$$

where $\dot{q}_\omega = \begin{bmatrix} \omega_B^T & \dot{\theta}^T \end{bmatrix}^T$, C collects the nonlinear velocity-dependent terms, D is the damping matrix, and f_D is the damping force mapped at the CoM. Note also that the external force f_{ext} has been added. This force is used as the driving input during the reflex phase. Thereby, the system passivity will be ensured, s.t.

$$\frac{1}{2}Mv_C^T v_C + \frac{1}{2}\dot{q}_\omega I \dot{q}_\omega^T = \frac{1}{2}\dot{q}_\omega^T(MJ^T J + I)\dot{q}_\omega \leq \int (f_C^{ref})^T J\dot{q}_\omega dt + \int f_{ext}^T J\dot{q}_\omega dt \tag{5.164}$$

holds. After the disturbance, a recovery phase control is invoked based on a simple PD feedback control law. We have

$$f_C^{ref} = -K_D(\dot{r}_C - \dot{r}_C^{des}) - K_P(r_C - r_C^{des}). \tag{5.165}$$

Here K_D and K_P denote the feedback gains. The CoM initial position, r_C^{init}, is recovered with the settings $r_C^{des} = r_C^{init}$ and $\dot{r}_C^{des} = 0$. The desired ZMP can be determined from the above f_C^{ref} in a straightforward manner, under the assumption of a negligible angular momentum. Finally, the CoP-in-BoS constraint is imposed on the desired ZMP thus determined.

The above balance controller has the advantage of simplicity and robustness stemming from passivity. There is no need of a precise dynamic model of the robot. The performance of the controller was experimentally examined in combination with the noniterative wrench distribution technique described in Section 5.10.1. The quasistatic and wrench distribution approaches used in the controller have been confirmed to yield robust whole-body postural variations in response to relatively weak disturbances applied at arbitrary parts of the body [63]. The robot was also capable of adaptation to uneven terrain [62]. The controller does not provide any means for reactive stepping under stronger disturbances, however.

The performance of the controller can be improved in two main aspects [114]. First, it is straightforward to add the capability of dealing with external moment–type disturbances. This can be done by determining a reference CRB moment derived from a PD feedback body-orientation control law, using suitable local coordinates (e.g. the quaternion notation). Second, instead of generalized inverse–based optimization, the WD problem could be handled with iterative optimization. In this way, the GRF/GRM/CoP interdependencies (cf. Section 5.8.1) can be resolved by involving soft constraints, using sequential or nonsequential optimization, e.g. as described in Section 5.14.4. In this case, there is no need to determine a desired ZMP, which leads to an advantage as clarified with the DCM generalized inverse method in Section 5.10.4. In [114], one of the soft constraints in the iterative optimization was selected to minimize the Euclidean norm of the contact forces. This, however, leads to equal distribution which inhibits reactive stepping, as already clarified.

5.13.2 Whole-Body Compliance With Multiple Contacts and Passivity

Multicontact balance control refers to the case when the robot contacts the environment with one or two hands or any other intermittent links, while in a single or double stance. A number of tasks exist that require multicontact balance control, e.g. painting a wall [13], cleaning a surface with one of the hands while supporting the weight with the other [135], sitting on a chair [90], ladder climbing [110,16,24,153], pivoting heavy objects [102,104], pushing heavy objects with shoulder–elbow or hip [103], and stair climbing with handrail support [80,83]. Multicontact balance control implies a closed-chain formulation with a single loop-closure link (the environment), whereby interdependent closed loops are formed (cf. Section 3.6).

A possible approach to design a balance controller for the case of multiple contacts is to make use of a dynamic model in terms of spatial (end-link) coordinates (cf. Section 4.228). As already discussed, the representation of dynamics in mixed quasicoordinates is quite convenient for balance control. The underlying velocity relation is derived from (2.131) as

$$\begin{bmatrix} \mathcal{V}_M \\ \mathcal{V} \end{bmatrix} = \begin{bmatrix} \boldsymbol{E} & \boldsymbol{0} \\ \mathbb{C}_C^T & \mathcal{J}_M \end{bmatrix} \begin{bmatrix} \mathcal{V}_M \\ \dot{\boldsymbol{\theta}} \end{bmatrix}. \tag{5.166}$$

The 4×4 block matrix on the r.h.s., denoted as $\boldsymbol{T}_M$, is used as a transformation matrix for the equation of motion in accordance with the transformation procedure described in Section 4.228 (see also [38]). The original transformation matrix, $\boldsymbol{T}_B$, is thereby replaced with $\boldsymbol{T}_M$. After the transformation, the equation of motion assumes the following form:

$$\mathcal{M}_M(\boldsymbol{q}) \begin{bmatrix} \dot{\mathcal{V}}_M \\ \dot{\mathcal{V}} \end{bmatrix} + \mathcal{C}_M(\boldsymbol{q}, \dot{\boldsymbol{q}}) \begin{bmatrix} \mathcal{V}_M \\ \mathcal{V} \end{bmatrix} + \boldsymbol{g}(\boldsymbol{q}) = \begin{bmatrix} -\mathbb{C}_C \\ \boldsymbol{E} \end{bmatrix} \mathcal{J}_M^{-T} \boldsymbol{\tau} + \boldsymbol{T}_M^{-T} \mathcal{Q}_{ext}. \tag{5.167}$$

The following remarks are due. First, note that to support a controller design, the nonlinear velocity-dependent term has been recast into a vector matrix form. Second, note that the gravity term is unaltered ($\boldsymbol{g}_M(\boldsymbol{q}) = \boldsymbol{g}(\boldsymbol{q})$) since, with mixed quasicoordinates, the CoM dynamics are decoupled from the rest. Third, observe that the transformation involves the inverse of $\boldsymbol{T}_M$, which implies that the Jacobian $\mathcal{J}_M$ is assumed square (nonredundant limbs) and nonsingular. Finally, note that an external force term has been added (the last term on the r.h.s.) to account for unexpected perturbations.

Furthermore, the last equation can be structured via the notation of constrained/unconstrained motion directions at the contact joints. Then, the first terms on the l.h.s. and the r.h.s. become

$$\mathcal{M}_M(\boldsymbol{q}_M) \begin{bmatrix} \dot{\mathcal{V}}_M \\ \dot{\overline{\mathcal{V}}}^c \\ \dot{\overline{\mathcal{V}}}^m \end{bmatrix} \quad \text{and} \quad \begin{bmatrix} -\mathbb{C}_C \mathcal{J}_M^{-T} \\ \mathcal{J}_{Mc}^{-T} \\ \mathcal{J}_{Mm}^{-T} \end{bmatrix} \boldsymbol{\tau} \equiv \begin{bmatrix} \mathcal{F}^M \\ \overline{\mathcal{F}}^c \\ \overline{\mathcal{F}}^m \end{bmatrix}, \tag{5.168}$$

respectively. This structuring supports the three basic control tasks: the CRB (i.e. momentum rate) motion control, the contact wrench force control (i.e. the wrench distribution problem), and the end-link motion control along the unconstrained motion directions.

The desired closed-loop behavior is determined as [38]

$$\mathcal{M}_M \begin{bmatrix} \Delta\dot{\mathcal{V}}_M \\ \boldsymbol{0} \\ \Delta\dot{\overline{\mathcal{V}}}^m \end{bmatrix} + \mathcal{C}_M \begin{bmatrix} \Delta\mathcal{V}_M \\ \boldsymbol{0} \\ \Delta\overline{\mathcal{V}}^m \end{bmatrix} = \boldsymbol{T}_M^{-T} \mathcal{Q}_{ext} - \begin{bmatrix} \mathcal{F}_u^M \\ \overline{\mathcal{F}}_u^c \\ \overline{\mathcal{F}}_u^m \end{bmatrix}, \tag{5.169}$$

where the subscript u stands for the control input and $\Delta(\circ)$ denotes an error term. The CRB ($\mathcal{F}_u^M$) and end-link motion ($\overline{\mathcal{F}}_u^m$) control inputs comprise PD feedback components. This implies that the motion of the CRB is controlled via a virtual linear/torsional spring-damper system. The end-link motion control input comprises also a desired feedforward (inertia force) component. The resulting end-link behavior is of impedance type. Note that the motion along the constrained motion directions is inhibited and, therefore, $\Delta\overline{\mathcal{V}}^c$ has been assumed zero.

The contact wrenches, $\overline{\mathcal{F}}_u^c$, can be determined from the spatial dynamics. To this end, first note that from (5.167) and (5.169), one obtains

$$\begin{bmatrix} -\mathbb{C}_C \\ \boldsymbol{E} \end{bmatrix} \mathcal{J}_M^{-T}\boldsymbol{\tau} = \boldsymbol{\mathcal{M}}_M \begin{bmatrix} \dot{\mathcal{V}}_M^{des} \\ \mathbf{0} \\ (\dot{\overline{\mathcal{V}}}^m)^{des} \end{bmatrix} + \boldsymbol{\mathcal{C}}_M \begin{bmatrix} \mathcal{V}_M^{des} \\ \mathbf{0} \\ (\overline{\mathcal{V}}^m)^{des} \end{bmatrix} + \boldsymbol{g} - \begin{bmatrix} \mathcal{F}_u^M \\ \overline{\mathcal{F}}_u^c \\ \overline{\mathcal{F}}_u^m \end{bmatrix}. \tag{5.170}$$

Next, the above equation is decomposed into two parts in accordance with the projector in the joint torque term. Then we have

$$-\mathbb{C}_C\left(\mathcal{J}_M^{-T}\boldsymbol{\tau}\right) = \boldsymbol{\mathcal{M}}_{M1} \begin{bmatrix} \dot{\mathcal{V}}_M^{des} \\ \mathbf{0} \\ (\dot{\overline{\mathcal{V}}}^m)^{des} \end{bmatrix} + \boldsymbol{\mathcal{C}}_{M1} \begin{bmatrix} \mathcal{V}_M^{des} \\ \mathbf{0} \\ (\overline{\mathcal{V}}^m)^{des} \end{bmatrix} + \begin{bmatrix} M\boldsymbol{a}_g \\ \mathbf{0} \end{bmatrix} - \mathcal{F}_u^M, \tag{5.171}$$

$$\mathcal{J}_M^{-T}\boldsymbol{\tau} = \boldsymbol{\mathcal{M}}_{M2} \begin{bmatrix} \dot{\mathcal{V}}_M^{des} \\ \mathbf{0} \\ (\dot{\overline{\mathcal{V}}}^m)^{des} \end{bmatrix} + \boldsymbol{\mathcal{C}}_{M2} \begin{bmatrix} \mathcal{V}_M^{des} \\ \mathbf{0} \\ (\overline{\mathcal{V}}^m)^{des} \end{bmatrix} - \mathcal{F}_u^*, \tag{5.172}$$

where $\mathcal{F}_u^* \equiv \left[(\overline{\mathcal{F}}_u^c)^T \quad (\overline{\mathcal{F}}_u^m)^T\right]^T$. To obtain a representation in terms of spatial dynamics, eliminate the joint torque from the last two equations, i.e.

$$\mathbb{C}_C\mathcal{F}_u^* = (\boldsymbol{\mathcal{M}}_{M1} + \mathbb{C}_C\boldsymbol{\mathcal{M}}_{M2}) \begin{bmatrix} \dot{\mathcal{V}}_M^{des} \\ \mathbf{0} \\ (\dot{\overline{\mathcal{V}}}^m)^{des} \end{bmatrix} + (\boldsymbol{\mathcal{C}}_{M1} + \mathbb{C}_C\boldsymbol{\mathcal{C}}_{M2}) \begin{bmatrix} \mathcal{V}_M^{des} \\ \mathbf{0} \\ (\overline{\mathcal{V}}^m)^{des} \end{bmatrix} + \begin{bmatrix} M\boldsymbol{a}_g \\ \mathbf{0} \end{bmatrix} - \mathcal{F}_u^M. \tag{5.173}$$

The driving force of the spatial dynamics, $\mathcal{F}_u^M$, is composed by PD feedback control terms, as already noted. The only unknown quantity is the contact wrench, $\overline{\mathcal{F}}_u^c$. It is easy to confirm that (5.173) is underdetermined w.r.t. the contact wrench. Hence, to solve the wrench distribution problem, optimization should be invoked.

The optimization approach suggested in [38] is based on soft constraints. The objective function minimizes the sum of three quadratic (penalty) terms determined as residuals w.r.t. $(\overline{\mathcal{F}}^c)^{des}$, $\overline{\mathcal{F}}_u^m$ and (5.173), over $\boldsymbol{x} = \mathcal{F}^*$. The constraints are composed of the $\mathfrak{F}_{FC}$, $\mathfrak{F}_{BoS}$, and $\mathfrak{F}_{TL}$ subdomains. To maintain the contacts, minimal vertical GRFs are required as an additional constraint. Note that the contact wrench penalty term is defined w.r.t. the desired values. In this way, the need of contact wrench measurement for balance can be avoided. The unique solution for the control joint torque[16] is derived from (5.172) as

$$\boldsymbol{\tau} = \mathcal{J}_M^T \left(\mathcal{M}_{M2} \begin{bmatrix} \dot{\mathcal{V}}_M^{des} \\ \mathbf{0} \\ (\dot{\bar{\mathcal{V}}}^m)^{des} \end{bmatrix} + \mathcal{C}_{M2} \begin{bmatrix} \mathcal{V}_M^{des} \\ \mathbf{0} \\ (\bar{\mathcal{V}}^m)^{des} \end{bmatrix} - \mathcal{F}_{opt}^* \right). \tag{5.174}$$

This expression is also used in the joint torque-limit constraint, $\mathcal{F}_{TL}$. From the above derivations, it is apparent that the optimization approach can be characterized as a nonsequential, spatial dynamics–based one.

The controller thus designed has the property of asymptotic stability and passivity. To show this, note first that under the assumption of maintaining the balancing contact wrenches, the dimension of the system can be reduced. This is done via the constraint elimination approach. The closed-loop system of reduced dimension becomes

$$\mathcal{M}_M^* \begin{bmatrix} \Delta\dot{\mathcal{V}}_M \\ \Delta\dot{\bar{\mathcal{V}}}^m \end{bmatrix} + \mathcal{C}_M^* \begin{bmatrix} \Delta\mathcal{V}_M \\ \Delta\bar{\mathcal{V}}^m \end{bmatrix} = \boldsymbol{T}^* \boldsymbol{T}_M^{-T} \mathcal{Q}_{ext} - \begin{bmatrix} \mathcal{F}_u^M \\ \bar{\mathcal{F}}_{opt}^m \end{bmatrix} \tag{5.175}$$

or

$$\mathcal{M}_M^* \begin{bmatrix} \Delta\dot{\mathcal{V}}_M \\ \Delta\dot{\bar{\mathcal{V}}}^m \end{bmatrix} + \mathcal{C}_M^* \begin{bmatrix} \Delta\mathcal{V}_M \\ \Delta\bar{\mathcal{V}}^m \end{bmatrix} + \begin{bmatrix} \mathcal{F}_u^M \\ \bar{\mathcal{F}}_u^m \end{bmatrix} = \boldsymbol{T}^* \boldsymbol{T}_M^{-T} \mathcal{Q}_{ext} - \begin{bmatrix} \mathbf{0} \\ (\bar{\mathcal{F}}^m)^{des} \end{bmatrix}, \tag{5.176}$$

where $\boldsymbol{T}^*$ is a transform that removes the middle (contact wrench) row from the equation of motion. Apparently, the difference term on the r.h.s. drives the error dynamics. Asymptotic stability in the sense of PD+ control [115] can be confirmed for the case of unconstrained motion, i.e. when the driving force is zero. Passivity can be confirmed for the regulation case, in accordance with [113].

The above controller has been implemented with a torque-controlled robot and shown to perform well under multicontacts, or when standing on soft ground and subjected to small external perturbations. Note that motion redundancy was not considered in the above derivations. In the implementation, the redundancy was resolved via a joint-space impedance component in the null space of the transposed Jacobian. Another problem is the assumption of nonsingular configurations. In [38], it was suggested to use the damped least-squares regularization method. This method leads to deteriorated tracking in the vicinity of a singular configuration, though (cf. also the discussion in Chapter 2).

5.13.3 Multicontact Motion/Force Control With Whole-Body Compliance

The multicontact balance controller described in Section 5.12.1 has the capability to react to weak disturbances. This is demonstrated with the same task (surface cleaning) as in the example in Section 5.12.1. While performing the task, the robot is subjected to two unexpected disturbances of 500 ms duration each. First, a pitch moment of magnitude 3 Nm is applied at the base link. This disturbance is accommodated by the torso rotation, as apparent from the

[16] Recall the assumption for a square Jacobian and, hence, no kinematic redundancy. This assumption will be dropped later on.

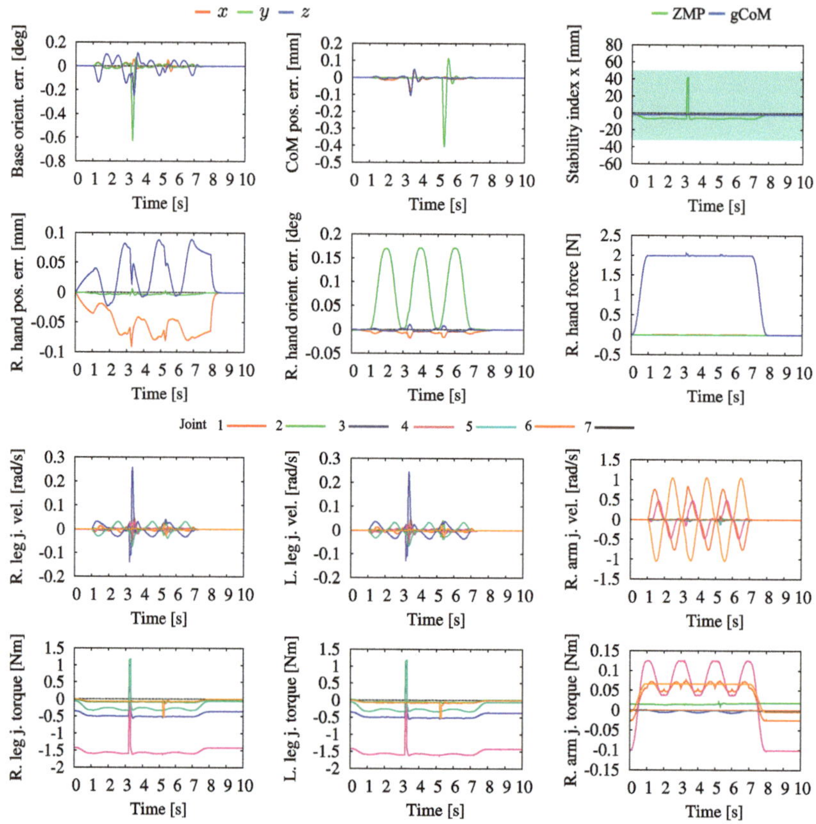

FIGURE 5.35 Multicontact motion/force control task as in Section 5.12.1. Two unexpected disturbances are applied; first, a pitch moment of magnitude 3 Nm at the base link and then, a lateral push of magnitude 5 N at the CoM. The disturbances are accommodated with a torso rotation and a CoM displacement, respectively. The deterioration of the hand motion/force task performance by the disturbances is negligible. R and L stand for "right" and "left," respectively.

base orientation error graph in Fig. 5.35. Next, a lateral push of magnitude 5 N is applied at the CoM. This disturbance is accommodated via the CoM displacement, as apparent from the CoM position error graph and the joint velocity and torque graphs in the same figure. Note that the right-hand position/orientation errors and force differ insignificantly from those obtained without disturbance (cf. Fig. 5.31). This demonstrates the decoupling capability of the controller. The simulation result is shown in animated form in Video 5.13-1 [137].

5.14 ITERATIVE OPTIMIZATION IN BALANCE CONTROL

A number of balance control methods have been developed throughout the years that make use of iterative convex optimization techniques. The main advantage is that inequality constraints can be directly included into the optimization task. At present, off-the-shelf QP solvers are preferably used.

The general form of a QP task formulation that includes both equality- and inequality-type constraints is given in (2.57). In the case of balance control, the objective is to minimize a sum of appropriately defined linear-quadratic functions, in a form similar to that used with noniterative optimization (e.g. as in (5.130)). A variety of task formulations for balance control have been implemented so far. There are two main concerns when formulating the iterative optimization task. One of them is the computational load. In this aspect, it is preferable to formulate a few smaller tasks than a large one, as already mentioned in Section 5.9. This approach is referred to as *sequential* optimization [60]. Examples of such optimization can be found in [13,14,88]. Note that sequential optimization can include mixed iterative/noniterative minimization tasks [142,167,38,37]. In this way, the computational burden can be further decreased due to the inherent time efficiency of the noniterative tasks.

The other main concern is the lack of a solution because of the presence of multiple contradicting constraints. There are three possible approaches to avoid such overconstraints. First, an equality-type constraint relaxation could be employed, as discussed in Section 5.9.2. Second, the so-called *slack variables* could be employed (cf. Section 2.60). The role of these variables is also a constraint relaxation, especially in the case of inequality-type constraints. Recall (from (2.61)) that the variables themselves are subjected to the minimization. Third, an order of priority could be introduced among the constraints. When compared to weight tuning, the prioritization approach yields the advantage of decoupled minimization tasks. In this way, stability can be increased, as reported in [18]. The prioritization approach, however, has also some drawbacks. Fixing the priorities may not always guarantee good results. Introducing variable priorities, on the other hand, is not that straightforward either (cf. the discussion in Section 2.8.3).

As a consequence of the ambiguity in weight tuning/priority assignment, it is difficult to prove the performance stability of an iterative optimization method. The tuning process could be cumbersome and special care could be needed to eventually obtain satisfactory performance. It should be noted, though, that despite this problem a number of studies have been based on the iterative optimization. Representative examples will be given in what follows.

5.14.1 A Brief Historical Overview

One of the first balance controllers based on iterative optimization was introduced in [151, 65] for a proactive single-leg balancing task. The objective function is formulated in terms of the joint angle increments. The joint-angle input trajectory is modified based on the result of the optimization. The optimization makes use of constraints $\mathcal{F}_{JT}$, $\mathcal{F}_{CoP}$, and $\mathcal{M}_{SM}$. Two components from $\mathcal{M}_{SM}$ are used to restrain the gCoM motion. Note that this constraint is quite conservative—it restricts the solution space of the optimizer excessively by not allowing the gCoM to move within the BoS. The three angular momentum components in $\mathcal{M}_{SM}$ are subjected to inequality-type constraints to avoid excessive accumulation of angular momentum.

The balance controller in [81] makes use of two types of control laws to ensure an appropriate reactive behavior in response to external disturbances applied while standing in a symmetric double-stance posture. In the case of a relatively small disturbance, a PD feedback control law is preferably used. The advantage of this controller is its suitability for real-time implementation. This controller, however, is inadequate in the case of larger disturbances. Such disturbances are dealt with an QP optimization controller. The objective function minimizes the cost of the joint acceleration, $c_{JA}|_{\ddot{\theta}^{ref}=0}$, and maximizes the CoM acceleration in the lateral direction, to avoid falls. The constraints are $\mathcal{F}_{BoS}$ and $\mathcal{M}_{AL}$, along with a PD-type restraint of gCoM motion, with components from $\mathcal{F}_{SM}$. There is also a specific constraint that makes use of symmetry relations within the closed-loop chain formed by the double stance. Human-like balance motion patterns can be achieved with this balance controller, including the use of an arm ("windmilling") motion.

The balance controller described in [144] is also based on the QP approach. The input trajectories are specified in terms of the joint angles, CoM/ZMP motion, and the vertical component of the GRF, in a consistent way. The task of the optimizer is to ensure stabilization by taking into account the ZMP/gCoM dynamics when reacting to relatively small, short-term disturbances. The resulting long-term deviation from the CoM input trajectory is compensated in a feedback manner with PD feedback gains. Effectively, the method can be regarded as an indirect ZMP manipulation approach. The joint variables are obtained from the output of the QP optimizer that uses the CoM inverse kinematics and other kinematic subtasks as constraints.

5.14.2 SOCP-Based Optimization

A family of balance control algorithms can be formulated as *second-order cone programming* (SOCP) problems, i.e. as a special class of convex optimization problems. Recall that the SOCP formulation (cf. (3.79)) is especially suitable when enforcing the friction cone constraints. In [117], examples of three convex optimization subtasks are given. First, the CoP-in-BoS constraint is defined as a convex minimization problem by requiring that $\|\boldsymbol{r}_p - \boldsymbol{r}_p^{des}\| \leq \epsilon_p$, ϵ_p being a user-defined constant threshold. Second, the gCoM-in-BoS constraint is defined in the form of a linear inequality obtained from the finite-difference approximation of the current gCoM. Third, the closed-loop kinematic relations are enforced as equality-type constraints. In this way, a dynamic environment (e.g. translating and/or rotating support) can be included into the model. The optimizer performed well in both proactive (a kicking motion) and reactive (support displacement) balance control tasks.

The SOCP approach seems quite promising. Different types of constraints can be mixed: nonlinear and linear, as well as equality- and inequality-type ones. There are problematic issues, though, similar to those pertinent to the QP solvers. The most important is the difficulty in proving the stability of the balance controller. There are a number of parameters that need to be set in a consistent way. The sequential approach above provides a nice solution. Other parameters, notably those related to the feedback-type controller performance (e.g. the ϵ_p error threshold in the first SOCP controller in this section and the PD gains for the CoM regulator in Section 5.14.1) can only be tuned in an intuitive way. Furthermore, the real-time implementation of an SOCP controller is still problematic. This problem is expected to be alleviated in the near future with the development of faster SOCP algorithms [6], in combination with high-performance computational hardware.

5.14.3 Iterative Contact Wrench Optimization

The contact wrench optimization problem can be cast as a QP task, with solution $\boldsymbol{x} = \overline{\mathcal{F}}^c$. The friction cone and the CoP inequality-type constraints can be directly included into the task formulation. The optimal contact wrench components, $\overline{\mathcal{F}}^c_{opt}$, will determine the best location of the CoPs together with appropriate reaction forces and a vertical moment at each foot. In this way, a foot slip, transitions from plane to line or point contacts via a foot roll and/or pitch rotations as well as torsional foot (yaw) rotation can be avoided.

A reasonable objective is the minimization of the ankle torques. The motivation behind this objective is threefold: (1) energy efficiency,[17] (2) minimization of the performance requirements on the ankle actuators [158], and (3) avoidance of foot roll [88]. A "generic" formulation of the problem is obtained from (5.115) as [158]

$$\min_{\boldsymbol{x}} \sum_{j\in\{r,l\}} \|\overline{\boldsymbol{\tau}}^c_j - (\boldsymbol{r}_{A_j} - \boldsymbol{r}_{P_j}) \times \overline{\boldsymbol{f}}^c_j\|^2. \tag{5.177}$$

The constraints are $\mathcal{F}_{SD}$, $\mathcal{F}_{FC}$, and $\mathcal{F}_{BoS}$, as defined in Section 5.9.2. On flat ground, an additional constraint should be included to restrict the vertical coordinate of the CoP to zero.

There are two problems associated with the above optimization. First, the formulation is *nonconvex* due to the vector cross-product. Second, there is no guarantee that the CoPs will be located away from the boundaries of the BoS, to avoid a foot roll. The first problem can be tackled by resorting to a point contact model, similar to that described in Section 5.10.1; see also [50,2,63,168]. In this case, the vector cross-product is determined from the position vectors of the point contacts. These are assumed constant during the optimization time interval. Thus a skew-symmetric matrix representation of the vector cross-product can be employed. The objective function can then be represented as a semidefinite quadratic form that renders the optimization problem convex, i.e.

$$\min_{\boldsymbol{x}} c_{CW}, \tag{5.178}$$

with c_{CW} determined as the relaxation of $\mathcal{F}_{CW}|_{(\overline{\mathcal{F}}^c)^{ref}=0}$. The same constraints as above are used. Recall that when the weight matrix $\boldsymbol{W}_{CW}$ is defined as in (5.133), the objective function assumes the form $c_{CW} = \|\boldsymbol{\tau}^c\|^2_{W_\kappa}$. This implies that the contact wrenches can be minimized implicitly, via the joint torque.

The second problem can be tackled by amending the objective function with the relaxation of the CoP tracking constraint:

$$\mathbb{C}_P \overline{\boldsymbol{\tau}}^c_t = \boldsymbol{r}^{ref}_P \tag{5.179}$$

to obtain

$$\min_{\boldsymbol{x}} c_{CW} + c_{CoP}. \tag{5.180}$$

[17] The equivalent of the metabolic cost/energy [3,15] in the field of biomechanics.

With (5.179), the tangential contact moments $\bar{\tau}_t^c$ are mapped via the $\mathbb{C}_P$ to obtain the current CoPs. The current CoPs are then penalized via the c_{CoP} at the desired CoPs, $\boldsymbol{r}_P^{ref}$. The usual, though somewhat conservative setting for $\boldsymbol{r}_P^{ref}$ is at the ground projection of the ankle joint. In this case, there is no need to use the CoP-in-BoS constraint, $\mathcal{F}_{BoS}$, anymore. There are also other choices for c_{CoP} that can avoid the conservative setting problem, e.g. to more directly penalize possible transitions to edge contacts [158].

In some cases, there may be no solution to the optimization task at all. This problem can be alleviated via the relaxation of the strict spatial dynamics equality constraint $\mathcal{F}_{SD}$ [142]. We have

$$\min_{\boldsymbol{x}} c_{CW} + c_{SD}. \tag{5.181}$$

With this formulation, there will always be a solution but the acceleration component in $\mathcal{F}_C^{ref}$ (i.e. the rate of change of the spatial momentum) cannot be guaranteed anymore.

5.14.4 Iterative Spatial Dynamics Optimization

It is possible to employ an iterative approach to optimize the contact wrenches *together with* the joint accelerations appearing in the spatial dynamics equation, (5.109), with the solution $\boldsymbol{x} = \begin{bmatrix} \ddot{\boldsymbol{\theta}}^T & (\bar{\mathcal{F}}^c)^T \end{bmatrix}^T$. Examples of sequential as well as nonsequential optimization task formulations will be presented below.

Sequential Approach

The following objective function is an example of a spatial dynamics–based optimization [88]:

$$\min_{\boldsymbol{x}} \sum_{j \in \{r,l\}} \left(\|\bar{\boldsymbol{f}}_j^c\|_{W_f}^2 + \|\bar{\boldsymbol{\tau}}_j^c\|_{W_\tau}^2 \right) + \|\dot{\boldsymbol{p}}(\boldsymbol{f}^c) - \dot{\boldsymbol{p}}^{ref}\|_{W_p}^2 + \|\dot{\boldsymbol{l}}_C(\bar{\mathcal{F}}^c) - \dot{\boldsymbol{l}}_C^{ref}\|_{W_l}^2. \tag{5.182}$$

There are four subtasks. The first two minimize the contact wrench components, i.e. the GRFs and the GRMs (or equivalently, the CoP location/ankle torques). The last two minimize the errors in the linear and angular momentum rate components. It is assumed that the reference momentum rate components, $\dot{\boldsymbol{p}}^{ref}$ and $\dot{\boldsymbol{l}}_C^{ref}$, are specified in an independent way, via a proportional-derivative CoM motion and proportional angular momentum feedback control, respectively. This poses a problem, however, since the two momentum rate components are coupled, as noted in Section 5.8.1. This is the reason why the above objective attempts to minimize the two components in an independent way. With regard to the GRF and GRM minimization subtasks, it should be noted that the former restricts the application field of the method since the foot loading/unloading will be rendered impossible, while the latter will render the formulation nonconvex, because of the GRF cross-product terms (cf. (5.115)).

To alleviate the inherent problems in the above optimization, the optimization task is reformulated as follows. A possible solution to the nonconvexity problem was presented in Section 5.14.3. Thereby, a contact model that approximates the plane contact with multipoint contacts was used. Here, another possibility will be explored that avoids the use of a multipoint contact model. To this end, the objective function will be reformulated to yield a sequential problem: first minimize the GRFs and then the GRMs [88]. The GRFs are determined by the following minimization task:

$$\min_{\boldsymbol{x}} \ \|\dot{\boldsymbol{p}}(\boldsymbol{f}^c) - \dot{\boldsymbol{p}}^{ref}\|^2 + \|\dot{\boldsymbol{l}}_C(\overline{\mathcal{F}}^c) - \dot{\boldsymbol{l}}_C^{ref}\|_{W_l}^2 + \sum_{j\in\{r,l\}} \|\bar{\boldsymbol{f}}_j^c\|_{W_f}^2 \tag{5.183}$$

under the $\mathfrak{F}_{FC}$ constraints. Note that unit weights are used for the linear momentum rate error subtask. In this way, the subtask is rendered high-priority. On the other hand, note that the linear momentum rate depends on the GRF. In order to avoid interference with the GRF minimization subtask (the third term), the weights should be set sufficiently small. In other words, the priority of the GRF minimization subtask should be set at the lowest level. Next, note that very small (approximately zero) angular momentum rate errors will yield very small ankle torques. Nevertheless, the angular momentum rate errors cannot be ignored in reality. This means that appropriate GRM/ankle torques have to be determined that will minimize the errors. Since the optimal GRFs are already known, the GRMs can be expressed with the linear relation:

$$\bar{\boldsymbol{\tau}}_j^c = \left[(\bar{\boldsymbol{f}}_{j\ opt}^c)^\times\right](\boldsymbol{r}_{P_j} - \boldsymbol{r}_{A_j}) + \boldsymbol{e}_z m_z. \tag{5.184}$$

Here $\boldsymbol{e}_z m_z$ accounts for the reaction torque around the normal. This equation renders the GRM minimization (or CoP allocation) problem convex. It is then straightforward to design an appropriate objective function [88].

The contact wrenches (the GRFs and GRMs) resulting from the above two sequential optimization subtasks are then used to obtain the *admissible* rate of change of the spatial momentum, $\dot{\mathcal{L}}_C^{ref}$. The final optimization subtask in the cascade can then be invoked, i.e.

$$\min_{\ddot{\theta}} c_{SM} + c_{JA}|_{\ddot{\theta}^{ref}=0}. \tag{5.185}$$

The c_{SM} objective stems from the relaxation of the $\mathfrak{F}_{SM}$ equality constraint. Apparently, this final optimization subtask falls also within the motion constraint domain. For example, as shown in [112], this objective can be used to maintain the initial posture of the upper body while standing on a narrow support, in the presence of disturbances.

Nonsequential Approach

A nonsequential spatial dynamics–based iterative optimization approach has been explored in [77,161]:

$$\min_{\boldsymbol{x}} \left(c_{CW} + c_{CoP} + c_{SM} + c_{EL} + c_{JA}|_{\ddot{\theta}^{ref}=\boldsymbol{0}} \right). \tag{5.186}$$

The force constraints are the equality-type $\mathfrak{F}_{SD}$ and the friction cone constraint $\mathfrak{F}_{FC}$. The inequality-type motion constraint, $\mathcal{M}_{AL}$, is used to enforce the joint limits. Since all constraints are formulated as components of the objective function, problems related to an overconstrained system can be avoided. This, however, is a compromise that leads to the problem of weight adjustment. As already mentioned, such an adjustment can be done, but only in an intuitive, nonrigorous way.

5.14.5 Complete Dynamics–Based Optimization

A number of multiobjective QP methods based on complete dynamics formulations have been developed in the past decade, mainly for the purpose of animating articulated figures, including humans [2,13,14,141,92,100,18,95,96]. In these works, the main aim is to generate a "realistically looking" *dynamic motion*. This goal cannot be achieved with kinematics- or quasistatics-based approaches. On the other hand, complete dynamics–based formulations in the field of humanoid robotics are still quite rare. There are a number of reasons for that. First of all, note that such formulations are useful for torque-controlled robots. Only a few such robots exist at present. The formulations are based on dynamic models that cannot be obtained in a straightforward manner. Indeed, the dynamic parameters from CAD data are imprecise since components like motors, wiring, connectors, and boards are not included. Thus, cumbersome identification procedures have to be used to obtain a precise dynamic model. Furthermore, dynamic models impose a relatively heavy computational burden. Therefore, it is quite challenging to ensure feedback control in real-time (typically, within 1 ms). Performing experiments with a torque-controlled robot is also quite demanding. Nevertheless, these hurdles can be overcome as will become apparent in what follows.

The optimal solution to the optimization task, based on the complete dynamics, is represented in the form of (5.113). The optimization task is formulated as a multiobjective function composed of penalty-type terms with or without a hierarchical structure for task prioritization.

Hierarchical Multiobjective Optimization With Hard Constraints

The problem of an overconstrained system can be tackled by introducing a hierarchical structure among the tasks. This approach, known as "hierarchical QP" (HQP), was discussed in Section 2.8.3. Recall also that a hierarchical structure can lessen the computational burden to some extent. The hierarchy defines the fixed task priorities. The constraints that need to be strictly enforced (hard constraints) are inequality-type, physical constraints. They include the joint angle and torque limits ($\mathcal{M}_{AL}$ and $\mathcal{F}_{TL}$), the friction constraints ($\mathcal{F}_{FC}$ and $\mathcal{F}_{FT}$), and the CoP-in-BoS constraint, $\mathcal{F}_{BoS}$. The notation pertinent to the general QP task formulation, (2.57), will be used below. The form of the solution vector is given in (5.113). The optimization task can be written as [40]

$$\begin{aligned} \min_{\boldsymbol{x},\boldsymbol{v},\boldsymbol{w}} \ & \|\boldsymbol{v}\|^2 + \|\boldsymbol{w}\|^2 \\ \text{s.t.} \quad & \boldsymbol{W}(\boldsymbol{A}\boldsymbol{x} + \boldsymbol{a}) = \boldsymbol{w}, \\ & \boldsymbol{V}(\boldsymbol{B}\boldsymbol{x} + \boldsymbol{b}) \le \boldsymbol{v}. \end{aligned} \tag{5.187}$$

Vectors $\boldsymbol{v}$ and $\boldsymbol{w}$ denote the slack variables that are used for constraint relaxation to avoid numerical instabilities. Recall that these variables are determined as an outcome from the optimization process; they do not need to be specified by the user. On the other hand, matrices $\boldsymbol{V}$ and $\boldsymbol{W}$ determining the relative weights for the subtasks are used-specified. Note that the above formulation is penalty-type. Strict task prioritization can be enforced with the help of the method introduced in Section 2.8.3 (see also [17,18]). First, denote one particular optimal solution at priority level r as $(\boldsymbol{x}_r^*,\ \boldsymbol{v}_r^*)$. All optimal solutions can then be expressed as

$$\boldsymbol{x} = \boldsymbol{x}_r^* + \boldsymbol{N}_r \boldsymbol{p}_{r+1} \tag{5.188}$$

$$\overline{A}_r x + \bar{a}_r \le v_r^*, \tag{5.189}$$
$$\dots$$
$$\overline{A}_1 x + \bar{a}_1 \le v_1^*.$$

Here N_r is a projector onto the compound null space of all higher-priority equality-type constraints, $\mathcal{N}(\overline{B}_r) \cap \ldots \cap \mathcal{N}(\overline{B}_1)$. Vector p_{r+1} is arbitrary; it parametrizes the compound null space. The overbar notation stands for a quantity restricted by the respective weight matrix, s.t. $\overline{A} = VA$, $\bar{a} = Va$, $\overline{B} = WB$, and $\bar{b} = Wb$. The null-space projectors can be computed via the singular-value decomposition SVD (cf. Section 2.7.1). The SVD procedure can be invoked in parallel to the QP task at level $r-1$. In the implementation reported in [40] it was observed that the SVD thus adds only an insignificant computational overhead.

With the above notation, the minimization task at level $r+1$ can be written as

$$\begin{aligned} \min_{v_{r+1},\, p_{r+1}} \quad & \|v_{r+1}\|^2 + \|\overline{B}_{r+1}(x_r^* + N_r p_{r+1}) + \bar{b}_{r+1}\|^2 \\ \text{s.t.} \quad & \overline{A}_{r+1}(x_r^* + N_r p_{r+1}) + \bar{a}_{r+1} \le v_{r+1}, \\ & \overline{A}_r(x_r^* + N_r p_{r+1}) + \bar{a}_r \le v_r^*, \\ & \dots \\ & \overline{A}_1(x_r^* + N_r p_{r+1}) + \bar{a}_1 \le v_1^*. \end{aligned} \tag{5.190}$$

Apparently, the priority solutions are obtained in a recursive fashion. With this formulation it can be guaranteed that the optimal solution will be found.

The above optimization method has been implemented with a torque-controlled humanoid robot of 25 DoFs [40]. Five priority levels were introduced that cover a large number of the constraints listed in Section 5.9.2. In the experiments, only 14 DoFs (no upper body) were used. With the complete dynamics formulation, about 5 ms were needed on an Intel Core i7-2600 CPU with 3.4 GHz. With the 14-DoF model, the $\mathcal{F}_{EoM}$ equality-type constraint was excluded. In this case, real-time performance (under 1 ms) has been achieved. Satisfactory performance has been reported for both proactive- and reactive-type tasks, despite the presence of the dynamic model inaccuracies.

Penalty-Based Multiobjective Optimization With Soft Constraints

Soft constraints always provide a solution, without numerical instabilities that sometimes may occur when using hard constraints. A penalty-type formulation can be employed to derive the optimal solution from the reference total force acting at the CoM, f_C^{ref} [22]. This force is obtained as the output of the DCM stabilizer in Section 5.6.3. The formulation is

$$\min_{x} \|\mathbb{C}_{cC_f} \overline{\mathcal{F}}^c - f_C^{ref}\|_{W_c} + \|\Delta r(\ddot{q}, \tau, \overline{\mathcal{F}}^c(C_\oplus))\|_{W_r}. \tag{5.191}$$

The relevant constraints are: $\mathcal{F}_{EoM}$, $\mathcal{F}_{FC}$, $\mathcal{F}_{TL}$, and $\mathcal{M}_{CL}$ (cf. Tables 5.1 and 5.2). The friction cone constraints are approximated by pyramids (cf. Section 3.1; see also e.g. [88]), vector $C_\oplus$ being composed of stacked C_k matrices (cf. (3.12)) that collect the pyramid basis vectors. Note that $C_\oplus$ parametrizes the contact wrenches. The two objectives in (5.191) minimize: (i) the

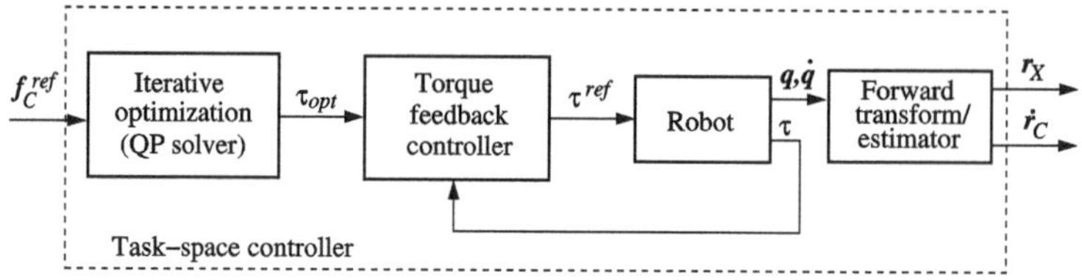

FIGURE 5.36 Block diagram of a task-space controller for the DCM stabilizer in Section 5.6.3. The task-space controller is designed for implementation with a torque-controlled robot. The QP solver uses the complete robot dynamics as a minimization constraint.

error of the total force acting at the CoM and (ii) the motion control errors along the unconstrained motion directions of the end links, as well as the self-motion control errors (the $\Delta \boldsymbol{r}$ term). The QP solver is part of the task-space controller appearing in the DCM-based balance controller in Fig. 5.15. The components of the task-space controller are shown in Fig. 5.36. Apparently, the QP solver determines the joint torque $\boldsymbol{\tau}_{opt}$ as a control input for the local joint-space controller.

Inverse Dynamics Plus Inverse Kinematics–Based Optimization Approach

A dynamic model of sufficient precision may not always be available. To cope with problems stemming from an imperfect dynamic model, one could use the information inherent to the (precise) inverse kinematics model. An inverse dynamics plus inverse kinematics–based optimization approach seems to be especially suitable for hydraulically driven humanoid robots with mixed force/velocity input commands [40,25]. In the implementation discussed in [25], two QP problems for the inverse dynamics and inverse kinematics are solved at each time step. The inverse dynamics solver ($\boldsymbol{x}$ as in (5.113)) is updated with the sensor states. The inverse kinematics solver ($\boldsymbol{x} = \dot{\boldsymbol{q}}_{int}$), on the other hand, is updated with states obtained via an integration, to avoid instabilities. This leads to accumulation of errors in the contact positions, though. The problem can be alleviated by employing a "leaky" integrator. The objective function (soft constraints), derived from the linear form $\boldsymbol{W}(\boldsymbol{A}\boldsymbol{x} - \boldsymbol{a})$, contains rows of mixed dynamic and kinematic objectives. The dynamic objectives stem from the relaxation of the $\mathcal{F}_{CoP}$ and $\mathcal{M}_{EL}$ constraints, as well as from additional dynamic objectives that include: (1) state tracking ($\boldsymbol{x} = \boldsymbol{x}^{ref}$), (2) weight distribution in accordance with (5.116) (only in the z-direction), and (3) suppression of sudden changes in the torque commands. The hard constraints used are: the equation of motion, the friction cone, and the CoP-in-BoS ($\mathcal{F}_{EoM}$, $\mathcal{F}_{FC}$, and $\mathcal{F}_{BoS}$, respectively). The kinematic objectives included are: (1) the end-link velocity tracking, (2) the direct joint-velocity tracking ($\dot{\boldsymbol{\theta}} = \dot{\boldsymbol{\theta}}^{ref}$), and (3) the suppression of sudden changes in the velocity commands. Further implementation details can be found in [25].

5.14.6 Mixed Iterative/Noniterative Optimization Approaches

One possibility to reduce the high computational burden inherent to iterative optimization based on the complete dynamics is to invoke a two-step sequential optimization scheme with mixed, iterative/noniterative optimization. The iterative part consists of spatial dynamics–based contact wrench optimization under specific constraints. The noniterative part solves the complete inverse dynamics problem via a suitable generalized inverse. In [142], for example, the iterative optimization task is defined as

$$\min_{\boldsymbol{x}} c_{SD} + c_{CW}|_{(\bar{\mathcal{F}}^c)^{ref}=0},$$

where $\boldsymbol{x} = \bar{\mathcal{F}}^c$. This task is solved under the inequality-type constraints $\mathcal{F}_{FC}$ and $\mathcal{F}_{BoS}$. Then, the equation of motion, $\mathcal{F}_{EoM}$, and the closed-chain motion constraint, $\mathcal{M}_{CC}$, are combined into one equation. We have

$$\begin{bmatrix} \boldsymbol{M}(\boldsymbol{q}) & -\boldsymbol{S} \\ \boldsymbol{J}_c & \boldsymbol{0} \end{bmatrix} \begin{bmatrix} \ddot{\boldsymbol{q}} \\ \boldsymbol{\tau} \end{bmatrix} = \begin{bmatrix} -\boldsymbol{c}(\boldsymbol{q}, \dot{\boldsymbol{q}}) - \boldsymbol{g}(\boldsymbol{q}) + \boldsymbol{J}_c^T \bar{\mathcal{F}}_{opt}^c \\ -\dot{\boldsymbol{J}}_c \dot{\boldsymbol{q}} \end{bmatrix}. \tag{5.192}$$

Here $\bar{\mathcal{F}}_{opt}^c$ are the contact wrenches obtained from the iterative optimization task. This equation is solved via a regularized (left) pseudoinverse for the joint acceleration and torque. Simple reactive tasks were examined in [142], with a torque-controlled robot (the Sarcos humanoid robot). The spatial dynamics objective was used mainly for the CoM PD feedback control. The desired rate of change of the angular momentum was set in a conservative way, to maintain the initial torso orientation while the robot was pushed on the back. Note however, that this approach does not allow for a sufficiently compliant response to the disturbance, e.g. to obtain a sufficiently large torso rotation as required under the hip strategy.

Using Generalized Acceleration Input Data Obtained From a Motion Capture System

Another example of a mixed, iterative contact wrench/noniterative complete dynamics optimization approach is reported in [167]. Desired generalized accelerations, $\ddot{\boldsymbol{q}}^{des}(t)$, obtained from a motion capture system, are used as the inputs. The iterative task is formulated as a penalty-type relaxation of the spatial dynamics equation expressed in terms of the base quasivelocity as follows:

$$\min_{\boldsymbol{f}} c_{SD} = \|\boldsymbol{\mathcal{A}}_{\overleftarrow{CB}} \ddot{\boldsymbol{q}}_B^{des} + \mathcal{C}_B - \mathbb{C}_{cB_f} \boldsymbol{f}\|^2,$$

subjected to the $\mathcal{F}_{FC}$ constraints. A point-contact model is used, the rectangular feet comprising four point contacts at the vertices. Vector $\boldsymbol{f}$ collects all reactions. The optimal $\boldsymbol{f}_{opt}$ is then used to derive optimal contact wrenches, $\bar{\mathcal{F}}_{opt}^c$, in a straightforward manner. Furthermore, the noniterative minimization task is defined as

$$\min_{\boldsymbol{\tau}} c_{JA} + c_{JT}|_{\boldsymbol{\tau}^{ref}=0},$$

subject to the complete dynamics equality constraint, $\mathcal{F}_{EoM}(\bar{\mathcal{F}}_{opt}^c)$, and the end-link motion equality constraints $\mathcal{M}_{EL}$. This task is solved via an appropriately defined generalized inverse.

Hierarchical Task Formulation With Decoupling

The mixed, iterative contact wrench/noniterative complete dynamics optimization approach can be further improved by introducing a hierarchical structure for the noniterative part, via the null-space projectors [38,37]. To this end, first rewrite the complete dynamics representation in mixed quasivelocity, as in (4.158), in the following quasistatic form:

$$\begin{bmatrix} \mathbf{0} \\ \mathbf{0} \\ \boldsymbol{\tau} \end{bmatrix} = \begin{bmatrix} \boldsymbol{g}_f \\ \mathbf{0} \\ \mathbf{0} \end{bmatrix} - \mathbb{J}^T \mathbb{F}. \tag{5.193}$$

The second term on the r.h.s. encodes the hierarchical structure [37]; we have

$$\mathbb{J}^T = \begin{bmatrix} \mathbb{J}_f^T \\ \mathbb{J}_m^T \\ \mathbb{J}_\tau^T \end{bmatrix} \equiv \begin{bmatrix} \boldsymbol{J}_c^T & \boldsymbol{N}_2^* \boldsymbol{J}_m^T & \boldsymbol{N}_3^* \boldsymbol{J}_q^T & \boldsymbol{N}_4^* \boldsymbol{S}^T \end{bmatrix}, \quad \mathbb{F} \equiv \begin{bmatrix} \overline{\mathcal{F}}^c \\ \overline{\mathcal{F}}^m \\ \mathcal{F}_C \\ \boldsymbol{\tau}_a \end{bmatrix}. \tag{5.194}$$

The $\boldsymbol{N}_{(\circ)}^* \in \Re^{(n+6)\times(n+6)}$ matrices denote the null-space projectors defined in Section 4.5.3. Their subscripts signify the priority level. It is important to emphasize that these null-space projectors ensure decoupling and thus guarantee the *asymptotic control stability* of the hierarchical system. The constraint ($\boldsymbol{J}_c$), mobility ($\boldsymbol{J}_m$), and spatial dynamics ($\boldsymbol{J}_q$) Jacobians are defined in (2.96), (2.95), and (4.99), respectively. The underactuation filtering matrix, $\boldsymbol{S}$, is defined in (2.102). The highest priority is assigned to the physical (contact) constraints. Motion in the mobility directions has higher priority than the CRB motion (used for balance control and to ensure a compliant whole-body response to an external disturbance). In this way, the object handling is not influenced by the CRB motion. The lowest priority is assigned to the posture control (e.g. in the form of a joint impedance) via the arbitrary torque vector $\boldsymbol{\tau}_a$. Furthermore, from the above two equations it is apparent that the following relations must hold:

$$\boldsymbol{g}_f - \mathbb{J}_f^T \mathbb{F} = \mathbf{0} = \mathbb{J}_m^T \mathbb{F}. \tag{5.195}$$

The iterative wrench distribution problem is then formulated simply as $\min_{\boldsymbol{x}} c_{CW}$, $\boldsymbol{x} = \overline{\mathcal{F}}^c$, under the constraints $\mathfrak{F}_{FC}$, $\mathfrak{F}_{FT}$, $\mathfrak{F}_{BoS}$, $\mathfrak{F}_{TL}$, and (5.195). The soft constraints c_{CW} require reference $(\overline{\mathcal{F}}^c)^{ref}$ to be specified as an input by a planner. The optimal contact wrenches are inserted, together with appropriate reference values for the rest of the $\mathbb{F}$ components, into the lower part of (5.193) to obtain the control torque as

$$\boldsymbol{\tau} = -\mathbb{J}_\tau^T \mathbb{F}.$$

It is important to note that no external force measurement is needed with this approach (cf. Section 4.5.3).

5.14.7 Computational Time Requirements

A comparison of the maximum computational times obtained with some of the iterative optimization approaches discussed in this chapter is given below. The following computational environment was used: a desktop PC with Intel Core i7 340 GHz CPU and 8 GB RAM,

TABLE 5.3 Comparison of the maximum computational times for sequential optimization

	Optimization			
	WD [μs]	FSD [μs]	JSID [μs]	Total [μs]
sequential WD → FSD	49	212	–	261
sequential WD → JSID	49	–	873	922

running Linux with the Real-Time PREEMPT patch. QuadProg++ [19] was used as the iterative solver.

The computational time requirements for the two sequential optimization approaches (cf. Fig. 5.22) are given in Table 5.3. In the case of nonsequential optimization, the formulation based on the FSD required 431 μs. The optimization based on the complete dynamics (CD) required 1728 μs, with variations of about ± 100 μs depending on the constraints.

For comparison, the two noniterative optimization approaches, DCM-GI-based WD and FSD optimization, required 1.62 μs and 2.12 μs, respectively.

References

[1] M. Abdallah, A. Goswami, A biomechanically motivated two-phase strategy for biped upright balance control, in: IEEE Int. Conf. on Robotics and Automation, Barcelona, Spain, 2005, pp. 1996–2001.
[2] Y. Abe, M. Da Silva, J. Popović, Multiobjective control with frictional contacts, in: ACM SIGGRAPH/Eurographics Symposium on Computer Animation, 2007, pp. 249–258.
[3] P.G. Adamczyk, S.H. Collins, A.D. Kuo, The advantages of a rolling foot in human walking, Journal of Experimental Biology 209 (2006) 3953–3963.
[4] Y. Aoustin, A.M. Formal'skii, On optimal swinging of the biped arms, in: IEEE/RSJ International Conference on Intelligent Robots and Systems, 2008, pp. 2922–2927.
[5] J.P. Aubin, Viability Theory, Birkhäuser, Basel, 2009.
[6] S.P. Boyd, B. Wegbreit, Fast computation of optimal contact forces, IEEE Transactions on Robotics 23 (2007) 1117–1132.
[7] T. Bretl, S. Lall, Testing static equilibrium for legged robots, IEEE Transactions on Robotics 24 (2008) 794–807.
[8] S. Caron, Q. Cuong Pham, Y. Nakamura, Leveraging cone double description for multi-contact stability of humanoids with applications to statics and dynamics, in: Robotics: Science and Systems XI. Robotics: Science and Systems Foundation, Rome, Italy, 2015, pp. 28–36.
[9] S. Caron, Q.-C. Pham, Y. Nakamura, ZMP support areas for multicontact mobility under frictional constraints, IEEE Transactions on Robotics 33 (2017) 67–80.
[10] Y. Choi, D. Kim, Y. Oh, B.-J. You, Posture/walking control for humanoid robot based on kinematic resolution of CoM Jacobian with embedded motion, IEEE Transactions on Robotics 23 (2007) 1285–1293.
[11] Y. Choi, D. Kim, B.-J. You, On the walking control for humanoid robot based on the kinematic resolution of CoM Jacobian with embedded motion, in: IEEE International Conference on Robotics and Automation, 2006, pp. 2655–2660.
[12] Y. Choi, B.-j. You, S.-r. Oh, On the stability of indirect ZMP controller for biped robot systems, in: IEEE/RSJ International Conference on Intelligent Robots and Systems, IROS, 2004, pp. 1966–1971.
[13] C. Collette, A. Micaelli, C. Andriot, P. Lemerle, Dynamic balance control of humanoids for multiple grasps and non coplanar frictional contacts, in: IEEE-RAS International Conference on Humanoid Robots, IEEE, 2007, pp. 81–88.
[14] C. Collette, A. Micaelli, C. Andriot, P. Lemerle, Robust balance optimization control of humanoid robots with multiple non coplanar grasps and frictional contacts, in: IEEE International Conference on Robotics and Automation, 2008, pp. 3187–3193.
[15] S.H. Collins, P.G. Adamczyk, A.D. Kuo, Dynamic arm swinging in human walking, Proceedings. Biological Sciences/The Royal Society 276 (2009) 3679–3688, arXiv:1010.2247.

[16] H. Dai, A. Valenzuela, R. Tedrake, Whole-body motion planning with centroidal dynamics and full kinematics, in: IEEE-RAS International Conference on Humanoid Robots, Madrid, Spain, 2014, pp. 295–302.
[17] M. De Lasa, A. Hertzmann, Prioritized optimization for task-space control, in: IEEE/RSJ International Conference on Intelligent Robots and Systems, 2009, pp. 5755–5762.
[18] M. De Lasa, I. Mordatch, A. Hertzmann, Feature-based locomotion controllers, ACM Transactions on Graphics 29 (2010) 131.
[19] L. Di Gaspero, QuadProg++: a C++ library for quadratic programming which implements the Goldfarb-Idnani active-set dual method, https://github.com/liuq/QuadProgpp, 2016.
[20] R.C. Dorf, R.H. Bishop, Modern Control Systems, 13 edition, Pearson, 2017.
[21] J. Englsberger, C. Ott, A. Albu-Schaffer, Three-dimensional bipedal walking control using divergent component of motion, in: IEEE/RSJ International Conference on Intelligent Robots and Systems, 2013, pp. 2600–2607.
[22] J. Englsberger, C. Ott, A. Albu-Schaffer, Three-dimensional bipedal walking control based on divergent component of motion, IEEE Transactions on Robotics 31 (2015) 355–368.
[23] J. Englsberger, C. Ott, M.A. Roa, A. Albu-Schaffer, G. Hirzinger, Bipedal walking control based on capture point dynamics, in: IEEE International Conference on Intelligent Robots and Systems, IEEE, 2011, pp. 4420–4427.
[24] S. Feng, E. Whitman, X. Xinjilefu, C.G. Atkeson, Optimization based full body control for the Atlas robot, in: IEEE-RAS International Conference on Humanoid Robots, Madrid, Spain, 2014, pp. 120–127.
[25] S. Feng, E. Whitman, X. Xinjilefu, C.G. Atkeson, Optimization-based full body control for the DARPA robotics challenge, Journal of Field Robotics 32 (2015) 293–312.
[26] Fujitsu, Miniature Humanoid Robot HOAP-2 Manual, 1st edition, Fujitsu Automation Co., Ltd, 2004 (in Japanese).
[27] J. Furusho, A. Sano, Sensor-based control of a nine-link biped, The International Journal of Robotics Research 9 (1990) 83–98.
[28] E.G. Gilbert, K.T. Tan, Linear systems with state and control constraints: the theory and application of maximal output admissible sets, IEEE Transactions on Automatic Control 36 (1991) 1008–1020.
[29] A. Goswami, Postural stability of biped robots and the foot-rotation indicator (FRI) point, The International Journal of Robotics Research 18 (1999) 523–533.
[30] A. Goswami, V. Kallem, Rate of change of angular momentum and balance maintenance of biped robots, in: IEEE International Conference on Robotics and Automation, New Orleans, LA, USA, 2004, pp. 3785–3790.
[31] R.J. Griffin, G. Wiedebach, S. Bertrand, A. Leonessa, J. Pratt, Straight-leg walking through underconstrained whole-body control, in: IEEE International Conference on Robotics and Automation, ICRA, 2018, pp. 5747–5757.
[32] J.-S. Gutmann, M. Fukuchi, M. Fujita, Stair climbing for humanoid robots using stereo vision, in: IEEE/RSJ International Conference on Intelligent Robots and Systems, 2004, pp. 1407–1413.
[33] T. Hamano, D. Nenchev, Posture stabilization at the BoS boundary, Robotic Life Support Laboratory, Tokyo City University, 2018 (Video clip), https://doi.org/10.1016/B978-0-12-804560-2.00012-2.
[34] K. Harada, K. Hauser, T. Bretl, J.-C. Latombe, Natural motion generation for humanoid robots, in: IEEE/RSJ International Conference on Intelligent Robots and Systems, 2006, pp. 833–839.
[35] K. Harada, S. Kajita, K. Kaneko, H. Hirukawa, Pushing manipulation by humanoid considering two-kinds of ZMPs, in: IEEE International Conference on Robotics and Automation, 2003, pp. 1627–1632.
[36] K. Harada, S. Kajita, H. Saito, M. Morisawa, F. Kanehiro, K. Fujiwara, K. Kaneko, H. Hirukawa, A humanoid robot carrying a heavy object, in: IEEE International Conference on Robotics and Automation, 2005, pp. 1712–1717.
[37] B. Henze, A. Dietrich, C. Ott, An approach to combine balancing with hierarchical whole-body control for legged humanoid robots, IEEE Robotics and Automation Letters 1 (2016) 700–707.
[38] B. Henze, M.A. Roa, C. Ott, Passivity-based whole-body balancing for torque-controlled humanoid robots in multi-contact scenarios, The International Journal of Robotics Research 35 (2016) 1522–1543.
[39] H. Herr, M. Popovic, Angular momentum in human walking, Journal of Experimental Biology 211 (2008) 467–481.
[40] A. Herzog, L. Righetti, F. Grimminger, P. Pastor, S. Schaal, Balancing experiments on a torque-controlled humanoid with hierarchical inverse dynamics, in: IEEE International Conference on Intelligent Robots and Systems, IEEE, 2014, pp. 981–988, arXiv:1305.2042.
[41] A. Herzog, N. Rotella, S. Mason, F. Grimminger, S. Schaal, L. Righetti, Momentum control with hierarchical inverse dynamics on a torque-controlled humanoid, Autonomous Robots 40 (2016) 473–491, arXiv:1410.7284v1.

[42] R. Hinata, D. Nenchev, Kicking motion generated with the relative angular momentum (RAM) balance controller, Robotic Life Support Laboratory, Tokyo City University, 2018 (Video clip), https://doi.org/10.1016/B978-0-12-804560-2.00012-2.
[43] R. Hinata, D. Nenchev, Upper-body twists generated with centroidal angular momentum conservation, Robotic Life Support Laboratory, Tokyo City University, 2018 (Video clip), https://doi.org/10.1016/B978-0-12-804560-2.00012-2.
[44] R. Hinata, D. Nenchev, Upper-body twists generated with coupling angular momentum conservation, Robotic Life Support Laboratory, Tokyo City University, 2018 (Video clip), https://doi.org/10.1016/B978-0-12-804560-2.00012-2.
[45] R. Hinata, D. Nenchev, Upper-body twists generated with the resolved momentum approach, Robotic Life Support Laboratory, Tokyo City University, 2018 (Video clip), https://doi.org/10.1016/B978-0-12-804560-2.00012-2.
[46] R. Hinata, D.N. Nenchev, Velocity-based contact stabilization of a rolling foot (convergent case), Robotic Life Support Laboratory, Tokyo City University, 2018 (Video clip), https://doi.org/10.1016/B978-0-12-804560-2.00012-2.
[47] R. Hinata, D.N. Nenchev, Velocity-based contact stabilization of a rolling foot (divergent case), Robotic Life Support Laboratory, Tokyo City University, 2018 (Video clip), https://doi.org/10.1016/B978-0-12-804560-2.00012-2.
[48] K. Hirai, M. Hirose, Y. Haikawa, T. Takenaka, The development of Honda humanoid robot, in: IEEE Int. Conf. on Robotics and Automation, Leuven, Belgium, 1998, pp. 1321–1326.
[49] K. Hirata, M. Fujita, Analysis of conditions for non-violation of constraints on linear discrete-time systems with exogenous inputs, in: 36th IEEE Conference on Decision and Control, IEEE, 1997, pp. 1477–1478.
[50] H. Hirukawa, S. Hattori, K. Harada, S. Kajita, K. Kaneko, F. Kanehiro, K. Fujiwara, M. Morisawa, A universal stability criterion of the foot contact of legged robots – adios ZMP, in: IEEE International Conference on Robotics and Automation, 2006, pp. 1976–1983.
[51] A.L. Hof, The 'extrapolated center of mass' concept suggests a simple control of balance in walking, Human Movement Science 27 (2008) 112–125.
[52] A.L. Hof, M.G.J. Gazendam, W.E. Sinke, The condition for dynamic stability, Journal of Biomechanics 38 (2005) 1–8.
[53] A. Hofmann, M. Popovic, H. Herr, Exploiting angular momentum to enhance bipedal center-of-mass control, in: IEEE Int. Conf. on Robotics and Automation, Kobe, Japan, 2009, pp. 4423–4429.
[54] M.H. Honarvar, M. Nakashima, A new measure for upright stability, Journal of Biomechanics 47 (2014) 560–567.
[55] M.A. Hopkins, D.W. Hong, A. Leonessa, Humanoid locomotion on uneven terrain using the time-varying divergent component of motion, in: IEEE-RAS International Conference on Humanoid Robots, 2014, pp. 266–272.
[56] F.B. Horak, S.M. Henry, A. Shumway-Cook, Postural perturbations: new insights for treatment of balance disorders, Physical Therapy 77 (1997) 517–533.
[57] M. Hosokawa, D.N. Nenchev, DCM-GI based contact transition control, Robotic Life Support Laboratory, Tokyo City University, 2018 (Video clip), https://doi.org/10.1016/B978-0-12-804560-2.00012-2.
[58] M. Hosokawa, D.N. Nenchev, T. Hamano, The DCM generalized inverse: efficient body-wrench distribution in multi-contact balance control, Advanced Robotics 32 (2018) 778–792.
[59] Q.H.Q. Huang, S. Kajita, N. Koyachi, K. Kaneko, K. Yokoi, H. Arai, K. Komoriya, K. Tanie, A high stability, smooth walking pattern for a biped robot, in: IEEE International Conference on Robotics and Automation, 1999, pp. 65–71.
[60] M. Hutter, H. Sommer, C. Gehring, M. Hoepflinger, M. Bloesch, R. Siegwart, Quadrupedal locomotion using hierarchical operational space control, The International Journal of Robotics Research 33 (2014) 1047–1062.
[61] F.E. Huxham, P.A. Goldie, A.E. Patla, Theoretical considerations in balance assessment, Australian Journal of Physiotherapy 47 (2001) 89–100.
[62] S.-H. Hyon, Compliant terrain adaptation for biped humanoids without measuring ground surface and contact forces, IEEE Transactions on Robotics 25 (2009) 171–178.
[63] S.-H. Hyon, J. Hale, G. Cheng, Full-body compliant human-humanoid interaction: balancing in the presence of unknown external forces, IEEE Transactions on Robotics 23 (2007) 884–898.
[64] K. Iqbal, Y.-c. Pai, Predicted region of stability for balance recovery: motion at the knee joint can improve termination of forward movement, Journal of Biomechanics 33 (2000) 1619–1627.

[65] S. Kagami, F. Kanehiro, Y. Tamiya, M. Inaba, H. Inoue, AutoBalancer: an online dynamic balance compensation scheme for humanoid robots, in: Int. Workshop Alg. Found. Robot, WAFR, 2000, pp. 329–339.

[66] S. Kagami, T. Kitagawa, K. Nishiwaki, T. Sugihara, M. Inaba, H. Inoue, A fast dynamically equilibrated walking trajectory generation method of humanoid robot, Autonomous Robots 12 (2002) 71–82.

[67] S. Kajita, H. Hirukawa, K. Harada, K. Yokoi, Introduction to Humanoid Robotics, Springer Verlag, Berlin, Heidelberg, 2014.

[68] S. Kajita, F. Kanehiro, K. Kaneko, K. Fujiwara, K. Harada, K. Yokoi, H. Hirukawa, Biped walking pattern generation by using preview control of zero-moment point, in: IEEE International Conference on Robotics and Automation, 2003, pp. 1620–1626.

[69] S. Kajita, F. Kanehiro, K. Kaneko, K. Fujiwara, K. Harada, K. Yokoi, H. Hirukawa, Resolved momentum control: humanoid motion planning based on the linear and angular momentum, in: IEEE/RSJ International Conference on Intelligent Robots and Systems, Las Vegas, Nevada, 2003, pp. 1644–1650.

[70] S. Kajita, M. Morisawa, K. Harada, K. Kaneko, F. Kanehiro, K. Fujiwara, H. Hirukawa, Biped walking pattern generator allowing auxiliary ZMP control, in: IEEE International Conference on Intelligent Robots and Systems, 2006, pp. 2993–2999.

[71] S. Kajita, M. Morisawa, K. Miura, S. Nakaoka, K. Harada, K. Kaneko, F. Kanehiro, K. Yokoi, Biped walking stabilization based on linear inverted pendulum tracking, in: IEEE/RSJ International Conference on Intelligent Robots and Systems, Taipei, Taiwan, 2010, pp. 4489–4496.

[72] S. Kajita, K. Tani, Study of dynamic walk control of a biped robot on rugged terrain – derivation and application of the linear inverted pendulum mode, Journal of Robotics and Mechatronics 5 (1993) 516–523.

[73] S. Kajita, K. Tani, Experimental study of biped dynamic walking, IEEE Control Systems Magazine 16 (1996) 13–19.

[74] S. Kajita, K. Yokoi, M. Saigo, K. Tanie, Balancing a humanoid robot using backdrive concerned torque control and direct angular momentum feedback, in: IEEE International Conference on Robotics and Automation, 2001, pp. 3376–3382.

[75] K. Kaneko, F. Kanehiro, S. Kajita, M. Morisawa, K. Fujiwara, K. Harada, H. Hirukawa, Motion suspension system for humanoids in case of emergency; real-time motion generation and judgment to suspend humanoid, in: IEEE/RSJ International Conference on Intelligent Robots and Systems, IEEE, 2006, pp. 5496–5503.

[76] A. Konno, T. Myojin, T. Matsumoto, T. Tsujita, M. Uchiyama, An impact dynamics model and sequential optimization to generate impact motions for a humanoid robot, The International Journal of Robotics Research 30 (2011) 1596–1608.

[77] T. Koolen, S. Bertrand, G. Thomas, T. de Boer, T. Wu, J. Smith, J. Englsberger, J. Pratt, Design of a momentum-based control framework and application to the humanoid robot Atlas, International Journal of Humanoid Robotics 13 (2016) 1650007.

[78] T. Koolen, T. de Boer, J. Rebula, A. Goswami, J. Pratt, Capturability-based analysis and control of legged locomotion, Part 1: theory and application to three simple gait models, The International Journal of Robotics Research 31 (2012) 1094–1113.

[79] T. Koolen, M. Posa, R. Tedrake, Balance control using center of mass height variation: limitations imposed by unilateral contact, in: IEEE-RAS 16th International Conference on Humanoid Robots (Humanoids), IEEE, 2016, pp. 8–15.

[80] K. Koyanagi, H. Hirukawa, S. Hattori, M. Morisawa, S. Nakaoka, K. Harada, S. Kajita, A pattern generator of humanoid robots walking on a rough terrain using a handrail, in: IEEE/RSJ International Conference on Intelligent Robots and Systems, 2008, pp. 2617–2622.

[81] S. Kudoh, T. Komura, K. Ikeuchi, The dynamic postural adjustment with the quadratic programming method, in: IEEE/RSJ International Conference on Intelligent Robots and System, 2002, pp. 2563–2568.

[82] S. Kudoh, T. Komura, K. Ikeuchi, Stepping motion for a human-like character to maintain balance against large perturbations, in: IEEE International Conference on Robotics and Automation, 2006, pp. 2661–2666.

[83] M. Kudruss, M. Naveau, O. Stasse, N. Mansard, C. Kirches, P. Soueres, K. Mombaur, Optimal control for whole-body motion generation using center-of-mass dynamics for predefined multi-contact configurations, in: IEEE-RAS 15th International Conference on Humanoid Robots (Humanoids), 2015, pp. 684–689.

[84] J.J. Kuffner, S. Kagami, K. Nishiwaki, M. Inaba, H. Inoue, Dynamically-stable motion planning for humanoid robots, Autonomous Robots 12 (2002) 105–118.

[85] S. Kuindersma, R. Deits, M. Fallon, A. Valenzuela, H. Dai, F. Permenter, T. Koolen, P. Marion, R. Tedrake, Optimization-based locomotion planning, estimation, and control design for the atlas humanoid robot, Autonomous Robots 40 (2016) 429–455.

[86] A. Kuo, An optimal control model for analyzing human postural balance, IEEE Transactions on Biomedical Engineering 42 (1995) 87–101.
[87] S.-H. Lee, A. Goswami, Ground reaction force control at each foot: a momentum-based humanoid balance controller for non-level and non-stationary ground, in: IEEE/RSJ International Conference on Intelligent Robots and Systems, IEEE, Taipei, Taiwan, 2010, pp. 3157–3162.
[88] S.-H.H. Lee, A. Goswami, A momentum-based balance controller for humanoid robots on non-level and non-stationary ground, Autonomous Robots 33 (2012) 399–414.
[89] Y. Lee, S. Hwang, J. Park, Balancing of humanoid robot using contact force/moment control by task-oriented whole body control framework, Autonomous Robots 40 (2016) 457–472.
[90] S. Lengagne, J. Vaillant, E. Yoshida, A. Kheddar, Generation of whole-body optimal dynamic multi-contact motions, The International Journal of Robotics Research 32 (2013) 1104–1119.
[91] C.W. Luchies, N.B. Alexander, A.B. Schultz, J. Ashton-Miller, Stepping responses of young and old adults to postural disturbances: kinematics, Journal of the American Geriatrics Society 42 (1994) 506–512.
[92] A. Macchietto, V. Zordan, C.R. Shelton, Momentum control for balance, ACM Transactions on Graphics 28 (2009) 80.
[93] B. Maki, W. Mcilroy, G. Fernie, Change-in-support reactions for balance recovery, IEEE Engineering in Medicine and Biology Magazine 22 (2003) 20–26.
[94] B.E. Maki, W.E. McIlroy, The role of limb movements in maintaining upright stance: the "change-in-support" strategy, Physical Therapy 77 (1997) 488–507.
[95] D. Mansour, A. Micaelli, P. Lemerle, A computational approach for push recovery in case of multiple noncoplanar contacts, in: IEEE/RSJ International Conference on Intelligent Robots and Systems, IEEE, San Francisco, CA, USA, 2011, pp. 3213–3220.
[96] D. Mansour, A. Micaelli, P. Lemerle, Humanoid push recovery control in case of multiple non-coplanar contacts, in: IEEE International Conference on Intelligent Robots and Systems, 2013, pp. 4137–4144.
[97] K. Mitobe, G. Capi, Y. Nasu, Control of walking robots based on manipulation of the zero moment point, Robotica 18 (2000) 651–657.
[98] K. Miura, M. Morisawa, F. Kanehiro, S. Kajita, K. Kaneko, K. Yokoi, Human-like walking with toe supporting for humanoids, in: IEEE/RSJ International Conference on Intelligent Robots and Systems, San Francisco, CA, USA, 2011, pp. 4428–4435.
[99] F. Miyazaki, S. Arimoto, A control theoretic study on dynamical biped locomotion, Journal of Dynamic Systems, Measurement, and Control 102 (1980) 233.
[100] I. Mordatch, M. de Lasa, A. Hertzmann, Robust physics-based locomotion using low-dimensional planning, ACM Transactions on Graphics 29 (2010) 1.
[101] M. Morisawa, K. Kaneko, F. Kanehiro, S. Kajita, K. Fujiwara, K. Harada, H. Hirukawa, Motion planning of emergency stop for humanoid robot by state space approach, in: IEEE International Conference on Intelligent Robots and Systems, 2006, pp. 2986–2992.
[102] M. Murooka, S. Noda, S. Nozawa, Y. Kakiuchi, K. Okada, M. Inaba, Achievement of pivoting large and heavy objects by life-sized humanoid robot based on online estimation control method of object state and manipulation force, Journal of the Robotics Society of Japan 32 (2014) 595 602.
[103] M. Murooka, S. Nozawa, Y. Kakiuchi, K. Okada, M. Inaba, Whole-body pushing manipulation with contact posture planning of large and heavy object for humanoid robot, in: IEEE International Conference on Robotics and Automation, 2015, pp. 5682–5689.
[104] M. Murooka, R. Ueda, S. Nozawa, Y. Kakiuchi, K. Okada, M. Inaba, Global planning of whole-body manipulation by humanoid robot based on transition graph of object motion and contact switching, Advanced Robotics 31 (2017) 322–340.
[105] M.P. Murray, A. Seireg, R.C. Scholz, Center of gravity, center of pressure, and supportive forces during human activities, Journal of Applied Physiology 23 (1967) 831–838.
[106] K. Nagasaka, M. Inaba, H. Inoue, Stabilization of dynamic walk on a humanoid using torso position compliance control, in: Proceedings of the 17th Annual Conference of the Robotics Society of Japan, 1999, pp. 1193–1194.
[107] T. Nakamura, D.N. Nenchev, Velocity-based posture stabilization on a balance board, Robotic Life Support Laboratory, Tokyo City University, 2018 (Video clip), https://doi.org/10.1016/B978-0-12-804560-2.00012-2.
[108] L.M. Nashner, G. McCollum, The organization of human postural movements: a formal basis and experimental synthesis, Behavioral and Brain Sciences 8 (1985) 135–150.

[109] E.S. Neo, K. Yokoi, S. Kajita, K. Tanie, Whole-body motion generation integrating operator's intention and robot's autonomy in controlling humanoid robots, IEEE Transactions on Robotics 23 (2007) 763–775.
[110] S. Noda, M. Murooka, S. Nozawa, Y. Kakiuchi, K. Okada, M. Inaba, Generating whole-body motion keep away from joint torque, contact force, contact moment limitations enabling steep climbing with a real humanoid robot, in: IEEE International Conference on Robotics and Automation, 2014, pp. 1775–1781.
[111] D.E. Orin, A. Goswami, Centroidal momentum matrix of a humanoid robot: structure and properties, in: IEEE/RSJ International Conference on Intelligent Robots and Systems, IROS, Nice, France, 2008, pp. 653–659.
[112] D.E. Orin, A. Goswami, S.H. Lee, Centroidal dynamics of a humanoid robot, Autonomous Robots 35 (2013) 161–176.
[113] C. Ott, A. Kugi, Y. Nakamura, Resolving the problem of non-integrability of nullspace velocities for compliance control of redundant manipulators by using semi-definite Lyapunov functions, in: IEEE International Conference on Robotics and Automation, 2008, pp. 1999–2004.
[114] C. Ott, M.A. Roa, G. Hirzinger, Posture and balance control for biped robots based on contact force optimization, in: IEEE-RAS International Conference on Humanoid Robots, Bled, Slovenia, 2011, pp. 26–33.
[115] B. Paden, R. Panja, Globally asymptotically stable 'PD+' controller for robot manipulators, International Journal of Control 47 (1988) 1697–1712.
[116] Y.-c. Pai, J. Patton, Center of mass velocity-position predictions for balance control, Journal of Biomechanics 30 (1997) 347–354.
[117] J. Park, J. Haan, F.C. Park, Convex optimization algorithms for active balancing of humanoid robots, IEEE Transactions on Robotics 23 (2007) 817–822.
[118] K. Park, P. Chang, J. Salisbury, A unified approach for local resolution of kinematic redundancy with inequality constraints and its application to nuclear power plant, in: International Conference on Robotics and Automation, IEEE, 1997, pp. 766–773.
[119] O.E.R. Ponce, Generation of the Whole-Body Motion for Humanoid Robots with the Complete Dynamics, Ph.D. thesis, Toulouse University, France, 2014.
[120] M. Popovic, A. Hofmann, H. Herr, Angular momentum regulation during human walking: biomechanics and control, in: IEEE International Conference on Robotics and Automation, 2004, pp. 2405–2411.
[121] M.B. Popovic, A. Goswami, H. Herr, Ground reference points in legged locomotion: definitions, biological trajectories and control implications, The International Journal of Robotics Research 24 (2005) 1013–1032.
[122] J. Pratt, J. Carff, S. Drakunov, A. Goswami, Capture point: a step toward humanoid push recovery, in: IEEE-RAS International Conference on Humanoid Robots, Genoa, Italy, 2006, pp. 200–207.
[123] Qiang Huang, K. Yokoi, S. Kajita, K. Kaneko, H. Arai, N. Koyachi, K. Tanie, Planning walking patterns for a biped robot, IEEE Transactions on Robotics and Automation 17 (2001) 280–289.
[124] O.E. Ramos, N. Mansard, P. Soueres, Whole-body motion integrating the capture point in the operational space inverse dynamics control, in: IEEE-RAS International Conference on Humanoid Robots, Madrid, Spain, 2014, pp. 707–712.
[125] J. Rebula, F. Cañnas, J. Pratt, A. Goswami, Learning capture points for humanoid push recovery, in: IEEE-RAS International Conference on Humanoid Robots, 2008, pp. 65–72.
[126] L. Righetti, J. Buchli, M. Mistry, M. Kalakrishnan, S. Schaal, Optimal distribution of contact forces with inverse-dynamics control, The International Journal of Robotics Research 32 (2013) 280–298.
[127] T. Saida, Y. Yokokohji, T. Yoshikawa, FSW (feasible solution of wrench) for multi-legged robots, in: IEEE International Conference on Robotics and Automation, 2003, pp. 3815–3820.
[128] A. Sano, J. Furusho, 3D dynamic walking of biped robot by controlling the angular momentum, Transactions of the Society of Instrument and Control Engineers 26 (1990) 459–466.
[129] A. Sano, J. Furusho, Realization of natural dynamic walking using the angular momentum information, in: IEEE International Conference on Robotics and Automation, Tsukuba, Japan, 1990, pp. 1476–1481.
[130] A. Sano, J. Furusho, Control of torque distribution for the BLR-G2 biped robot, in: Fifth International Conference on Advanced Robotics, vol. 1, IEEE, 1991, pp. 729–734.
[131] P. Sardain, G. Bessonnet, Forces acting on a biped robot. Center of pressure—zero moment point, IEEE Transactions on Systems, Man and Cybernetics. Part A. Systems and Humans 34 (2004) 630–637.
[132] T. Sato, S. Sakaino, E. Ohashi, K. Ohnishi, Walking trajectory planning on stairs using virtual slope for biped robots, IEEE Transactions on Industrial Electronics 58 (2011) 1385–1396.
[133] L. Sentis, Synthesis and Control of Whole-Body Behaviors in Humanoid Systems, Ph.D. thesis, Stanford University, 2007.

[134] L. Sentis, O. Khatib, Synthesis of whole-body behaviors through hierarchical control of behavioral primitives, International Journal of Humanoid Robotics 02 (2005) 505–518.
[135] L. Sentis, O. Khatib, Compliant control of multicontact and center-of-mass behaviors in humanoid robots, IEEE Transactions on Robotics 26 (2010) 483–501.
[136] M. Shibuya, T. Suzuki, K. Ohnishi, Trajectory planning of biped robot using linear pendulum mode for double support phase, in: IECON Proceedings (Industrial Electronics Conference), 2006, pp. 4094–4099.
[137] T. Shibuya, D. Nenchev, Inclined surface cleaning with motion/force control under external disturbances, Robotic Life Support Laboratory, Tokyo City University, 2018 (Video clip), https://doi.org/10.1016/B978-0-12-804560-2.00012-2.
[138] T. Shibuya, S. Sakaguchi, D. Nenchev, Inclined surface cleaning with motion/force control, Robotic Life Support Laboratory, Tokyo City University, 2018 (Video clip), https://doi.org/10.1016/B978-0-12-804560-2.00012-2.
[139] A. Shumway-Cook, F.B. Horak, Vestibular rehabilitation: an exercise approach to managing symptoms of vestibular dysfunction, Seminars in Hearing 10 (1989) 196–209.
[140] B. Siciliano, L. Sciavicco, L. Villani, G. Oriolo, Robotics: Modelling, Planning and Control, Springer-Verlag, London, 2009.
[141] M. da Silva, Y. Abe, J. Popović, Interactive simulation of stylized human locomotion, ACM Transactions on Graphics 27 (2008) 1.
[142] B.J. Stephens, C.G. Atkeson, Dynamic balance force control for compliant humanoid robots, in: IEEE/RSJ International Conference on Intelligent Robots and Systems, 2010, pp. 1248–1255.
[143] T. Sugihara, Standing stabilizability and stepping maneuver in planar bipedalism based on the best COM-ZMP regulator, in: IEEE International Conference on Robotics and Automation, Kobe, Japan, 2009, pp. 1966–1971.
[144] T. Sugihara, Y. Nakamura, Whole-body cooperative balancing of humanoid robot using COG Jacobian, in: IEEE/RSJ Int. Conf. on Intelligent Robots and System, Lausanne, Switzerland, 2002, pp. 2575–2580.
[145] T. Sugihara, Y. Nakamura, H. Inoue, Real-time humanoid motion generation through ZMP manipulation based on inverted pendulum control, in: IEEE Int. Conf. on Robotics and Automation, Washington, DC, 2002, pp. 1404–1409.
[146] A. Takanishi, M. Ishida, Y. Yamazaki, I. Kato, The realization of dynamic walking by the biped walking robot WL-10RD, Journal of the Robotics Society of Japan 3 (1985) 325–336.
[147] A. Takanishi, H.-o. Lim, M. Tsuda, I. Kato, Realization of dynamic biped walking stabilized by trunk motion on a sagittally uneven surface, in: IEEE International Workshop on Intelligent Robots and Systems, 1990, pp. 323–330.
[148] A. Takanishi, M. Tochizawa, H. Karaki, I. Kato, Dynamic biped walking stabilized with optimal trunk and waist motion, in: IEEE/RSJ International Workshop on Intelligent Robots and Systems, 1989, pp. 187–192.
[149] T. Takenaka, T. Matsumoto, T. Yoshiike, Real time motion generation and control for biped robot – 1st report: walking gait pattern generation, in: IEEE/RSJ International Conference on Intelligent Robots and Systems, 2009, pp. 1084–1091.
[150] T. Takenaka, T. Matsumoto, T. Yoshiike, T. Hasegawa, S. Shirokura, H. Kaneko, A. Orita, Real time motion generation and control for biped robot – 4th report: integrated balance control, in: IEEE/RSJ Int. Conf. on Intelligent Robots and System, 2009, pp. 1601–1608.
[151] Y. Tamiya, M. Inaba, H. Inoue, Realtime balance compensation for dynamic motion of full-body humanoid standing on one leg, Journal of the Robotics Society of Japan 17 (1999) 112–118.
[152] T. Tanaka, T. Takubo, K. Inoue, T. Arai, Emergent stop for humanoid robots, in: IEEE/RSJ International Conference on Intelligent Robots and Systems, IEEE, 2006, pp. 3970–3975.
[153] J. Vaillant, A. Kheddar, H. Audren, F. Keith, S. Brossette, A. Escande, K. Bouyarmane, K. Kaneko, M. Morisawa, P. Gergondet, E. Yoshida, S. Kajita, F. Kanehiro, Multi-contact vertical ladder climbing with an HRP-2 humanoid, Autonomous Robots 40 (2016) 561–580.
[154] M. Vukobratović, Contribution to the study of anthropomorphic systems, Kybernetika 8 (1972) 404–418.
[155] M. Vukobratovic, B. Borovac, Zero-moment point — thirty five years of its life, International Journal of Humanoid Robotics 01 (2004) 157–173.
[156] M. Vukobratović, A.A. Frank, D. Juricić, On the stability of biped locomotion, IEEE Transactions on Bio-Medical Engineering 17 (1970) 25–36.
[157] M. Vukobratovic, D. Juricic, Contribution to the synthesis of biped gait, IEEE Transactions on Biomedical Engineering BME-16 (1969) 1–6.

[158] P.M. Wensing, G. Bin Hammam, B. Dariush, D.E. Orin, Optimizing foot centers of pressure through force distribution in a humanoid robot, International Journal of Humanoid Robotics 10 (2013) 1350027.
[159] P.-B. Wieber, On the stability of walking systems, in: Third IARP International Workshop on Humanoid and Human Friendly Robotics, Tsukuba, Japan, 2002, pp. 1–7.
[160] P.-B. Wieber, Viability and predictive control for safe locomotion, in: IEEE/RSJ International Conference on Intelligent Robots and Systems, 2008, pp. 1103–1108.
[161] G. Wiedebach, S. Bertrand, T. Wu, L. Fiorio, S. McCrory, R. Griffin, F. Nori, J. Pratt, Walking on partial footholds including line contacts with the humanoid robot Atlas, in: IEEE-RAS International Conference on Humanoid Robots, 2016, pp. 1312–1319.
[162] D. Winter, A.E. Patla, J.S. Frank, Assessment of balance control in humans, Medical Progress Through Technology 16 (1990) 31–51.
[163] A. Witkin, M. Kass, Spacetime constraints, ACM SIGGRAPH Computer Graphics 22 (1988) 159–168.
[164] J. Yamaguchi, A. Takanishi, I. Kato, Development of a biped walking robot compensating for three-axis moment by trunk motion, in: IEEE/RSJ International Conference on Intelligent Robots and Systems, 1993, pp. 561–566.
[165] K. Yamamoto, Control strategy switching for humanoid robots based on maximal output admissible set, Robotics and Autonomous Systems 81 (2016) 17–32.
[166] R. Yang, Y.-Y. Kuen, Z. Li, Stabilization of a 2-DOF spherical pendulum on X-Y table, in: IEEE International Conference on Control Applications, 2000, pp. 724–729.
[167] Yu Zheng, K. Yamane, Human motion tracking control with strict contact force constraints for floating-base humanoid robots, in: IEEE-RAS International Conference on Humanoid Robots, IEEE, Atlanta, GA, USA, 2013, pp. 34–41.
[168] Y. Zheng, C.-M. Chew, Fast equilibrium test and force distribution for multicontact robotic systems, Journal of Mechanisms and Robotics 2 (2010) 021001.
[169] Y.F. Zheng, J. Shen, Gait synthesis for the SD-2 biped robot to climb sloping surface, IEEE Transactions on Robotics and Automation 6 (1990) 86–96.

CHAPTER

6

Cooperative Object Manipulation and Control

6.1 INTRODUCTION

The problem of multiarm and multirobot cooperation is similar to the problem of multifinger grasping. The main difference is in the modeling of the contacts between the fingertips (or the hands) and the object.

When a hand grasps an object, the contacts established between the fingers and the object are generally unilateral. As discussed in Section 2.9.3, the transmission of velocities and angular velocities (twists) between the fingers and the object is represented by velocity transform bases $\mathbb{B}_m$. The transmission of forces and moments (wrenches), on the other hand, is represented by constraint bases $\mathbb{B}_c$.

In the discussion about multiarm and multirobot cooperation, it is assumed that the robot hands grasp the object firmly. Hence, each robot hand can apply a 6-DoF wrench to the object. Therefore, the respective velocity transform basis $\mathbb{B}_m$ becomes a zero matrix, while the constraint basis $\mathbb{B}_c$ becomes a 6D identity matrix.

In multifinger grasping and multiarm and multirobot cooperation, the grasped object may deform. Since the aim in the following discussion, however, is to better understand the basics of grasping and cooperation, it will be assumed that the object does not deform.

The chapter is structured as follows. Section 6.2 discusses multifinger grasping. The ability of a grasp to resist external wrenches applied to the grasped object deeply depends on the type of grasping. The grasping ability can be characterized with the help of the *form closure* and *force closure* concepts. These concepts make use of the grasp and hand Jacobian matrices and of the kinematics relationship between the fingertips and the object. At the end of the section, simple examples for better understanding will be presented.

Section 6.3 discusses the problem of multiarm cooperation. There are two major approaches: master-slave and symmetry-type cooperation. When p robot arms grasp an object, the arms can apply $6p$-dimensional wrenches to the object in total. Among the $6p$-dimensional wrenches, a 6D wrench is used to control the external wrench. The remaining $6(p-1)$-dimensional wrenches are used to control the internal wrenches. The methods used to control external and internal wrenches are discussed in this section as well.

In Section 6.4, multirobot cooperation is discussed. It is worth noting that when multiple wheeled robots cooperate, the so-called "leader-follower"-type cooperation is widely used. This type of cooperation, however, is not suitable when multiple legged robots cooperate

Humanoid Robots
https://doi.org/10.1016/B978-0-12-804560-2.00013-4

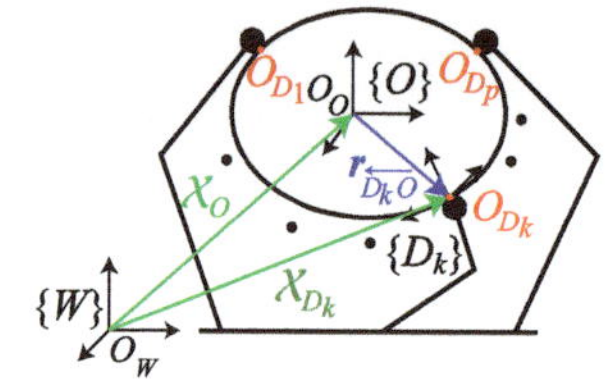

FIGURE 6.1 The object is grasped by p robot fingers.

since there will be a time lag between the motion of the leader and that of the followers. The time lag may cause the robots to fall down. In multirobot cooperation, the so-called "symmetry"-type cooperation is more suitable; it can be expected that all robots move synchronously, and hence, no internal wrenches will be generated in the ideal case. Therefore, the focus in Section 6.4 is on the symmetry-type cooperation.

The problem of dynamic object manipulation is highlighted in Section 6.5.

6.2 MULTIFINGER GRASPING

This section discusses mainly the ability of a robotic hand to resist external wrenches applied to the grasped object. The ability of the hand to resist the external wrenches is characterized by two concepts: the *form closure* and the *force closure*. The terms *form closure* and *force closure* originally appeared in [20] and were used to represent constraint conditions. Partial form and force closure properties of grasping were discussed in [2].

Several different types of finger contacts have been modeled so far. Some examples from the taxonomy of contacts are: (1) lack of contact (six freedoms), (2) point contact without friction (five freedoms), (3) line contact without friction (four freedoms), (4) point contact with friction (three freedoms), (5) planar contact without friction (three freedoms), (6) soft finger (two freedoms), (7) line contact with friction (one freedom), and (8) planar contact with friction (0 freedoms) [15]. Among the above finger contact models, three contact models, (2) point contact without friction, (4) point contact with friction, and (6) soft finger, are most common [21]. This section deals with these three contact models to analyze closure properties.

6.2.1 Grasp Matrix and Hand Jacobian Matrix

Motion constraints through contacts were presented in Section 2.9. Suppose that p robot fingers grasp an object as illustrated in Fig. 6.1. Let $\{O\}$, $\{D_k\}$, and $\{W\}$ be coordinate frames fixed on the object, at the kth contact point between a finger and the object, and the inertial coordinate frame, respectively. The axes of frame $\{D_k\}$ are defined so that the z-axis is normal to the contact surface and points to the interior of the object. The other two axes are orthogonal and lie in the tangential plane (see Fig. 6.2). The origin of frame $\{D_k\}$, O_{D_k}, is placed at the contact point in case of a point contact, or at the center of pressure (CoP) in the case of a plane or line contact. The terms O_O and O_W denote the origins of coordinate frames $\{O\}$ and $\{W\}$, respectively. In general, O_O can be any point on the object. The preferable choice, however, is

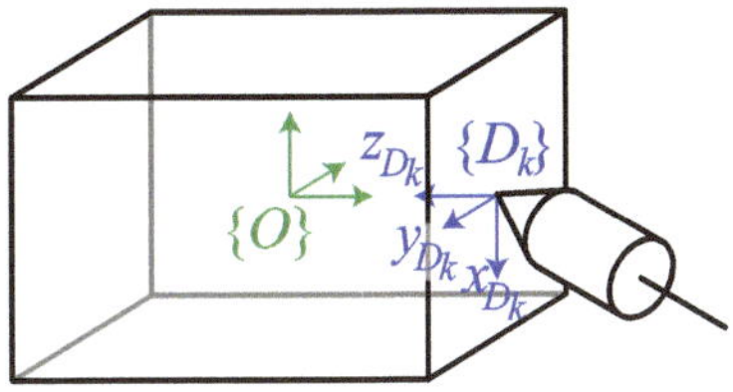

FIGURE 6.2 Contact coordinate frame $\{D_k\}$.

at the center of mass (CoM) since in this case, a force applied along $\boldsymbol{r}_{\overleftarrow{D_k O}}$ (see Fig. 6.1) will not produce a moment about the CoM. Thus, the equation of motion will be simple. Note that in the above figures and throughout this chapter, the j subscript will be used to indicate the number of the finger (or arm).[1] The terms $\mathcal{X}_O$ and $\mathcal{X}_{D_k}$ denote the 6D position (position and orientation) of the object at O_O and O_{D_k}, respectively, with respect to $\{W\}$.

The number of motion constraints at each contact joint is denoted as c_j (< 6) as defined in Section 2.9. The case of $c_j = 6$ will be described in Section 6.3; $\boldsymbol{r}_{\overleftarrow{D_k O}}$ is a 3D vector pointing from O_O to O_{D_k} with respect to $\{W\}$.

Let $\mathcal{V}_O = \begin{bmatrix} \boldsymbol{v}_O^T & \boldsymbol{\omega}_O^T \end{bmatrix}^T$ and ${}^{D_k}\mathcal{V}_{D_k}^O = \begin{bmatrix} {}^{D_k}\boldsymbol{v}_{D_k}^T & {}^{D_k}\boldsymbol{\omega}_{D_k}^T \end{bmatrix}^T$ be the translational and angular velocities (twists) (see Section 2.4) of the object at O_O with respect to $\{W\}$ and at contact point $\mathcal{X}_{D_k}$ with respect to $\{D_k\}$, respectively. The relationship between $\mathcal{V}_O$ and ${}^{D_k}\mathcal{V}_{D_k}^O$ is given by

$$ {}^{D_k}\mathcal{V}_{D_k}^O = {}^{D_k}\mathbb{X}_{W_{\overleftarrow{D_k O}}} \mathcal{V}_O, \tag{6.1} $$

where

$$ {}^{D_k}\mathbb{X}_{W_{\overleftarrow{D_k O}}} = \begin{bmatrix} \boldsymbol{R}_{D_k}^T & 0_3 \\ 0_3 & \boldsymbol{R}_{D_k}^T \end{bmatrix} \begin{bmatrix} \boldsymbol{E}_3 & -\left[\boldsymbol{r}_{\overleftarrow{D_k O}}^{\times}\right] \\ 0_3 & \boldsymbol{E}_3 \end{bmatrix} = \begin{bmatrix} \boldsymbol{R}_{D_k}^T & -\boldsymbol{R}_{D_k}^T\left[\boldsymbol{r}_{\overleftarrow{D_k O}}^{\times}\right] \\ 0_3 & \boldsymbol{R}_{D_k}^T \end{bmatrix} \in \Re^{6\times 6} \tag{6.2} $$

is the twist transform matrix from $\mathcal{V}_O$ to ${}^{D_k}\mathcal{V}_{D_k}^O$; $\boldsymbol{R}_{D_k}$ is the rotation matrix that represents the orientation of $\{D_k\}$ with respect to $\{W\}$, ${}^{D_k}\mathbb{X}$ is called *partial grasp matrix* [24].

Let ${}^{D_k}\mathcal{V}_{D_k}^l$ be the twist of the finger at contact point $\mathcal{X}_{D_k}$ with respect to $\{D_k\}$; ${}^{D_k}\mathcal{V}_{D_k}^l$ is given as follows:

$$ {}^{D_k}\mathcal{V}_{D_k}^l = {}^{D_k}\boldsymbol{J}(\boldsymbol{\theta}_{D_k})\dot{\boldsymbol{\theta}}_{D_k}, \tag{6.3} $$

where ${}^{D_k}\boldsymbol{J}$ is the finger Jacobian matrix defined in $\{D_k\}$ and $\boldsymbol{\theta}_{D_k}$ is the joint angle vector of the finger in contact with the object at O_{D_k}.

Contact joint models were described in Section 2.9.3. The velocities of the object and the finger in contact are constrained at O_{D_k}. The motion constraint of the object and the finger at

[1] In other chapters of this work (e.g. in Chapter 2) the j subscript is used to indicate l (left) or r (right).

contact point O_{D_k} is given by

$$ {}^{D_k}\mathbb{B}_c^T \left({}^{D_k}\mathcal{V}_{D_k}^l - {}^{D_k}\mathcal{V}_{D_k}^O \right) = \mathbf{0}_{c_j}, \tag{6.4} $$

where ${}^{D_k}\mathbb{B}_c \in \Re^{6\times c_j}$ is the constraint basis at contact point O_{D_k} with respect to $\{D_k\}$. See Section 2.10.1 for motion analysis of closed loop chains.

The twists of the object and the finger, ${}^{D_k}\mathcal{V}_{D_k}^O$ and ${}^{D_k}\mathcal{V}_{D_k}^l$, for $k = 1, \cdots, p$, are combined into ${}^C\mathcal{V}_C^O$ and ${}^C\mathcal{V}_C^l$, respectively, as follows:

$$ {}^C\mathcal{V}_C^O = \begin{bmatrix} {}^{D_1}\mathcal{V}_{D_1}^O \\ \vdots \\ {}^{D_p}\mathcal{V}_{D_p}^O \end{bmatrix}, \quad {}^C\mathcal{V}_C^l = \begin{bmatrix} {}^{D_1}\mathcal{V}_{D_1}^l \\ \vdots \\ {}^{D_p}\mathcal{V}_{D_p}^l \end{bmatrix}. \tag{6.5} $$

Here ${}^C\mathcal{V}_C^O$ and ${}^C\mathcal{V}_C^l$ are given by

$$ {}^C\mathcal{V}_C^O = {}^C\mathbb{X}\mathcal{V}_O, \tag{6.6} $$

$$ {}^C\mathcal{V}_C^l = {}^C\boldsymbol{J}(\boldsymbol{\theta}_C)\dot{\boldsymbol{\theta}}_C, \tag{6.7} $$

where

$$ {}^C\mathbb{X} = \begin{bmatrix} {}^{D_1}\mathbb{X} \\ \vdots \\ {}^{D_p}\mathbb{X} \end{bmatrix}, \quad {}^C\boldsymbol{J} = \begin{bmatrix} {}^{D_1}\boldsymbol{J} & & \mathbf{0} \\ & \ddots & \\ \mathbf{0} & & {}^{D_p}\boldsymbol{J} \end{bmatrix}, \quad \boldsymbol{\theta}_C = \begin{bmatrix} \boldsymbol{\theta}_{D_1} \\ \vdots \\ \boldsymbol{\theta}_{D_p} \end{bmatrix}. $$

The total motion constraints are described as follows:

$$ {}^C\mathbb{B}_c^T \left({}^C\mathcal{V}_C^l - {}^C\mathcal{V}_C^O \right) = \mathbf{0}_c, \tag{6.8} $$

where

$$ c = \sum_{j=1}^{p} c_j \quad \text{and} \quad {}^C\mathbb{B}_c^T = \begin{bmatrix} {}^{D_1}\mathbb{B}_c^T & & \mathbf{0} \\ & \ddots & \\ \mathbf{0} & & {}^{D_p}\mathbb{B}_c^T \end{bmatrix} \in \Re^{c\times 6p}. $$

Eq. (6.8) can be rewritten in a matrix form as

$$ \begin{bmatrix} -{}^C\boldsymbol{G}^T & {}^C\boldsymbol{J} \end{bmatrix} \begin{bmatrix} \mathcal{V}_O \\ \dot{\boldsymbol{\theta}}_C \end{bmatrix} = \mathbf{0}_c, \tag{6.9} $$

$$ {}^C\boldsymbol{G}^T \equiv {}^C\mathbb{B}_c^{T\,C}\mathbb{X} \in \Re^{c\times 6}, $$

$$ {}^C\boldsymbol{J} \equiv {}^C\mathbb{B}_c^{T\,C}\boldsymbol{J} \in \Re^{c\times n_{\theta_C}}, $$

where n_{θ_c} is the dimension of $\boldsymbol{\theta}_C$; ${}^C\boldsymbol{G}$ and ${}^C\boldsymbol{J}$ are called *grasp matrix* and *hand Jacobian*, respectively [24].

6.2.2 Statics of Grasping

Eq. (6.9) is rewritten as follows:

$$ {}^{C}\boldsymbol{G}^{T}\mathcal{V}_{O} = {}^{C}\boldsymbol{J}\dot{\boldsymbol{\theta}}_{C} = {}^{C}\mathbb{B}_{c}^{T\,C}\mathcal{V}_{C}^{l} \in \Re^{c}. \tag{6.10}$$

Here ${}^{C}\mathbb{B}_{c}^{T}(\boldsymbol{q}_{C})^{C}\mathcal{V}_{C}^{l}$ represents a vector of restricted velocities at the contact points.

Let ${}^{D_k}\mathcal{F}_{D_k}^{l}$ be the forces and moments (wrenches) applied to the object at contact point $\mathcal{X}_{D_k}$ by the jth finger with respect to $\{D_k\}$; $\mathcal{F}_{O}$ is the net object wrench at O_O with respect to $\{W\}$, resulting from the finger wrenches. Note that the wrenches applied by the fingers are restricted by the contact types given in (3.15).

The wrenches applied to each contact point by the fingers are stacked as follows:

$$ {}^{C}\mathcal{F}_{C}^{l} = \begin{bmatrix} {}^{D_1}\mathcal{F}_{D_1}^{l} \\ \vdots \\ {}^{D_p}\mathcal{F}_{D_p}^{l} \end{bmatrix}.$$

The instantaneous power defined in the object coordinate frame and the one defined in the contact coordinate frame must be the same. Therefore, using (6.10), the following relationship is obtained:

$$\begin{aligned} \mathcal{V}_{O}^{T}\mathcal{F}_{O} &= \left({}^{C}\mathbb{B}_{c}^{T\,C}\mathcal{V}_{C}^{l}\right)^{T}\left({}^{C}\mathbb{B}_{c}^{T\,C}\mathcal{F}_{C}^{l}\right) \\ &= \left({}^{C}\boldsymbol{G}^{T}\mathcal{V}_{O}\right)^{T}\left({}^{C}\mathbb{B}_{c}^{T\,C}\mathcal{F}_{C}^{l}\right) \\ &= \mathcal{V}_{O}^{T\,C}\boldsymbol{G}\left({}^{C}\mathbb{B}_{c}^{T\,C}\mathcal{F}_{C}^{l}\right). \end{aligned} \tag{6.11}$$

From (6.11), the force generated at O_O with respect to $\{W\}$ is given as follows:

$$ \mathcal{F}_{O} = {}^{C}\boldsymbol{G}\left({}^{C}\mathbb{B}_{c}^{T\,C}\mathcal{F}_{C}^{l}\right). \tag{6.12}$$

6.2.3 Constraint Types

Various grasp types are illustrated in Fig. 6.3. In the case of Fig. 6.3A, if the constraint points are locked, the object cannot move even infinitesimally. Such a grasp type is referred to as the *form closure* (more precisely the *first-order form closure*). A form-closed grasp is also referred to as the *power grasp* or *enveloping grasp*. In the case of Fig. 6.3B, the object cannot move much but can infinitesimally rotate around the barycenter even if the constraint points are locked, because the contact points do not constrain the object in the tangential directions. This grasp type is a *form closure* of higher order. The order of the *form closure* depends on the curvature of the contact surface of the object or the hand. It is easily understood that the object can rotate around the barycenter in the case of Fig. 6.3C, even if the constraint points are locked. However, if the friction forces are considered, the grasp illustrated in Fig. 6.3C can keep the balance. Hence such a grasp is characterized as a *force closure*. Note that all the form-closure grasps are also force closure.

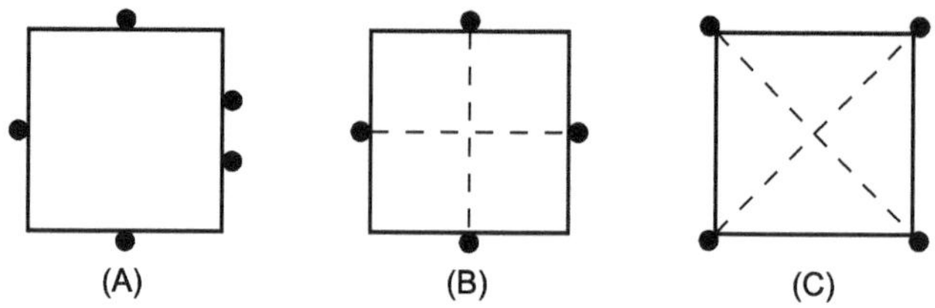

FIGURE 6.3 Various grasp types. The square denotes the grasped object. The ● symbols are the grasping points. (A) First order form-closure. (B) Second order form-closure. (C) Non-form-closure.

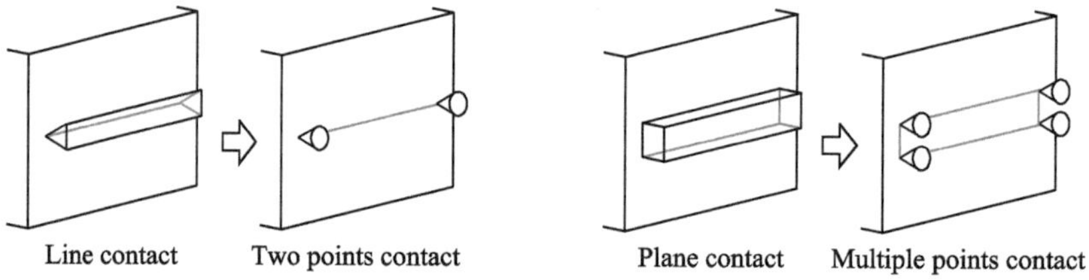

FIGURE 6.4 Approximation of a line contact and a plane contact.

In order to analyze the constraints imposed by the fingers, the line contacts and the plane contacts will be replaced by corresponding multipoint contacts, as illustrated in Fig. 6.4. Let ψ_j be the gap between the object and the jth fingertip; ψ_j will be a function of the object's position and orientation, $\mathcal{X}_O$, and the joint coordinates of the jth finger, $\boldsymbol{\theta}_{D_k}$. If ψ_j is negative, the jth fingertip penetrates into the object. If penetration of a fingertip into the object is not permitted, inequality constraint conditions have to be defined at each contact point, as follows:

$$\psi_j\left(\mathcal{X}_O, \boldsymbol{\theta}_{D_k}\right) \geq 0. \tag{6.13}$$

6.2.4 Form Closure

There are several definitions for form closure. In [2], form closure was defined as

A set of contact constraints is defined as the form closure if, for all object motions, at least one contact constraint is violated.

Reference [24] defined the form closure using the gap function given in (6.13). Suppose the following equation holds for $\mathcal{X}_O$ and $\boldsymbol{\theta}_{D_k}$:

$$\psi_j\left(\mathcal{X}_O, \boldsymbol{\theta}_{D_k}\right) = 0 \quad \forall j = 1, \ldots, c, \tag{6.14}$$

where c is the number of constraints. The definition of the form closure in [24] is

A grasp $(\mathcal{X}_O, \boldsymbol{\theta}_{D_k})$ has a form closure if and only if the following implication holds:

$$\boldsymbol{\psi}\left(\mathcal{X}_O + d\mathcal{X}_O, \boldsymbol{\theta}_{D_k}\right) \geq \mathbf{0} \quad \Rightarrow \quad d\mathcal{X}_O = \mathbf{0} \tag{6.15}$$

where $\boldsymbol{\psi}$ is the c-dimensional vector of gap functions with jth component equal to $\psi_j\left(\mathcal{X}_O, \boldsymbol{\theta}_{D_k}\right)$.

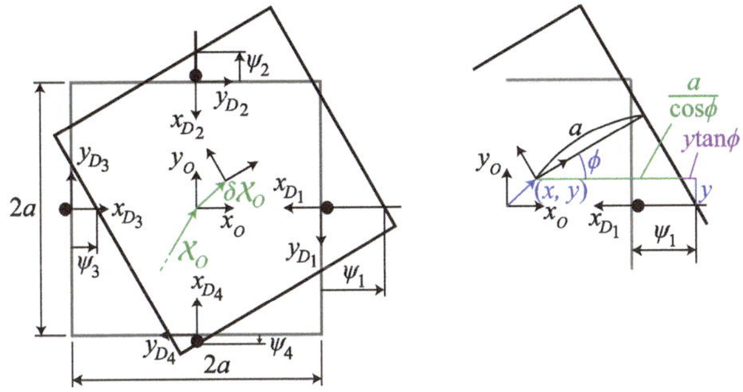

FIGURE 6.5 Constraint analysis. The 6D object position/orientation $\mathcal{X}_O$ moves to $\mathcal{X}_O + \delta\mathcal{X}_O$, and the finger tips do not move.

If the first-order approximation of $\boldsymbol{\psi}\left(\mathcal{X}_O + d\mathcal{X}_O, \boldsymbol{\theta}_{D_k}\right)$ is considered, the definition of the form closure (6.15) can be written as

$$\boldsymbol{\psi}\left(\mathcal{X}_O + d\mathcal{X}_O, \boldsymbol{\theta}_{D_k}\right) \simeq \frac{\partial \boldsymbol{\psi}\left(\mathcal{X}_O, \boldsymbol{\theta}_{D_k}\right)}{\partial \mathcal{X}_O} d\mathcal{X}_O \geq \mathbf{0} \quad \Rightarrow \quad d\mathcal{X}_O = \mathbf{0}. \tag{6.16}$$

If a grasp satisfies (6.16), the grasp is *first-order form-closed.*

Case Study

Using the planar case illustrated in Fig. 6.3B, the form closure is analyzed as a case study. Suppose a square object is constrained at $(x, y) = (a, 0)$, $(0, a)$, $(-a, 0)$, and $(0, -a)$. The initial position and orientation of the object is $(x_0, y_0, \phi_0) = (0, 0, 0)$. When the object moves to (x, y, ϕ), some constraints may be violated and some constraints may break as illustrated in Fig. 6.5, if the constraint points are locked. Gap functions ψ_j are defined as gaps between the constraint position and the edge of the object along the x_{D_k}-axes.

As illustrated in Fig. 6.5, point $(a, -y_0, 0)$ on the right edge of the object moves to $(a - \psi_1, 0, \phi)$ when the object moves from $(0, 0, 0)$ to (x, y, ϕ). This transformation is given as

$$\begin{bmatrix} a - \psi_1 \\ 0 \end{bmatrix} = \begin{bmatrix} \cos\phi & -\sin\phi \\ \sin\phi & \cos\phi \end{bmatrix} \begin{bmatrix} a \\ -y_a \end{bmatrix} + \begin{bmatrix} x \\ y \end{bmatrix}. \tag{6.17}$$

By eliminating y_a from (6.17), the gap function ψ_1 is obtained. In the same manner, ψ_j for $j = 2, 3, 4$ are obtained. Finally, the constraint condition is given as

$$\begin{bmatrix} \psi_1(x, y, \phi) \\ \psi_2(x, y, \phi) \\ \psi_3(x, y, \phi) \\ \psi_4(x, y, \phi) \end{bmatrix} = \begin{bmatrix} -x - y\tan\phi - \dfrac{a}{\cos\phi} + a \\ x\tan\phi - y - \dfrac{a}{\cos\phi} + a \\ x + y\tan\phi - \dfrac{a}{\cos\phi} + a \\ -x\tan\phi + y - \dfrac{a}{\cos\phi} + a \end{bmatrix} \geq \begin{bmatrix} 0 \\ 0 \\ 0 \\ 0 \end{bmatrix}. \tag{6.18}$$

It is obvious that $\psi_j(0,0,0)=0$ for $j=1,\ldots,4$. The first-order approximation of (6.18) around $\psi_j(x,y,\phi)=(0,0,0)$ is

$$\begin{bmatrix}\psi_1(\delta x,\delta y,\delta\psi)\\ \psi_2(\delta x,\delta y,\delta\psi)\\ \psi_3(\delta x,\delta y,\delta\psi)\\ \psi_4(\delta x,\delta y,\delta\psi)\end{bmatrix} \simeq \boldsymbol{\psi}(0,0,0)+\left\{\frac{\partial\boldsymbol{\psi}}{\partial x}\delta x+\frac{\partial\boldsymbol{\psi}}{\partial y}\delta y+\frac{\partial\boldsymbol{\psi}}{\partial\phi}\delta\phi\right\}\Bigg|_{(0,0,0)} = \begin{bmatrix}-\delta x\\ -\delta y\\ \delta x\\ \delta y\end{bmatrix} \geq \begin{bmatrix}0\\0\\0\\0\end{bmatrix}. \tag{6.19}$$

The solutions to (6.19) include $(\delta x,\delta y,\delta\phi)=(0,0,\delta\phi)$. Since the grasp illustrated in Fig. 6.3B allows for infinitesimal rotations $\delta\phi$, it follows that the grasp cannot be first-order form-closed.

The second-order approximation of (6.18) around $\psi_j(x,y,\phi)=(0,0,0)$ becomes as follows:

$$\begin{bmatrix}\psi_1(\delta x,\delta y,\delta\psi)\\ \psi_2(\delta x,\delta y,\delta\psi)\\ \psi_3(\delta x,\delta y,\delta\psi)\\ \psi_4(\delta x,\delta y,\delta\psi)\end{bmatrix} \simeq \begin{bmatrix}-\delta x-\delta y\delta\phi-\frac{a}{2}(\delta\phi)^2\\ -\delta y+\delta x\delta\phi-\frac{a}{2}(\delta\phi)^2\\ \delta x+\delta y\delta\phi-\frac{a}{2}(\delta\phi)^2\\ \delta y-\delta x\delta\phi-\frac{a}{2}(\delta\phi)^2\end{bmatrix} \geq \begin{bmatrix}0\\0\\0\\0\end{bmatrix}. \tag{6.20}$$

The only solution to (6.20) is $(\delta x,\delta y,\delta\phi)=(0,0,0)$. Therefore, the grasp illustrated in Fig. 6.3B is second-order form-closed.

6.2.5 Force Closure

In Section 6.2.4, frictionless constraints were assumed. However, because of friction forces, a grasp may be maintained for all object motions even if the grasp is not form-closed. Such grasps are referred to as the *force closure* or *force-closed*. The main difference between a *form closure* and a *force closure* is the consideration of the friction at the grasping points.

Contact joints with friction were discussed in Section 2.9.3. The cases of point contact with friction and a soft-finger contact were formulated in (3.9) and (3.11), respectively. A box grasped by two fingers and the respective friction cones are illustrated in Fig. 6.6. The half-angle of the friction cone is given by $\tan^{-1}\mu$, $\mu>0$ being the static friction coefficient (cf. Section 2.9.3).

Definition of Force Closure

The definition of *force closure*, according to [16], is as follows:

A grasp is a force closure if, [for] any given external wrench $\mathcal{F}_e$ ($\mathcal{F}_e\in\Re^6$ in 3D grasping, while $\mathcal{F}_e\in\Re^3$ in 2D grasping) applied to the object, there exist contact forces $({}^C\mathbb{B}_c^T\,{}^C\mathcal{F}_C^l)\in FC$ such that

$${}^C\boldsymbol{G}\left({}^C\mathbb{B}_c^T\,{}^C\mathcal{F}_C^l\right)=-\mathcal{F}_e. \tag{6.21}$$

Here FC is a set of forces within the friction cones defined as $FC=\left\{f_{D_k}\in FC_{D_k};\ j=1,\cdots,p\right\}$. See (3.9) and (3.11) for FC_{D_k}. The l.h.s. of (6.21) is equivalent to $\mathcal{F}_O$ calculated in (6.12).

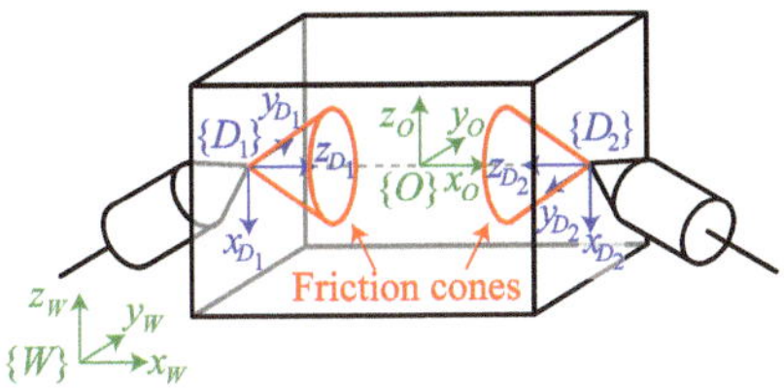

FIGURE 6.6 Friction cones when two soft fingers grasp a box.

Case Study

Suppose two soft fingers grasp a box as illustrated in Fig. 6.6 (this example is the same as Example 5.2 in [16]). It is clear that the grasp illustrated in Fig. 6.6 is not form-closed. The force closure of this example is examined in this case study.

The rotation matrices $\boldsymbol{R}_{D_1}$ and $\boldsymbol{R}_{D_1}$ are

$$\boldsymbol{R}_{D_1} = \begin{bmatrix} 0 & 0 & 1 \\ 0 & 1 & 0 \\ -1 & 0 & 0 \end{bmatrix}, \qquad \boldsymbol{R}_{D_2} = \begin{bmatrix} 0 & 0 & -1 \\ 0 & -1 & 0 \\ -1 & 0 & 0 \end{bmatrix}. \tag{6.22}$$

The position vectors $\boldsymbol{r}_{D_1}$ and $\boldsymbol{r}_{D_2}$ are given as $\boldsymbol{r}_{D_1} = \begin{bmatrix} -r & 0 & 0 \end{bmatrix}^T$ and $\boldsymbol{r}_{D_2} = \begin{bmatrix} r & 0 & 0 \end{bmatrix}^T$, respectively.

Here ${}^{D_1}\mathbb{X}$ and ${}^{D_2}\mathbb{X}$ are calculated from (6.2) as

$${}^{D_1}\mathbb{X} = \begin{bmatrix} 0 & 0 & -1 & 0 & -r & 0 \\ 0 & 1 & 0 & 0 & 0 & -r \\ 1 & 0 & 0 & 0 & 0 & 0 \\ 0 & 0 & 0 & 0 & 0 & -1 \\ 0 & 0 & 0 & 0 & 1 & 0 \\ 0 & 0 & 0 & 1 & 0 & 0 \end{bmatrix}, \qquad {}^{D_2}\mathbb{X} = \begin{bmatrix} 0 & 0 & -1 & 0 & r & 0 \\ 0 & -1 & 0 & 0 & 0 & -r \\ -1 & 0 & 0 & 0 & 0 & 0 \\ 0 & 0 & 0 & 0 & 0 & -1 \\ 0 & 0 & 0 & 0 & -1 & 0 \\ 0 & 0 & 0 & -1 & 0 & 0 \end{bmatrix}. \tag{6.23}$$

The constraint bases ${}^{D_1}\mathbb{B}_c$ and ${}^{D_2}\mathbb{B}_c$ are given as

$${}^{D_1}\mathbb{B}_c = {}^{D_2}\mathbb{B}_c = \begin{bmatrix} 1 & 0 & 0 & 0 \\ 0 & 1 & 0 & 0 \\ 0 & 0 & 1 & 0 \\ 0 & 0 & 0 & 0 \\ 0 & 0 & 0 & 0 \\ 0 & 0 & 0 & 1 \end{bmatrix}. \tag{6.24}$$

The transpose of the grasp matrix can be calculated from (6.23) and (6.24) as

$$
{}^{C}\boldsymbol{G}^{T} = \begin{bmatrix} {}^{D_1}\mathbb{B}_c^{T}\, {}^{D_1}\mathbb{X} \\ {}^{D_2}\mathbb{B}_c^{T}\, {}^{D_2}\mathbb{X} \end{bmatrix} = \left[\begin{array}{cccccc} 0 & 0 & -1 & 0 & -r & 0 \\ 0 & 1 & 0 & 0 & 0 & -r \\ 1 & 0 & 0 & 0 & 0 & 0 \\ 0 & 0 & 0 & 1 & 0 & 0 \\ \hline 0 & 0 & -1 & 0 & r & 0 \\ 0 & -1 & 0 & 0 & 0 & -r \\ -1 & 0 & 0 & 0 & 0 & 0 \\ 0 & 0 & 0 & -1 & 0 & 0 \end{array}\right]. \tag{6.25}
$$

The l.h.s. of (6.21) becomes

$$
{}^{C}\boldsymbol{G}\left({}^{C}\mathbb{B}_c^{T}\, {}^{C}\mathcal{F}_C^{l}\right) = {}^{C}\boldsymbol{G}\left[\begin{array}{c} {}^{D_1}\overline{\mathcal{F}}_{D_1} \\ \hline {}^{D_2}\overline{\mathcal{F}}_{D_2} \end{array}\right] = {}^{C}\boldsymbol{G}\left[\begin{array}{c} {}^{D_1}f_{D_{1x}} \\ {}^{D_1}f_{D_{1y}} \\ {}^{D_1}f_{D_{1z}} \\ {}^{D_1}m_{D_{1z}} \\ \hline {}^{D_2}f_{D_{2x}} \\ {}^{D_2}f_{D_{2y}} \\ {}^{D_2}f_{D_{2z}} \\ {}^{D_2}m_{D_{2z}} \end{array}\right] = \begin{bmatrix} {}^{D_1}f_{D_{1z}} - {}^{D_2}f_{D_{2z}} \\ {}^{D_1}f_{D_{1y}} - {}^{D_2}f_{D_{2y}} \\ -{}^{D_1}f_{D_{1x}} - {}^{D_2}f_{D_{2x}} \\ {}^{D_1}m_{D_{1z}} - {}^{D_2}m_{D_{2z}} \\ -r\left({}^{D_1}f_{D_{1x}} - {}^{D_2}f_{D_{2x}}\right) \\ -r\left({}^{D_1}f_{D_{1y}} + {}^{D_2}f_{D_{2y}}\right) \end{bmatrix}. \tag{6.26}
$$

Therefore, if contact forces ${}^{D_1}\overline{\mathcal{F}}_{D_1} \in FC_{D_1}$ and ${}^{D_2}\overline{\mathcal{F}}_{D_2} \in FC_{D_2}$ exist for a given external wrench $\mathcal{F}_e$, such that

$$
{}^{C}\boldsymbol{G}\left[\begin{array}{c} {}^{D_1}\overline{\mathcal{F}}_{D_1} \\ \hline {}^{D_2}\overline{\mathcal{F}}_{D_2} \end{array}\right] = -\mathcal{F}_e, \tag{6.27}
$$

the grasp is force-closed (but not form-closed). The friction cones, given in (3.11), are rewritten here as

$$
\left\{FC_{D_1} : \sqrt{{}^{D_1}f_{D_{1x}}^2 + {}^{D_1}f_{D_{1y}}^2} \le \mu\, {}^{D_1}f_{D_{1z}},\ {}^{D_1}f_{D_{1z}} \ge 0,\ |{}^{D_1}m_{D_{1z}}| \le \gamma\, {}^{D_1}f_{D_{1z}}\right\},
$$
$$
\left\{FC_{D_2} : \sqrt{{}^{D_2}f_{D_{2x}}^2 + {}^{D_2}f_{D_{2y}}^2} \le \mu\, {}^{D_2}f_{D_{2z}},\ {}^{D_2}f_{D_{2z}} \ge 0,\ |{}^{D_2}m_{D_{2z}}| \le \gamma\, {}^{D_2}f_{D_{2z}}\right\}.
$$

If point contacts with friction are assumed instead of soft-finger contacts, as in the above case, the rotation of the object around z_{D_k} will be unconstrained. In this case, constant bases ${}^{D_1}\boldsymbol{C}_{D_1}$ and ${}^{D_2}\boldsymbol{C}_{D_2}$ are given as

$$
{}^{D_1}\mathbb{B}_c = {}^{D_2}\mathbb{B}_c = \begin{bmatrix} 1 & 0 & 0 \\ 0 & 1 & 0 \\ 0 & 0 & 1 \\ 0 & 0 & 0 \\ 0 & 0 & 0 \\ 0 & 0 & 0 \end{bmatrix}. \tag{6.28}
$$

The transpose of the grasp matrix, obtained from (6.23) and (6.28), can be calculated as

$$ {}^{C}\boldsymbol{G}^{T} = \begin{bmatrix} {}^{D_1}\mathbb{B}_{C}^{T}\,{}^{D_1}\mathbb{X} \\ {}^{D_2}\mathbb{B}_{C}^{T}\,{}^{D_2}\mathbb{X} \end{bmatrix} = \left[\begin{array}{cccccc} 0 & 0 & -1 & 0 & -r & 0 \\ 0 & 1 & 0 & 0 & 0 & -r \\ 1 & 0 & 0 & 0 & 0 & 0 \\ \hline 0 & 0 & -1 & 0 & r & 0 \\ 0 & -1 & 0 & 0 & 0 & -r \\ -1 & 0 & 0 & 0 & 0 & 0 \end{array}\right]. \tag{6.29} $$

The l.h.s. of (6.21) becomes

$$ {}^{C}\boldsymbol{G}\left({}^{C}\mathbb{B}_{C}^{T}\,{}^{C}\mathcal{F}_{C}^{I}\right) = {}^{C}\boldsymbol{G}\left[\begin{array}{c} {}^{D_1}\overline{\mathcal{F}}_{D_1} \\ \hline {}^{D_2}\overline{\mathcal{F}}_{D_2} \end{array}\right] = {}^{C}\boldsymbol{G}\left[\begin{array}{c} {}^{D_1}f_{D_{1x}} \\ {}^{D_1}f_{D_{1y}} \\ {}^{D_1}f_{D_{1z}} \\ \hline {}^{D_2}f_{D_{2x}} \\ {}^{D_2}f_{D_{2y}} \\ {}^{D_2}f_{D_{2z}} \end{array}\right] = \begin{bmatrix} {}^{D_1}f_{D_{1z}} - {}^{D_2}f_{D_{2z}} \\ {}^{D_1}f_{D_{1y}} - {}^{D_2}f_{D_{2y}} \\ -{}^{D_1}f_{D_{1x}} - {}^{D_2}f_{D_{2x}} \\ 0 \\ -r\left({}^{D_1}f_{D_{1x}} - {}^{D_2}f_{D_{2x}}\right) \\ -r\left({}^{D_1}f_{D_{1y}} + {}^{D_2}f_{D_{2y}}\right) \end{bmatrix}. \tag{6.30} $$

It is clear from (6.30) that the external moment around x_O cannot be controlled, neither by ${}^{D_1}\overline{\mathcal{F}}_{D_1}$ nor by ${}^{D_2}\overline{\mathcal{F}}_{D_2}$.

From the above discussion it should be apparent that the grasp illustrated in Fig. 6.6 is not force-closed, when the contact model is a point contact with friction (cf. (3.9)). On the other hand, when a soft-finger contact model is assumed, the grasp will be force-closed. It should be also obvious that, if a contact model without friction is assumed, the grasp cannot be force-closed.

6.3 MULTIARM OBJECT MANIPULATION CONTROL METHODS

6.3.1 Background of Multiarm Object Manipulation

In order to achieve an object manipulation with a dual-arm manipulator system without applying an excessive force to the object, the master-slave control approach was proposed in the pioneering work [17]. According to the approach, one of the robot arms behaves as a master and controls the position of the object, while the other arm behaves as a slave and controls the internal wrenches imposed on the object. Other control approaches for cooperating manipulators, studied throughout the years, are the hybrid position/force [5,33,27,18] and the impedance [22,26,19,3,6,14] control methods.

With the focus on internal wrench control, virtual constructs have been introduced such as the *virtual linkage* [28] and the *virtual sticks* [27].

Another related field of research is the cooperative object transportation by multiple mobile robots. In this field, the leader-follower-type control schemes have been studied enthusiastically [13,7]. The leader-follower-type control between two humanoid robots was discussed in [30]. Studies on cooperation between a humanoid robot and a human have also been carried out [32,1]. Symmetry (i.e. not leader-follower)-type cooperation among multiple humanoid robots was proposed in [29,31].

6.3.2 Kinematics and Statics of Multiarm Cooperation

Let $(\boldsymbol{r}, \boldsymbol{R})$ denote the 6D position of a rigid body. A *small displacement* of the body in 6D can be expressed as $(\delta\boldsymbol{r}, \delta\boldsymbol{R})$, where

$$\delta\boldsymbol{r} = \boldsymbol{r}' - \boldsymbol{r}, \tag{6.31}$$

$$\delta\boldsymbol{R} = \boldsymbol{R}'\boldsymbol{R}^T. \tag{6.32}$$

Here $(\boldsymbol{r}', \boldsymbol{R}')$ is the 6D position of the body after the displacement. The angular part of the displacement can also be expressed by a vector as follows:

$$\delta\boldsymbol{\phi} = (\ln\ \delta\boldsymbol{R})^{\vee}, \tag{6.33}$$

where $(\ln\ \delta\boldsymbol{R}) = [\delta\boldsymbol{\phi}^{\times}]$ [8]. The $()^{\vee}$ operator extracts a vector from a skew-symmetric matrix, i.e.

$$\left[\delta\boldsymbol{\phi}^{\times}\right]^{\vee} = \begin{bmatrix} 0 & -\delta\phi_z & \delta\phi_y \\ \delta\phi_z & 0 & -\delta\phi_x \\ -\delta\phi_y & \delta\phi_x & 0 \end{bmatrix}^{\vee} \equiv \begin{bmatrix} \delta\phi_x \\ \delta\phi_y \\ \delta\phi_z \end{bmatrix}. \tag{6.34}$$

The angular displacement vector $\delta\boldsymbol{\phi}$ can then be obtained as [8,16]

$$\delta\boldsymbol{\phi} = (\ln\ \delta\boldsymbol{R})^{\vee} = \begin{cases} \boldsymbol{0} & \text{if } \delta\boldsymbol{R} = \boldsymbol{E}_3 \\ |\delta\phi| \dfrac{\boldsymbol{t}}{\|\boldsymbol{t}\|} & \text{otherwise,} \end{cases} \tag{6.35}$$

where

$$\delta\boldsymbol{R} = \begin{bmatrix} r_{11} & r_{12} & r_{13} \\ r_{21} & r_{22} & r_{23} \\ r_{31} & r_{32} & r_{33} \end{bmatrix}, \quad \boldsymbol{t} = \begin{bmatrix} r_{32} - r_{23} \\ r_{13} - r_{31} \\ r_{21} - r_{12} \end{bmatrix}, \quad |\delta\phi| = \text{atan2}\left(\|\boldsymbol{t}\|, r_{11} + r_{22} + r_{33} - 1\right).$$

Note that $\delta\phi$ is defined here in $-\pi < \delta\phi < \pi$, and $\boldsymbol{t}$ includes the sign of $\delta\phi$. Using the above notation, in what follows the small displacement of the rigid body in 6D will be expressed as

$$\delta\mathcal{X} \equiv \begin{bmatrix} \delta\boldsymbol{r} \\ \delta\boldsymbol{\phi} \end{bmatrix}. \tag{6.36}$$

Suppose p robot arms grasp an object, as illustrated in Fig. 6.7. As clarified in Section 6.1, it is assumed that the robot hands grasp the object firmly whereby each hand impresses a

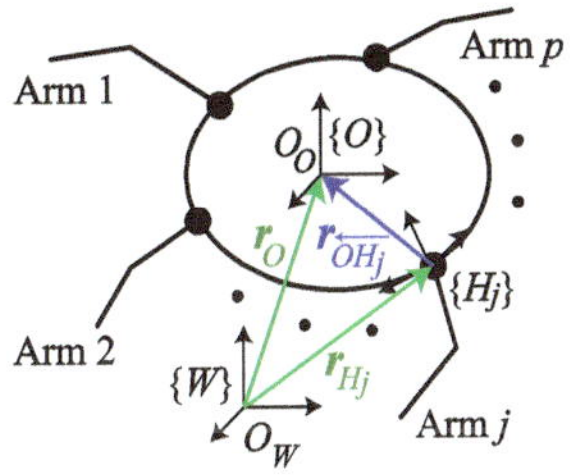

FIGURE 6.7 *p* robot arms that grasp an object.

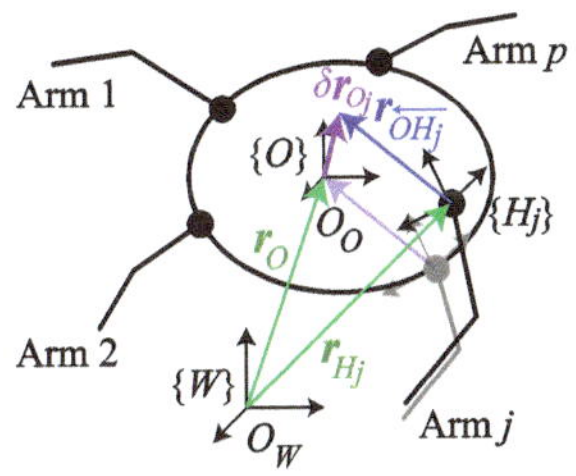

FIGURE 6.8 Virtual displacement at the tip of the virtual stick.

wrench bilaterally. Let $\{O\}$ and $\{H_j\}$ be coordinate frames fixed on the object and at the hand of Arm j, respectively, and $\{W\}$ be the inertial coordinate frame; O_O and O_W denote the origins of the coordinate frames $\{O\}$ and $\{W\}$.

Furthermore, let $(\boldsymbol{r}_{H_j}, \boldsymbol{R}_{H_j})$ and $(\boldsymbol{r}_O, \boldsymbol{R}_O)$ denote the 6D position of Arm j at the grasping point and of the grasped object, respectively. Vector $^{H_j}\boldsymbol{r}_{\overleftarrow{OH_j}}$ is the 3D vector pointing from grasping point $\boldsymbol{r}_{H_j}$ to O_O, w.r.t. the hand coordinate frame $\{H_j\}$. When the grasped object is rigid and does not deform (as assumed here), the vector $^{H_j}\boldsymbol{r}_{\overleftarrow{OH_j}}$ will be constant in coordinate frame $\{H_j\}$. Vector $^{H_j}\boldsymbol{r}_{\overleftarrow{OH_j}}$ is represented as $\boldsymbol{r}_{\overleftarrow{OH_j}} = \boldsymbol{R}_{H_j}{}^{H_j}\boldsymbol{r}_{\overleftarrow{OH_j}}$ in the inertial coordinate frame $\{W\}$. Note that $^{H_j}\boldsymbol{r}_{\overleftarrow{OH_j}}$ can be considered a stick fixed in the coordinate frame $\{H_j\}$. This vector was named the *virtual stick* in [27].

The position of the tip and the orientation of $\boldsymbol{r}_{\overleftarrow{OH_j}}$ are represented by $\boldsymbol{r}_{O_j}$ $(=\boldsymbol{r}_{H_j} + \boldsymbol{r}_{\overleftarrow{OH_j}})$ and $\boldsymbol{R}_{O_j}$, respectively. Since $^{H_j}\boldsymbol{r}_{\overleftarrow{OH_j}}$ is constant in coordinate frame $\{H_j\}$, $\boldsymbol{r}_{O_j}$ is equal to $\boldsymbol{r}_O$, and $\boldsymbol{R}_{O_j}$ is equal to $\boldsymbol{R}_O$. However, if a virtual displacement is considered, $\boldsymbol{r}_{O_j}$ and $\boldsymbol{R}_{O_j}$ will not always be equal to $\boldsymbol{r}_O$ and $\boldsymbol{R}_O$, respectively (see Fig. 6.8).

Let $\boldsymbol{q}_j$ be the joint angle vector of Arm j. A small displacement of the hand of Arm j can be expressed as

$$\delta\mathcal{X}_{O_j} = \begin{bmatrix} \delta\boldsymbol{r}_{O_j} \\ \delta\boldsymbol{\phi}_{O_j} \end{bmatrix} = \boldsymbol{J}_{O_j}\delta\boldsymbol{q}_j, \tag{6.37}$$

where $\boldsymbol{J}_{O_j}$ is the Jacobian matrix of Arm j.

6.3.3 Force and Moment Applied to the Object

Let ${}^{H_j}\mathcal{F}_{H_j}$ and $\mathcal{F}_{O_j}$ be the wrenches applied to the grasped object by Arm j at $\mathcal{X}_{H_j}$ with respect to $\{H_j\}$ and at O_O with respect to $\{W\}$, accordingly. The relationship between ${}^{H_j}\mathcal{F}_{H_j}$ and $\mathcal{F}_{O_j}$ is given as

$$\mathcal{F}_{O_j} = {}^{O}\mathbb{X}_{H_j}^{T}\, {}^{H_j}\mathcal{F}_{H_j}, \tag{6.38}$$

where ${}^{O}\mathbb{X}_{H_j}^{T}$ is the wrench transform matrix from $\{H_j\}$ to $\{O\}$ given in (2.7). The wrench ${}^{H_j}\mathcal{F}_{H_j}$ is calculated from the possible force ${}^{H_j}\overline{\mathcal{F}}_{H_j} \in \Re^{6\times c_j}$ (cf. (3.15)) as

$${}^{H_j}\mathcal{F}_{H_j} = {}^{H_j}\mathbb{B}_c\, {}^{H_j}\overline{\mathcal{F}}_{H_j}. \tag{6.39}$$

Here ${}^{H_j}\mathbb{B}_c$ is the constraint basis at the contact point with respect to $\{H_j\}$ (see Section 2.9.3). Since the robot hands firmly grasp the object, ${}^{H_j}\mathbb{B}_c$ becomes a 6D identity matrix (${}^{H_j}\mathbb{B}_c = \boldsymbol{E}_6$).

When p robot arms grasp an object, the resultant wrench is given as

$$\begin{aligned} \mathcal{F}_O^A &= \sum_{j=1}^{p} \mathcal{F}_{O_j} = \boldsymbol{E}_{6\times 6p}\mathcal{F}_O, \\ \boldsymbol{E}_{6\times 6p} &\equiv \begin{bmatrix} \boldsymbol{E}_6 & \cdots & \boldsymbol{E}_6 \end{bmatrix} \in \Re^{6\times 6p}, \\ \mathcal{F}_O &\equiv \begin{bmatrix} \mathcal{F}_{O_1}^T & \cdots & \mathcal{F}_{O_p}^T \end{bmatrix}^T \in \Re^{6p}, \end{aligned} \tag{6.40}$$

where $\mathcal{F}_O^A$ is the wrench that affects the movement of the object. The wrench $\mathcal{F}_O$ is obtained from the (least-squares) solution of (6.40), i.e.

$$\mathcal{F}_O = \left(\boldsymbol{E}_{6\times 6p}\right)^{\#} \mathcal{F}_O^A + \boldsymbol{N}\left(\boldsymbol{E}_{6\times 6p}\right) \mathcal{F}_O^a, \tag{6.41}$$

where $\left(\boldsymbol{E}_{6\times 6p}\right)^{\#} \in \Re^{6p\times 6}$ is a generalized inverse matrix of $\boldsymbol{E}_{6\times 6p}$,

$$\boldsymbol{N}\left(\boldsymbol{E}_{6\times 6p}\right) = \boldsymbol{E}_{6p} - \left(\boldsymbol{E}_{6\times 6p}\right)^{\#} \boldsymbol{E}_{6\times 6p}$$

is a projection matrix onto the null space of $\boldsymbol{E}_{6\times 6p}$, and

$$\mathcal{F}_O^a = \begin{bmatrix} \left(\mathcal{F}_{O_1}^a\right)^T & \cdots & \left(\mathcal{F}_{O_p}^a\right)^T \end{bmatrix}^T \in \Re^{6p}$$

is an arbitrary wrench.

The reduced-form representation of (6.41) can be written as

$$\begin{aligned} \mathcal{F}_O &= \left(\boldsymbol{E}_{6\times 6p}\right)^{\#} \mathcal{F}_O^A + \boldsymbol{V}\mathcal{F}_O^I = \boldsymbol{U}\mathcal{F}_O^{AI}, \\ \boldsymbol{U} &\equiv \begin{bmatrix} \left(\boldsymbol{E}_{6\times 6p}\right)^{\#} & \boldsymbol{V} \end{bmatrix} \in \Re^{6p\times 6p}, \\ \mathcal{F}_O^{AI} &\equiv \begin{bmatrix} \mathcal{F}_O^A \\ \mathcal{F}_O^I \end{bmatrix} \in \Re^{6p}, \end{aligned} \tag{6.42}$$

where $\boldsymbol{V} \in \Re^{6p \times 6(p-1)}$ is a matrix that satisfies

$$\boldsymbol{E}_{6\times 6p}\boldsymbol{V} = \boldsymbol{0}_{6\times 6(p-1)}, \tag{6.43}$$

and $\boldsymbol{0}_{6\times 6(p-1)}$ is the $6 \times 6(p-1)$ zero matrix; $\mathcal{F}_O^I$ is a set of wrenches defined as

$$\mathcal{F}_O^I = \left[\left(\mathcal{F}_{O_1}^I\right)^T \quad \cdots \quad \left(\mathcal{F}_{O_{p-1}}^I\right)^T \right]^T \in \Re^{6(p-1)}, \tag{6.44}$$

and $\mathcal{F}_{O_j}^I$ corresponds to the jth internal wrench. The motion of the object is unaffected by the internal wrenches by definition.

Case Study

The two expressions (6.41) and (6.42) for $\mathcal{F}_O$ are compared taking $p = 2$ as a case study. The pseudoinverse is used as a generalized inverse to compare the two expressions.

First, $\mathcal{F}_O$ is derived from (6.41) as

$$\begin{aligned} \mathcal{F}_O &= \frac{1}{2}\begin{bmatrix} \boldsymbol{E}_6 \\ \boldsymbol{E}_6 \end{bmatrix} \mathcal{F}_O^A + \frac{1}{2}\begin{bmatrix} \boldsymbol{E}_6 & -\boldsymbol{E}_6 \\ -\boldsymbol{E}_6 & \boldsymbol{E}_6 \end{bmatrix} \mathcal{F}_O^a \\ &= \frac{1}{2}\begin{bmatrix} \boldsymbol{E}_6 \\ \boldsymbol{E}_6 \end{bmatrix} \mathcal{F}_O^A + \frac{1}{2}\begin{bmatrix} \boldsymbol{E}_6 \\ -\boldsymbol{E}_6 \end{bmatrix} \mathcal{F}_{O_1}^a + \frac{1}{2}\begin{bmatrix} -\boldsymbol{E}_6 \\ \boldsymbol{E}_6 \end{bmatrix} \mathcal{F}_{O_2}^a, \end{aligned} \tag{6.45}$$

where $\mathcal{F}_{O_1}^a$ and $\mathcal{F}_{O_2}^a$ are arbitrary 6D wrenches. It should be apparent that the second and third terms on the r.h.s. are redundant expressions.

Next, $\mathcal{F}_O$ is derived using (6.42). The general form of matrix $\boldsymbol{V}$ that satisfies (6.43) is given as

$$\boldsymbol{V} = \xi \begin{bmatrix} \boldsymbol{E}_6 \\ -\boldsymbol{E}_6 \end{bmatrix}, \tag{6.46}$$

where ξ is an arbitrary scalar. Substituting (6.46) into (6.42), $\mathcal{F}_O$ becomes

$$\mathcal{F}_O = \frac{1}{2}\begin{bmatrix} \boldsymbol{E}_6 \\ \boldsymbol{E}_6 \end{bmatrix} \mathcal{F}_O^A + \xi \begin{bmatrix} \boldsymbol{E}_6 \\ -\boldsymbol{E}_6 \end{bmatrix} \mathcal{F}_O^{Ia}, \tag{6.47}$$

where $\mathcal{F}_O^{Ia}$ is an arbitrary 6D wrench.

Comparing (6.47) with (6.45), it becomes apparent that (6.42) is a reduced-form representation of (6.41). The main advantages of (6.42) are that $\boldsymbol{U}$ is a square matrix of size $6p$ and that $\boldsymbol{V}$ is specified so that $\boldsymbol{U}$ is invertible.

6.3.4 Load Distribution

Consider the following matrix:

$$\left(\boldsymbol{E}_{6\times 6p}\right)^{\#} = \begin{bmatrix} \left(1-\sum_{j=2}^{p}\lambda_j\right)\boldsymbol{E}_6 \\ \lambda_2\boldsymbol{E}_6 \\ \cdots \\ \lambda_p\boldsymbol{E}_6 \end{bmatrix}, \tag{6.48}$$

where $0 \leq \lambda_j \leq 1$ and $0 \leq \sum_{j=2}^{p}\lambda_j \leq 1$. It can be verified that the above matrix satisfies the definitive equation of a generalized inverse matrix, i.e. $\boldsymbol{E}_{6\times 6p}\left(\boldsymbol{E}_{6\times 6p}\right)^{\#}\boldsymbol{E}_{6\times 6p} = \boldsymbol{E}_{6\times 6p}$. Substituting (6.48) into (6.42), the net wrench $\mathcal{F}_O$, that should be applied at point O_O by the p robot arms, can be obtained as

$$\mathcal{F}_O = \begin{bmatrix} \mathcal{F}_{O_1} \\ \mathcal{F}_{O_2} \\ \vdots \\ \mathcal{F}_{O_p} \end{bmatrix} = \begin{bmatrix} \left(1-\sum_{j=2}^{p}\lambda_j\right)\boldsymbol{E}_6 \\ \lambda_2\boldsymbol{E}_6 \\ \cdots \\ \lambda_p\boldsymbol{E}_6 \end{bmatrix} \mathcal{F}_O^A + \boldsymbol{V}\mathcal{F}_O^I. \tag{6.49}$$

From this equation it is apparent that the applied force $\mathcal{F}_O^A$ can be distributed to the p robot arms in a desirable way, by setting the λ_js in an appropriate way. For example, a higher load can be distributed to the robot arms that have high-power actuators; λ_j is called the *load sharing coefficient* [25].

It is easily confirmed that $\mathcal{F}_O$ calculated from (6.49) satisfies (6.40). When λ_j is set as $\lambda_j = 1/p$ $(j = 1 \sim p)$, $\boldsymbol{E}_{6\times 6p}^{\#}$ becomes the pseudoinverse matrix, and the applied force $\mathcal{F}_O^A$ will be equally distributed to each robot arm.

6.3.5 Control of the External and Internal Wrenches

Suppose that p robot arms grasp an object and each arm applies a 6D wrench on the object. The p robot arms apply $6p$-dimensional wrenches on the object in total. In other words, the p robot arms can control $6p$-dimensional wrenches generated at the object. A 6D wrench out of the $6p$ wrenches is used to balance the applied wrench $\mathcal{F}_O^A$ defined in (6.40). The rest of the wrenches constitute a composite $6(p-1)$-dimensional wrench that is used to control the internal wrench $\mathcal{F}_O^I$ appearing in (6.42). From the relationship presented in (6.43) it is apparent that the term $\boldsymbol{V}\mathcal{F}_O^I$ in (6.42) exists in the null space of $\boldsymbol{E}_{6\times 6p}$. Hence, the $\boldsymbol{V}\mathcal{F}_O^I$ component will be annihilated when $\mathcal{F}_O$ is premultiplied by $\boldsymbol{E}_{6\times 6p}$, as shown in (6.40).

Suppose that p robot arms grasp an object as illustrated in Fig. 6.9. The number of two-combinations from p robot arms is given by ${}_pC_2 = p(p-1)/2$. Therefore, as illustrated in Fig. 6.9, $6 \times {}_pC_2 = 3p(p-1)$-dimensional internal wrenches can be considered, if an internal wrench is defined between each pair of robot arms.

However, as described above, only $6(p-1)$-dimensional internal wrenches can be controlled by p robot arms. If two robot arms grasp an object, p equals to two and hence $6 \times {}_pC_2 = 6(p-1) = 6$. In this case, the dimension of the internal wrenches is equal to the

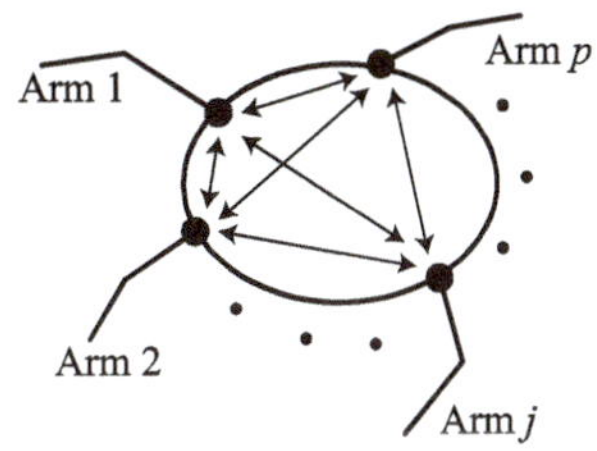

FIGURE 6.9 Internal forces defined between two-combinations from p robot arms.

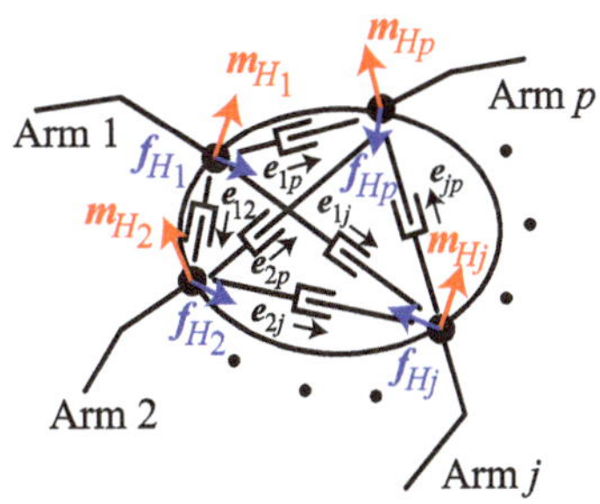

FIGURE 6.10 Concept of a virtual linkage.

controllable dimension. However, when the number of robot arms is more than two, the dimension of the internal wrenches defined for all two-combinations from p (> 2) robot arms is greater than the controllable dimension, which is expressed as $6 \times {}_pC_2 > 6(p-1)$. In such a case, $6(p-1)$-dimensional internal wrenches can be directly controlled, and the remaining $6({}_pC_2-(p-1))$-dimensional internal wrenches are indirectly (implicitly) controlled. Hence $6(p-1)$-dimensional internal wrenches must be chosen to be directly controlled.

Virtual Linkage [28] (see also Section 3.5.2)

The virtual linkage was defined in [28] as: *"A virtual linkage associated with an n-grasp manipulation task is a $6(p-1)$-degree-of-freedom mechanism whose actuated joints characterize the object's internal forces and moments."*

Suppose p robot arms grasp an object as illustrated in Fig. 6.10. Robot Arm j applies force $\boldsymbol{f}_{H_j} \in \Re^3$ and moment $\boldsymbol{m}_{H_j} \in \Re^3$. The net wrench applied at O_O is given as

$$\mathcal{F}_O^A = \begin{bmatrix} \boldsymbol{f}_O^A \\ \boldsymbol{m}_O^A \end{bmatrix} = \begin{bmatrix} \mathbb{T}_f^T & \mathbb{T}_m^T \end{bmatrix} \begin{bmatrix} \boldsymbol{f}_H \\ \boldsymbol{m}_H \end{bmatrix}, \tag{6.50}$$

$$\mathbb{T}_f^T = \begin{bmatrix} \boldsymbol{E}_3 & \cdots & \boldsymbol{E}_3 \\ -[\boldsymbol{r}_1^\times] & \cdots & -[\boldsymbol{r}_p^\times] \end{bmatrix} \in \Re^{6\times 3p}, \quad \mathbb{T}_m^T = \begin{bmatrix} \boldsymbol{0}_3 & \cdots & \boldsymbol{0}_3 \\ \boldsymbol{E}_3 & \cdots & \boldsymbol{E}_3 \end{bmatrix} \in \Re^{6\times 3p}, \tag{6.51}$$

$$\boldsymbol{f}_H = \begin{bmatrix} \boldsymbol{f}_{H_1} \\ \vdots \\ \boldsymbol{f}_{H_p} \end{bmatrix} \in \Re^{3p}, \quad \boldsymbol{m}_H = \begin{bmatrix} \boldsymbol{m}_{H_1} \\ \vdots \\ \boldsymbol{m}_{H_p} \end{bmatrix} \in \Re^{3p}. \tag{6.52}$$

Virtual linkages are introduced between all two-combinations of the p robot arms, as illustrated in Fig. 6.10. Let $\boldsymbol{r}_{H_j}$ be the 3D position vector of the grasping point of Arm j. Note that $\boldsymbol{r}_{H_j}$ can be extracted from the first three components of $\mathcal{X}_{H_j}$.

Unit vectors $\boldsymbol{e}_{ij}$ are defined between $\boldsymbol{r}_{H_i}$ and $\boldsymbol{r}_{H_j}$ as follows:

$$\boldsymbol{e}_{ij} = \frac{\boldsymbol{r}_{H_j} - \boldsymbol{r}_{H_i}}{\|\boldsymbol{r}_{H_j} - \boldsymbol{r}_{H_i}\|}, \quad \boldsymbol{e}_{ji} = -\boldsymbol{e}_{ij}. \tag{6.53}$$

The internal force between $\boldsymbol{r}_{H_i}$ and $\boldsymbol{r}_{H_j}$ is defined as

$$\boldsymbol{f}^I_{H_{ij}} = f^{int}_{ij} \boldsymbol{e}_{ij}. \tag{6.54}$$

Here f^{int}_{ij} $(= f_{ji} \geq 0)$ is the magnitude of the internal force, whereas $\boldsymbol{e}_{ij}$ is its direction. If the magnitudes of the internal forces $\boldsymbol{f}^{int} = \begin{bmatrix} \cdots & f^{int}_{ij} & \cdots \end{bmatrix}^T \in \Re^{{}_pC_2}$ are specified, the force $\boldsymbol{f}_H$ (defined in (6.52)) that realizes the desired internal forces is given as

$$\boldsymbol{f}^I_H = \boldsymbol{V}_L \boldsymbol{f}^{int}. \tag{6.55}$$

Here $\boldsymbol{V}_L \in \Re^{3p \times {}_pC_2}$ is a matrix that relates $\boldsymbol{f}^{int} \in \Re^{{}_pC_2}$ to $\boldsymbol{f}^I_H \in \Re^{3p}$.

Using (6.50) and (6.55), the relationship between resultant/internal forces and applied wrenches is given by

$$\begin{bmatrix} \boldsymbol{f}^A_O \\ \boldsymbol{m}^A_O \\ \boldsymbol{f}^{int} \\ \boldsymbol{m}_H \end{bmatrix} = \mathbb{G} \begin{bmatrix} \boldsymbol{f}_H \\ \boldsymbol{m}_H \end{bmatrix}, \quad \mathbb{G} = \begin{bmatrix} \mathbb{T}^T_f & \mathbb{T}^T_m \\ \boldsymbol{V}^+_L & \boldsymbol{0}_{{}_pC_2 \times 3p} \\ \boldsymbol{0}_{3p} & \boldsymbol{E}_{3p} \end{bmatrix} \in \Re^{({}_pC_2+3p+6)\times 6p}, \tag{6.56}$$

where $\boldsymbol{V}^+_L$ is the Moore–Penrose pseudoinverse matrix of $\boldsymbol{V}_L$. The matrix $\mathbb{G}$ is referred to as the *grasp description matrix* [28]. Note that the matrix $\mathbb{G}$ defined here is different from the *grasp matrix* [16] that indicates convexity conditions.

When the desired resultant wrench, the magnitudes of the internal forces, and the moments at the grasp points are given, the wrenches which should be applied to the grasp points are calculated as follows:

$$\begin{bmatrix} \boldsymbol{f}_H \\ \boldsymbol{m}_H \end{bmatrix} = \mathbb{G}^+ \begin{bmatrix} \boldsymbol{f}^A_O \\ \boldsymbol{m}^A_O \\ \boldsymbol{f}^{int} \\ \boldsymbol{m}_H \end{bmatrix}. \tag{6.57}$$

Consider as an example the case when three robot arms grasp an object, as illustrated in Fig. 6.11. The internal forces are obtained from (6.55) as follows:

$$\begin{bmatrix} \boldsymbol{f}^I_{H_1} \\ \boldsymbol{f}^I_{H_2} \\ \boldsymbol{f}^I_{H_3} \end{bmatrix} = \begin{bmatrix} \boldsymbol{e}_{12} & \boldsymbol{0}_{3\times 1} & \boldsymbol{e}_{13} \\ \boldsymbol{e}_{21} & \boldsymbol{e}_{23} & \boldsymbol{0}_{3\times 1} \\ \boldsymbol{0}_{3\times 1} & \boldsymbol{e}_{32} & \boldsymbol{e}_{31} \end{bmatrix} \begin{bmatrix} f^{int}_{12} \\ f^{int}_{23} \\ f^{int}_{31} \end{bmatrix} = \boldsymbol{V}_L \boldsymbol{f}^{int}, \tag{6.58}$$

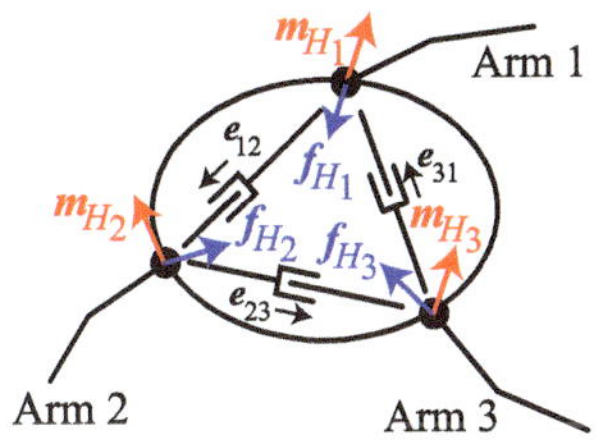

FIGURE 6.11 Three-grasp virtual linkages.

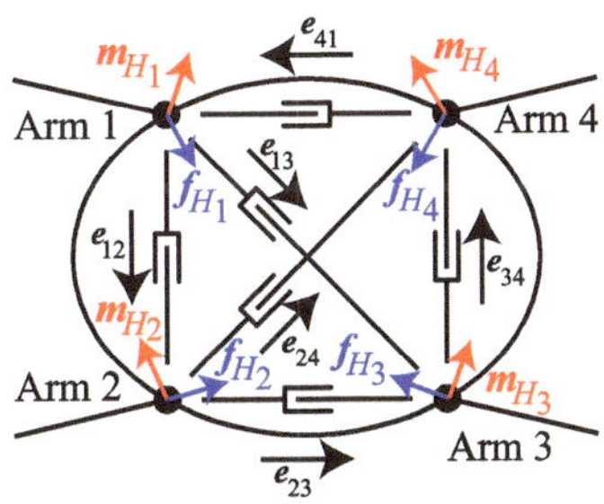

FIGURE 6.12 Four-grasp virtual linkages.

where $V_L \in \Re^{9\times 3}$. When the three grasping points form an equilateral triangle, the unit vectors between each two grasping points are defined as follows:

$$e_{12} = \begin{bmatrix} -\frac{1}{2} \\ -\frac{\sqrt{3}}{2} \\ 0 \end{bmatrix}, \quad e_{23} = \begin{bmatrix} 1 \\ 0 \\ 0 \end{bmatrix}, \quad e_{31} = \begin{bmatrix} -\frac{1}{2} \\ \frac{\sqrt{3}}{2} \\ 0 \end{bmatrix}. \tag{6.59}$$

When the unit vectors e_{ij} are determined as above, the rank of V_L is three and there is no rank deficiency.

Next, suppose that four arms grasp an object, as illustrated in Fig. 6.12. The internal forces obtained from (6.55) are

$$\begin{bmatrix} f^I_{H_1} \\ f^I_{H_2} \\ f^I_{H_3} \\ f^I_{H_4} \end{bmatrix} = \begin{bmatrix} e_{12} & 0_{3\times1} & 0_{3\times1} & e_{14} & e_{13} & 0_{3\times1} \\ e_{21} & e_{23} & 0_{3\times1} & 0_{3\times1} & 0_{3\times1} & e_{24} \\ 0_{3\times1} & e_{32} & e_{34} & 0_{3\times1} & e_{31} & 0_{3\times1} \\ 0_{3\times1} & 0_{3\times1} & e_{43} & e_{41} & 0_{3\times1} & e_{42} \end{bmatrix} \begin{bmatrix} f^{int}_{12} \\ f^{int}_{23} \\ f^{int}_{34} \\ f^{int}_{41} \\ f^{int}_{13} \\ f^{int}_{24} \end{bmatrix} = V_L f^{int}, \tag{6.60}$$

where $V_L \in \Re^{12\times 6}$. When the four grasping points form a square, the unit vectors between each two grasping points are defined as follows:

$$e_{12} = \begin{bmatrix} 0 \\ -1 \\ 0 \end{bmatrix}, \quad e_{23} = \begin{bmatrix} 1 \\ 0 \\ 0 \end{bmatrix}, \quad e_{34} = \begin{bmatrix} 0 \\ 1 \\ 0 \end{bmatrix}, \quad e_{41} = \begin{bmatrix} -1 \\ 0 \\ 0 \end{bmatrix}, \quad e_{13} = \frac{1}{\sqrt{2}}\begin{bmatrix} -1 \\ -1 \\ 0 \end{bmatrix}, \quad e_{24} = \frac{1}{\sqrt{2}}\begin{bmatrix} 1 \\ 1 \\ 0 \end{bmatrix}. \tag{6.61}$$

When the unit vectors e_{ij} are determined as above, the rank of V_L is just five, which implies rank deficiency. In general, V_L is not always of full rank, hence $\mathbb{G}^+$ in (6.57) is not always available.

The problems of the virtual linkage model are summarized as follows:

- There are no criteria how to determine the desired magnitudes of internal forces f^{int}. Note that there is no guarantee that the internal forces defined between each two pairs of grasp points will be always independent.
- In the case of dependent internal forces, the rank of grasp description matrix $\mathbb{G}$ may be deficient, as explained above (also mentioned in [28]).
- In the virtual linkage model, only the internal force along e_{ij} is controlled. Therefore, unexpected internal force/moment along other directions may be generated.

Virtual Stick [25,27]

The virtual stick model [27] can deal with 6D internal wrenches between two grasp points. When two robot arms grasp an object ($p = 2$), V is given by (6.46). By setting the arbitrary scalar as $\xi = -1$, V is obtained as [27]

$$V = \begin{bmatrix} -E_6 \\ E_6 \end{bmatrix}. \tag{6.62}$$

With this matrix, the external/internal wrenches can be calculated as follows:

$$\begin{bmatrix} \mathcal{F}_O^A \\ \mathcal{F}_O^I \end{bmatrix} = \left[(E_{6\times 12})^+ \quad V \right]^{-1} \mathcal{F}_O, \tag{6.63}$$

where $(\circ)^+$ denotes the (right) Moore–Penrose pseudoinverse matrix. Substituting (6.62) into (6.63), the internal wrench is obtained as

$$\mathcal{F}_O^I = \frac{1}{2}\left(\mathcal{F}_{O_2} - \mathcal{F}_{O_1}\right). \tag{6.64}$$

The above equation indicates that the internal wrench is defined as the difference between the forces generated by robot arms 1 and 2. This equation clearly presents the physical meaning of the internal force.

When the number of robot arms is more than two ($p > 2$), however, it is difficult to intuitively understand what the meaning of the internal wrench $\mathcal{F}_O^I$ obtained from (6.63) is. Note that this case is not discussed in [25,27].

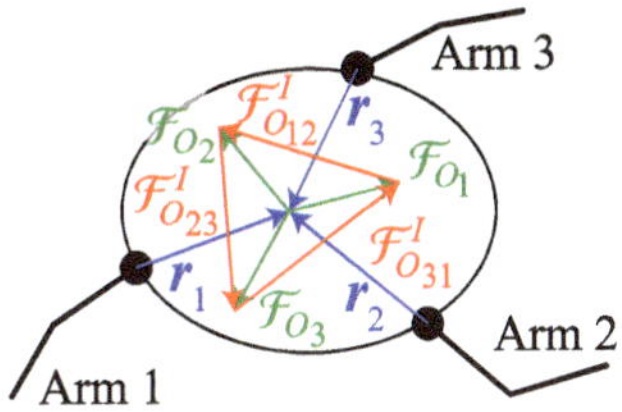

FIGURE 6.13 Internal forces generated among three robot arms.

A way of defining the internal force for the case $p > 2$ was proposed in [31]. To avoid the explicit appearance of $\boldsymbol{V}$, instead of (6.63) use

$$\begin{bmatrix} \mathcal{F}_O^A \\ \mathcal{F}_O^I \end{bmatrix} = \begin{bmatrix} \boldsymbol{E}_{6\times 6p} \\ \boldsymbol{Q} \end{bmatrix} \mathcal{F}_O. \tag{6.65}$$

The internal force can be specified via matrix $\boldsymbol{Q}$. To this end, note that matrix $\boldsymbol{U}$ in (6.42) can be obtained as

$$\boldsymbol{U} = \begin{bmatrix} \boldsymbol{E}_{6\times 6p} \\ \boldsymbol{Q} \end{bmatrix}^{-1}. \tag{6.66}$$

The method proposed in [31] is explained here through an example. Consider that three robot arms grasp an object, as illustrated in Fig. 6.13. Arm j applies the 6D wrench $\mathcal{F}_{O_j}$ to the object. The internal force between each pair of arms is not uniquely determined. However, the internal force between arms i and j can be expressed as (cf. (6.64))

$$\mathcal{F}_{O_{ij}}^I = \frac{1}{2}\left(\mathcal{F}_{O_j} - \mathcal{F}_{O_i}\right). \tag{6.67}$$

Note that the above equation can also be derived from (6.63) with an appropriate matrix $\boldsymbol{V}$ that satisfies (6.43). Note that $\mathcal{F}_{O_{ij}}^I$, defined in (6.67), is an internal force. Furthermore, note that the sum of all $\mathcal{F}_{O_{ij}}^I$ is also an internal force, s.t. $\mathcal{F}_O^A$ will be unaffected by this force. In [31], the internal forces were defined as appropriate combinations of $\mathcal{F}_{O_{ij}}^I$.

Cooperation Among Three Robot Arms

In the case of cooperation among three robot arms, the number of two-combinations is ${}_3C_2 = 3(3-1)/2 = 3$. This means that three internal wrenches can be defined. They are denoted as $\mathcal{F}_{O_{12}}^I$, $\mathcal{F}_{O_{23}}^I$, and $\mathcal{F}_{O_{31}}^I$ (see Fig. 6.13). Note, however, that the DoFs available to control the internal wrenches are just $6(p-1) = 6(3-1) = 12$. This means that only two out of the three internal wrenches can be explicitly controlled. In other words, only two of the internal wrenches are independent. The third internal wrench can be expressed as a sum of the other two, i.e. $\mathcal{F}_{O_{31}}^I = -(\mathcal{F}_{O_{12}}^I + \mathcal{F}_{O_{23}}^I)$.

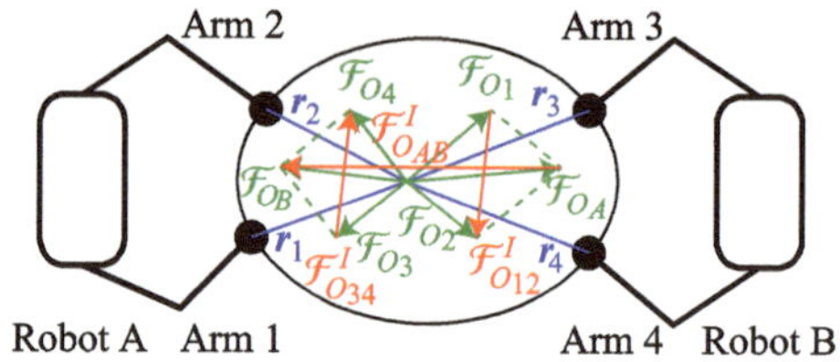

FIGURE 6.14 Cooperation between two dual-arm robots.

Suppose that $\mathcal{F}^I_{O_{12}}$ and $\mathcal{F}^I_{O_{23}}$ are chosen as the internal wrenches for explicit control; $\mathcal{F}^I_O$ can then be obtained as follows:

$$\mathcal{F}^I_O = \begin{bmatrix} \mathcal{F}^I_{O_{12}} \\ \mathcal{F}^I_{O_{23}} \end{bmatrix} = \boldsymbol{Q} \begin{bmatrix} \mathcal{F}_{O_1} \\ \mathcal{F}_{O_2} \\ \mathcal{F}_{O_3} \end{bmatrix}, \tag{6.68}$$

$$\boldsymbol{Q} = \frac{1}{2} \begin{bmatrix} -\boldsymbol{E}_6 & \boldsymbol{E}_6 & \boldsymbol{0}_6 \\ \boldsymbol{0}_6 & -\boldsymbol{E}_6 & \boldsymbol{E}_6 \end{bmatrix}.$$

Substituting (6.40) and (6.68) into (6.65), the following equation is obtained:

$$\begin{bmatrix} \mathcal{F}^A_O \\ \mathcal{F}^I_{O_{12}} \\ \mathcal{F}^I_{O_{23}} \end{bmatrix} = \begin{bmatrix} \boldsymbol{E}_{6\times 6p} \\ \boldsymbol{Q} \end{bmatrix} \begin{bmatrix} \mathcal{F}_{O_1} \\ \mathcal{F}_{O_2} \\ \mathcal{F}_{O_3} \end{bmatrix} = \begin{bmatrix} \boldsymbol{E}_6 & \boldsymbol{E}_6 & \boldsymbol{E}_6 \\ -\frac{1}{2}\boldsymbol{E}_6 & \frac{1}{2}\boldsymbol{E}_6 & \boldsymbol{0}_6 \\ \boldsymbol{0}_6 & -\frac{1}{2}\boldsymbol{E}_6 & \frac{1}{2}\boldsymbol{E}_6 \end{bmatrix} \begin{bmatrix} \mathcal{F}_{O_1} \\ \mathcal{F}_{O_2} \\ \mathcal{F}_{O_3} \end{bmatrix}. \tag{6.69}$$

Grasp wrenches $\mathcal{F}_{O_j}$, that can be used to achieve the desired $\mathcal{F}^A$, $\mathcal{F}^I_{O_{12}}$, and $\mathcal{F}^I_{O_{23}}$, are calculated as follows:

$$\begin{bmatrix} \mathcal{F}_{O_1} \\ \mathcal{F}_{O_2} \\ \mathcal{F}_{O_3} \end{bmatrix} = \begin{bmatrix} \boldsymbol{E}_{6\times 6p} \\ \boldsymbol{Q} \end{bmatrix}^{-1} \begin{bmatrix} \mathcal{F}^A_O \\ \mathcal{F}^I_{O_{12}} \\ \mathcal{F}^I_{O_{23}} \end{bmatrix} = \frac{1}{3} \begin{bmatrix} \boldsymbol{E}_6 & -4\boldsymbol{E}_6 & -2\boldsymbol{E}_6 \\ \boldsymbol{E}_6 & 2\boldsymbol{E}_6 & -2\boldsymbol{E}_6 \\ \boldsymbol{E}_6 & 2\boldsymbol{E}_6 & 4\boldsymbol{E}_6 \end{bmatrix} \begin{bmatrix} \mathcal{F}^A_O \\ \mathcal{F}^I_{O_{12}} \\ \mathcal{F}^I_{O_{23}} \end{bmatrix} = \begin{bmatrix} \boldsymbol{E}^+_{6\times 6p} & \boldsymbol{V} \end{bmatrix} \begin{bmatrix} \mathcal{F}^A_O \\ \mathcal{F}^I_O \end{bmatrix}, \tag{6.70}$$

$$\mathcal{F}^I_O = \begin{bmatrix} \mathcal{F}^I_{O_{12}} \\ \mathcal{F}^I_{O_{23}} \end{bmatrix}, \quad \boldsymbol{E}^+_{6\times 6p} = \frac{1}{3} \begin{bmatrix} \boldsymbol{E}_6 \\ \boldsymbol{E}_6 \\ \boldsymbol{E}_6 \end{bmatrix}, \quad \boldsymbol{V} = \frac{1}{3} \begin{bmatrix} -4\boldsymbol{E}_6 & -2\boldsymbol{E}_6 \\ 2\boldsymbol{E}_6 & -2\boldsymbol{E}_6 \\ 2\boldsymbol{E}_6 & 4\boldsymbol{E}_6 \end{bmatrix}.$$

Cooperation Between Two Humanoid Robots

Suppose two humanoid robots grasp an object with their hands. As illustrated in Fig. 6.14, the robot arms are numbered from 1 to 4. Since there are four robot arms ($p = 4$), the number of two-combinations is ${}_4C_2 = 4(4-1)/2 = 6$. Hence, there are six internal wrenches. They are denoted as $\mathcal{F}^I_{O_{12}}$, $\mathcal{F}^I_{O_{23}}$, $\mathcal{F}^I_{O_{34}}$, $\mathcal{F}^I_{O_{41}}$, $\mathcal{F}^I_{O_{13}}$, and $\mathcal{F}^I_{O_{24}}$. Note, however, that there will be only three independent internal wrenches (since $p-1 = 4-1$).

Generally, the internal forces of importance in this case will be: (1) internal force $\mathcal{F}^I_{O_{AB}}$ between the two robots, (2) internal force $\mathcal{F}^I_{O_{12}}$ between arms 1 and 2, and (3) internal force $\mathcal{F}^I_{O_{34}}$ between Arms 3 and 4 (see Fig. 6.14).

Robot A applies two wrenches, $\mathcal{F}_{O_1}$ and $\mathcal{F}_{O_2}$, to point O_O by using Arms 1 and 2. The resultant wrench is $\mathcal{F}_{O_A} = \mathcal{F}_{O_1} + \mathcal{F}_{O_2}$. In the same way, Robot B applies the resultant force, $\mathcal{F}_{O_B} = \mathcal{F}_{O_3} + \mathcal{F}_{O_4}$, to point O_O by its Arms 3 and 4. Then, the internal force generated between the two robots can be defined as

$$\begin{aligned} \mathcal{F}^I_{O_{AB}} = \frac{1}{2}(\mathcal{F}_{O_B} - \mathcal{F}_{O_A}) &= \frac{1}{2}((\mathcal{F}_{O_3} + \mathcal{F}_{O_4}) - (\mathcal{F}_{O_1} + \mathcal{F}_{O_2})) \\ &= \mathcal{F}^I_{O_{13}} + \mathcal{F}^I_{O_{24}}. \end{aligned} \tag{6.71}$$

Note that the expression in the last row is the sum of internal forces defined in (6.67).

Internal wrench $\mathcal{F}^I_O$ can be defined as follows:

$$\mathcal{F}^I_O = \begin{bmatrix} \mathcal{F}^I_{O_{12}} \\ \mathcal{F}^I_{O_{34}} \\ \mathcal{F}^I_{O_{AB}} \end{bmatrix} = \boldsymbol{Q} \begin{bmatrix} \mathcal{F}_{O_1} \\ \mathcal{F}_{O_2} \\ \mathcal{F}_{O_3} \\ \mathcal{F}_{O_4} \end{bmatrix}, \tag{6.72}$$

$$\boldsymbol{Q} = \frac{1}{2} \begin{bmatrix} -\boldsymbol{E}_6 & \boldsymbol{E}_6 & \boldsymbol{0}_6 & \boldsymbol{0}_6 \\ \boldsymbol{0}_6 & \boldsymbol{0}_6 & -\boldsymbol{E}_6 & \boldsymbol{E}_6 \\ -\boldsymbol{E}_6 & -\boldsymbol{E}_6 & \boldsymbol{E}_6 & \boldsymbol{E}_6 \end{bmatrix}.$$

Substituting (6.40) and (6.72) into (6.65), the following equation is obtained:

$$\begin{bmatrix} \mathcal{F}^A_O \\ \mathcal{F}^I_{O_{12}} \\ \mathcal{F}^I_{O_{34}} \\ \mathcal{F}^I_{O_{AB}} \end{bmatrix} = \begin{bmatrix} \boldsymbol{E}_{6\times 6p} \\ \boldsymbol{Q} \end{bmatrix} \begin{bmatrix} \mathcal{F}_{O_1} \\ \mathcal{F}_{O_2} \\ \mathcal{F}_{O_3} \\ \mathcal{F}_{O_4} \end{bmatrix} = \begin{bmatrix} \boldsymbol{E}_6 & \boldsymbol{E}_6 & \boldsymbol{E}_6 & \boldsymbol{E}_6 \\ -\frac{1}{2}\boldsymbol{E}_6 & \frac{1}{2}\boldsymbol{E}_6 & \boldsymbol{0}_6 & \boldsymbol{0}_6 \\ \boldsymbol{0}_6 & \boldsymbol{0}_6 & -\frac{1}{2}\boldsymbol{E}_6 & \frac{1}{2}\boldsymbol{E}_6 \\ -\frac{1}{2}\boldsymbol{E}_6 & -\frac{1}{2}\boldsymbol{E}_6 & \frac{1}{2}\boldsymbol{E}_6 & \frac{1}{2}\boldsymbol{E}_6 \end{bmatrix} \begin{bmatrix} \mathcal{F}_{O_1} \\ \mathcal{F}_{O_2} \\ \mathcal{F}_{O_3} \\ \mathcal{F}_{O_4} \end{bmatrix}. \tag{6.73}$$

Given the desired wrench $(\mathcal{F}^A)^{des}$ to be impressed on the object and the desired internal forces $\mathcal{F}^I_{O_{12}}$, $\mathcal{F}^I_{O_{34}}$, and $\mathcal{F}^I_{O_{AB}}$, the desired hand wrenches $\mathcal{F}_{O_j}$ to be applied at point O_o are calculated as follows:

$$\begin{bmatrix} \mathcal{F}_{O_1} \\ \mathcal{F}_{O_2} \\ \mathcal{F}_{O_3} \\ \mathcal{F}_{O_4} \end{bmatrix} = \begin{bmatrix} \frac{1}{4}\boldsymbol{E}_6 & -\boldsymbol{E}_6 & \boldsymbol{0}_6 & -\frac{1}{2}\boldsymbol{E}_6 \\ \frac{1}{4}\boldsymbol{E}_6 & \boldsymbol{E}_6 & \boldsymbol{0}_6 & -\frac{1}{2}\boldsymbol{E}_6 \\ \frac{1}{4}\boldsymbol{E}_6 & \boldsymbol{0}_6 & -\boldsymbol{E}_6 & \frac{1}{2}\boldsymbol{E}_6 \\ \frac{1}{4}\boldsymbol{E}_6 & \boldsymbol{0}_6 & \boldsymbol{E}_6 & \frac{1}{2}\boldsymbol{E}_6 \end{bmatrix} \begin{bmatrix} \mathcal{F}^A_O \\ \mathcal{F}^I_{O_{12}} \\ \mathcal{F}^I_{O_{34}} \\ \mathcal{F}^I_{O_{AB}} \end{bmatrix} = \begin{bmatrix} \boldsymbol{E}^+_{6\times 6p} & \boldsymbol{V} \end{bmatrix} \begin{bmatrix} \mathcal{F}^A_O \\ \mathcal{F}^I_O \end{bmatrix}, \tag{6.74}$$

$$\mathcal{F}^I_O = \begin{bmatrix} \mathcal{F}^I_{O_{12}} \\ \mathcal{F}^I_{O_{34}} \\ \mathcal{F}^I_{O_{AB}} \end{bmatrix}, \quad \boldsymbol{E}^+_{6\times 6p} = \frac{1}{4} \begin{bmatrix} \boldsymbol{E}_6 \\ \boldsymbol{E}_6 \\ \boldsymbol{E}_6 \\ \boldsymbol{E}_6 \end{bmatrix}, \quad \boldsymbol{V} = \begin{bmatrix} -\boldsymbol{E}_6 & \boldsymbol{0}_6 & -\frac{1}{2}\boldsymbol{E}_6 \\ \boldsymbol{E}_6 & \boldsymbol{0}_6 & -\frac{1}{2}\boldsymbol{E}_6 \\ \boldsymbol{0}_6 & -\boldsymbol{E}_6 & \frac{1}{2}\boldsymbol{E}_6 \\ \boldsymbol{0}_6 & \boldsymbol{E}_6 & \frac{1}{2}\boldsymbol{E}_6 \end{bmatrix}.$$

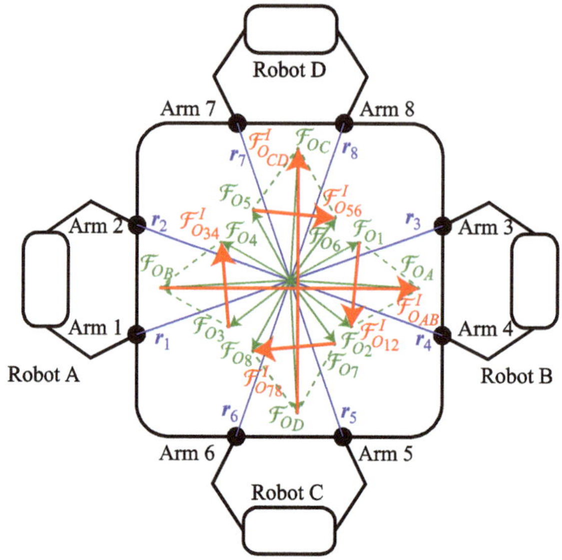

FIGURE 6.15 Cooperation among four humanoid robots.

The above wrenches are just particular solutions of the underdetermined wrench distribution problem. Other solutions can also be considered by choosing appropriate sums of internal forces (6.67).

Cooperation Among Four Humanoid Robots

The last example to be discussed is about cooperation among four humanoid robots. As illustrated in Fig. 6.15, the robot arms are numbered from 1 to 8. Since there are eight robot arms ($p = 8$), the number of two-combinations is ${}_8C_2 = 8(8-1)/2 = 28$. Thus, there will be 28 internal wrenches. Among them only seven ($p - 1 = 8 - 1$) will be independent, however. The remaining 21 wrenches will be dependent.

Generally, the important internal wrenches in this case will be: (1) internal force between two pairs of robots, e.g. between robots A and B and robots C and D, $\mathcal{F}^I_{O_{(AB)(CD)}}$; (2) the internal forces between the pair of robots facing each other, i.e. $\mathcal{F}^I_{O_{AB}}$ and $\mathcal{F}^I_{O_{CD}}$; and (3) internal forces $\mathcal{F}^I_{O_{12}}$, $\mathcal{F}^I_{O_{34}}$, $\mathcal{F}^I_{O_{56}}$, and $\mathcal{F}^I_{O_{78}}$ between the arms of each robot (see Fig. 6.15).

Robot A applies two wrenches $\mathcal{F}_{O_1}$ and $\mathcal{F}_{O_2}$ at point O_O by using Arms 1 and 2. The resultant wrench is given as $\mathcal{F}_{O_A} = \mathcal{F}_{O_1} + \mathcal{F}_{O_2}$. In the same way, Robots B, C, and D apply resultant wrenches $\mathcal{F}_{O_B} = \mathcal{F}_{O_3} + \mathcal{F}_{O_4}$, $\mathcal{F}_{O_C} = \mathcal{F}_{O_5} + \mathcal{F}_{O_6}$, and $\mathcal{F}_{O_D} = \mathcal{F}_{O_7} + \mathcal{F}_{O_8}$, respectively.

The internal force between the two pairs of robots, A–B and C–D, can be defined as follows:

$$
\begin{aligned}
\mathcal{F}^I_{O_{(AB)(CD)}} &= \frac{1}{2}\left(\mathcal{F}_{O_C} + \mathcal{F}_{O_D}\right) - \frac{1}{2}\left(\mathcal{F}_{O_A} + \mathcal{F}_{O_B}\right) \\
&= \frac{1}{2}\left(-\mathcal{F}_{O_1} - \mathcal{F}_{O_2} - \mathcal{F}_{O_3} - \mathcal{F}_{O_4} + \mathcal{F}_{O_5} + \mathcal{F}_{O_6} + \mathcal{F}_{O_7} + \mathcal{F}_{O_8}\right). \qquad (6.75)
\end{aligned}
$$

The internal forces generated between the two pairs of robots, facing each other, can be expressed as

$$\mathcal{F}^I_{O_{AB}} = \frac{1}{2}\left(\mathcal{F}_{O_B} - \mathcal{F}_{O_A}\right) = \frac{1}{2}\left(-\mathcal{F}_{O_1} - \mathcal{F}_{O_2} + \mathcal{F}_{O_3} + \mathcal{F}_{O_4}\right), \tag{6.76}$$

$$\mathcal{F}^I_{O_{CD}} = \frac{1}{2}\left(\mathcal{F}_{O_D} - \mathcal{F}_{O_C}\right) = \frac{1}{2}\left(-\mathcal{F}_{O_5} - \mathcal{F}_{O_6} + \mathcal{F}_{O_7} + \mathcal{F}_{O_8}\right). \tag{6.77}$$

Furthermore, $\mathcal{F}^I_{O_{(AB)(CD)}}$, $\mathcal{F}^I_{O_{AB}}$, and $\mathcal{F}^I_{O_{CD}}$ can be represented as sums of internal forces (cf. (6.67)), i.e.

$$\begin{aligned}
\mathcal{F}^I_{O_{(AB)(CD)}} &= \mathcal{F}^I_{O_{15}} + \mathcal{F}^I_{O_{26}} + \mathcal{F}^I_{O_{37}} + \mathcal{F}^I_{O_{48}}, \\
\mathcal{F}^I_{O_{AB}} &= \mathcal{F}^I_{O_{13}} + \mathcal{F}^I_{O_{24}}, \\
\mathcal{F}^I_{O_{CD}} &= \mathcal{F}^I_{O_{57}} + \mathcal{F}^I_{O_{68}},
\end{aligned}$$

respectively.

Internal forces $\mathcal{F}^I_O$ can be defined as follows:

$$\mathcal{F}^I_O = \boldsymbol{Q}\mathcal{F}_O, \tag{6.78}$$

$$\mathcal{F}^I_O = \left[\mathcal{F}^I_{O_{(AB)(CD)}} \quad \mathcal{F}^I_{O_{AB}} \quad \mathcal{F}^I_{O_{CD}} \quad \mathcal{F}^I_{O_{12}} \quad \mathcal{F}^I_{O_{34}} \quad \mathcal{F}^I_{O_{56}} \quad \mathcal{F}^I_{O_{78}}\right]^T,$$

$$\mathcal{F}_O = \left[\mathcal{F}_{O_1} \quad \mathcal{F}_{O_2} \quad \mathcal{F}_{O_3} \quad \mathcal{F}_{O_4} \quad \mathcal{F}_{O_5} \quad \mathcal{F}_{O_6} \quad \mathcal{F}_{O_7} \quad \mathcal{F}_{O_8}\right]^T,$$

$$\boldsymbol{Q} = \frac{1}{2}\begin{bmatrix}
-\boldsymbol{E}_6 & -\boldsymbol{E}_6 & -\boldsymbol{E}_6 & -\boldsymbol{E}_6 & \boldsymbol{E}_6 & \boldsymbol{E}_6 & \boldsymbol{E}_6 & \boldsymbol{E}_6 \\
-\boldsymbol{E}_6 & -\boldsymbol{E}_6 & \boldsymbol{E}_6 & \boldsymbol{E}_6 & \boldsymbol{0}_6 & \boldsymbol{0}_6 & \boldsymbol{0}_6 & \boldsymbol{0}_6 \\
\boldsymbol{0}_6 & \boldsymbol{0}_6 & \boldsymbol{0}_6 & \boldsymbol{0}_6 & -\boldsymbol{E}_6 & -\boldsymbol{E}_6 & \boldsymbol{E}_6 & \boldsymbol{E}_6 \\
-\boldsymbol{E}_6 & \boldsymbol{E}_6 & \boldsymbol{0}_6 & \boldsymbol{0}_6 & \boldsymbol{0}_6 & \boldsymbol{0}_6 & \boldsymbol{0}_6 & \boldsymbol{0}_6 \\
\boldsymbol{0}_6 & \boldsymbol{0}_6 & -\boldsymbol{E}_6 & \boldsymbol{E}_6 & \boldsymbol{0}_6 & \boldsymbol{0}_6 & \boldsymbol{0}_6 & \boldsymbol{0}_6 \\
\boldsymbol{0}_6 & \boldsymbol{0}_6 & \boldsymbol{0}_6 & \boldsymbol{0}_6 & -\boldsymbol{E}_6 & \boldsymbol{E}_6 & \boldsymbol{0}_6 & \boldsymbol{0}_6 \\
\boldsymbol{0}_6 & \boldsymbol{0}_6 & \boldsymbol{0}_6 & \boldsymbol{0}_6 & \boldsymbol{0}_6 & \boldsymbol{0}_6 & -\boldsymbol{E}_6 & \boldsymbol{E}_6
\end{bmatrix}.$$

Substituting (6.40) and (6.78) into (6.65), the following equation is obtained:

$$\begin{aligned}
\begin{bmatrix} \mathcal{F}^A_O \\ \mathcal{F}^I_O \end{bmatrix} &= \begin{bmatrix} \boldsymbol{E}_{6\times 6p} \\ \boldsymbol{Q} \end{bmatrix} \mathcal{F}_O \\
&= \begin{bmatrix}
\boldsymbol{E}_6 & \boldsymbol{E}_6 & \boldsymbol{E}_6 & \boldsymbol{E}_6 & \boldsymbol{E}_6 & \boldsymbol{E}_6 & \boldsymbol{E}_6 & \boldsymbol{E}_6 \\
-\frac{1}{2}\boldsymbol{E}_6 & -\frac{1}{2}\boldsymbol{E}_6 & -\frac{1}{2}\boldsymbol{E}_6 & -\frac{1}{2}\boldsymbol{E}_6 & \frac{1}{2}\boldsymbol{E}_6 & \frac{1}{2}\boldsymbol{E}_6 & \frac{1}{2}\boldsymbol{E}_6 & \frac{1}{2}\boldsymbol{E}_6 \\
-\frac{1}{2}\boldsymbol{E}_6 & -\frac{1}{2}\boldsymbol{E}_6 & \frac{1}{2}\boldsymbol{E}_6 & \frac{1}{2}\boldsymbol{E}_6 & \boldsymbol{0}_6 & \boldsymbol{0}_6 & \boldsymbol{0}_6 & \boldsymbol{0}_6 \\
\boldsymbol{0}_6 & \boldsymbol{0}_6 & \boldsymbol{0}_6 & \boldsymbol{0}_6 & -\frac{1}{2}\boldsymbol{E}_6 & -\frac{1}{2}\boldsymbol{E}_6 & \frac{1}{2}\boldsymbol{E}_6 & \frac{1}{2}\boldsymbol{E}_6 \\
-\frac{1}{2}\boldsymbol{E}_6 & \frac{1}{2}\boldsymbol{E}_6 & \boldsymbol{0}_6 & \boldsymbol{0}_6 & \boldsymbol{0}_6 & \boldsymbol{0}_6 & \boldsymbol{0}_6 & \boldsymbol{0}_6 \\
\boldsymbol{0}_6 & \boldsymbol{0}_6 & -\frac{1}{2}\boldsymbol{E}_6 & \frac{1}{2}\boldsymbol{E}_6 & \boldsymbol{0}_6 & \boldsymbol{0}_6 & \boldsymbol{0}_6 & \boldsymbol{0}_6 \\
\boldsymbol{0}_6 & \boldsymbol{0}_6 & \boldsymbol{0}_6 & \boldsymbol{0}_6 & -\frac{1}{2}\boldsymbol{E}_6 & \frac{1}{2}\boldsymbol{E}_6 & \boldsymbol{0}_6 & \boldsymbol{0}_6 \\
\boldsymbol{0}_6 & \boldsymbol{0}_6 & \boldsymbol{0}_6 & \boldsymbol{0}_6 & \boldsymbol{0}_6 & \boldsymbol{0}_6 & -\frac{1}{2}\boldsymbol{E}_6 & \frac{1}{2}\boldsymbol{E}_6
\end{bmatrix}
\begin{bmatrix} \mathcal{F}_{O_1} \\ \mathcal{F}_{O_2} \\ \mathcal{F}_{O_3} \\ \mathcal{F}_{O_4} \\ \mathcal{F}_{O_5} \\ \mathcal{F}_{O_6} \\ \mathcal{F}_{O_7} \\ \mathcal{F}_{O_8} \end{bmatrix}.
\end{aligned} \tag{6.79}$$

Given the desired wrench to be applied, $(\mathcal{F}_O^A)^{des}$, and the desired internal forces, $\mathcal{F}_O^{I,des}$, the desired wrench to be applied by Arm j at point O_O is calculated as

$$\begin{bmatrix} \mathcal{F}_{O_1} \\ \mathcal{F}_{O_2} \\ \mathcal{F}_{O_3} \\ \mathcal{F}_{O_4} \\ \mathcal{F}_{O_5} \\ \mathcal{F}_{O_6} \\ \mathcal{F}_{O_7} \\ \mathcal{F}_{O_8} \end{bmatrix} = \begin{bmatrix} \frac{1}{8}\boldsymbol{E}_6 & -\frac{1}{4}\boldsymbol{E}_6 & -\frac{1}{2}\boldsymbol{E}_6 & \boldsymbol{0}_6 & -\boldsymbol{E}_6 & \boldsymbol{0}_6 & \boldsymbol{0}_6 & \boldsymbol{0}_6 \\ \frac{1}{8}\boldsymbol{E}_6 & -\frac{1}{4}\boldsymbol{E}_6 & -\frac{1}{2}\boldsymbol{E}_6 & \boldsymbol{0}_6 & \boldsymbol{E}_6 & \boldsymbol{0}_6 & \boldsymbol{0}_6 & \boldsymbol{0}_6 \\ \frac{1}{8}\boldsymbol{E}_6 & -\frac{1}{4}\boldsymbol{E}_6 & \frac{1}{2}\boldsymbol{E}_6 & \boldsymbol{0}_6 & \boldsymbol{0}_6 & -\boldsymbol{E}_6 & \boldsymbol{0}_6 & \boldsymbol{0}_6 \\ \frac{1}{8}\boldsymbol{E}_6 & -\frac{1}{4}\boldsymbol{E}_6 & \frac{1}{2}\boldsymbol{E}_6 & \boldsymbol{0}_6 & \boldsymbol{0}_6 & \boldsymbol{E}_6 & \boldsymbol{0}_6 & \boldsymbol{0}_6 \\ \frac{1}{8}\boldsymbol{E}_6 & \frac{1}{4}\boldsymbol{E}_6 & \boldsymbol{0}_6 & -\frac{1}{2}\boldsymbol{E}_6 & \boldsymbol{0}_6 & \boldsymbol{0}_6 & -\boldsymbol{E}_6 & \boldsymbol{0}_6 \\ \frac{1}{8}\boldsymbol{E}_6 & \frac{1}{4}\boldsymbol{E}_6 & \boldsymbol{0}_6 & -\frac{1}{2}\boldsymbol{E}_6 & \boldsymbol{0}_6 & \boldsymbol{0}_6 & \boldsymbol{E}_6 & \boldsymbol{0}_6 \\ \frac{1}{8}\boldsymbol{E}_6 & \frac{1}{4}\boldsymbol{E}_6 & \boldsymbol{0}_6 & \frac{1}{2}\boldsymbol{E}_6 & \boldsymbol{0}_6 & \boldsymbol{0}_6 & \boldsymbol{0}_6 & -\boldsymbol{E}_6 \\ \frac{1}{8}\boldsymbol{E}_6 & \frac{1}{4}\boldsymbol{E}_6 & \boldsymbol{0}_6 & \frac{1}{2}\boldsymbol{E}_6 & \boldsymbol{0}_6 & \boldsymbol{0}_6 & \boldsymbol{0}_6 & \boldsymbol{E}_6 \end{bmatrix} \begin{bmatrix} \mathcal{F}_O^A \\ \mathcal{F}_{O(AB)(CD)}^I \\ \mathcal{F}_{O_{AB}}^I \\ \mathcal{F}_{O_{CD}}^I \\ \mathcal{F}_{O_{12}}^I \\ \mathcal{F}_{O_{34}}^I \\ \mathcal{F}_{O_{56}}^I \\ \mathcal{F}_{O_{78}}^I \end{bmatrix}. \tag{6.80}$$

The above wrenches are just particular solutions of the underdetermined wrench distribution problem. Other solutions can also be considered by choosing appropriate sums of internal forces.

6.3.6 Hybrid Position/Force Control

Cooperation among multiple robot arms based on force control was discussed in Section 6.3.5. This section discusses a position/force hybrid control approach for the case of multirobot cooperation. The deformation of the object and possible slippage between the object and the hands are not considered here. As discussed in Section 6.3.2, the position and orientation, $\boldsymbol{r}_{O_j}$ and $\boldsymbol{R}_{O_j}$ (see Fig. 6.7), generally correspond to those of the object at O_O, i.e. $\boldsymbol{r}_O$ and $\boldsymbol{R}_O$.

Suppose that p robot arms grasp an object, applying thereby the net wrench $\mathcal{F}_O$. The relationship between $\mathcal{F}_O$ and the resultant wrenches, i.e. external wrench $\mathcal{F}_O^A$ and internal wrench $\mathcal{F}_O^I$, is calculated from (6.42). Making use of the principle of virtual work, the following equation can be obtained:

$$\begin{bmatrix} (\mathcal{F}_O^A)^T & (\mathcal{F}_O^I)^T \end{bmatrix} \begin{bmatrix} \delta\mathcal{X}_O^A \\ \delta\mathcal{X}_O^I \end{bmatrix} = \mathcal{F}_O^T \delta\mathcal{X}_O, \tag{6.81}$$

$$\delta\mathcal{X}_O^I \equiv \begin{bmatrix} \delta\mathcal{X}_{O_1}^I \\ \vdots \\ \delta\mathcal{X}_{O_{(p-1)}}^I \end{bmatrix} \in \Re^{6(p-1)}, \quad \delta\mathcal{X}_O \equiv \begin{bmatrix} \delta\mathcal{X}_{O_1} \\ \vdots \\ \delta\mathcal{X}_{O_p} \end{bmatrix} \in \Re^{6p}.$$

Here $\delta\mathcal{X}_O^A$ is the small displacement of the object, $\delta\mathcal{X}_O^I$ is the stacked relative displacements caused by internal force $\mathcal{F}_O^I$, and $\delta\mathcal{X}_O$ are stacked displacement vectors $\delta\mathcal{X}_{O_j}$ defined in (6.37).

From (6.42) and (6.81), the small 6D displacement caused by external wrench $\mathcal{F}_O^A$ and internal wrench $\mathcal{F}_O^I$ is given as

$$\delta\mathcal{X}_O^{AI} = \boldsymbol{U}^T \delta\mathcal{X}_O, \tag{6.82}$$

$$\delta\mathcal{X}_O^{AI} \equiv \begin{bmatrix} \delta\mathcal{X}_O^A \\ \delta\mathcal{X}_O^I \end{bmatrix},$$

matrix $\boldsymbol{U}$ being defined in (6.66).

Position Controller

Given the desired position $\boldsymbol{r}_O^{des}$ and orientation $\boldsymbol{R}_O^{des}$ of the object, the position error is calculated from

$$\left(\delta\mathcal{X}_O^A\right)_{\mathcal{X}} = \begin{bmatrix} \boldsymbol{r}_O^{des} - \boldsymbol{r}_O \\ \left(\ln \boldsymbol{R}_O^T \boldsymbol{R}_O^{des}\right)^{\vee} \end{bmatrix}. \tag{6.83}$$

The internal wrench can also be controlled, by determining appropriate desired relative positions among the robot arms. If the desired relative position along the lth internal wrench, $\mathcal{F}_{O_l}^I$ $(1 \leq l \leq p-1)$, is given, the error in the relative position is calculated as follows:

$$\left(\delta\mathcal{X}_{O_l}^I\right)_{\mathcal{X}} = \begin{bmatrix} \boldsymbol{r}_{O_l}^{des} - \boldsymbol{r}_{O_l} \\ \left(\ln \boldsymbol{R}_{O_l}^T \boldsymbol{R}_{O_l}^{des}\right)^{\vee} \end{bmatrix}. \tag{6.84}$$

The superscript *des* denotes a desired state.

Given $(\delta\mathcal{X}_{O_l}^I)_{\mathcal{X}} = \boldsymbol{0}_6$, the corresponding robot arms are controlled to maintain the kinematic constraints; hence $\mathcal{T}_{O_l}^I$ is controlled to be zero.

Position-Based Force Controller

The reference position for force control is given by

$$\left(\delta\mathcal{X}_O^{AI}\right)_{\mathcal{F}} = \boldsymbol{K}_{\mathcal{F}}^{AI} \left\{ \left(\mathcal{F}_O^{AI}\right)^{des} - \left(\mathcal{F}_O^{AI}\right) \right\}, \tag{6.85}$$

where $\boldsymbol{K}_{\mathcal{F}}^{AI}$ is a diagonal gain matrix and $\mathcal{F}_O^{AI}$ is transformed from $\mathcal{F}_O$ using (6.65).

Hybrid Position/Force Controller

In most cases of multirobot cooperation it is desirable to control the position of the object and the internal wrenches distributed among the robots. The following position/force hybrid force controller can be employed in this case:

$$\left(\delta\mathcal{X}_O^{AI}\right)_{hybrid} = (\boldsymbol{E}_{6n} - \boldsymbol{S}) \left(\delta\mathcal{X}_O^{AI}\right)_{\mathcal{X}} + \boldsymbol{S} \left(\delta\mathcal{X}_O^{AI}\right)_{\mathcal{F}}, \tag{6.86}$$

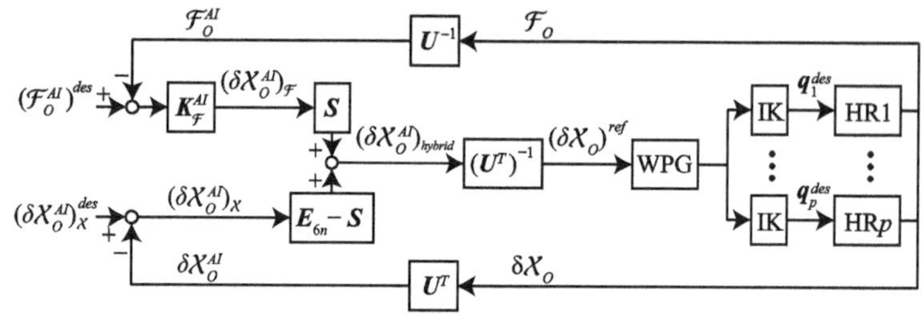

FIGURE 6.16 Position/force hybrid controller for multirobot cooperation. WPG, IK, and HRj denote walking pattern generator, inverse kinematics, and jth humanoid robot, respectively.

where $\boldsymbol{S}$ is a selection matrix. The reference position obtained from the above controller can be used to determine the reference tip positions of the virtual sticks, i.e.

$$\begin{bmatrix} \delta\mathcal{X}_{O_1}^{ref} \\ \vdots \\ \delta\mathcal{X}_{O_p}^{ref} \end{bmatrix} = \left(\boldsymbol{U}^T\right)^{-1} \left(\delta\mathcal{X}_O^{AI}\right)_{hybrid}. \tag{6.87}$$

The desired joint angles of the jth robot arm can then be obtained via the instantaneous kinematics relations (cf. (6.37)) as follows:

$$\boldsymbol{q}_j^{des} = \boldsymbol{J}_{O_j}^{+} \delta\mathcal{X}_{O_j}^{ref} + \boldsymbol{q}_j. \tag{6.88}$$

The block diagram of the position/force controller is illustrated in Fig. 6.16.

6.4 COOPERATION BETWEEN MULTIPLE HUMANOIDS

6.4.1 On-Line Footstep Planning

This section introduces a footstep planning method for cooperative object manipulation, as proposed in [30]. The concept is shown in Fig. 6.17. The planning of the footsteps is done for each robot separately. After grasping the object, each robot calculates the distances between the ground projection of the mid-point between the hands and the origins of the foot positions ($\boldsymbol{d}_1$ and $\boldsymbol{d}_2$, as illustrated in Fig. 6.17). When the robots start to move the object, the reference foot positions $\boldsymbol{d}_1^{des}$ and $\boldsymbol{d}_2^{des}$ are continuously calculated. When the error between the reference and the current foot position exceeds a specified threshold, the robot starts to walk. Once the reference foot position is decided, the reference zero-moment point (ZMP) trajectory is obtained by interpolating the discrete foot positions between the current and the reference ones. The reference CoM trajectory is computed using the preview control theory [9]. The robots stop walking when their foot position errors become lower than the threshold. The error includes the position error, e_{P}, and the attitude error in the yaw direction, e_{A}. These

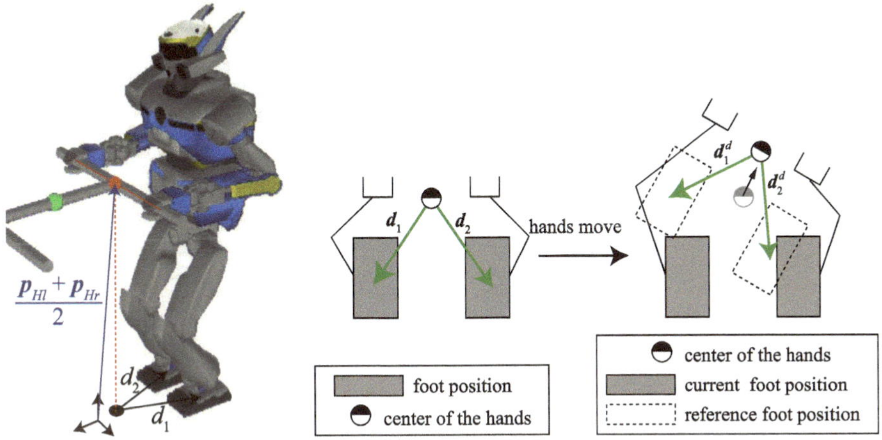

FIGURE 6.17 Concept of on-line footstep planning.

errors are defined as

$$e_{pk} = ||\boldsymbol{d}_k^{des} - \boldsymbol{d}_k||,$$
$$e_{Ak} = |\psi_k^{des} - \psi_k|,$$

where $k = 1, 2$ denotes the foot number and $\boldsymbol{d}_k^{des}$ and ψ_k^{des} are the desired position and yaw angle, respectively, and e_{pk} and e_{Ak} are checked after every step to make a decision about initializing another step or terminating the stepping.

The relative position error between the robots can be successfully compensated since the internal force of the system is controlled by stretching or shrinking the arms and since the desired foot position of the robot is calculated from the mid-point between the hands.

6.4.2 Coordinated Movement of Hands and Feet

Suppose the reference position $(\delta\mathcal{X}_O^{AI})_{hybrid}^{ref}$ is given by the hybrid controller (6.86). The corresponding 6D positions of the virtual sticks for the two hands of the humanoid robot, $\delta\mathcal{X}_{O_j}^{ref}$ and $\delta\mathcal{X}_{O_{j+1}}^{ref}$, are calculated from (6.87).

As illustrated in Fig. 6.18, a humanoid robot walks swaying its CoM. However, the hand position must be controlled independently of the swaying motion.

The positions of the hands, ${}^{F_{st}}\delta\mathcal{X}_{H_l}^{ref}$ and ${}^{F_{st}}\delta\mathcal{X}_{H_r}^{ref}$, are defined w.r.t. the stance foot coordinate frame $\{F_{st}\}$. These positions are calculated from the positions of the virtual sticks $\delta\mathcal{X}_{O_j}^{ref}$ and $\delta\mathcal{X}_{O_{j+1}}^{ref}$.

Then, $\boldsymbol{d}_1$ and $\boldsymbol{d}_2$ can be calculated by using ${}^{F_{st}}\delta\mathcal{X}_{H_l}^{ref}$ and ${}^{F_{st}}\delta\mathcal{X}_{H_r}^{ref}$, as discussed in Section 6.4.1. In the case when e_{pk} or e_{Ak} exceeds the previously specified threshold, the on-line walking pattern generator defines the desired CoM and the trajectory of the swing foot.

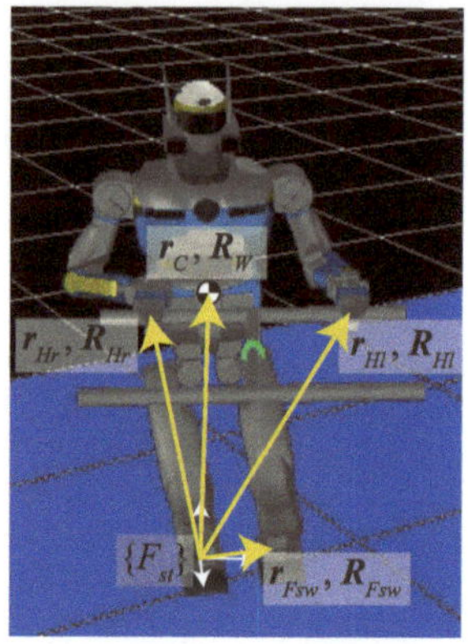

FIGURE 6.18 Position control of the hands, swing foot, and CoM.

Hence, the constraints can be summarized as: (1) the positions of the hands, (2) the CoM, and (3) the position of the swing foot.

The instantaneous kinematics equation (6.88) is therefore rewritten using the stacked Jacobian matrix and the stacked error vectors as follows:

$$\boldsymbol{q}_j^{des} = {}^{F_{st}}\underline{\boldsymbol{J}}_j^{+}\, {}^{F_{st}}\underline{\delta\mathcal{X}}_j^{ref} + \boldsymbol{q}_j, \tag{6.89}$$

where

$${}^{F_{st}}\underline{\boldsymbol{J}}_j = \begin{bmatrix} {}^{F_{st}}\boldsymbol{J}_C \\ {}^{F_{st}}\boldsymbol{J}_{F_{sw}} \\ {}^{F_{st}}\boldsymbol{J}_{H_l} \\ {}^{F_{st}}\boldsymbol{J}_{H_r} \end{bmatrix},$$

$${}^{F_{st}}\underline{\delta\mathcal{X}}_j^{ref} = \begin{bmatrix} {}^{F_{st}}\delta\mathcal{X}_C^{ref} \\ {}^{F_{st}}\delta\mathcal{X}_{F_{sw}}^{ref} \\ {}^{F_{st}}\delta\mathcal{X}_{H_l}^{ref} \\ {}^{F_{st}}\delta\mathcal{X}_{H_r}^{ref} \end{bmatrix},$$

$${}^{F_{st}}\delta\mathcal{X}_C^{ref} = \begin{bmatrix} {}^{F_{st}}\boldsymbol{r}_C^{des} - {}^{F_{st}}\boldsymbol{r}_C \\ (\ln {}^{F_{st}}\boldsymbol{R}_C^T\, {}^{F_{st}}\boldsymbol{R}_C^{des})^{\vee} \end{bmatrix},$$

$${}^{F_{st}}\delta\mathcal{X}_{F_{sw}}^{ref} = \begin{bmatrix} {}^{F_{st}}\boldsymbol{r}_{F_{sw}}^{des} - {}^{F_{st}}\boldsymbol{r}_{F_{sw}} \\ \left(\ln {}^{F_{st}}\boldsymbol{R}_{F_{sw}}^T\, {}^{F_{st}}\boldsymbol{R}_{F_{sw}}^{des}\right)^{\vee} \end{bmatrix},$$

$${}^{F_{st}}\delta\mathcal{X}_{H_e}^{ref} = \begin{bmatrix} {}^{F_{st}}\boldsymbol{r}_{H_e}^{des} - {}^{F_{st}}\boldsymbol{r}_{H_e} \\ \left(\ln {}^{F_{st}}\boldsymbol{R}_{H_e}^T\, {}^{F_{st}}\boldsymbol{R}_{H_e}^{des}\right)^{\vee} \end{bmatrix}, \quad (e = l \text{ or } r).$$

The control variables are listed below.

$\boldsymbol{r}_C$	position vector of the CoM
$\boldsymbol{R}_C$	rotation matrix of the CoM (actually of the waist)
$\boldsymbol{r}_{F_{sw}}$	position vector of the swing foot
$\boldsymbol{R}_{F_{sw}}$	rotation matrix of the swing foot
$\boldsymbol{r}_{F_{H_e}}$	position vector of the hand ($e = l$ or r)
$\boldsymbol{R}_{F_{H_e}}$	rotation matrix of the hand ($e = l$ or r)
$\boldsymbol{q}_j$	joint angular velocity vector of the whole body
$\boldsymbol{J}_C$	analytical Jacobian matrix that relates $\delta\boldsymbol{q}$ to $\delta\mathcal{X}_C$
$\boldsymbol{J}_{F_{sw}}$	analytical Jacobian matrix that relates $\delta\boldsymbol{q}$ to $\delta\mathcal{X}_{F_{sw}}$
$\boldsymbol{J}_{H_e}$	analytical Jacobian matrix that relates $\delta\boldsymbol{q}$ to $\delta\mathcal{X}_{F_{H_e}}$ ($e = l$ or r)

The trailing superscript des represents the desired state. The trailing superscript ref denotes the difference between the desired and the current states.

The meaning of the leading superscript F_{st} ($st = l$ or r) is that the vector or the matrix is defined w.r.t. the stance foot coordinate frame $\{F_{st}\}$ ($st = l$ or r). Since the left and right foot alternately play the role of the stance foot, two different sets of Jacobian matrices and reference vectors, i.e. (${}^{F_l}\underline{\boldsymbol{J}}_j$, ${}^{F_l}\underline{\delta\mathcal{X}}_j$) and (${}^{F_r}\underline{\boldsymbol{J}}_j$, ${}^{F_r}\underline{\delta\mathcal{X}}_j$), have to be used alternately. When the robot is in the double-support phase, the inverse kinematics are solved by letting ${}^{F_{st}}\delta\mathcal{X}^{ref}_{F_{sw}} = \boldsymbol{0}_6$.

6.4.3 Leader-Follower- and Symmetry-Type Cooperation

The multiarm object manipulation control method discussed in Section 6.3 can be applied to the problem of cooperating humanoid robots. Multirobot cooperation is classified into (a) leader-follower-type cooperation and (b) symmetry-type cooperation. The concepts of these two types of cooperation are visualized in Fig. 6.19.

In leader-follower-type cooperation, one of the robots behaves as a leader while the remaining robots behave as followers. The leader robot controls the position and orientation of the grasped object while the followers control the internal wrenches. Generally, only the gravitational force of the grasped object is distributed to all robots. The motion of the object along the other five axes is controlled by the leader robot.

In symmetry-type cooperation, on the other hand, all robots equally control the motion of the grasped object and the internal forces.

Implementations of leader-follower-type cooperation and symmetry-type cooperation are presented in the following subsections.

6.4.4 Leader-Follower-Type Cooperative Object Manipulation

This section introduces a method of a leader-follower-type cooperative object manipulation, as proposed in [30]. The method applies to humanoid robots comprising gear drives with high reduction ratios and hardware joint-position servos. Most of the actual humanoid

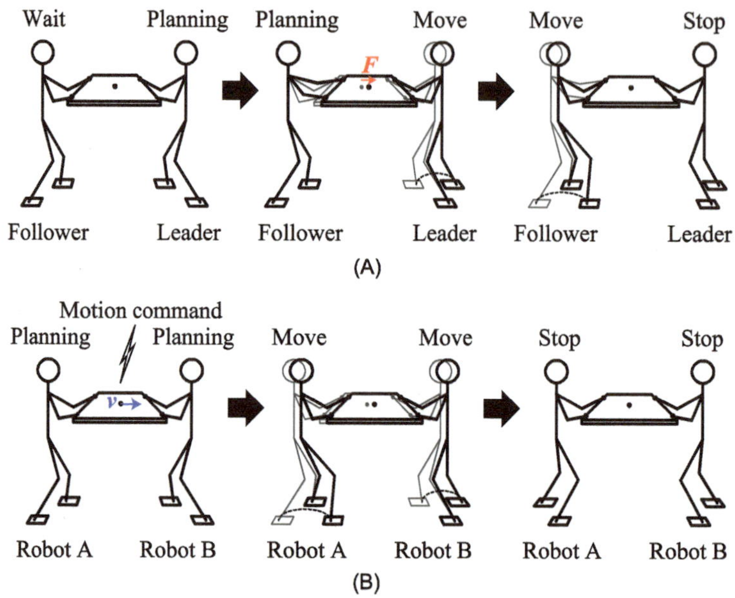

FIGURE 6.19 Two types of multirobot cooperation. (A) Concept of leader-follower-type cooperation. (B) Concept of symmetry-type cooperation.

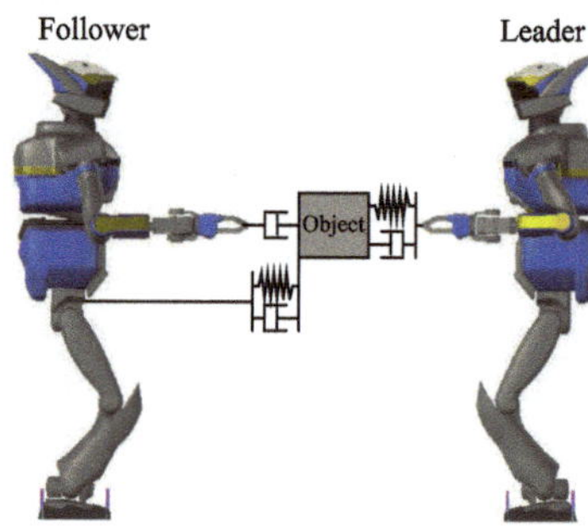

FIGURE 6.20 Concept of a leader-follower-type cooperation.

robots fall within this category. They accept control commands specified in terms of joint positions.

Concept of a Leader-Follower-Type Cooperative Object Manipulation

Generally, the leader robot controls the position (or motion) of the carried object and the follower robots control the internal force. The concept of leader-follower-type cooperation proposed in [30] is illustrated in Fig. 6.20. The leader robot walks in the direction commanded to carry the object. The motion of its arms is determined by an impedance control law. The motion of the arms of the follower robots, on the other hand, is determined by a damping con-

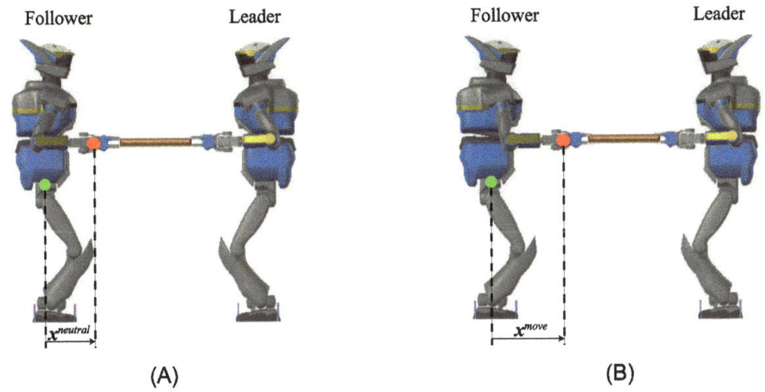

FIGURE 6.21 Time lag between the movements of the leader and the follower robots. (A) Neutral hand position. (B) Hand position after the leader moves.

trol law. The walking gaits of the robots are generated on-line, assuming virtual springs and dampers between the object and the robots, as illustrated in Fig. 6.20. Thereby, the distance between the grasped object and the follower robots is controlled by the generated walking gait.

When the leader robot moves the grasped object, the hands of the follower robot are stretched under the damping control. Let $\mathcal{X}^{neutral}$ and $\mathcal{X}^{move}$ be the neutral hand position of the follower robot and the hand position after the leader robot moves, respectively, as illustrated in Fig. 6.21. The difference between $\mathcal{X}^{neutral}$ and $\mathcal{X}^{move}$ is defined by $\Delta\mathcal{X} = \mathcal{X}^{move} - \mathcal{X}^{neutral}$. A PD controller is designed to generate a step that controls $\Delta\mathcal{X}$ to zero as follows:

$$\mathcal{X}^{step} = \boldsymbol{K}_{Pstep}\Delta\mathcal{X} + \boldsymbol{K}_{Dstep}\Delta\dot{\mathcal{X}}. \tag{6.90}$$

Here $\mathcal{X}^{step}$ is the next desired footprint with respect to the current stance foot; $\boldsymbol{K}_{Pstep}$ and $\boldsymbol{K}_{Dstep}$ act as a virtual spring and damper, respectively, as illustrated in Fig. 6.20. The reference ZMP is computed in accordance with the walking command. The CoM trajectory can then be calculated with the help of the preview control approach [9].

Experiment of Object Transportation

Results from the leader-follower-type cooperative object manipulation experiment reported in [30] are presented in this section. The goal was to carry the object in the direction of the x-axis (see Fig. 6.22A for the coordinate frame). The commanded object positions along the y- and z-axes were constant.

In the experiment, the leader robot first walked backward, thus moving the grasped object in its direction. Then, the direction of the movement was changed and the leader robot moved the object in the direction of the follower robot, while walking forward. Snapshots from the movements of the leader and follower robots are presented in Fig. 6.22. More detailed data, such as hand positions and forces generated during the experiment, are given in [30].

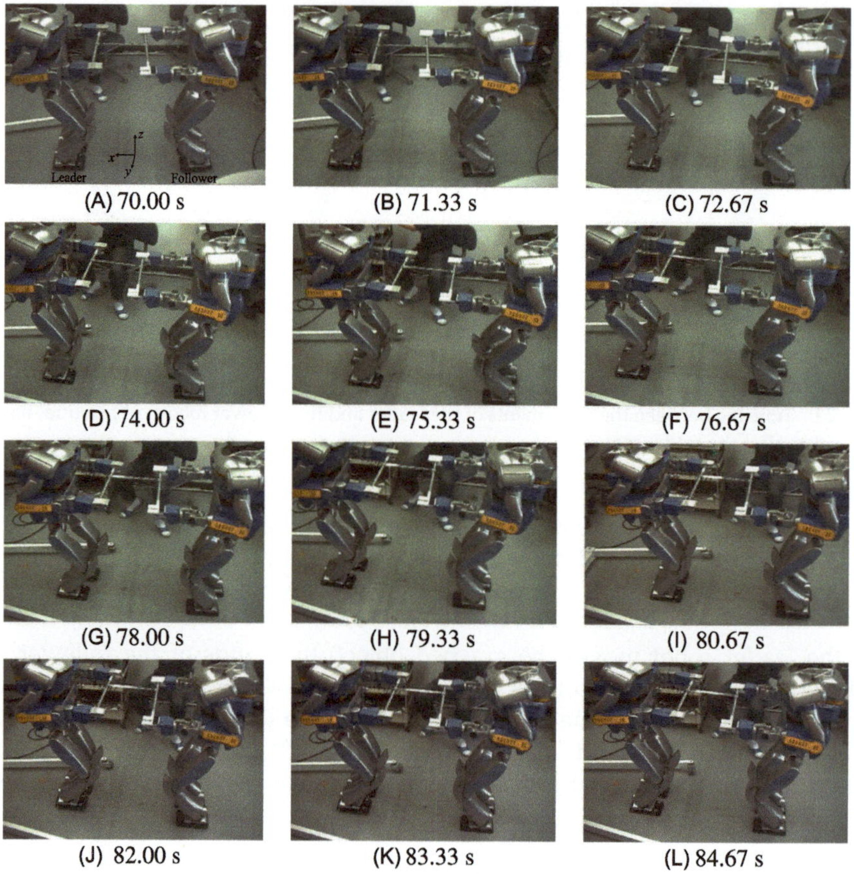

FIGURE 6.22 Cooperative object transportation by two humanoid robots [30].

6.4.5 Symmetry-Type Cooperative Object Manipulation

The multiarm object manipulation control method introduced in Section 6.3 is applied to control the cooperation between multiple humanoid robots. In what follows, cooperative object manipulation between two humanoid robots will be discussed.

Simulation of Symmetry-Type Cooperation

An example of symmetry-type cooperation between two humanoid robots is illustrated in Fig. 6.14. Given the desired applied wrench, $\mathcal{F}^{A,des}$, and the desired internal wrenches, $\mathcal{F}^{I}_{O_{12}}$, $\mathcal{F}^{I}_{O_{34}}$, and $\mathcal{F}^{I}_{O_{AB}}$, the desired hand wrenches $\mathcal{F}^{des}_{O_j}$ can be calculated from (6.74).

In the simulation experiment for symmetry-type cooperation, two models of the humanoid robot HRP-2 [11] were used. OpenHRP-3 [10] was used for the dynamic simulation.

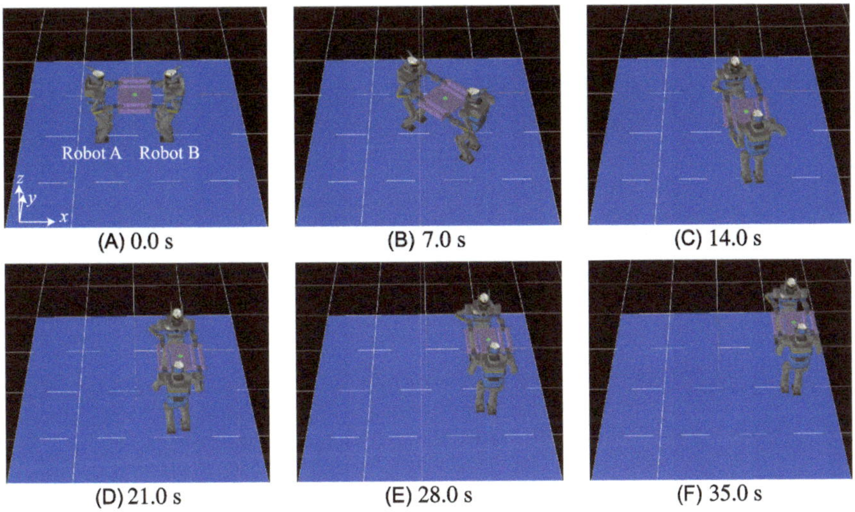

FIGURE 6.23 Snapshots from on-line operation [29].

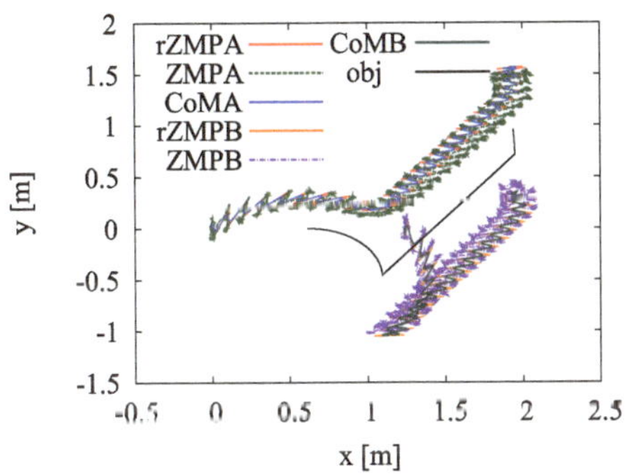

FIGURE 6.24 Result of on-line operation [29].

Video 6.4-1 [12] shows the simulation results from the cooperative motion of four humanoid robots.

Simulation Results

The snapshots in Fig. 6.23 are from the movement of two humanoid robots during the simulation of cooperative object transportation. Fig. 6.24 shows the trajectory of the reference ZMP (rZMP), ZMP, CoM, and the object. Fig. 6.25 shows the internal wrench $\mathcal{F}^I_{O_{AB}}$ generated during the simulation. This wrench is defined in (6.71). When the internal wrench is not controlled, spikes of the internal wrench are observed especially in the x-direction, as shown in Fig. 6.25. The spikes of the internal wrench are well reduced by applying force control.

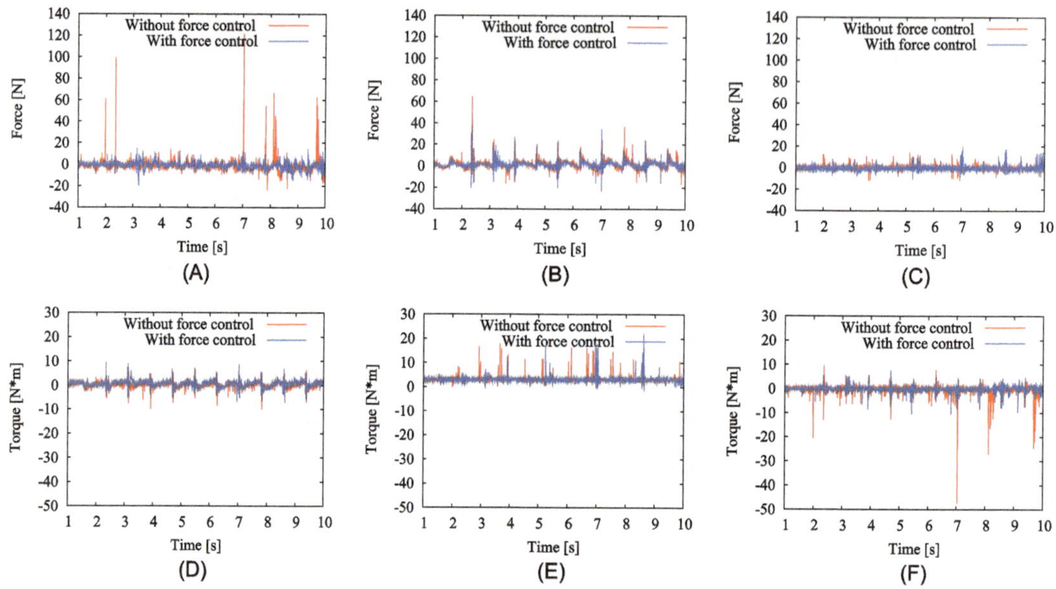

FIGURE 6.25 Internal wrench generated at the object [29]. (A) Internal force along the x-axis. (B) Internal force along the y-axis. (C) Internal force along the z-axis. (D) Internal moment around the x-axis. (E) Internal moment around the y-axis. (F) Internal moment around the z-axis.

6.4.6 Comparison Between Leader-Follower-Type and Symmetry-Type Cooperation

A comparison between the symmetry-type cooperation and the leader-follower-type cooperation based on a dynamic simulation is discussed in this section. In the simulation, the same constant reference velocity of 0.1 m/s is given for the object along the x-axis. Figs. 6.26 and 6.27 show the snapshots of the dynamic simulation of the leader-follower-type cooperation and the symmetry-type cooperation, respectively. The ZMP trajectories during the simulation are plotted in Fig. 6.28.

In the simulation of the leader-follower-type cooperation, the leader robot lost the balance because of the time lag between the leader and the follower robot, as shown in Fig. 6.26C. The leader robot moved first; then the follower robot moved, but the follower robot could not respond quickly enough. The ZMP in the x-axis went to the front fringe of the support polygon at $t = 7.2$ s, as shown in Fig. 6.28A.

In the symmetry-type cooperation, the two robots stably transported the object, as shown in Fig. 6.27. The ZMP always stayed near the center of the support polygon, as shown in Fig. 6.28B.

The results show that the symmetry-type cooperation is more stable for biped robots than the leader-follower-type cooperation. The advantages and disadvantages of each type of cooperation are summarized as follows.

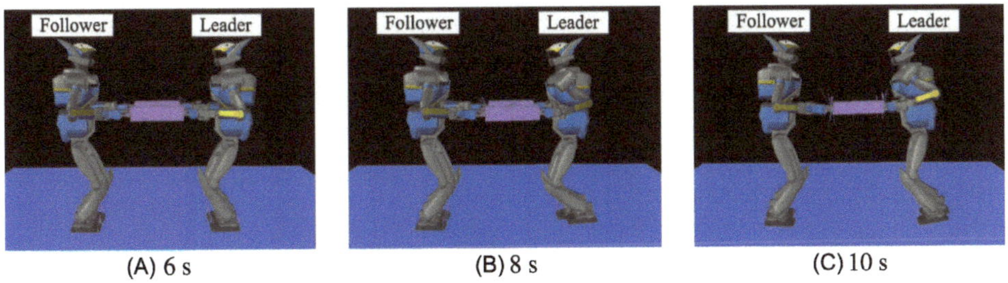

FIGURE 6.26 Simulation of the leader-follower-type cooperation.

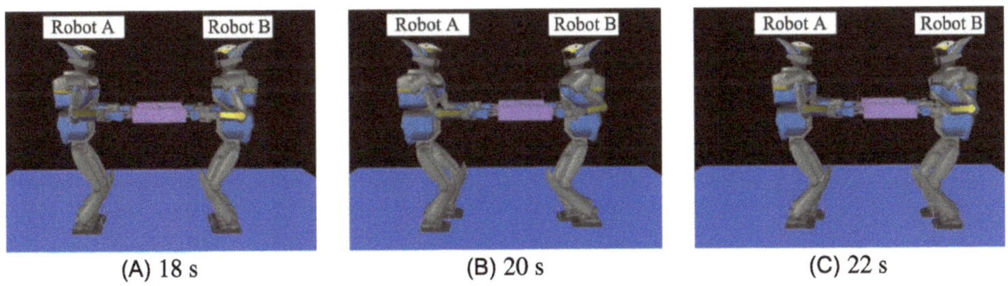

FIGURE 6.27 Simulation of the symmetry-type cooperation.

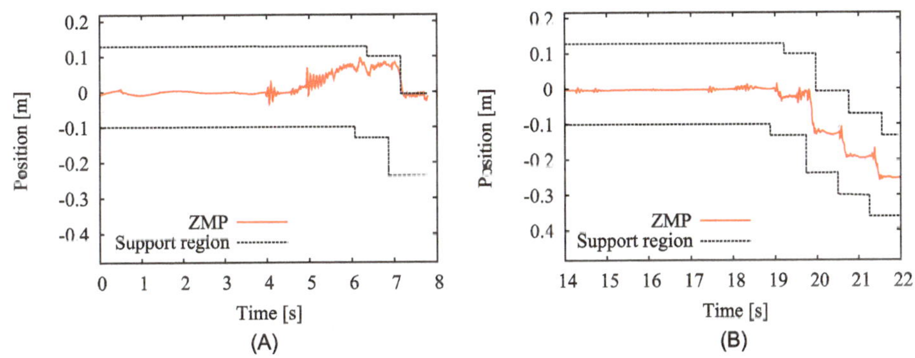

FIGURE 6.28 ZMP trajectory along the x-direction. (A) ZMP trajectory of the leader robot in the leader-follower-type cooperation. (B) ZMP trajectory of the robot B in the symmetry-type cooperation.

Leader-follower-type cooperation

Advantage: The control of the robots is decentralized. Therefore, there is no need of synchronization among the robots.

Disadvantage: The follower robots move based on the internal force generated at the object. Therefore, there is a time lag between the movements of the leader robot and those of the follower robots. The time lag may cause one or more of the robots to fall down.

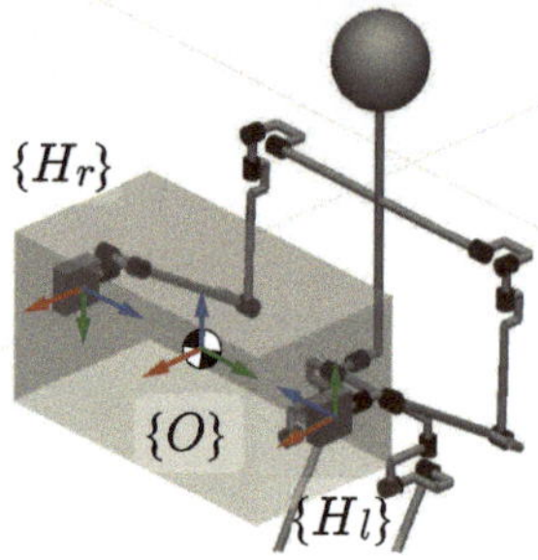

FIGURE 6.29 Object and hand coordinate frames. The contact joints at the hands impose unilateral planar constraints.

Symmetry-type cooperation

Advantage: All robots move synchronously. Ideally, there is no time lag among the robots. Thus, the risk of a robot to fall down is lower than that during leader-follower-type cooperation.

Disadvantage: The control of the robots is centralized. Therefore, synchronization among the robots is needed.

6.5 DUAL-ARM DYNAMIC OBJECT MANIPULATION CONTROL

Most tasks in the field of cooperative multifinger, dual-arm, and multirobot object manipulation are not time-critical. Such tasks can be handled with control methods based on the kinematic and kinetostatic relationships introduced in the previous sections of this chapter. There are, however, time-critical tasks where the speed of manipulation matters. Such type of tasks require a dynamics-based object manipulation control method. The control approach to be introduced in the remainder of this section is based on the kinematic, kinetostatic, and dynamic models developed in the previous chapters of this work. The approach is general in the sense that unilateral constraints at the hand contact joints can be handled, in addition to the bilateral ones used in the previous sections.

6.5.1 Equation of Motion of the Object

Suppose a humanoid robot grasps a rigid-body object with its hands s.t. *unilateral* plane contacts are established, as shown in Fig. 6.29. The coordinate frames of the object and the hands contacts, $\{O\}$ and $\{H_j\}$, respectively, are also shown in the figure. The equation of motion of the object is written in compact form as

$$\mathbb{M}_O \dot{\mathcal{V}}_O + \mathcal{C}_O + \mathcal{G}_O = \mathcal{F}_O. \tag{6.91}$$

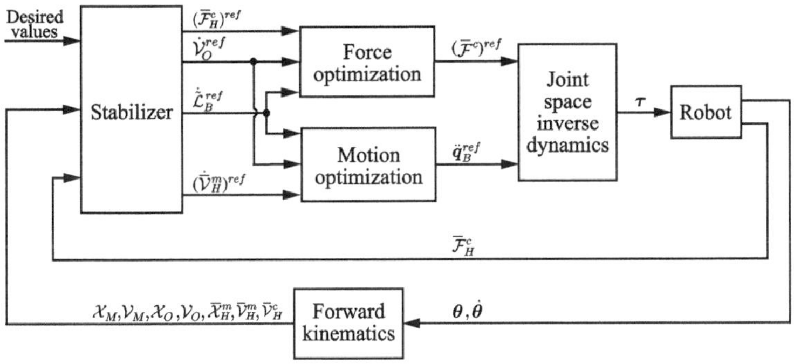

FIGURE 6.30 The motion/force controller based on the closed-chain formulation (cf. Fig. 5.29) is augmented with $\dot{\mathcal{V}}_O^{ref}$, the reference spatial acceleration for the desired object motion.

The expanded-form representation is

$$\begin{bmatrix} M_O \boldsymbol{E}_3 & \boldsymbol{0} \\ \boldsymbol{0} & \boldsymbol{I}_O \end{bmatrix} \dot{\mathcal{V}}_O + \begin{bmatrix} \boldsymbol{0} \\ \boldsymbol{\omega}_O \times \boldsymbol{I}_O \boldsymbol{\omega}_O \end{bmatrix} + \begin{bmatrix} M_O \boldsymbol{a}_g \\ \boldsymbol{0} \end{bmatrix} = -\mathbb{C}_{cO}(\boldsymbol{q}_H) \begin{bmatrix} \overline{\mathcal{F}}^c_{H_r} \\ \overline{\mathcal{F}}^c_{H_l} \end{bmatrix}. \tag{6.92}$$

Here M_O, $\boldsymbol{I}_O$ denote the mass and inertia tensor w.r.t. the CoM of the object, $\boldsymbol{\omega}_O$ is its angular velocity; $\mathbb{C}_{cO}(\boldsymbol{q}_H)$ is the contact map for the hands. Note the minus sign in the term on the r.h.s. It appears since the hand contact wrenches balance the *impressed* net wrench on the object, $\mathcal{F}_O$.

6.5.2 Controller

Dynamic object manipulation capabilities can be ensured with the dynamic motion/force controller introduced in Section 5.12.1. To this end, the controller in Fig. 5.29 is augmented as follows (see Fig. 6.30). First, the reference spatial acceleration for the object motion is defined as

$$\dot{\mathcal{V}}_O^{ref} = \begin{bmatrix} \dot{\boldsymbol{v}}_O^{ref} \\ \dot{\boldsymbol{\omega}}_O^{ref} \end{bmatrix} = \begin{bmatrix} \dot{\boldsymbol{v}}_O^{des} + K_{vO} \dot{\boldsymbol{e}}_{pO} + K_{pO} \boldsymbol{e}_{pO} \\ \dot{\boldsymbol{\omega}}_O^{des} + K_{\omega O} \boldsymbol{e}_{\omega O} + K_{oO} \boldsymbol{e}_{oO} \end{bmatrix}. \tag{6.93}$$

Here $\dot{\boldsymbol{v}}_O^{des}$ and $\dot{\boldsymbol{\omega}}_O^{des}$ are the desired linear and angular accelerations of the object, $\boldsymbol{e}_{\omega_O} = \boldsymbol{\omega}_O^{des} - \boldsymbol{\omega}_O$ denotes the error in the angular velocity, and $\boldsymbol{e}_{p_O}$ and $\boldsymbol{e}_{o_O}$ are the position and orientation errors, the latter defined by the preferable parametrization. The $K_{(\circ)}$ quantities stand for PD feedback gains. It is assumed here that the current 6D position and twist of the object are known, either calculated via the forward kinematics or measured.

Second, in the null-space prioritization scheme of the *motion optimization component* (see Fig. 5.30), an additional level is added in the hierarchy to account for the motion constraints in the independent closed loop formed by the hands and the object and for the motion of the object. This is shown in Fig. 6.31. Altogether, there are now four subtasks of priority.

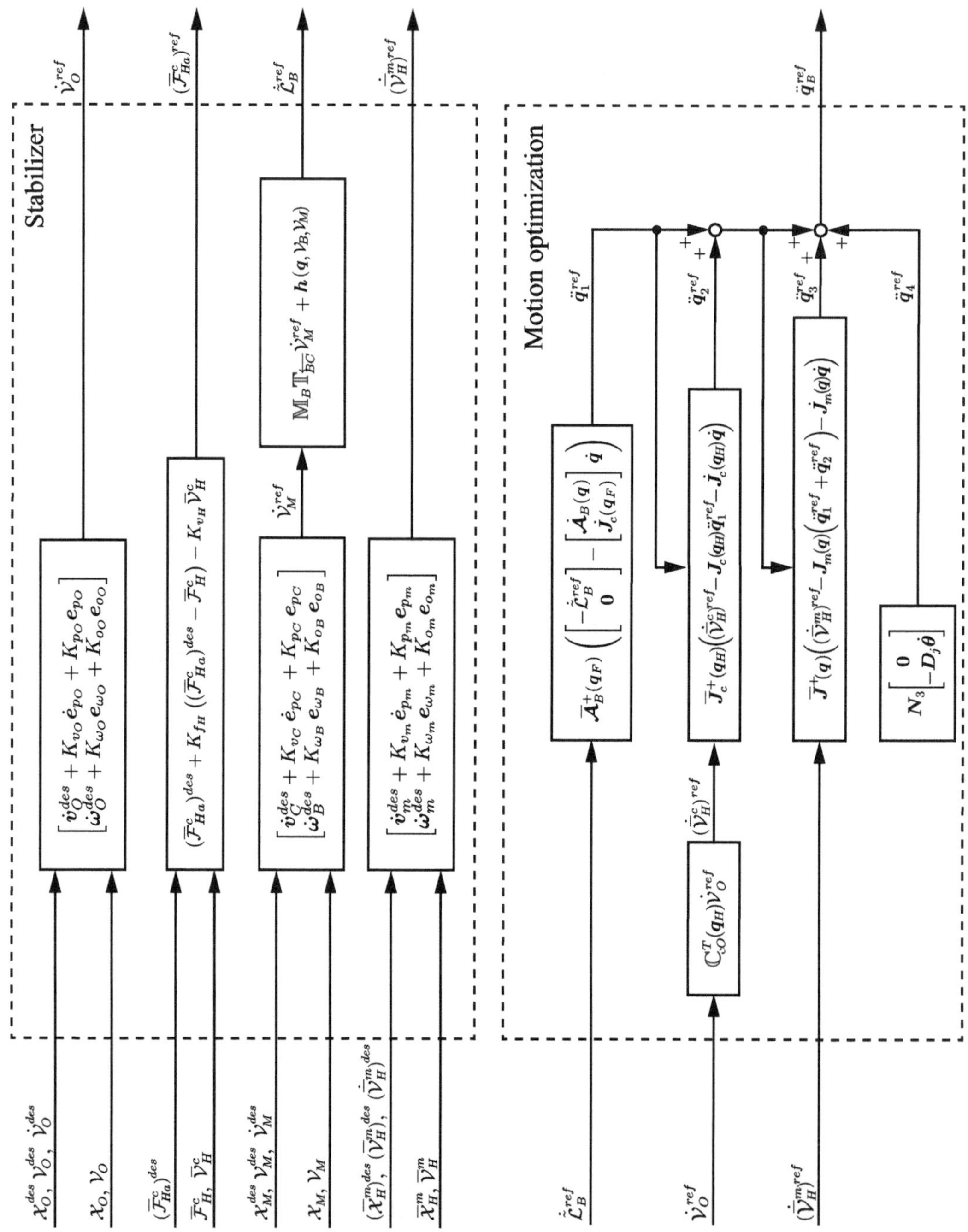

FIGURE 6.31 The equations used in the Stabilizer and Motion optimization blocks in Fig. 6.30.

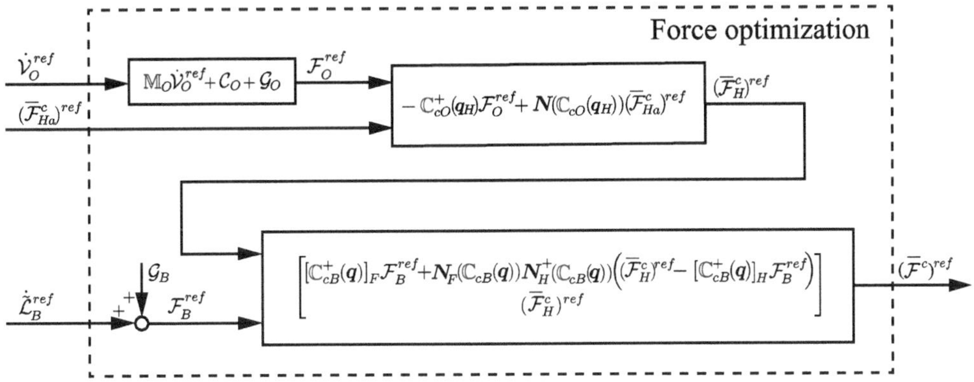

FIGURE 6.32 The equations used in the Force optimization block in Fig. 6.30.

The highest priority is for the balance control subtask ($\ddot{q}_1$), the second highest for the object motion ($\ddot{q}_2$), the third for the motion in the unconstrained (mobility) directions ($\ddot{q}_3$), and the fourth for the self-motion ($\ddot{q}_4$), in the form of joint-space damping in this case.

Third, the *wrench distribution problem for the hands* is solved in accordance with (3.66). The conventional force/moment control law

$$(\overline{\mathcal{F}}^c_{Ha})^{ref} = (\overline{\mathcal{F}}^c_{Ha})^{des} + K_{fH}\left((\overline{\mathcal{F}}^c_{Ha})^{des} - \overline{\mathcal{F}}^c_H\right) - K_{vH}\overline{\mathcal{V}}^c_H, \tag{6.94}$$

which also appears in the controller in Fig. 5.29, is used here to squeeze the object with a desired force determined by $(\overline{\mathcal{F}}^c_{Ha})^{des}$. This is done within the null space of the object contact map, $\mathcal{N}(\mathbb{C}_{cO})$, as apparent from Fig. 6.32. In this way, an appropriate internal wrench control can be ensured to enforce the contact wrench cone constraints.

The performance of the controller is examined with a dynamic object motion task simulated with a small-size humanoid robot.[2] An object of mass $M_O = 0.5$ kg and inertia tensor $\boldsymbol{I}_O = \text{diag}(1.92, 0.833, 1.92) \times 10^{-3}$ kgm^2 is grasped by the robot, lifted a bit, then shaken strongly, first in the x-, then in the y- and z-directions, followed by three consecutive rotations around the same axes and in the same order. The results are shown in graphical form in Fig. 6.33 and in Video 6.5-1 [23].

The desired position, velocity, orientation and angular velocity of the object are displayed in the upper row in Fig. 6.33. From the arm joint velocity graphs it can be confirmed that the motion is quite fast, indeed. The hand force graphs are presented in the local coordinate frames. They show that an internal (squeezing) force of 10 N is applied in the z-direction that

[2] The parameters of the model were derived from a HOAP-2 robot [4] (see Section A.2).

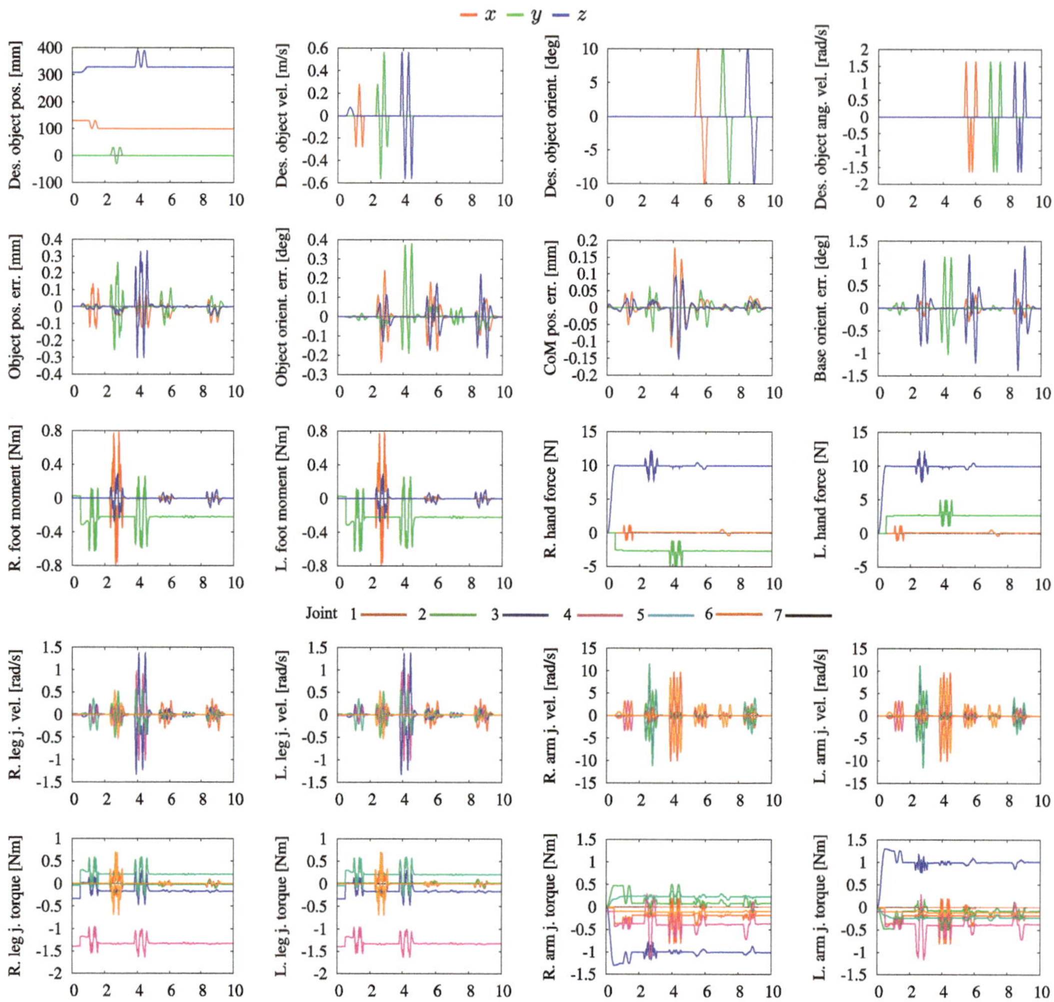

FIGURE 6.33 Results from a dynamic object-motion simulation task. The object is grasped, lifted a bit, then shaken strongly in the x-, y- and z-directions, and finally, rotated around these axes. The internal (squeezing) force prevents the object from slipping. The hand force graphs are presented in the local (sensor) frames. All other workspace related quantities are shown w.r.t. the world frame.

penetrates the object. In the world coordinate frame, this is the y-direction. From the error graphs for the object position/orientation, the CoM and the base rotation (the second row in Fig. 6.33) it is apparent that the performance is quite satisfactory since the errors are relatively small.

It should be noted that, in practice, the inertia tensor of the object may be unknown. The object mass can be estimated with measurements from the force/torque sensors in the wrists. The object dynamics (6.91) can then be reduced by ignoring the inertia and nonlinear wrench components. The performance of the controller is quite robust w.r.t. the object dynamics reduced in this way.

References

[1] D.J. Agravante, A. Cherubini, A. Bussy, P. Gergondet, A. Kheddar, Collaborative human-humanoid carrying using vision and haptic sensing, in: Proceedings of IEEE International Conference on Robotics and Automation, 2014, pp. 607–612.

[2] A. Bicchi, On the closure properties of robotic grasping, The International Journal of Robotics Research 14 (1995) 319–334.

[3] F. Caccavale, P. Chiacchio, A. Marino, L. Villani, Six-dof impedance control of dual-arm cooperative manipulators, IEEE/ASME Transactions on Mechatronics 13 (2008) 576–586.

[4] Fujitsu, Miniature Humanoid Robot HOAP-2 Manual, 1st edition, Fujitsu Automation Co., Ltd, 2004 (in Japanese).

[5] S. Hayati, Hybrid position/force control of multi-arm cooperating robotics, in: Proceedings of IEEE International Conference on Robotics and Automation, 1986, pp. 82–89.

[6] D. Heck, D. Kostić, A. Denasi, H. Nijmeijer, Internal and external force-based impedance control for cooperative manipulation, in: Proceedings of European Control Conference, 2013, pp. 2299–2304.

[7] Y. Hirata, T. Sawada, Z.-D. Wang, K. Kosuge, Leader-follower type motion control algorithm of multiple mobile robots with dual manipulators for handling a single object in coordination, in: Proceedings of IEEE International Conference on Intelligent Mechatronics and Automation, 2004, pp. 362–367.

[8] S. Kajita, H. Hirukawa, K. Harada, K. Yokoi, Introduction to Humanoid Robotics, Springer Tracts in Advanced Robotics (STAR), Springer, 2014.

[9] S. Kajita, F. Kanehiro, K. Kaneko, K. Fujiwara, K. Harada, K. Yokoi, H. Hirukawa, Biped walking pattern generation by using preview control of zero-moment point, in: Proceedings of IEEE International Conference on Robotics and Automation, 2003, pp. 1620–1626.

[10] F. Kanehiro, K. Fujiwara, S. Kajita, K. Yokoi, K. Kaneko, H. Hirukawa, Y. Nakamura, K. Yamane, Open architecture humanoid robotics platform, in: 2002 IEEE International Conference on Robotics and Automation, IEEE, 2002, pp. 24–30.

[11] K. Kaneko, F. Kanehiro, S. Kajita, H. Hirukawa, T. Kawasaki, M. Hirata, K. Akachi, T. Isozumi, Humanoid robot HRP-2, in: IEEE International Conference on Robotics and Automation, Proceedings, vol. 2, ICRA '04, 2004, 2004.

[12] A. Konno, Cooperative object handling by four humanoid robots, 2016 (Video clip), https://doi.org/10.1016/B978-0-12-804560-2.00013-4.

[13] Y. Kume, Y. Hirata, Z.-D. Wang, K. Kosuge, Decentralized control of multiple mobile manipulators handling a single object in coordination, in: Proceedings of IEEE/RSJ International Conference on Intelligent Robots and Systems, 2002, pp. 2758–2763.

[14] Z. Li, S. Deng, C.Y. Su, T. Chai, C. Yang, J. Fu, Decentralized control of multiple cooperative manipulators with impedance interaction using fuzzy systems, in: Proceedings of IEEE International Conference on Information and Automation, 2014, pp. 665–670.

[15] M.T. Mason, Mechanics of Robotic Manipulation, Intelligent Robotics and Autonomous Agents Series, A Bradford Book, 2001.

[16] R.M. Murray, Z. Li, S.S. Sastry, A Mathematical Introduction to Robotic Manipulation, CRC Press, 1994.

[17] E. Nakano, S. Ozaki, T. Isida, I. Kato, Cooperational control of the anthropomorphous manipulator 'MELARM', in: Proceedings of 4th International Symposium of Industrial Robots, 1974, pp. 251–260.

[18] P. Pagilla, M. Tomizuka, Hybrid force/motion control of two arms carrying an object, in: Proceedings of American Control Conference, 1994, pp. 195–199.

[19] R. Rastegari, S.A.A. Moosavian, Multiple impedance control of cooperative manipulators using virtual object grasp, in: 2006 IEEE Conference on Computer Aided Control System Design, 2006 IEEE International Conference on Control Applications, 2006 IEEE International Symposium on Intelligent Control, IEEE, 2006, pp. 2872–2877.

[20] F. Reuleaux, The Kinematics of Machinery, Macmillan, New York, 1875. Republished by Dover, New York, 1963.

[21] J.K. Salisbury, B. Roth, Kinematic and force analysis of articulated mechanical hands, Journal of Mechanisms, Transmissions, and Automation in Design 105 (1983) 35–41.
[22] S. Schneider, R. Cannon, Object impedance control for cooperative manipulation: theory and experimental results, IEEE Transactions on Robotics and Automation 8 (1992) 383–394.
[23] T. Shibuya, S. Sakaguchi, D. Nenchev, Dynamic object shaking, Robotic Life Support Laboratory, Tokyo City University, 2018 (Video clip), https://doi.org/10.1016/B978-0-12-804560-2.00013-4.
[24] B. Siciliano, O. Khatib (Eds.), Springer Handbook of Robotics, Springer, Berlin, Germany, 2008, pp. 671–700, Chapter 28.
[25] B. Siciliano, O. Khatib (Eds.), Springer Handbook of Robotics, Springer, Berlin, Germany, 2008, pp. 701–718, Chapter 29.
[26] J. Szewczyk, F. Plumet, P. Bidaud, Planning and controlling cooperating robots through distributed impedance, Journal of Robotic Systems 19 (2002) 283–297.
[27] M. Uchiyama, P. Dauchez, Symmetric kinematic formulation and non-master/slave coordinated control of two-arm robots, Advanced Robotics 7 (1993) 361–383.
[28] D. Williams, O. Khatib, The virtual linkage: a model for internal forces in multi-grasp manipulation, in: 1993 IEEE International Conference on Robotics and Automation, 1993, pp. 1025–1030.
[29] M.-H. Wu, A. Konno, S. Ogawa, S. Komizunai, Symmetry cooperative object transportation by multiple humanoid robots, in: Proceedings of IEEE International Conference on Robotics and Automation, Hong Kong, China, 2014, pp. 3446–3451.
[30] M.-H. Wu, A. Konno, M. Uchiyama, Cooperative object transportation by multiple humanoid robots, in: Proceedings of IEEE/SICE International Symposium on System Integration, Kyoto, Japan, 2011, pp. 779–784.
[31] M.-H. Wu, S. Ogawa, A. Konno, Symmetry position/force hybrid control for cooperative object transportation using multiple humanoid robots, Advanced Robotics 30 (2016) 131–149.
[32] K. Yokoyama, H. Handa, T. Isozumi, Y. Fukase, K. Kaneko, F. Kanehiro, Y. Kawai, F. Tomita, H. Hirukawa, Cooperative works by a human and a humanoid robot, in: Proceedings of IEEE International Conference on Robotics and Automation, 2003, pp. 2985–2991.
[33] T. Yoshikawa, X.Z. Zheng, Coordinated dynamic control for multiple robot manipulators handling an object, in: Proceedings of Fifth International Conference on Advanced Robotics, 1991, pp. 579–584.

CHAPTER

7

Motion Generation and Control: Selected Topics With Applications

7.1 OVERVIEW

Humanoid robots are expected to perform a broad variety of tasks, similar to humans. Basically, two large groups of periodic and aperiodic motion tasks can be distinguished. The most prominent example of a periodic motion task is walking on a regular (flat) ground. This is an indispensable task for a humanoid robot. Other examples of periodic motion tasks are ascending/descending a staircase and running. Historically, the initial focus in the humanoid robotics research field was on gait generation and walking control [114]. This was a very challenging task for more than 25 years, until the announcement of the P2 humanoid prototype by Honda Motor Co. in 1996 [39]. The recent state of the art in periodic gait generation and control is based on the capture point (CP) concept (see Section 5.5). An implementation that stabilizes the unstable component of motion during walking has been described in [105]. In this chapter, the implementation described in [22] is explained and applied to gait generation and walking control on sand [60].

The class of "aperiodic" [9] or "nongaited" [35] motion tasks includes such tasks as fetching/manipulating objects within cluttered environments, performing a job with a hand-held tool, sitting/lying down and standing up, negotiating highly irregular terrain (e.g. a free-climbing robot [12]), crawling through narrow spaces, and climbing a ladder. These types of tasks require a somewhat different approach than those used in periodic motion generation tasks. In the pioneering work [70], motion generation for aperiodic tasks was based on the given initial and target postures. The process is accomplished in two phases. First, a statically stable collision-free path is computed via a modified version of the Rapidly exploring Random Trees (RRTs) algorithm [72]. Second, the computed path is smoothed and transformed into a dynamically stable whole-body trajectory with the help of the AutoBalancer [55]. This approach can be characterized as time consuming and prone to robustness issues, though. Heuristics have to be employed during the search, and hence there is no guarantee that an existing solution will be found.

The roots of motion generation for humanoid robots are found in the closely related field of physics-based human character animation methods. These methods became widely adopted recently. Already at an early stage, though, it was clarified [119] that to obtain a "realistic" and smooth motion of the animated character for reaching a prespecified goal, two problems have to be addressed. First, the initial state has to be determined carefully to ensure the

Humanoid Robots
https://doi.org/10.1016/B978-0-12-804560-2.00014-6

appropriateness of the generated motion trajectory. Second, since the character is underactuated, its successful propulsion requires a force (reaction wrench) input trajectory. The force trajectory cannot be generated in an independent fashion from the motion trajectory; they are intertwined since the impressed/reaction wrenches depend upon the acceleration of the CoM and the angular accelerations of the body segments (cf. Section 5.8.1). As a consequence, it became apparent that simulation methods based on initial-value problems are not appropriate; the above objectives require the solution of a constrained two-point boundary problem. Furthermore, it has been also argued that the movement "realism" can be further improved by introducing joint torque minimization into the algorithm. This implies that, in addition to the motion/force task constraints, physical constraints also need to be specified. The former determine the task completely; they freely propagate forward/backward in time. The latter are used to limit the "muscle" forces (i.e. the actuated joint torque) and to determine an appropriate contact (i.e. passive-joint) behavior. An optimal, physically feasible solution trajectory can then be found with the help of an appropriate constraint optimization method, e.g. one of those described in Chapter 5. It is important to note that the character of the solution is global, but time consuming and, hence, only suitable for off-line calculations. To summarize, for motion generation and control of complex tasks as those mentioned above, the robot should be equipped with the following two general capabilities: (1) multicontact planning [12,24] and (2) whole-body motion generation consistent with the constraints [70,35,33].

There are a few ways of improving the efficacy of the process of motion generation for aperiodic motion tasks. First, discrete sets of contact states can be added to the motion plan [12]. The contact states are determined from a coarse model/map of the environment constructed in advance. The process is referred to as contact planning [24]. For each given contact state, a set of statically stable postures is obtained.[1] This process is referred to as the posture generation [15]. Posture generation is accomplished with the help of a constraint-based inverse kinematics and statics solver [10,11]. Recall that, as discussed in Chapter 3, for a given set of contacts there is in general an infinite number of statically stable postures [12–14]. There are also an infinite number of motion/force trajectories that connect the generated keyframes in accordance with the kinematic/dynamic relationships and the various constraints. A dynamically feasible trajectory can be generated via interpolation [70,33]. This approach is referred to as the "contact-before-motion" approach [35]. It is important to note that because of the variety of tasks and environments envisioned, the contact planning process is nontrivial. For example, it is not sufficient to plan one contact state at a time since the current contact state determines all future states [12]. A number of methods have been developed so far for contact planning purposes. Interested readers are referred to [35,12,23,11,24,15,109]. It should be noted that, although the contact state–based posture/trajectory generation approach is faster than the algorithms based on RRTs, it is still quite time demanding due to the involvement of iterative optimization procedures. It is also important to note that the method is not straightforward since there is no guarantee that the preplanned contact states will facilitate the generation of smooth, "naturally looking" constraint-consistent movements. Additional contact states/keyframes may be have to be inserted to support the process of motion generation, as already noted.

[1] Such postures are sometimes called "keyframes" [33], a term borrowed from the field of animation.

Another way of increasing the time efficiency in motion generation is by making use of the reduced dynamics representations in terms of end-link and CRB coordinates. Note that the posture-based motion generation process implies that the optimization runs over the joint space, involving all of the DoFs. The number of variables can be drastically decreased by resorting to one of the reduced forms of the equation of motion (cf. Section 4.13). In this case, the preplanned contact states are connected via trajectories determined from the CRB dynamic relations. As a result, one obtains constraint-consistent CRB and end-link motion/force trajectories that can be resolved for the joint torques via the inverse dynamics [69]. Another possibility for improvement is to divide the trajectory generation process into two phases. First, a parametrized path is generated that connects the contact states. Then, a constraint-consistent trajectory is obtained by determining the appropriate timing along the path [33]. This approach also has the advantage of facilitating time-optimal motion generation [97,98].

The time efficiency of motion generation can also be increased by making use of motion synergies or primitives [30,34]. These are motion patterns in joint-space that are controlled with fewer variables than the number of DoFs. An approach to synergy-based motion generation based on an inverse kinematics solution with task prioritization is introduced in Section 7.5. Synergy-based motion generation for reactive balance control with planar models is discussed in Section 7.6. Reactive synergies obtained with a whole-body model are addressed in Section 7.7. The final section in this chapter deals with motion generation for impacts.

7.2 ICP-BASED GAIT GENERATION AND WALKING CONTROL

7.2.1 CP-Based Walking Control

This section introduces a gait generation and walking control approach based on the CP concept [22]. The following expressions for the extrapolated CoM (xCoM) and the CP dynamics represented in 2D space will be employed (cf. also (5.55) and (5.56)):

$$\boldsymbol{r}_{ex}(t) = \boldsymbol{r}_g(t) + \frac{\dot{\boldsymbol{r}}_g(t)}{\omega}, \tag{7.1}$$

$$\dot{\boldsymbol{r}}_{ex} = \omega\left(\boldsymbol{r}_{ex} - \boldsymbol{r}_p\right). \tag{7.2}$$

Here $\boldsymbol{r}_{ex}$, $\boldsymbol{r}_g$, and $\boldsymbol{r}_p$ are 2D vectors denoting the xCoM, the gCoM, and the CoP, respectively (see Section 5.3). Eqs. (7.1) and (7.2) are rewritten in the frequency domain as follows:

$$\boldsymbol{r}_g(s) = \frac{1}{1 + \frac{1}{\omega}s}\boldsymbol{r}_{ex}(s), \tag{7.3}$$

$$\boldsymbol{r}_{ex}(s) = \frac{1}{1 - \frac{1}{\omega}s}\boldsymbol{r}_p(s). \tag{7.4}$$

Eq. (7.3) represents the first-order open-loop CoM dynamics taking the CP as the input, while (7.4) represents the first-order open-loop CP dynamics taking the CoP as the input.

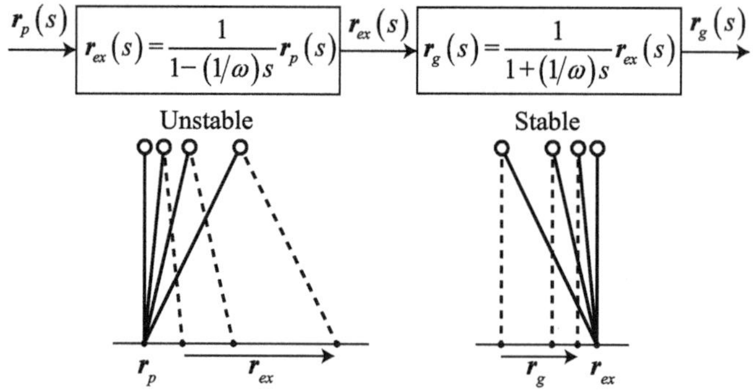

FIGURE 7.1 Coupled dynamics of the CP and the CoM.

The system (7.3) is stable because the real part of the characteristic root, $s = -\omega$, is negative. On the contrary, the system (7.4) is unstable because the real part of the characteristic root, $s = \omega$, is positive. Fig. 7.1 illustrates the coupled dynamics of the CP and CoM. In order to stabilize the CP dynamics, the following CP controller was proposed in [22] (cf. also (5.57)):

$$\boldsymbol{r}_p^{ref} = \frac{1}{1 - e^{\omega dT}} \boldsymbol{r}_{ex}^{des} - \frac{e^{\omega dT}}{1 - e^{\omega dT}} \boldsymbol{r}_{ex}. \tag{7.5}$$

Substituting $\boldsymbol{r}_p^{ref}$ from (7.5) into $\boldsymbol{r}_p$ in (7.2), the following system is obtained:

$$\dot{\boldsymbol{r}}_{ex} = \frac{\omega}{1 - e^{\omega dT}} \left(\boldsymbol{r}_{ex} - \boldsymbol{r}_{ex}^{des} \right). \tag{7.6}$$

The frequency-domain representation can be written as

$$\boldsymbol{r}_{ex}(s) = \frac{1}{1 - \frac{1 - e^{\omega dT}}{\omega} s} \boldsymbol{r}_{ex}^{des}(s). \tag{7.7}$$

Note that dT is always positive; hence $e^{\omega dT}$ is greater than 1, and consequently the real part of the characteristic root, $s = \omega / \left(1 - e^{\omega dT}\right)$, is negative. Therefore, the CP controller (7.5) stabilizes the system. The block diagram of the stabilized CP and CoM dynamic system is illustrated in Fig. 7.2.

7.2.2 CP-Based Gait Generation

Let T_{step} be the specified duration of the step. The gait generation sequence is based on relative time, i.e. $0 \le t \le T_{step}$. The sequence comprises the following phases:

(a) Planning of foot prints: The foot prints are designed with the step length L_x, step width L_y, and step direction L_θ (see Fig. 7.3). The step width L_y is generally set to the natural

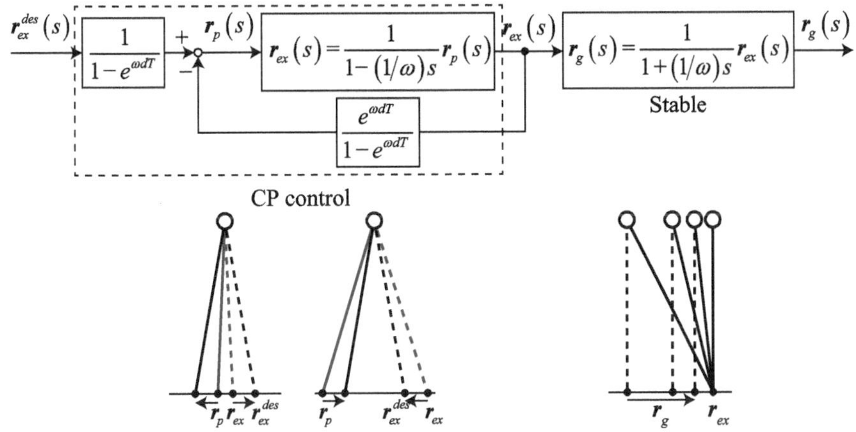

FIGURE 7.2 CP stabilization control.

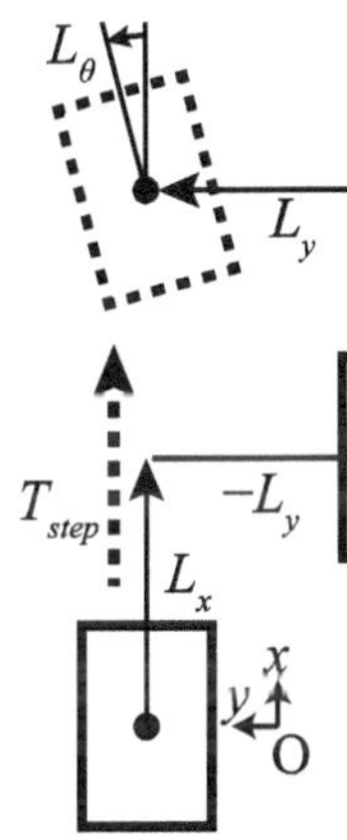

FIGURE 7.3 Step parameters.

length between the left and right legs. The step length L_x and step direction L_θ are heuristically designed depending on the dimensions of the humanoid robot and the desired pathway.

(b) Planning of the CPs at the end of each step (eos): The desired CP at the end of the ith step $\boldsymbol{r}^i_{ex,eos}$ is located at the ground projections of the corresponding ankle positions (see Fig. 7.4A). The CP at the beginning of the ith step $\boldsymbol{r}^i_{ex0}$ is determined based on the initial position and velocity vectors of the gCoM. The reference CoP at the ith step $\boldsymbol{r}^{ref,i}_{p0}$ is calculated based on $\boldsymbol{r}^i_{ex,eos}$ and $\boldsymbol{r}^i_{ex0}$. However, there is no guarantee that the calculated reference CoP, $\boldsymbol{r}^{ref,i}_{p0}$, will lie within the BoS. When the calculated reference CoP lies outside the BoS, it is modified, as explained in Phase (d). Depending on the parameter setting for the desired walking pattern, it might happen that $\boldsymbol{r}^{ref,i}_{p0}$ frequently lies outside the BoS. This problem can be alleviated by

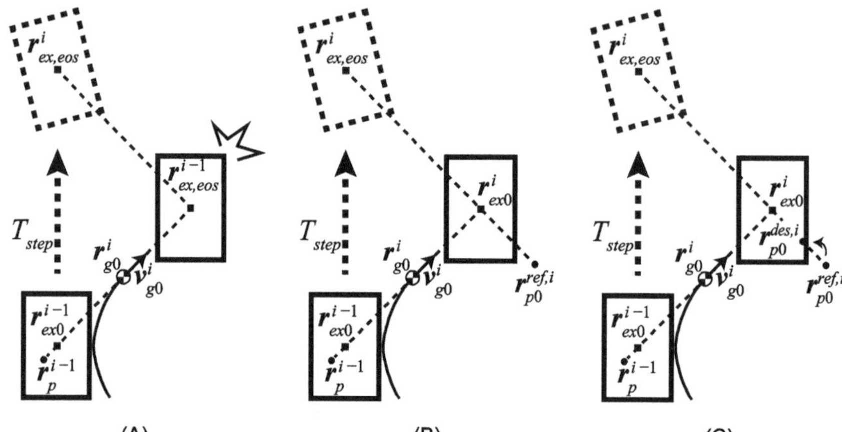

FIGURE 7.4 Design components for walking pattern generation based on the CP: (A) Planning of footprints and CPs at the end of each step (eos). (B) Calculation of the reference CoP $\boldsymbol{r}_p^{ref}$ using (7.5). (C) When $\boldsymbol{r}_p^{ref}$ is located outside the BoS, it is shifted within the BoS.

modifying $\boldsymbol{r}^i_{ex,eos}$ with a heuristically designed constant offset that is added to the preplanned ankle position, such that the resulting CoP $\boldsymbol{r}_{p0}^{ref,i}$ always lies within the BoS for the desired walking pattern. In this way, the frequent modification of the reference CoPs $\boldsymbol{r}_{p0}^{ref,i}$ can be avoided.

(c) Calculation of the reference CoP: Let $\boldsymbol{r}^i_{g0}$ and $\boldsymbol{v}^i_{g0}$ be the 2D initial position and velocity vectors of the gCoM, respectively, in the ith step (see Fig. 7.4A). Note that $\boldsymbol{r}^i_{g0}$ and $\boldsymbol{v}^i_{g0}$ are equal to the terminal position and velocity of the gCoM in the $(i-1)$th step. Substituting $\boldsymbol{r}^i_{g0}$ and $\boldsymbol{v}^i_{g0}$ into (7.1), the initial CP at the ith step $\boldsymbol{r}^i_{ex0}$ is calculated as follows:

$$\boldsymbol{r}^i_{ex0} = \boldsymbol{r}^i_{g0} + \frac{\boldsymbol{v}^i_{g0}}{\omega}. \tag{7.8}$$

The reference CoP at the ith step $\boldsymbol{r}_{p0}^{ref,i}$ is calculated by substituting $\boldsymbol{r}^{des}_{ex} = \boldsymbol{r}^i_{ex,eos}$, $\boldsymbol{r}_{ex} = \boldsymbol{r}^i_{ex0}$, and $dT = T_{step}$ into (7.5). Thus

$$\boldsymbol{r}_{p0}^{ref,i} = \frac{1}{1-e^{\omega T_{step}}}\boldsymbol{r}^i_{ex,eos} - \frac{e^{\omega T_{step}}}{1-e^{\omega T_{step}}}\boldsymbol{r}^i_{ex0}. \tag{7.9}$$

(d) Calculation of the desired CoP for step initialization: When the reference CoP $\boldsymbol{r}_{p0}^{ref,i}$ (calculated from (7.9)) lies within the BoS, the desired CoP for step initialization is determined as $\boldsymbol{r}_p^{des,i} = \boldsymbol{r}_{p0}^{ref,i}$. However, $\boldsymbol{r}_{p0}^{ref,i}$ may not always lie within the BoS (see Fig. 7.4B). In this case, $\boldsymbol{r}_{p0}^{ref,i}$ is shifted to the nearest point on the footprint boundary along the line $\boldsymbol{r}^i_{ex,eos} - \boldsymbol{r}^i_{ex0}$

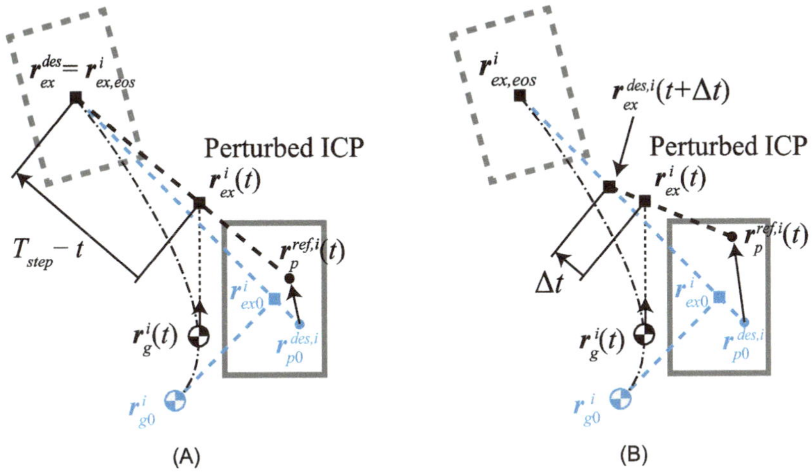

FIGURE 7.5 Concept of two CP controllers. (A) The CP end-of-step (CPS) controller. (B) The CP Tracking (CPT) controller.

(see Fig. 7.4C). If the line does not intersect the ith footprint, the line is slightly tilted to do so. The desired CoP $r_{p0}^{des,i}$ is then determined from the intersection point.

7.2.3 ICP Controller

The desired CoP $r_p^{des,i}$ at the ith step is obtained from the Phases (a)–(d) presented in Section 7.2.2. The desired CP trajectory is obtained by solving (7.2) with $r_{ex} = r_{ex0}^{i}$ and $r_p = r_{p0}^{des,i}$ as

$$r_{ex}^{des,i}(t) = e^{\omega t}\left(r_{ex0}^{i} - r_{p0}^{des,i}\right) + r_{p0}^{des,i}, \tag{7.10}$$

which is a 2D extension of (5.50); $r_{ex}^{des,i}(t)$ $(0 \le t \le T_{step})$ is referred to as the *instantaneous* CP (ICP) trajectory [67]. The ICP trajectory heads toward $r_{ex,eos}^{i}$. However, in reality the ICP trajectory is perturbed by modeling errors and/or disturbances, or for other reasons. In the CP control method proposed in [22], two different approaches were verified: the CP end-of-step (CPS) control and the CP Tracking (CPT) control. The concepts of the CPS and CPT controllers are illustrated in Fig. 7.5.

CP end-of-step (CPS) controller: Let T_{step} be the specified duration of a step. As illustrated in Fig. 7.5A, the ICP trajectory is controlled so that $r_{ex}^{i}(t)$ reaches $r_{ex,eos}^{i}$ at $t = T_{step}$. Therefore, (7.5) has to be modified as follows:

$$r_p^{ref,i}(t) = \frac{1}{1 - e^{\omega(T_{step}-t)}} r_{ex,eos}^{i} - \frac{e^{\omega(T_{step}-t)}}{1 - e^{\omega(T_{step}-t)}} r_{ex}^{i}(t) \quad \left(0 < t \le T_{step}\right). \tag{7.11}$$

Note that the CPS controller does not depend on future step positions but only on the final CP at the ith step $r_{ex,eos}^{i}$ and the current CP $r_{ex}^{i}(t)$. Hence reactive gait planning is possible.

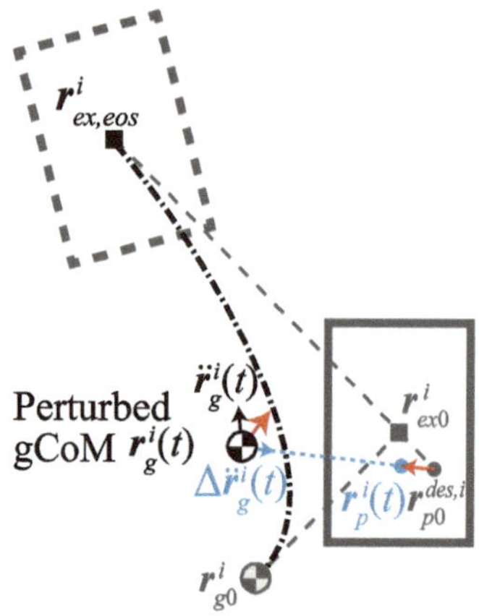

FIGURE 7.6 Concept of CP-based gait generation and ZMP control.

CP tracking (CPT) controller: Let Δt be the control period. As illustrated in Fig. 7.5B, the CPT controller controls the current ICP $\boldsymbol{r}^i_{ex}(t)$ toward the desired ICP at the next sampling period $\boldsymbol{r}^{des,i}_{ex}(t+\Delta t)$. Since $\boldsymbol{r}^{des,i}_{ex}(t+\Delta t)$ is used as the desired ICP, (7.5) is rewritten as

$$\boldsymbol{r}^{ref,i}_p(t) = \frac{1}{1-e^{\omega\Delta t}}\boldsymbol{r}^{des,i}_{ex}(t+\Delta t) - \frac{e^{\omega\Delta t}}{1-e^{\omega\Delta t}}\boldsymbol{r}^i_{ex}(t) \quad \left(0 < t \leq T_{step}\right). \tag{7.12}$$

In the simulation in [22], Δt was set at 0.05 s, while T_{step} was set at 0.8 s.

Modification of the reference CoP: The reference CoP $\boldsymbol{r}^{ref,i}_p(t)$ calculated from (7.11) or (7.12) may not always lie within the BoS. When $\boldsymbol{r}^{ref,i}_p(t)$ is outside the BoS, it should be shifted within the BoS (cf. Phase (d) in Section 7.2.2). The desired CoP $\boldsymbol{r}^{des,i}_p(t)$ is then set at the modified reference CoP.

7.2.4 CP-Based Gait Generation and ZMP Control

The main disadvantage of the well-known gait generation approach based on preview control [56] is that the preview controller depends on future reference CoPs to generate the motion. Hence, the preview controller can not modify the generated gait to react to an external disturbance. CP-based gait generation overcomes this problem.

Both the CPS and the CPT controller calculate the current ICP $\boldsymbol{r}^i_{ex}(t)$ at every control cycle. The reference CoP $\boldsymbol{r}^i_p(t)$ is given based on the difference between the desired and current ICPs. The CoP is controlled so that the current CP moves toward the desired CP.

In [60], the CP dynamics were used just to generate the reference gCoM trajectory, and a conventional ZMP controller was used for stabilization. The main advantage of this method is that a well-established ZMP controller can be used while reactively generating the gait pattern. The concept of an on-line gait generation and stabilizing control used in [60] is illustrated in Fig. 7.6.

Note that the calculation of the desired CoP in [60] is different from the CPS or CPT controllers. The initial ICP $\boldsymbol{r}^{i}_{ex0}$ and the reference CoP $\boldsymbol{r}^{ref,i}_{p0}$ at the ith step are calculated by (7.8) and (7.9), respectively. The reference CoP is modified so that the CoP lies within BoS, and the desired CoP is determined (if the reference CoP lies within BoS, the desired CoP is set at the reference CoP).

The reference CoP $\boldsymbol{r}^{ref,i}_{p}$ corresponds to $\bar{\boldsymbol{u}}$ in Fig. 5.12. As shown in Fig. 5.12, the ZMP manipulator receives the desired CoP $\boldsymbol{r}^{des,i}_{p0}$ and produces the desired CoM trajectory $\left(\boldsymbol{r}^{des,i}_{g}(t), \dot{\boldsymbol{r}}^{des,i}_{g}(t)\right)$ (see Fig. 7.4C).

The reference gCoM trajectory is calculated from the explicit solution of the gCoM dynamics (5.11), i.e.

$$\begin{bmatrix} \boldsymbol{r}^{i}_{g}(t) \\ \dot{\boldsymbol{r}}^{i}_{g}(t) \end{bmatrix} = \begin{bmatrix} \cosh(\omega t) & 0 & \frac{1}{\omega}\sinh(\omega t) & 0 \\ 0 & \cosh(\omega t) & 0 & \frac{1}{\omega}\sinh(\omega t) \\ \omega\sinh(\omega t) & 0 & \cosh(\omega t) & 0 \\ 0 & \omega\sinh(\omega t) & 0 & \cosh(\omega t) \end{bmatrix} \begin{bmatrix} \boldsymbol{r}^{i}_{g0} \\ \boldsymbol{v}^{i}_{g0} \end{bmatrix} + \begin{bmatrix} 1-\cosh(\omega t) & 0 \\ 0 & 1-\cosh(\omega t) \\ -\omega\sinh(\omega t) & 0 \\ 0 & -\omega\sinh(\omega t) \end{bmatrix} \boldsymbol{r}^{des,i}_{p0}, \tag{7.13}$$

where the hyperbolic functions are defined as follows:

$$\cosh(\omega t) = \frac{e^{\omega t} + e^{-\omega t}}{2}, \qquad \sinh(\omega t) = \frac{e^{\omega t} - e^{-\omega t}}{2}.$$

The deviation of the gCoM from the desired trajectory can be controlled by any well-established ZMP controller. In [60], the torso position compliance control (TPCC) approach [85] (see also Section 5.4.5) was used in the stabilizing controller.

7.3 BIPED WALK ON SAND

7.3.1 Landing Position Control for Walking on Sand

The difficulty of walking on loose soil for a humanoid robot is that the soil under the supporting foot easily deforms because of the weight of the robot. Hence, if the CoP goes to the fringe of the BoS, the soil under the CoP will be depressed, and consequently the robot body will tilt. Most of the stabilizing controllers control the net CoP (the ZMP), however the shift of the CoP may cause sinking on loose soil.

In order to compensate the tilt of the robot body caused by sinking and sliding on soil, a landing position control approach was proposed in [60]. According to this approach, when the deviation of the gCoM from the desired position exceeds a previously specified threshold $\delta\boldsymbol{r}^{th}_{g}$, the deviation is compensated by shifting the landing position. The shift of the landing

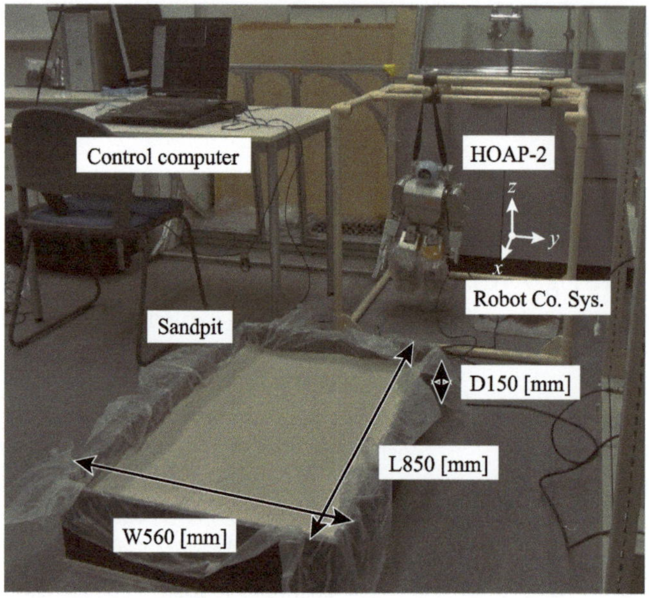

FIGURE 7.7 Experimental setup for biped walking on sand.

position at the next step is calculated by

$$\Delta \boldsymbol{r}_{k,j+1} = \boldsymbol{K}_L \max\left(\delta \boldsymbol{r}_{g,j}(t)\right) + \boldsymbol{\lambda} \Delta \boldsymbol{r}_{k,j} \quad (0 \le t \le T_{step}), \tag{7.14}$$

where $\boldsymbol{r}_{k,j}$ $k \in \{Fr, Fl\}$ is the 2D position vector of the ground projection of the supporting foot ankle position at the jth step, $\max(\delta \boldsymbol{r}_{g,j})$ is the maximum amount of deviation of the gCoM from the desired position during the jth step, $\boldsymbol{K}_L$ is a diagonal gain matrix, and $\boldsymbol{\lambda}$ is a diagonal forgetting factor matrix. The first term on the r.h.s. compensates the deviation of the gCoM, while the second term contributes to the reduction of the accumulated deviation.

7.3.2 Experiments of Walking on Sand

Fig. 7.7 shows the experimental setup. A small-size humanoid robot HOAP-2 manufactured by Fujitsu Automation Ltd. was used in the experiments. The height of the robot is 500 mm and the weight is 7 kg. The sampling period was set at 1 ms. The legs were covered with plastic sheet for dust proofing. The dimensions of the experimental sand pool were 850 × 560 × 150 mm (length × width × depth). Silica sand was used with a particle diameter of about 0.01 ∼ 0.04 mm.

The foot prints were designed as described in Section 7.2.1, with the parameters $L_x =$ 15 mm, $L_y = 78$ mm, and $L_\theta = 0°$ (see Fig. 7.3). The robot walked 28 steps forward and eight steps in place, and then halted. Based on the foot prints, the CPs at the end of each step $\boldsymbol{r}^i_{ex,eos}$ were designed. The duration of the step, T_{step}, was set at 0.4 s. The threshold for the landing position control was set as $\delta \boldsymbol{r}_g^{th} = [2.0 \;\; 6.0]^T$ (mm).

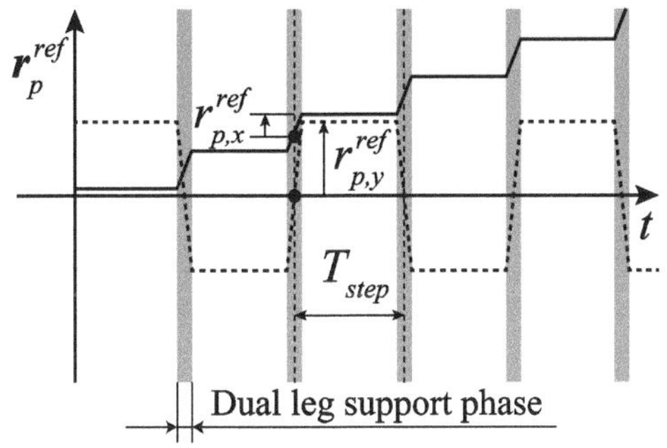

FIGURE 7.8 Design of reference ZMP trajectories.

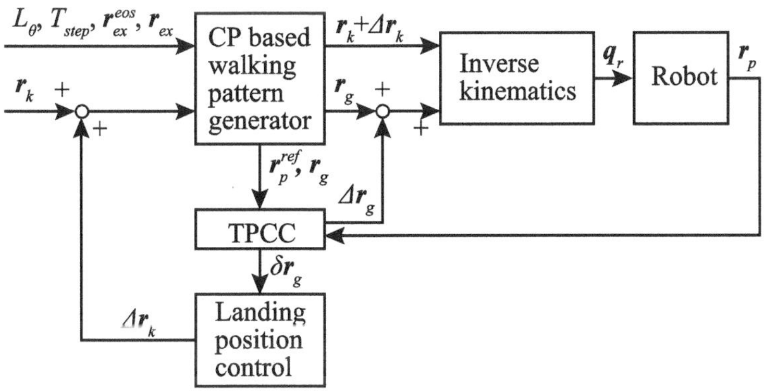

FIGURE 7.9 Block diagram of the control system of biped walking on sand.

The gait was generated on-line in the way described in Section 7.2.2. The designed reference CoP is plotted in Fig. 7.8. The simplified TPCC relations (5.43) and (5.44) were applied to the robot to stabilize its walking. Experiments with and without the landing position control in (7.14) were performed to verify the effectiveness. With landing position control, the forgetting factor λ was set to 0.7 for both the x- and the y-direction. The block diagram of the control system is shown in Fig. 7.9.

Fig. 7.10 shows the experimental results of walking on a flat floor (on a metal plate) without landing position control ($\Delta \boldsymbol{r}_k = \boldsymbol{0}$ in Fig. 7.10C). In biped walking, the humanoid robot laterally sways its body; hence lateral swaying of the CoM is seen in Fig. 7.10C. The lateral and rolling motions were kept in the constant range and the robot could stably walk, step, and stop.

Fig. 7.11 shows the experimental results of walking on sand without landing position control. As seen in Fig. 7.11B, approximately after the 10th step (about 5 s), the pitch angle rapidly increased (the robot tilted backward) and at the 18th step the robot fell down backward, despite the TPCC stabilizing approach. The same experiments were performed several times; however, in all experiments the robot could not stably walk on sand without landing position control.

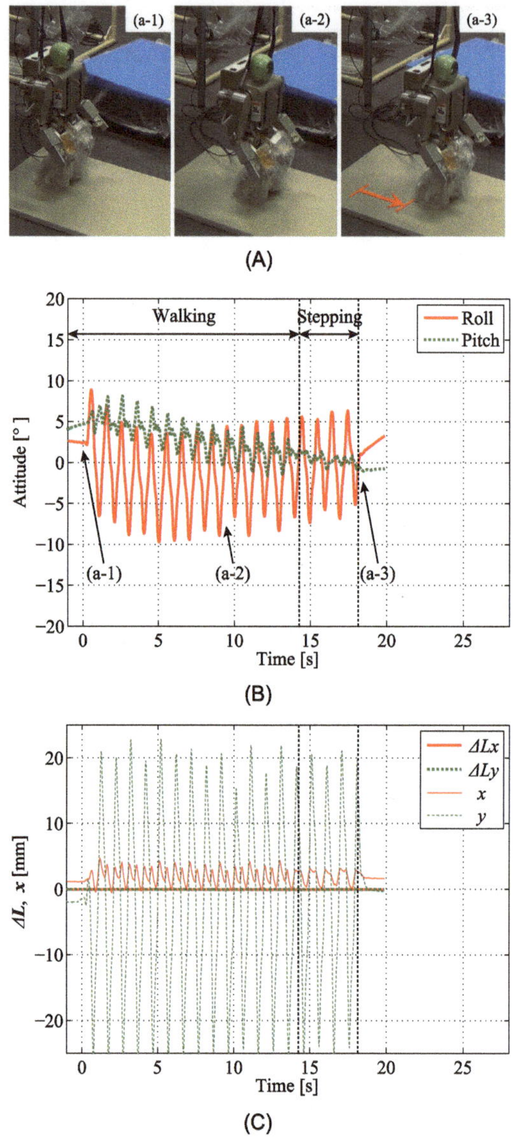

FIGURE 7.10 Walking on a floor without the landing position control. (A) Snapshots of the experiment. (B) Roll and pitch angles. (C) CoM trajectory and modification of the landing position.

Fig. 7.12 shows the experimental results of walking on sand with landing position control. As seen in Fig. 7.12C, the desired deviation of the landing position $\Delta \boldsymbol{r}_k$ was calculated and then applied to the robot. As seen in Fig. 7.12B, the landing position control successfully compensated the deviation of the gCoM; consequently the divergence of the roll and pitch angles was well suppressed.

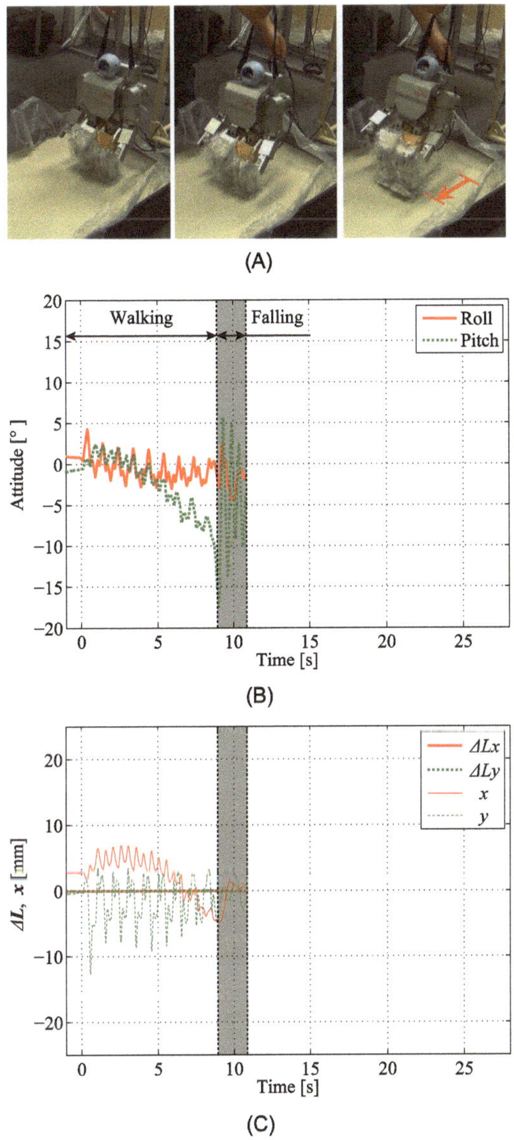

FIGURE 7.11 Walking on sand without the landing position control. (A) Snapshots of the experiment. (B) Roll and pitch angles. (C) CoM trajectory and modification of the landing position.

In order to verify whether the landing position control is effective even on a flat floor, an experiment of walking on a metal plate with landing position control was performed. Fig. 7.13 shows the results. Compared with the results shown in Fig. 7.10, it seems that the landing position control contributes to suppress the lateral swaying motion. Because of the elasticity of the roll joints at the ankles and hips, the robot tends to tilt inside (toward the

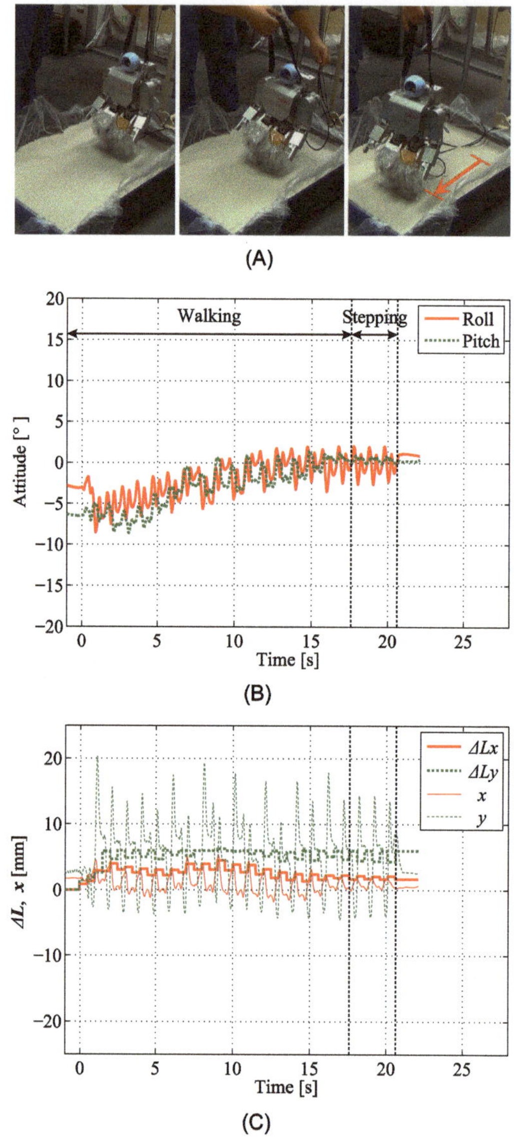

FIGURE 7.12 Walking on sand with landing position control. (A) Snapshots of the experiment. (B) Roll and pitch angles. (C) CoM trajectory and modification of the landing position.

swing leg). As a result, the swing foot tends to land slightly before the expected landing time, and hence *kicks* the floor. This *kicking* motion enlarges the rolling of the body, as seen in Fig. 7.10B, which ideally should be zero. It is assumed that the orbital energy is decreased by changing the landing position, and hence the *kicking* effect is suppressed.

The experimental results are shown in Video 7.3-1 [62].

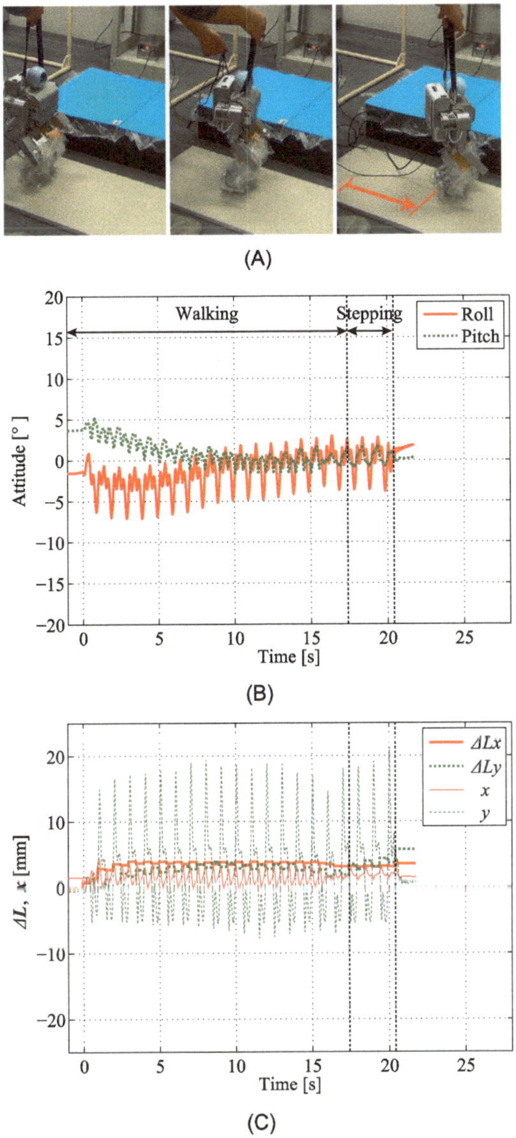

FIGURE 7.13 Walking on a hard floor with landing position control. (A) Snapshots of the experiment. (B) Roll and pitch angles. (C) CoM trajectory and modification of the landing position.

7.3.3 Summary and Discussion

This section presented a way to realize biped walking of humanoid robots on loose soil. The CP-based gait generation, TPCC, and landing position control were implemented with the humanoid robot HOAP-2 walking on sand. Those controllers are easy to implement and

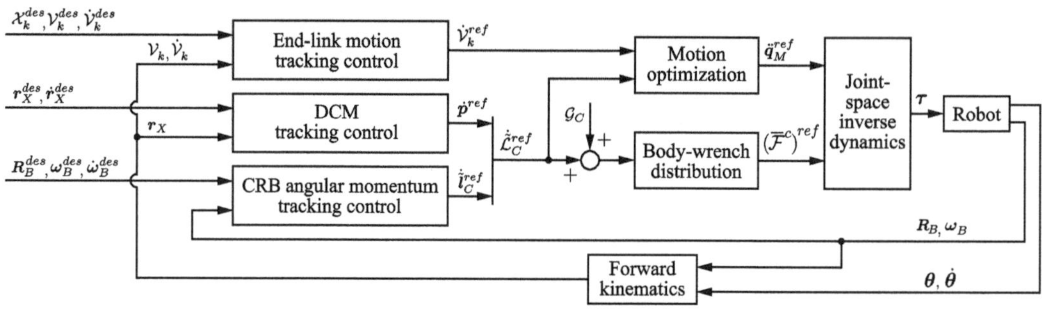

FIGURE 7.14 Block diagram of the VRP/DCM dynamic gait generation and walking controller for irregular terrain based on the VRP-GI body wrench distribution approach. The structure is identical with that of the motion/force controller in Fig. 5.29.

are effective for biped walking on loose soil, such as sand, as experimentally validated in Section 7.3.2.

7.4 GAIT GENERATION FOR IRREGULAR TERRAIN AND VRP-GI-BASED WALKING CONTROL

The ICP-based dynamic gait generation and walking control method described in Section 7.2 is limited to flat-ground environments. The method can be adopted to gait generation for irregular terrain in 3D by employing the DCM analysis based on the eCMP and VRP concepts (cf. Section 5.6.2). In this section, an implementation of two gait generation approaches, referred to as the "Continuous Double-Support" (CDS) and "Heel-to-Toe" (HT) gait generation, [20,21] will be described. These approaches ensure continuous GRFs that can be computed in real-time. The HT approach facilitates the so-called toe-off gait, whereby the support foot is allowed to roll around the BoS boundary at the toe. Such a motion is useful to avoid the stretched-knee kinematic singularity that occurs in a gait with a larger step width or while ascending/descending a staircase with a larger step height.

The controller shown in Fig. 7.14 will be used as the walking controller. Note that the controller's structure is identical with that of the motion/force controller in Fig. 5.29. The VRP generalized inverse (VRP-GI) (cf. Section 5.10.5) will be used to distribute the body wrench during the double-stance (DS) phase and to facilitate smooth transitions between the single-stance (SS) and DS phases. The employment of the VRP-GI is computationally efficient; the wrench distribution problem can be solved much faster [44] than with a "conventional" approach based on iterative optimization (see e.g. [66,116]).

7.4.1 Continuous Double-Support (CDS) Gait Generation

Recall that the ICP gait generation method in Section 7.2 was based on control inputs in terms of the desired CoPs $\boldsymbol{r}_P^{des,i}$ that remain fixed within the foot BoS during the step. The same approach is used in CDS-based gait generation. Instead of desired CoPs, the control inputs are determined now as desired VRPs, i.e.

$$\boldsymbol{r}_{vrp}^{des,i} = \boldsymbol{r}_a^{des,i} + \begin{bmatrix} 0 & 0 & \bar{z}_{vrp} \end{bmatrix}^T + \boldsymbol{R}_a^{des,i} \begin{bmatrix} 0 & \bar{y}_{vrp} & 0 \end{bmatrix}^T. \tag{7.15}$$

Here $\boldsymbol{r}_a^{des,i}$ are the ankle positions at the preplanned footprints. The last term on the r.h.s. is related to the setting of the dynamic stability margin in the envisioned VRP-GI body wrench distribution approach. As explained in Section 5.10.4, to minimize the generated foot moments (which is desirable from the viewpoint of walking control) the dynamic stability margin should be set between the ankle joints. But in this case, a zero weight will be generated by (5.123), which leads to rank deficiency of the VRP-GI. To avoid this, the desired VRP should not pass exactly above the ankles but nearby. To this end, a small offset $\bar{y}_{vrp} > 0$ is introduced; $\boldsymbol{R}_a^{des,i}$ denotes the rotation matrix for the desired foot orientation.

Next, the desired DCM is determined by reverse-order recursion, i.e.

$$\boldsymbol{r}_X^{des,i} = e^{-\frac{T_{step}}{T_X}} \left(\boldsymbol{r}_X^{des,i+1} - \boldsymbol{r}_{vrp}^{des,i} \right) + \boldsymbol{r}_{vrp}^{des,i}, \tag{7.16}$$

where $T_X = 1/\bar{\omega}_X = \sqrt{\bar{z}_{vrp}/g}$ is the time constant of the DCM dynamics. The reference DCM trajectories can then be obtained as

$$\boldsymbol{r}_X^{des,i}(t) = e^{\frac{t-T_{step}}{T_X}} \left(\boldsymbol{r}_X^{des,i+1} - \boldsymbol{r}_{vrp}^{des,i} \right) + \boldsymbol{r}_{vrp}^{des,i}, \tag{7.17}$$

$$\dot{\boldsymbol{r}}_X^{des,i}(t) = \frac{1}{T_X} \left(\boldsymbol{r}_X^{des,i}(t) - \boldsymbol{r}_{vrp}^{des,i} \right), \tag{7.18}$$

for $0 \le t \le T_{step}$.

Assuming a stationary initial state at Step $i = 0$, s.t. $\boldsymbol{r}_X^{des,0} = \boldsymbol{r}_C(0)$, the initial desired VRP is calculated from (7.16) as

$$\boldsymbol{r}_{vrp}^{des,0} = \frac{1}{1 - e^{-\frac{T_{step}}{T_X}}} \boldsymbol{r}_C(0) + \frac{1}{1 - e^{\frac{T_{step}}{T_X}}} \boldsymbol{r}_X^{des,1}. \tag{7.19}$$

The above relations are valid for a gait whereby the swing and support foot exchange instantaneously. They can be used in gait generation but the resulting trajectories will be discontinuous. To alleviate this problem, a DS phase is introduced. Denote by T_{DS} the DS time interval. To make use of the above notation, T_{DS} is divided into subintervals, i.e. $T_{DS} = T_{DS}^{init} + T_{DS}^{end}$; $T_{DS}^{init} = \alpha_{DS}^{init} T_{DS}$ and $T_{DS}^{end} = (1 - \alpha_{DS}^{init}) T_{DS}$ stand for the time intervals before and after the (imaginary) instantaneous transition, and $0 \le \alpha_{DS}^{init} \le 1$ is constant. In the following simulations $\alpha_{DS}^{init} = 0.5$ will be used, as suggested in [20,21]. The desired DCM states at the end and the beginning of the DS phase can then be determined as

$$\boldsymbol{r}_{X,DS^{end}}^{des,i} = e^{\frac{T_{DS}^{end}}{T_X}} \left(\boldsymbol{r}_X^{des,i} - \boldsymbol{r}_{vrp}^{des,i} \right) + \boldsymbol{r}_{vrp}^{des,i}, \tag{7.20}$$

$$\dot{\boldsymbol{r}}_{X,DS^{end}}^{des,i} = \frac{1}{T_X} \left(\boldsymbol{r}_{X,DS^{end}}^{des,i} - \boldsymbol{r}_{vrp}^{des,i} \right) \tag{7.21}$$

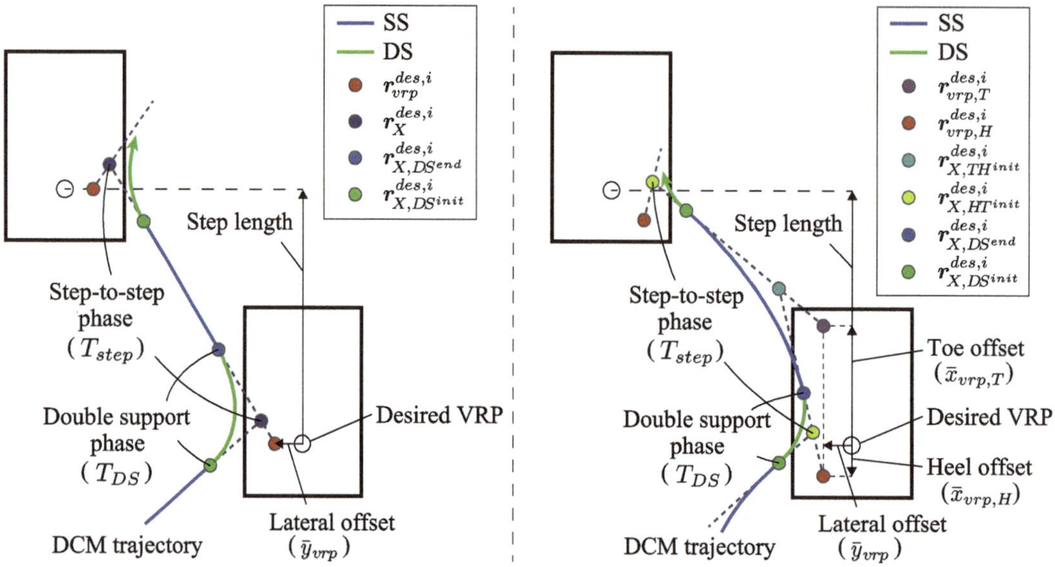

FIGURE 7.15 DCM trajectory generation methods [20]. Left (A): Continuous double-support (CDS). Right (B): Heel-to-toe (HT).

and

$$r_{X,DS^{init}}^{des,i} = e^{-\frac{T_{DS}^{init}}{T_X}}\left(r_X^{des,i} - r_{vrp}^{des,i-1}\right) + r_{vrp}^{des,i-1}, \tag{7.22}$$

$$\dot{r}_{X,DS^{init}}^{des,i} = \frac{1}{T_X}\left(r_{X,DS^{init}}^{des,i} - r_{vrp}^{des,i-1}\right), \tag{7.23}$$

respectively. The above two boundary states are connected via a fifth-order spline to generate the desired DCM trajectories during the DS phase. The CDS gait generation procedure is outlined in Fig. 7.15A.

7.4.2 Heel-to-Toe (HT) Gait Generation

To generate a toe-off–type gait, instead of the desired VRP $r_{vrp}^{des,i}$ used in the CDS approach, two desired VRPs will be employed: $r_{vrp,H}^{des,i}$ and $r_{vrp,T}^{des,i}$ for the heel and toe, respectively. They are obtained with the help of offsets $\bar{x}_{vrp,H}$ and $\bar{x}_{vrp,T}$, respectively (cf. Fig. 7.15B), i.e.

$$r_{vrp,H}^{des,i} = r_a^{des,i} + \begin{bmatrix}0 & 0 & \bar{z}_{vrp}\end{bmatrix}^T + R_a^{des,i}\begin{bmatrix}\bar{x}_{vrp,H} & \bar{y}_{vrp} & 0\end{bmatrix}^T, \tag{7.24}$$

$$r_{vrp,T}^{des,i} = r_a^{des,i} + \begin{bmatrix}0 & 0 & \bar{z}_{vrp}\end{bmatrix}^T + R_a^{des,i}\begin{bmatrix}\bar{x}_{vrp,T} & \bar{y}_{vrp} & 0\end{bmatrix}^T. \tag{7.25}$$

The desired DCMs can then be determined by the following reverse-order recursions:

$$r_{X,TH^{init}}^{des,i} = e^{-\frac{T_{TH}}{T_X}}\left(r_{X,HT^{init}}^{des,i+1} - r_{vrp,T}^{des,i}\right) + r_{vrp,T}^{des,i}, \tag{7.26}$$

$$\boldsymbol{r}^{des,i}_{X,HT^{init}} = e^{-\frac{T_{HT}}{T_X}}\left(\boldsymbol{r}^{des,i}_{X,TH^{init}} - \boldsymbol{r}^{des,i}_{vrp,H}\right) + \boldsymbol{r}^{des,i}_{vrp,H}, \tag{7.27}$$

where $T_{HT} = \alpha_{HT} T_{step}$ and $T_{TH} = (1 - \alpha_{HT}) T_{step}$, $0 \leq \alpha_{HT} \leq 1$.

As in the CDS algorithm, assume a stationary initial state at Step $i = 0$. The initial desired VRP can then be calculated as

$$\boldsymbol{r}^{des,0}_{vrp,T} = \frac{1}{1 - e^{-\frac{T_{TH}}{T_X}}}\boldsymbol{r}_C(0) + \frac{1}{1 - e^{\frac{T_{TH}}{T_X}}}\boldsymbol{r}^{des,1}_{X,HT^{init}}. \tag{7.28}$$

The desired DCM states at the end and the beginning of the DS phase can then be determined as

$$\boldsymbol{r}^{des,i}_{X,DS^{end}} = e^{\frac{T^{end}_{DS}}{T_X}}\left(\boldsymbol{r}^{des,i}_{X,HT^{init}} - \boldsymbol{r}^{des,i}_{vrp,H}\right) + \boldsymbol{r}^{des,i}_{vrp,H}, \tag{7.29}$$

$$\dot{\boldsymbol{r}}^{des,i}_{X,DS^{end}} = \frac{1}{T_X}\left(\boldsymbol{r}^{des,i}_{X,DS^{end}} - \boldsymbol{r}^{des,i}_{vrp,H}\right) \tag{7.30}$$

and

$$\boldsymbol{r}^{des,i}_{X,DS^{init}} = e^{-\frac{T^{init}_{DS}}{T_X}}\left(\boldsymbol{r}^{des,i}_{X,HT^{init}} - \boldsymbol{r}^{des,i-1}_{vrp,T}\right) + \boldsymbol{r}^{des,i-1}_{vrp,T}, \tag{7.31}$$

$$\dot{\boldsymbol{r}}^{des,i}_{X,DS^{init}} = \frac{1}{T_X}\left(\boldsymbol{r}^{des,i}_{X,DS^{init}} - \boldsymbol{r}^{des,i-1}_{vrp,T}\right), \tag{7.32}$$

respectively. The above two boundary states are connected via a fifth-order spline to generate the desired DCM trajectories during the DS phase. The IIT gait generation procedure is outlined in Fig. 7.15B.

7.4.3 Simulation

A small-sized humanoid robot model with parameters similar to those of the HOAP-2 robot [25] was used in the Choreonoid environment [87]. For the numbering of the joints and other relevant data see Section A.1. The results from three simulations with the HT algorithm will be presented. The following parameters were used: $T_{step} = 0.5$ s, $T_{DS} = 0.1$ s, $\bar{y}_{vrp} = 15$ mm, $\bar{x}_{vrp,T} = 25$ mm, $\bar{x}_{vrp,H} = -15$ mm, $\alpha_{DS} = \alpha_{HT} = 0.5$, and the step length and height were 100 and 20 mm, respectively. The desired rotation of the trunk (the base link) is the initial (zero) one. The feedback gains for the CRB trajectory were $K_X = 300 = K_{OB}$, $K_{\omega B} = 50$. The P and D feedback gains for the swing foot trajectory tracking were set at 3000 and 500, respectively. The arm motion was controlled by joint damping only. The damping was set at 100 for each arm joint.

The first simulation demonstrates walking on a flat floor, without toe-off. The results are shown in Video 7.4-1 [81] and Fig. 7.16. Apparently, the generated dynamic gait and the controller ensure a stable performance.

The second simulation demonstrates the ascending/descending of a staircase. The height of the stair steps was 20 mm. The maximum height of the swing foot trajectory was set at 10 mm above the ground. The results are shown in Video 7.4-2 [82] and Fig. 7.17. Although there is a slight difference in the CoP fluctuation while ascending and descending, overall the walking is quite stable.

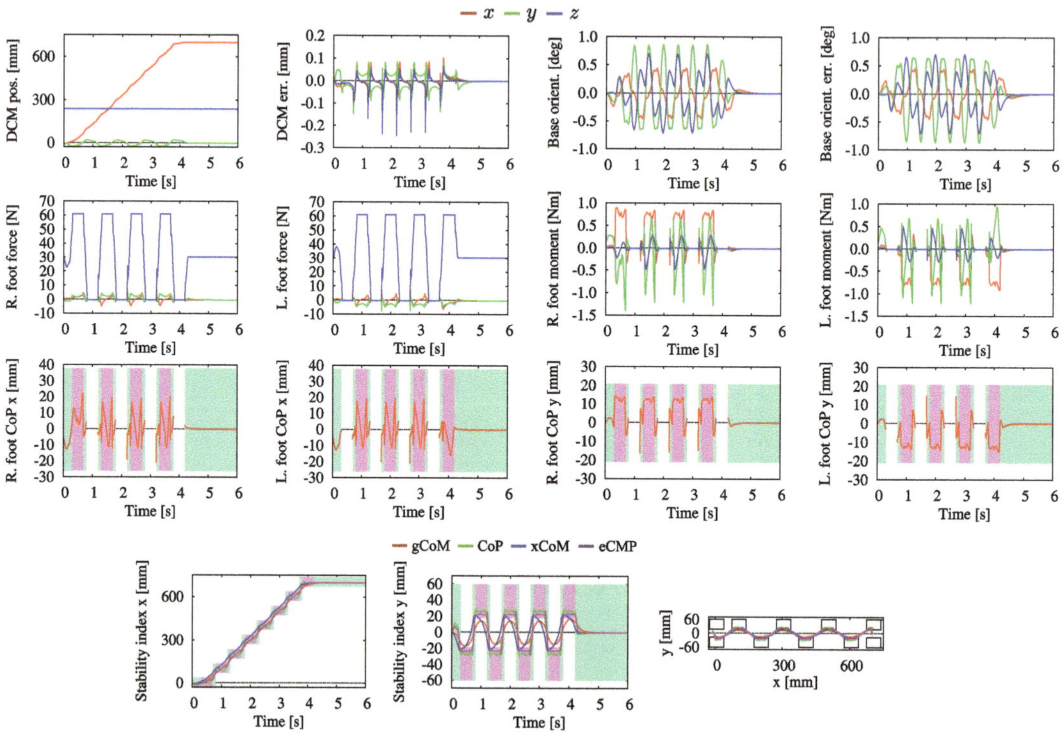

FIGURE 7.16 Dynamic straight walking on a flat floor. The light-blue/pink areas signify the double/single-support phases, respectively.

The third simulation is for walking on a flat floor with toe-off. The result is shown in Video 7.4-3 [83]. This simulation also demonstrates very stable dynamic walking.

7.5 SYNERGY-BASED MOTION GENERATION

Research on motor control in the field of neurophysiology has revealed that determining the appropriate muscle activation to achieve a given motion objective is a fairly complex task. The reasons, as pointed out in [19], are the high dimensionality of the search space and the nonlinear and dynamic nature of the transformation between the muscle activity and the movement. It has been hypothesized that this task is solved by the central nervous system (CNS) resorting to *muscle synergies* or muscle activation patterns. A muscle synergy induces a respective *movement synergy* (also called a "kinematic synergy," a "motion pattern" or a "motion primitive"), i.e. a variation in the posture of a human or an animal whereby the movement in the body segments is coordinated in a specific way [71]. The above hypothesis has been confirmed via experimental results [19]. The results have shown that the motor control in animals and humans alike is organized in a modular way, s.t. various behaviors are determined either by linear combinations of the same synergies, or by specific synergies. In this way, the load on the CNS can be significantly decreased.

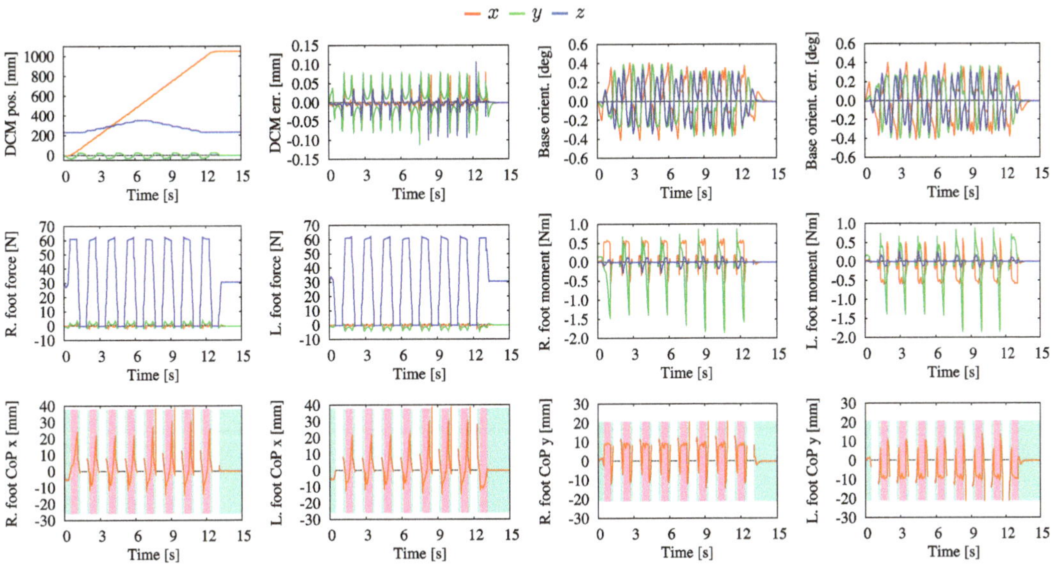

FIGURE 7.17 Ascending/descending a staircase. The light-blue/pink areas signify the double/single-support phases, respectively.

The term "synergy" was introduced in the field of robotics in the pioneering work [74], with regard to motion analysis/generation in the presence of kinematic redundancy. As discussed in Chapter 2, a kinematically redundant manipulator can achieve its end-link motion task with an infinite number of movements. To arrive at the desired unique movement, one or more additional subtasks have to be specified. It is preferable that the subtasks are specified in the lower-dimensional subspace of the redundant DoFs, rather than in the higher-dimensional joint space. Each redundant DoF yields a specific synergy that determines a movement pattern. A desirable complex movement pattern can be obtained by linear combinations of the synergies.

Kinematic synergies are useful not only in redundancy resolution, but also in dealing with kinematically singular manipulator configurations. As revealed by the analysis based on the Singularity-Consistent method [89] (cf. Section 2.5), *instantaneous* motion patterns emerge at such configurations. Furthermore, kinematic synergies have also been shown to be useful in specifying the manipulation tasks [8]. In the field of computer-based animation of articulated figures, the synergetic choreography approach in [120] deserves to be mentioned.

Kinematic synergies have also been employed in balance control of humanoid robots. With the method described in [29], postural variations that resemble the ankle and hip strategies used by humans in response to a perturbation can be generated. A discussion on kinematic synergies in relation to balance control, including the formal definition of a kinematic synergy as a *one-parameter* family of curves in joint space (i.e. a manifold), can be found in [31,32].

In the remaining part of this section, the focus will be on synergy-based motion generation for a humanoid robot. Synergy-based balance control will be discussed in Section 7.6 and Section 7.7.

TABLE 7.1 Primitive motion synergies generated with a single command

	Primitive motion synergy	$\mathcal{V}_M$	$\bar{\mathcal{V}}^m$	$\dot{\theta}_u$
P1	limb self-motion (internal-link motion with stationary CoM/base)	$=0$	$=0$	$\neq 0$
P2	end-link motion with stationary CoM/base	$=0$	$\neq 0$	$=0$
P3	CoM/base motion (fixed end links, no self-motion)	$\neq 0$	$=0$	$=0$

TABLE 7.2 Combinations of synergies generated with multiple commands

	Combinations of synergies	$\mathcal{V}_M$	$\bar{\mathcal{V}}^m$	$\dot{\theta}_u$
C1	whole-body motion with stationary CoM/base	$=0$	$\neq 0$	$\neq 0$
C2	whole-body self-motion (internal-link and CoM/base motion)	$\neq 0$	$=0$	$\neq 0$
C3	end-link and CoM/base motion	$\neq 0$	$\neq 0$	$=0$
C4	whole-body motion	$\neq 0$	$\neq 0$	$\neq 0$

7.5.1 Primitive Motion Synergies

The constraint-consistent instantaneous-motion inverse kinematics solutions for a humanoid robot, expressed in terms of the base and the mixed quasivelocity, are given in (2.98) and (2.133), respectively. The latter solution will be implemented below for the synergy-based motion generation. Recall that this solution has an advantage that enables the robot to attain a completely erected posture, as shown with the example in Section 2.11.4.

Each of the three components of solution (2.133) determines a *primitive kinematic synergy*. This is apparent from the fact that the components are parametrized by lower-dimensional inputs. The input in the first component, $\dot{\boldsymbol{\theta}}^c$, is the mixed-quasivelocity twist $\mathcal{V}_M$. This component determines the CoM/base-link movement synergy. The second component, $\dot{\boldsymbol{\theta}}^m$, specifies a synergy for the end-link movements along the unconstrained motion directions via the input $\bar{\mathcal{V}}^m$. The third component, $\dot{\boldsymbol{\theta}}^n$, determines the self-motion synergy. The three synergies are called primitive since they are generated by a single command input, as shown in Table 7.1.

7.5.2 Combinations of Primitive Synergies

As already pointed out, primitive motion synergies can be combined to obtain complex movement patterns. The primitive synergies in Table 7.1 are combined to obtain the four movement patterns shown in Table 7.2. The synergies and their combinations can be related to typical motion tasks. For example, the three stationary-CoM/base motion synergies/patterns (P1, P2, and C1, obtained with $\mathcal{V}_M = \mathbf{0}$) are useful for dual-stance (DS) erected or sitting postures, whereby only the arms move. When the hands are immobilized (e.g. when the hands are placed on a table s.t. $\bar{\mathcal{V}}_H^m = \mathbf{0}$), the limb self-motion synergy P1 can be used to adjust the arm configurations, e.g. to apply optimal contact forces while standing up. A movement pattern for standing up can be realized with the CoM/base-motion synergy P3 or the whole-body self-motion pattern C2.

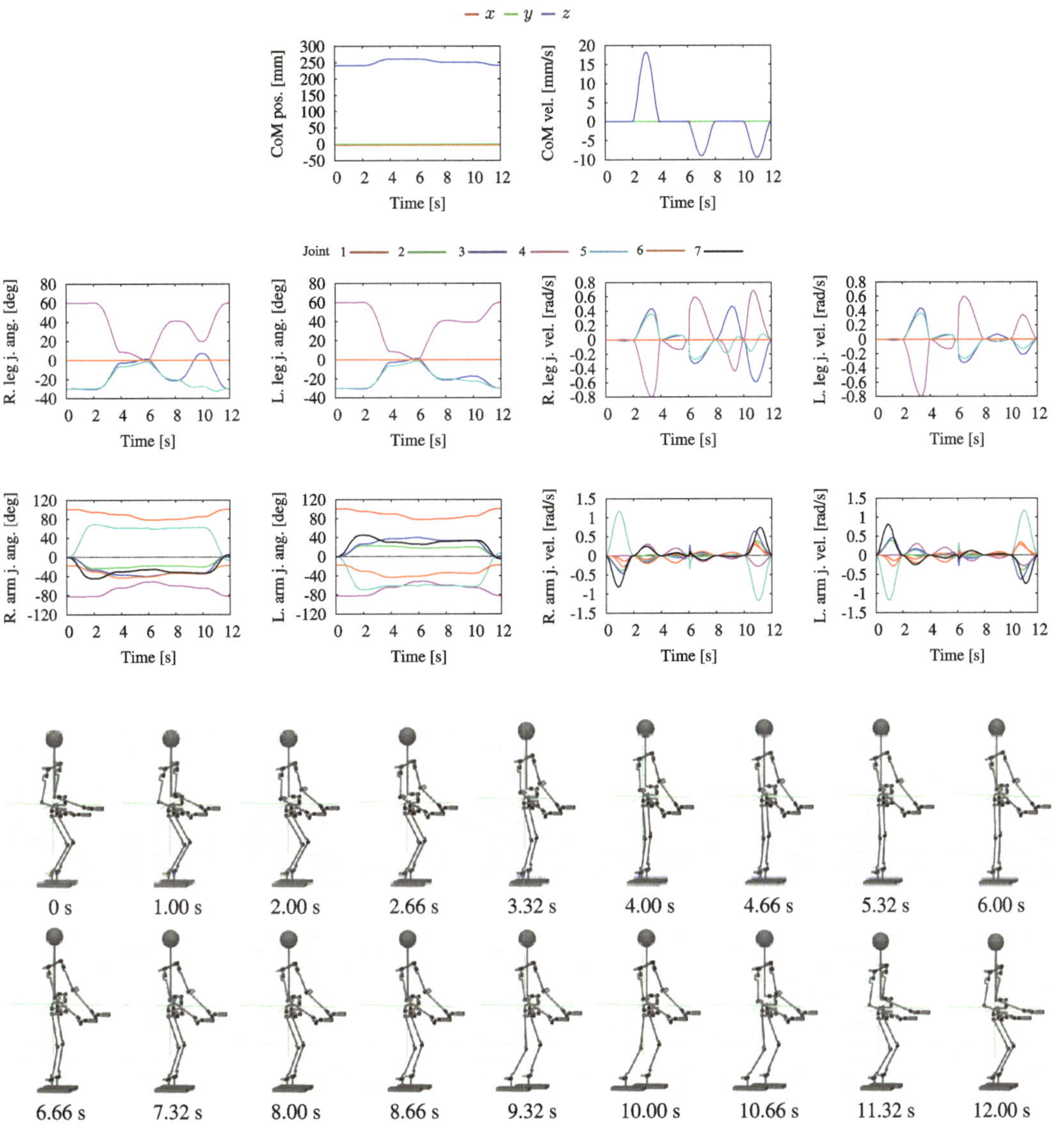

FIGURE 7.18 Synergy-based motion generation. The position and velocity of the CoM are shown in the upper-row graphs. The joint angles and their rates for the legs and arms are shown in the middle and lower rows, respectively. Snapshots from the animated motion are shown at the bottom.

Video 7.5-1 [106] demonstrates the movements of a small-sized humanoid robot[2] generated with primitive motion synergies and their combinations. The CoM position/velocity and the joint angle/angular velocity graphs are shown in Fig. 7.18. Snapshots from the sim-

[2] The parameters of the model were derived from a HOAP-2 robot [25] (see Section A.2).

ulation are shown at the bottom of the same figure. The following synergies/patterns are performed:

1. $0 \sim 2$ s arm self-motion (P1);
2. $2 \sim 4$ s CoM/base upward motion (P3);
3. $4 \sim 6$ s hands sliding along an invisible horizontal surface, with stationary CoM/base (C3);
4. $6 \sim 8$ s CoM/base downward motion (P3);
5. $8 \sim 10$ s right-leg sliding motion with stationary CoM/base (P2);
6. $10 \sim 12$ s whole-body motion for returning to the initial posture (C4).

Note that the contacts at the feet/hands have been set as frictionless to allow the robot to adjust the postures of its limbs by sliding on the closure-link surfaces (i.e. the floor surface for the feet and the invisible table surface for the hands) in the desirable way. Note also that while the hands slide during Phase 3, the posture becomes fully erected. This is also apparent from the knee-joint graphs in Fig. 7.18 (drawn in purple color). No instabilities are observed at this posture because the mixed quasivelocity notation was employed. This would not be the case if the notation in terms of the base quasivelocity were used. The motion in the final, sixth phase is composed of all previous synergies, but executed in the reverse direction. As apparent from the graphs, smooth movements can be generated with this method.

7.5.3 Multiple Synergies Generated With a Single Command Input

There are special cases when multiple synergies can be invoked with a single command input. Assume a humanoid robot with redundant arms, the hands being constrained to keep their positions on a table. The initial posture of the robot is with slightly bent knees. The CoM/base upward motion command (P3) first invokes upward CoM/base motion, as expected. When the legs become fully extended, the base cannot move anymore upward. But the CoM can do so, by (automatically) invoking self-motion in the arms. This type of behavior can be seen in Video 7.5-2 [127].

7.6 SYNERGY-BASED REACTIVE BALANCE CONTROL WITH PLANAR MODELS

In Section 5.13 three balance controllers were introduced capable of accommodating external disturbances via a whole-body compliance. It was noted, though, that such controllers can accommodate relatively weak disturbances only. This limitation stems from the constant PD feedback terms in the CRB spatial dynamics control component; the error bounds cannot be set sufficiently low to allow for larger postural variations. Note that a larger postural variation is needed to accommodate a larger disturbance. The reactive balance controller should ensure sufficiently large postural variations without deteriorating the balance. As will be shown in what follows, motion synergies can be used to generate such postural variations.

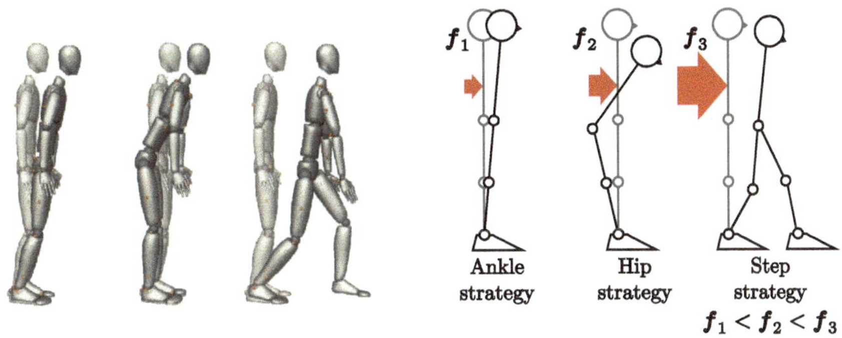

FIGURE 7.19 Sagittal-plane ankle, hip, and step reactive strategies in response to an external force acting on the back. The ankle strategy is initialized by a weak disturbance that is accommodated by motion in the ankle joints and displacement of the CoM within the BoS. The hip strategy is initialized by a disturbance of a mid-level magnitude; it is realized by bending the upper body with prevailing motion in the hip joints. Sometimes, this strategy is characterized as an antiphase excursion in the ankle and hip joints. The stepping strategy is initialized by a strong disturbance leading to a change in the BoS (reactive stepping).

7.6.1 Motion Synergies for Balance Control Used by Humans

Researchers in the fields of biomechanics and physical therapy have paid attention to the balance recovery strategies used by humans to assess balance disorders [102,118,117,42]. For the purpose of analysis, most often simple models on the sagittal and lateral planes (cf. Fig. 2.1) have been adopted. A brief overview of the results will be presented below.

Sagittal Plane

When a healthy person standing upright is subjected to an external disturbance generated by horizontal perturbations in the support surface, he or she reacts either with the ankle or the hip synergy [102,42]. When the support surface is sufficiently large, postural balance is achieved with the ankle synergy. Thereby, the disturbance accommodation and the equilibrium restoration are done via movements in the ankle joints mainly. The hip synergy, on the other hand, is invoked when the person stands on a support surface shortened in relation to the foot length. This synergy produces a horizontal shear force against the support surface, with little or no motion in the ankles but with a predominant motion in the hips. In addition to the ankle and hip strategies, a third one, the so-called *stepping or stumbling strategy*, was also identified. This strategy is invoked when certain boundary values (in position and/or velocity) during the hip strategy are exceeded. The three reactive strategies are illustrated in Fig. 7.19.

Lateral Plane

A conceptual model for investigating the role of the CNS in the performance of swaying movements within the lateral plane[3] was proposed in [53]. A four-link model was used to

[3] Also referred to as a frontal or coronal plane.

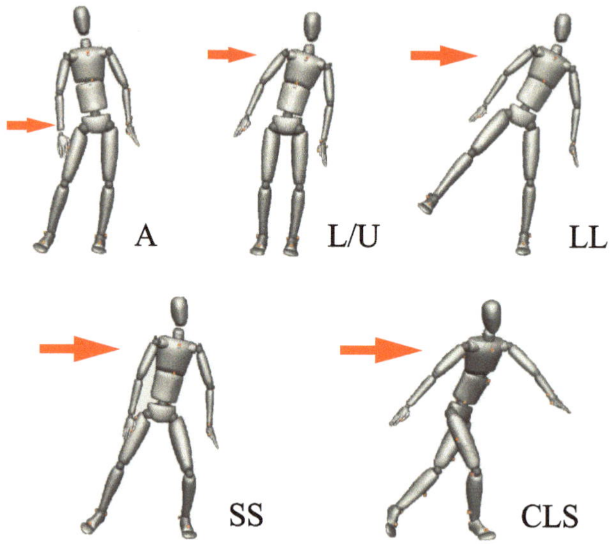

FIGURE 7.20 Lateral-plane reactive response patterns. *A*, Ankle. *L/U*, Load/Unload. *LL*, Lift-Leg. *SS*, Side-Step. *CLS*, Cross-Leg-Step. Patterns A, L/U, SS, and CLS are known from studies in the field of biomechanics [100,79,96]. Pattern LL was identified in [126]; it also appears in a humanoid robot study [41].

derive parallel-leg motion patterns with predominant motion in the ankle/hip joints. A more detailed study [100] came to the conclusion that the lateral-plane ankle strategy is used when the disturbance is relatively small (e.g. when applied at the pelvis). With a larger disturbance, acting e.g. at the shoulder, the response is with the so-called *load/unload synergy*. Also, it was mentioned that a step could be invoked to maintain the balance, when the disturbance grows even larger. In [79] and [96], two main protective-step patterns were identified: the loaded side-step and the unloaded crossover-step strategies in response to invariant disturbances. In [126] it was observed that when the disturbance invoking the load/unload synergy grows larger, the contact at the unloaded foot is lost, resulting in the so-called *lift-leg* synergy. This synergy was also obtained with a humanoid robot model [41]. The lateral plane balance strategies confirmed in [126] are shown in Fig. 7.20.

The reaction/recovery motion patterns used by humans have been examined experimentally via simulations. Some of the patterns have been implemented with real robots. Fairly complex full-body models have been introduced at an early stage. Solutions that can accommodate not only weak, but also medium disturbances, were obtained with the linear [29] and quadratic [68] programming approaches. Most studies have been based on simple models, though, mainly in the sagittal plane [103,6,122]. In the remaining part of this section, the focus will be on the implementation of synergies that do *not require a change in the BoS*. Such synergies are used to accommodate a disturbance of medium magnitude. Synergies that are typically used to respond to a larger disturbance require a change in the BoS (e.g. as the sagittal-plane step strategy or the lateral-plane lift-leg strategy). These types of synergies will be discussed in Section 7.7.5.

7.6.2 RNS-Based Reactive Synergies

Given a statically stable posture with a sufficiently large stability margin, an infinite set of *large* postural variations for disturbance accommodation can be generated via a spatial momentum conservation approach, e.g. the Reaction Null-Space (RNS) method (cf. Section 4.9). Such variations, henceforth referred to as *RNS synergies*, are reactionless; they can ensure balance stability while conserving the initial (zero) spatial momentum. The conservation implies that a motion will be generated that keeps the CoM and the centroidal-moment point (CMP) fixed throughout the motion synergy. This is a somewhat conservative approach, but the balance controller is simple and yields robust performance, also with a position-controlled robot as will be shown below. In the following discussion, models composed of the legs and the trunk will be employed. Whole-body models that include arm motion will be discussed in Section 7.7. Furthermore, it will be assumed that the direction of the external disturbance does not pass through the CoM. This means that the disturbance can be accommodated mainly via rotations in the trunk/legs, without inducing a large variation in the CoM.

7.6.3 Sagittal-Plane Ankle/Hip Synergies

The ankle synergy is used to accommodate a relatively small disturbance. Thus, it can be realized with one of the whole-body compliance control approaches discussed in Chapter 5, e.g. the passivity-based gravity compensation (cf. Section 5.13.1, see also [48]). The ankle synergy can also be invoked in a straightforward manner with a simple linearized IP-on-foot model and an LQR controller [103]. The realization of the hip synergy, on the other hand, is somewhat more challenging. The ultimate goal is to design a balance controller that can invoke any of these two synergies depending on the magnitude of the disturbance. It is also desirable to accommodate possible variations in the disturbance, via smooth transitions between the two synergies.

The simplest model for the hip synergy is the double-IP-on-foot model with foot/leg/torso segments (cf. Fig. 7.21B). With this model, the hip synergy can be generated based on the RNS method, whereby the foot is designated as the root link of the floating-base system [92, 91]. The balance stability can be preserved under a disturbance *force* applied horizontally at the torso, provided the leg/torso force accommodating movements do not impose horizontal force and moment components at the foot. The foot will then remain stationary, s.t. $v_{Bx} = 0 = \omega_B$. In this case, the momentum conservation condition assumes the simple form $\boldsymbol{J}_{\overleftarrow{CB}_x}\dot{\boldsymbol{\theta}} = 0$, where

$$\boldsymbol{J}_{\overleftarrow{CB}_x} = \begin{bmatrix} z_g & k_m l_{g2}\cos(\theta_1 + \theta_2) \end{bmatrix} \tag{7.33}$$

is the horizontal (x) component of the CoM Jacobian, $k_m = M_2/M$, $M = M_1 + M_2$ denoting the total mass. The joint velocity vector is composed of the two joint rates (ankle and hip). Solving for the joint rates, one obtains the following set of *reactionless joint velocities*:

$$\{\dot{\boldsymbol{\theta}}_{rl}\} = \{b\boldsymbol{n}\}, \tag{7.34}$$

where b is an arbitrary scalar that parametrizes the hip synergy; $\boldsymbol{n} \in \Re^2$ is a vector from the kernel of $\boldsymbol{J}_{\overleftarrow{CB}_x}$, i.e.

$$\boldsymbol{n} = \begin{bmatrix} n_1 \\ n_2 \end{bmatrix} = \begin{bmatrix} -M_2 l_{g2} \cos(\theta_1 + \theta_2) \\ (M_1 l_{g1} + M_2 l_1) \cos\theta_1 + M_2 l_{g2} \cos(\theta_1 + \theta_2) \end{bmatrix}. \tag{7.35}$$

This relation generates a vector field in the joint space that specifies the synergy.

The following relation for the joint rates in the hip synergy can then be obtained:

$$\dot{\theta}_{1h}^{ref} = \frac{n_1}{n_2} \dot{\theta}_{2h}^{ref}. \tag{7.36}$$

Here $\dot{\theta}_{ih}^{ref}$, $i = 1, 2$, are reference joint rates for the hip strategy.

Next, consider the IP-on-foot model for the ankle synergy shown in Fig. 7.21A. The equation of motion is written as

$$M l_g^2 \ddot{\theta}_1 - M g x_g = \tau_1 - D_a \dot{\theta}_1 - K_a \theta_1. \tag{7.37}$$

Here l_g is the distance from the ankle to the CoM, $x_g = l_g \sin\theta_1$ is the CoM ground projection, g is the gravity acceleration, and D_a and K_a are virtual damper/spring constants. Linearizing around the upright posture, the ankle torque can be replaced as $\tau_1 \approx -M g x_p$ (cf. (4.8)). The reference joint acceleration for the ankle can then be obtained as

$$\ddot{\theta}_{1a}^{ref} = \frac{1}{M l_g^2} \left(M g (x_g - x_p) - D_a \dot{\theta}_1 - K_a \theta_1 \right). \tag{7.38}$$

The same relation can be used to obtain the reference joint acceleration for the hip joint in the hip synergy, under the simplified assumption that the torso segment represents a single IP, whereby the leg is instantaneously immobilized. Thus,

$$\ddot{\theta}_{2h}^{ref} = \frac{1}{M_2 l_{g2}^2} \left(M_2 g (x_g - x_p) - D_h \dot{\theta}_2 - K_h \theta_2 \right). \tag{7.39}$$

Here C_h and K_h are virtual damper/spring constants for the torso; the rest of the parameters should be clear from Fig. 7.21. The joint accelerations are integrated to obtain the reference joint angle rates to be used as control inputs in the following equations.

The above ankle and hip synergies can be combined by considering the displacement of the CoM during the two synergies. During the ankle synergy, the gCoM is displaced within the BoS. During the hip strategy, the CoM is displaced only in the vertical direction, whereas its ground projection remains stationary (cf. Fig. 7.21B). The aim of the transition between the ankle (A) and hip (H) synergies is twofold. First, it should be ensured that hip motion is initialized in a timely fashion before x_g reaches the BoS boundary (A–H transition). Second, the gCoM should move back swiftly to the position of the erected posture, after the disturbance disappears. In order to ensure such movements, make use of

$$\dot{x}_g = \boldsymbol{J}_{\overleftarrow{CB}_x} \dot{\boldsymbol{\theta}} \tag{7.40}$$

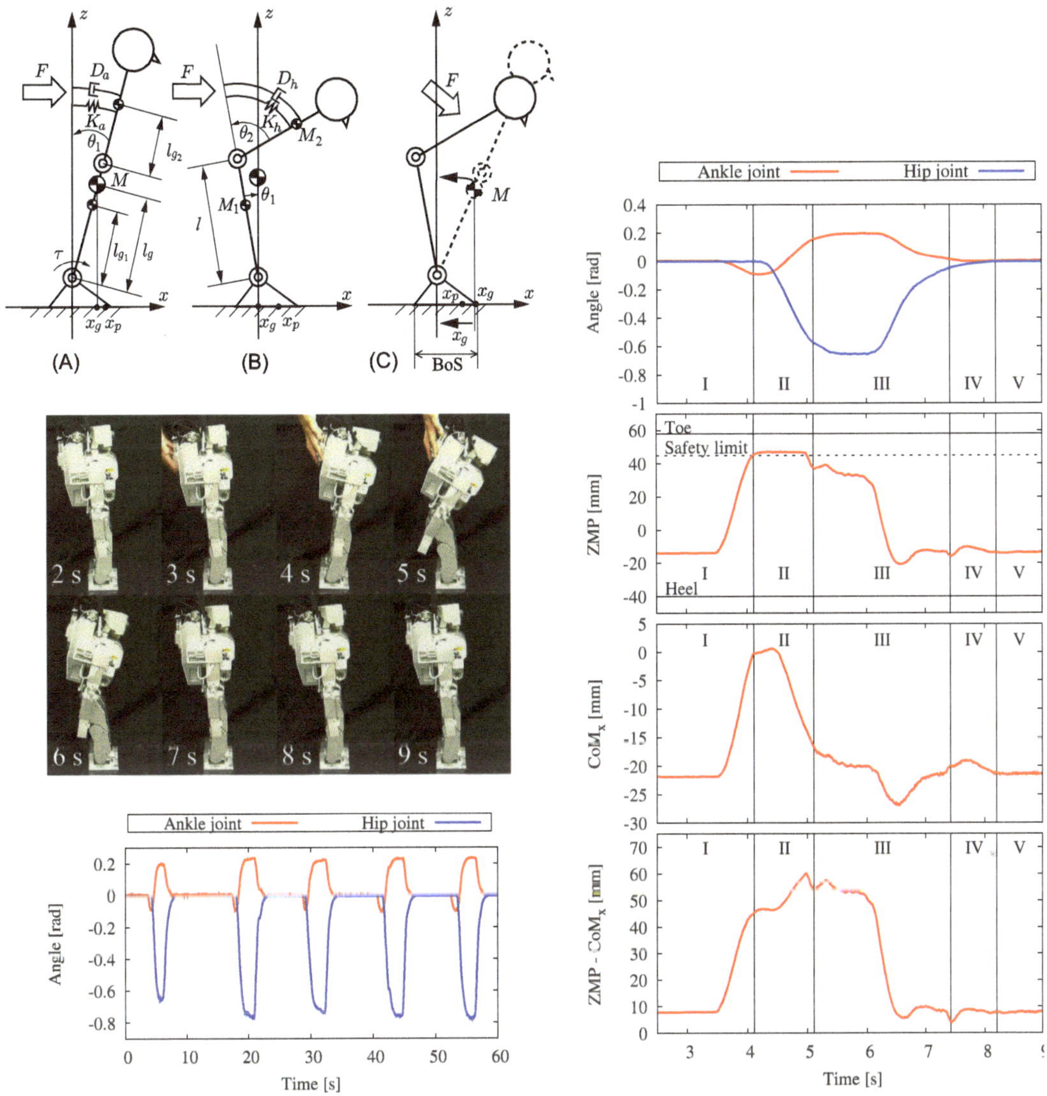

FIGURE 7.21 Ankle-hip reaction synergies within the sagittal plane [57]. Left-top: models for the ankle (A) and hip (B) synergies and the transition (C). Left-middle: snapshots from the experiment. Left-bottom: joint angle data for five consecutive disturbances; Right: detailed data from the first disturbance cycle. The motion phases are enumerated according to Table 7.3; $K_{(\circ)}/D_{(\circ)}$ are the virtual spring/damper variables.

with solution

$$\dot{\boldsymbol{\theta}} = \boldsymbol{J}^{+}_{\overleftarrow{CB}_x} \dot{x}_g + b\boldsymbol{n}. \tag{7.41}$$

Consider first the A–H transition phase. With the help of (7.33) and (7.35), (7.41) can be expanded in the following form:

TABLE 7.3 Motion phases and variables

Phase	Synergy	θ_1	θ_2	Variable of transition
I	Ankle (A)	(7.38)	–	–
II	A–H transition	(7.43) (7.44)	(7.39)	x_p
		$x_g^{des}: x_g^{lim} \overset{spline}{\longrightarrow} x_g^{init}$	$k_w: 0 \overset{spline}{\longrightarrow} 1$	
III	Hip (H)	(7.36)	(7.39)	–
IV	H–A transition	(7.38)	(7.45)	x_p, z_g
V (I′)	Ankle	(7.38)	–	–

$$\begin{bmatrix} \dot{\theta}_1 \\ \dot{\theta}_2 \end{bmatrix} = \frac{\dot{x}_g^{ref}}{z_g^2 + (k_m l_{g2} \cos(\theta_1 + \theta_2))^2} \begin{bmatrix} z_g \\ k_m l_{g2} \cos(\theta_1 + \theta_2) \end{bmatrix} + \boldsymbol{bn}. \tag{7.42}$$

The reference ankle joint rate can then be derived as

$$\dot{\theta}_1^{ref} = \frac{z_g}{z_g^2 + (k_m l_g \cos(\theta_1 + \theta_2))^2} \dot{x}_g^{ref} + k_w \frac{n_1}{n_2} \dot{\theta}_2^{ref}. \tag{7.43}$$

The first component on the r.h.s. depends on the motion rate of the gCoM as a component of the ankle synergy. The second component, on the other hand, is an RNS hip synergy component (cf. (7.36)). The superposition of the two components is realized via a fifth-order spline, i.e. $0 \leq k_w(t) \leq 1$. The reference gCoM rate is calculated from the simple feedback law

$$\dot{x}_g^{ref} = k_{pc} \left(x_g^{des} - x_g \right), \tag{7.44}$$

k_{pc} denoting a P feedback gain; x_g^{des} is also determined as a fifth-order spline, to ensure smooth transition between x_g^{lim} and $x_g^{init} (= 0)$, where x_g^{lim} is the gCoM at the time instant when $x_p = x_p^{lim}$, where x_p^{lim} is a safety limit, set in the vicinity of the BoS boundary.

The H–A transition is initialized after the disturbance disappears. To this end, the values of the ZMP (x_p) and the vertical CoM coordinate (z_g) are monitored. Note that the desired posture at the end of the transition equals the initial (erect) one ($\theta_2 = 0$). With proper values, the transition will be initialized when θ_2 is close to zero. Hence, a regulator-type feedback controller can be employed, i.e.

$$\dot{\theta}_2^{ref} = -k_{p\theta} \theta_2, \tag{7.45}$$

$k_{p\theta}$ denoting a feedback gain. At the end, four distinct phases of motion are obtained, as summarized in Table 7.3.

Experimental data obtained with a miniature humanoid robot HOAP-2 [25] (see Section A.1) are shown in Fig. 7.21. The robot is subjected to unknown (variable) continuous disturbance forces applied five times in the horizontal direction, by pushing its back with a hand during a time interval of 60 s. The time span for the A–H transition was set to 1 s, within the spline functions for the variables k_w and x_g^{des}. The ZMP safety limit was set as $x_p^{lim} = 45$ mm. The H–A transition was initialized when the following condition was met: $264.5 < z_g < 266.5$ mm and $x_p < 0$. The joint angle time history is shown in the bottom-left of Fig. 7.21. Detailed data graphs from the first push are shown on the right side of the figure. The experimental data demonstrate that the ankle/hip synergies with transitions can be realized in a stable manner with a simple velocity-based balance controller on a position-controlled humanoid robot. The experimental results are shown in Video 7.6-1 [94].

It is desirable to extend the controller capabilities in two ways [6]. First, the foot could be allowed to roll. Second, the system parameters should be determined in a rigorous way by taking into account the system state and the stability analysis methods discussed in the previous sections.

The simple double-IP-on-foot model has also been employed in [103] to design an integral controller that accommodates disturbances via the ankle and hip synergies, achieved with an LQR controller and a gCoM regulator, respectively. This is in contrast to the RNS synergy approach where the hip synergy is generated in a simpler way, using the feedforward control only. Another way of generating the hip synergy is by employing a centroidal moment component, as with the RWP model. Motion control during the reaction (or reflex) phase can then be based on spatial dynamics control of the type (5.136). An implementation of this approach with a four-link model (a foot, a two-segment leg with a knee joint, and a torso) is described in [1]. Since the model is planar, the reference momentum rate can be designed with scalar components $\dot{p}_x^{ref} = k_p(x_p - x_g)$ and $\dot{l}^{ref} = -k_\omega l$. The optimal quasivelocity, obtained e.g. via the pseudoinverse, can then be inserted into the joint-space inverse dynamics (JSID) solution to derive the joint torque. During the recovery phase, on the other hand, the movements in the three segments can be generated via a potential energy maximization. This approach yields an erected (stretched-leg) upright posture.

7.6.4 Lateral Plane Ankle, Load/Unload and Lift-Leg Synergies

There are relatively few studies that address the implementation of postural balance synergies within the lateral plane with a humanoid robot. In [16], a two-stage balance control strategy was proposed and implemented with a miniature position-controlled robot. The lateral-plane ankle strategy was realized via the ZMP feedback control during the first stage. During the second stage, a combined ankle-hip strategy was applied to deal with the larger disturbances. However, the proposed models did not refer to an in-depth investigation of the human postural balance strategies. In [41], a control approach based on the ZMP/CoM dynamics and the upper-body angular momentum was adopted, resulting in motions wherein the swing foot was used for balance. No particular reaction patterns were envisioned as control input, though, but the robot behavior was found to resemble that of the human participants.

A more detailed study on the implementation of the lateral-plane ankle and load/unload synergies with follow-up transitions to the lift-leg synergy was presented in [126]. The synergies were realized with the help of the simplified multilink models shown in Fig. 7.22.

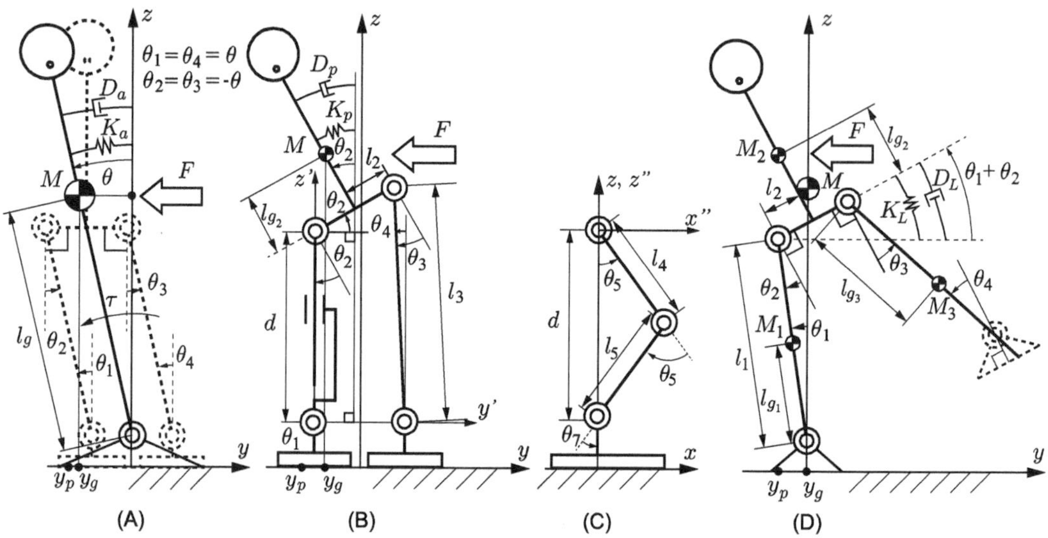

FIGURE 7.22 Models for the three strategies in the lateral plane. (A) Ankle. (B), (C) Load/unload. (D) Lift-leg. The slider joint in the right leg of (B) is a virtual joint—an equivalent of the real leg subchain involving the three pitch joints: hip (θ_5), knee (θ_6), and ankle (θ_7), as shown in (C).

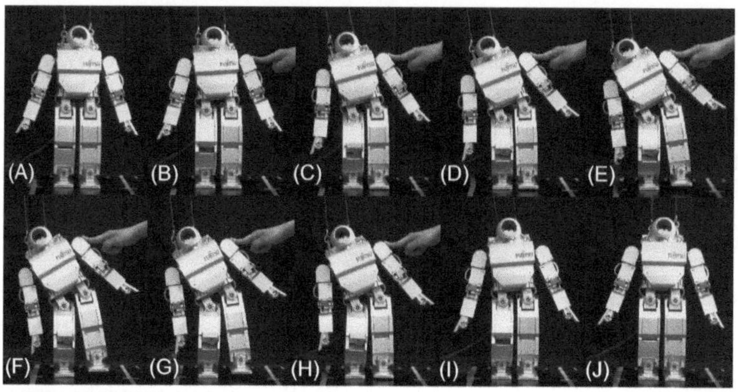

FIGURE 7.23 Snapshots from Load/Unload and Lift-leg integration, when the disturbance was a continuous force. (A) and (B): Load/Unload strategy. (C)–(H): Lift-leg strategy. (I) and (J): Load/Unload strategy.

Snapshots from the load-unload/lift-leg motion patterns are shown in Fig. 7.23. The ankle/lift-leg synergy motion is shown in Video 7.6-2 [125]. As apparent from the data, the desired motion patterns could be achieved. It should be noted, though, that the implementation was not that straightforward since, first, the models had to be exchanged and, second, the control laws comprised difficult to tune nonlinear spring coefficients and feedback gains. As will be shown in Section 7.7.2, there is an alternative, preferable approach for generating and controlling the lateral-plane reactive synergies with a whole-body model in 3D.

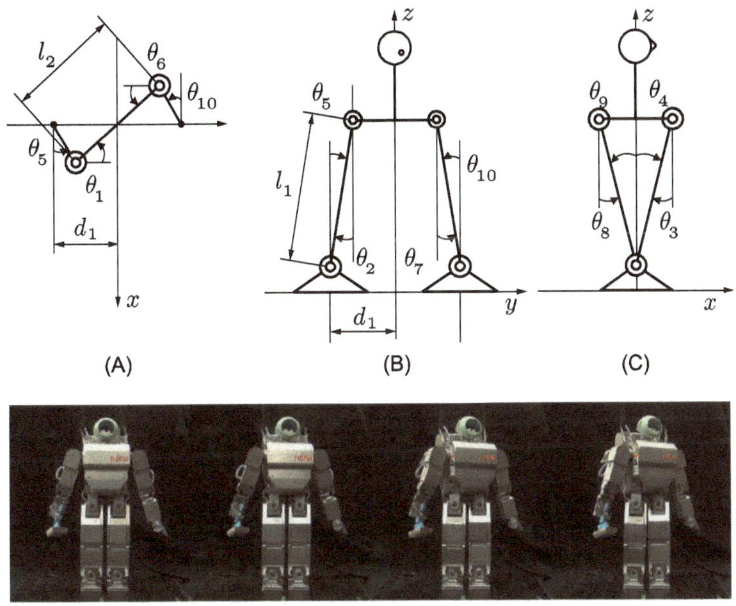

FIGURE 7.24 Transverse-plane twist strategy. A simple model is used to generate the synergy. (A) Top view. (B) Front view. (C) Side view. The meaning of the joint angles is as follows: hip roll/yaw/pitch: $\theta_1/\theta_2/\theta_3$ and $\theta_6/\theta_7/\theta_8$ for right and left, respectively; ankle pitch/yaw: θ_4/θ_5 and θ_9/θ_{10} for right and left, respectively.

7.6.5 Transverse-Plane Twist Synergy

The transverse-plane twist synergy is used to accommodate a sagittal-plane disturbance force applied from the back (e.g. at the right shoulder), by rotating the upper body around the vertical. Such rotation can be generated via a rolling motion in the pelvis and/or in a special waist roll joint, if available. The focus here is on the pelvis roll. This synergy can be obtained with the simplified two-leg model shown in Fig. 7.24. The following angular relations can be validated:

$$\begin{aligned}
\theta_4 &= \theta_8 = -\theta_9 = -\theta_3, \\
\theta_5 &= \theta_7 = -\theta_{10} = -\theta_2, \\
\theta_6 &= \theta_1.
\end{aligned}$$

Here $\theta_1/\theta_2/\theta_3$ and $\theta_6/\theta_7/\theta_8$ stand for the roll/yaw/pitch in the right and left hip, respectively; θ_4/θ_5 and θ_9/θ_{10} denote the pitch/yaw in the right and left ankle, respectively. Furthermore, from Fig. 7.24B and C one obtains

$$\begin{aligned}
l_2 \sin\theta_5 + d_1 \cos\theta_1 &= d_1, \\
l_2 \sin\theta_4 + d_2 \sin\theta_1 &= 0,
\end{aligned}$$

respectively. Thus,

$$\theta_2 = \arcsin\left(\frac{d_1 - d_1 \cos\theta_1}{l_1}\right),$$

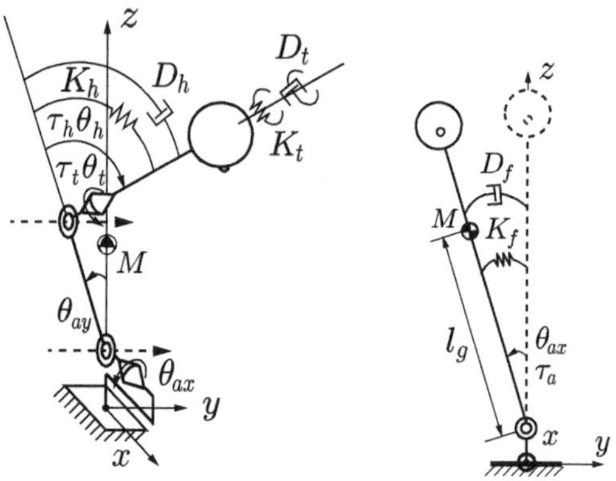

FIGURE 7.25 Simple models for an ankle-hip-twist strategy superposition. Left (A): four-joint 3D model for the hip-twist and lateral-ankle synergies. Right (B): a single-joint planar model for the lateral-ankle synergy. The virtual spring/damper variables are denoted as $K_{(\circ)}/D_{(\circ)}$, respectively.

$$\theta_3 = -\arcsin\left(\frac{d_1}{l_1}\sin\theta_1\right).$$

Apparently, all angles can be expressed as functions of θ_1. This angle parametrizes the twist synergy. Snapshots from the realization of the twist strategy are shown in the lower part of Fig. 7.24.

7.6.6 Complex Reactive Synergies Obtained by Superposition of Simple Ones

The ultimate goal in synergy-based balance control is the design of a controller that can ensure responses to a variety of disturbances applied from different directions. This implies that the appropriate reaction/recovery synergy has to be invoked depending on the magnitude and direction of the disturbance. Two approaches will be introduced below. First, in the remainder of this subsection, the superposition of the simple in-plane synergies discussed so far will be explored. The goal is to obtain a set of out-of-plane synergies that can handle disturbances from various directions. The second approach, to be introduced in Section 7.7, is based on synergy generation with whole-body models.

The sagittal-plane ankle/hip, lateral-plane ankle, and transverse-plane twist synergies can be combined with the help of the simple model shown in Fig. 7.25A. The model comprises two 2-DoF joints: a pitch-yaw ankle joint and a pitch-roll hip joint. A separate model for ankle-yaw is shown in Fig. 7.25B. This model is used in the lateral-plane ankle synergy. Virtual spring/dampers are attached to the joints, with the exception of the ankle-pitch joint. This joint is the dependent joint in the hip synergy. The hip-roll joint is used in the twist synergy that was explained in Section 7.6.5; τ_a, τ_h, and τ_t are derived from the external disturbances for yaw, pitch, and roll measured by a force/torque sensor embedded into the pelvis of the experimental robot.[4] These inputs are inserted into the forward dynamic relations of the two

[4] A HOAP-2 robot was modified to accommodate the force/torque sensor.

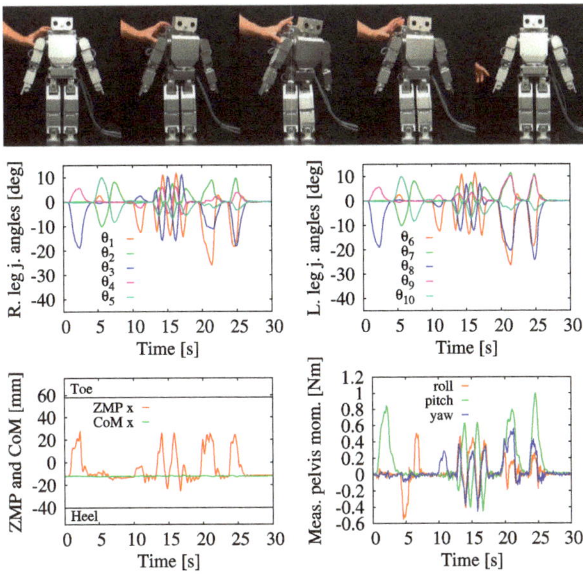

FIGURE 7.26 Superposition of the ankle-hip-twist synergies. The moment graphs in the bottom-right plot present data obtained from a force/torque sensor embedded into the pelvis of a HOAP-2 robot. The leg joint angles are numbered in increasing order from the hip toward the ankle. The knee joint is not included into the model.

simple models, to obtain the joint accelerations, and the velocities and positions after integration. These are then mapped to the joints of the real robot. As apparent from Video 7.6-3 [93] and the graphs in Fig. 7.26, the robot responds swiftly to perturbations from various directions, resulting in a very compliant behavior. Note that the hip synergy is realized via the RNS approach, as described in Section 7.6.3. In this way, the ZMP excursion within the sagittal plane can be minimized.

The lateral-plane ankle synergy is the only lateral plane synergy realized in this experiment. It could be possible to implement the lateral-plane load/unload and lift-leg synergies, but as already noted, the handling of the transitions between numerous models and the respective synergies is not that straightforward. It would be much more preferable to generate the complex reactive motion patterns with a single whole-body model. Such approach will be introduced in Section 7.7.

7.6.7 Summary and Discussion

Simple models can be employed to obtain reactive synergies for accommodating unknown external disturbances of *medium magnitude* that do not require a change in the BoS. The synergies are derived from those used by humans to maintain their balance. It is possible to superimpose the in-plane synergies to obtain fairly complex whole-body motion patterns in response to disturbances acting from various directions. Implementations with a position-controlled humanoid robot have demonstrated satisfactory results, including transitions between the synergies. However, the tuning of the parameters for the transitions is not straightforward. An alternative and more straightforward approach for generating complex out-of-plane reactive synergies with a single whole-body model will be introduced in Section 7.7.

7.7 REACTIVE SYNERGIES OBTAINED WITH A WHOLE-BODY MODEL

Reactive synergies that can cope with various types of external disturbances of medium magnitude can be generated with the whole-body models described in this work. To obtain a compliant response during the reflex phase, relatively low feedback gains for the CRB trajectories are used. After the disturbance disappears, the feedback gains are switched to higher values to ensure the swift recovery of the initial posture. This approach is known as *gain scheduling* (see e.g. [73,58]). Alternatively, the initial posture can be recovered with CRB trajectory tracking, as described in Section 5.8.7.

In this section, four balance control approaches will be explored, based on:

- a simple dynamic torque controller;
- a DCM-GI-based wrench distribution implementation for the load/unload and lift-leg strategies;
- a compliant-body response;
- an impact accommodation via angular momentum damping from the RNS (cf. Section 5.11.2);
- a reactive change-of-BoS (reactive stepping).

It is important to distinguish between continuous and impulsive types of disturbances. The discussion in Section 7.6 was based on the former type exclusively. In this section, it will be shown how to implement the balance control methods for both types of disturbances.

All simulations in this section were performed with a small-size humanoid robot model with parameters similar to those of the HOAP-2 robot [25] in the Choreonoid environment [87]. For the numbering of the robot joints and other relevant data, see Section A.1.

7.7.1 Reactive Synergies Generated With a Simple Dynamic Torque Controller

A whole-body reactive motion in 3D can be generated and controlled with the simple control law (5.128). Thereby, the linear component of the reference rate of change of the spatial momentum will be obtained from the DCM stabilizer (cf. (5.76)), while the angular one comprises just a damping term. The controller block diagram is the same as that shown in Fig. 5.25. The only difference is the setting of the desired values.

During the reactive (reflex) phase ($T_0 \leq t < T_{rec}$), admittance motion control will be employed. This can be done in a straightforward manner by setting the desired DCM at the current CoM position $\left(\boldsymbol{r}_X^{des}(t) = \boldsymbol{r}_C(t)\right)$ in the DCM stabilizer. The reference DCM can then be obtained from (5.76) and the definition of the DCM as

$$\dot{\boldsymbol{r}}_X^{ref}(t) = (1 - T_X k_X)\dot{\boldsymbol{r}}_C(t), \tag{7.46}$$

where $T_X = 1/\bar{\omega}_X$ is the time constant of the DCM dynamics.

When the external disturbance disappears (at T_{rec}), the recovery phase is initialized. During this phase, the desired DCM trajectory is determined as

$$\boldsymbol{r}_X^{des}(t) = \boldsymbol{r}_C(t) + (1 - e^{-\frac{t-T_{rec}}{T_X}})(\boldsymbol{r}_X^s(t) - \boldsymbol{r}_C(t)). \tag{7.47}$$

Here $\boldsymbol{r}_X^s(t)$, $T_{rec} \le t \le T_f$, is obtained via a fifth-order spline with the nonstationary initial state $\boldsymbol{r}_X^s(T_{rec})$ and the stationary final state $\boldsymbol{r}_C(T_f) = \boldsymbol{r}_C^{init}$, $\dot{\boldsymbol{r}}_C(T_f) = \boldsymbol{0}$, $\boldsymbol{r}_C^{init}$ denoting the initial CoM position (before the appearance of the disturbance); $\boldsymbol{r}_X^{des}(t)$ and its time derivative are substituted into (5.76) to obtain $\dot{\boldsymbol{r}}_X^{ref}$ for the recovery phase. Note that in this way, a smooth transition from the reflex to the recovery phase will also be ensured.

As an example, consider the case when an "unknown" external pushing force is applied on the back of a standing upright humanoid robot horizontally within the sagittal plane. The magnitude of the disturbance is gradually increased from zero to the maximum of 6 Nm via a fifth-order spline, within $2 \sim 4$ s. The magnitude stays constant within $4 \sim 12$ s and then decreases gradually within $12 \sim 14$ s. The application point is at (0, −0.050, 0.145) m w.r.t. the base coordinate frame, i.e. close to the right shoulder. This implies that the robot has to react with a combination of the hip and twist synergies. There are no particular models for the synergies, though. The reflex movement patterns are obtained by just setting the feedback gains to mimic the virtual spring/dampers. The simulation result is shown in Video 7.7-1 [50]. Apparently, the robot reacts as desired, in a compliant way and without loosing balance. After the disturbance disappears, a swift recovery movement is performed, also in a stable way. The stable performance during the two phases can be confirmed from the graphs in Fig. 7.27.

Furthermore, when the magnitude of the disturbance is set initially at a smaller value, the sagittal-plane ankle synergy will be invoked first. An increase in the magnitude of the disturbance during the ankle strategy yields a smooth transition to the hip-twist synergy. The results from the simulation can be seen in Video 7.7-2 [49].

7.7.2 The Load/Unload and Lift-Leg Strategies Revisited

The lateral plane lift-leg synergy described in Section 7.6.4 can be realized with a whole-body dynamic model under the DCM generalized inverse distribution control approach described in Section 5.10.4. The control method is examined via a simulation with the model and controller parameters set as in the examples described in Section 5.10.4. The reactive balance control task demonstrates the performance in the presence of an "unknown" disturbance force impulse, specified by a force vector of (0, −20, −20) N, applied from between $3 \sim 5$ s at the left hip of the robot. Accommodation of this disturbance requires a variation in the CoM height. Note that the DCM control approach admits such variations. The $\bar{z}_{vrp}$ constant in the DCM controller (cf. Section 5.6.2) is set at the initial CoM height (242 mm), which yields $\bar{\omega}_X = 0.169$. The force impulse results first in the loading/unloading of the right/left foot, then in lifting the left leg. Thus, this task demonstrates transitions from double to single and back to double stance (DS→SS→DS).

Video 7.7-3 [43] presents the result in animated form; the graphs are displayed in Fig. 7.28. The gray/pink areas in Fig. 7.28 signify the load/unload (DS) and lift-leg (SS) strategy, respectively. The left/right column displays the results from a simulation whereby the dynamic stability area has been set as large/small, respectively. From Fig. 7.28A and B it is seen that

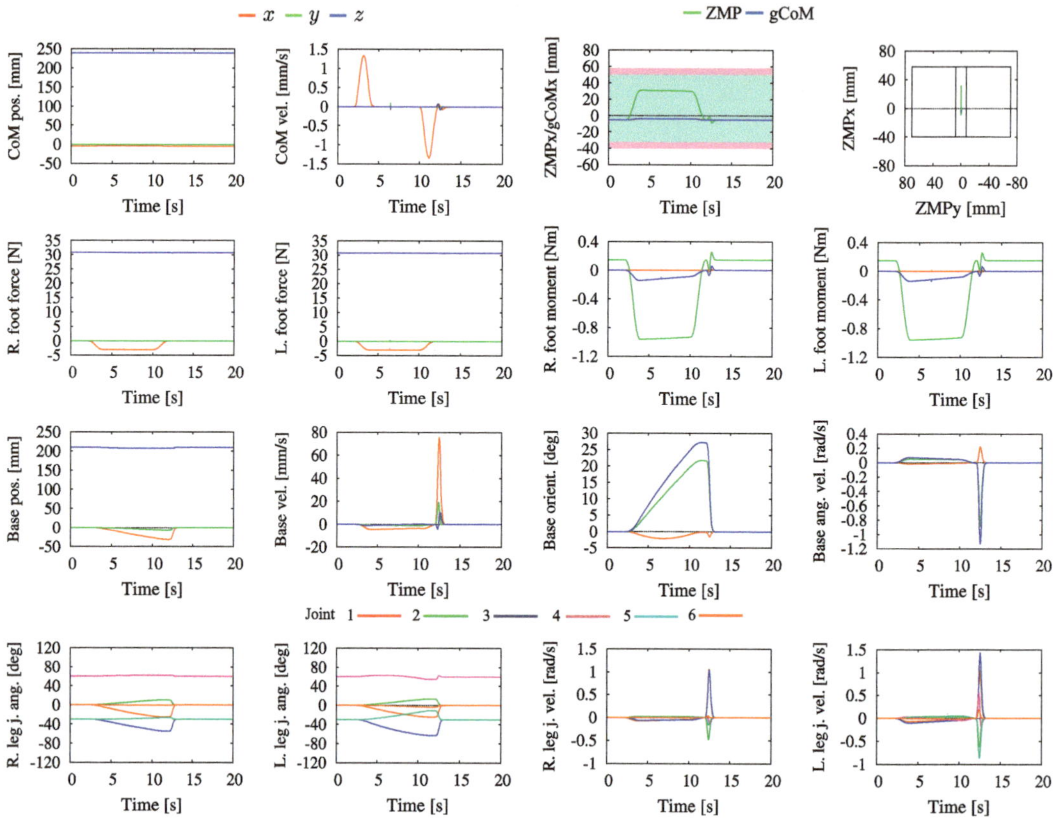

FIGURE 7.27 Simulation data for the hip-twist reflex synergy followed by a fast recovery motion. The magnitude of the disturbance is gradually increased within 2 ∼ 4 s, stays constant at 6 Nm within 4 ∼ 12 s, and then decreases gradually within 12 ∼ 14 s. The robot recovers thereafter its initial posture, swiftly and in a stable way.

the disturbance component in the z-direction yields a significant variation in the vertical CoM position. The reaction forces in Fig. 7.28C–F demonstrate the continuous asymmetric force distribution capability of the DCM-GI. The CoP graphs are shown in Fig. 7.28G and H. From the results it is apparent that enlarging the dynamic stability area beyond the ankle-joint positions is possible. However, as seen from Fig. 7.28G, the left foot CoP (green line) almost reaches the limit of the BoS. The foot roll can be avoided, though, since the robot switches from DS to SS in a stable manner. The joint velocity graphs for the four limbs are shown in Fig. 7.28I–L. They demonstrate that the transitions are performed in a continuous way, without excessive accelerations.

7.7.3 Compliant-Body Response

A compliant body–type of behavior in response to arbitrary external wrenches was demonstrated in Section 7.6.6. The motion was obtained from the superposition of synergies derived

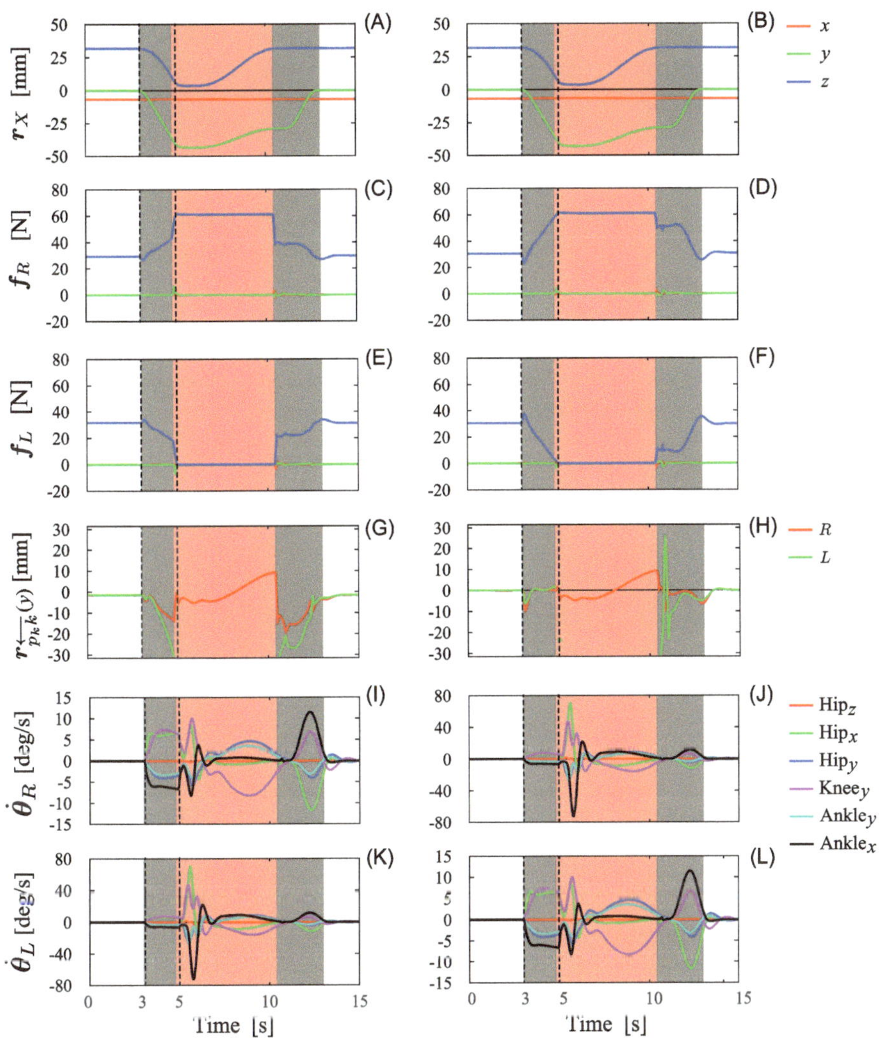

FIGURE 7.28 Reactive task under an unknown disturbance acting at the base/hip. The graphs on the l.h.s./r.h.s. display the results from a simulation where by the dynamic stability area has been set as large/small, respectively. The gray/pink areas signify the load/unload and the lift-leg strategies, respectively. The time span of the disturbance is signified with dotted vertical lines.

from multiple simple models. In this section, it will be shown that the same type of response can be obtained with a CRB trajectory tracking controller based on the constraint-consistent joint acceleration (5.141), i.e.

$$\ddot{\boldsymbol{\theta}} = \mathcal{J}_{cM}^{+}\left((\dot{\bar{\mathcal{V}}}^{c}) - \mathbb{C}_{cC}^{T}\dot{\mathcal{V}}_{M}^{ref} - \dot{\mathcal{J}}_{cM}\dot{\boldsymbol{\theta}} - \dot{\mathbb{C}}_{cC}^{T}\mathcal{V}_{M}\right) + \boldsymbol{N}(\mathcal{J}_{cM})\,\ddot{\boldsymbol{\theta}}_{u}. \tag{7.48}$$

The acceleration in the constrained motion directions, $\dot{\mathcal{V}}^c$, is set at zero since the robot stands on firm ground. The control input for the CRB trajectory tracking is defined in accordance with the results in Section 5.8.7 as

$$\dot{\mathcal{V}}_M^{ref} = \mathbb{M}_C^{-1} \mathcal{F}_B^{ext} + \begin{bmatrix} K_{vC} \dot{e}_{pC} \\ K_{\omega B} e_{\omega B} \end{bmatrix} + \begin{bmatrix} K_{pC} e_{pC} \\ K_{oB} e_{oB} \end{bmatrix}. \tag{7.49}$$

The feedback gains $K_{(\circ)}$ and the errors $e_{(\circ)}$ are defined in Section 5.8.7; $\mathcal{F}_B^{ext}$ is the external wrench measured by the force/torque sensor mounted at the pelvis (cf. Section 7.6.6). This wrench is mapped by the inverse of the CRB (locked) inertia to obtain a feedforward control component in terms of CRB acceleration. The control input $\ddot{\theta}_u$ determines the motion of the arms. In the following experiment, the arms will be kept motionless.

The experiment was the same as that described in Section 7.6.6: arbitrary wrenches were generated by slightly pushing the upper body in various directions. Since the force/moment measurements were noisy, a threshold of 1 N/Nm was used to filter out the noise. Note that a low-pass filter should be avoided due to the time delay; the robot's response would not be sufficiently compliant. The P and D feedback gains for the CRB trajectory tracking were set at 300 and 100, respectively, for each, position and orientation. The control time interval was 2.5 ms. The result is shown in Video 7.7-4 [51]. The graphs for the measured force/moment, the displacement errors for the CoM and the base rotation, and the joint angles in the legs are displayed in Fig. 7.29.

7.7.4 Impact Accommodation With Angular Momentum Damping From the RNS

In general, an impact can be accommodated either with the linear or the angular component of the spatial momentum, or with both. When the xCoM is located in the vicinity of the BoS boundary, even a weak impact could invoke a critical postural state leading to a foot roll (State D in Fig. 5.3). As clarified in Section 5.3.3, a swift action would then be required to restore the balance. Note that at a critical state, it would be impossible to accommodate the impact via the CoM motion alone; either an angular momentum–based accommodation should be invoked or a reactive step taken. Below, the focus is on the angular momentum–based accommodation with trunk and arm rotations. A reactive step–type response will be discussed in Section 7.7.5.

In Section 5.8.3 it was shown that the *relative angular momentum/velocity* (RAM/V) balance controller can be used to achieve fast proactive torso rotations without deteriorating the balance. The movements of the arms, though not under direct control, have been shown to play an important role. The method can be adopted in a straightforward manner to generate and control a whole-body reflexive motion of a humanoid robot subjected to an impact. To accommodate the impact, angular momentum damping will be injected into the system. This can be done with the *relative angular acceleration* (RAA) controller derived from the second-order formulation of the method (cf. Section 5.11.2).

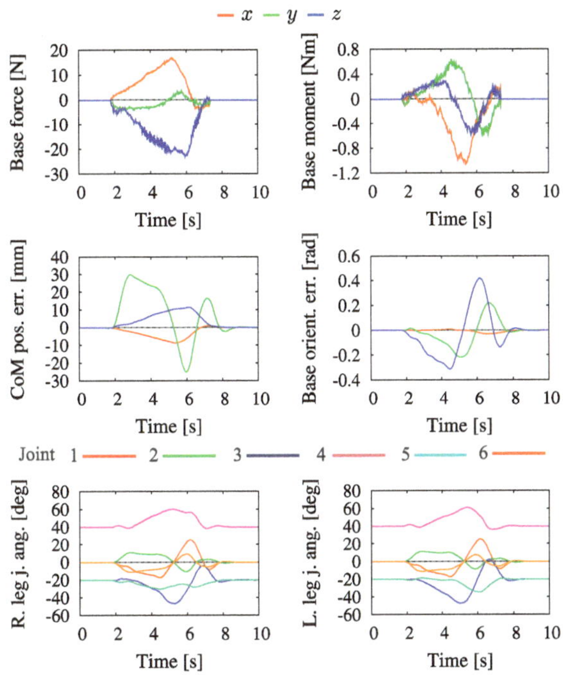

FIGURE 7.29 Compliant-body response experiment. The measured base force/moment is used as a feedforward control input for the CRB acceleration. The errors denote deviations from the initial configuration.

The constraint-consistent joint acceleration given in (5.142) comprises three *independent* control inputs: the inertia CoM acceleration, $\dot{\boldsymbol{v}}_{C_I}^{ref}$, the *system* angular acceleration, $\dot{\boldsymbol{\omega}}_C^{ref}$, and the angular acceleration of the base link, $\dot{\boldsymbol{\omega}}_B^{ref}$. Recall that when the mixed quasivelocity is employed, the CoM acceleration component will be completely decoupled from the angular acceleration one. This implies that $\dot{\boldsymbol{v}}_{C_I}^{ref}$ can be designed independently from the two reference angular accelerations.

The conservation of either the system or the coupling angular momentum has been pointed out (in Section 5.11.2) as a possible goal in the design of the angular accelerations. Note, however, that in the case of an impact at a critical state, the system angular momentum conservation would be of no help to reverse the motion in the rolling foot or feet. The only possibility to stabilize the robot at such a state is to *invoke a coupling angular momentum conservation*. This implies an RNS-based design for the angular accelerations, s.t. (cf. Section 5.11.2)

$$\dot{\boldsymbol{\omega}}_C^{ref} = \dot{\boldsymbol{\omega}}_B^{ref} - D_\omega \boldsymbol{J}_\omega \dot{\boldsymbol{\theta}}. \tag{7.50}$$

The capability of the RNS-based postural stabilization at a critical state (a foot roll) invoked by an impact will be demonstrated with the following simulation. The humanoid robot is placed on a flat ground in a symmetric posture, the feet being aligned. The robot is leaning forward s.t. the gCoM is in the vicinity of the BoS boundary (the toe area). The initial posture is stabilized with the asymptotic trajectory tracking control approach, using

TABLE 7.4 Gain scheduling for the base-link rotation PD feedback gains (impact expected)

Phase	Preimpact		Impact	Postimpact I	Postimpact II
Time (s)	$0 \sim 0.9$	$0.9 \sim 1.0$	$1.0 \sim 1.05$	$1.05 \sim 1.25$	$1.25 \sim$
K_{oB} (P-gain)	300	$300 \sim 0.01$	0.01	$0.01 \sim 30$	30
$K_{\omega B}$ (D-gain)	50	$50 \sim 0.001$	0.001	$0.001 \sim 5$	5

the RAA controller given in (5.142). The CoM motion is regulated toward the initial position. The respective PD feedback gains are set at relatively high values ($K_{pC} = 300$, $K_{vC} = 50$). The damping gain, used in (7.50) to ensure the RNS-based motion generation, is set at $D_\omega = 100$.

The disturbance accommodation during the impact phase and the stabilization of the posture during the postimpact phase are done with a prevailing upper body rotation (in the trunk and the arms), as already explained. The disturbance is applied within the sagittal plane, in the horizontal direction from behind. The application point is around the neck, the exact coordinates (in the base-link frame) are (0, 0, 145) (mm). The impulse of the disturbance is of magnitude 5.5 N, applied for 50 ms. Since the disturbance direction has no lateral component, it can be expected that the impulse will be accommodated with a base-link rotation in the pitch direction mainly. In general, $\dot{\boldsymbol{\omega}}_B^{ref}$ could be designed as a conventional feedforward plus PD feedback control law. In the special case of a disturbance accommodation task, though, there is no need for a feedforward term. Only PD feedback control can be used to ensure the compliant behavior at impact and the fast recovery after the impact. It will be sufficient to regulate the base-link angular deviation toward the initial posture during all three phases (i.e. preimpact, impact, and postimpact).

Anticipatory-Type Impact Accommodation

To accommodate the disturbance during the reflex (impact) phase and to stabilize the posture in the postimpact phase, a gain scheduling approach is employed (see e.g. [73,58]). The set of the PD gains for the base rotation is given in Table 7.4. Initially, high gains are used to ensure the desired base orientation (i.e. the initial orientation). Just before the impact, the gains are lowered significantly. After the impact, the gains are switched to higher values to ensure swift recovery. Note that the lowering of the gains *before* the impact corresponds to an *anticipatory* type of behavior.

The results of the simulation are shown in an animated form in Video 7.7-5 [36] and in graphic form in Fig. 7.30. From the graphs it is seen that the impact was successfully accommodated, mainly with a base angular deviation in the pitch direction, as expected. The displacement of the CoM was insignificant. During the accommodation period, the net CoP arrived at the BoS boundary. From the foot angular velocity graphs it is seen that the feet were about to roll. The robot was able to avoid this critical state, however, since the RNS joint acceleration component generated an appropriate swift arm motion. After that, the posture was stabilized to the initial one.

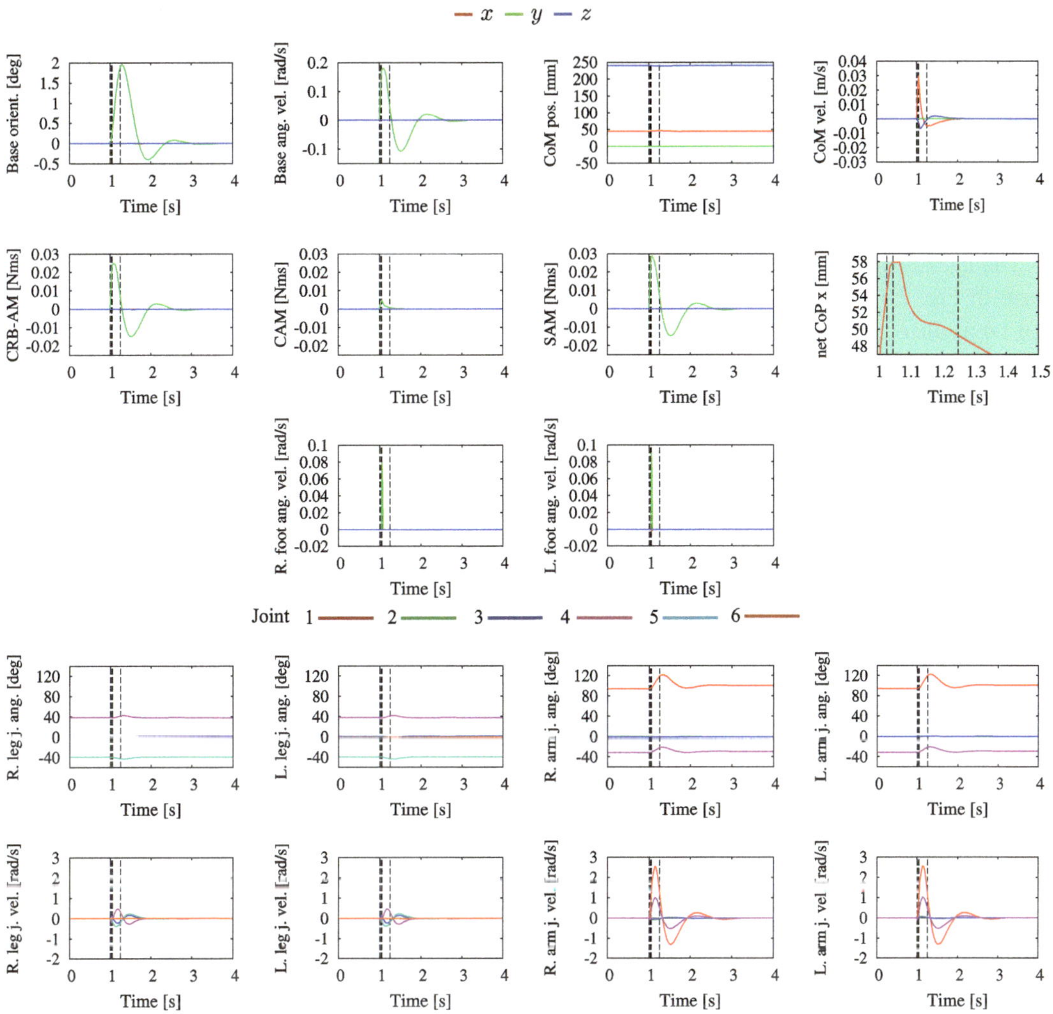

FIGURE 7.30 Simulation result for reflexive impulse accommodation with anticipatory type of behavior (the PD feedback gains for base rotation are lowered *before* the impact). A foot roll is successfully suppressed via the arm motion generated with the RNS approach. The vertical dashed lines signify the time instants in Table 7.4.

Nonanticipatory-Type Impact Accommodation

The same simulation was repeated with a gain scheduling policy corresponding to a *nonanticipatory* type of behavior. The set of the PD feedback gains for the base rotation is given in Table 7.5. Note that the initial high gains are maintained during the first 20 ms after the impact onset (i.e. Impact I phase). The gains are then lowered to obtain a compliant response during the remaining 30 ms of the impact (i.e. Impact II phase) and the Postimpact I phase. In the final phase, Postimpact II, the gains are switched again to higher values to ensure swift recovery of the initial posture.

TABLE 7.5 Gain scheduling for the base-link rotation PD feedback gains (impact unexpected).

Phase	Preimpact	Impact I	Impact II	Postimpact I	Postimpact II
Time (s)	$0 \sim 1.0$	$1.0 \sim 1.03$	$1.03 \sim 1.05$	$1.05 \sim 1.25$	$1.25 \sim$
K_{oB} (P-gain)	300		$300 \sim 0.01$	$0.01 \sim 30$	30
$K_{\omega B}$ (D-gain)	50		$50 \sim 0.001$	$0.001 \sim 5$	5

The results of the simulation are shown in animated form in Video 7.7-6 [37] and in graphic form in Fig. 7.31. The main difference with the previous experiment is that a foot roll could not be avoided. Both feet started rolling since the gains were high at impact onset, and the base link could not accommodate the impact with pitch roll as much as in the previous simulation. Nevertheless, the RNS-based generated motion response in the arms was able to ensure the recovery of the plane contacts at the feet and that of the posture stability. Note that no provision was made in the controller for the contact transitions at the feet. This clearly demonstrates the robustness of the controller w.r.t. the model discrepancies.

7.7.5 Reactive Stepping

The so-called *change-of-BoS* strategy is used by humans in response to unexpected postural perturbations with a relatively large magnitude. The most prominent example is the reactive stepping (or stumbling) [102,75,76]. Other examples include strategies performed mainly in the lateral plane, known as the "loaded side step" and the "unloaded crossover step" [96], as well as strategies that involve hand support, the so-called "reach-and-grasp" strategies [76]. In this section, the focus will be on the reactive stepping.

From the field of biomechanics it is known that when a relatively large disturbance is applied to the human body, one or more steps may be needed to restore the balance with an upright standing posture [2]. The decision on the number of steps to be taken can be based on the outcome of the so-called "N-step capturability" analysis [67]. The analysis uses the CP theory applied to the 3D LIP, the 3D LIP-on-foot, and the 3D RWP models described in Sections 4.4.2 and 4.4.3. The state at which the balance can eventually be restored after taking one or more steps is referred to as the *captured state*. Being a generalization of the CP approach, the N-step capturability framework is suitable for motion generation and balance control that can endow a humanoid robot with the ability to arrive at a captured state without falling, by taking N or fewer steps [99].

In what follows, an implementation of the N-step capturability framework for balance control during reactive stepping will be described. The following assumptions are made for clarity and simplicity:

- The reactive stepping strategy is implemented on a flat ground.
- The initial state is stationary, the feet being aligned.
- Multiple steps can be taken, each step is completed in time T_{step}.
- The DCM trajectory is in the horizontal plane (the LIP model applies).
- The swing leg path is a circular arc with maximum height l_h.

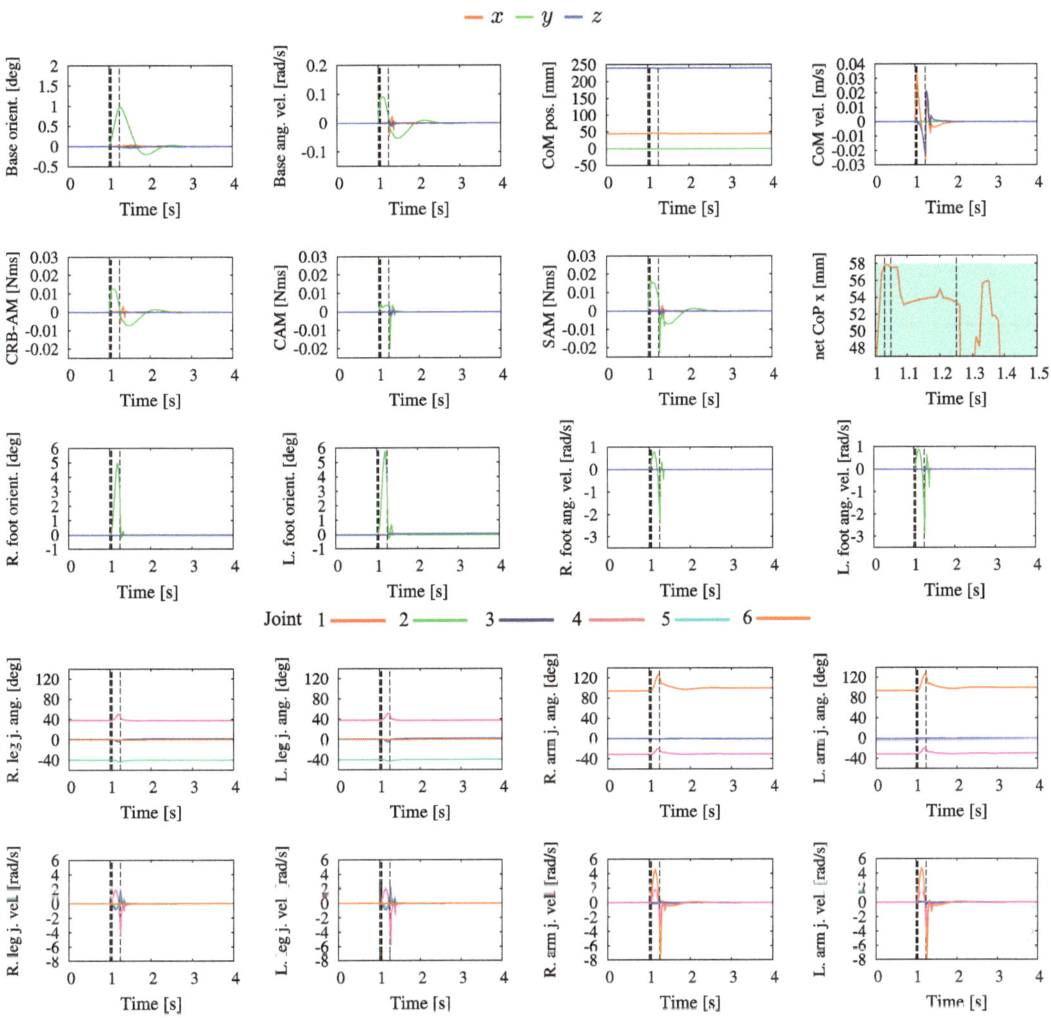

FIGURE 7.31 Simulation result for reflexive impulse accommodation with nonanticipatory type of behavior with high-gain impact. A foot roll is unavoidable but the plane contacts at the feet and the stability can be recovered via the arm motion generated with the RNS approach. The vertical dashed lines signify the time instants in Table 7.5.

The disturbance is in the form of an impulse expressed as the force/time product $\boldsymbol{f}_{ext}\Delta t_{imp}$, Δt_{imp} denoting the time duration of the impulse. The disturbance induces a change in momentum, i.e. $M\Delta\boldsymbol{v}_C = \boldsymbol{f}_{ext}\Delta t_{imp}$. Apparently, the change in the CoM velocity is determined by the direction and magnitude of the applied force. The xCoM/ICP trajectory $\boldsymbol{r}_{ex}(t)$ may leave the BoS (e.g. as shown with $\boldsymbol{r}_{ex}\left(t_{imp}^{end}\right)$ in Fig. 7.32A), resulting in a foot roll and eventually in a fall. One possibility to alleviate this problem is to make use of the centroidal moment, as clarified in Chapter 5. The linear reaction-wheel pendulum model (LRWP) described in Section 5.6.1 can be employed. The 2D version of the xCoM dynamics (5.61) is

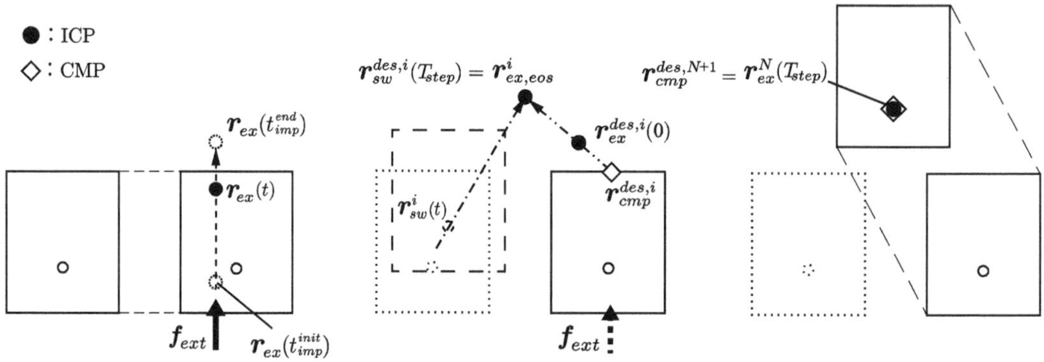

FIGURE 7.32 Reactive stepping. Left (A): Impact phase. Middle (B): Stepping phase. Right (C): Recovery phase.

written as

$$\dot{\boldsymbol{r}}_{ex}(t) = \omega(\boldsymbol{r}_{ex}(t) - \boldsymbol{r}_{cmp}(t)). \tag{7.51}$$

Recall that ω is the natural angular frequency of the LIP dynamics. The above equation can also be obtained by projecting the DCM dynamics (5.72) onto the ground surface. The solution is given as

$$\boldsymbol{r}_{ex}(t) = e^{\omega t}\left(\boldsymbol{r}_{ex}(0) - \boldsymbol{r}_{cmp}(t)\right) + \boldsymbol{r}_{cmp}(t). \tag{7.52}$$

As already noted, $\boldsymbol{r}_{ex}(t)$ is also referred to as the *instantaneous* CP (ICP) [67].

Another possibility to avoid a foot roll is to initialize reactive stepping. The xCoM location after the impact, $\boldsymbol{r}_{ex}\left(t_{imp}^{end}\right)$, is used to assess the capturability, i.e. whether zero, one, or more steps will be needed to arrive at a desired captured state $\boldsymbol{r}_{ex}^{des}$. In the case of a nonzero step capturability, as assumed here, $\boldsymbol{r}_{ex}^{des,i}(0) = \boldsymbol{r}_{ex}\left(t_{imp}^{end}\right)$ is located outside the BoS (cf. Fig. 7.32B).

The reactive stepping algorithm comprises four phases: preimpact, impact, stepping, and recovery. The desired DCM/ICP trajectories are generated with the help of (7.51) and (7.52) as follows.

Impact Phase

The impact phase duration is $\Delta t_{imp} = t_{imp}^{end} - t_{imp}^{init}$. During this time interval, the impact is accommodated with an appropriate CoM translation and a base rotation. The xCoM trajectory is designed to yield an admittance type of behavior, i.e.

$$\dot{\boldsymbol{r}}_{ex}^{des}(t) = \dot{\boldsymbol{r}}_g(t), \quad \boldsymbol{r}_{ex}^{des}(t) = \boldsymbol{r}_g(t), \tag{7.53}$$

where $t_{imp}^{init} \le t \le t_{imp}^{end}$. A rotary damper is used for the base-link rotation, the respective PD gains being designed with the help of a gain scheduling approach. The ICP trajectory is depicted in Fig. 7.32A.

Stepping Phase

In general, N reactive steps are assumed. The stepping is initialized at the end of the impact, t_{imp}^{end}. For each step, relative time is used, s.t. $0 \leq t \leq T_{step}$. Since the foot roll problem is tackled with reactive stepping, a zero centroidal moment is admissible. In this case, a constant CMP can be employed. The desired ICP trajectory for Step i is obtained from (7.51) and (7.52) as

$$\boldsymbol{r}_{ex}^{des,\,i}(t) = e^{\omega t}\left(\boldsymbol{r}_{ex}^{i}(0) - \boldsymbol{r}_{cmp}^{des,\,i}\right) + \boldsymbol{r}_{cmp}^{des,\,i}, \tag{7.54}$$

$$\dot{\boldsymbol{r}}_{ex}^{des,\,i}(t) = \omega\left(\boldsymbol{r}_{ex}^{des,\,i}(t) - \boldsymbol{r}_{cmp}^{des,\,i}\right), \tag{7.55}$$

respectively. Here CMP $\boldsymbol{r}_{cmp}^{des,\,i}$ denotes the constant desired CMP. Its location is set at the BoS boundary, at the crossing point with the line determined by $\boldsymbol{r}_{ex}^{des,\,i}(0)$ and the desired CP at the end of the step, $\boldsymbol{r}_{ex,eos}^{i}$. The latter should be set in a way that minimizes the step length. This situation is depicted in Fig. 7.32B; $\boldsymbol{r}_{sw}(t)$ is the ground projection of the swing leg path.

Recovery Phase

After completing the N steps, at $t = T_{rec}$ the ICP is within the BoS. The robot state is captured with

$$\boldsymbol{r}_{ex}(T_{rec}) = \boldsymbol{r}_{ex}^{N}(T_{step}) = \boldsymbol{r}_{cmp}^{des,\,N+1}. \tag{7.56}$$

This situation is depicted in Fig. 7.32C. Note that motion in the upper body (trunk and arms) is admissible, when generated from within the null space of the xCoM stabilization task.

The walking controller shown in Fig. 7.14 will be used. The reference contact wrenches are determined via the VRP-GI (cf. Section 5.10.5). The linear component of the reference rate of change of the spatial momentum is obtained from the DCM stabilizer (cf. (5.76)). The angular component (i.e. the rotation of the trunk/base link) is specified as a regulator with a damping term (obtained from (5.106) by removing the feedforward term). The motion of the arms is determined by a joint damping.

Simulation

The results from a simulation of reactive stepping under the above control approach are shown in Video 7.7-7 [52]. The impact force $\boldsymbol{f}_{ext} = (30,\ 0)$ N was applied at $t_{imp}^{init} = 1.0$ s for $\Delta t_{imp} = 0.1$ s. This resulted in $\boldsymbol{r}_{ex}(0) = (70.8,\ -25.2)$ mm and $N = 1$. The stepping time duration was preset at $T_{step} = 0.2$ s. The CP was calculated as $\boldsymbol{r}_{ex}^{des,\,N} = \boldsymbol{r}_{ex}^{des}(T_{rec}) = (124.9, 0.0)$ mm, where $T_{rec} = t_{imp}^{end} + T_{step} = 1.1 + 0.2 = 1.3$ s. Feedback was used only during the initial (preimpact) and final stabilization. During the impact and the stepping, only feedforward was applied. This implies an anticipatory type of behavior. The joint damping gain for the arm motion was set at $K_{D_h} = 100$. The stability margins for the VRP-GI-based wrench distribution at the initial and final DS postures were set to pass under the ankles. The maximum step height was set at 10 mm.

The data graphs for the mixed quasicoordinate and quasivelocity, the states of the end links, the GRFs and GRMs at the feet, the DCM error, the gCoM, the net CoP, and the xCoM are shown in Fig. 7.33. The CoP coordinates are presented in the local frames, the light-blue

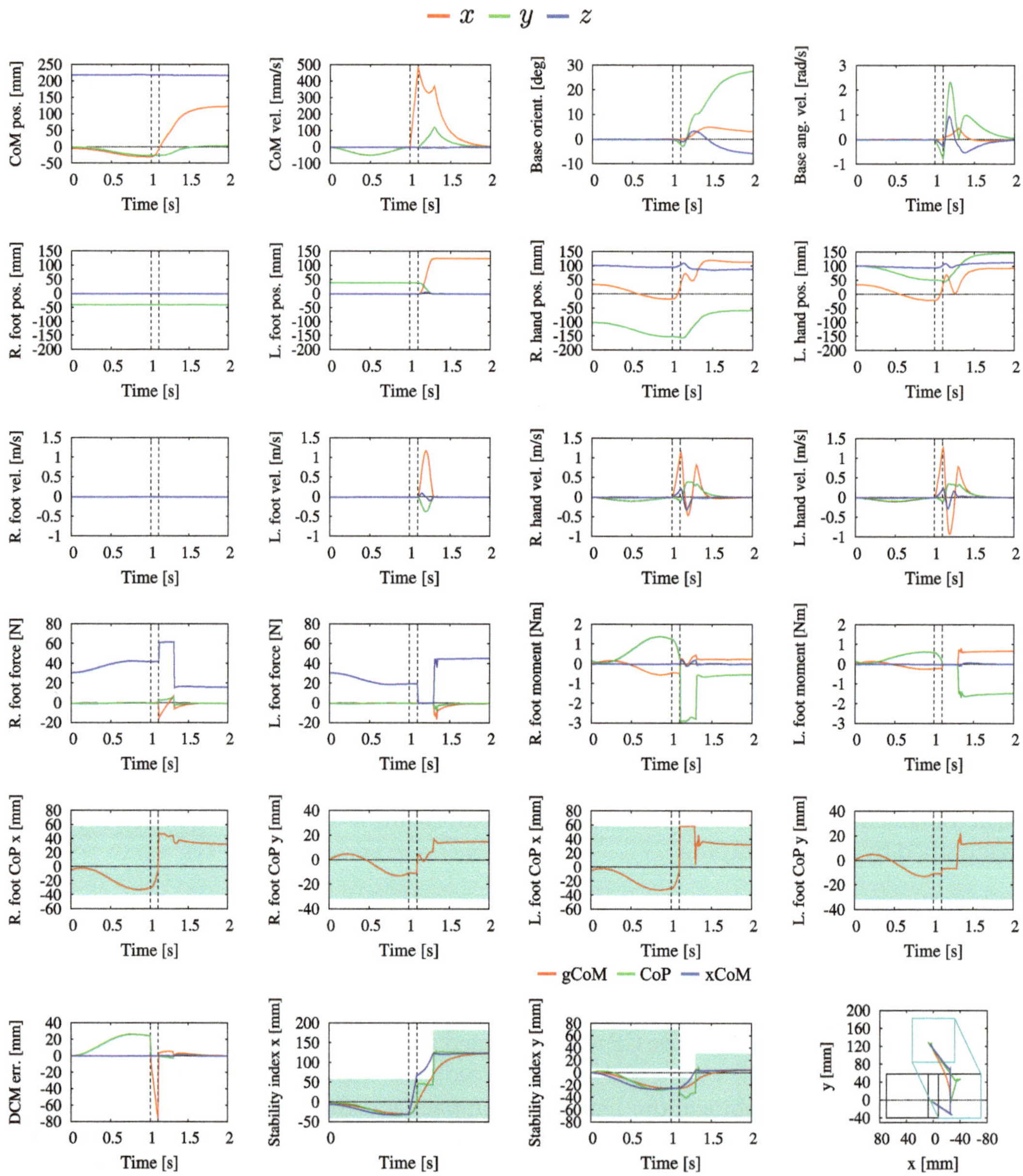

FIGURE 7.33 Data graphs for a simulation of reactive stepping. Shown are the states of the CoM, the rotation of the base link and the end links, the GRFs and GRMs at the feet, the DCM error, the gCoM, the net CoP and the xCoM. The CoP coordinates are presented in the local frames, the light-blue area denotes the BoS.

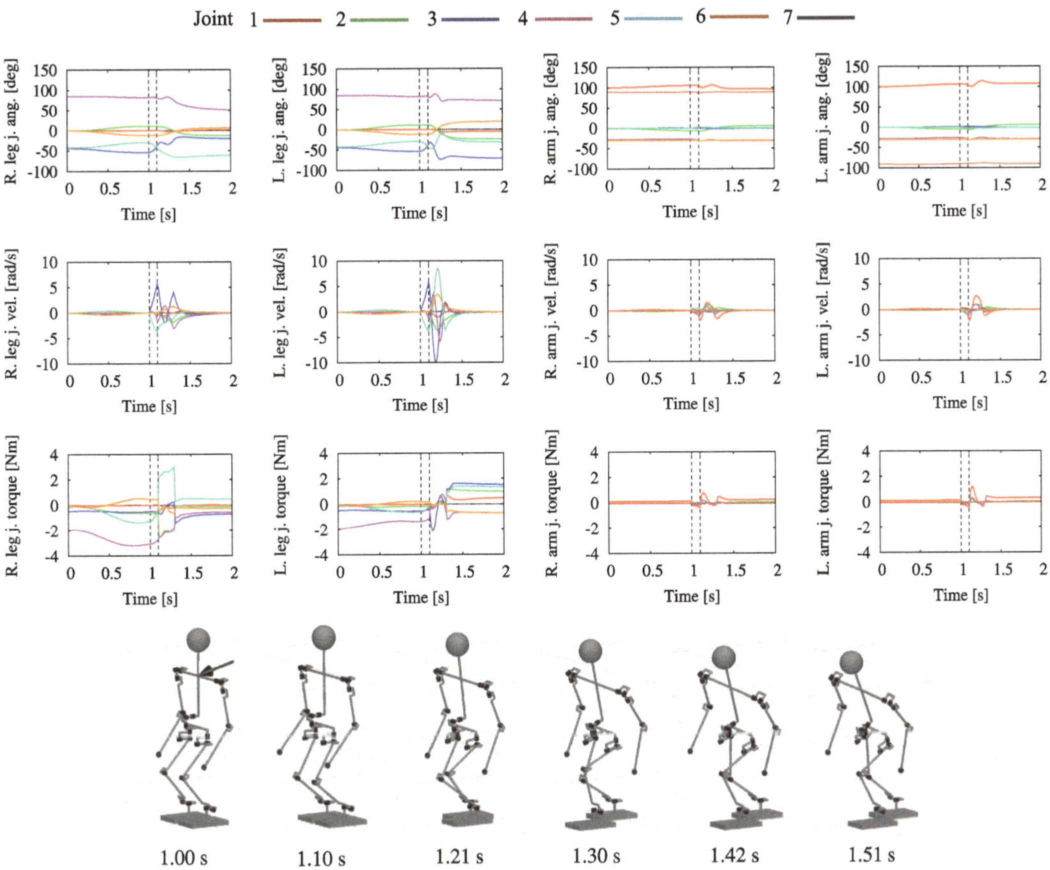

FIGURE 7.34 Data graphs for a simulation of reactive stepping. Shown are the graphs for the variations in the joint angles, rates, and torques in time. Snapshots from the animated motion as shown at the bottom of the figure.

area denotes the BoS. Furthermore, the data graphs for the joint-space quantities (joint angles, rates, and torques) are shown in Fig. 7.34. From the graphs, a stable stepping response can be confirmed.

With the above approach, there is some ambiguity in the parameter setting. The stepping parameters depend on the torque and velocity constraints. In the case of constraint violation, the stepping time could be increased and/or a smaller step could be taken, at the expense of a smaller dynamic stability margin. As an alternative, the motion planning and generation processes can be formulated as a recursive optimization task (a QP task) that includes the above constraints, as suggested in [27].

It should be noted that the impact applied was relatively large; it was impossible to accommodate the impact with a synergetic motion, neither with the ankle nor with the hip/twist synergies or combinations thereof. In Section 7.7.6, an RNS-based accommodation approach will be explored as an alternative.

TABLE 7.6 Gain scheduling and snapshots for accommodating a large impact without stepping

Phase	Preimpact		Impact	Postimpact I	Postimpact II	
Time [s]	0 ~ 0.4	0.4 ~ 0.5	1.0 ~ 1.1	1.1 ~ 1.6	1.6 ~ 3.0	3.0 ~
K_{oB} (P-gain)	300	300 ~ 0.01	0.01		0.01 ~ 300	300
$K_{\omega B}$ (D-gain)	50	50 ~ 0.001	0.001	0.001 ~ 5	5 ~ 50	50

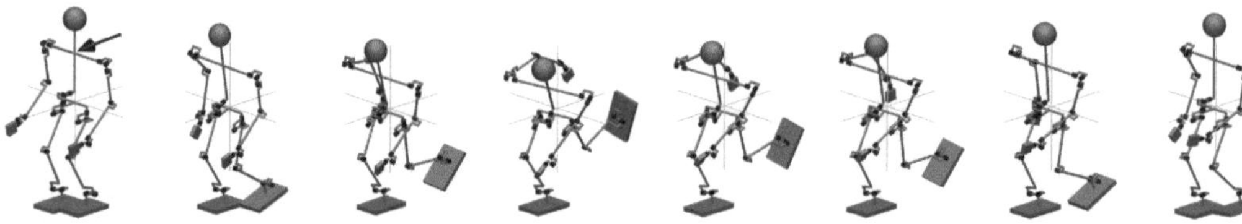

0.50 s 0.70 s 0.86 s 1.16 s 1.66 s 1.70 s 1.83 s 2.83 s

7.7.6 Accommodating a Large Impact Without Stepping

The main idea behind accommodating a large impact without stepping is to make use of the whole-body motion generation and control approach with angular momentum damping from the RNS, when the robot is in a single stance. In this way, the motion in the free leg can contribute, together with that in the upper torso and the arm motion, to the dissipation of the energy of the impact. The *relative angular acceleration* control framework from Section 7.7.4 is suitable to implement the RNS balance controller. During the reflex phase, an anticipatory type of behavior is assumed, i.e. the virtual-spring gain for the impact accommodating base rotation is preset at a relatively low value *before* the impact. The gain is also held at that value during the phases Impact and Postimpact I.

The performance of the designed controller is examined with a simulation. The initial xCoM and the impact parameters were set as in the reactive stepping simulation in Section 7.7.5. [5] The initial posture is an SS one, the left foot being lifted off the floor. The timing of the impact and the gain scheduling setting are apparent from Table 7.6. The resulting motion can be seen in Video 7.7-8 [38].

The graphs from the simulation are displayed in Figs. 7.35 and 7.36. The graphs in Fig. 7.35 are for the CoM position and velocity, the base-link orientation and the angular velocity, the end-link positions/orientations and the respective velocities/angular velocities, the CRB, the coupling and the system angular momentum (abbreviated as CRB-AM, CAM, and SAM, respectively), the respective rates of change, and, finally, the CoP graphs. In Fig. 7.36, graphs for the joints-space quantities (joint angles, rates, and torques) are shown. From the snapshots, the video, and the graphs it is apparent that the large impact has been accommodated successfully. The impact induced a rotational disturbance in the support foot, but only instantaneously. The state was recovered swiftly with the relative angular acceleration control law. The energy of the impact was dissipated in the postimpact phases, s.t. the robot came to rest without disturbing the stance foot state significantly.

[5] Recall that the impact was such that it could not be accommodated with any of the known synergies.

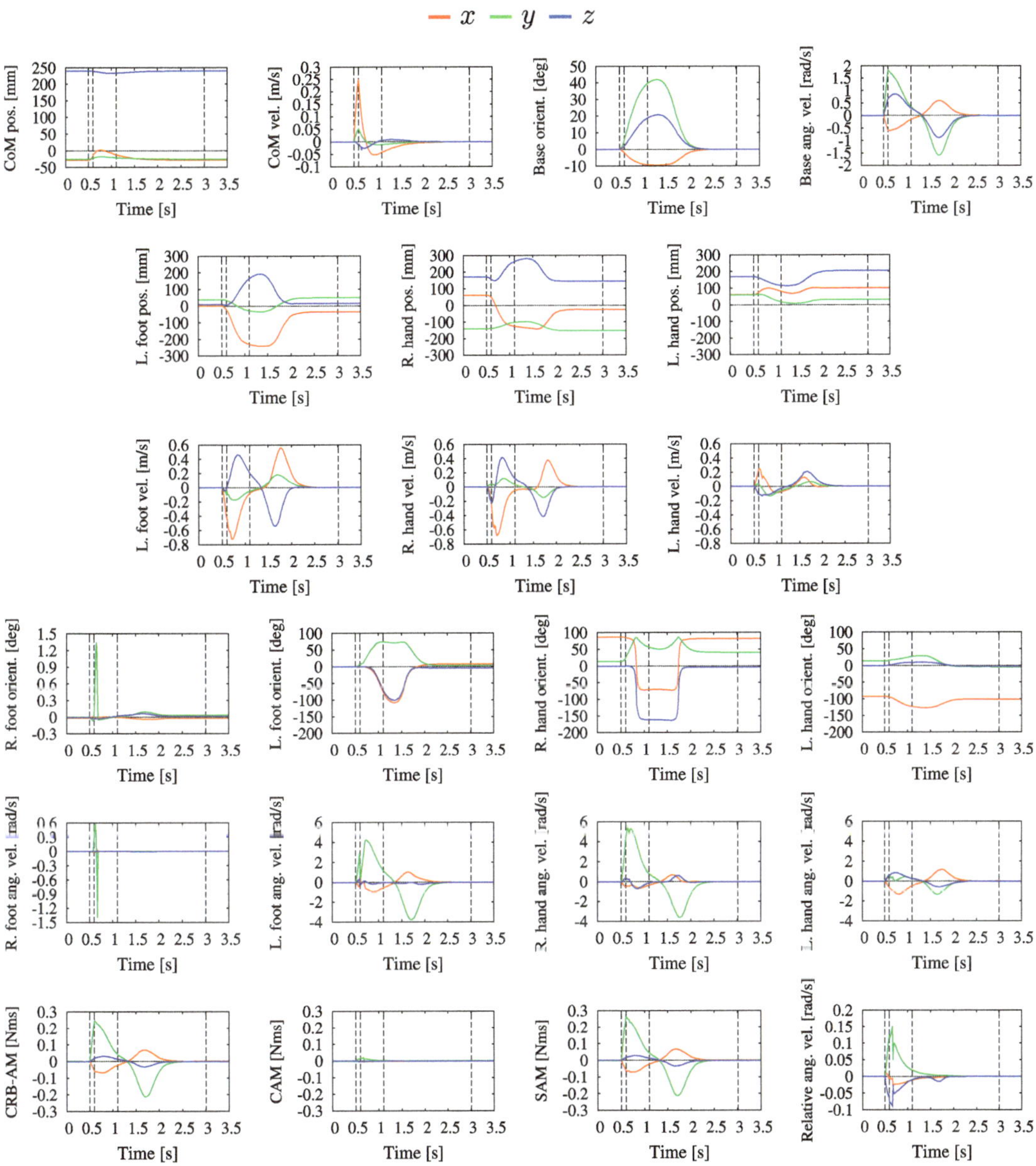

FIGURE 7.35 Accommodation of a large impact on the back without stepping. The graphs are for the CoM position and velocity, the base-link orientation and angular velocity, the end-link positions/orientations and the respective velocities/angular velocities, the CRB, the coupling and the system angular momentum (abbreviated as CRB-AM, CAM and SAM, respectively), the reference rates of change of the angular momenta, and finally, the CoP. The light-blue areas stand for the BoS.

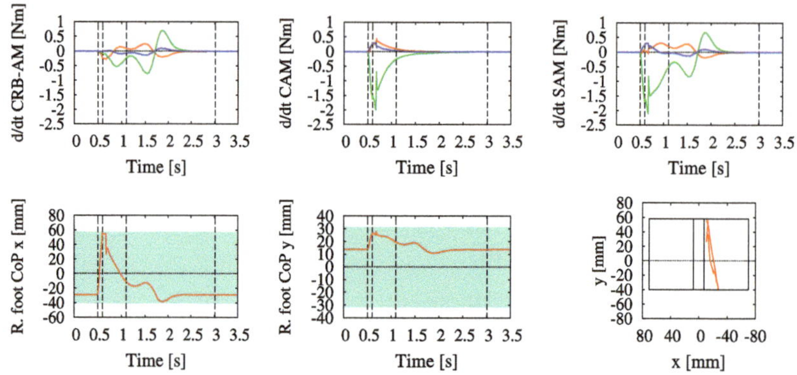

FIGURE 7.35 (Continued.)

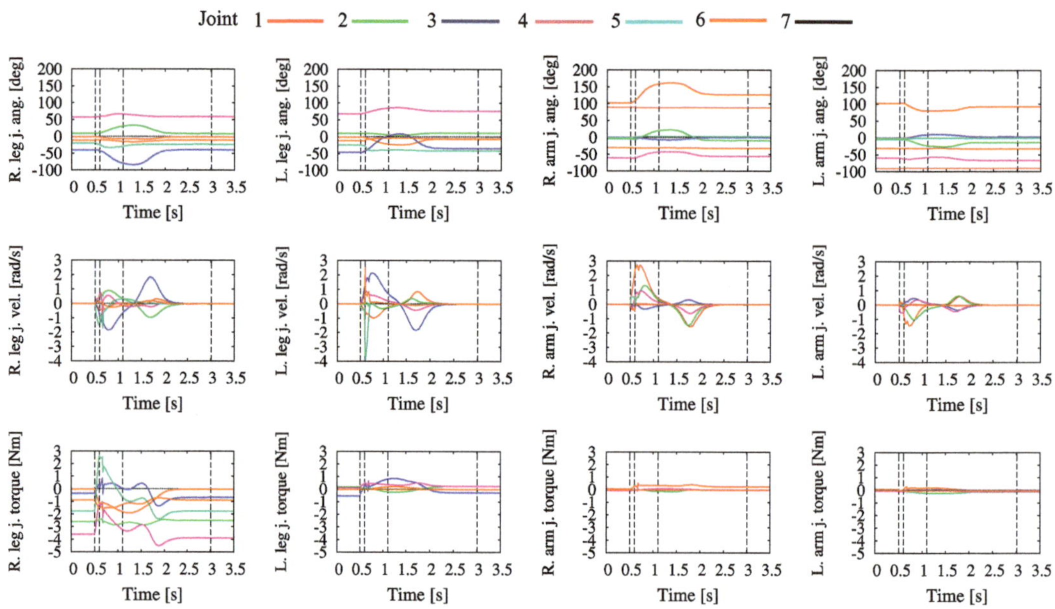

FIGURE 7.36 Accommodation of a large impact on the back without stepping. The graphs are for the joint-space quantities (joint angles, rates and torques).

From the above example it becomes apparent that with the RNS balance control approach it is possible to accommodate relatively large impacts without taking a step. Up to now, such type of impacts would require taking one or more steps, according to the N-step capturability theory [67]. The above result means that the theory can be further developed in the direction of increasing the zero-step capturability region via RNS-based motion generation and control.

7.8 IMPACT MOTION GENERATION

An impact motion is defined here as a motion that generates impulsive forces when contacting objects from the environment, whereby the magnitude of the force exerted on the object is not limited by the actuator torque constraints, as is the case for forces generated under static conditions.

7.8.1 Historical Background

When a fixed-base robot arm collides with a hard environment, there is a possibility of control destabilization due to a flip-flop–type transition between the free and constrained spaces. A humanoid robot that comprises a floating base may fall down if a large impulsive force is applied to its body. These examples show that the dynamics of impacts play an important role in modeling and control. Indeed, a considerable number of studies have been conducted in this field over the past few decades.

An impact dynamics model suitable for fixed-base robots was discussed in [128]. In order to avoid an excessive impulse force at collision, numerous control approaches have been proposed so far. They include the optimum approach velocity for force-controlled contacts [86, 59], an impact control scheme for stable hard-on-hard contact [110], a discontinuous control approach for transition between a noncontact and contact motion [80], a unified control strategy capable of achieving stable contact within both hard and soft environments [77], and a sensor-referenced control method using positive acceleration feedback and a switching control strategy [107]. In order to analyze and evaluate the effects of impacts on fixed-base robots, some measures have been proposed such as the *virtual mass* [5], the *dynamic impact measure*, and the *generalized impact measure* [115].

In the field of biped locomotion, impacts occur at collisions between the swing leg and the floor at the end of the swing phase. This type of collision is referred to as the *heel strike*. The effect of such impacts on the stability of biped locomotion was discussed in [45,46]. In [28], biped locomotion was modeled as a nonlinear system with impulse effects. The stability of the gait was analyzed in [28,88] using a Poincaré map. The toe-off action at the end of the double support phase was modeled in [17] by an impulsive foot actuator. In [84], the contact phase after the impact was analyzed using a five-link humanoid robot model. Strategies for tripping caused by an unexpected impact at the swing leg were addressed in [95]. Impacts occurring at touchdown after jumping were analyzed in [3]. Stabilization after a hard impact was studied in [121,101].

Free-floating space robots also belong to the class of floating-base robots. In the case of a collision with a free-floating object, the state of the robot may become completely uncontrollable. The *Reaction Null Space*–based control method has been proposed to deal with such situations. This control approach can minimize the impulsive reaction wrench and the disturbance on the floating satellite base [124,90].

Most works have tried to overcome the problems caused by impacts. On the other hand, some trials have been made to harness impacts. Note that, when a robot statically applies a force to the environment, the magnitude of the force is bounded by the maximum torque of the actuators. When the applied force is in the form of an impulse, its magnitude is not

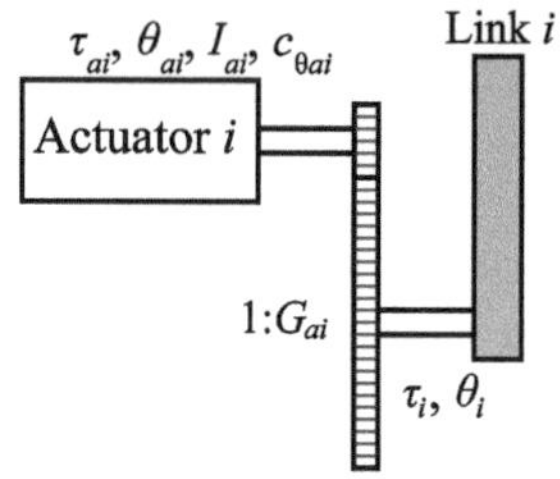

FIGURE 7.37 The model of a robot joint driven by an actuator with a reduction gear train.

bounded in such a way. Some applications that utilize an impulsive force have been presented so far, such as hammering a nail with a 3-DOF manipulator [108], with a rigid-link manipulator [104], and with a flexible-link manipulator [54], or sawing a wooden plate [104].

If a humanoid robot could do heavy work utilizing impulsive forces as well as a human does, the application field of humanoid robots would be drastically enlarged. Some attempts of using impulsive forces by humanoid robots have been reported so far, e.g. power lifting, diving, and gymnastics [7], pushing a wall and turning a valve [47], drum-beating [63], dynamic lifting [4], breaking woods with karate chop [78,64], and kicking a soccer ball [18].

It is important to note that when a legged robot applies an impulsive force to the environment, maintaining the balance is of utmost importance. A few works have addressed this problem, e.g. the *Adios Zero Moment Point* approach [40] (cf. Section 5.8.5).

7.8.2 Considering the Effects of the Reduction Gear Train

Most humanoid robots developed so far are equipped with gear trains with high reduction ratios in the joints. Such gears are generally characterized by a low joint backdrivability stemming from the friction in the gear transmission and the high effective rotor inertia. Recall that the friction and inertia moments at the actuator side are multiplied by the square of the reduction ratio, when seen from the output side. A model of a joint equipped with a high-reduction gear train is illustrated in Fig. 7.37.

Let $\tau_{ai}, \theta_{ai}, I_{ai}$, and c_{ai} be the torque, rotation angle, rotor inertia, and viscosity coefficient of the ith actuator and its gear (before reduction), respectively. Furthermore, let τ_i and θ_i be the torque and the rotation angle of the output side (after reduction). Suppose the reduction ratio is $1/G_{ai}$; G_{ai} takes a negative value when the output shaft rotates in the direction opposite to that of the motor shaft.

The equation of motion at the actuator side can be written as follows:

$$\boldsymbol{\tau}_a = \boldsymbol{I}_a \ddot{\boldsymbol{\theta}}_a + \boldsymbol{c}_a \dot{\boldsymbol{\theta}} + \boldsymbol{G}_a^{-1} \boldsymbol{\tau}, \tag{7.57}$$

where

$$\boldsymbol{\tau}_a = \begin{bmatrix} \tau_{a1} \\ \vdots \\ \tau_{an} \end{bmatrix}, \quad \boldsymbol{\theta}_a = \begin{bmatrix} \theta_{a1} \\ \vdots \\ \theta_{an} \end{bmatrix}, \quad \boldsymbol{I}_a = \begin{bmatrix} I_{a1} & & \boldsymbol{0} \\ & \ddots & \\ \boldsymbol{0} & & I_{an} \end{bmatrix},$$

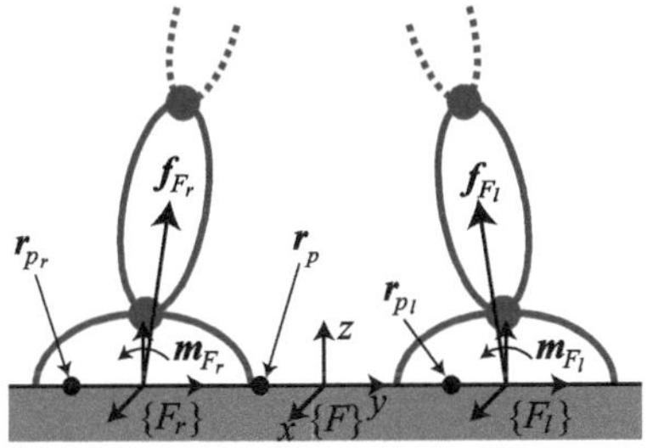

FIGURE 7.38 Reaction force and moment.

$$\boldsymbol{c}_a = \begin{bmatrix} c_{a1} & & \mathbf{0} \\ & \ddots & \\ \mathbf{0} & & c_{an} \end{bmatrix}, \quad \boldsymbol{G}_a = \begin{bmatrix} G_{a1} & & \mathbf{0} \\ & \ddots & \\ \mathbf{0} & & G_{an} \end{bmatrix},$$

and $\boldsymbol{\tau}$ is defined in (3.71). The equation of motion (7.57) can be rewritten as follows:

$$\boldsymbol{\tau} = \boldsymbol{G}\boldsymbol{\tau}_a - \boldsymbol{G}_a \boldsymbol{I}_a \boldsymbol{G}_a \ddot{\boldsymbol{\theta}} - \boldsymbol{G}_a \boldsymbol{c}_a \boldsymbol{G}_a \dot{\boldsymbol{\theta}} \quad (\because \boldsymbol{\theta}_a = \boldsymbol{G}_a \boldsymbol{\theta}). \tag{7.58}$$

Substituting (7.58) into (4.155), the equation of motion is rewritten as follows:

$$\begin{bmatrix} \mathbb{M}_B & \boldsymbol{H}_{BB} \\ \boldsymbol{H}_{BB}^T & \boldsymbol{M}_{\theta B_G} \end{bmatrix} \begin{bmatrix} \dot{\mathcal{V}}_B \\ \ddot{\theta} \end{bmatrix} + \begin{bmatrix} \mathcal{C}_B \\ \boldsymbol{c}_{\theta B_G} \end{bmatrix} + \begin{bmatrix} \mathcal{G}_B \\ \boldsymbol{g}_\theta \end{bmatrix} = \begin{bmatrix} \mathbf{0} \\ \boldsymbol{G}\boldsymbol{\tau}_a \end{bmatrix} + \begin{bmatrix} \mathbb{C}_{cB} \\ \mathcal{J}_{cB}^T \end{bmatrix} \overline{\mathcal{F}}^c, \tag{7.59}$$

where

$$\boldsymbol{M}_{\theta B_G} = \boldsymbol{M}_{\theta B} + \boldsymbol{G}_a \boldsymbol{I}_a \boldsymbol{G}_a, \quad \boldsymbol{c}_{\theta B_G} = \boldsymbol{c}_{\theta B} + \boldsymbol{G}_a \boldsymbol{c}_a \boldsymbol{G}_a \dot{\boldsymbol{\theta}}. \tag{7.60}$$

Here $\overline{\mathcal{F}}^c$ denotes the contact (reaction) forces defined in (3.61). For the definitions of $\mathbb{C}_c$ and $\mathcal{J}_{cB}$ refer to (2.74), (2.75), and (2.81).

7.8.3 Ground Reaction Force and Moment

Let $\boldsymbol{f}_{F_j} \in \Re^3$ and $\boldsymbol{m}_{F_j} \in \Re^3$ be the GRF and GRM measured at the origin of the foot coordinate system $\{F_j\}$, $j \in \{r, l\}$, as illustrated in Fig. 7.38. The wrench $\mathcal{F}_{F_j}$ can be written as (cf. (3.57))

$$\mathcal{F}_{F_j} = \begin{bmatrix} \boldsymbol{f}_{F_j} \\ \boldsymbol{m}_{F_j} \end{bmatrix} \triangleq \mathbb{B}_{cF_j} \overline{\mathcal{F}}_{F_j}^c. \tag{7.61}$$

Let $\boldsymbol{r}_{F_r}, \boldsymbol{r}_{F_l}$, and $\boldsymbol{r}_F$ denote the origins of coordinate systems $\{F_r\}, \{F_l\}$, and $\{F\}$, respectively. In double-support stance, the net reaction wrench is given as follows (cf. (3.57)):

$$\mathcal{F}_F = \begin{bmatrix} \boldsymbol{f}_F \\ \boldsymbol{m}_F \end{bmatrix} = \begin{bmatrix} \boldsymbol{f}_{F_r} + \boldsymbol{f}_{F_l} \\ \boldsymbol{m}_{F_r} - [\boldsymbol{r}_{\overleftarrow{FF_r}}^{\times}]\boldsymbol{f}_{F_r} + \boldsymbol{m}_{F_l} - [\boldsymbol{r}_{\overleftarrow{FF_l}}^{\times}]\boldsymbol{f}_{F_l} \end{bmatrix}, \tag{7.62}$$

where $\boldsymbol{r}_{\overleftarrow{FF_j}} = \boldsymbol{r}_F - \boldsymbol{r}_{F_j}$. Usually, a humanoid robot is equipped with force/torque sensors mounted at the ankles or pressure sensors embedded into the soles [85,65]. The sensors detect the wrenches $\mathcal{F}_{FS_j}$ with respect to the sensor coordinate frame $\{FS_j\}$. The wrench at $\boldsymbol{r}_{F_j}$ is obtained from the sensor readings as follows:

$$\mathcal{F}_{F_j} = {}^W\mathbb{X}^T_{FS_j\overleftarrow{F_jFS_j}}\, {}^{FS_j}\mathcal{F}_{FS_j}, \tag{7.63}$$

where ${}^W\mathbb{X}^T_{FS_j\overleftarrow{F_jFS_j}}$ is a wrench coordinate transform matrix from the force sensor coordinate frame $\{FS_j\}$ to the world coordinate frame $\{W\}$ (cf. (2.7)).

The net CoP $\boldsymbol{r}_p$, which is equivalent to the ZMP [111,112] on a flat ground, can be detected by the ankle force/torque sensors or the foot sole pressure sensors, in general. The wrench at $\boldsymbol{r}_{F_j}$ is expressed as a function of the ZMP as follows:

$$\mathcal{F}_{F_j} = \begin{bmatrix} \boldsymbol{f}_{F_j} \\ \boldsymbol{m}_{F_j} \end{bmatrix} = \begin{bmatrix} \boldsymbol{f}_{Pj} \\ -[\boldsymbol{r}^\times_{\overleftarrow{F_jP_j}}]\boldsymbol{f}_{P_j} \end{bmatrix}, \tag{7.64}$$

where $\boldsymbol{r}_{\overleftarrow{F_jP_j}} = \boldsymbol{r}_{F_j} - \boldsymbol{r}_{P_j}$, and $\boldsymbol{f}_{P_j} \in \Re^3$ is the GRF at the CoP.

Note that the z-components of $\boldsymbol{r}_{P_j}$ and $\boldsymbol{r}_{F_j}$ are zero because they are on the ground. The relationship between the CoP and $\mathcal{F}_{F_j}$ is obtained as follows:

$$r_{Pjx} = r_{Fjx} - \frac{m_{Fjy}}{f_{Fjz}}, \qquad r_{Pjy} = r_{Fjy} + \frac{m_{Fjx}}{f_{Fjz}}, \tag{7.65}$$

where the subscripts x and y indicate components of the respective vector quantities.

7.8.4 Dynamic Effects Caused by Impacts

Suppose an impulsive force stemming from an impact is applied to a point $\boldsymbol{r}_I$ of a humanoid robot at time t_0, as illustrated in Fig. 7.39. The dynamic equation at the moment of the impact is given as follows:

$$\begin{aligned}
&\begin{bmatrix} \mathbb{M}_B & \boldsymbol{H}_{BB} \\ \boldsymbol{H}^T_{BB} & \boldsymbol{M}_{\theta BG} \end{bmatrix}\begin{bmatrix} \dot{\mathcal{V}}_B \\ \ddot{\boldsymbol{\theta}} \end{bmatrix} + \begin{bmatrix} \mathcal{C}_B \\ \boldsymbol{c}_{\theta BG} \end{bmatrix} + \begin{bmatrix} \mathcal{G}_B \\ \boldsymbol{g}_\theta \end{bmatrix} \\
&= \begin{bmatrix} \boldsymbol{0} \\ \boldsymbol{G}\boldsymbol{\tau}_a \end{bmatrix} + \begin{bmatrix} \mathbb{C}_{cB}(\boldsymbol{q}) \\ \mathcal{J}^T_{cB}(\boldsymbol{q}) \end{bmatrix}\left(\mathcal{F}^c + \overline{\mathcal{I}}^c\delta(t-t_0)\right) + \begin{bmatrix} \mathbb{C}_{cB}(\boldsymbol{q}_I) \\ \mathcal{J}^T_{cB}(\boldsymbol{q}_I) \end{bmatrix}\overline{\mathcal{I}}\delta(t-t_0),
\end{aligned} \tag{7.66}$$

where $\delta(t-t_0)$ is the Dirac delta function, and hence $\overline{\mathcal{I}}\delta(t-t_0)$ represents the impulsive force applied at $t = t_0$. Note that $\overline{\mathcal{I}}$ does not include moment components; $\overline{\mathcal{I}}^c\delta(t-t_0)$ is the reaction impulse induced by $\overline{\mathcal{I}}\delta(t-t_0)$ at the constraint points such as feet (see Fig. 7.39) and $\boldsymbol{q}_I$ is the generalized coordinate vector with respect to $\boldsymbol{r}_I$.

During the short period from $t_0 - \Delta t/2$ to $t_0 + \Delta t/2$, it is assumed that the inertia matrices are quasiconstant, i.e. $\dot{\mathbb{M}}_B = \boldsymbol{0}$, $\dot{\boldsymbol{H}}_{BB} = \boldsymbol{0}$, and $\dot{\boldsymbol{M}}_{\theta BG} = \boldsymbol{0}$. Therefore, from the definitions of $\mathcal{C}_B$ and $\boldsymbol{c}_{\theta B}$ given in (4.125) and (4.145), respectively, the velocity-dependent terms are zero, i.e.

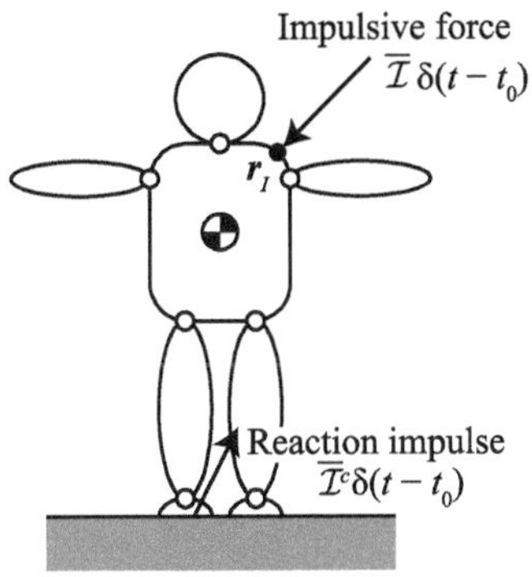

FIGURE 7.39 Impulsive force applied to a humanoid robot.

$\mathcal{C}_B = \mathbf{0}$ and $\boldsymbol{c}_{\theta B} = \mathbf{0}$. Furthermore, it can be assumed that the gravitational terms $\mathcal{G}_B$ and $\boldsymbol{g}_\theta$, the actuator torque $\boldsymbol{\tau}_a$, and the constraint wrench $\overline{\mathcal{F}}^c$ are also quasiconstant during the short period from $t_0 - \Delta t/2$ to $t_0 + \Delta t/2$.

Integrating both sides of (7.66), the following equation is obtained:

$$\begin{aligned}
\begin{bmatrix} \mathbb{M}_B & \boldsymbol{H}_{BB} \\ \boldsymbol{H}_{BB}^T & \boldsymbol{M}_{\theta BG} \end{bmatrix} \begin{bmatrix} \Delta\mathcal{V}_B \\ \Delta\dot{\boldsymbol{\theta}} \end{bmatrix} &= \int_{t_0-\frac{\Delta t}{2}}^{t_0+\frac{\Delta t}{2}} \left\{ -\begin{bmatrix} \mathbf{0} \\ \mathbf{G}_a\boldsymbol{c}_a\mathbf{G}_a\dot{\boldsymbol{\theta}} \end{bmatrix} - \begin{bmatrix} \mathcal{G}_B \\ \boldsymbol{g}_\theta \end{bmatrix} + \begin{bmatrix} \mathbf{0} \\ \boldsymbol{G}\boldsymbol{\tau}_a \end{bmatrix} + \begin{bmatrix} \mathbb{C}_{cB}(\boldsymbol{q}) \\ \mathcal{J}_{cB}^T(\boldsymbol{q}) \end{bmatrix} \overline{\mathcal{F}}^c \right\} dt \\
&+ \int_{t_0-\frac{\Delta t}{2}}^{t_0+\frac{\Delta t}{2}} \left\{ \begin{bmatrix} \mathbb{C}_{cB}(\boldsymbol{q}) \\ \mathcal{J}_{cB}^T(\boldsymbol{q}) \end{bmatrix} \overline{\mathcal{I}}^c \delta(t-t_0) + \begin{bmatrix} \mathbb{C}_{cB}(\boldsymbol{q}_I) \\ \mathcal{J}_{cB}^T(\boldsymbol{q}_I) \end{bmatrix} \overline{\mathcal{I}}\delta(t-t_0) \right\} dt \\
&= -\begin{bmatrix} \mathbf{0} \\ \boldsymbol{G}_a\boldsymbol{c}_a\boldsymbol{G}_a \Big[\boldsymbol{\theta}(t)\Big]_{t_0-\frac{\Delta t}{2}}^{t_0+\frac{\Delta t}{2}} \end{bmatrix} + \left\{ -\begin{bmatrix} \mathcal{G}_B \\ \boldsymbol{g}_\theta \end{bmatrix} + \begin{bmatrix} \mathbf{0} \\ \boldsymbol{G}\boldsymbol{\tau}_a \end{bmatrix} + \begin{bmatrix} \mathbb{C}_{cB}(\boldsymbol{q}) \\ \mathcal{J}_{cB}^T(\boldsymbol{q}) \end{bmatrix} \overline{\mathcal{F}}^c \right\} \Delta t \\
&+ \left\{ \begin{bmatrix} \mathbb{C}_{cB}(\boldsymbol{q}) \\ \mathcal{J}_{cB}^T(\boldsymbol{q}) \end{bmatrix} \overline{\mathcal{I}}^c + \begin{bmatrix} \mathbb{C}_{cB}(\boldsymbol{q}_I) \\ \mathcal{J}_{cB}^T(\boldsymbol{q}_I) \end{bmatrix} \overline{\mathcal{I}} \right\} \int_{t_0-\frac{\Delta t}{2}}^{t_0+\frac{\Delta t}{2}} \delta(t-t_0)dt \\
&= \begin{bmatrix} \mathbb{C}_{cB}(\boldsymbol{q}) \\ \mathcal{J}_{cB}^T(\boldsymbol{q}) \end{bmatrix} \overline{\mathcal{I}}^c + \begin{bmatrix} \mathbb{C}_{cB}(\boldsymbol{q}_I) \\ \mathcal{J}_{cB}^T(\boldsymbol{q}_I) \end{bmatrix} \overline{\mathcal{I}},
\end{aligned} \tag{7.67}$$

because

$$\int_{t_0-\frac{\Delta t}{2}}^{t_0+\frac{\Delta t}{2}} \delta(t-t_0)dt = 1, \quad \lim_{\Delta t \to 0} \Big[\boldsymbol{\theta}(t)\Big]_{t_0-\frac{\Delta t}{2}}^{t_0+\frac{\Delta t}{2}} = \mathbf{0},$$

and, since $\mathcal{G}_B$, $\boldsymbol{g}_\theta$, $\boldsymbol{\tau}_a$, and $\overline{\mathcal{F}}^c$ are assumed to be quasiconstant during the short period, the terms $\mathcal{G}_B\Delta t$, $\boldsymbol{g}_\theta\Delta t$, $\boldsymbol{\tau}_a\Delta t$, and $\overline{\mathcal{F}}^c\Delta t$ vanish as $\Delta t \to 0$ [128]. The differences $\Delta\mathcal{V}_B$ and $\Delta\dot{\boldsymbol{\theta}}$ are defined as follows:

$$\Delta\mathcal{V}_B = \mathcal{V}_B^+ - \mathcal{V}_B^-, \quad \Delta\dot{\boldsymbol{\theta}} = \dot{\boldsymbol{\theta}}^+ - \dot{\boldsymbol{\theta}}^-,$$

where $\mathcal{V}_B^+$ and $\dot{\boldsymbol{\theta}}^+$ are the base and the joint velocity after the impact, while $\mathcal{V}_B^-$ and $\dot{\boldsymbol{\theta}}^-$ are the velocities before the impact, respectively.

The differences in the velocities are obtained from (7.67) as

$$\begin{bmatrix} \Delta\mathcal{V}_B \\ \Delta\dot{\boldsymbol{\theta}} \end{bmatrix} = \begin{bmatrix} \boldsymbol{X}_{11} & \boldsymbol{X}_{12} \\ \boldsymbol{X}_{21} & \boldsymbol{X}_{22} \end{bmatrix} \left\{ \begin{bmatrix} \mathbb{C}_{cB}(\boldsymbol{q}) \\ \mathcal{J}_{cB}^T(\boldsymbol{q}) \end{bmatrix} \overline{\mathcal{I}}^{\,c} + \begin{bmatrix} \mathbb{C}_{cB}(\boldsymbol{q}_I) \\ \mathcal{J}_{cB}^T(\boldsymbol{q}_I) \end{bmatrix} \overline{\mathcal{I}} \right\}, \tag{7.68}$$

where

$$\boldsymbol{X}_{11} = \left(\mathbb{M}_B - \boldsymbol{H}_{BB}\boldsymbol{M}_{\theta BG}^{-1}\boldsymbol{H}_{BB}^T\right)^{-1},$$
$$\boldsymbol{X}_{12} = -\left(\mathbb{M}_B - \boldsymbol{H}_{BB}\boldsymbol{M}_{\theta BG}^{-1}\boldsymbol{H}_{BB}^T\right)^{-1}\boldsymbol{H}_{BB}\boldsymbol{M}_{\theta BG}^{-1},$$
$$\boldsymbol{X}_{21} = -\left(\boldsymbol{M}_{G\theta_B} - \boldsymbol{H}_{BB}^T\mathbb{M}_B^{-1}\boldsymbol{H}_{BB}\right)^{-1}\boldsymbol{H}_{BB}^T\mathbb{M}_B^{-1},$$
$$\boldsymbol{X}_{22} = \left(\boldsymbol{M}_{G\theta_B} - \boldsymbol{H}_{BB}^T\mathbb{M}_B^{-1}\boldsymbol{H}_{BB}\right)^{-1}.$$

As discussed in Section 7.8.2, most humanoid robots are equipped with high-reduction gear trains in their joints. When the gear reduction ratios (the diagonal elements of $\boldsymbol{G}_a$) are large enough, the diagonal elements of $\boldsymbol{M}_{G\theta_B}$ become quite large (see (7.60)). The inverse of $\boldsymbol{M}_{\theta BG}$ can then be approximated as $\boldsymbol{M}_{\theta BG}^{-1} \simeq \boldsymbol{0}_{n\times n}$. In this case, the submatrices $\boldsymbol{X}_{11}$, $\boldsymbol{X}_{12}$, $\boldsymbol{X}_{21}$, and $\boldsymbol{X}_{22}$ can be approximated as follows:

$$\boldsymbol{X}_{11} \simeq \mathbb{M}_B^{-1}, \quad \boldsymbol{X}_{12} \simeq \boldsymbol{0}, \quad \boldsymbol{X}_{21} \simeq \boldsymbol{0}, \quad \boldsymbol{X}_{22} \simeq \boldsymbol{0}. \tag{7.69}$$

Substituting (7.69) into (7.68), the following equation is obtained:

$$\begin{bmatrix} \Delta\mathcal{V}_B \\ \Delta\dot{\boldsymbol{\theta}} \end{bmatrix} \simeq \begin{bmatrix} \mathbb{M}_B^{-1}\left\{\mathbb{C}_{cB}(\boldsymbol{q})\overline{\mathcal{I}}^{\,c} + \mathbb{C}_{cB}(\boldsymbol{q}_I)\overline{\mathcal{I}}\right\} \\ \boldsymbol{0} \end{bmatrix}. \tag{7.70}$$

The above equation indicates that a humanoid robot with high reduction ratios in the joints (and hence, low backdrivability) behaves as a CRB during the impulse.

7.8.5 Virtual Mass

Fig. 7.40 visualizes the concept of a *virtual mass* [5], also referred to as the *equivalent mass*. The virtual mass is a point mass which is equivalent to the mass of the robot reflected at the contact point. The virtual mass of a fixed-base robot can be determined as the ratio of the impulsive force to the resultant acceleration [5]. It should be noted that the original definition of the virtual mass does not consider the friction in the joints. The virtual mass concept was adopted in free-floating space robots with joint resistance, by introducing the *joint resistance factor* [123].

In this section, the virtual mass concept is discussed considering the effects of the reduction gear train. Unilateral point contacts are assumed at the collision points between the robot and

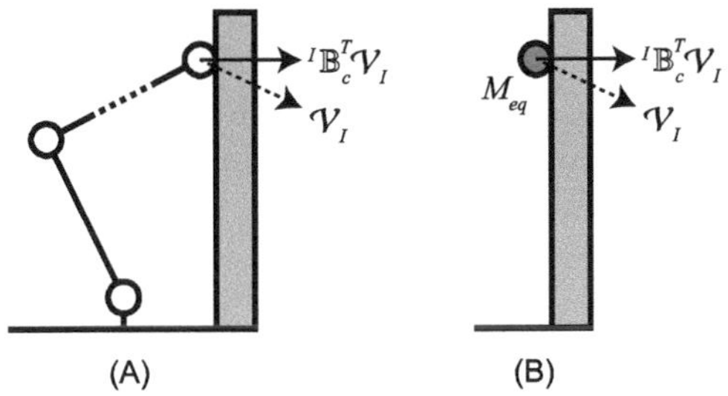

FIGURE 7.40 Equivalent point mass. (A) Robot manipulator. (B) Equivalent point mass.

the environment. Hence, the collision does not produce an impulsive moment. Under this assumption, the constraint basis ${}^{I}\mathbb{B}_c$ can be written as

$$ {}^{I}\mathbb{B}_c = \begin{bmatrix} \boldsymbol{e}_{nI} \\ \boldsymbol{0}_{3\times 1} \end{bmatrix} \in \Re^{6\times 1}, \tag{7.71} $$

where $\boldsymbol{e}_{nI} \in \Re^3$ is the unit vector normal to the constraint surface. As illustrated in Fig. 7.40, the component of $\mathcal{V}_I$ that is normal to the constraint surface is given by ${}^{I}\mathbb{B}_c^T\mathcal{V}_I$.

The difference between the velocity at the impact point before the impact, $\mathcal{V}_I^-$, and that after the impact, $\mathcal{V}_I^+$, is expressed with respect to the world coordinate system as follows:

$$ \Delta\mathcal{V}_I = \mathbb{T}_{\overleftarrow{BI}}\Delta\mathcal{V}_B + \boldsymbol{J}(\boldsymbol{\theta}_I)\Delta\dot{\boldsymbol{\theta}}. \tag{7.72} $$

Substituting (7.70) into (7.72), $\Delta\mathcal{V}_I$ is rewritten as follows:

$$ \begin{aligned} \Delta\mathcal{V}_I &= \mathbb{T}_{\overleftarrow{BI}}\mathbb{M}_B^{-1}\mathbb{C}_{cB}(\boldsymbol{q})\overline{\mathcal{I}}^{\,c} + \mathbb{T}_{\overleftarrow{BI}}\mathbb{M}_B^{-1}\mathbb{C}_{cB}(\boldsymbol{q}_I)\overline{\mathcal{I}}^{\,c} \\ &= \left\{ \mathbb{T}_{\overleftarrow{BI}}\mathbb{M}_B^{-1}\mathbb{C}_{cB}(\boldsymbol{q})\frac{\overline{\mathcal{I}}^{\,c}}{\overline{\mathcal{I}}} + \mathbb{T}_{\overleftarrow{BI}}\mathbb{M}_B^{-1}\mathbb{C}_{cB}(\boldsymbol{q}_I) \right\}\overline{\mathcal{I}}. \end{aligned} \tag{7.73} $$

If a point mass M_{eq} collides with an object from the environment, the generated impulse would be

$$ M_{eq}\,{}^{I}\mathbb{B}_c^T\Delta\mathcal{V}_I = \overline{\mathcal{I}}. \tag{7.74} $$

Comparing (7.74) with (7.73), the equivalent mass is given as follows:

$$ M_{eq} = \frac{1}{{}^{I}\mathbb{B}_c^T\mathbb{T}_{\overleftarrow{BI}}\mathbb{M}_B^{-1}\mathbb{C}_{cB}(\boldsymbol{q})\dfrac{\overline{\mathcal{I}}^{\,c}}{\overline{\mathcal{I}}} + {}^{I}\mathbb{B}_c^T\mathbb{T}_{\overleftarrow{BI}}\mathbb{M}_B^{-1}\mathbb{C}_{cB}(\boldsymbol{q}_I)}. \tag{7.75} $$

The equivalent mass can be utilized to generate the whole-body motion that maximizes/minimizes the impulsive force. An example of motion generation using the equivalent mass is given in Section 7.8.

7.8.6 CoP Displacement Induced by the Impulsive Force

Suppose that only the feet are constrained, with plane contacts on a flat floor (i.e. the hands are completely free). If the humanoid robot can keep the plane contacts after the impact without slipping, the base velocity difference $\Delta\mathcal{V}_B$ will be zero. The upper part of (7.70) can then be rewritten as

$$\mathbb{C}_{cB}(\boldsymbol{q}_F)\overline{\mathcal{I}}_F^{\,c} + \mathbb{C}_{cB}(\boldsymbol{q}_I)\overline{\mathcal{I}} = \mathbf{0}. \tag{7.76}$$

The reactive impulse can be calculated from (7.76) and the definition of $\mathbb{C}_{cB}$ (cf. (2.74)) as

$$\mathcal{I}_F^c = {}^I\mathbb{B}_c(\boldsymbol{q}_F)\overline{\mathcal{I}}_F^{\,c} = -\mathbb{T}_{\overleftarrow{FB}}^T\mathbb{C}_{cB}\overline{\mathcal{I}}. \tag{7.77}$$

The reactive impulse can be expressed as

$$\mathcal{I}_F^c \triangleq \begin{bmatrix}\left(\boldsymbol{f}_F^{\mathcal{I}c}\right)^T & \left(\boldsymbol{m}_F^{\mathcal{I}c}\right)^T\end{bmatrix}\Delta t = \begin{bmatrix} f_{Fx}^{\mathcal{I}c} & f_{Fy}^{\mathcal{I}c} & f_{Fz}^{\mathcal{I}c} & m_{Fx}^{\mathcal{I}c} & m_{Fy}^{\mathcal{I}c} & m_{Fz}^{\mathcal{I}c}\end{bmatrix}^T \Delta t. \tag{7.78}$$

From the relationship between $\boldsymbol{f}_F^{\mathcal{I}c}$ and $\boldsymbol{m}_F^{\mathcal{I}c}$ given in (7.77), the CoP displacement $\Delta\boldsymbol{r}_p$ induced by the impact $\overline{\mathcal{I}}$ is estimated as

$$\Delta r_{p_x} = -\frac{m_{Fy}^{\mathcal{I}c}}{f_{Fz}^{\mathcal{I}c}}, \quad \Delta r_{p_y} = \frac{m_{Fx}^{\mathcal{I}c}}{f_{Fz}^{\mathcal{I}c}}. \tag{7.79}$$

Note that there is no consideration for the dimension of the supporting polygon in the calculation of (7.79), hence the obtained CoP is the *imaginary ZMP* [113] (also known as the *Foot Rotation Indicator* (FRI) [26]).

7.8.7 Optimization Problems for Impact Motion Generation

The impact force (the acting force) generated during the impact motion should be as large as possible and should not lead to balance deterioration after the impact. Thus, the objective of the impact motion is to maximize (i) the impact force and (ii) the stability margin in the balance control.

In the impact motion generation approach presented in [64], the initial and final positions were not explicitly specified as boundary conditions. Instead, the position and velocity at the impact were determined by solving an optimization problem. Then, decelerating the motion at impact forwards and backwards in time, the initial and final positions were determined. The concept of this type of motion generation is illustrated in Fig. 7.41. This approach comprises the following five steps:

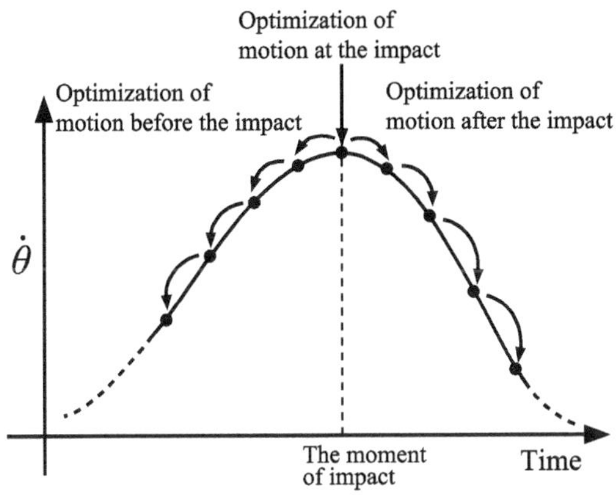

FIGURE 7.41 Conceptual sketch of the impact motion generation approach in [64].

Step 1: $\boldsymbol{\theta}(t)$ and $\dot{\boldsymbol{\theta}}(t)$ at the moment of impact $t = t_I$ are determined so that the objective function J_I is minimized.
Step 2: $\dot{\boldsymbol{\theta}}$ at $t = t_I - \Delta t$ is determined so that the objective function J_{baI} is minimized.
Step 3: Step 2 is iterated with $t = t - \Delta t$ until $\dot{\boldsymbol{\theta}}(t)^T\dot{\boldsymbol{\theta}}(t)$ becomes zero.
Step 4: $\dot{\boldsymbol{\theta}}$ at $t = t_I + \Delta t$ is determined so that the objective function J_{baI} (the same function used in Step 2) is minimized.
Step 5: Step 4 is iterated with $t = t + \Delta t$ until $\dot{\boldsymbol{\theta}}(t)^T\dot{\boldsymbol{\theta}}(t)$ becomes zero.

Here $\boldsymbol{\theta}(t)$ and $\dot{\boldsymbol{\theta}}(t)$ denote the joint angle and angular velocity vectors, respectively; Δt stands for the time step.

As a result of the optimization before/after the impact, the joint velocity vector $\dot{\boldsymbol{\theta}}(t)$ that minimizes the given objective function J_{baI} will be obtained. The joint angle vector $\boldsymbol{\theta}(t)$ and the joint acceleration $\ddot{\boldsymbol{\theta}}(t)$ are calculated as follows:

$$\boldsymbol{\theta}(t) = \begin{cases} \boldsymbol{\theta}(t+\Delta t) - \dot{\boldsymbol{\theta}}(t)\Delta t & \text{(for the motion before the impact)} \\ \boldsymbol{\theta}(t-\Delta t) + \dot{\boldsymbol{\theta}}(t)\Delta t & \text{(for the motion after the impact)} \end{cases} \tag{7.80}$$

$$\ddot{\boldsymbol{\theta}}(t) = \begin{cases} \dfrac{\dot{\boldsymbol{\theta}}(t+\Delta t) - \dot{\boldsymbol{\theta}}(t)}{\Delta t} & \text{(for the motion before the impact)} \\ \dfrac{\dot{\boldsymbol{\theta}}(t) - \dot{\boldsymbol{\theta}}(t-\Delta t)}{\Delta t} & \text{(for the motion after the impact).} \end{cases} \tag{7.81}$$

Furthermore, limits on the joint angles and their time derivatives are imposed in the form of the following inequality constraints:

$$\theta_{i,\min} \le \theta_i(t) \le \theta_{i,\max}, \tag{7.82}$$
$$-\dot{\theta}_{i,\max} \le \dot{\theta}_i(t) \le \dot{\theta}_{i,\max}, \tag{7.83}$$
$$-\ddot{\theta}_{i,\max} \le \ddot{\theta}_i(t) \le \ddot{\theta}_{i,\max}. \tag{7.84}$$

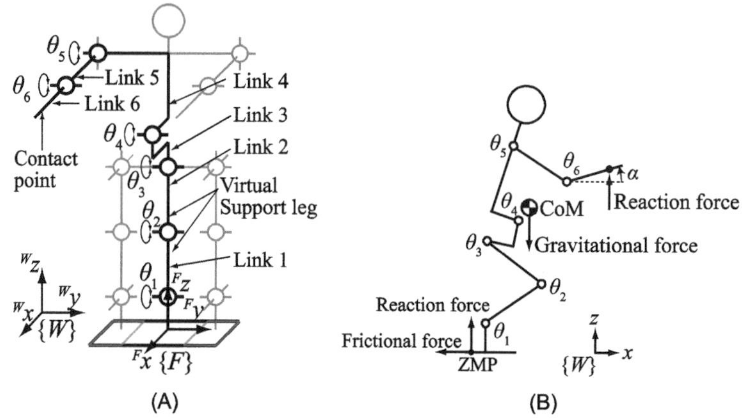

FIGURE 7.42 (A) A simplified model of the humanoid robot HOAP-2. (B) Side view of the model.

Here $\theta_i(t)$ denotes the angle of the ith joint at t, $\dot{\theta}_i(t)$ and $\ddot{\theta}_i(t)$ are the first and second time derivatives. Substituting (7.81) into (7.84), the constraints for the joint rates are obtained as

$$\begin{cases} \dot{\theta}_i(t+\Delta t) - \ddot{\theta}_{i,\max}\Delta t \le \dot{\theta}_i(t) \le \dot{\theta}_i(t+\Delta t) + \ddot{\theta}_{i,\max}\Delta t & \text{(for the motion before the impact)} \\ \dot{\theta}_i(t-\Delta t) - \ddot{\theta}_{i,\max}\Delta t \le \dot{\theta}_i(t) \le \dot{\theta}_i(t-\Delta t) + \ddot{\theta}_{i,\max}\Delta t & \text{(for the motion after the impact).} \end{cases} \tag{7.85}$$

In the case the joint acceleration given in (7.81) can be ensured during the entire motion, it can be said that the generated motion is continuous. Such a motion can be obtained by constraining the joint velocity as in (7.85). The maximum joint acceleration can be estimated from (7.59) as

$$\ddot{\boldsymbol{\theta}}_{\max} = \boldsymbol{M}_{G\theta}^{-1}\left(\boldsymbol{G}\boldsymbol{\tau}_{a,\max} - \boldsymbol{c}_{\theta B_G} + \boldsymbol{g}_\theta\right), \tag{7.86}$$

for $\dot{\boldsymbol{v}}_B = \boldsymbol{0}$ and $\boldsymbol{F}_{hi} = \boldsymbol{0}$. Note that the maximum joint acceleration depends on the arm dynamics. In order to obtain an accurate $\ddot{\boldsymbol{\theta}}_{\max}(t)$, these dynamics have to be solved at every iteration during optimization. This approach is not realistic, however, due to the computational cost. This problem can be alleviated in case the minimum value of $\ddot{\boldsymbol{\theta}}_{\max}(t)$ is estimated in advance.

7.8.8 A Case Study: Karate Chop Motion Generation

A Simplified Model of the Humanoid Robot HOAP-2

The humanoid robot HOAP-2 (Fujitsu Automation) comprising 25 DoFs is chosen as the test platform for the Karate chop experiments. For the impact motion generation task, a simplified, planar 6-DoF model is used, as shown in Fig. 7.42. The simplified model consists of six links and six joints. The left and right legs of the real robot are combined into a virtual support leg in the simplified model, as illustrated in Fig. 7.42A. Therefore, the masses of Links 1 and 2 are the sums of the masses of the corresponding left and right leg links, respectively. The masses of each link of the simplified model are listed in Table 7.7.

TABLE 7.7 Mass distribution of the HOAP-2 model

	Link 1	Link 2	Link 3	Link 4	Link 5	Link 6
Mass (kg)	0.57	0.88	0.92	2.45	0.62	0.17

TABLE 7.8 Joint limitations

i	1	2	3	4	5	6
$\theta_{i,\min}$ (°)	−60	−130	−70	0	−90	−115
$\theta_{i,\max}$ (°)	60	0	80	90	90	0

The joint limits of the simplified model are specified in accordance with those of the real robot, as listed in Table 7.8. The maximum joint velocity for each joint is set at π rad/s, which is the actual limit of the actuators used in the real robot.

Performance Index for Impact Force Evaluation

First, recall that the objective of the impact motion is to maximize the impact force (acting force) and the stability margin (cf. Section 7.8.7). When a link of the robot collides with an object from the environment the generated impulse can be obtained as a difference between the pre- and postimpact momenta. Using the equivalent mass relation (7.74), the preimpact momentum can be expressed as $M_{eq}{}^{I}\mathbb{B}_c^T\mathcal{V}_I^+$, where ${}^{I}\mathbb{B}_c$ is the constraint basis at the instantaneous contact joint established at the point of collision (cf. (7.71)). This momentum can be used as a performance index for the impact force minimization goal. Note that ${}^{I}\mathbb{B}_c^T\mathcal{V}_I^+$ denotes the component of $\mathcal{V}_I^+$ along the normal to the contact surfaces at the collision point (see Fig. 7.40B).

Performance Index for Stability Margin Evaluation

The dynamic equation of a humanoid robot, as given in (4.155) will be employed in the following derivations. Recall that the spatial dynamics relationship (4.130) stems from the upper part of (4.155) and the expression for the *system spatial momentum* (SSM). This relationship can be rewritten with respect to the foot coordinate system $\{F\}$ as

$$\frac{d}{dt}\mathcal{L}_F + \mathcal{G}_F = \mathcal{F}_F, \tag{7.87}$$

where

$$\frac{d}{dt}\mathcal{L}_F = \frac{d}{dt}\left({}^{F}\mathbb{X}^T_{B_{\overleftarrow{FB}}}\mathcal{L}_B(\boldsymbol{q},\dot{\boldsymbol{q}}_B)\right) \equiv \begin{bmatrix}\dot{\boldsymbol{p}}\\ \dot{\boldsymbol{l}}_F\end{bmatrix}, \quad \mathcal{G}_F = M\begin{bmatrix}-\boldsymbol{E}_3\\ [\boldsymbol{r}^{\times}_{\overleftarrow{FC}}]\end{bmatrix}\boldsymbol{a}_g, \quad \mathcal{F}_F = \begin{bmatrix}\boldsymbol{f}_F\\ \boldsymbol{m}_F\end{bmatrix},$$

and M is the total mass of the robot, $\boldsymbol{a}_g = \begin{bmatrix}0 & 0 & -g\end{bmatrix}^T$ is the gravitational acceleration vector, $\boldsymbol{r}_{\overleftarrow{FC}}$ is the position of the origin of $\{F\}$ w.r.t. the CoM. f_{pz} and m_{Fy} are calculated from (7.87) as

$$f_{Fz} = \dot{p}_z + Mg, \quad m_{Fy} = \dot{l}_{Fy} + \left(\boldsymbol{r}_{\overleftarrow{FC}}\right)_x Mg. \tag{7.88}$$

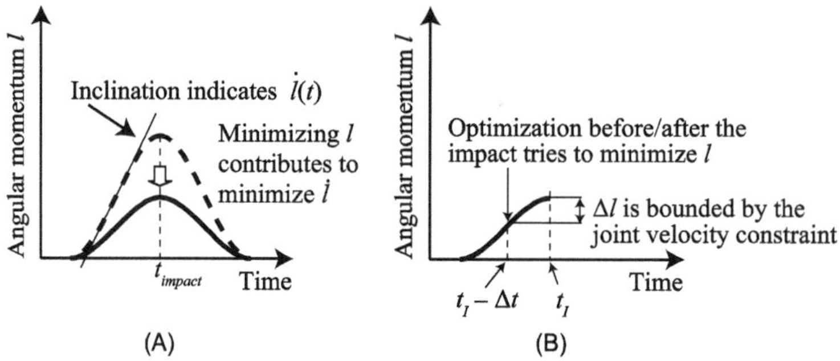

FIGURE 7.43 Minimizing angular momentum l in: (A) optimization at the impact and (B) optimization before/after the impact.

Substituting (7.88) into (7.65), the x-coordinate of the CoP is expressed using the time derivative of SSM as follows:

$$\left(\boldsymbol{r}_{\overleftarrow{pF}}\right)_x = -\frac{\dot{l}_{Fy} + \left(\boldsymbol{r}_{\overleftarrow{FC}}\right)_x Mg}{Mg + \dot{p}_z}. \tag{7.89}$$

In general, the rate of change of the linear momentum along the z-axis, $\dot{p}_z$, can be assumed much smaller than the gravitational force Mg. From the above relation it follows then that the rate of change of the angular momentum around the y-axis, $\dot{\boldsymbol{l}}_{Fy}$, is the crucial parameter that will dominantly affect the x-coordinate of the CoP.

In the impact motion generation approach discussed here, it is assumed that the robot is stationary at the initial ($t = 0$) and final ($t = T$) postures. Hence, $\boldsymbol{p}(0) = \boldsymbol{p}(T) = \boldsymbol{0}$ and $\boldsymbol{l}_F(0) = \boldsymbol{l}_F(T) = \boldsymbol{0}$. Furthermore, in the optimization process, the angular momentum is constrained to take its maximum value at the impact and to be monotonically reduced before/after the impact, as illustrated in Fig. 7.41. That is why it can be assumed that minimizing $\boldsymbol{l}_F^2$ will contribute to the minimization of $\dot{\boldsymbol{l}}_F$, as illustrated in Fig. 7.43A. On the contrary, during the optimization before/after the impact, the minimization of $\boldsymbol{l}_F^2$ at $t = t_I \pm \Delta t$ yields an increase in $\dot{\boldsymbol{l}}_F$ (cf. Fig. 7.43B). Note, however, that $\dot{\boldsymbol{l}}$ will be bounded by the joint velocity constraints given in (7.85).

From the reasons described above, it is possible to employ the quantity l_{Fy}^2 as a performance index for the stability margin evaluation goal, instead of the rate of change of the angular momentum, $\dot{l}_{Fy}$.

Optimization of the Posture and Velocity at the Impact

A constrained optimization problem is solved to find the optimal posture and velocity of the robot at an instant when the hand hits the wooden board in the Karate chop experiments. The Sequential Quadratic Programming (SQP) method is used to solve the constrained optimization problem. This method defines the objective function and the constraints as nonlinear functions of the design parameters. In this work, the `fmincon()` function of the numerical computing language MATLAB (MathWorks Inc.) was used to solve the SQP problem.

The objective function J_A is defined as follows:

$$J_A = w_A \left(1 - \frac{M_{eq}{}^I\mathbb{B}_c^T \mathcal{V}_I^+}{(M_{eq}{}^I\mathbb{B}_c^T \mathcal{V}_I^+)_{\max}}\right)^2 + (1 - w_A)\left(\frac{l_{Fy}}{l_{Fy,\max}}\right)^2. \tag{7.90}$$

The first term on the r.h.s. in (7.90) is minimized when $M_{eq}{}^I\mathbb{B}_c^T \mathcal{V}_I^+$ is maximized. The second term on the r.h.s., on the other hand, is minimized when l_{Fy} is minimized; w_A ($0 \le w_A \le 1$) is a weighting factor.

The maximum values $(M_{eq}{}^I\mathbb{B}_c^T \mathcal{V}_I^+)_{\max}$ and $l_{Fy,\max}$ are estimated in advance by solving the constrained optimization problem with objective functions $J_{A,MV} = -M_{eq}{}^I\mathbb{B}_c^T \mathcal{V}_I^+$ and $J_{A,l_{Fy}} = -l_{Fy}$, respectively. Minimizing $-M_{eq}{}^I\mathbb{B}_c^T \mathcal{V}_I^+$ or $-l_{Fy}$ yields the maximization of $M_{eq}{}^I\mathbb{B}_c^T \mathcal{V}_I^+$ or l_{Fy}, respectively.

The optimization problem is to find an optimal joint angle $\boldsymbol{\theta}$ and joint angular velocity $\dot{\boldsymbol{\theta}}$ at the impact under the following conditions:

$$\begin{aligned}
\text{minimize} \quad & J_A(\boldsymbol{\theta}, \dot{\boldsymbol{\theta}}), \\
\text{subject to} \quad & \theta_{i,\min} + \theta_m \le \theta_i(t) \le \theta_{i,\max} - \theta_m, \\
& -\dot{\theta}_{i,\max} \le \dot{\theta}_i(t) \le \dot{\theta}_{i,\max}, \\
& \alpha_{\min} \le \alpha(\boldsymbol{\theta}) \le \alpha_{\max}, \\
& \left(r_{\overleftarrow{pF}}\right)_{x,\min} \le \left(r_{\overleftarrow{pF}}(\boldsymbol{\theta})\right)_x \le \left(r_{\overleftarrow{pF}}\right)_{x,\max},
\end{aligned} \tag{7.91}$$

where α is the angle between the wooden board and the hand (Link 6) and θ_m is the safety margin obtained from the mechanical limits of the joint.

In this case study, the hand position was not included in the constraints in the optimization process. The hand position was calculated from the obtained joint angle $\boldsymbol{\theta}$. The wooden board was then placed at the calculated hand position for the experiments. It should not be too difficult, though, to employ an equality constraint for the hand position.

Furthermore, the initial posture was assigned in a heuristic way. In cases when the optimization did not converge, other initial postures were explored, heuristically. The optimization process may also get stuck into a local minimum. Therefore, several initial postures have to be explored to obtain the posture that yields the best J_A value.

Optimization of the Velocity Before/After the Impact

As illustrated in Fig. 7.41, the robot motion is obtained step by step decelerating the joint velocities from the moment of the impact. Two sets of constrained optimization problems are solved to find the optimal velocities of the robot at $t - \Delta t$ for the swing before the impact and at $t + \Delta t$ for the swing after the impact. The optimizations are iteratively invoked to generate the whole swing motion before and after the impact.

The objective function J_{BC} is defined as follows:

$$J_{BC} = w_{BC}\hat{\dot{\boldsymbol{\theta}}}^T\hat{\dot{\boldsymbol{\theta}}} + (1 - w_{BC})\left(\frac{\dot{l}_{Fy}}{\dot{l}_{Fy,\max}}\right)^2, \tag{7.92}$$

where w_{BC} $(0 \le w_{BC} \le 1)$ is a weighting factor. The normalized joint velocity $\hat{\dot{\boldsymbol{\theta}}}$ is defined as follows:

$$\hat{\dot{\boldsymbol{\theta}}} = \begin{bmatrix} \hat{\dot{\theta}}_1 & \hat{\dot{\theta}}_2 & \cdots & \hat{\dot{\theta}}_n \end{bmatrix}^T, \quad \hat{\dot{\theta}}_i = \frac{\dot{\theta}_i}{\dot{\theta}_{i,\max}}. \tag{7.93}$$

The optimization problem is to find the optimal joint velocities $\dot{\boldsymbol{\theta}}$ at $t - \Delta t$ and at $t + \Delta t$ under the following conditions:

$$\begin{aligned} \text{minimize} \quad & \begin{cases} J_A(\boldsymbol{\theta}, \dot{\boldsymbol{\theta}}) & (k \le k') \\ J_{BC}(\dot{\boldsymbol{\theta}}) & (k > k') \end{cases} \\ \text{subject to} \quad & \theta_{i,\min} \le \theta_i(t) \le \theta_{i,\max}, \\ & -\dot{\theta}_{i,\max} \le \dot{\theta}_i(t) \le \dot{\theta}_{i,\max}, \\ & -\ddot{\theta}_{i,\max} \le \ddot{\theta}_i(t) \le \ddot{\theta}_{i,\max}, \\ & \dot{l}_{Fy,\min} \le \dot{l}_{Fy} \le \dot{l}_{Fy,\max}, \end{aligned} \tag{7.94}$$

where k denotes the iteration number and k' is an arbitrary positive integer. When the humanoid robot follows exactly the generated motion, the hand will hit the wooden board exactly at the expected impact time. However, it is not realistic to expect the humanoid robot to follow the generated motion without any time delay. Therefore, the objective function J_A is invoked instead of J_{BC} during a small period around the estimated time of the impact. The period $\pm k'\Delta t$ corresponds to the time margin for the impact. Since the same objective function J_A is used for $k \le k'$, the joint velocity is not decelerated during the time interval when $k \le k'$.

The boundary condition for the joint acceleration is replaced by the boundary condition for the joint velocity, as given in (7.85). Therefore, the current joint velocity will be bounded by the joint velocity from the previous step plus/minus $\ddot{\theta}_{i,\max}\Delta t$. This boundary condition avoids the disparity of $\dot{\theta}_i$ at the optimization points. Since the joint velocity is minimized as J_{BC} is minimized, the joint velocity tends to be near the lower bound of (7.85) when $\dot{\theta}_i$ is positive. Consequently, the joint velocity will be gradually decelerated.

It should be noted, however, that there is no guarantee that $\dot{\boldsymbol{\theta}}$ will always converge to zero within the joint angle limits. If $\dot{\boldsymbol{\theta}}$ does not converge, $\ddot{\boldsymbol{\theta}}_{\max}$ is slightly increased and the optimization process is restarted.

As a result of the optimization, the joint velocity $\dot{\boldsymbol{\theta}}$ is obtained. Then, the current joint angle vector $\boldsymbol{\theta}$ is calculated from $\dot{\boldsymbol{\theta}}$ and the joint angle vector at the previous step, as given in (7.80). Note that the obtained joint trajectory after the impact (Optimization C) is just a reference trajectory, and hence, it does not guarantee stability after the impact. When the applied momentum $M_{eq}{}^I\mathbb{B}_c^T\mathcal{V}_I^+$ is completely transformed into the form of the applied impact, the impact will be maximized. Using the estimation of the maximum impact force, the net CoP/ZMP displacement can be estimated for the worst case, via (7.79).

TABLE 7.9 Conditions for the motion optimization

Parameters	Values
Δt (ms)	5
w_A, w_{BC}	0.03, 0.5
θ_m (rad)	0.523
$\alpha_{min}, \alpha_{max}$ (°)	−5, 5
$\left(r_{\overleftarrow{pF}}\right)_{x,min}, \left(r_{\overleftarrow{pF}}\right)_{x,max}$ (m)	0.010, 0.030
$\dot{l}_{Fy,min}, \dot{l}_{Fy,max}$ (Nm)	−0.20, 0.20
k'	5

TABLE 7.10 Success rate of the board breaking motion

Motion	Thickness (mm)	Trial	Success	Failure	Success rate (%)
Hand-made motion	3	10	10	0	100
	5	10	1	9	10
	7	5	0	5	0
Optimized motion	3	3	3	0	100
	5	10	9	1	90
	7	5	0	5	0

7.8.9 Experimental Verification of the Generated Impact Motion

Experiments on breaking artificial-wood boards using a hand-made motion and an optimized motion approach will be described in what follows. The parameters used to calculate the optimized motion are listed in Table 7.9. The optimized joint angles are obtained every 5 ms. Since the sampling time in the HOAP-2 humanoid robot controller is 1 ms, it is necessary to interpolate the optimized joint angles before using them as control inputs. In [78], a stabilizing control approach was proposed for the postimpact phase. This approach will not be employed in the experiments, though, since the differences between the hand-made motion and the optimized one will be better exhibited then.

Three artificial-wood boards of various thickness are used in the experiments. The success rates of breaking them are presented in Table 7.10. As apparent from the table, the success rate in the case of the hand-made motion for a board of 5 mm thickness was 10%, while the success rate for the optimized motion for the same board was 90%. Apparently, the success rate was drastically improved by using the proposed method.

The normal component of the velocity of the hand and the angular momentum in the y (pitch)-direction, calculated from the designed motion, are plotted in Fig. 7.44A and B, respectively (note that these are not experimental results). As shown in Fig. 7.44A, the hand's velocity of the optimized motion reaches a peak at the impact, while in the hand-made motion the hand's velocity reaches a peak before the impact. Furthermore, the angular momentum in the y-direction is kept at a minimum in the optimized motion. In the hand-made motion, on the other hand, vibration is observed that may cause the humanoid robot to fall.

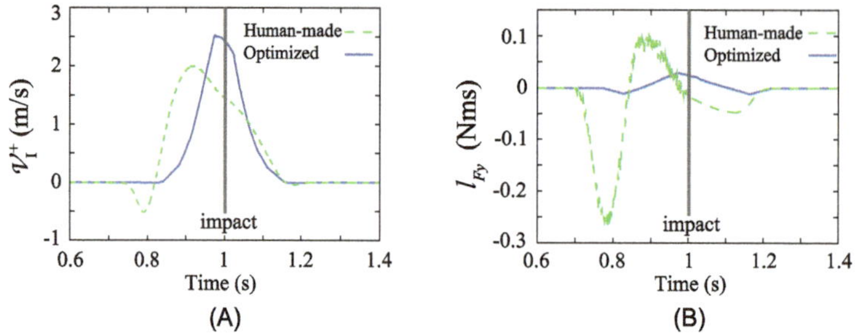

FIGURE 7.44 Comparison between the hand-made motion and optimized motion in: (A) velocity of the hand $\mathcal{V}_I^+$ and (B) angular momentum l_{Fy}.

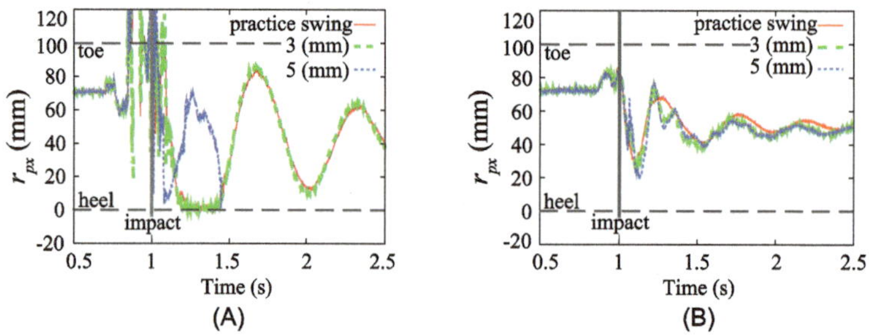

FIGURE 7.45 CoP along the x-axis at: (A) hand-made motion and (B) optimized motion.

In the experiments, the humanoid robot HOAP-2 was placed on a force plate manufactured by KISTLER Corp. to measure the net CoP. The CoP trajectories for the hand-made and the optimized motion are plotted in Fig. 7.45A and B, respectively. Since the humanoid robot slightly jumped up during the hand-made motion, the CoP plotted in Fig. 7.45A exceeds the supporting polygon at some points. These points represent unreliable data.

At the moment of the impact, the robot contacts the environment with its hand and feet. Therefore, at the moment of the impact, the CoP is not on the ground but on the supporting polygon formed by the hand and the feet. However, this is an instantaneous phenomenon as the board is immediately broken after the impact. Note that the CoP value plotted in Fig. 7.45A and B is measured with the force plate. Thus, strictly speaking, the CoPs in Fig. 7.45A and B at the moment of impact are incorrect.

In the hand-made motion, the CoP drastically varies after the impact, while in the optimized motion the CoP stays near the center of the support polygon. As mentioned above, the stabilizing control in [78] was not applied in these experiments. Therefore, the differences between the two generation approaches can be clearly seen. The CoP does not drastically vary in the optimized motion, even at the impact. The reason is that the board is successfully broken, and hence, the reaction force is not very large.

The values of $M_{eq}{}^I\mathbb{B}_c^T\mathcal{V}_I^+$ at the impact are presented in Table 7.11 for the cases of the hand-made and optimized motion. The snapshots of the experiments are presented in Figs. 7.46 and 7.47, Video 7.8-1 [61] was taken during the experiments.

TABLE 7.11 Estimated momentum

	M_{eq} (kg)	$\mathcal{V}_I^+$ (m/s)	$M_{eq}{}^I\mathbb{B}_c^T\mathcal{V}_I^+$ (Ns)	l_{Fy} (Nms)
Hand-made motion	0.097	1.46	0.141	−0.0158
Optimized motion	0.102	2.43	0.249	0.0245

(A) 0 s (B) 14/30 s (C) 17/30 s (D) 20/30 s

FIGURE 7.46 Snapshots of the hand-made motion. The thickness of the board was 5 mm. Snapshot (D) shows that the robot failed to break the board. The success rate was 10%.

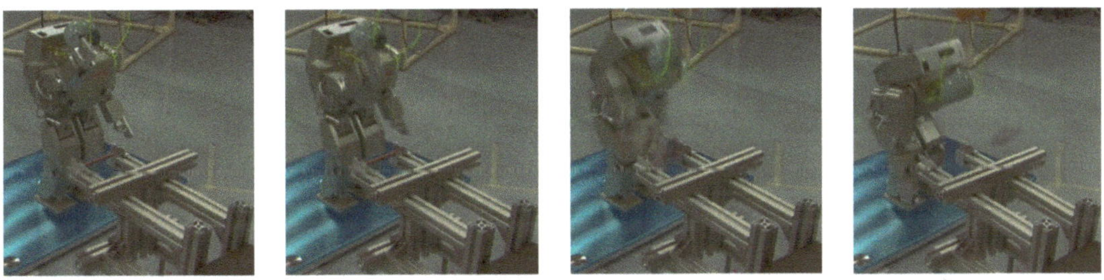

(A) 0 s (B) 14/30 s (C) 17/30 s (D) 20/30 s

FIGURE 7.47 Snapshots of the optimized motion. The thickness of the board was 5 mm. Snapshot (D) shows that the robot succeeded to break the board. The success rate was 90 (%).

References

[1] M. Abdallah, A. Goswami, A biomechanically motivated two-phase strategy for biped upright balance control, in: IEEE Int. Conf. on Robotics and Automation, Barcelona, Spain, 2005, pp. 1996–2001.

[2] Z. Aftab, T. Robert, P.-B. Wieber, Predicting multiple step placements for human balance recovery tasks, Journal of Biomechanics 45 (2012) 2804–2809.

[3] Y. Aoustin, A. Formalskii, Upward jump of a biped, International Journal of Humanoid Robotics 10 (2013) 1350032.

[4] H. Arisumi, J.-R. Chardonnet, A. Kheddar, K. Yokoi, Dynamic lifting motion of humanoid robots, in: Proceedings of IEEE International Conference on Robotics and Automation, Rome, Italy, 2007, pp. 2661–2667.

[5] H. Asada, K. Ogawa, On the dynamic analysis of a manipulator and its end effector interacting with the environment, in: Proceedings of IEEE International Conference on Robotics and Automation, 1987, pp. 751–756.

[6] D. Asmar, B. Jalgha, A. Fakih, Humanoid fall avoidance using a mixture of strategies, International Journal of Humanoid Robotics 09 (2012) 1250002.

[7] J.E. Bobrow, B. Martin, G. Sohl, E.C. Wang, F.C. Park, J. Kim, Optimal robot motions for physical criteria, Journal of Robotic Systems 18 (2001) 785–795.

[8] G. Borghesan, E. Aertbelien, J. De Schutter, Constraint- and synergy-based specification of manipulation tasks, in: 2014 IEEE International Conference on Robotics and Automation, ICRA, IEEE, 2014, pp. 397–402.

[9] K. Bouyarmane, A. Escande, F. Lamiraux, A. Kheddar, Potential field guide for humanoid multicontacts acyclic motion planning, in: IEEE International Conference on Robotics and Automation, Kobe, Japan, 2009, pp. 1165–1170.

[10] K. Bouyarmane, A. Kheddar, Static multi-contact inverse problem for multiple humanoid robots and manipulated objects, in: IEEE-RAS International Conference on Humanoid Robots, Nashville, TN, USA, 2010, pp. 8–13.

[11] K. Bouyarmane, A. Kheddar, Humanoid robot locomotion and manipulation step planning, Advanced Robotics 26 (2012) 1099–1126.

[12] T. Bretl, Motion planning of multi-limbed robots subject to equilibrium constraints: the free-climbing robot problem, The International Journal of Robotics Research 25 (2006) 317–342.

[13] T. Bretl, S. Lall, A fast and adaptive test of static equilibrium for legged robots, in: IEEE International Conference on Robotics and Automation, IEEE, 2006, pp. 1109–1116.

[14] T. Bretl, S. Lall, Testing static equilibrium for legged robots, IEEE Transactions on Robotics 24 (2008) 794–807.

[15] S. Brossette, A. Escande, G. Duchemin, B. Chrétien, A. Kheddar, Humanoid posture generation on non-Euclidean manifolds, in: IEEE-RAS International Conference on Humanoid Robots, 2015, pp. 352–358.

[16] A. Carmona, L. Molina-Tanco, M. Azuaga, J.a. Rodriguez, F. Sandoval, Online absorption of mediolateral balance disturbances for a small humanoid robot using accelerometer and force-sensor feedback, in: 2007 IEEE/ASME International Conference on Advanced Intelligent Mechatronics, IEEE, Tenerife, Spain, 2007, pp. 1–6.

[17] J.H. Choi, J. Grizzle, Planar bipedal robot with impulsive foot action, in: IEEE Conf. on Decision and Control, Atlantis, Paradise Island, Bahamas, 2004, pp. 296–302.

[18] R. Cisneros, K. Yokoi, E. Yoshida, Impulsive pedipulation of a spherical object for reaching a 3D goal position, in: IEEE-RAS International Conference on Humanoid Robots, Atlanta, GA, USA, 2013, pp. 154–160.

[19] A. D'Avella, E. Bizzi, Shared and specific muscle synergies in natural motor behaviors, Proceedings of the National Academy of Sciences 102 (2005) 3076–3081.

[20] J. Englsberger, T. Koolen, S. Bertrand, J. Pratt, C. Ott, A. Albu-Schaffer, Trajectory generation for continuous leg forces during double support and heel-to-toe shift based on divergent component of motion, in: IEEE/RSJ International Conference on Intelligent Robots and Systems, 2014, pp. 4022–4029.

[21] J. Englsberger, C. Ott, A. Albu-Schaffer, Three-dimensional bipedal walking control based on divergent component of motion, IEEE Transactions on Robotics 31 (2015) 355–368.

[22] J. Englsberger, C. Ott, M.A. Roa, A. Albu-Schaffer, G. Hirzinger, Bipedal walking control based on capture point dynamics, in: IEEE International Conference on Intelligent Robots and Systems, IEEE, 2011, pp. 4420–4427.

[23] A. Escande, A. Kheddar, S. Miossec, Planning support contact-points for humanoid robots and experiments on HRP-2, in: IEEE International Conference on Intelligent Robots and Systems, 2006, pp. 2974–2979.

[24] A. Escande, A. Kheddar, S. Miossec, Planning contact points for humanoid robots, Robotics and Autonomous Systems 61 (2013) 428–442.

[25] Fujitsu, Miniature Humanoid Robot HOAP-2 Manual, 1st edition, Fujitsu Automation Co., Ltd, 2004 (in Japanese).

[26] A. Goswami, Postural stability of biped robots and the foot-rotation indicator (fri) point, The International Journal of Robotics Research 18 (1999) 523–533.

[27] R.J. Griffin, A. Leonessa, A. Asbeck, Disturbance compensation and step optimization for push recovery, in: IEEE/RSJ International Conference on Intelligent Robots and Systems, IROS, IEEE, 2016, pp. 5385–5390.

[28] J. Grizzle, G. Abba, F. Plestan, Asymptotically stable walking for biped robots: analysis via systems with impulse effects, IEEE Transactions on Automatic Control 46 (2001) 51–64.

[29] M. Guihard, P. Gorce, Dynamic control of bipeds using ankle and hip strategies, in: IEEE/RSJ Int. Conf. on Intelligent Robots and Systems, Lausanne, Switzerland, 2002, pp. 2587–2592.

[30] K. Harada, K. Hauser, T. Bretl, J.-C. Latombe, Natural motion generation for humanoid robots, in: IEEE/RSJ International Conference on Intelligent Robots and Systems, 2006, pp. 833–839.

[31] H. Hauser, G. Neumann, A.J. Ijspeert, W. Maass, Biologically inspired kinematic synergies provide a new paradigm for balance control of humanoid robots, in: IEEE-RAS International Conference on Humanoid Robots, Pittsburgh, PA, 2007, pp. 73–80.

[32] H. Hauser, G. Neumann, A.J. Ijspeert, W. Maass, Biologically inspired kinematic synergies enable linear balance control of a humanoid robot, Biological Cybernetics 104 (2011) 235–249.

[33] K. Hauser, Fast interpolation and time-optimization with contact, The International Journal of Robotics Research 33 (2014) 1231–1250.

[34] K. Hauser, T. Bretl, K. Harada, J.-C. Latombe, Using motion primitives in probabilistic sample-based planning for humanoid robots, in: S. Akella, N.M. Amato, W.H. Huan, B. Mishra (Eds.), Algorithmic Foundation of Robotics VII, in: Springer Tracts in Advanced Robotics, vol. 47, Springer Berlin Heidelberg, Berlin, Heidelberg, 2008, pp. 507–522.

[35] K. Hauser, T. Bretl, J.C. Latombe, Non-gaited humanoid locomotion planning, in: IEEE-RAS International Conference on Humanoid Robots, 2005, pp. 7–12.

[36] R. Hinata, D. Nenchev, Anticipatory reflexive behavior in response to an impulsive force applied while leaning forward, Robotic Life Support Laboratory, Tokyo City University, 2018 (Video clip), https://doi.org/10.1016/B978-0-12-804560-2.00014-6.

[37] R. Hinata, D. Nenchev, Nonanticipatory reflexive behavior in response to an impulsive force applied while leaning forward, Robotic Life Support Laboratory, Tokyo City University, 2018 (Video clip), https://doi.org/10.1016/B978-0-12-804560-2.00014-6.

[38] R. Hinata, D.N. Nenchev, Accommodating a large impulsive force without stepping, Robotic Life Support Laboratory, Tokyo City University, 2018 (Video clip), https://doi.org/10.1016/B978-0-12-804560-2.00014-6.

[39] K. Hirai, M. Hirose, Y. Haikawa, T. Takenaka, The development of Honda humanoid robot, in: IEEE Int. Conf. on Robotics and Automation, Leuven, Belgium, 1998, pp. 1321–1326.

[40] H. Hirukawa, S. Hattori, K. Harada, S. Kajita, K. Kaneko, F. Kanehiro, K. Fujiwara, M. Morisawa, A universal stability criterion of the foot contact of legged robots – adios ZMP, in: IEEE International Conference on Robotics and Automation, 2006, pp. 1976–1983.

[41] A. Hofmann, M. Popovic, H. Herr, Exploiting angular momentum to enhance bipedal center-of-mass control, in: IEEE Int. Conf. on Robotics and Automation, Kobe, Japan, 2009, pp. 4423–4429.

[42] F.B. Horak, S.M. Henry, A. Shumway-Cook, Postural perturbations: new insights for treatment of balance disorders, Physical Therapy 77 (1997) 517–533.

[43] M. Hosokawa, T. Hamano, D. Nenchev, Lift-leg reactive synergy with DCM-GI based control, Robotic Life Support Laboratory, Tokyo City University, 2018 (Video clip), https://doi.org/10.1016/B978-0-12-804560-2.00014-6.

[44] M. Hosokawa, D.N. Nenchev, T. Hamano, The DCM generalized inverse: efficient body-wrench distribution in multi-contact balance control, Advanced Robotics 32 (2018) 778–792.

[45] Y. Hürmüzlü, G.D. Moskowitz, Bipedal locomotion stabilized by impact and switching: I. Two- and three-dimensional, three-element models, Dynamics and Stability of Systems 2 (1987) 73–96.

[46] Y. Hürmüzlü, G.D. Moskowitz, Bipedal locomotion stabilized by impact and switching: II. Structural stability analysis of a four-element bipedal locomotion model, Dynamics and Stability of Systems 2 (1987) 97–112.

[47] Y. Hwang, A. Konno, M. Uchiyama, Whole body cooperative tasks and static stability evaluations for a humanoid robot, in: Proceedings of IEEE/RSJ International Conference on Intelligent Robots and Systems, 2003, pp. 1901–1906.

[48] S.-H. Hyon, R. Osu, Y. Otaka, Integration of multi-level postural balancing on humanoid robots, in: IEEE International Conference on Robotics and Automation, Kobe, Japan, 2009, pp. 1549–1556.

[49] R. Iizuka, D.N. Nenchev, Reflex/recovery motion with ankle/hip/twist synergies, Robotic Life Support Laboratory, Tokyo City University, 2018 (Video clip), https://doi.org/10.1016/B978-0-12-804560-2.00014-6.

[50] R. Iizuka, D.N. Nenchev, Reflex/recovery motion with hip/twist synergies, Robotic Life Support Laboratory, Tokyo City University, 2018 (Video clip), https://doi.org/10.1016/B978-0-12-804560-2.00014-6.

[51] C. Inamura, T. Nakamura, D.N. Nenchev, Compliant whole–body balance control, Robotic Life Support Laboratory, Tokyo City University, 2018 (Video clip), https://doi.org/10.1016/B978-0-12-804560-2.00014-6

[52] R. Iizuka, D.N. Nenchev, Reactive stepping, Robotic Life Support Laboratory, Tokyo City University, 2018 (Video clip), https://doi.org/10.1016/B978-0-12-804560-2.00014-6.

[53] K. Iqbal, H. Hemami, S. Simon, Stability and control of a frontal four-link biped system, IEEE Transactions on Biomedical Engineering 40 (1993) 1007–1018.

[54] T. Izumi, Y. Kitaka, Control of a hitting velocity and direction for a hammering robot using a flexible link, Journal of the Robotics Society of Japan 11 (1993) 436–443 (in Japanese).

[55] S. Kagami, F. Kanehiro, Y. Tamiya, M. Inaba, H. Inoue, AutoBalancer: an online dynamic balance compensation scheme for humanoid robots, in: Int. Workshop Alg. Found. Robot, WAFR, 2000, pp. 329–339.

[56] S. Kajita, F. Kanehiro, K. Kaneko, K. Fujiwara, K. Harada, K. Yokoi, H. Hirukawa, Biped walking pattern generation by using preview control of zero-moment point, in: IEEE International Conference on Robotics and Automation, 2003, pp. 1620–1626.
[57] Y. Kanamiya, S. Ota, D. Sato, Ankle and hip balance control strategies with transitions, in: IEEE International Conference on Robotics and Automation, Anchorage, AK, USA, 2010, pp. 3446–3451.
[58] H. Khalil, Nonlinear Systems, 3rd edition, Prentice Hall, Upper Saddle River, New Jersey, USA, 2002.
[59] K. Kitagaki, M. Uchiyama, Optimal approach velocity of end-effector to the environment, in: Proceedings of IEEE International Conference on Robotics and Automation, 1992, pp. 1928–1934.
[60] S. Komizunai, Y. Onuki, M.H. Wu, T. Tsujita, A. Konno, A walking stabilization control for biped robots on loose soil, Journal of the Robotics Society of Japan 35 (2017) 548–556.
[61] A. Konno, Karate chop experiments with the HOAP-2 robot, 2015 (Video clip), https://doi.org/10.1016/B978-0-12-804560-2.00014-6.
[62] A. Konno, HOAP-2 walking on sand, 2016 (Video clip), https://doi.org/10.1016/B978-0-12-804560-2.00014-6.
[63] A. Konno, T. Matsumoto, Y. Ishida, D. Sato, M. Uchiyama, Drum beating a martial art Bojutsu performed by a humanoid robot, in: A.C. de Pina Filho (Ed.), Humanoid Robots: New Developments, Advanced Robotic Systems International and I-Tech Education and Publishing, ISBN 978-3-902613-00-4, 2007, pp. 521–530, Chapter 29.
[64] A. Konno, T. Myojin, T. Matsumoto, T. Tsujita, M. Uchiyama, An impact dynamics model and sequential optimization to generate impact motions for a humanoid robot, The International Journal of Robotics Research 30 (2011) 1596–1608.
[65] A. Konno, Y. Tanida, K. Abe, M. Uchiyama, A plantar h-slit force sensor for humanoid robots to detect the reaction forces, in: Proceedings of IEEE/RSJ International Conference on Intelligent Robots and Systems, 2005, pp. 1470–1475.
[66] T. Koolen, S. Bertrand, G. Thomas, T. de Boer, T. Wu, J. Smith, J. Englsberger, J. Pratt, Design of a momentum-based control framework and application to the humanoid robot Atlas, International Journal of Humanoid Robotics 13 (2016) 1650007.
[67] T. Koolen, T. de Boer, J. Rebula, A. Goswami, J. Pratt, Capturability-based analysis and control of legged locomotion, Part 1: theory and application to three simple gait models, The International Journal of Robotics Research 31 (2012) 1094–1113.
[68] S. Kudoh, T. Komura, K. Ikeuchi, The dynamic postural adjustment with the quadratic programming method, in: IEEE/RSJ International Conference on Intelligent Robots and System, 2002, pp. 2563–2568.
[69] M. Kudruss, M. Naveau, O. Stasse, N. Mansard, C. Kirches, P. Soueres, K. Mombaur, Optimal control for whole-body motion generation using center-of-mass dynamics for predefined multi-contact configurations, in: IEEE-RAS 15th International Conference on Humanoid Robots (Humanoids), 2015, pp. 684–689.
[70] J.J. Kuffner, S. Kagami, K. Nishiwaki, M. Inaba, H. Inoue, Dynamically-stable motion planning for humanoid robots, Autonomous Robots 12 (2002) 105–118.
[71] M.L. Latash, Synergy, Oxford University Press, 2008.
[72] S.M. LaValle, J.J. Kuffner Jr., Rapidly-exploring random trees: progress and prospects, in: Workshop on the Algorithmic Foundations of Robotics, Hanover, New Hampshire, 2000, pp. 293–308.
[73] D.J. Leith, W.E. Leithead, Survey of gain-scheduling analysis and design, International Journal of Control 73 (2000) 1001–1025.
[74] A. Liegeois, Automatic supervisory control of the configuration and behavior of multibody mechanisms, IEEE Transactions on Systems, Man and Cybernetics 7 (1977) 868–871.
[75] C.W. Luchies, N.B. Alexander, A.B. Schultz, J. Ashton-Miller, Stepping responses of young and old adults to postural disturbances: kinematics, Journal of the American Geriatrics Society 42 (1994) 506–512.
[76] B. Maki, W. Mcilroy, G. Fernie, Change-in-support reactions for balance recovery, IEEE Engineering in Medicine and Biology Magazine 22 (2003) 20–26.
[77] N. Mandal, S. Payandeh, Control strategies for robotic contact tasks: an experimental study, The International Journal of Robotics Research 12 (1995) 67–92.
[78] T. Matsumoto, A. Konno, L. Gou, M. Uchiyama, A humanoid robot that breaks wooden boards applying impulsive force, in: Proceedings of IEEE/RSJ International Conference on Intelligent Robots and Systems, 2006, pp. 5919–5924.
[79] M.-L. Mille, M.E. Johnson, K.M. Martinez, M.W. Rogers, Age dependent differences in lateral balance recovery through protective stepping, Clinical Biomechanics 20 (2005) 607–616.

[80] J.K. Mills, D.M. Lokhorst, Control of robotic manipulators during general task execution: a discontinuous control approach, The International Journal of Robotics Research 12 (1993) 146–163.

[81] S. Miyahara, D.N. Nenchev, VRP-GI based dynamic walk on a flat ground, Robotic Life Support Laboratory, Tokyo City University, 2018 (Video clip), https://doi.org/10.1016/B978-0-12-804560-2.00014-6.

[82] S. Miyahara, D.N. Nenchev, VRP-GI based dynamic walk on a staircase, Robotic Life Support Laboratory, Tokyo City University, 2018 (Video clip), https://doi.org/10.1016/B978-0-12-804560-2.00014-6.

[83] S. Miyahara, D.N. Nenchev, VRP-GI based dynamic walk with toe-off, Robotic Life Support Laboratory, Tokyo City University, 2018 (Video clip), https://doi.org/10.1016/B978-0-12-804560-2.00014-6.

[84] X. Mu, Q. Wu, On impact dynamics and contact events for biped robots via impact effects, IEEE Transactions on Systems, Man and Cybernetics. Part B. Cybernetics 36 (2006) 1364–1372.

[85] K. Nagasaka, M. Inaba, H. Inoue, Stabilization of dynamic walk on a humanoid using torso position compliance control, in: Proceedings of Annual Conference of Robotics Society of Japan, 1999, pp. 1193–1194 (in Japanese).

[86] K. Nagata, T. Ogasawara, T. Omata, Optimum velocity vector of articulated robot for soft bumping, Journal of the SICE 26 (1990) 435–442 (in Japanese).

[87] S. Nakaoka, Choreonoid: extensible virtual robot environment built on an integrated GUI framework, in: IEEE/SICE International Symposium on System Integration, SII, 2012, pp. 79–85.

[88] T. Narukawa, M. Takahashi, K. Yoshida, Stability analysis of a simple active biped robot with a torso on level ground based on passive walking mechanisms, in: M. Hackel (Ed.), Humanoid Robots, Human-Like Machines, I-Tech Education and Publishing, 2007, pp. 163–174.

[89] D. Nenchev, Y. Tsumaki, M. Uchiyama, Singularity-consistent parameterization of robot motion and control, The International Journal of Robotics Research 19 (2000) 159–182.

[90] D. Nenchev, K. Yoshida, Impact analysis and post-impact motion control issues of a free-floating space robot subject to a force impulse, IEEE Transactions on Robotics and Automation 15 (1999) 548–557.

[91] D.N. Nenchev, A. Nishio, Ankle and hip strategies for balance recovery of a biped subjected to an impact, Robotica 26 (2008) 643–653.

[92] A. Nishio, K. Takahashi, D. Nenchev, Balance control of a humanoid robot based on the reaction null-space method, in: IEEE/RSJ Int. Conf. on Intelligent Robots and Systems, Beijing, China, 2006, pp. 1996–2001.

[93] S. Onuma, D. Nenchev, Reactive motion in 3D obtained by super-positioning of simple synergies, Robotic Life Support Laboratory, Tokyo City University, 2015 (Video clip), https://doi.org/10.1016/B978-0-12-804560-2.00014-6.

[94] S. Ota, S. Onuma, D. Nenchev, Ankle-hip reactive synergies with transitions, Robotic Life Support Laboratory, Tokyo City University, 2011 (Video clip), https://doi.org/10.1016/B978-0-12-804560-2.00014-6.

[95] H.W. Park, A. Ramezani, J.W. Grizzle, A finite-state machine for accommodating unexpected large ground-height variations in bipedal robot walking, IEEE Transactions on Robotics 29 (2013) 331–345.

[96] J.L. Patton, M.J. Hilliard, K. Martinez, M.-L. Mille, M.W. Rogers, A simple model of stability limits applied to sidestepping in young, elderly and elderly fallers, in: IEEE EMBS Annual International Conference, New York City, NY, 2006, pp. 3305–3308.

[97] Q.-C. Pham, Y. Nakamura, Time-optimal path parameterization for critically dynamic motions of humanoid robots, in: 2012 12th IEEE-RAS International Conference on Humanoid Robots, IEEE, Osaka, Japan, 2012, pp. 165–170.

[98] Q.-C. Pham, O. Stasse, Time-optimal path parameterization for redundantly actuated robots: a numerical integration approach, IEEE/ASME Transactions on Mechatronics 20 (2015) 3257–3263.

[99] J. Pratt, T. Koolen, T. de Boer, J. Rebula, S. Cotton, J. Carff, M. Johnson, P. Neuhaus, Capturability-based analysis and control of legged locomotion, Part 2: application to M2V2, a lower-body humanoid, The International Journal of Robotics Research 31 (2012) 1117–1133.

[100] S. Rietdyk, A. Patla, D. Winter, M. Ishac, C. Little, Balance recovery from medio-lateral perturbations of the upper body during standing, Journal of Biomechanics 32 (1999) 1149–1158.

[101] M. Rijnen, E. de Mooij, S. Traversaro, F. Nori, N. van de Wouw, A. Saccon, H. Nijmeijer, Control of humanoid robot motions with impacts: numerical experiments with reference spreading control, in: IEEE International Conference on Robotics and Automation, ICRA, 2017, pp. 4102–4107.

[102] A. Shumway-Cook, F.B. Horak, Vestibular rehabilitation: an exercise approach to managing symptoms of vestibular dysfunction, Seminars in Hearing 10 (1989) 196–209.

[103] B.J. Stephens, Integral control of humanoid balance, in: IEEE/RSJ International Conference on Intelligent Robots and Systems, San Diego, CA, USA, 2007, pp. 4020–4027.
[104] K. Takase, Task execution by robot hand, Journal of the SICE 29 (1990) 213–219 (in Japanese).
[105] T. Takenaka, T. Matsumoto, T. Yoshiike, Real time motion generation and control for biped robot – 1st report: walking gait pattern generation, in: IEEE/RSJ International Conference on Intelligent Robots and Systems, 2009, pp. 1084–1091.
[106] K. Tamoto, R. Yui, R. Hinata, D.N. Nenchev, Motion synergy generation with multiple commands, Robotic Life Support Laboratory, Tokyo City University, 2017 (Video clip), https://doi.org/10.1016/B978-0-12-804560-2.00014-6.
[107] T.-J. Tarn, Y. Wu, N. Xi, A. Isidori, Force regulation and contact transition control, IEEE Control Systems 16 (1996) 32–40.
[108] M. Uchiyama, A Control Algorithm Constitution Method for Artificial Arm and Dynamic Control Modes, University of Tokyo Press, 1975 (in Japanese).
[109] J. Vaillant, A. Kheddar, H. Audren, F. Keith, S. Brossette, A. Escande, K. Bouyarmane, K. Kaneko, M. Morisawa, P. Gergondet, E. Yoshida, S. Kajita, F. Kanehiro, Multi-contact vertical ladder climbing with an HRP-2 humanoid, Autonomous Robots 40 (2016) 561–580.
[110] R. Volpe, P. Khosla, A theoretical and experimental investigation of impact control for manipulators, The International Journal of Robotics Research 12 (1993) 351–365.
[111] M. Vukobratović, Contribution to the study of anthropomorphic systems, Kybernetika 8 (1972) 404–418.
[112] M. Vukobratović, Legged Locomotion Robots and Anthropomorphic Mechanisms, Mihailo Pupin Institute, Belgrade, 1975.
[113] M. Vukobratović, B. Borovac, D. Šurdilović, Zero-moment point – proper interpretation and new applications, in: Proceedings of the IEEE/RAS International Conference on Humanoid Robots, 2001, pp. 237–244.
[114] M. Vukobratovic, D. Juricic, Contribution to the synthesis of biped gait, IEEE Transactions on Biomedical Engineering BME-16 (1969) 1–6.
[115] I.D. Walker, Impact configurations and measures for kinematically redundant and multiple armed robot system, IEEE Transactions on Robotics and Automation 10 (1994) 670–683.
[116] G. Wiedebach, S. Bertrand, T. Wu, L. Fiorio, S. McCrory, R. Griffin, F. Nori, J. Pratt, Walking on partial footholds including line contacts with the humanoid robot Atlas, in: IEEE-RAS International Conference on Humanoid Robots, 2016, pp. 1312–1319.
[117] D. Winter, Human balance and posture control during standing and walking, Gait & Posture 3 (1995) 193–214.
[118] D. Winter, A.E. Patla, J.S. Frank, Assessment of balance control in humans, Medical Progress Through Technology 16 (1990) 31–51.
[119] A. Witkin, M. Kass, Spacetime constraints, ACM SIGGRAPH Computer Graphics 22 (1988) 159–168.
[120] K. Yamane, Y. Nakamura, Synergetic CG choreography through constraining and deconstraining at will, in: IEEE International Conference on Robotics and Automation, Washington, DC, 2002, pp. 855–862.
[121] S.-J. Yi, B.-t. Zhang, D. Hong, D.D. Lee, Active stabilization of a humanoid robot for impact motions with unknown reaction forces, in: IEEE/RSJ International Conference on Intelligent Robots and Systems, IEEE, Vilamoura, Algarve, Portugal, 2012, pp. 4034–4039.
[122] S.-J. Yi, B.-T. Zhang, D. Hong, D.D. Lee, Whole-body balancing walk controller for position controlled humanoid robots, International Journal of Humanoid Robotics 13 (2016) 1650011.
[123] K. Yoshida, R. Kurazume, N. Sashida, Y. Umetani, Modeling of collision dynamics for space free-floating links with extended generalized inertia tensor, in: Proceedings of IEEE International Conference on Robotics and Automation, 1992, pp. 899–904.
[124] K. Yoshida, D.N. Nenchev, Space robot impact analysis and satellite-base impulse minimization using reaction null-space, in: Proceedings of IEEE International Conference on Robotics and Automation, 1995, pp. 1271–1277.
[125] Y. Yoshida, D.N. Nenchev, Ankle/Lift-leg lateral-plane reactive synergy from simple model, Robotic Life Support Laboratory, Tokyo City University, 2014 (Video clip), https://doi.org/10.1016/B978-0-12-804560-2.00014-6.
[126] Y. Yoshida, K. Takeuchi, Y. Miyamoto, D. Sato, D.N. Nenchev, Postural balance strategies in response to disturbances in the frontal plane and their implementation with a humanoid robot, IEEE Transactions on Systems, Man, and Cybernetics: Systems 44 (2014) 692–704.
[127] R. Yui, R. Hinata, D.N. Nenchev, Motion synergy generation with single command, Robotic Life Support Laboratory, Tokyo City University, 2018 (Video clip), https://doi.org/10.1016/B978-0-12-804560-2.00014-6.
[128] Y.-F. Zheng, H. Hemami, Mathematical modeling of a robot collision with its environment, Journal of Robotic Systems 2 (1985) 289–307.

CHAPTER

8

Simulation

8.1 OVERVIEW

By using a robot simulator, the time and cost for developing a controller of a humanoid robot can be reduced. At present, the employment of a robot simulator is indispensable for research on humanoid robots. This is the reason this chapter has been included at the end of this work. Robot simulators receive reference values (position, velocity, acceleration, torque) for the joints of a virtual robot. These quantities are generated by a controller designed by the user. In this way, the behavior of the robot can be simulated in a virtual world. The modeling and control methods described in the previous chapters can be tested beforehand in a robot simulator, without conducting experiments with the actual humanoid robot. Thus, it is highly beneficial to know how to use a robot simulator.

In order to control a humanoid robot, the user should prepare a control program for the intended task. First of all, the user should investigate thoroughly whether the control program for the task has already been developed and whether it is available to download from a server or not. If the source code is already available, it is possible to significantly shorten the development period, even if some modification is required. If a novel algorithm is proposed or there is no existing program, the development will proceed according to Fig. 8.1.

As a first step of the development process, a model or an algorithm for control is considered. The previous chapters of this book focus on this problem. In addition to the performance

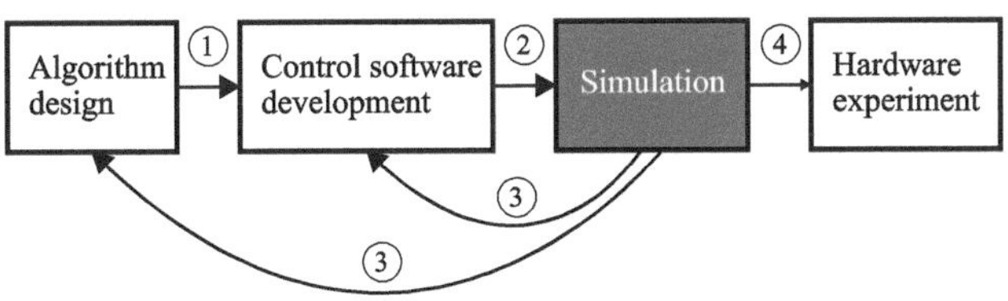

FIGURE 8.1 Common development flow of a robot controller. First, a model or an algorithm for control is considered. Second, whichever is chosen is implemented as a controller program. Third, the control program is verified in a robot simulator. If problems are found in the controller implementation or the algorithm at this stage, they should be reviewed. Finally, the controller is evaluated using an actual humanoid robot.

Humanoid Robots
https://doi.org/10.1016/B978-0-12-804560-2.00015-8

of the algorithm, another factor for efficient controller development is how easy it can be implemented.

The second step is to implement the model or the algorithm as a controller program. In many cases, at this phase the user would consider hardware-specific limitations, e.g. joint angle limits and joint torque limits. It is desirable to develop an universal controller by generalizing the hardware-specific problems, so that the controller can be used for various robots. Therefore, the hardware-specific information should not be hard-coded into the program. It is also desirable to prepare files that describe the structure of the robot. There are specific formats or programming languages for this purpose, such as the SDF (Simulator Description Format), the URDF (Universal Robotic Description Format), the YAML (YAML Ain't a Markup Language), and the VRML (Virtual Reality Modeling Language). The files can then be easily imported by the simulator program.

As a third step, the control program is to be evaluated in the robot simulator. Note that, when using a human-sized humanoid robot, the evaluation of the developed controller can be a tough job since the process of conducting experiments is time consuming and costly. Also, there is the risk of breaking the robot. The process of simulation is beneficial even when a small-size humanoid robot is used. For that reason the user can test the controller under development and obtain the status of the robot that is computed during the simulation. Even when the robot does not have the means for measuring the status, the user can analyze the status in the simulator and in this way debug the controller. Any problems found at this stage of the implementation of the controller algorithm should be reviewed.

At the final step, the developed control program is evaluated using the actual humanoid robot. This phase is indispensable since, despite a successful test result, the robot simulator cannot guarantee a robot behavior that matches completely that of the actual robot. There is always a possibility that, when using the actual robot, a phenomenon may occur that did not occur at the simulation stage. Thus, experiments with the actual robot have to be carried out with great attention.

This chapter is structured as follows. Section 8.2 gives an outline and introduces some well-known robot simulators. In Section 8.3, the structure of a generic robot simulator is described and shown in flowcharts. In Section 8.4 it is explained how to use Simscape Multibody in a step-by-step manner, so that the reader can conduct a simple simulation.

8.2 ROBOT SIMULATORS

A simulator plays a key role in mechanical design, control, motion planning, and so on. The main components of a robot simulator are: a model loader, a physics engine based on multibody dynamics, contact models, and a visualization interface. In many robot simulators, users can access the functions of the simulator using a programming language such as Python, C/C++, or MATLAB. Recent robot simulators also comprise a Graphical User Interface (GUI) for controlling the simulation. With a dynamics-based simulation of a humanoid robot, it is possible to evaluate postural stability, estimate the forces between the robot and the environment, check the actuator torques generated to perform the desired motion, and so on.

The user programs the controller, which communicates with the simulator, using a programming language. The controller is tested in the simulator and is modified, if necessary, according to the simulation results. Several robot simulators, e.g. Gazebo with the Robot Operating System (ROS) and Choreonoid, can transparently handle actual and virtual robots. In other words, a controller developed in the robot simulator for a target humanoid robot can be used in the actual robot without changing the source code.

In some robot simulators, it is possible to emulate external sensors mounted on the robot, e.g. a camera, a depth sensor, or force/torque sensors. The virtual sensors are helpful to develop controllers for object handling, mapping and localization algorithms, etc. A large number of robot simulators have been developed so far [30,31,10]. Some well-known simulators are listed below.

Gazebo: Gazebo is an open-source software for robot simulation running on Linux [35,18] which is commonly used with the (ROS) [52,1]. This simulator was used in the Defense Advanced Research Projects Agency (DARPA) Virtual Robotics Challenge (VRC) [2]. The core libraries for the physics engine in Gazebo are Open Dynamics Engine [48], Bullet [7], Simbody [56], and DART [11]. The Gazebo framework can easily switch between these physics engines. Since Gazebo communicates via ROS messages, the developed controllers can be used on actual machines without modifying the source code [54]. A number of control packages for the ROS environment have been developed in the community that can be used with Gazebo. For example, MoveIt! is a well-known package for motion planning [8,43].

Choreonoid: Choreonoid is an open-source robot simulator running on Linux and Windows. It includes a choreography function [44]. The core library for the physics engine in Choreonoid is the AIST engine developed by the National Institute of Advanced Industrial Science and Technology (AIST) in Japan [45]. The library was developed for the open architecture Humanoid Robotics Platform (OpenHRP) [32]. In addition to the AIST engine, the Open Dynamics Engine [48], Bullet [7], and PhysX [51] can also be used with Choreonoid. The functions of Choreonoid can be extended by plug-in modules. For example, graspPlugin can solve the grasp planning, trajectory planning, and task planning problems [24]. The ROS plug-in provides the communication interface with ROS. This simulator will be used in the Tunnel disaster response and recovery competition during the World Robot Challenge 2018 [62].

Vortex Studio: Vortex Studio, developed by CM Labs Simulations, is an interactive real-time multibody dynamics simulation software [67]. Unlike a nonreal-time multibody dynamics tool that focuses on motion generation in a virtual world only, it is a tool for physical simulation that requires human interactive manipulation input in real-time. The full-function version of this software is for profit, but Vortex Studio Essentials with restricted functionality can be used for free.

V-Rep: V-Rep is a commercial cross-platform robot simulator and runs on Windows, macOS, and Linux [53,65]. An unlimited educational version is available for free. Robot controllers for V-Rep can be written in C/C++, Python, Java, Lua, MATLAB (Mathworks, Inc.) [41], or Octave [21]. The core libraries for the physics engine in V-Rep are Open Dynamics Engine [48], Bullet [7], Vortex Dynamics [67], and Netwon Dynamics [47]. V-Rep also has an application programming interface (API) to communicate with ROS.

Webots: Webots is a commercial cross-platform robot simulator that runs on Windows, macOS, and Linux [68]. Webots can import robot models written in VRML formats and maps for virtual environment from OpenStreetMap [49] or Google Maps [22]. The humanoid robot models of Atlas [3], DARwIn-OP [25], HOAP-2 [27], Nao [23], etc. are included in Webots. The robot controller for Webots can be written in C/C++, Python, Java, or MATLAB. The core library for the physics engine in Webots is an extended version of the Open Dynamics Engine [48]. Webots also has an interface API to communicate with ROS.

FROST: FROST is an open-source MATLAB toolkit for developing model-based controllers and planning algorithms for dynamics of legged locomotion [26,17]. FROST is an integrated software for modeling, optimization, and simulation of robotics systems focusing on the hybrid dynamical system and virtual constraints. By using FROST with a Nonlinear Programming (NLP) solver and Mathematica (Wolfram Research, Inc.) [40], a motion planning problem can be solved as a nonlinear constrained optimization problem.

Simscape Multibody: Although the simulators described above are very sophisticated, it takes time to learn how to use the simulator, and it is not the shortest route to validate algorithms focusing on the modeling and control described in this book. Simscape Multibody [57] is a commercial multibody dynamics–based simulation software and a toolbox for Simulink [59] (Mathworks, Inc.). The advantage of Simscape Multibody is that simulation models can be written easily with Simulink-like block diagrams. Also, a number of useful MATLAB/Simulink functions and blocks can be utilized. The core libraries for the Simscape Multibody physics engine are commercial and implemented in MATLAB/Simulink. In addition, a connection with ROS is available, and it is possible to use various ROS functions via the Robotics System Toolbox. The usage of Simscape Multibody will be described in Section 8.4.

8.3 STRUCTURE OF A ROBOT SIMULATOR

As mentioned in Section 8.2, a robot simulator consists of a model loader, a physics engine, and a visualization program (cf. Fig. 8.2). The physics engine, or dynamics-based simulation program, can compute the behavior of the humanoid robot based on the multibody dynamics. In order to compute the behavior, the physics engine has to include components for rigid-body physics, collision detection, and contact physics [42]. Several physical engines include a model loader and visualization functions.

The following explanation is given according to Figs. 8.3 and 8.4. Figs. 8.3 and 8.4 show the flowcharts of inverse dynamics–based simulation and forward dynamics–based simulation, respectively. In the beginning of the simulation, the robot's structure and the simulation parameters are loaded. In the simulation loop, the first step is to check the contact state between the robot and the environment and to compute the contact force. As a second step, the robot controller computes the input joint acceleration for the inverse dynamics–based simulation or the input joint torques. In the third step, the dynamics simulator computes the motion of the

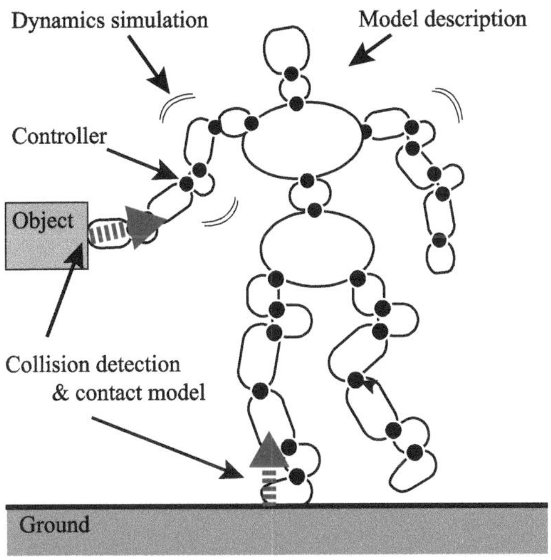

FIGURE 8.2 Required components for a robot simulator. The core libraries of the robot simulator are a model loader, a physics engine, and a visualization program. The physics engine can compute the behavior of the humanoid robot based on multibody dynamics. In order to compute the behavior, the physics engine has to include components for rigid-body physics, collision detection, and contact physics.

robot based on the input. Finally, the computed motion is visualized in the form of computer graphics. The details of each part are explained below.

Model loader: In the beginning of the simulation, the model loader parses the robot model, which includes the kinematic and dynamic parameters, and sends this information to the physics engine. There are various formats for describing the structure of a robot. YAML and VRML formats are supported by Choreonoid [5]. The Simulator Description Format (SDF) and Universal Robotic Description Format (URDF) are supported by Gazebo [38,64]. Simscape Multibody supports URDF, as explained in Section 8.4.1. In addition, the simulation parameters, e.g. the sampling time, the solver, and the contact model parameters, are loaded.

Collision detection & contact model: The collision detection program computes the interaction between the robot and the environment, or the self-collision of the robot [66,13]. By using this program, collision positions at the robot's links and the respective reaction forces can be computed based on the specific contact force model [16]. These values are substituted into the equation of motion of the robot. An implementation example of a collision detection procedure and a contact model is explained in Section 8.4.4.

Robot controller: The robot controller computes the joint-space reference values like joint torques, positions, velocities, and/or accelerations. In the case the controller computes reference joint torques, the robot motion is obtained by a forward dynamics–based simulation. This type of controller is mostly used in humanoid robots with torque-controlled actuators [12,28,60].

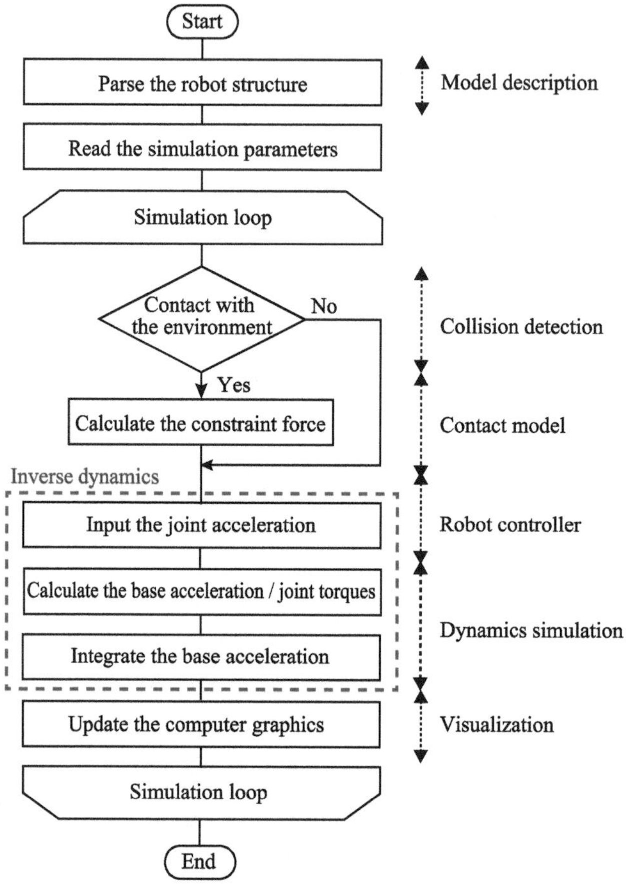

FIGURE 8.3 Flowchart of a robot simulation based on the inverse dynamics. In the beginning of the simulation, the robot's structure and the simulation parameters are loaded. In the simulation loop, the first step is to check the contact state between the robot and the environment and to compute the contact force. As a second step, the robot controller computes the input joint accelerations. In the third step, the dynamics simulator computes the motion of the robot based on the input. Finally, the computed motion is visualized in the form of computer graphics.

In the case the controller computes the joint positions, velocities, or accelerations, joint-servo controllers and joint models are required in the simulator. Fig. 8.5 shows an example of a joint-servo controller and a joint model implemented in a forward dynamics–based simulation. The joint model consists of an actuator, a driver circuit, and a reduction gear. The robot controller computes the reference joint angles while the servo controller controls the actual joint angle, according to the reference in Fig. 8.5. This type of controller is mostly used in a humanoid robot with position- or velocity-controlled actuators. Many motion-controlled humanoid robots have Harmonic drive reduction gears [33,36,50,55]. Models of such gears can be found in [34] for example. Several small-size humanoids have small, smart actuators that consist of a microcon-

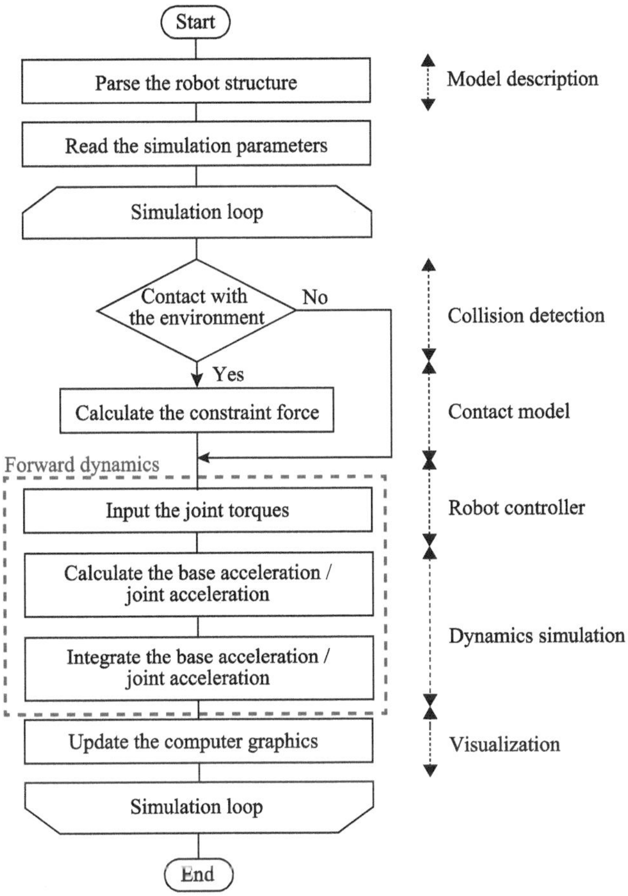

FIGURE 8.4 Flowchart of a robot simulation based on the forward dynamics. In the beginning of the simulation, the robot's structure and the simulation parameters are loaded. In the simulation loop, the first step is to check the contact state between the robot and the environment and to compute the contact force. As a second step, the robot controller computes the input joint torques. In the third step, the dynamics simulator computes the motion of the robot based on the input. Finally, the computed motion is visualized in the form of computer graphics.

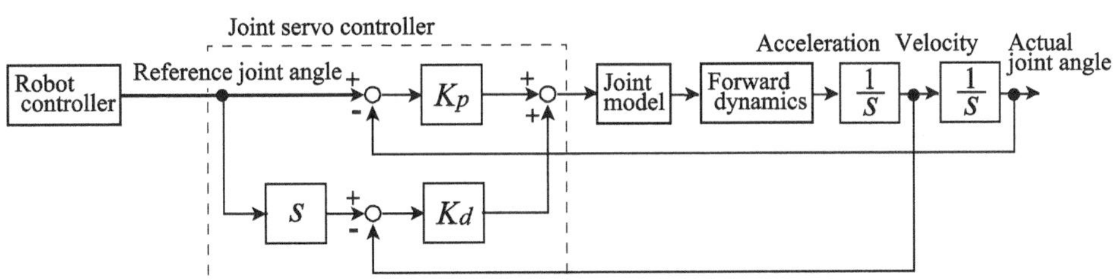

FIGURE 8.5 An example of a joint-servo controller for forward dynamics–based simulation. The joint model consists of an actuator, a driver circuit, and a reduction gear. The robot controller computes the reference joint angles and the servo controller controls the joint angle according to the reference.

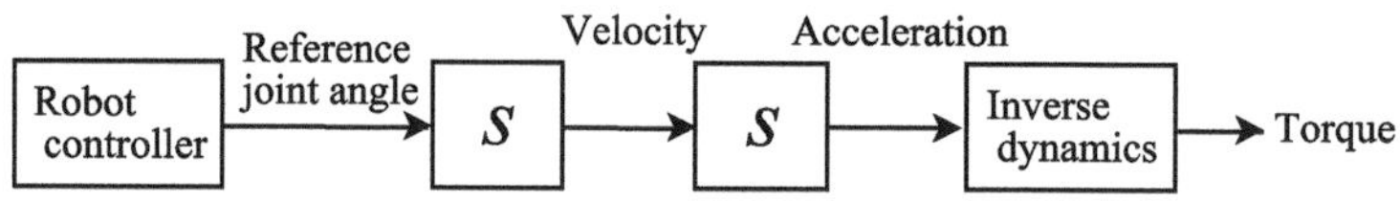

FIGURE 8.6 An example of a joint-servo controller for inverse dynamics–based simulation. The robot controller computes the reference joint angles. The second time derivative of the reference angles (acceleration) is inserted into the inverse dynamics simulator.

troller, a motor, and a reduction gear [4,25]. A well-known smart actuator is the Dynamixel AX-12, which is modeled in [37]. Also, models of the actuator and the driver are widely used, not only in the field of robotics but also in many other fields. In humanoid robots, brushless DC motors and drivers are mostly used because of their high durability. Respective models can be found in [6] for example.

In some cases, the trajectories of the joint angles are given by the robot controller and the required joint torques are computed by inverse dynamics–based simulation. In such cases, the user obtains information about the required torque and the postural stability of the robot, based on the assumption that all actuators track perfectly the reference trajectories. Fig. 8.6 shows an example of a joint-servo controller for inverse dynamics–based simulation. The robot controller computes the reference joint angles and supplies the second derivatives of the reference angles to the dynamics simulator, as shown in Fig. 8.6. In Section 8.4, it is explained how to conduct this type of simulation using MATLAB/Simulink.

Dynamics-based simulation: In inverse dynamics–based simulation, the joint acceleration is the input and the joint torque is the output. On the other hand, in forward dynamics–based simulation, the joint torque is the input and the joint acceleration is the output. The details of dynamics-based simulation are explained in Chapter 4. Generally, the complexity of computing the joint and base acceleration from (4.1) is $O(N^3)$, where N denotes the DoF in the forward dynamics–based simulation [15]. In order to accelerate the development of the controller, the robot simulation needs to be performed in a short time. Several high-speed computation methods have been developed. For example, forward dynamics based on propagation methods have computational complexity of $O(N)$ [15]. By utilizing parallel computing, $O(\log(N))$ algorithms have been developed [14,70].

By solving the equation of motion considering the contact force, accelerations in the joints and the base of the robot can be obtained. By integrating the joint and base accelerations numerically, joint velocities, angles, and the whole-body motion can be obtained.

Visualization: Visualization is an important part of the robot simulator. A humanoid robot usually comprises a number of joints and links. The numerous plots of their states are not easy to understand. Three-dimensional computer graphics can help the user to know what will happen when executing the designed motion. For the display of computer graphics, there are useful libraries. OpenGL [46], for example, is one of the graphics APIs for rendering, texture mapping, etc. Being a cross-language and a cross-platform, it is widely used in robot simulators. OpenGL is a low-level library that is usually assessed via toolkits compatible with Open Inventor [69], such as GLUT [20], GLFW [19] or Coin3D [9].

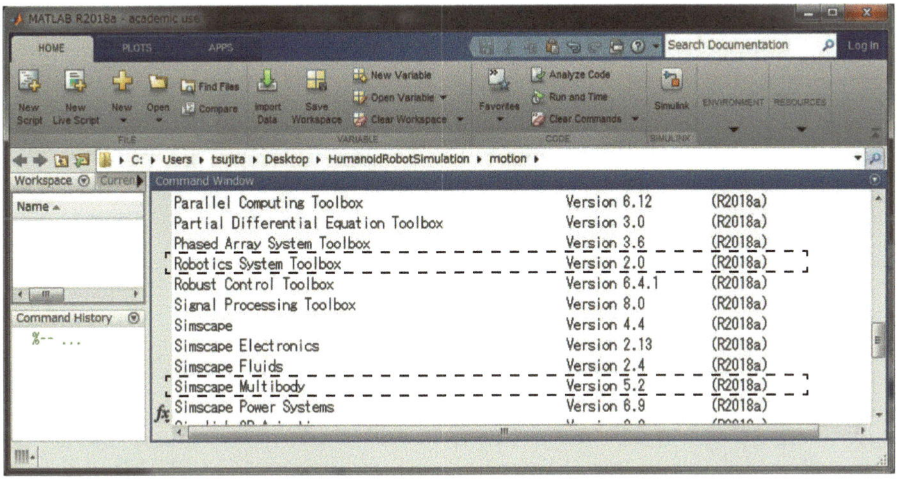

FIGURE 8.7 An example of a list with licensed components. This figure shows the output in the command window of MATLAB, generated by the "ver" command. The components in the dotted rectangles should be displayed.

8.4 DYNAMICS SIMULATION USING MATLAB/SIMULINK

In this section, it is explained how to perform dynamics-based simulation using MATLAB/Simulink with Simscape Multibody and the Robotics System Toolbox. Simscape Multibody and the Robotics System Toolbox are used to perform dynamics-based simulation and to generate the motion of the humanoid robot, respectively. To check whether Simscape Multibody and the Robotics System Toolbox are available, enter the following command in the MATLAB command window:

```
>>ver
```

As a result, the components shown in Fig. 8.7 should be displayed.

8.4.1 Generating a Robot Tree Model for Simulink

In general, it is possible to describe the structure of a robot with Simscape alone. Since the structure of a humanoid robot is quite complex, this would be quite a tedious task, though. It is much more preferable to make use of the robot CAD files or the Unified Robot Description Format (URDF) file. In what follows, it will be assumed that either is available. In the case the robot is designed by the user, the following subsection will be helpful. In the case a commercial robot that can be used with ROS is at hand, the subsection named "Using a URDF file" will be helpful.

Using CAD Files

Simscape Multibody can import the structure of the robot from its CAD files via Simscape Multibody Link [58]. At the time being, Simscape Multibody Link supports typical mid-range CAD software packages like Solidworks (Dassault Systemes SolidWorks Corp.), Creo (PTC), and Autodesk Inventor (Autodesk Inc.). Simscape Multibody can import XML (eXtensible Markup Language) files that contain the structure of a robot. In order to export an XML file from the CAD files, the Simscape Multibody Link plug-in for the respective CAD software package must be downloaded and installed. With an appropriate maintenance contract for Simscape Multibody, it should be possible to download the plug-in from the Mathworks site [58]. The method of installing the plug-in for each CAD software package is explained in the documentation of the Simscape Multibody Link plug-in [29]. Here, a brief explanation is given for the following environment.

Operating system	**Windows 7 x64 professional**
CAD software	SolidWorks 2016 x64
Simulation software	MATLAB/Simulink R2018a x64

Step 1. Download [58] and save the archived file smlink.r2018a.win64.zip and the MATLAB file install_addon.m in a convenient folder. If the version of MATLAB is different, read the name of the ZIP file as appropriate.

Step 2. Right-click on the MATLAB icon and run with administrator privileges.

Step 3. In the command window of MATLAB, add the folder containing the downloaded ZIP file to the MATLAB path with the "addpath" command and install the add-on with the "install_addon" command:

```
>>addpath('D:\tmp')
>>install_addon('smlink.r2018a.win64.zip')
```

D:\tmp is the folder which contains the ZIP file.

Step 4. In order to connect via the Simscape Multibody Link plug-in, MATLAB must be registered as an automation server. In the command window of MATLAB, type the following command:

```
>>regmatlabserver
```

Step 5. To enable the Simscape Multibody Link plug-in for SolidWorks, enter the following command in the command window:

```
>>smlink_linksw
```

After successfully registering the plug-in, start SolidWorks.

Step 6. Open a dialog box by selecting the "Add-Ins" in the tools menu of SolidWorks. Select the "Simscape Multibody Link" check box, as shown in Fig. 8.8, and click on the "OK" button.

The abovementioned procedure completes the installation of the Simscape Multibody Link plug-in. The plug-in is ready now to export the robot model from SolidWorks.

Next, it is explained how to assemble a robot with SolidWorks. In the SolidWorks assembly file, set a mate entity so that it is recognized as a joint in Simscape. An example of a hinge joint between two links is shown in Fig. 8.9. In the figure, the "Concentric" mate is set around the axis of the hinge, and the "Coincident" mate is set on the front surface of Link 1 and the back surface of Link 2. After these settings, Link 2 will be able to rotate freely in the SolidWorks assembly window. Also, the joint will be recognized in Simscape Multibody Link. Repeat this

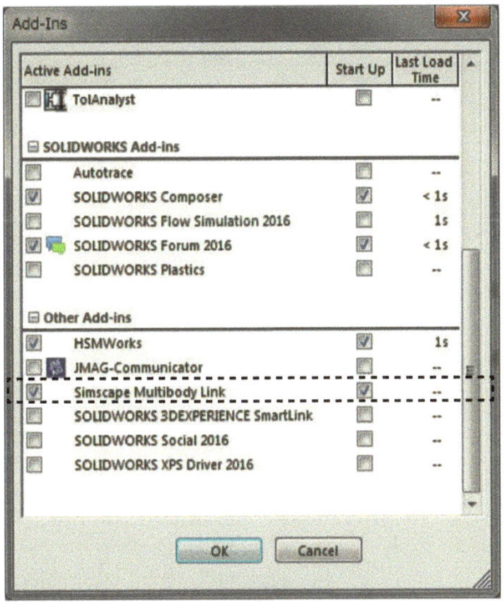

FIGURE 8.8 An example of the "Add-Ins" dialog box in SolidWorks. In order to be able to communicate with Simulink, check the "Simscape Multibody Link" box in the dotted rectangle.

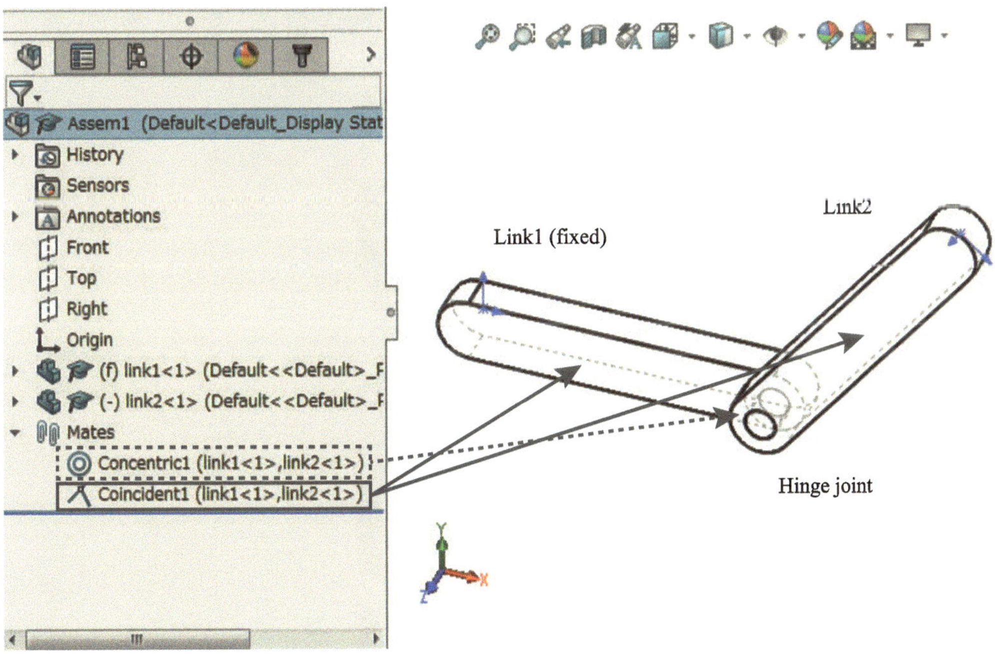

FIGURE 8.9 Example of a hinge joint setting in SolidWorks. In order to recognize the two parts as a hinge joint in Simscape, the "Concentric1" mate is set around the axis of the hinge, and the "Coincident1" mate is set on the front surface of Link 1 and the back surface of Link 2.

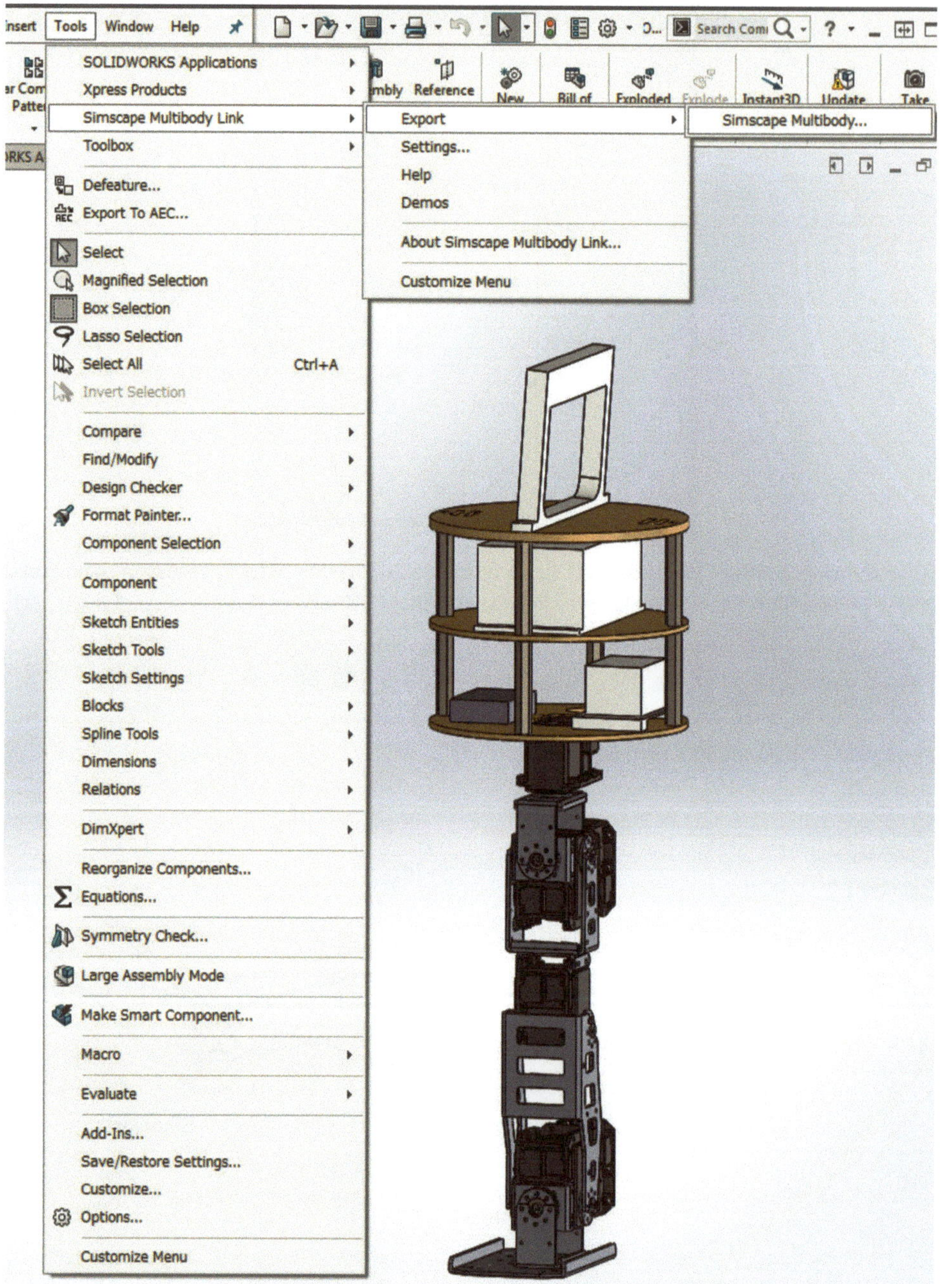

FIGURE 8.10 Design of a one-legged robot [61] using SolidWorks. The materials of all parts of this robot are specified for configuring the mass, the CoM, and the inertia.

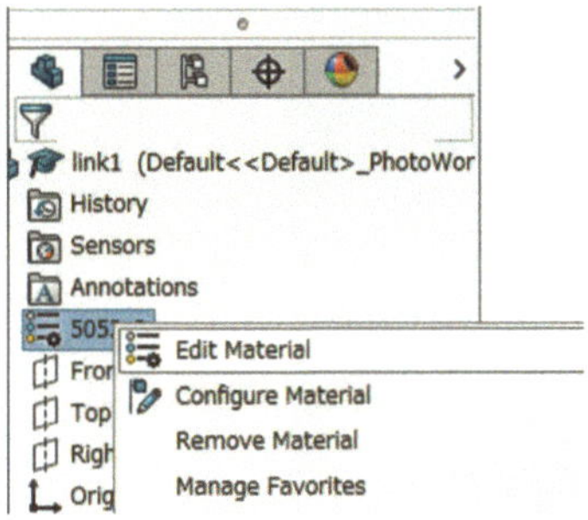

FIGURE 8.11 Menu for editing the material in the "FutureManager Design Tree." By selecting the "Edit Material," the "Material setting" window (Fig. 8.12) pops up.

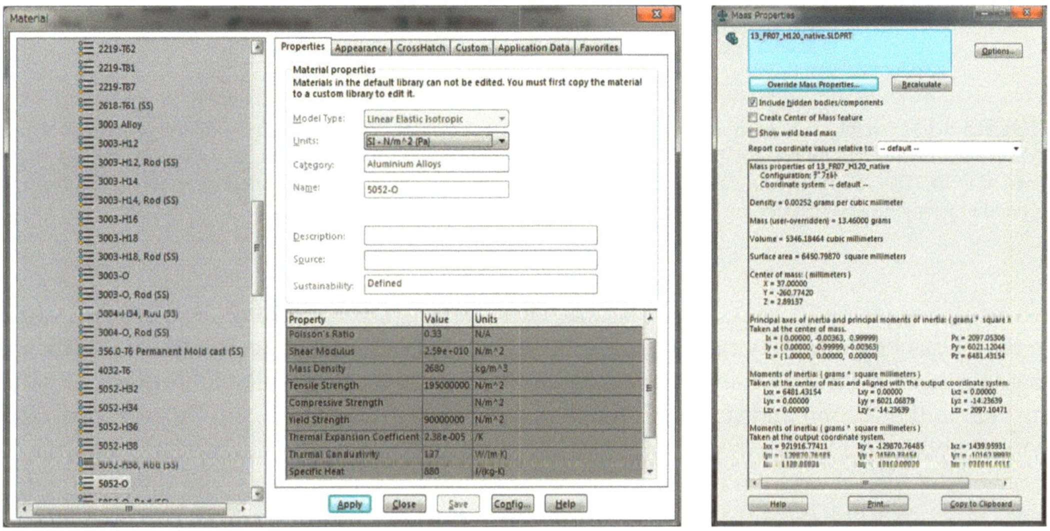

(A) (B)

FIGURE 8.12 Setting the mass properties of each rigid-body part. (A) Left: The "Material" window for specifying the material of a rigid-body part. The material can be chosen from a list. (B) Right: The "Mass Properties" window shows the CoM, the principal axes of inertia, the principal moments of inertia, and the moments of inertia of a rigid-body part. Note that there are other components, such as actuators, whose density is not constant. In this case, override the mass properties by clicking on the "Override Mass Properties" button.

procedure for each humanoid robot joint and set all joints to rotate freely. Note that, in the case the robot comprises other types of joints, it is necessary to set the mates according to the respective joint type [39].

Fig. 8.10 shows an example of a robot design [61]. To determine the mass, center of gravity position, and inertia tensor of a given part, it is necessary to select the "Edit Material" in the "FutureManager Design Tree" at the left end of the window, as shown in Fig. 8.11. The material density of each rigid-body part should be specified as close as possible to the actual material density (cf. Fig. 8.12A). Note that there are other components, such as actuators,

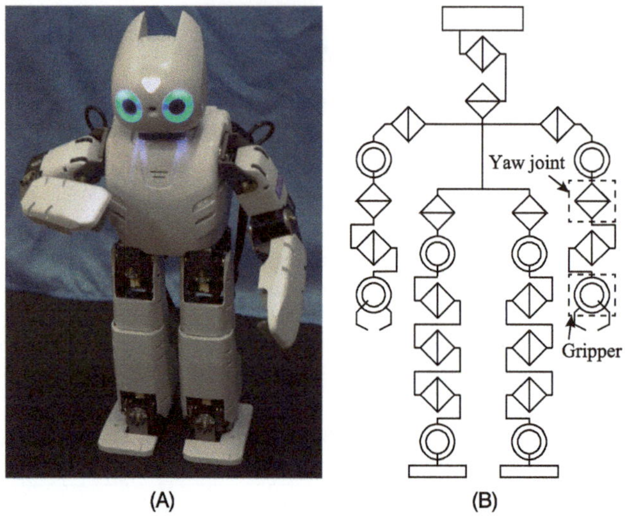

FIGURE 8.13 Small-size humanoid robot ROBOTIS-OP2. Its URDF file, used in this example, is available for free. (A) Left: overview of ROBOTIS-OP2 (basic arm model) having the same mechanics as DARwIn-OP [25]. (B) Right: DoFs of ROBOTIS-OP2 including yaw joints and grippers. The URDF files are written based on this structure. The appended yaw joint and gripper are in dotted rectangles.

whose density is not constant. In this case, click on the "Override Mass Properties" button in the "Mass Properties" window (cf. Fig. 8.12B) and specify the value. In order to check the mass properties, click on the "Mass Properties" button which is available from the "Evaluate" tab in the "CommandManager" tool bar. As shown in Fig. 8.12B, the "Mass Properties" window pops up and shows the center of mass, the principal axes of the inertia, the principal moments of inertia, and the moments of inertia of the rigid-body part. With this setting, the mass properties will be sent to the Simscape tree to be described later.

The next step is to generate the XML file and the STEP files via the Simscape Multibody Link plug-in for importing the structure and the shape of the designed robot. To this end, open Simscape Multibody Link by selecting the "Simscape Multibody Link" → the "Export" → the "Simscape Multibody" from the "Tools" menu in the menu bar of the SolidWorks assembly window, as shown in Fig. 8.10, and specify a name for saving. The saved XML and STEP files describe the structure of the robot and the shape of the links, respectively. The robot model is now ready to be loaded with Simscape Multibody.

Using the URDF File

Some robots comprise a URDF file that is written in XML. Simscape can parse the structure and mass properties of a robot described by the respective URDF file. The URDF files of commonly used robots are readily available from the ROS package.

In this section, the robot model of ROBOTIS-OP2 is used as an example. ROBOTIS-OP2 is a small-size humanoid robot (cf. Fig. 8.13A) having the same mechanics as DARwIn-OP [25],

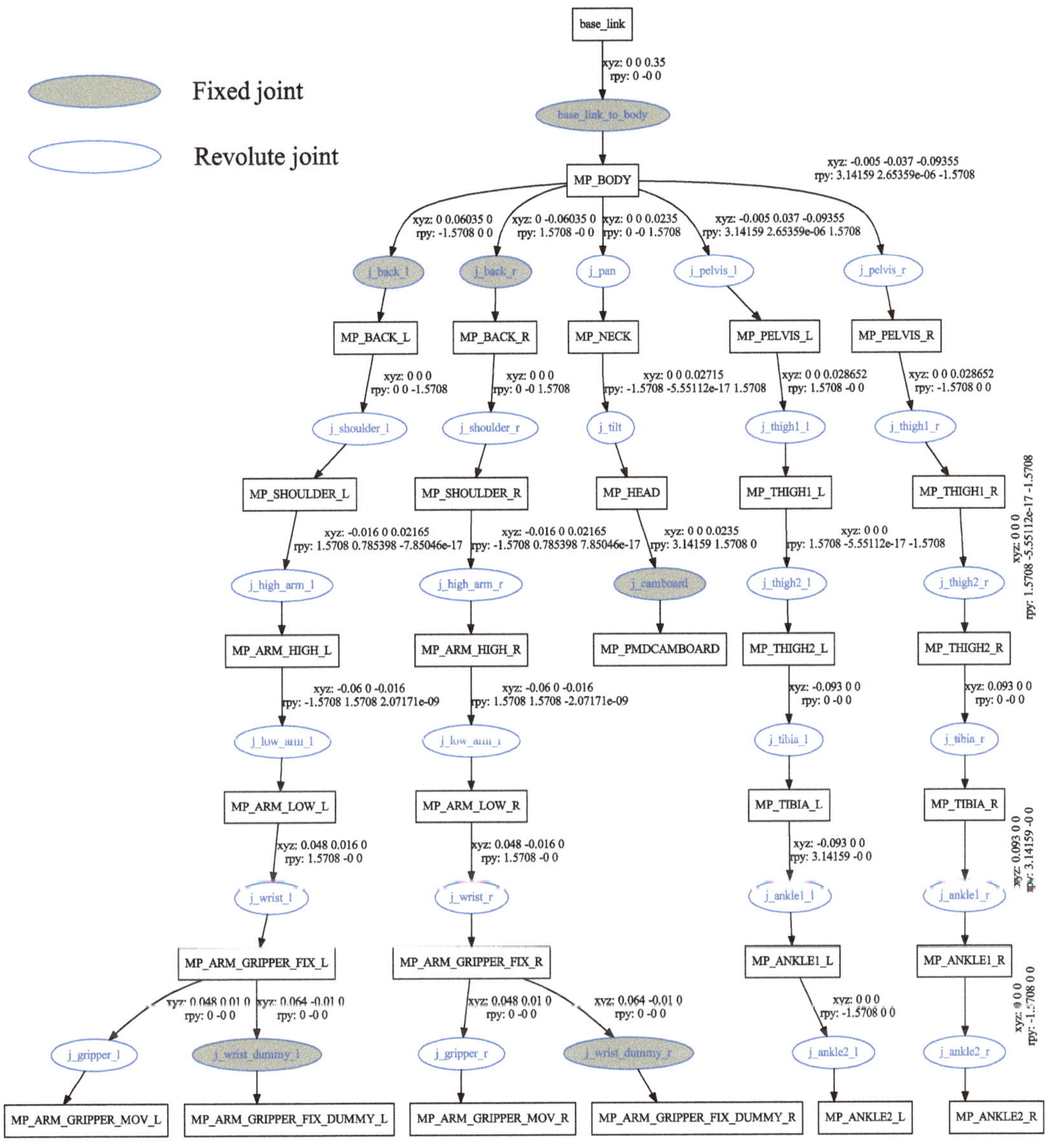

FIGURE 8.14 Structure of ROBOTIS-OP2. Fixed and revolute joints are denoted by gray and white ovals, respectively. A body that forms a joint is denoted by a white rectangle. An arrow expresses the connection between a joint and the body. The numerical value next to the arrow indicates the relationship for translation and orientation between the joint and the body.

with an enhanced embedded computer. Fig. 8.13B shows the DoFs of ROBOTIS-OP2, including yaw joints and grippers. The URDF files distributed on GitHub [63] are written based on this structure. Fig. 8.14 shows the connection between the links and the joints based on

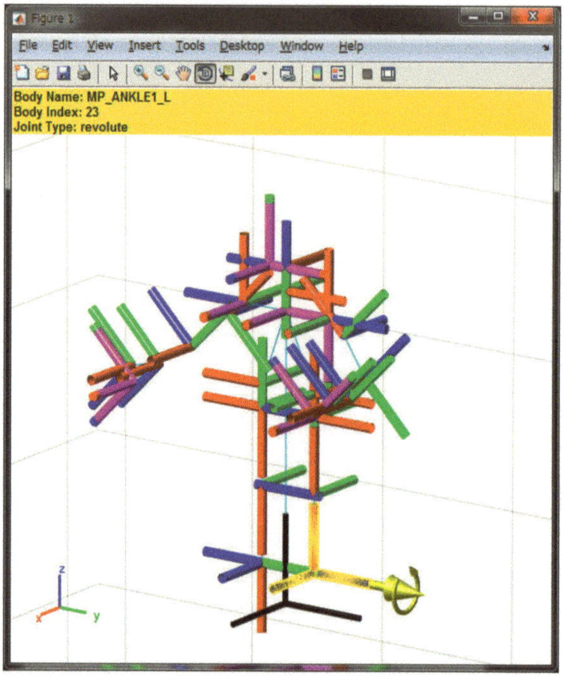

FIGURE 8.15 Visualization of the frames of ROBOTIS-OP2. By using the Robotics System Toolbox, the defined frames in the URDF files can be visualized as shown in this figure. Fixed joint frames are shown in magenta. Movable joint frames have three colors. The red, green, and blue cylinders denote the x-, y-, and z-axes, respectively. When selecting a movable joint frame, the frame is highlighted in yellow and the definition of the rotation axis is expressed by an arrow. The name of the frame is displayed in the upper part of the window.

the URDF file. Fixed and revolute joints are denoted by gray and white ovals, respectively. A body which is connected to a joint is denoted by a white rectangle. An arrow expresses the connection between a joint and the body. The numerical value next to the arrow indicates the relationship of translation/orientation between the joint and the body.

The joint frames of the robot can be visualized as shown in Fig. 8.15, by typing the following command (included in the Robotics System Toolbox):

```
>>cd 'location_of_URDF_files';
>>op2=importrobot('robotis_op.urdf');
>>op2.show('visuals','off')
```

where the "importrobot" command imports the rigid-body tree from the URDF file.

The colors of the x-, y-, and z-axes are according to the RGB convention; the fixed joints are shown in magenta. When selecting a movable joint frame, the frame is highlighted in yellow and the definition of the rotation axis is expressed by an arrow. The name of the frame is displayed in the upper part of the window.

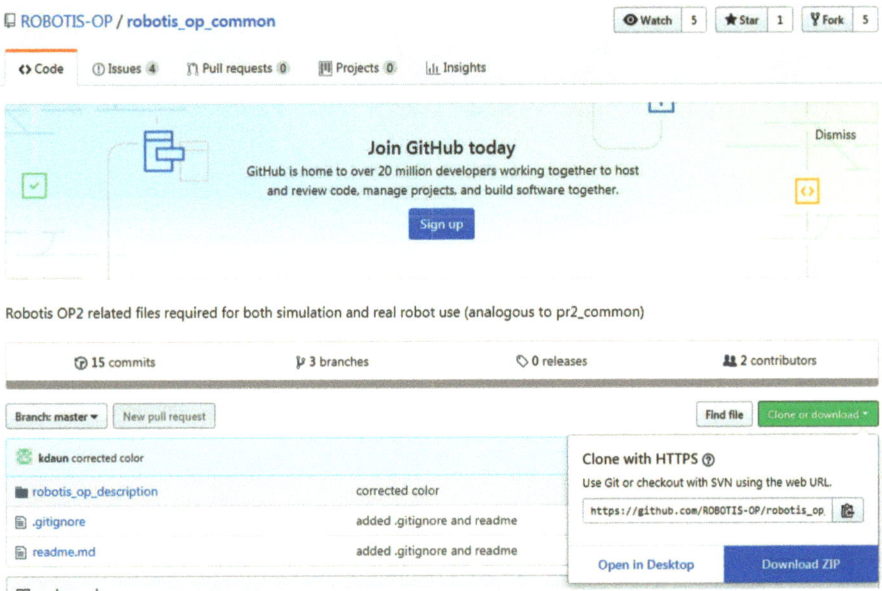

FIGURE 8.16 A URDF file and mesh files distributed on GitHub [63]. Click on the "Clone or download" button, and then click on the "Download ZIP." The URDF file and the mesh files can be obtained by extracting the downloaded file.

8.4.2 Generating the Simulink Model

In the case of an XML file, exported through the Simscape Multibody Link plug-in and imported into Simscape, type the following commands in the MATLAB window:

```
>>cd 'location_of_exported_files'
>>smimport('robot.xml')
```

In the case of a URDF file imported into Simscape, type the following commands:

```
>>cd 'location_of_URDF_file'
>>smimport('robot.urdf')
```

The details of using the URDF file of ROBOTIS-OP2 are explained below. From GitHub [63], download the ZIP file containing the URDF file that includes the robot's structure and mass properties and the mesh file that describes the appearance of each part. Click on the "Clone or download" button; then click on the "Download ZIP" as shown in Fig. 8.16 and extract the downloaded file. Place the URDF file and the STL files on the folder tree, as shown in Fig. 8.17. In order to be able to read the mesh file in Mechanics Explorer (the viewer of Simscape), set the path to "D:\MATLAB\HRS\RobotModels\robotis_op_description\meshes\" as in this example. The "Set Path" button is available from the "HOME" tab in the MATLAB main window. Next, type the following commands in the MATLAB window:

```
>>cd D:\MATLAB\HRS\RobotModels\robotis_op_description\robots\
>>smimport('robotis_op.urdf')
```

```
Program folder (ex. D:/MATLAB/HRS/)
├──── RobotModels/
│      └──── robotis_op_description/
│             ├──── robots/
│             │      └────robots_op.urdf
│             └──── meshes/
│                    └────STL files (ankle1_l.STL, ···, tibia_r.STL)
├────op2_motion.m: MATLAB script for generating a motion.
├────op2_simulation.slx: Simulink diagram for dynamics-based simulation.
└────MyLibs.slx: Simulink block library.
```

FIGURE 8.17 Example of a folder tree. Place the URDF file and the STL files downloaded from GitHub [63] on this folder tree. The RobotModels folder contains specific files for the robot. The MATLAB script file named op2_motion.m is a program file for generation of motion (cf. Section 8.4.6). The op2_simulation.slx and MyLibs.slx are simulink diagrams for dynamics-based simulation (cf. Section 8.4.2).

TABLE 8.1 Definition of the Simscape blocks used in the Simscape tree shown in Fig. 8.18. These blocks describe the mechanical properties and the simulation configuration

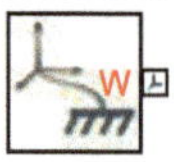	World frame	This block represents the inertial reference frame.
	Mechanism configuration	This block sets mechanical and simulation parameters that apply to an entire machine.
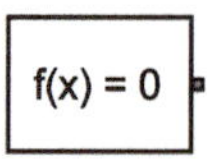	Solver configuration	This block defines the solver settings to use for simulation.
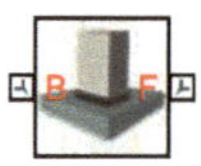	Weld joint	This block represents a fixed joint between two frames.
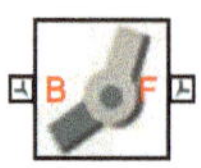	Revolute joint	This block represents a revolute joint acting between two frames.
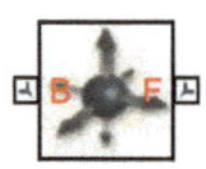	6-DOF joint	This block represents a 6-DoF joint acting between two frames.

The structure of ROBOTIS-OP2 imported into Simscape is shown in Fig. 8.18. The model tree is divided into right leg, left leg, neck, right hand, and left hand. The areas surrounded by a square box are grouped by the "area" button at the left end of the Simulink editor. The definition of each block is described in Table 8.1.

It can be seen that the base link (the "base_link") is fixed to the world frame (the inertial frame) and that five branches originate at the torso in Fig. 8.18. The upper two branches are the lower limbs, the central branch is the head, and the lower two branches are the upper

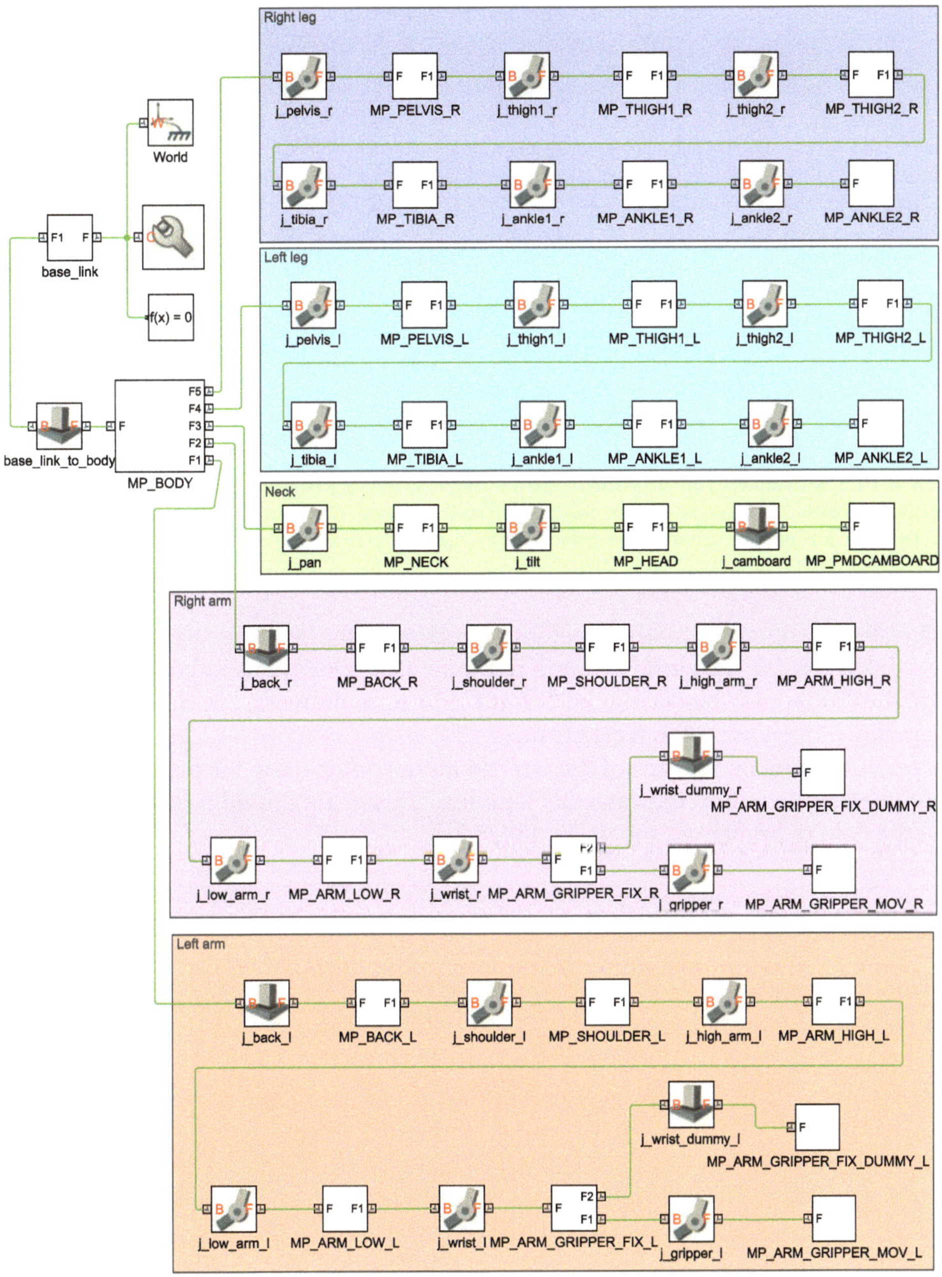

FIGURE 8.18 The Simscape tree of ROBOTIS-OP2. The model tree is divided into right leg, left leg, neck, right hand, and left hand. The areas surrounded by a square box are grouped by the "area" button at the left side of the Simulink editor.

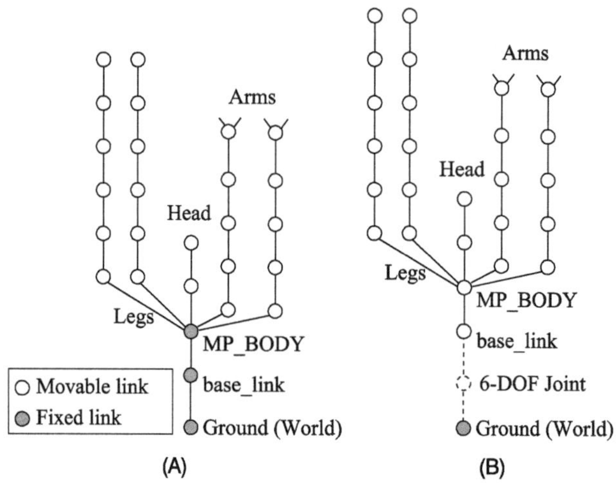

FIGURE 8.19 Tree connectivity structure of the kinematic chain of ROBOTIS-OP2. (A) Left: The original tree connectivity structure. The link named base_link is fixed on the ground. (B) Right: A free joint of six DoFs is attached between the base_link and the ground. This modification changes the fixed model to a free-floating model.

limbs. Fig. 8.19A shows the tree connectivity structure of the Simscape tree. In this state, since the base link is fixed to the world frame, a free joint of six DoFs (a rigid-body joint) is attached between the "base_link" block and the "World" block, as shown in Fig. 8.19B. In this way, the torso can move freely w.r.t. the world frame.

By double clicking on the icon of the wrench mark and opening the property of "Mechanical Configuration," confirm the value of "Gravity." This value should be [0 0 − 9.8] as is often used in the field of humanoid robots (cf. Fig. 8.21).

8.4.3 Joint Mode Configuration

The default state after import is such that the input mode of each joint is unspecified. Double click on the joint block and specify the joint input mode as shown in Fig. 8.22. In order to calculate the joint torque from the given angular acceleration of a joint using the inverse dynamics, users need to configure the "Actuation" item in Fig. 8.22. Set the "Automatically Computed" in the "Torque" configuration of the "Actuation" item as shown in Fig. 8.22A. In addition, set the "Provided by Input" in the "Motion" configuration. In order to output the computed torque from this block, check the "Actuator Torque" box in the "Sensing" item.

On the other hand, in order to calculate the joint acceleration from the given torque of a joint by the forward dynamics, set the "Provided by Input" in the "Torque" configuration of the "Actuation" item as shown in Fig. 8.22B. In addition, set the "Automatically Computed" in the "Motion" configuration. In order to output the computed joint acceleration from this block, check the "Acceleration" box in the "Sensing" item. If the angular velocity and/or the angle are needed (calculated by integrating the computed acceleration), also check the "Velocity" box and/or the "Position" box in the "Sensing" item.

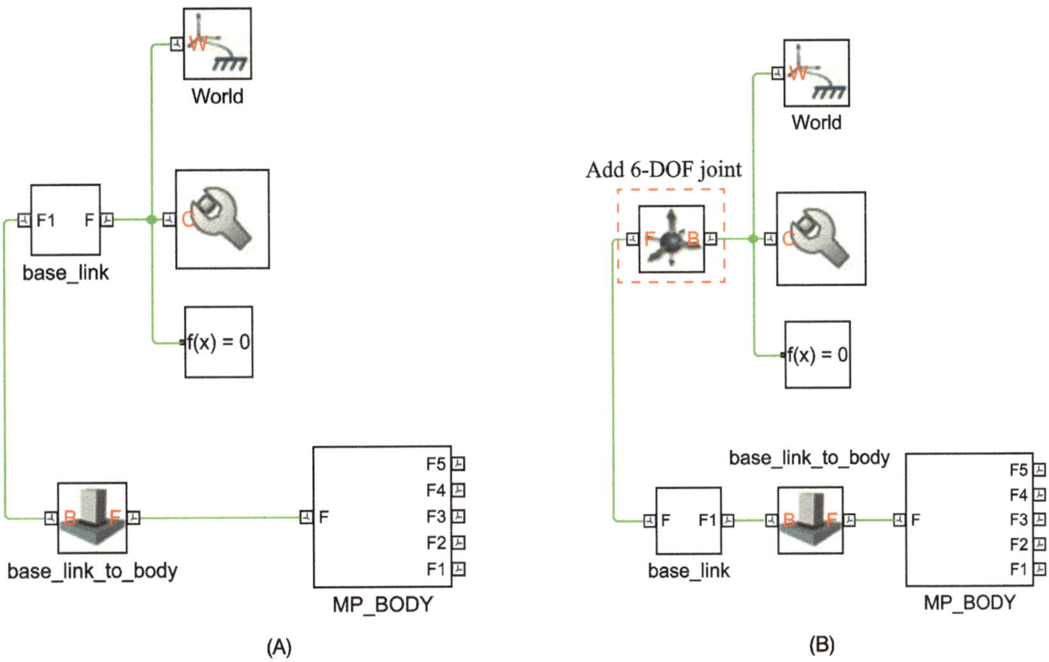

FIGURE 8.20 (A) Left: The original diagram of the Simscape tree between the "World" block and the "MP_BODY" block. Since the "base_link" block is connected directly to the "World" block, the "base_link" is fixed. (B) Right: A free joint block named "6-DOF joint" is attached between the "base_link" block and the "World" block.

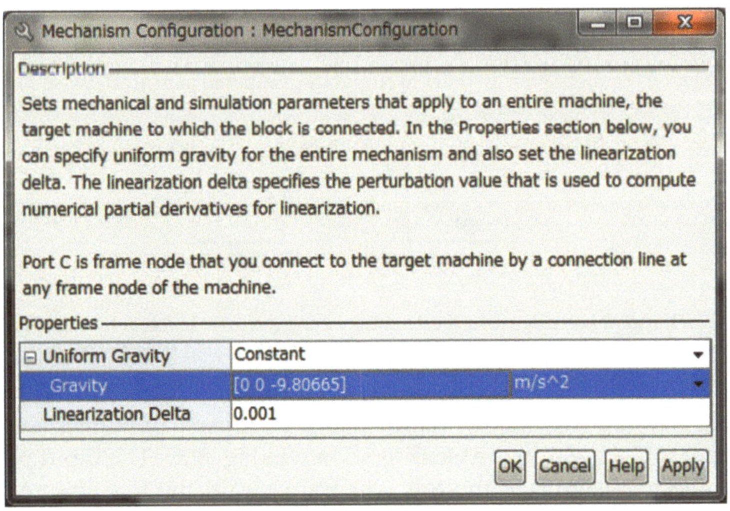

FIGURE 8.21 The "Mechanism Configuration" window. By double clicking on the icon of the wrench mark in the Simscape diagram shown in Fig. 8.20, this window pops up and confirms the value of "Gravity." This value should be [0 0 -9.8], as shown in this figure.

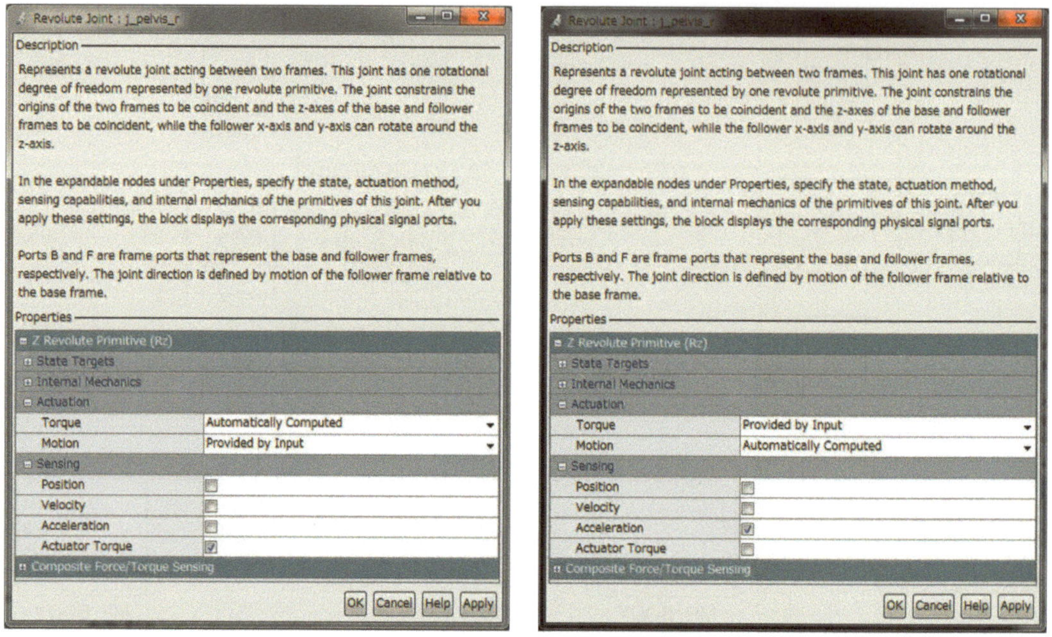

FIGURE 8.22 Joint configuration for a revolute joint. (A) Left: In order to calculate the joint torque from the given angular acceleration to the joint using the inverse dynamics, users need to configure the "Actuation" item in this property window. Set the "Automatically Computed" in the "Torque" configuration of the "Actuation" item as shown in this figure. In addition, set the "Provided by Input" in the "Motion" configuration. In order to output the computed torque from this block, check the "Actuator Torque" box in the "Sensing" item. (B) Right: In order to calculate the joint acceleration from the given torque of a joint by the forward dynamics, set the "Provided by Input" in the "Torque" configuration of the "Actuation" item as shown in this figure. In addition, set the "Automatically Computed" in the "Motion" configuration. In order to output the computed joint acceleration from this block, check the "Acceleration" box in the "Sensing" item.

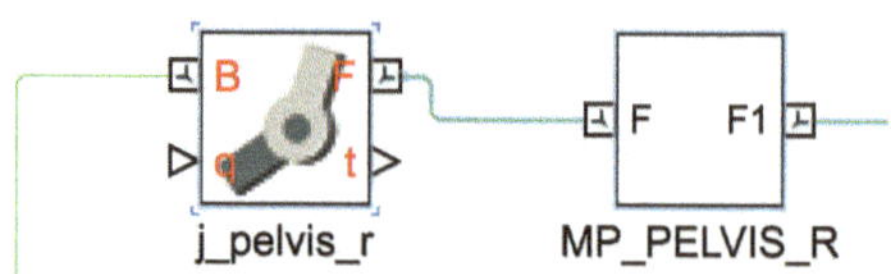

FIGURE 8.23 When the joint is configured for calculation using the inverse dynamics, the input port "q" and the output port "t" appear in the block. The input port is connected to the desired trajectory signal.

When the joint is configured for calculation using the inverse dynamics, the input port "q" and the output port "t" appear in the block as shown in Fig. 8.23. The input port is connected to the desired trajectory signal of the follower frame w.r.t. the base frame along the joint primitive axis.

Fig. 8.24 shows an example of how to calculate the torque using a sinusoidal trajectory as the target trajectory in a joint. The "Sine Wave" block that generates the desired trajectory of the joint is a standard source block in Simulink. However, it cannot be directly connected to

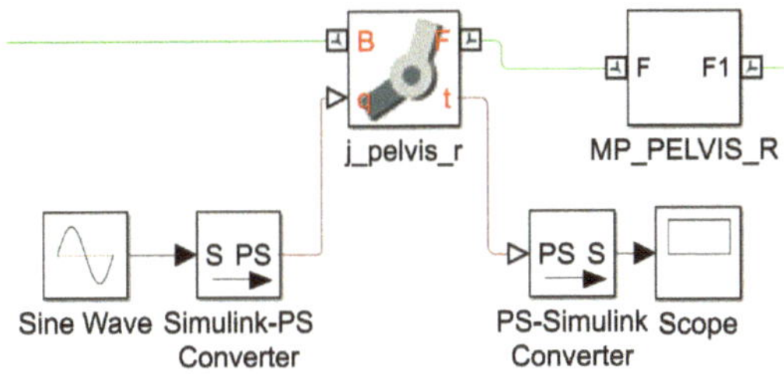

FIGURE 8.24 An example of desired trajectory input. The "Sine Wave" block that generates the desired trajectory of the joint is a standard source block in Simulink, but it cannot be directly connected to the Simscape block. Therefore, the source block is connected via the "Simulink-PS Converter" block to the signal input "q" of the joint.

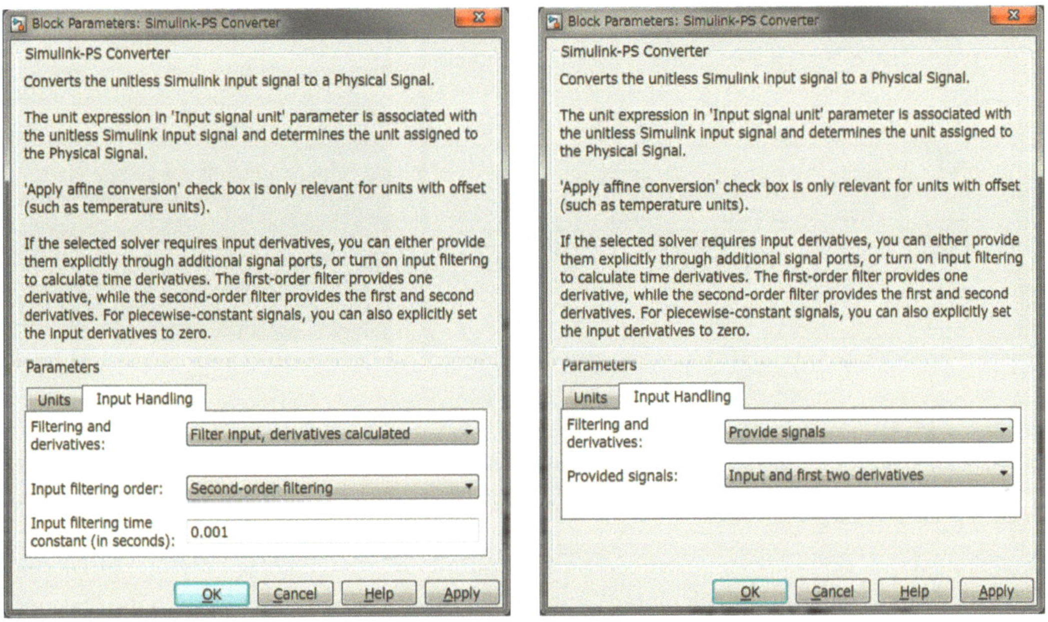

(A) (B)

FIGURE 8.25 By double clicking on the "Simulink-PS Converter" block, the calculation method for angular acceleration can be configured. (A) Left: By setting the "Filtering and derivatives:" item to the "Filter input, derivatives calculated" and the "Input filtering order" item to the "Second-order filtering," the input trajectory is filtered and its second derivative is calculated in the "Simulink-PS Converter" block for the simulation using the inverse dynamics. (B) Right: Alternatively, the calculation method for acceleration can be specified by setting the "Filtering and derivatives" item to the "Provide signals."

the Simscape block. Therefore, the output of the source block is connected via the "Simulink-PS Converter" block to the "q" input of the joint.

In order to calculate the joint torque using the inverse dynamics, it is also necessary to obtain the angular acceleration from the signal input "q." The calculation method of the angular acceleration can be configured by double clicking on the "Simulink-PS Converter" block as shown in Fig. 8.25. As shown in Fig. 8.25A, by setting the "Filtering and derivatives:" item to

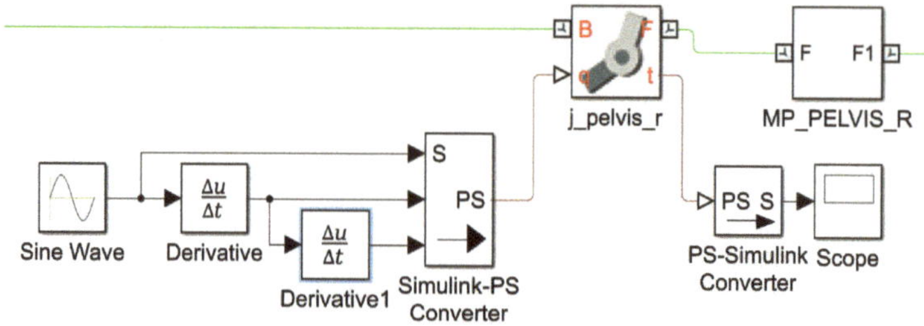

FIGURE 8.26 An example of a user-defined calculation method for angular acceleration. The angular acceleration and velocity are derived by using the two "Derivative" blocks (included in the Simulink standard blocks).

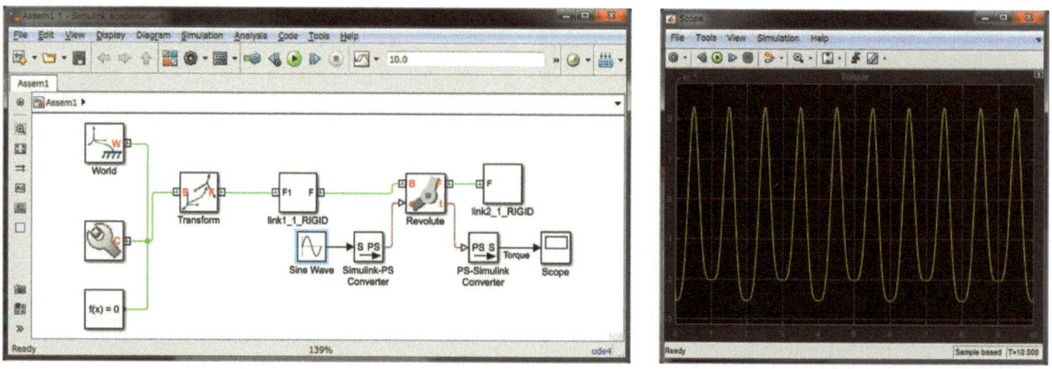

(A) (B)

FIGURE 8.27 Inverse dynamics–based simulation of the two-link arm. (A) Left: The Simulink window for editing the diagram and starting the simulation. The simulation is initialized by clicking on the play icon (▷) in this window. (B) Right: The scope window, displaying the computed joint torque, is connected to the output port of the joint block via the "PS-Simulink Converter" block.

the "Filter input, derivatives calculated" and the "Input filtering order" item to the "Second-order filtering", the input trajectory is filtered and its second derivative is calculated in the "Simulink-PS Converter" block. Alternatively, the calculation method for acceleration can be specified by setting the "Filtering and derivatives" item to the "Provide signals" as shown in Fig. 8.25B. When an item in the "Provide signals" is set to "Input and first two derivatives," inputs for the first-order and second-order differentiation appear in the input signal of the "Simulink-PS converter" block. By doing this, the calculation method of acceleration can be set voluntarily as shown in Fig. 8.26. In addition, it is also possible to input the speed and position into the "Simulink-PS Converter" block, after motion generation in terms of acceleration and then integrating. To observe or handle the computed torque for the input acceleration from Simulink, the "PS-Simulink Converter" block is used. In the example shown in Fig. 8.26, the computed torque is input into the "Scope" block. On the other hand, to input torque, simply convert it with the "Simulink-PS Converter" block without differentiation.

Fig. 8.27 shows an example of inverse dynamics calculation of the two link arm shown in Fig. 8.9. Here, simulation is performed assuming that Link 1 is fixed. By the "smim-

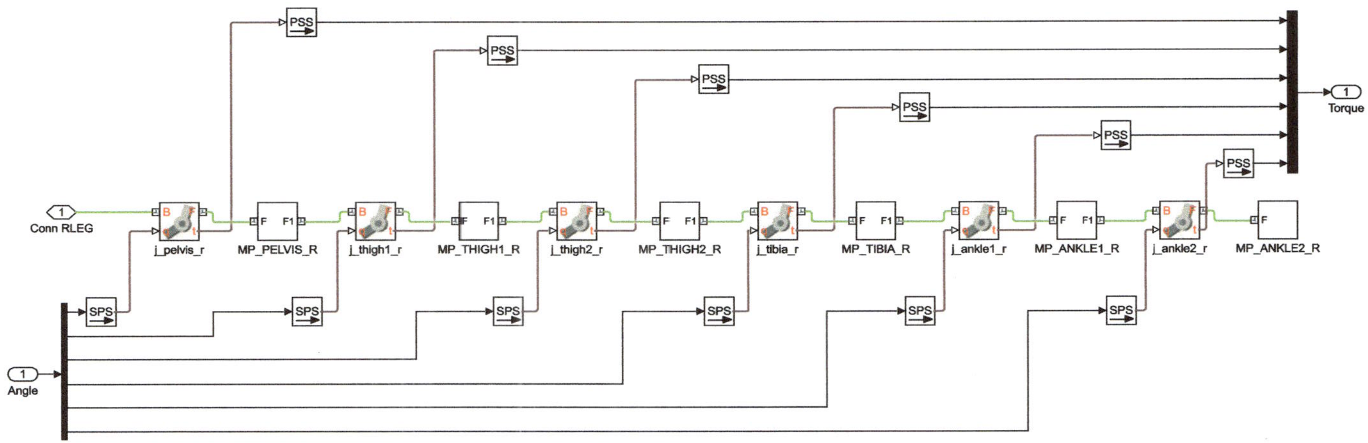

FIGURE 8.28 Right leg tree for the inverse dynamics calculation. This subsystem is the "Right leg" area shown in Fig. 8.18. It can be created by a right click on the area and selecting the "Create Subsystem from Area" as shown in Fig. 8.29. This subsystem uses a reference joint angle vector as input and generates a computed torque vector.

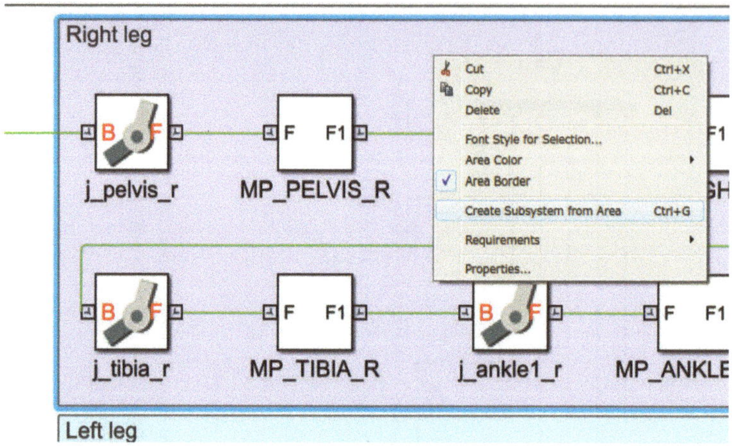

FIGURE 8.29 Creating a subsystem from an area. A subsystem can be created by a right click on the area and selecting "Create Subsystem from Area."

port" command, the "Revolute" joint block is generated between the "link1_1_RIGID" and "link2_1_RIGID" blocks, and a desired joint trajectory is specified for the joint block with a sinusoidal waveform as shown in Fig. 8.9A. The "Scope" block is connected to the output port of the joint block via the "PS-Simulink Converter" block. By clicking on the play icon (▷) in the Simulink window, the simulation starts and the computed torque is plotted as shown in Fig. 8.9B.

Inverse dynamics–based simulation of a whole-body motion of a humanoid robot can be performed by setting the abovementioned procedures for all joints. Fig. 8.28 is a subsystem created from the "Right leg" area shown in Fig. 8.18. It represents an example for inverse dynamics calculation. This subsystem can be created by a right click on the area and selecting "Create Subsystem from Area" as shown in Fig. 8.29. This subsystem uses a reference joint angle vector as an input and generates a computed torque vector.

In order to increase the readability of the diagram, it is helpful to create subsystems for all areas as shown in Fig. 8.30. All subsystems have an input port for the reference joint angle vector and an output port for the computed torque vector. In this example, the reference joint angle vector will be generated by the MATLAB script and saved as a workspace variable. The variable is then imported into Simulink by using the "From Workspace" block, as shown in Fig. 8.31. In order to prevent unintended movement, set the "Form output after final data value by" item to "Holding final value" as shown in Fig. 8.32. The variable is a time-series object comprising information for the time and the 24 joint angles. Since the number of inputs of the arms, the neck, and the legs are five, two, and six, respectively, the composite joint angle vector is decomposed by the "Demux" block, by setting its parameters as shown in Fig. 8.33. In order to observe the computed torque at all joints, the outputs of the subsystems are connected to the "Scope" block (included in the Simulink standard blocks).

The minimum setting to perform the inverse dynamics simulation is finalized by the abovementioned procedure. However, there will be no reaction force from the floor. Thus,

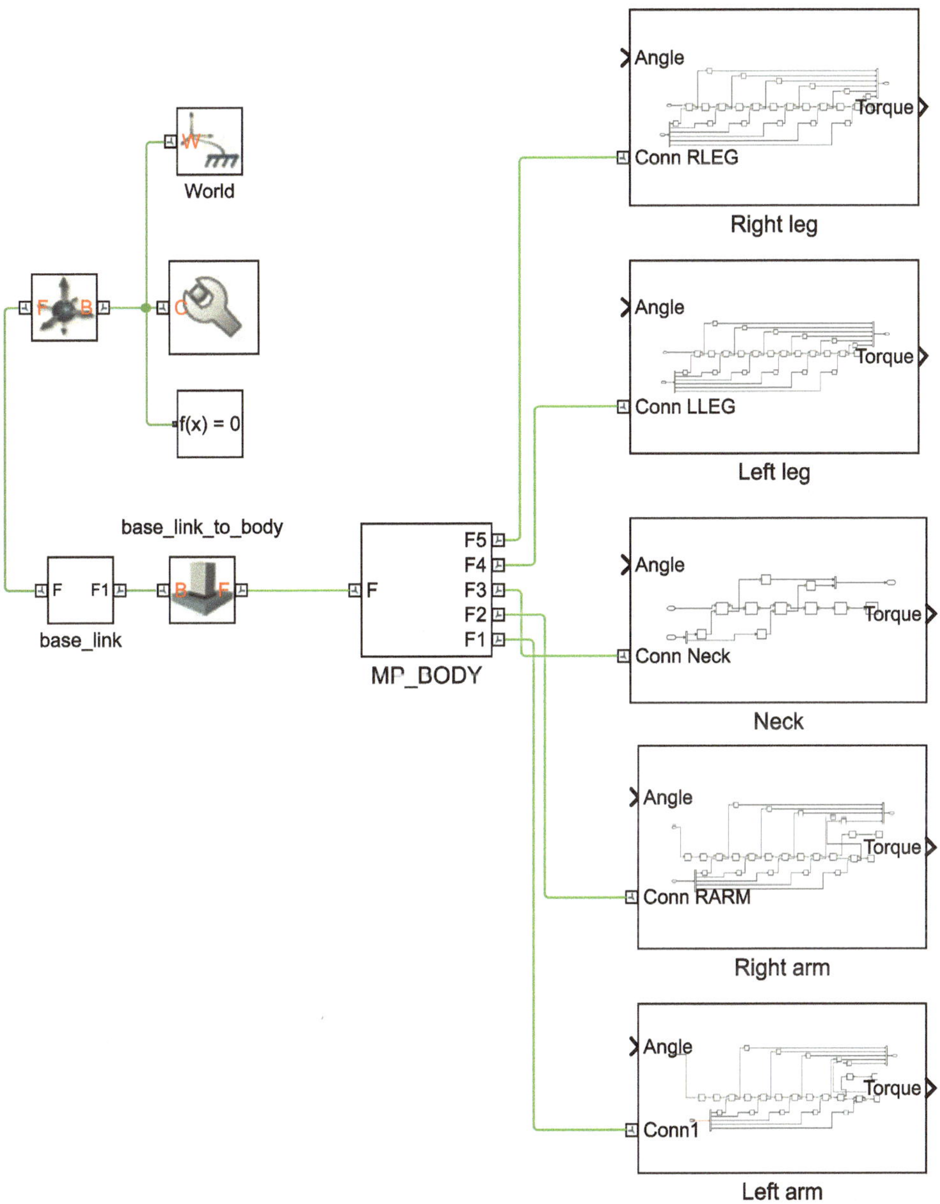

FIGURE 8.30 A Simscape tree with five subsystems. In order to increase the readability of the complicated diagram of the humanoid robot, it is helpful to create subsystems for the branches of the kinematic chain. All subsystems have an input port for the reference joint angle vector and an output port for the computed torque vector.

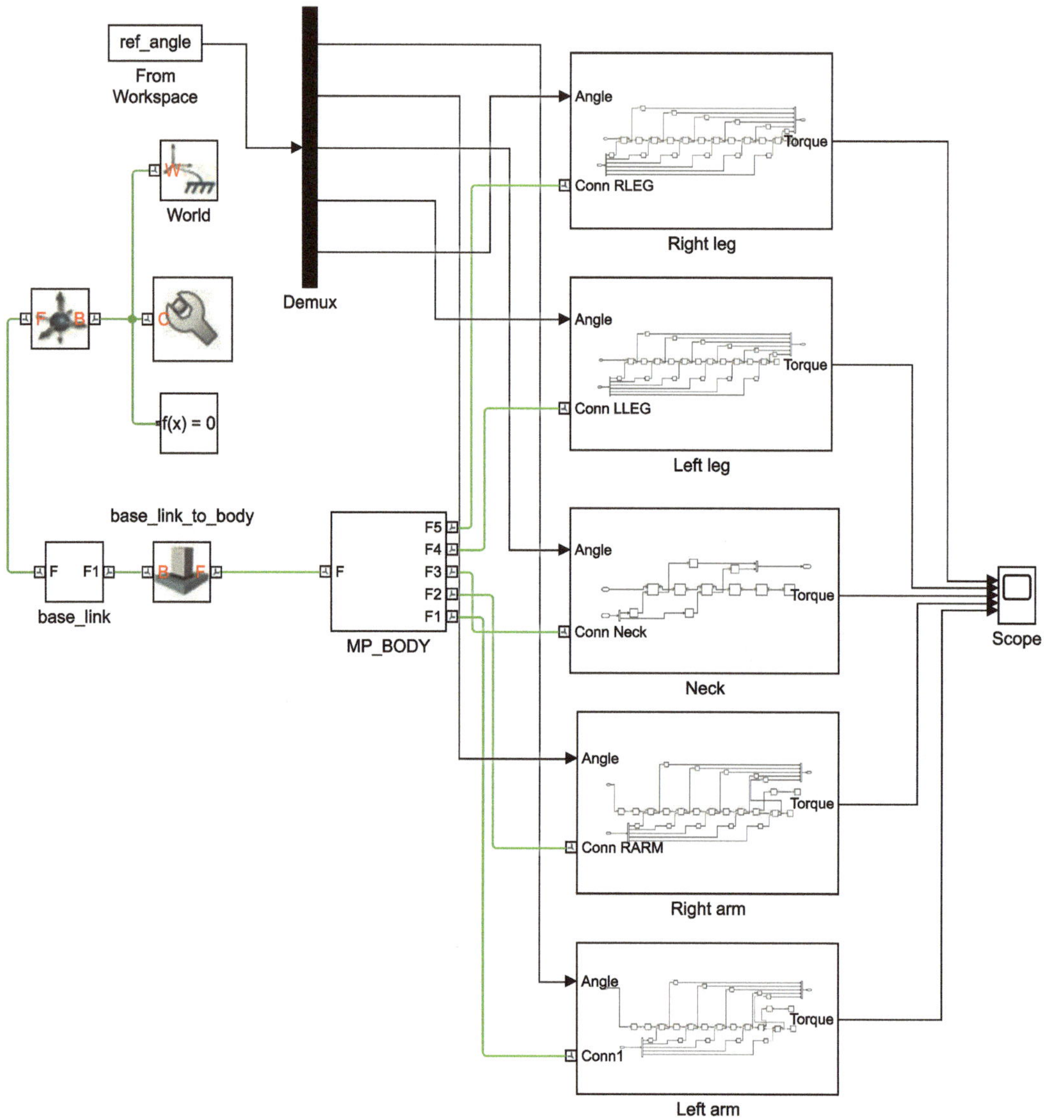

FIGURE 8.31 The joint reference is an input from the MATLAB workspace. The reference joint angle vector will be generated by the MATLAB script and saved as a workspace variable. The variable is then imported into Simulink by using the "From Workspace" block. The variable is a time-series object comprising information for the time and all of the 24 joint angles. The composite joint angle vector is then decomposed by the "Demux" block.

when the simulation is initialized, the robot will fall down. In order to check the settings so far, set "Gravity" to zero. This item can be configured by double clicking on the icon of the wrench mark in Fig. 8.31. Then try a simulation with reference joint angles.

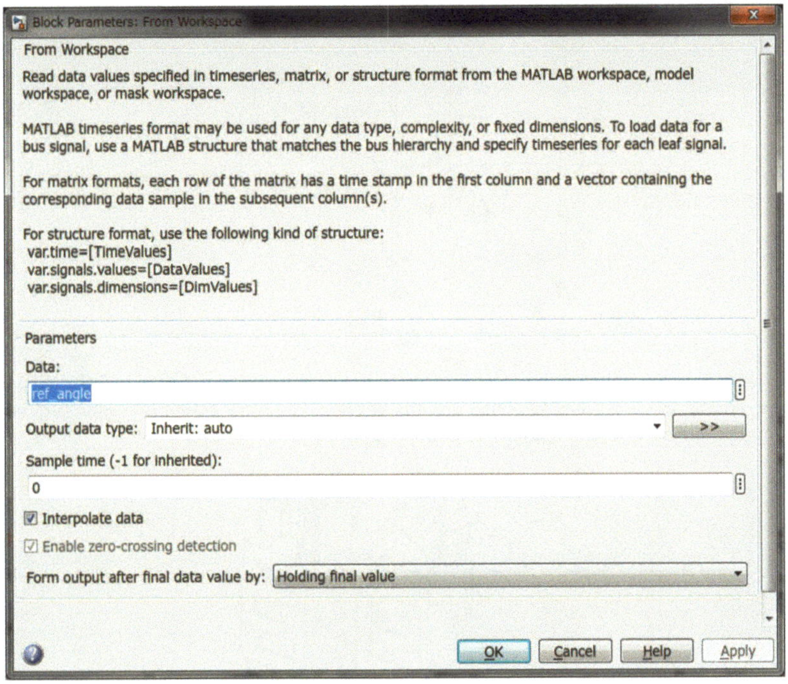

FIGURE 8.32 Property of the "Form Workspace" block. In order to prevent unintended movement, set the "Form output after final data value by" item to "Holding final value," as shown in this figure.

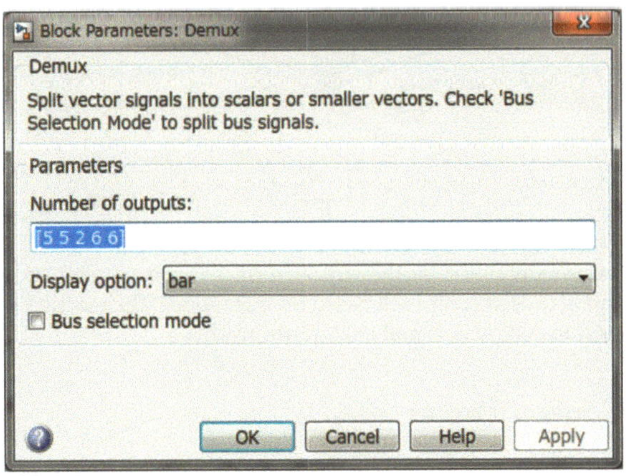

FIGURE 8.33 Parameters of the "Demux" block. The composite reference joint angle vector is decomposed by the "Demux" block into angles for the legs, the neck and the arms. Since the number of inputs of the arms, the neck, and the legs are five, two, and six, respectively, set the "Number of outputs:" item to [5 5 2 6 6].

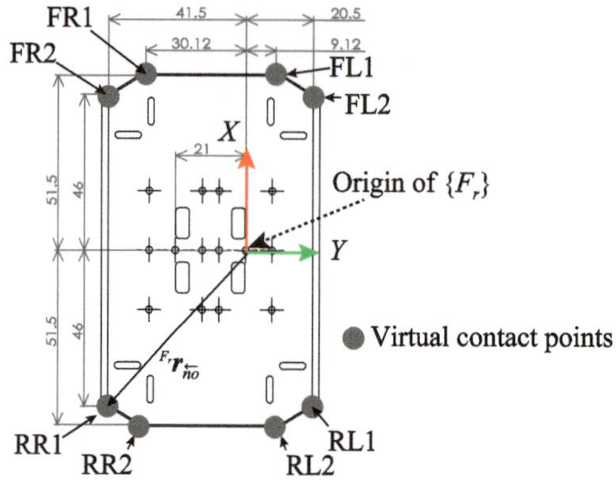

FIGURE 8.34 Definition of contact points on the sole (a view of the sole from below). Eight virtual contact points are placed per sole so that the robot has 16 contact points in total. The force acting on each point is calculated by a spring-damper model. The friction cones are as shown in Fig. 8.35.

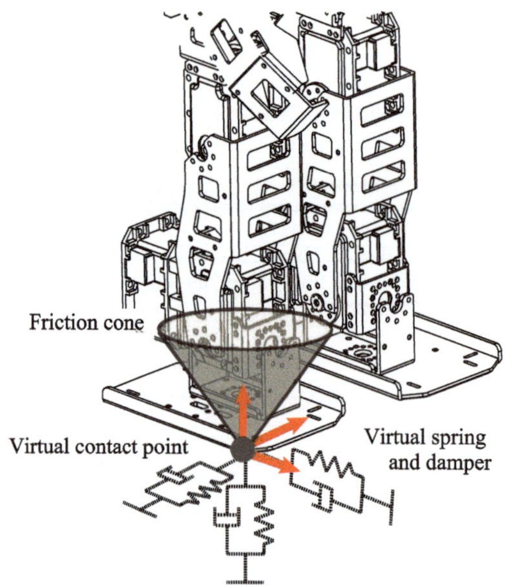

FIGURE 8.35 Contact force model based on the virtual spring-damper model and the friction cone. The force acting at each point shown in Fig. 8.34 is computed based on this model. The reaction force in the normal direction and the tangential forces are expressed by (8.1) and (8.2), respectively.

8.4.4 Modeling of Contact Forces

Assuming the robot is in contact with the floor only at the soles, the contact forces between the soles and the floor are modeled based on the sample point contact model described in Section 3.3.2. As shown in Fig. 8.34, eight virtual contact points are placed at each sole. Thus, the robot comprises 16 contact points in total. The force acting at each point is calculated by a spring-damper model. The friction cones are shown in Fig. 8.35. Here, the surface of the horizontal floor is assumed to be at zero ground level. Therefore, when the height of the nth virtual contact point z_n is negative, the point will penetrate the floor and a reaction force will be generated based on the depth and speed of the penetration.

Using the spring-damper model, the reaction force in the normal direction is expressed by the following equations:

$$f_{k_{nz}} = \begin{cases} -K_z \Delta z_n - D_z \dot{z}_n & (z_n < 0), \\ 0 & (z \geq 0), \end{cases} \tag{8.1}$$

where K_z and D_z denote the spring constant and the damping coefficient of the virtual spring and damper, respectively; Δz_n stands for the penetration depth of the point w.r.t. the floor surface. The tangential forces can be obtained from (3.9) and (3.10) as

$$f_{k_{nt}} = \begin{cases} -K_t \Delta t_n - D_t \dot{t}_n & \left(z_n < 0,\ \sqrt{f_{k_{nx}}^2 + f_{k_{ny}}^2} \leq \mu_k f_{k_{nz}}\right), \\ -\mu_k f_{k_{nz}} \frac{v_{k_{nt}}}{\sqrt{v_{k_{nx}}^2 + v_{k_{ny}}^2}} & \left(z_n < 0,\ \sqrt{f_{k_{nx}}^2 + f_{k_{ny}}^2} > \mu_k f_{k_{nz}}\right), \\ 0 & (z_n \geq 0), \end{cases} \tag{8.2}$$

where $t \in \{x, y\}$, and μ_k (> 0) denotes the constant static friction coefficient. Fig. 8.36 shows a Simulink diagram based on this contact force model. This diagram is repeatedly used as many times as the number of the virtual contact points. Therefore, it is recommend to save it as a library (My_Libs.slx) in the same folder as op2_simulation.slx, as shown in Fig. 8.17. The Physical Modeling Connection (PMC) port named "External Force" is connected to the virtual contact point (cf. Fig. 8.37) and the "Transform Sensor" block generates the position of the point. The "External Force and Torque" block applies the computed reaction force to the point.

The bottom area, surrounded by a rectangle, displays the calculation method of the normal force expressed in (8.1). The "Saturation" block is connected as shown in Fig. 8.36, in order to avoid generating attractive forces; Fig. 8.38 shows the property of the "Saturation" block. As shown in the figure, set the "Upper limit" item to "inf" and the "Lower limit" item to "0." The middle area, surrounded by a rectangle, displays the calculation method of the tangential force as given in the first row of (8.2). The output of the "Sample and Hold" block follows the input when the point does not penetrate the floor, or it slides on it when $\Delta t_{i,n}$ is zero. Otherwise, the output is on hold, i.e. when the point penetrates the floor and sticks onto it. Then, $\Delta t_{i,n}$ is computed. The "Sample and Hold" block is included in the Simscape Power Systems toolbox. If this toolbox is not available, a block that has the same function as the "Sample and Hold" block can be made, as shown in Fig. 8.39. This block can be used instead of the "Sample and Hold" block. The MATLAB code for this block is as follows.

Listing 8.1: MATLAB function for the block that has the same function as the "Sample and Hold" block.

```
function y = fcn(In,h,S)
if S>0
    y=In;
else
    y=h;
end
```

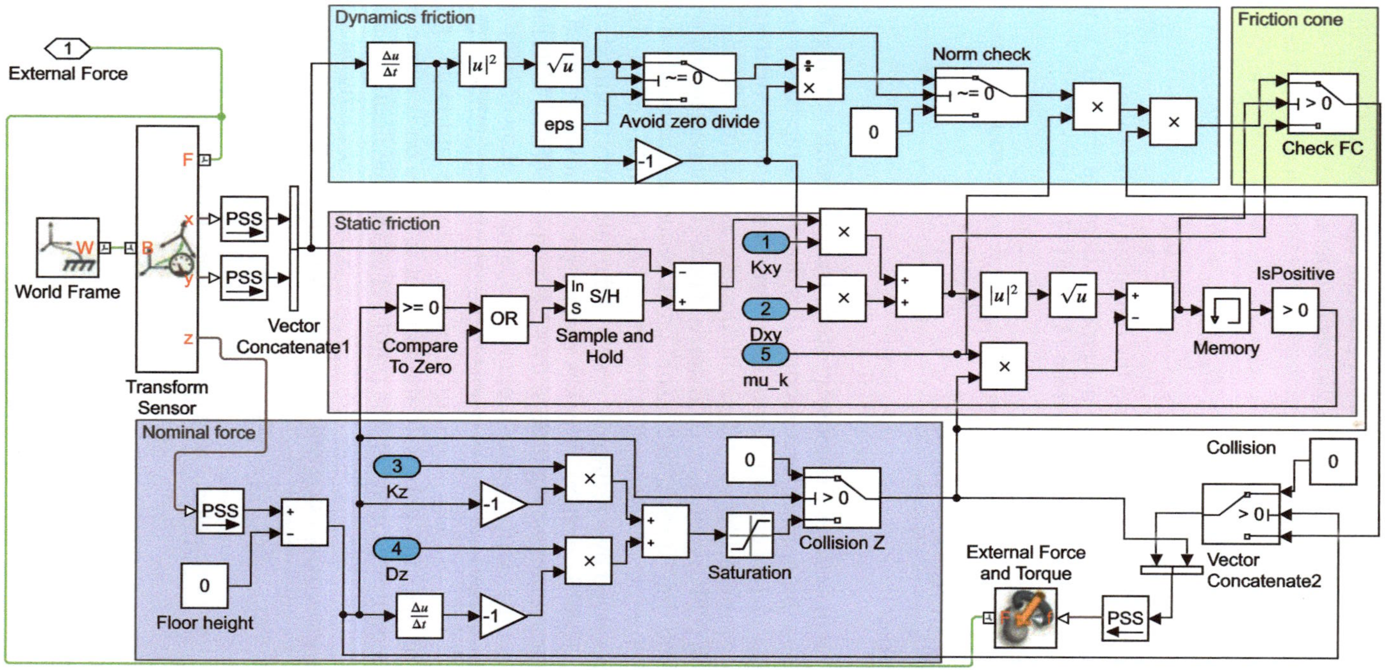

FIGURE 8.36 Modeling of the contact forces in Simulink based on (3.9) and (3.10). The bottom area, surrounded by a rectangle, displays the calculation method of the nominal force expressed in (8.1). The middle area, surrounded by a rectangle, displays the calculation method of the tangential force, as given in the first row of (8.2). The left-top area, surrounded by a rectangle, displays the calculation method of the tangential force as given in the second row of (8.2). The right-top area, surrounded by a rectangle, switches the tangential force based on the friction cone.

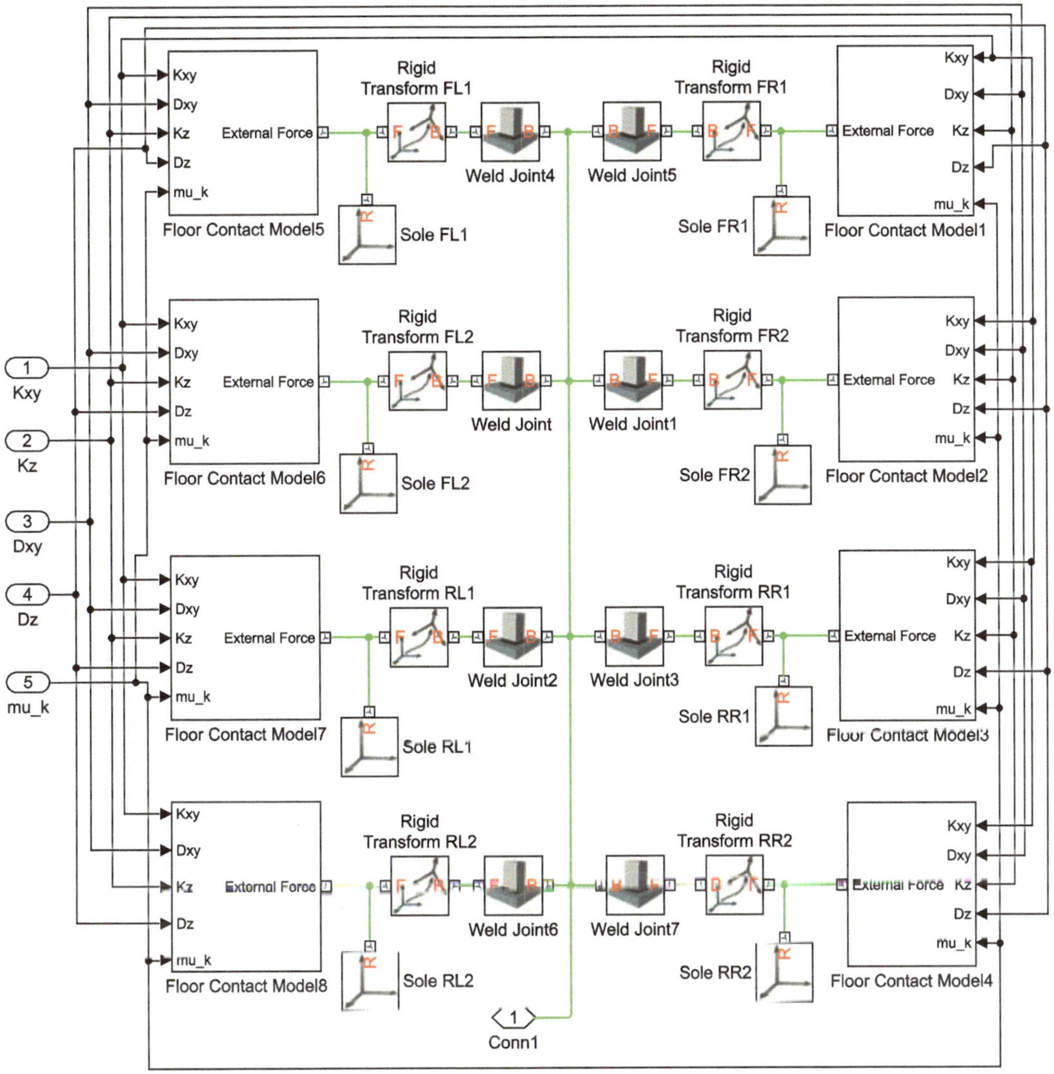

FIGURE 8.37 Definition of the virtual contact points on the right sole. The sum of the floor reaction forces given in the diagram shown in Fig. 8.36 is transmitted to the leg via the origin of the sole shown in Fig. 8.34.

The left-top area, surrounded by a rectangle, displays the calculation method for the tangential force expressed in the second row of (8.2). The right-top area, surrounded by a rectangle, switches the tangential force based on the friction cone. The computed force is an input into the "External Force and Torque" block via the "PS-Simulink Converter" block. In order to apply the force to the point, check the "Force" item in the property of the "External Force and Torque" block. Since (8.1) and (8.2) denote forces at the virtual point in the world frame, set the "Force Resolution Frame" item to "World," as shown in Fig. 8.40.

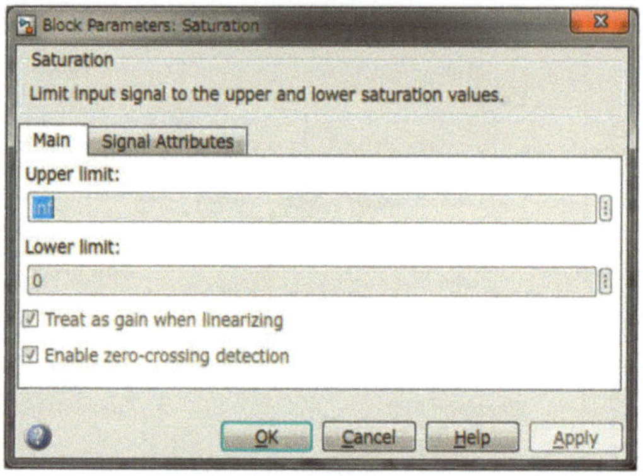

FIGURE 8.38 In order to avoid generating attractive forces, the "Saturation" block is connected in Fig. 8.37. This figure shows the property of the "Saturation" block. As shown in this figure, set the "Upper limit:" item to "inf" and the "Lower limit:" item to "0."

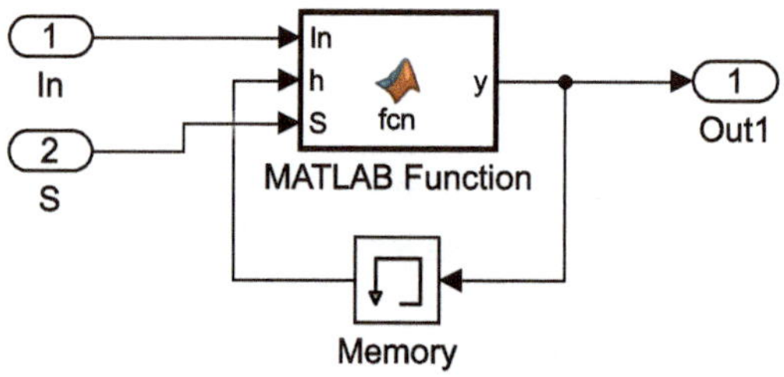

FIGURE 8.39 A block that has the same function as the "Sample and Hold" block.

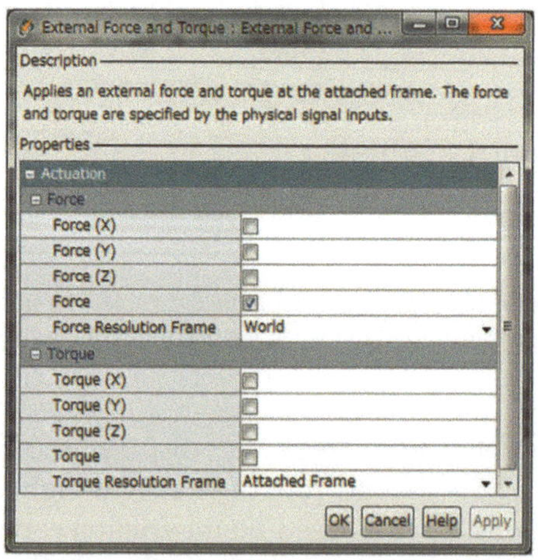

FIGURE 8.40 Property of the "External Force and Torque" block. In order to apply the force to the point, check the "Force" item and set the "Force Resolution Frame" item to "World" (since (8.1) and (8.2) are forces at the virtual point in the world frame).

TABLE 8.2 Offset parameters for the "Rigid Transform" blocks for the right sole. These offsets are measured as shown in the drawing in Fig. 8.34

Rigid Transform	Offset (mm)	Rigid Transform	Offset (mm)
FL1	[51.5 9.12 0]	FR1	[51.5 −30.12 0]
FL2	[46 20.5 0]	FR2	[46 −41.5 0]
RL1	[−46 20.5 0]	RR1	[−46 −41.5 0]
RL2	[−51.5 9.12 0]	RR2	[−51.5 −30.12 0]

Next, the sum of the floor reaction forces, given in the diagram shown in Fig. 8.36, is transmitted to the leg via the origin of the sole shown in Fig. 8.34. The force $\boldsymbol{f}_k$ and moment $\boldsymbol{m}_k$ at the origin of the sole in frame $\{k\}$ $(k \in \{F_r, F_l\})$ can be denoted as follows:

$$\boldsymbol{f}_k = \sum_{n=1}^{N} {}^k\boldsymbol{R}_W \boldsymbol{f}_{k_n}, \tag{8.3}$$

$$\boldsymbol{m}_k = \sum_{n=1}^{N} {}^k\boldsymbol{r}_{\overleftarrow{no}} \times {}^k\boldsymbol{R}_W \boldsymbol{f}_{k_n}, \tag{8.4}$$

where ${}^k\boldsymbol{R}_W \in \Re^{3\times3}$ denotes the rotation matrix that transforms vectors from the world frame to the foot frame $\{k\}$; $\boldsymbol{f}_{k_n} = \left[f_{k_{nx}}\ f_{k_{ny}}\ f_{k_{nz}}\right]^T$, N, and ${}^k\boldsymbol{r}_{\overleftarrow{no}}$ are the floor reaction force denoted in (8.1) and (8.2), the number of the contact points, and the position vector of the nth contact point in $\{k\}$, respectively. The diagram shown in Fig. 8.37 describes the geometrical relationships between the foot origin and the virtual contact points. The PMC port named "Conn1" is connected at the foot origin and the subsystem named "Floor Contact Model" is the library block shown in Fig. 8.36. The "Rigid Transform" block stands for ${}^k\boldsymbol{r}_{\overleftarrow{no}}$. The "Floor Contact Model" block outputs the tangential force components of $\boldsymbol{f}_k$ based on the spring constant "Kxy" and the damping coefficient "Dxy." In addition, the block outputs the nominal force of $\boldsymbol{f}_k$ based on the sprint constant "Kz," the damping coefficient "Dz," and the friction coefficient "mu_k." The values of the spring constants, the damping coefficients, and the friction coefficient are specified by the Simulink standard input ports. As shown in Fig. 8.41, the position of the virtual contact point can be set in the "Offset" item. The "Offset" values for all the points on the right sole are shown in Table 8.2. The offsets are measured as shown in the drawing in Fig. 8.34. Also, set the left sole to be symmetrical to the right one.

In order to connect the PMC port "Conn1" to the origin of the right foot frame, it is needed to make the right foot frame. Fig. 8.42 shows the geometrical relationship between the ankle and the foot frames. The origin of the foot frame is located at the projection of the origin of the ankle frame on the plantar surface. The directions of the x-, y-, and z-axes of the foot frame point forward, to the left, and upward.

Fig. 8.43 shows a diagram describing the kinematic relationships for the right sole. As shown in Fig. 8.28, the "MP_ANKLE2_R" body is at the end of the tree originally, and a PMC port named "Connection Port" is added to the "Reference Frame" of the "MP_ANKLE2_R" body. In order to attach the foot frames, the "Rigid Transform" block named "For rotation" is connected to the "Connection Port" that rotates from the reference frame of the "MP_ANKLE2_R" to the frame parallel to the foot frames. In the "Rigid Transform" block, the rotation

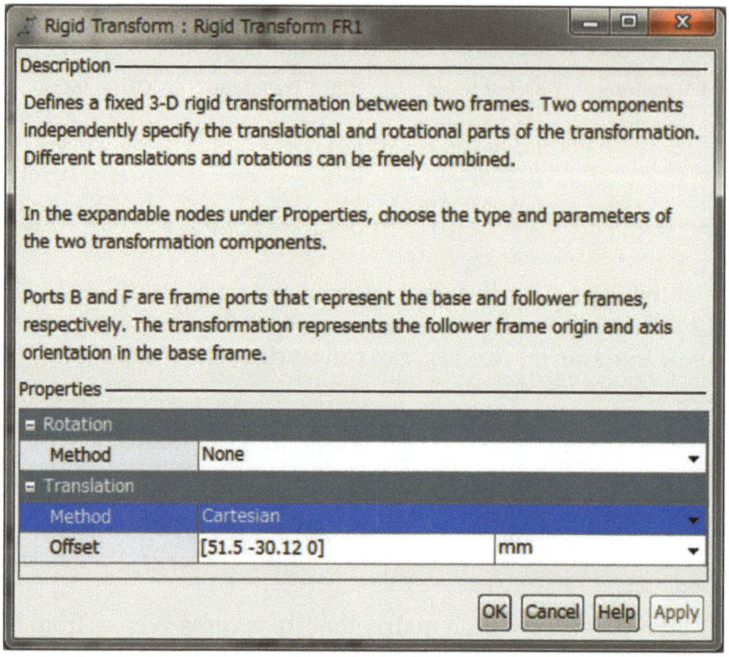

FIGURE 8.41 Property of the "Rigid Transform" block. The position of the virtual contact point can be set in the "Offset" item. The "Offset" values for all the points on the right sole are shown in Table 8.2.

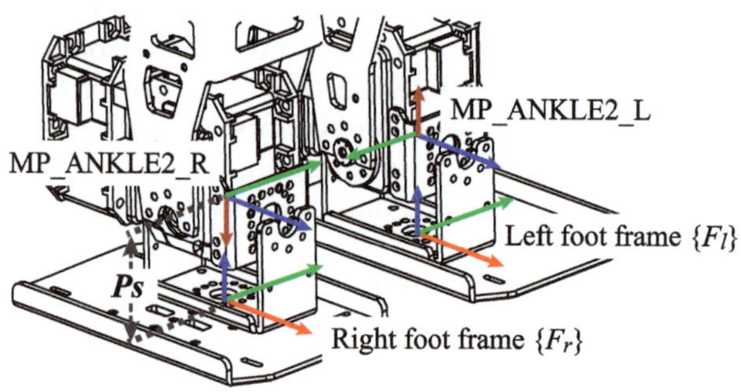

FIGURE 8.42 Geometrical relationship between the ankle and the foot frames. The origin of the foot frame is located at the projection of the origin of the ankle frame on the plantar surface. The directions of the *x*-, *y*-, and *z*-axes of the foot frame point forward, to the left, and upward.

can be applied by choosing a rotation method from the "Aligned Axes," the "Standard Axes," the "Arbitrary Axis," the "Rotation Sequence," and the "Rotation Matrix" methods. In this case, using the "Aligned Axes," it is easy to specify the rotation between the two frames. As shown in Fig. 8.42, the direction along the z-axis of the reference frame of the "MP_ANKLE2_R" is the same as the direction along the x-axis, and the directions along the z-axis of the two frames are the same. Fig. 8.44 shows the property of the "Rigid Transform" block named "For rotation" based on this geometrical relationship. The distance $\boldsymbol{P}_s$ between the

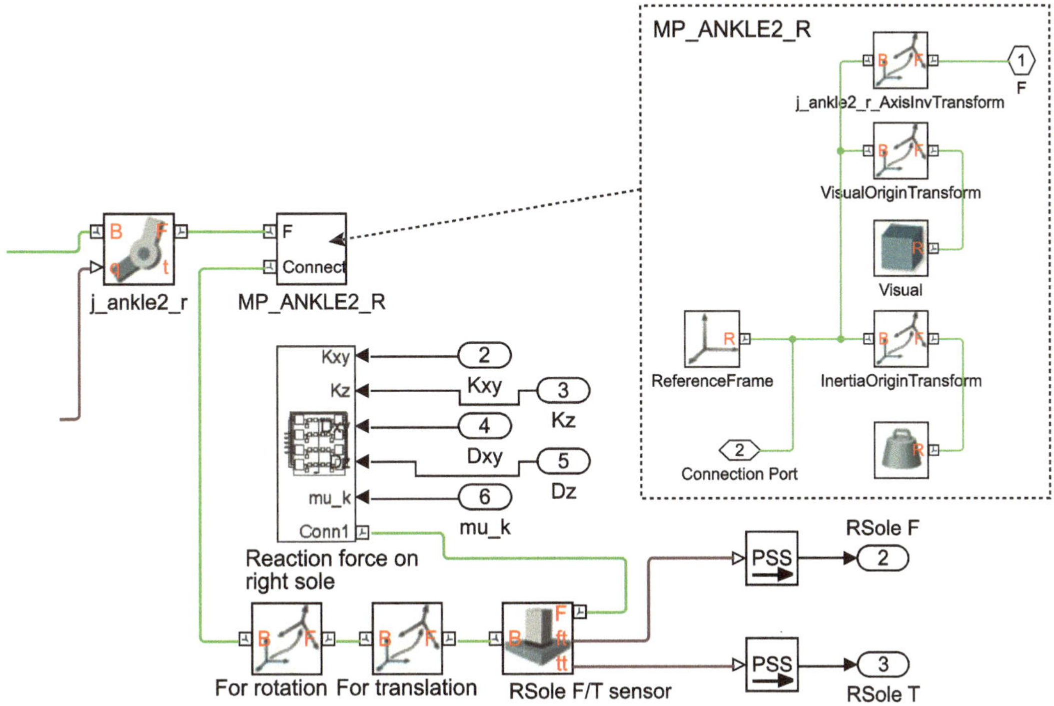

FIGURE 8.43 Making the right foot frame and an F/T (force and torque) sensor. A PMC port named "Connection Port" is added to "Reference Frame" of the "MP_ANKLE2_R" body. In order to attach the foot frame, the "Rigid Transform" block named "For rotation" is connected to "Connection Port" that rotates from the reference frame of "MP_ANKLE2_R" to the frame parallel to the foot frame. The distance between the reference frame of "MP_ANKLE2_R" and the right foot frame is applied by the "Rigid Transform" block named "For translation." In order to measure the force and the moment applied to the sole, a virtual F/T sensor is mounted at the origin of the foot frame. This virtual sensor can be implemented with the "Weld Joint" block that is a 0-DoF joint.

reference frame of the "MP_ANKLE2_R" and the right foot frame is $[0\ 0\ -33.5]^T$ (mm) and can be applied to the "Rigid Transform" block named "For translation" block in the same way as in Fig. 8.41.

In order to measure the force and the moment applied to the sole, a virtual F/T (force and torque) sensor is mounted on the origin of the foot frame. The measured force and moment will be used in Section 8.4.5 for computing the zero-moment point (ZMP). The virtual sensor can be implemented by the "Weld Joint" block, which is a 0-DoF joint. By checking the "Total Force" and the "Total Torque" in the property of the "Weld Joint," as shown in Fig. 8.45, the block measures these values. The measured values are connected to the Simulink standard output port via the "PS-Simulink Converter" block. The follower port of the "Weld Joint" is connected to the PMC port named "Conn1" of the subsystem named "Reaction force on right sole," which is explained in Fig. 8.37. Set these configurations to the left sole as well. The "Constant" blocks used for setting the parameters of the contact force model are placed at the top level of the hierarchical diagram shown in Fig. 8.46. By clicking on these blocks, the parameters of all the contact points can be set easily at the same time. In addition, a solid model

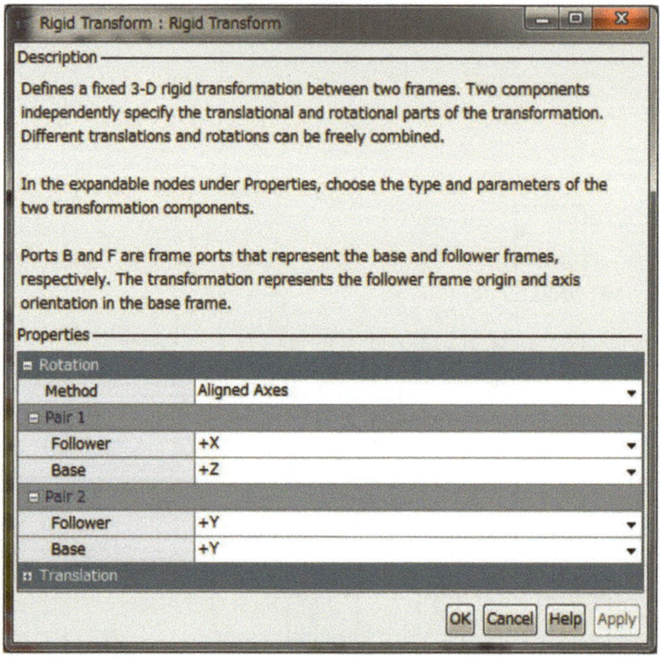

FIGURE 8.44 Property of the "Rigid Transform" block named "For rotation." In the "Rigid Transform" block, the rotation can be applied by choosing a rotation method from the "Aligned Axes," the "Standard Axes," the "Arbitrary Axis," the "Rotation Sequence," and the "Rotation Matrix" methods. In this case, using "Aligned Axes," it is easy to specify the rotation between the two frames.

is added to the top level diagram in the hierarchy for visualizing the floor. The "Solid" block named "Floor" in Fig. 8.46 is a large thin brick connected to the world fame. The property of the "Solid" block named "Floor" is shown in Fig. 8.47.

8.4.5 Computing the ZMP

The ZMP of a multilink system can be computed from the time derivative of the linear and angular momenta of the robot, as described in Section 4.4.5. When the robot is in contact with the floor surface only at its soles, the ZMP can be obtained based on the law of action and reaction from the force and moment measured by the F/T sensors mounted on the soles as follows:

$$r_{p_x} = -\frac{m_y}{f_z} = -\frac{m_{ry} + m_{ly}}{f_{rz} + f_{lz}}, \tag{8.5}$$

$$r_{p_y} = \frac{m_x}{f_z} = \frac{m_{rx} + m_{lx}}{f_{rz} + f_{lz}}, \tag{8.6}$$

where r_{p_t}, f_z, and m_j ($t \in \{x, y\}$) denote the position of the ZMP, the total force in the z-direction, and the moment around the j-axis in the world frame, respectively; f_{jz} and m_{jt} ($t \in \{x, y\}$, $j \in \{r, l\}$) are respectively the force in the z-direction and the moment around the

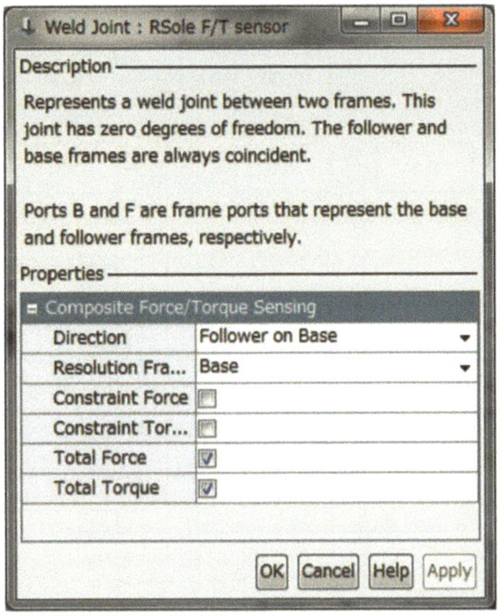

FIGURE 8.45 Property of the "Weld Joint" block used as a virtual F/T (force and torque) sensor. The virtual sensor can be implemented with the "Weld Joint" block which is a 0-DoF joint. By checking "Total Force" and "Total Torque" in the property of "Weld Joint," as shown in this figure, the block outputs these values to the Simulink standard output ports named "RSole F" and "RSole T."

z-axis in the world frame measured by each sensor. We have

$$ {}^{w}\boldsymbol{f}_k = {}^{w}\boldsymbol{R}_k \boldsymbol{f}_k, \tag{8.7}$$

$$ {}^{w}\boldsymbol{m}_k = {}^{w}\boldsymbol{R}_k \boldsymbol{m}_k + \boldsymbol{r}_k \times {}^{w}\boldsymbol{f}_k, \tag{8.8}$$

where ${}^{w}\boldsymbol{f}_k$ and ${}^{w}\boldsymbol{m}_k$ are the measured force and moment in the world frame, respectively; $\boldsymbol{r}_k$ and ${}^{w}\boldsymbol{R}_k$ denote respectively the position vector and the rotation matrix of the foot frame in the world frame.

Fig. 8.48 shows the Simulink implementation of (8.7) and (8.8) for the sensor mounted on the right sole. In order to obtain $\boldsymbol{r}_k$ and ${}^{w}\boldsymbol{R}_k$, the "Transform Sensor" block is connected to the virtual sensor block named "RSole F/T sensor." Although (8.7) and (8.8) can be implemented with a Simulink diagram, the "MATLAB Function" block is used for describing them simply. Input the measured force $\boldsymbol{f}_k$, the moment $\boldsymbol{m}_k$, $\boldsymbol{r}_k$, and ${}^{w}\boldsymbol{R}_k$ into the "MATLAB Function" block. The MATLAB code for this block is as follows.

Listing 8.2: MATLAB function for computing the forces and moments in the world frame.

```
function ft = fcn(mf,mt,wRm,wp)
f=wRm*mf;
t=wRm*mt+cross(wp,f);
ft=[f',t']';
end
```

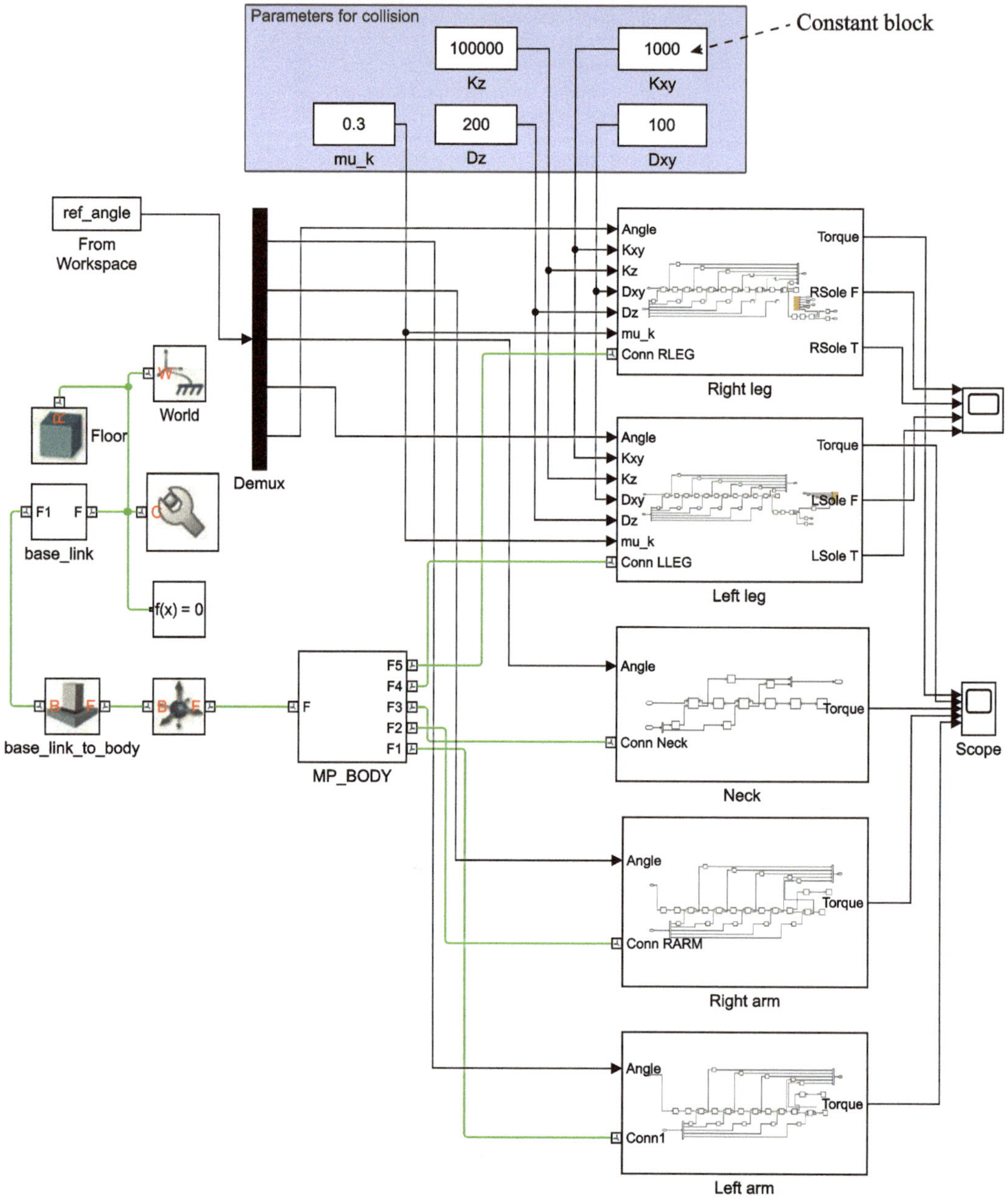

FIGURE 8.46 The top level in the hierarchical diagram for dynamics-based simulation with a humanoid robot. The "Constant" blocks for setting the parameters of the contact force model are placed at the top level of the hierarchical diagram. By clicking on these blocks, the parameters of all the contact points can be set easily at the same time. The "Solid" block named "Floor" is a large thin brick for visualizing the floor.

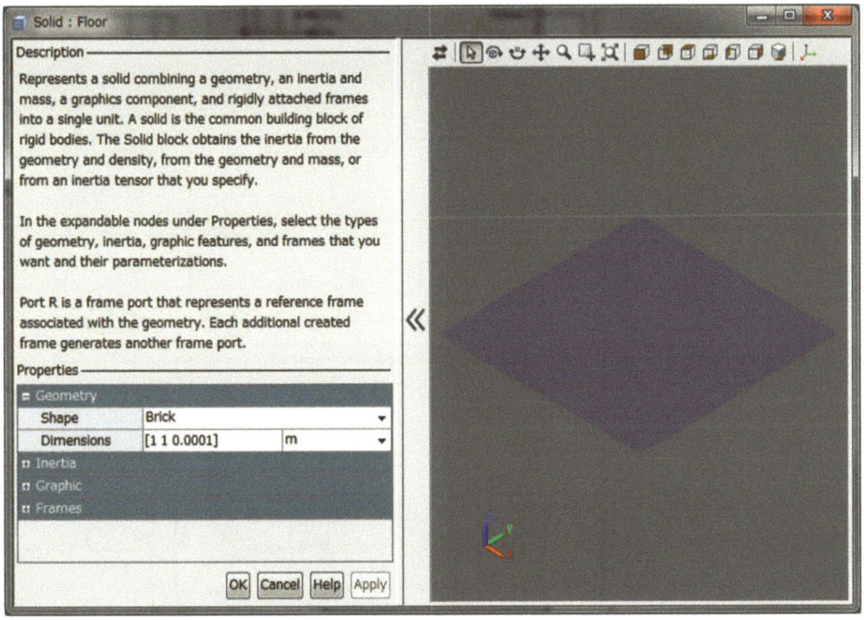

FIGURE 8.47 Property of the "Solid" block named "Floor." This solid model is added to the top level of the hierarchical diagram for visualizing the floor.

The output of the block is a vector which consists of forces and moments computed in the world frame. This output is connected to the Simulink standard output port named "FTw." Likewise, the force and moment in the world frame are also obtained from the sensor mounted on the left sole. For computing the ZMP, input the two vectors into the "MATLAB Function" block as shown in Fig. 8.49. The MATLAB code for this block based on (8.5) and (8.6) is as follows.

Listing 8.3: MATLAB function for computing the position of the ZMP.

```
function pzmp = fcn(FTr,FTl)
pzmp=zeros(3,1);
pzmp(1)=-(FTr(5)+FTl(5))/(FTr(3)+FTl(3));
pzmp(2)=(FTr(4)+FTl(4))/(FTr(3)+FTl(3));
pzmp(3)=0;
end
```

The computed position of the ZMP is sent to a subsystem named "ZMP marker" which displays the position by a sphere in Mechanics Explore. At the time of writing, since there is no function to draw a sphere as a computer graphics feature in Mechanics Explore, the sphere is handled as an object and moved by the "Cartesian joint" as shown in Fig. 8.50. This joint has three translational DoFs represented by three prismatic primitives. The "Cartesian joint" moves the sphere based on a reference position input from the Simulink standard port named "ZMPxyz." Set the items "Force" to the "Automatically Computed" and the items "Motion"

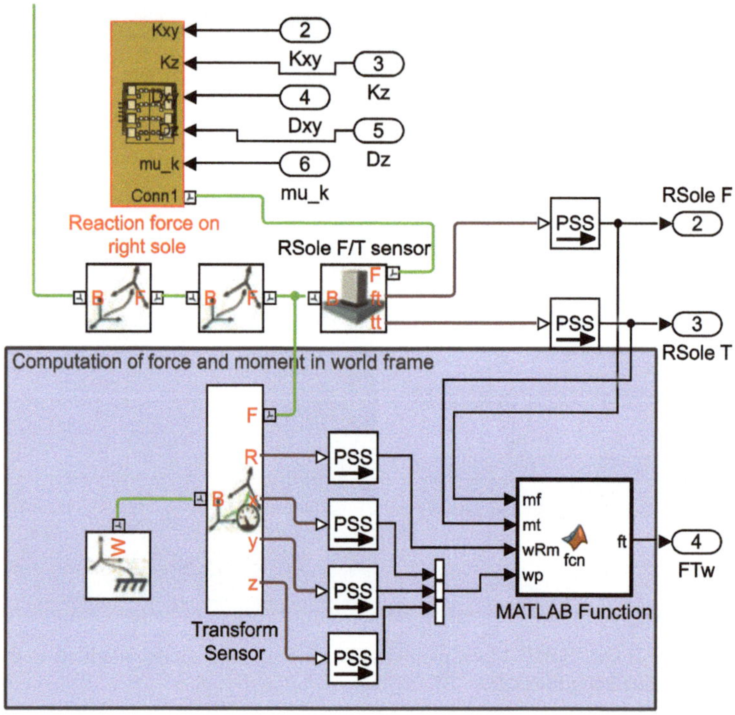

FIGURE 8.48 Computation of the force and moment in the world frame from the measured force and moment in the foot frame. In order to obtain $\boldsymbol{r}_k$ and ${}^{w}\boldsymbol{R}_k$, the "Transform Sensor" block is connected to the virtual sensor block named "RSole F/T sensor." The measured force $\boldsymbol{f}_k$ and moment $\boldsymbol{m}_k$, as well as $\boldsymbol{r}_k$ and ${}^{w}\boldsymbol{R}_k$, are connected to the inputs of the "MATLAB Function" block. The output of the block is a vector which consists of forces and moments computed in the world frame.

to the "Provided by Input" in the "X," the "Y," and the "Z" as shown in Fig. 8.51. Fig. 8.52 shows the property of the "Solid" block named "ZMP Marker." Its mass property can have any value.

Since the "MATLAB Function" block in Fig. 8.49 outputs NaN (Not a Number) when the sole has no contact with the floor, the "MATLAB Function" block in Fig. 8.50 handles the error by setting the position of the marker far from the origin as follows.

Listing 8.4: MATLAB function for error handling.

```
function y = fcn(u)
if sum(isnan(u))>0
    y=zeros(3,1);
    y(3)=-1000;
else
    y=u;
end
end
```

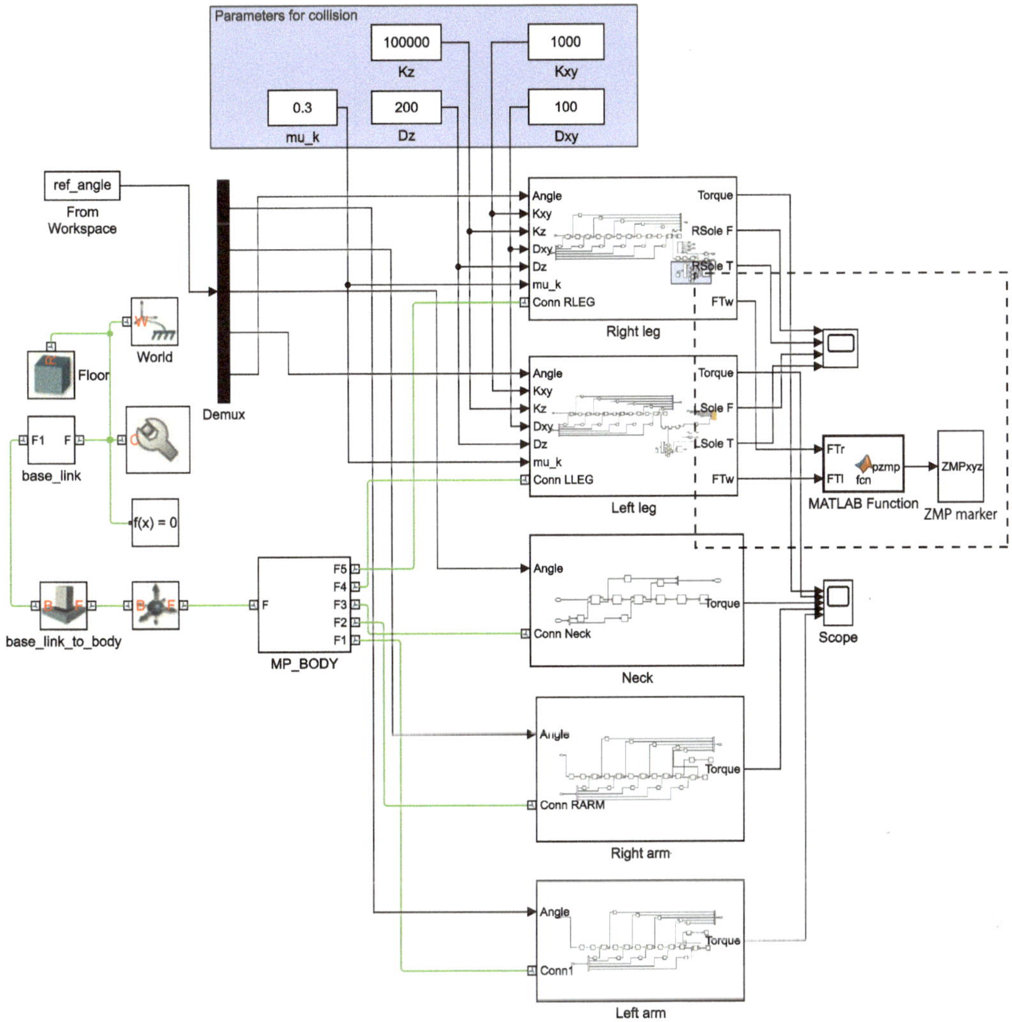

FIGURE 8.49 The top level in the hierarchical diagram with the ZMP marker. The "MATLAB Function" block shown in Fig. 8.48 outputs a vector which consists of forces and moments in the world frame. Such quantities are also obtained from the sensor mounted on the left sole. For computing the ZMP, input the two vectors into the "MATLAB Function" block as shown in this figure.

8.4.6 Motion Design

In this section, it is explained how to design an off-line motion with the MATLAB script utilizing Robotics System Toolbox. The reference trajectory of the left gripper of ROBOTIS-OP2 in the world frame is generated with a fifth-order polynomial interpolation and the time-series reference joint angles are computed by inverse kinematics. The MATLAB script is shown in Fig. 8.53 and is saved as op2_motion.m in the same folder with the Simulink files as shown in Fig. 8.17.

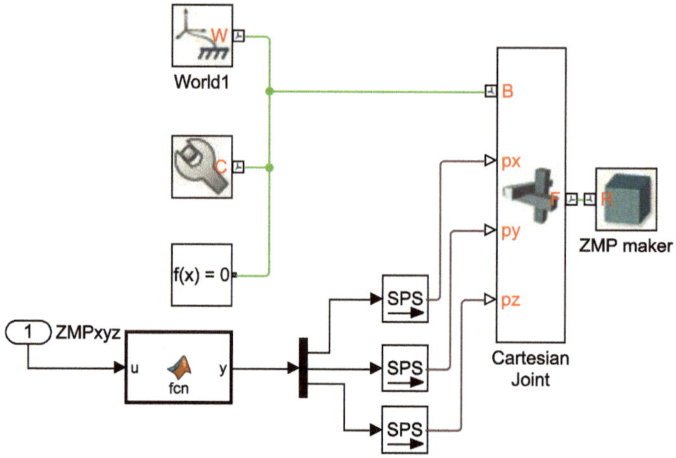

FIGURE 8.50 Subsystem for the ZMP marker. The computed position of the ZMP is sent to this subsystem which displays the position by a sphere in Mechanics Explore. The sphere is handled as an object and moved by the "Cartesian joint" as shown in this figure.

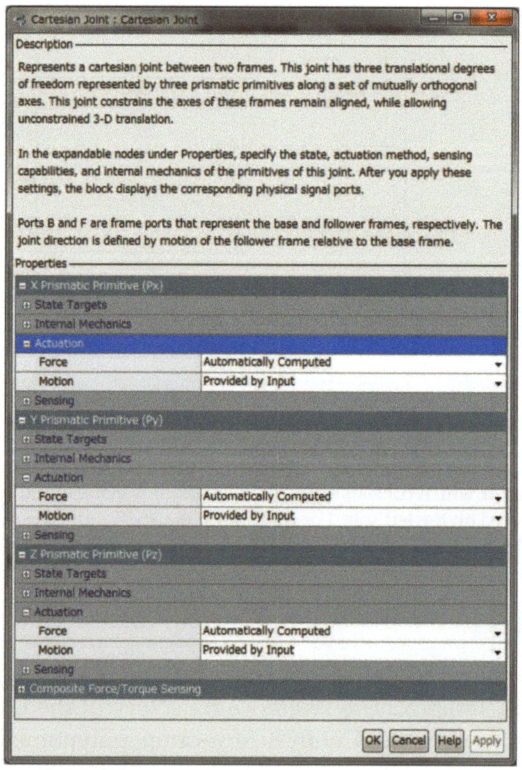

FIGURE 8.51 Property of the "Cartesian joint" block. This joint has three translational DoFs represented by three prismatic primitives.

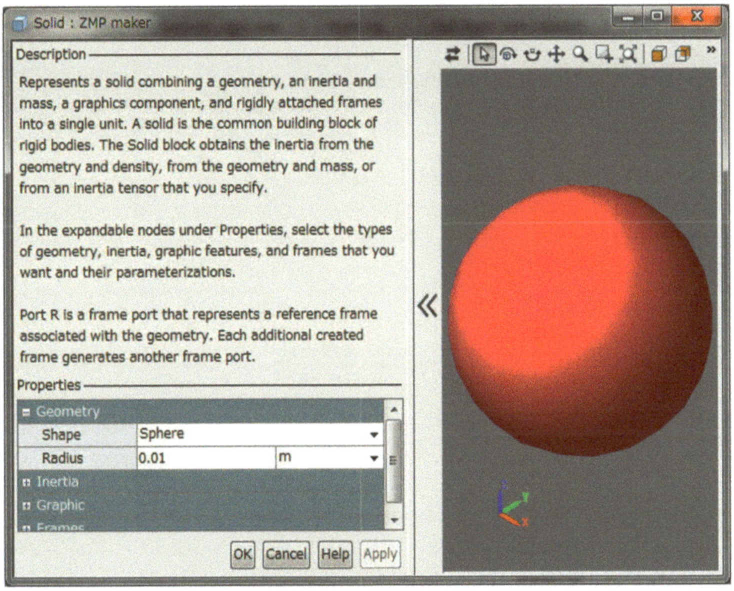

FIGURE 8.52 Property of the "Solid" block named "ZMP Marker."

In line 3, the URDF files of ROBOTIS-OP2 are parsed and stored in the Workspace memory. In line 11, the coordinate frame of the left gripper named "MP_ARM_GRIPPER_FIX_DUMMY_L" (cf. Fig. 8.14) is specified as the end effector. In order to generate the reference trajectory of the end effector with a fifth-order polynomial interpolation, the coefficients for the interpolation are prepared with the code from lines 15 to 29. The solver for the inverse kinematics is initialized in line 32. The first three elements of the variable "weight" in line 33 correspond to the weights on the tolerances for the orientation for the desired pose. The last three elements correspond to the weights on the tolerances for the x-y-z-position of the gripper for the desired pose. Form lines 35 to 46, for each sampling time, the trajectory is calculated and the inverse kinematics solver computes the joint angles. From lines 49 to 52, the stationary motion is attached before the motion generated in the section from lines 35 to 46. In line 55, the reference joint angles are stored in the Workspace as a time-series matrix, and they will be available in Simulink.

8.4.7 Simulation

After executing the MATLAB script explained in Section 8.4.6, the Simulink diagram (op2_simulation.slx) is ready to be executed. Fig. 8.54 shows an example of a simulation condition. The type of the solver and its time step should be carefully chosen based on the intended situation. In this example, the solver is the Runge–Kutta method and the step size is 1 ms. By clicking on the play icon (▷) on the Simulink window, the simulation starts. Figs. 8.55 and 8.56 show snapshots of the simulation and the result of the simulation, respectively. In the animation of the simulation, the ZMP marker moves in accordance with the designed

```
clearvars;
addpath('.\RobotModels\robotis_op_description\meshes');
op2=importrobot('.\RobotModels\robotis_op_description\robots\robotis_op.urdf');
q0=homeConfiguration(op2);

%% Generating the end-effector trajectory
SamplingTime=0.01;
tf=2;
t = (0:SamplingTime:tf)'; % Time
count = length(t);
endEffector = 'MP_ARM_GRIPPER_FIX_DUMMY_L';
T=op2.getTransform(q0,endEffector);
p=T(1:3,4);
%coefficients for 5th order polynomial interpolation
xs=0;
dxs=0;
ddxs=0;
xf=0.03;
dxf=0;
ddxf=0;
a0 = xs;
a1 = dxs;
a2 = ddxs/2.0;
a3 = (20.0*xf - 20.0*xs - (8.0*dxf + 12.0*dxs)*tf - ...
    (3.0*ddxs - ddxf)*tf.^2.0)/(2.0*tf.^3.0);
a4 = (30.0*xs - 30.0*xf + (14.0*dxf + 16.0*dxs)*tf + ...
    (3.0*ddxs - 2.0*ddxf)*tf.^2.0)/(2.0*tf.^4.0);
a5 = (12.0*xf - 12.0*xs - (6.0*dxf + 6.0*dxs)*tf - ...
    (ddxs - ddxf)*tf.^2.0)/(2.0*tf.^5.0);

%% Inverse kinematics
ik = robotics.InverseKinematics('RigidBodyTree', op2);
weights = [0 0 0 1 1 1];
qInitial=q0;
for i = 1:count
    ti=SamplingTime*(i-1);
    %5th order polynomial interpolation
    pi=a0+a1*ti+a2*ti^2+a3*ti^3+a4*ti^4+a5*ti^5;
    T(1:3,4)=  [p(1)+pi,p(2)-pi,p(3)]';
    % Solving inverse kinematics
    [qSol,solInfo] = ik(endEffector,T,weights,qInitial);
    % Store the configuration
    ref_q(i,:) = [qSol.JointPosition];
    % Start from prior solution
    qInitial = qSol;
end

%% Generating motion to be stationary for half second.
t_stop=0.5;
t_init = (0:SamplingTime:(t_stop-SamplingTime))'; % Time
len = length(t_init);
ref_q_init(1:len,:)=ones(len,1)*ref_q(1,:);

%% Making timeseries matrix for Simulink
ref_angle=timeseries([ref_q_init;ref_q],[t_init;(t+t_stop)]);
```

FIGURE 8.53 Matlab program code for moving the left arm. The reference trajectory of the gripper (MP_ARM_GRIPPER_FIX_DUMMY_L) is generated in the world frame with fifth order polynomial interpolation. The time-series reference joint angles are computed with the inverse kinematics.

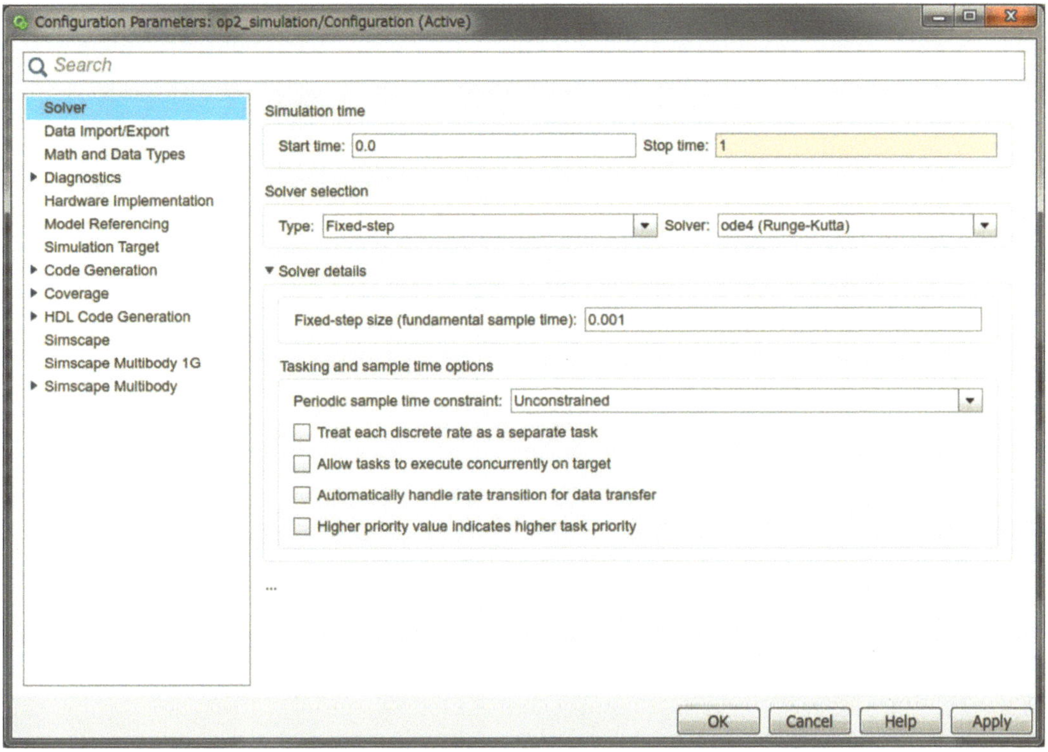

FIGURE 8.54 Configuration parameters for a dynamics-based simulation. The type of the solver and its time step should be carefully chosen based on the intended situation. In this example, the solver is the Runge–Kutta method and the step size is 1 ms.

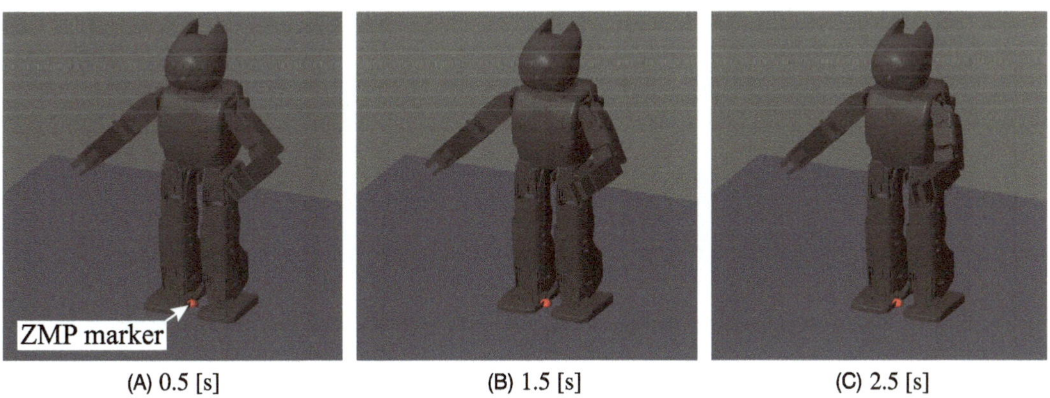

FIGURE 8.55 Visualization of dynamics-based simulation. As the left gripper moves to the right front, the ZMP marker also moves to the right front.

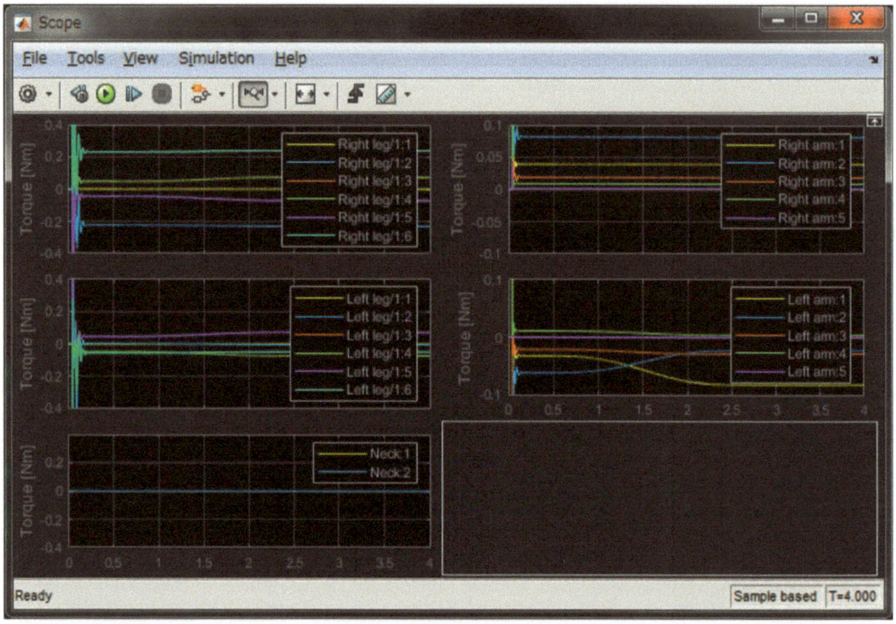

(A)

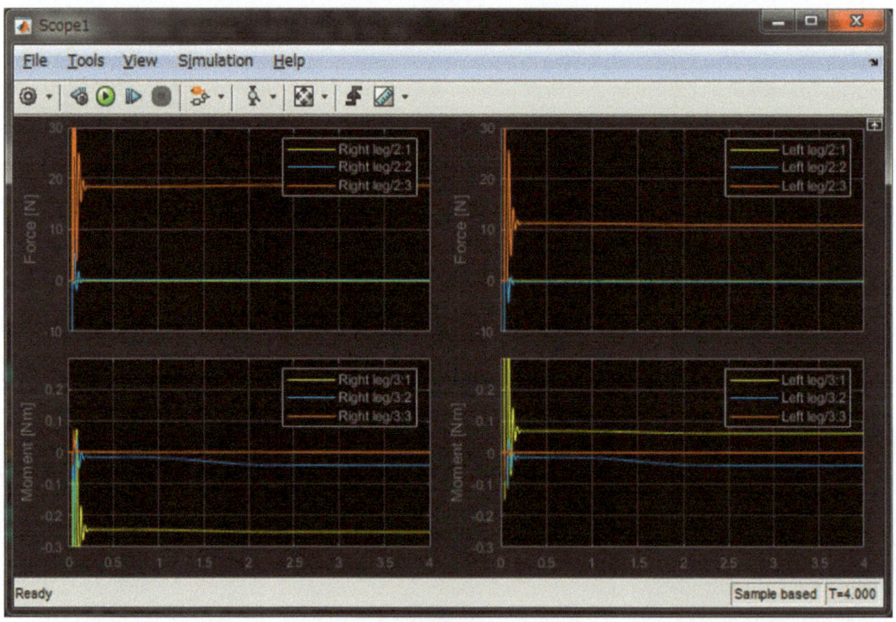

(B)

FIGURE 8.56 Simulation results plotted by the "Scope" blocks. (A) Above: Relationship between the time and the joint torques. (B) Below: Relationship between the time and the force/moment. The measured values oscillate in all the plots immediately after the start of the simulation. The robot is dropped from a certain height at the beginning of the simulation. That is why the values oscillate.

motion. As the left gripper moves to the right front, the ZMP marker also moves to the right front. As shown in Fig. 8.56, the measured values oscillate in all the plots immediately after the start of the simulation. The robot is dropped from a certain height at the beginning of the simulation. That is why the values oscillate. As shown in the plot "Left arm" in Fig. 8.56A, the motor torques are changing from 0.5 to 2.5 s.

References

[1] About ROS, http://www.ros.org/about-ros/.

[2] C.E. Agüero, N. Koenig, I. Chen, H. Boyer, S. Peters, J. Hsu, B. Gerkey, S. Paepcke, J.L. Rivero, J. Manzo, E. Krotkov, G. Pratt, Inside the virtual robotics challenge: simulating real-time robotic disaster response, IEEE Transactions on Automation Science and Engineering 12 (2015) 494–506.

[3] Atlas, Boston dynamics, https://www.bostondynamics.com/atlas.

[4] BIOROID GP, http://www.robotis.us/robotis-gp/.

[5] Body model, Choreonoid 1.6 documentation, http://choreonoid.org/en/manuals/1.6/handling-models/bodymodel.html.

[6] Brushless DC Motor, Simscape documentation, https://jp.mathworks.com/help/physmod/elec/examples/brushless-dc-motor.html.

[7] Bullet, http://bulletphysics.org/.

[8] S. Chitta, I. Sucan, S. Cousins, MoveIt! [ROS topics], IEEE Robotics & Automation Magazine 19 (2012) 18–19.

[9] Coin3D, https://bitbucket.org/Coin3D/coin/wiki/Home.

[10] D. Cook, A. Vardy, R. Lewis, A survey of AUV and robot simulators for multi-vehicle operations, in: IEEE/OES Autonomous Underwater Vehicles, 2015, pp. 1–8.

[11] DART, http://dartsim.github.io/.

[12] J. Englsberger, A. Werner, C. Ott, B. Henze, M.A. Roa, G. Garofalo, R. Burger, A. Beyer, O. Eiberger, K. Schmid, A. Albu-Schaffer, Overview of the torque-controlled humanoid robot TORO, in: IEEE-RAS International Conference on Humanoid Robots, 2014, pp. 916–924.

[13] C. Ericson, Real-Time Collision Detection, CRC Press, 2004.

[14] R. Featherstone, A divide-and-conquer articulated-body algorithm for parallel O(log(n)) calculation of rigid-body dynamics. Part 1: basic algorithm, The International Journal of Robotics Research 18 (1999) 876–892.

[15] R. Featherstone, Rigid Body Dynamics Algorithms, Springer, 2008.

[16] P. Flores, H.M. Lankarani, Contact Force Models for Multibody Dynamics, Springer, 2016.

[17] FROST: fast robot optimization and simulation toolkit on GitHub, https://github.com/ayonga/frost-dev.

[18] Gazebo, http://gazebosim.org/.

[19] GLFW, http://www.glfw.org/.

[20] GLUT– The OpenGL Utility Toolkit, https://www.opengl.org/resources/libraries/glut/.

[21] GNU Octave, https://www.gnu.org/software/octave/.

[22] Google Maps, https://www.google.com/maps/.

[23] D. Gouaillier, V. Hugel, P. Blazevic, C. Kilner, J. Monceaux, P. Lafourcade, B. Marnier, J. Serre, B. Maisonnier, Mechatronic design of NAO humanoid, in: IEEE International Conference on Robotics and Automation, 2009, pp. 769–774.

[24] graspPlugin, http://www.hlab.sys.es.osaka-u.ac.jp/grasp/en.

[25] I. Ha, Y. Tamura, H. Asama, J. Han, D.W. Hong, Development of open humanoid platform DARwIn-OP, in: SICE Annual Conference 2011, 2011, pp. 2178–2181.

[26] A. Hereid, A.D. Ames, FROST *: fast robot optimization and simulation toolkit, in: IEEE/RSJ International Conference on Intelligent Robots and Systems, 2017, pp. 719–726.

[27] HOAP-2, Fujitsu, https://en.wikipedia.org/wiki/HOAP.

[28] S.H. Hyon, D. Suewaka, Y. Torii, N. Oku, H. Ishida, Development of a fast torque-controlled hydraulic humanoid robot that can balance compliantly, in: IEEE-RAS International Conference on Humanoid Robots, 2015, pp. 576–581.

[29] Install the Simscape Multibody Link plug-in, https://jp.mathworks.com/help/physmod/sm/ug/installing-and-linking-simmechanics-link-software.html?lang=en.

[30] S. Ivaldi, J. Peters, V. Padois, F. Nori, Tools for simulating humanoid robot dynamics: a survey based on user feedback, in: IEEE-RAS International Conference on Humanoid Robots, 2014, pp. 842–849.
[31] S. Ivaldi, B. Ugurlu, Free simulation software and library, in: Humanoid Robotics: A Reference, Springer, 2018, Chapter 35.
[32] F. Kanehiro, H. Hirukawa, S. Kajita, OpenHRP: open architecture humanoid robotics platform, The International Journal of Robotics Research 23 (2004) 155–165.
[33] K. Kaneko, F. Kanehiro, M. Morisawa, K. Akachi, G. Miyamori, A. Hayashi, N. Kanehira, Humanoid robot HRP-4 – humanoid robotics platform with lightweight and slim body, in: IEEE International Conference on Intelligent Robots and Systems, 2011, pp. 4400–4407.
[34] C.W. Kennedy, J.P. Desai, Modeling and control of the Mitsubishi PA-10 robot arm harmonic drive system, IEEE/ASME Transactions on Mechatronics 10 (2005) 263–274.
[35] N. Koenig, A. Howard, Design and use paradigms for Gazebo, an open-source multi-robot simulator, in: IEEE/RSJ International Conference on Intelligent Robots and Systems, 2004, pp. 2149–2154.
[36] K. Kojima, T. Karasawa, T. Kozuki, E. Kuroiwa, S. Yukizaki, S. Iwaishi, T. Ishikawa, R. Koyama, S. Noda, F. Sugai, S. Nozawa, Y. Kakiuchi, K. Okada, M. Inaba, Development of life-sized high-power humanoid robot JAXON for real-world use, in: IEEE-RAS International Conference on Humanoid Robots, 2015, pp. 838–843.
[37] J.L. Lima, J.A. Gonçalves, P.G. Costa, A.P. Moreira, Humanoid realistic simulator: the servomotor joint modeling, in: International Conference on Informatics in Control, Automation and Robotics, 2009, pp. 396–400.
[38] Make a model, Gazebo tutorial, http://gazebosim.org/tutorials?tut=build_model&cat=build_robot.
[39] Mates and joints, https://jp.mathworks.com/help/physmod/smlink/ref/mates-and-joints.html.
[40] Mathematica, https://www.wolfram.com/mathematica/index.ja.html?footer=lang.
[41] MATLAB, https://jp.mathworks.com/products/matlab.html.
[42] I. Millington, Game Physics Engine Development: How to Build a Robust Commercial-Grade Physics Engine for Your Game, second edition, CRC Press, 2010.
[43] MoveIt!, https://moveit.ros.org/.
[44] S. Nakaoka, Choreonoid: extensible virtual robot environment built on an integrated GUI framework, in: IEEE/SICE International Symposium on System Integration, 2012, pp. 79–85.
[45] S. Nakaoka, S. Hattori, F. Kanehiro, S. Kajita, H. Hirukawa, Constraint-based dynamics simulator for humanoid robots with shock absorbing mechanisms, in: IEEE/RSJ International Conference on Intelligent Robots and Systems, 2007, pp. 3641–3647.
[46] J. Neider, T. Davis, M. Woo, OpenGL Programming Guide, Addison Wesley, 1993.
[47] Newton dynamics, http://newtondynamics.com/forum/newton.php.
[48] ODE (open dynamics engine), http://www.ode.org/.
[49] OpenStreetMap, http://www.openstreetmap.org/.
[50] I.W. Park, J.Y. Kim, J. Lee, J.H. Oh, Mechanical design of humanoid robot platform KHR-3 (KAIST humanoid robot – 3: HUBO), in: IEEE-RAS International Conference on Humanoid Robots, 2005, pp. 321–326.
[51] PhysX, https://developer.nvidia.com/gameworks-physx-overview.
[52] M. Quigley, K. Conley, B. Gerkey, J. Faust, T. Foote, J. Leibs, E. Berger, R. Wheeler, A. Mg, ROS: an open-source robot operating system, in: IEEE International Conference on Robotics and Automation Workshop on Open Source Robotics, 2009.
[53] E. Rohmer, S.P. Singh, M. Freese, V-REP: a versatile and scalable robot simulation framework, in: 2013 IEEE International Conference on Intelligent Robots and Systems, 2013, pp. 1321–1326.
[54] ROS control, http://gazebosim.org/tutorials/?tut=ros_control.
[55] S. Shirata, A. Konno, M. Uchiyama, Development of a light-weight biped humanoid robot, in: IEEE/RSJ International Conference on Intelligent Robots and Systems, 2004, pp. 148–153.
[56] Simbody, https://simtk.org/home/simbody/.
[57] Simscape Multibody, https://jp.mathworks.com/products/simmechanics.html.
[58] Simscape Multibody Link plug-in, https://jp.mathworks.com/help/physmod/smlink/index.html.
[59] Simulink, https://jp.mathworks.com/products/simulink.html.
[60] Toyota unveils third generation humanoid robot T-HR3, http://corporatenews.pressroom.toyota.com/releases/toyota+unveils+third+generation+humanoid+robot+thr3.htm.
[61] T. Tsujita, O. Altangerel, S. Abiko, A. Konno, Analysis of drop test using a one-legged robot toward parachute landing by a humanoid robot, in: IEEE International Conference on Robotics and Biomimetics, 2017, pp. 221–226.
[62] Tunnel disaster response and recovery challenge, http://worldrobotsummit.org/en/wrc2018/disaster/.

[63] A URDF file and Mesh files of ROBOTIS-OP2 on GitHub, https://github.com/ROBOTIS-OP/robotis_op_common/tree/master/robotis_op_description.
[64] URDF in Gazebo, Gazebo tutorial, http://gazebosim.org/tutorials/?tut=ros_urdf.
[65] V-Rep, http://www.coppeliarobotics.com/index.html.
[66] G. Van Den Bergen, Collision Detection in Interactive 3D Environments, Elsevier, 2003.
[67] Vortex Studio, https://www.cm-labs.com/vortex-studio/.
[68] Webots, https://www.cyberbotics.com/#webots.
[69] J. Wernecke, The Inventor Mentor, Addison Wesley, 1994.
[70] K. Yamane, Simulating and Generating Motions of Human Figures, Springer-Verlag, Berlin, Heidelberg, 2004.

APPENDIX

A

Appendix

Two data sets for models derived from a small-size humanoid robot HOAP-2 [1] are presented below. One of the models resembles the original with arms comprising four degrees of freedom (DoFs). In the other model, three more joints are added to the arms (wrist joints) to obtain kinematically redundant arms that can perform a number of application tasks. The data sets are in YAML format. They can be downloaded and directly used in the Chorenoid simulation environment (cf. Chapter 8).

As a common rule, the numbering of the joints is in agreement with the tree structure of the robot. The structure is determined by the limbs extending from the base link, as shown in Fig. 2.1. Thus, the joints of each limb are numbered in increasing order, starting from the joint proximal to the base link. Note that in the upper-body part there is a torso joint. This joint does not appear in the models, i.e. the base link is constituted of the pelvis and the trunk.

A.1 MODEL PARAMETERS FOR A SMALL-SIZE HUMANOID ROBOT WITH 4-DOF ARMS

The naming and numbering of the joints is shown in Fig. A.1. Link lengths are given in Fig. A.2 and Fig. A.3. The coordinate frames are shown in Fig. A.4. The data file for the inertia and other parameters can be downloaded from [2].

A.2 MODEL PARAMETERS FOR A SMALL-SIZE HUMANOID ROBOT WITH 7-DOF ARMS

The naming and numbering of the joints is shown in Fig. A.5. Link lengths are given in Fig. A.6 and Fig. A.7. The coordinate frames are shown in Fig. A.8. The data file for the inertia and other parameters can be downloaded from [3].

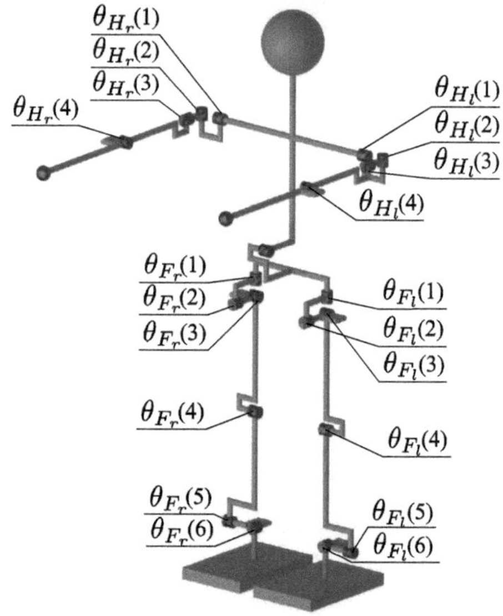

FIGURE A.1 Joint angle naming and numbering for the model with 4-DoF arms.

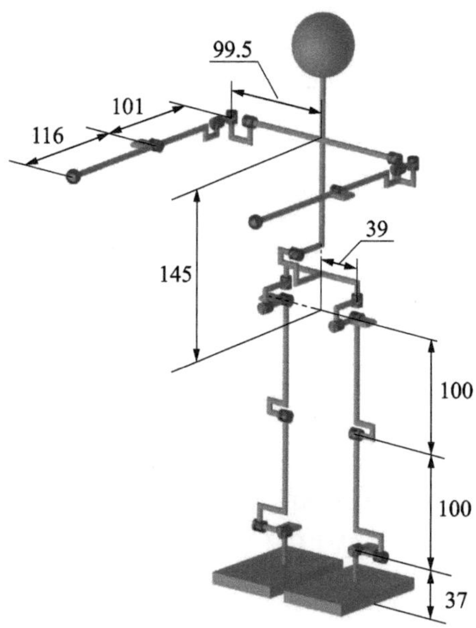

FIGURE A.2 Link lengths (in mm) for the model with 4-DoF arms.

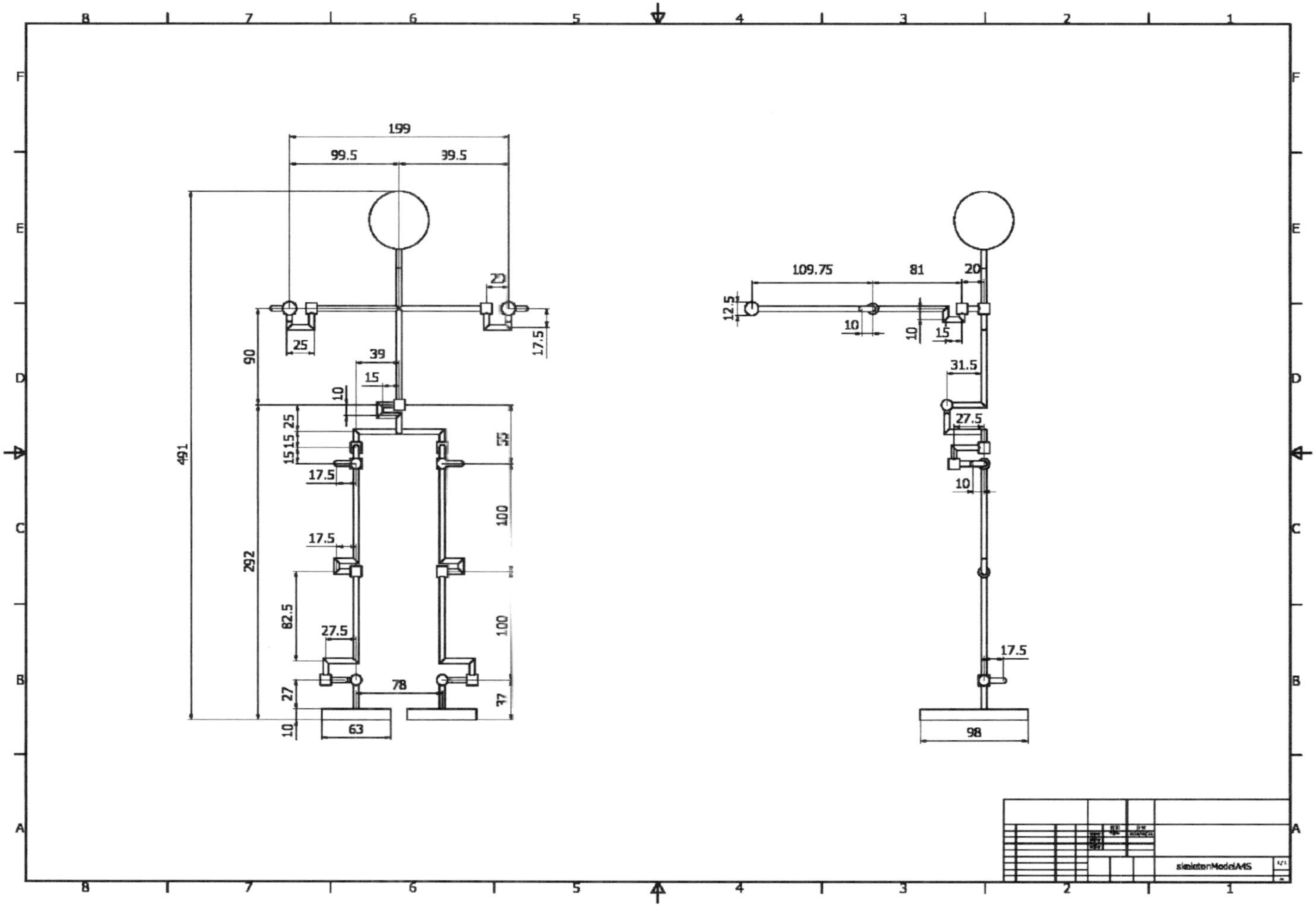

FIGURE A.3 Distances (in mm) from CAD data for the model with 4-DoF arms.

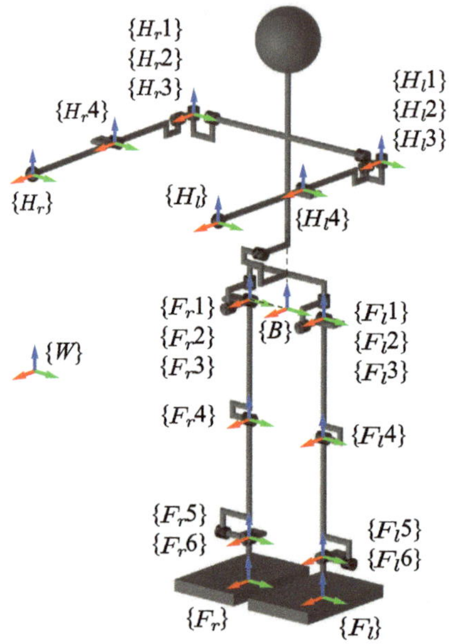

FIGURE A.4 Coordinate frames of the model with 4-DoF arms.

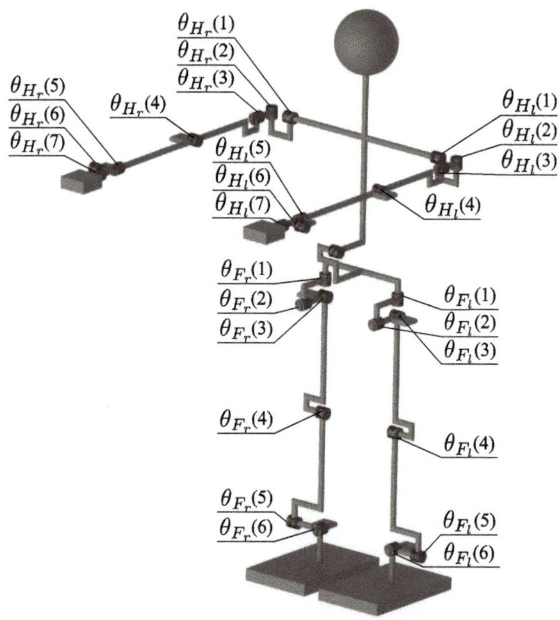

FIGURE A.5 Joint angle naming and numbering for the model with 7-DoF arms.

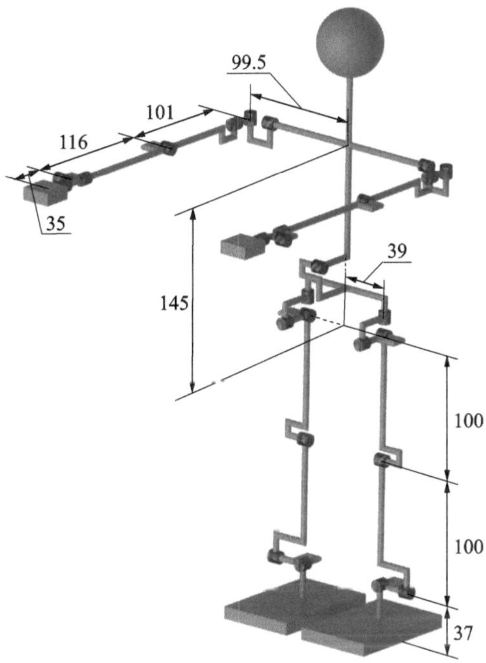

FIGURE A.6 Link lengths (in mm) for the model with 7-DoF arms.

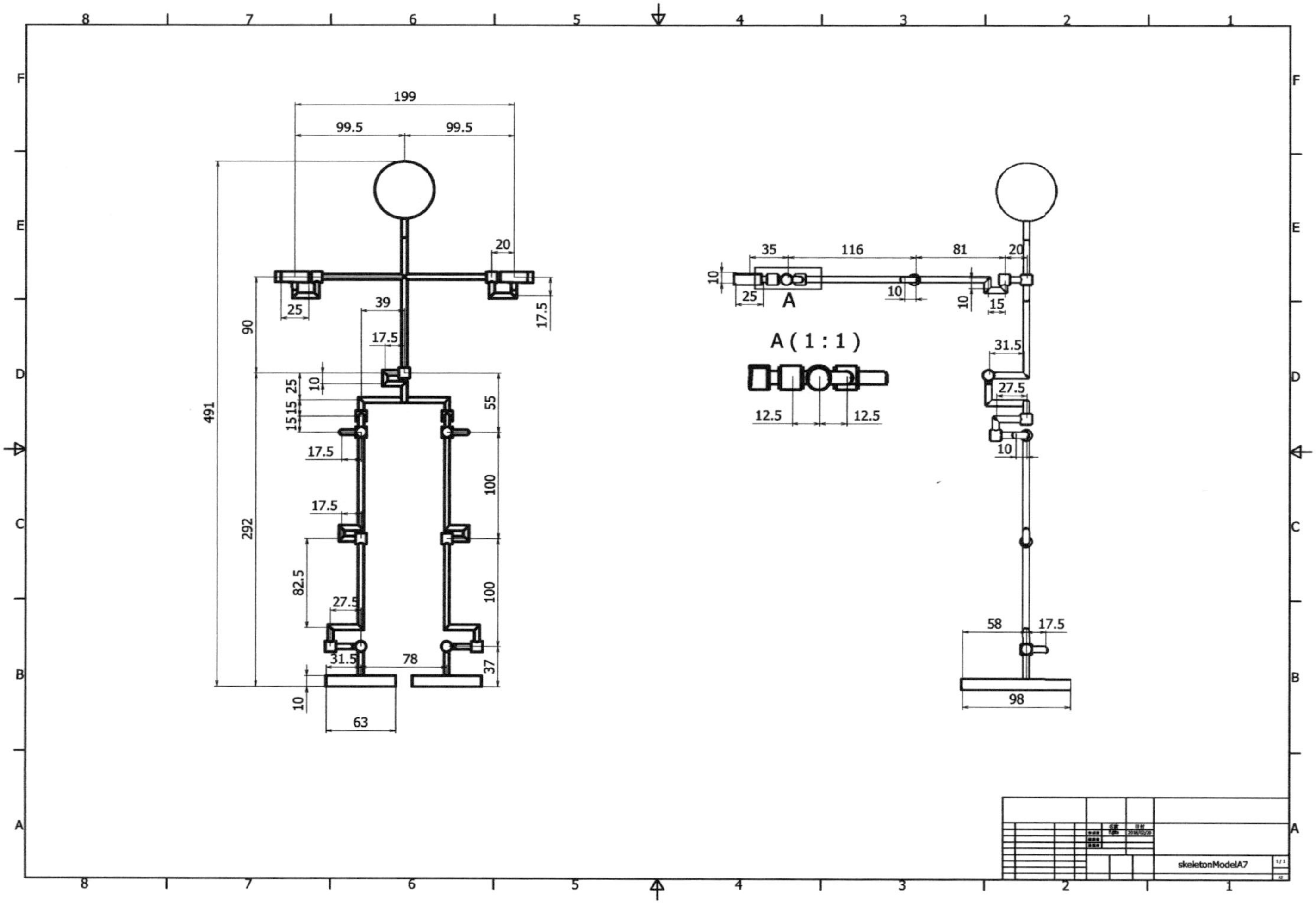

FIGURE A.7 Distances (in mm) from CAD data for the model with 7-DoF arms.

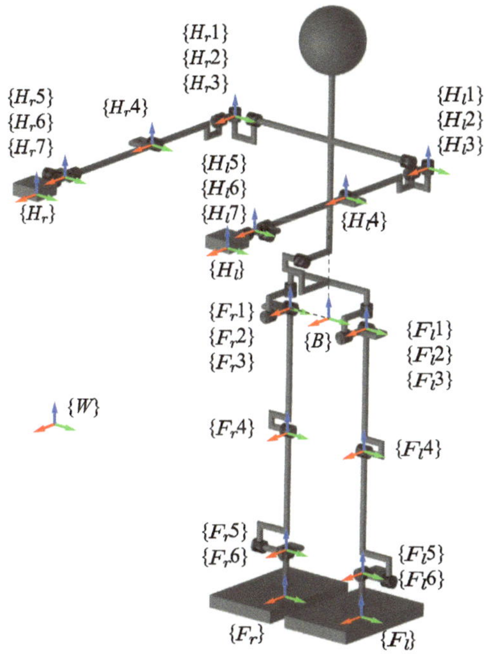

FIGURE A.8 Coordinate frames of the model with 7-DoF arms.

References

[1] Fujitsu, Miniature Humanoid Robot HOAP-2 Manual, 1st edition, Fujitsu Automation Co., Ltd, 2004 (in Japanese).

[2] RLS@TCU, Data set for a small-size humanoid robot model with four-DoF arms, Robotic Life Support Laboratory, Tokyo City University, 2018, https://doi.org/10.1016/B978-0-12-804560-2.00016-X.

[3] RLS@TCU, Data set for a small-size humanoid robot model with seven-DoF arms, Robotic Life Support Laboratory, Tokyo City University, 2018, https://doi.org/10.1016/B978-0-12-804560-2.00016-X.

Index

D

Q

R

FSC
www.fsc.org
MIX
Papier | Fördert
gute Waldnutzung
FSC® C083411

Kolibri 360 GmbH
Ferdinand-Jühlke-Straße 7
99095 Erfurt, Deutschland
produktsicherheit@kolibri360.de